1) lin in Y
2)

4) lod test
5)

7) lod test

*An Introduction
to Differential Equations
and Their Applications*
SECOND EDITION

An Introduction to Differential Equations and Their Applications

SECOND EDITION

Stephen L. Campbell
North Carolina State University

Wadsworth Publishing Company
Belmont, California
A Division of Wadsworth, Inc.

Mathematics Editors: Barbara Holland, Anne Scanlan-Rohrer
Editorial Assistant: Leslie With
Designer: Andrew H. Ogus
Production Editor: Harold Humphrey
Print Buyer: Karen Hunt
New Technical Illustrations: Judith Ogus/Random Arts
Compositor: Asco Trade Typesetting Ltd., Hong Kong
Cover: Andrew H. Ogus
Signing Representative: Charlie Delmar

Cover art and spine detail: Charles Sheeler, *Aerial Gyrations*, 1953. Oil on canvas. $23\frac{5}{8}'' \times 18\frac{5}{8}''$ (60.0 × 47.3 cm). San Francisco Museum of Modern Art, Mrs. Manfred Bransten Special Fund Purchase.

© 1990 by Wadsworth, Inc. All rights reserved. No part of this book may be reproduced, stored in a retrieval system, or transcribed, in any form or by any means, electronic, mechanical, photocopying, recording, or otherwise, without the prior written permission of the publisher, Wadsworth Publishing Company, Belmont, California 94002, a division of Wadsworth, Inc.
© 1986 by Longman, Inc.

Printed in the United States of America   85

2 3 4 5 6 7 8 9 10—94 93 92 91

Library of Congress Cataloging-in-Publication Data

Campbell, S. L. (Stephen La Vern)
   An introduction to differential equations and their applications/Stephen L. Campbell.—2nd ed.
       p.   cm.
   Includes index.
   ISBN 0-534-09468-6
   1. Differential equations.   I. Title.
QA371.C1919   1990
515'.35—dc20

89-31680
CIP

Contents

1   *Introduction*   1

2   *First-Order Equations*   13

        2.1   Existence and Uniqueness   *13*
        2.2   Direction Fields   *18*
        2.3   First-Order Linear Equations   *26*
        2.4   Bernoulli Equations   *35*
        2.5   Separable Equations   *37*
        2.6   Exact Equations   *43*
        2.7   Integrating Factors   *49*
        2.8   Homogeneous Equations   *53*
        2.9   Heat and Decay Problems   *56*

- 2.10 Mixing and Flow Problems *69*
- 2.11 Orthogonal Trajectories *79*
- 2.12 Circuits *82*
- 2.13 Mechanics and Chemical Reactions *91*

3 Higher-Order Differential Equations 98

- 3.1 Introduction and Basic Theory *98*
- 3.2 Two Helpful Substitutions *100*
- 3.3 Applications to Mechanics *105*
- 3.4 Second-Order Linear Equations *111*
- 3.5 Linear Independence and the Wronskian *120*
- 3.6 Operator Notation *126*
- 3.7 Reduction of Order *131*
- 3.8 Homogeneous Linear Constant-Coefficient Equations (Second-Order) *135*
- 3.9 nth-Order Linear Differential Equations *140*
- 3.10 Homogeneous Linear Constant-Coefficient Equations (nth-Order) *146*
- 3.11 Euler's Equation *149*
- 3.12 Undetermined Coefficients (Second-Order) *154*
- 3.13 Undetermined Coefficients (nth-Order) *165*
- 3.14 Variation of Parameters (Second-Order) *171*
- 3.15 Variation of Parameters (nth-Order) *178*
- 3.16 Mechanical Vibrations *180*
- 3.17 Linear Electrical Circuits *200*
- 3.18 Phase Angles, Complex Arithmetic, and Phasors *203*

4 The Laplace Transform 209

- 4.1 Definition and Basic Properties *209*
- 4.2 Initial-Value Problems *217*
- 4.3 Discontinuous Forcing Functions *227*

- 4.4 Periodic Functions *237*
- 4.5 Impulses and Distributions *242*
- 4.6 Convolution Integrals *246*
- 4.7 Transfer Functions *252*

5 *Series Solution of Linear Equations* 257

- 5.1 Introduction *257*
- 5.2 Review of Power Series *258*
- 5.3 Solution at an Ordinary Point (Theory) *267*
- 5.4 Solution at an Ordinary Point (Taylor-Series Method) *270*
- 5.5 Solution at an Ordinary Point (Undetermined Coefficients) *275*
- 5.6 Legendre Polynomials *284*
- 5.7 Regular Singular Points (Frobenius Method) *287*
- 5.8 Bessel Functions *302*

6 *Linear Systems of Differential Equations* 309

- 6.1 Introduction *309*
- 6.2 Elimination Methods *318*
- 6.3 Solution by Laplace Transform *328*
- 6.4 Mixing Problems *331*
- 6.5 Mechanical Systems *341*
- 6.6 Multiloop Circuits *350*
- 6.7 Matrices and Vectors *353*
- 6.8 Determinants and Linear Independence *368*
- 6.9 Differential Equations: Basic Theory *374*
- 6.10 Homogeneous Systems with Constant Coefficients Using Eigenvectors *380*
- 6.11 Nonhomogeneous Systems (Undetermined Coefficients) *407*
- 6.12 The Matrix Exponential *415*
- 6.13 Fundamental Solution Matrices *424*

7 Difference Equations 432

- 7.1 Introduction *432*
- 7.2 First-Order Linear Difference Equations *437*
- 7.3 Second-Order Homogeneous Difference Equations *440*
- 7.4 Nonhomogeneous Difference Equations *445*
- 7.5 Applications *450*

8 Numerical Methods 459

- 8.1 Introduction *459*
- 8.2 Euler's Method *462*
- 8.3 An Analysis of Euler's Method *467*
- 8.4 Second-Order Methods *476*
- 8.5 Fourth-Order Runge–Kutta *481*
- 8.6 Multistep Methods *483*
- 8.7 Systems *486*

9 Qualitative Analysis of Nonlinear Equations in the Plane 491

- 9.1 Introduction *491*
- 9.2 The Phase Plane *494*
- 9.3 Linear Systems *498*
- 9.4 Equilibria of Nonlinear Systems *503*
- 9.5 Periodic Solutions *508*
- 9.6 Population Models *512*
- 9.7 Nonlinear Circuits *519*
- 9.8 Mechanical Systems *522*

Appendix A	*Complex Numbers*	527
Appendix B	*Review of Partial Fractions*	533
Appendix C	*Existence and Uniqueness*	538
	Solutions to Selected Exercises	548
	Index	594

Preface

The Philosophy Behind This Book. This book was written with the student in mind. However, it is not a cookbook. The careful development of not only techniques and theory but also applications and the geometry of differential equations will provide the student with the balanced background needed to go on in his or her chosen field, be it mathematics, engineering, the sciences, or something else.

In an introductory calculus course a student must learn not only how to manipulate the notation but also how to interpret the derivative geometrically (tangent lines) and physically (velocities, accelerations). The same is true with differential equations. Rather than treat many applications in a cursory manner ("here is the equation for nerve transmission in left-tentacled squids"), we have taken three major applications and carried them throughout. They are *mechanics, mixing problems,* and *electrical circuits.* The applications are formulated initially as nonlinear problems. This not only provides interesting applications of our nonlinear techniques, but also helps the student appreciate the true role—and the limitations—of linearity. Additional applications that are discussed are *heat, radioactive decay, orthogonal trajectories, chemical reactions, population models,* and *economics.*

In writing this text, I have tried to follow certain basic principles:

"Everything should be made as simple as possible, but not simpler";

"Do one thing at a time"; and
"Spend extra time where the student has difficulty."

Thus, I devote more space then usual to Heaviside (unit-step) functions, for example, because my experience has shown that the topic is difficult for many students, even though it is almost trivial mathematically. Also, when given the choice between alternative approaches, I have chosen the one that best explains to the student how the concepts are used in practice. For example, the method of undetermined coefficients is developed using the roots of the characteristic equation because these roots (poles of the transfer function) are the basis of so many design procedures in the engineering disciplines. Another example is Chapter 6 on systems, where I have tried to avoid special techniques that are appropriate for only two-dimensional problems.

My experience is that many students need to review partial fractions. Accordingly, there is a review appendix (Appendix B). Also, some students may not have had the needed material on complex numbers, so there is an appendix on them (Appendix A). Similarily, Section 6.7 may be largely review for some students.

I have tried to make the text as flexible as possible. To avoid having to cover only part of a section, I put the topics that are most often deleted into separate sections. Thus, there is a wide variation in the length of sections, but the book is easily adaptable to a variety of syllabi.

Changes in the Second Edition. The second edition has benefited from the comments of students and faculty using the first edition, along with additional reviewing and editing. The spirit and approach of the first edition remain unchanged. However, there has been a lot of fine tuning. The major changes are:

— Over 400 exercises and word problems have been added. The majority of these are aimed at helping students develop their basic techniques and understanding of key concepts.

— The chapters, and some sections, have been reordered to move the more frequently covered material earlier.

— A discussion of biological and economic models has been added to the exercises of Sections 2.9, 2.10, 2.13, 6.4, and 7.5.

— The discussion and examples on taking Laplace and inverse Laplace transforms have been expanded in Chapter 4.

— Appendix C on existence and uniqueness of solutions has been added.

— A new solutions manual with more discussion and worked out examples has been written.

— All of the exercises and examples have been reworked to insure accuracy.

Organization. Certain features of the organization need comment. Sections 2.4, 2.7, and 2.8 are optional. Any of the applications in Sections 2.9 through 2.13 may be omitted. Similarly, Sections 3.2 and 3.3 may be omitted. However, when studying the mechanics applications, the students should study Sections 2.12, 3.16, 6.5, and 9.8 in that order. For circuits, Section 2.12 should precede

3.17, 6.6, and 9.7. Sections 3.6, 3.9, 3.10, 3.13, and 3.15 may be omitted from Chapter 3 if only second-order equations are being covered. If nth-order equations are being covered, the second-order equations serve to motivate and explain the corresponding nth-order section. However, Sections 3.6 and 3.10 are needed if elimination methods are to be used in Chapter 6. Section 3.15 may be omitted. Section 3.11 is used to motivate the Frobenius method in Chapter 5, but it may be omitted. If Chapter 4 (Laplace transform) is not covered, then Section 6.3 and the solution of Example 6.4.2 should be omitted. Chapter 7 on difference equations is not required for either Chapter 5 or Chapter 8. However, if Chapter 7 is covered, it might be desirable to do so before Chapter 5 (series) and Section 8.3 (analysis of Euler's method). Chapter 8 on numerical methods can be covered at any point after, or even during, Chapter 2. (This was done by some users of the first edition.) Section 8.3 may be omitted. Section 8.3 does use difference equations, but may be covered without them by omitting a few details in the error derivations.

Acknowledgements. Many people have had input into this book, including an excellent group of reviewers, users of the first edition, and an experienced editorial staff. However, two groups have especially influenced me: the many students to whom I have taught differential equations and the colleagues with whom I have worked in differential equations, numerical analysis, and electrical and mechanical engineering.

Finally, the book could not have been written without the support and understanding of my wife, Gail, and our two sons, Matthew and Eric.

Stephen L. Campbell

1

Introduction

Differential equations are found in many areas of mathematics, science, and engineering. Often, students taking a first course in differential equations have already seen simple examples of differential equations in their mathematics, physics, chemistry, or engineering courses. Besides explaining what a differential equation is to those who have not encountered them before, this first chapter discusses on an intuitive level the type of information present in a differential equation.

A *differential equation* is an equation relating an unknown function or functions to one or more of its derivatives. Examples are

$$\frac{dx(t)}{dt} = \cos(x(t)), \tag{1}$$

$$\frac{\partial x(t, u)}{\partial t} = \frac{\partial^2 x(t, u)}{\partial u^2}, \tag{2}$$

$$\frac{d^2 y}{dx^2} = y \frac{dy}{dx}, \tag{3}$$

$$\frac{\partial^2 u}{\partial x^2} + \frac{\partial^2 u}{\partial y^2} = u^2. \tag{4}$$

The derivatives tell us which variables are independent and which are dependent. Thus the notation d^2y/dx^2 and dy/dx in (3) tell us that in this equation, y is the dependent variable and x the independent one. In (2), $\partial x/\partial t$, $\partial^2 x/\partial u^2$ tell us that x is the dependent and u, t are the independent variables. If there are several unknown functions, then there will usually be several equations (or one vector equation). We will consider such *systems* later in Chapter 6. If there are several independent variables and partial derivatives with respect to them, as in (2), where x is a function of t and u, or (4), where u is a function of x and y, then we have a *partial differential equation*.

If there is only one independent variable such as t in (1) or x in (3), the differential equation is called an *ordinary differential equation*. This book is concerned almost entirely with ordinary differential equations. (The exceptions occur in Section 2.6 and Chapter 7.) We assume that we are dealing with ordinary differential equations unless stated otherwise.

In a typical application, physical laws often lead to a differential equation. As a simple example, which will be discussed more carefully later, Newton's law of cooling (which is actually an approximate relationship and not a physical law) says that under certain conditions,

> The rate of temperature loss is proportional to the difference between the surface temperature of the object and the ambient (surrounding) temperature.

If $T(t)$ is the temperature of the object at time t, and $S(t)$ is the temperature of the surroundings at time t, then this law is expressed by

$$\frac{dT}{dt} = k(T - S). \tag{5}$$

Here k is a proportionality constant determined by the physical characteristics of the object and the surroundings. The function S is considered to be a known or measurable quantity possibly dependent on the time t. The function $T(t)$ is unknown. Often we wish to find $T(t)$ by "solving" the differential equation.

As another example, suppose an object of mass m is x units away from a point b at time t and the basic law we are concerned with is

> The force on the object is inversely proportional to the square of the distance from b.

Since force is mass times acceleration (if the mass is constant) and acceleration is the second derivative of position, the force would then be $m\, d^2x/dt^2$. Thus the physical law is expressed by

$$m\frac{d^2x}{dt^2} = \frac{k}{x^2}, \tag{6}$$

where k is the constant of proportionality.

We set up the differential equations such as (5) and (6), in order to obtain certain information. In our problems, we shall usually wish to determine the unknown function, such as the temperature $T(t)$ in (5) or the position $x(t)$ in (6). There are three general approaches for analyzing differential equations. In practice, solving a given problem usually involves aspects of all three approaches. The approaches are:

1. qualitative,
2. numerical,
3. analytical.

Using the *qualitative* approach, we determine the behavior of the solutions without actually getting a formula for them. This approach is somewhat similar to the curve-sketching process in introductory calculus where we sketch curves by drawing maxima, minima, concavity changes, etc. Qualitative ideas will be discussed in Section 2.2 and in Chapter 9.

Using the *numerical* approach, we compute estimates for the values of the unknown function at certain values of the independent variable. Numerical methods are extremely important and for many difficult problems are the only practical approach. We will discuss numerical methods in Chapter 8. The safe and effective use of numerical methods requires an understanding of the basic properties of differential equations and their solutions.

Most of this book is devoted to developing *analytical* procedures, that is, obtaining explicit and implicit formulas for the solutions of various ordinary differential equations. We shall present a sufficient number of applications to enable the reader to understand how differential equations are used and develop some feeling for the physical information they convey.

As every calculus student knows, the process of differentiating is much easier than that of antidifferentiating. The derivative may be calculated in a straightforward manner. Antidifferentiating can be much more difficult. A small change can alter the integration method and may even make integration in terms of known functions impossible. For example,

$$\int \frac{1}{x^2 + 1} dx = \tan^{-1} x + C$$

is a standard formula, whereas

$$\int \frac{1}{x^2 - 1} dx$$

can be evaluated by partial fractions to yield

$$\tfrac{1}{2} \ln|x - 1| - \tfrac{1}{2} \ln|x + 1| + C.$$

As another example: $\int x e^{x^2} dx$ can be evaluated by substitution to yield $\tfrac{1}{2} e^{x^2} + C$, while $\int e^{x^2}\,dx$ cannot be expressed in terms of any of the usual functions taught in introductory calculus although the integral does exist.

Differential equations present even more difficult problems. Since Eq. (6) involves a derivative of x, d^2x/dt^2, we should expect to have to antidifferentiate something with respect to t. But we do not know d^2x/dt^2 in terms of t! We know it only in terms of x. Thus in solving differential equations we should be prepared for all the usual problems of antidifferentiation compounded by some new ones. We shall have to develop different techniques for different types of differential equations.

The *order* of a differential equation is the order of the highest derivative of the unknown function (dependent variable) that appears in the equation. For example, $dT/dt = k(T - S)$ is a first-order equation, whereas

$$m\frac{d^2x}{dt^2} = \frac{k}{x^2}$$

is a second-order equation. Most of the equations we shall deal with will be of first- or second-order.

An nth-order differential equation is *linear* if it can be written in the form

$$a_n(x)\frac{d^n y}{dx^n} + a_{n-1}(x)\frac{d^{n-1} y}{dx^{n-1}} + \cdots + a_1(x)\frac{dy}{dx} + a_0(x)y = f(x), \quad (7)$$

with a_n not identically zero. The $a_i(x)$ are known functions of x called *coefficients*. An equation that is not linear is called *nonlinear*. Whether or not an equation is linear is quite important, so let us examine the definition in some detail. An equation being linear places a restriction on what happens to the dependent variable [the unknown function y in formula (7)]. If the equation is linear, the only operations that may be performed on the dependent variable are:

1. Differentiation of the dependent variable
2. Multiplication of the dependent variable and its derivatives by functions of the *independent* variable only
3. Setting equal sums of the terms satisfying 2 and a function only of the independent variable

For first-order equations, (7) takes the form

$$a_1(x)\frac{dy}{dx} + a_0(x)y = f(x), \quad a_1(x) \not\equiv 0,$$

and for second-order equations (7) assumes the form

$$a_2(x)\frac{d^2y}{dx^2} + a_1(x)\frac{dy}{dx} + a_0(x)y = f(x), \quad a_2(x) \not\equiv 0.$$

For example,

$$x^2 \frac{d^3 y}{dx^3} + y = \sin x$$

is a third-order linear equation, while

$$\frac{d^2 y}{dx^2} + y \frac{dy}{dx} + y^3 = 0$$

is a second-order nonlinear equation because of the presence of both the $y\, dy/dx$ and the y^3 terms. The term $y\, dy/dx$ consists of a derivative of the dependent variable multiplied by a nonconstant function, y, of the dependent variable. The y^3 term also violates rule 2. The equation

$$\frac{d^2 y}{dx^2} + \sin x \frac{dy}{dx} = y \cos x + \frac{d^4 y}{dx^4}$$

is a fourth-order linear equation since it can be written

$$\frac{d^4 y}{dx^4} - \frac{d^2 y}{dx^2} - (\sin x)\frac{dy}{dx} + (\cos x) y = 0,$$

which is (7) with $n = 4$, $a_4 = 1$, $a_3 = 0$, $a_2 = -1$, $a_1 = -\sin x$, $a_0 = \cos x$, and $f(x) = 0$. The equation

$$\frac{d^2 y}{dx^2} + \sin x \frac{dy}{dx} = y \cos x + \left(\frac{dy}{dx}\right)^4$$

is a second-order nonlinear equation. The term $(dy/dx)^4$ is not allowed under rules 1 through 3. As a final example,

$$\frac{dy}{dx} = \frac{y}{x}$$

is a first-order linear equation, while

$$\frac{dy}{dx} = \frac{x}{y}$$

is a first-order nonlinear equation. On the other hand,

$$\frac{dx}{dy} = \frac{x}{y}$$

is linear, since dx/dy tells us that now x is the dependent variable and y the independent variable.

A function $y(x)$ is a *solution* of a differential equation if, when substituted into the differential equation, the resulting equality is true for *all* values of x in the domain of $y(x)$.

Example 1.1.1   Verify that $y(x) = \sin x + x^2$ is a solution of the second-order linear equation

$$\frac{d^2 y}{dx^2} + y = x^2 + 2.$$

Solution

We need to check whether

$$[\sin x + x^2]'' + [\sin x + x^2] = x^2 + 2$$

for all x. Performing the indicated differentiations gives

$$(-\sin x + 2) + \sin x + x^2 = x^2 + 2$$

or $2 + x^2 = x^2 + 2$, which is true for all x. ∎

The formula $y(x) = \sin x + x^2$ in Example 1.1.1 gave the solution y *explicitly* in terms of the independent variable x. Sometimes we get solutions defined *implicitly*.

Example 1.1.2

The equation $x = e^y + y$ implicitly defines y as a function of x. Verify that this implicitly defined function is a solution of the differential equation

$$\frac{dy}{dx} = \frac{1}{x - y + 1}. \tag{8}$$

Solution

Suppose $x = e^y + y$. Then x and y are related. Differentiating $x = e^y + y$ with respect to x gives

$$1 = e^y \frac{dy}{dx} + \frac{dy}{dx}.$$

Solving for dy/dx, we find

$$\frac{dy}{dx} = \frac{1}{e^y + 1}. \tag{9}$$

Substituting (9) and $x = e^y + y$ into (8) gives

$$\frac{1}{e^y + 1} = \frac{1}{[e^y + y] - y + 1}$$

or

$$\frac{1}{e^y + 1} = \frac{1}{e^y + 1}$$

which is true. ∎

If the solution of a differential equation contains (a) unknown or arbitrary constants, (b) is a solution for every value of the constants, and (c) satisfies the requirement that every solution of the differential equation is of this form, the solution is called the *general solution* of the differential equation. If a solution does not contain unknown constants, it is called a *particular solution*.

Example 1.1.3

Verify that $y = x^6 + c_1 x + c_2$ with c_1, c_2 arbitrary constants is the general solution of

$$y'' = 30x^4. \tag{10}$$

Solution

Since $(x^6 + c_1 x + c_2)'' = 30x^4$, it follows that $x^6 + c_1 x + c_2$ is a solution of (10) for any value of the constants c_1, c_2. On the other hand, if we antidifferentiate (10) twice, we find

$$y' = \int y''\, dx = \int 30x^4\, dx = 6x^5 + c_1$$

and

$$y = \int y'\, dx = \int (6x^5 + c_1)\, dx = x^6 + c_1 x + c_2,$$

where c_1, c_2 are arbitrary constants. Thus $y = x^6 + c_1 x + c_2$ gives all solutions and is the general solution. ∎

Note that $y = x^6 + c_1 x + 3$ is not the general solution of (10) even though it has an arbitrary constant, since it does not include the (particular) solution $y = x^6 + 2x + 4$.

One often knows the kinds of functions in the solution, but not the actual solution. This information may have been derived as part of a more general technique or it may be based on experience. Our task then is to determine a solution utilizing this information. Techniques based on this approach are developed in Chapter 3. For now we concentrate on determining a solution when we are given its "form."

Example 1.1.4

Determine the values of the constant r, if any, so that $y = \cos rx$ will be a solution of the second-order linear differential equation

$$\frac{d^2 y}{dx^2} + 9y = 0. \tag{11}$$

Solution

Substitute $y = \cos rx$ into the differential equation (11):

$$\frac{d^2(\cos rx)}{dx^2} + 9 \cos rx = 0.$$

Differentiating $\cos rx$ twice gives

$$-r^2 \cos rx + 9 \cos rx = 0. \tag{12}$$

Since (11) is a differential equation, (12) is to hold for all values of x. In particular, it holds for x for which $\cos rx \neq 0$. Dividing (12) by $\cos rx$ yields

$$-r^2 + 9 = 0 \quad \text{or} \quad r^2 = 9.$$

Thus $y = \cos 3x$ is a solution. [Note that $r = -3$ gives $\cos(-3x)$, which equals $\cos 3x$.] ∎

Example 1.1.5 Determine the values, if any, of the constants a, b so that the function $y = ax + b$ is a solution of the first-order linear equation

$$\frac{dy}{dx} + 3y = x. \tag{13}$$

Solution Substitute $y = ax + b$ into Eq. (13). This gives

$$\frac{d(ax+b)}{dx} + 3(ax+b) = x$$

or

$$a + 3ax + 3b = x. \tag{14}$$

One way to proceed is to note that (14) is $(a + 3b) + (3a)x = 0 + 1 \cdot x$. Equating coefficients of powers of x gives

$$a + 3b = 0, \qquad 3a = 1.$$

Thus $a = \frac{1}{3}$, $b = -\frac{1}{9}$, and $y = \frac{1}{3}x - \frac{1}{9}$. ∎

There will usually be many solutions to a differential equation. Often we want to find that solution which satisfies certain additional information. Frequently this information involves values of the unknown function and its derivatives at specific points, such as

$$y(0) = 0, \qquad y'(1) = 2. \tag{15}$$

If we are given values of the unknown function or its derivatives at different values of the independent variable, such as $x = 0$ and $x = 1$ in (15), we have a *boundary-value problem*, and the specified values are called *boundary conditions*. Boundary-value problems are very important, but we will not deal with them in this book. If we are given values of the unknown function and/or its derivatives at a single value of the independent variable, the problem is called an *initial-value problem*, and the specified values are called *initial conditions*. For example,

$$y'' = \sin y, \qquad y(1) = 3, \qquad y'(1) = 36$$

and

$$\frac{dy}{dx} = y + x, \qquad y(0) = 5$$

are initial-value problems, while

$$y'' = \sin y, \qquad y(0) = 0, \qquad y(1) = 0$$

and

$$\frac{dy}{dx} = y + x, \qquad y(0) - y(1) = 0$$

are boundary-value problems.

Example 1.1.6 Verify that $y = ce^{3x} - 1$, c an arbitrary constant, is a solution of

$$\frac{dy}{dx} - 3y = 3. \tag{16}$$

Then find the value of c so that y satisfies the initial condition

$$y(1) = 2. \tag{17}$$

Solution First we verify that $y = ce^{3x} - 1$ is a solution of Eq. (16). Substituting $y = ce^{3x} - 1$ into (16) gives

$$\frac{d(ce^{3x} - 1)}{dx} - 3(ce^{3x} - 1) = 3ce^{3x} - 3ce^{3x} + 3 = 3,$$

as desired. Now substitute $y = ce^{3x} - 1$ into the initial condition $2 = y(1)$ to get

$$2 = y(1) = ce^3 - 1.$$

Thus $ce^3 = 3$ or $c = 3e^{-3}$. A solution of the differential equation (16) satisfying the initial condition (17) is $y = 3e^{-3}e^{3x} - 1$. ∎

Not every function $f(x)$ has an antiderivative that can be expressed in terms of functions we are familiar with. The fundamental theorem of calculus provides an antiderivative if $f(x)$ is continuous.

If $f(x)$ is continuous on the interval $[a, b]$, $a < b$ and $a \le c \le b$, then

$$F(x) = \int_c^x f(s)\, ds.$$

is an antiderivative of $f(x)$. That is, $F'(x) = f(x)$.

For example,

$$\ln x = \int_1^x s^{-1}\, ds \quad \text{and} \quad \frac{d \ln x}{dx} = x^{-1}.$$

Since $y'(x) = f(x)$ is a simple differential equation, it follows that not every differential equation will have solutions given by an explicit formula involving familiar functions. As will be shown later, the solutions of these differential equations can sometimes be expressed using integrals or series (Chapter 5).

Example 1.1.7

Verify that $y(x) = e^{-x^3} \int_1^x e^{s^3}\, ds$ is a solution of the initial-value problem

$$\frac{dy}{dx} + 3x^2 y = 1, \qquad y(1) = 0. \tag{18}$$

Solution

First check whether $y(x) = e^{-x^3} \int_1^x e^{s^3}\, ds$ is a solution. Substituting $y(x)$ into the left-hand side of the differential equation (18) gives

$$\underbrace{\frac{d}{dx}\left(e^{-x^3} \int_1^x e^{s^3}\, ds\right)} + 3x^2 \left(e^{-x^3} \int_1^x e^{s^3}\, ds\right),$$

which, by the product rule for differentiation, is

$$\underbrace{\frac{d}{dx}(e^{-x^3}) \int_1^x e^{s^3}\, ds + e^{-x^3} \frac{d}{dx}\left(\int_1^x e^{s^3}\, ds\right)} + 3x^2 e^{-x^3} \int_1^x e^{s^3}\, ds$$

$$= \underbrace{-3x^2 e^{-x^3} \int_1^x e^{s^3}\, ds + e^{-x^3} e^{x^3}} + 3x^2 e^{-x^3} \int_1^x e^{s^3}\, ds$$

$$= e^{-x^3} e^{x^3} = 1.$$

Thus $y(x)$ is a solution. Next we verify that $y(x)$ satisfies the initial condition,

$$y(1) = e^{-1} \int_1^1 e^{s^3}\, ds = e^{-1} 0 = 0. \quad \blacksquare$$

Exercises

In Exercises 1 through 12, state whether the equation is an ordinary or partial differential equation. If it is an ordinary differential equation, state whether it is linear or nonlinear, and give its order.

1. $\dfrac{dx}{dt} + t^2 x = \sin t$

2. $\dfrac{\partial u}{\partial t} = \dfrac{\partial^2 u}{\partial x^2} + \dfrac{\partial^2 u}{\partial y^2}$

3. $\dfrac{dh}{dt} + \sin ht = t$

4. $x\dfrac{dx}{dt} + \cos\left(\dfrac{d^2 x}{dt^2}\right) = x$

5. $\dfrac{dy}{dx}\dfrac{d^2 y}{dx^2} = y$

6. $\dfrac{\partial x}{\partial t} = 3\sin\left(\dfrac{\partial^2 x}{\partial y^2}\right)$

7. $\dfrac{dy}{dx} = y + x$

8. $\dfrac{dy}{dx} = \dfrac{1}{y + x}$

9. $\dfrac{dx}{dy} = \dfrac{x}{y^2}$

10. $\dfrac{d^2 x}{dy^2} = \dfrac{x}{y+1}$

11. $\dfrac{dy}{dx} = \dfrac{x}{y^2}$

12. $\dfrac{d^2 y}{dx^2} = \left(\dfrac{dy}{dx}\right)^2$

In Exercises 13 through 22, verify that the indicated function is a solution of the differential equation for each value of the constants c and find the value of the constants such that the solution satisfies the initial condition(s). State whether or not the equation is linear.

13. $y' - 2y = 0, \qquad y = ce^{2x}, \qquad y(1) = 1$

14. $\dfrac{dy}{dx} + 2y = x, \qquad y = ce^{-2x} + \dfrac{x}{2} - \dfrac{1}{4}, \qquad y(0) = 1$

15. $y'' + y = 0, \qquad y = c_1 \sin x + c_2 \cos x,$
    $y(\pi) = 2, \qquad y'(\pi) = 3$

16. $\dfrac{d^2 y}{dx^2} + 4y = 4, \qquad y = 1 + c \sin 2x, \qquad y'(0) = 1$

17. $y\dfrac{dy}{dx} = 1, \qquad y = (2x + c)^{1/2}, \qquad y(0) = 4$

18. $y' - 2xy = 0, \qquad y = ce^{x^2}, \qquad y(0) = 3$

19. $y' + 4x^3 y = x$, $\quad y = e^{-x^4} \int_1^x s e^{s^4}\, ds + c e^{-x^4}$,
$y(1) = 2$

20. $y' + 2xy = 1$, $\quad y = e^{-x^2} \int_2^x e^{s^2}\, ds + c e^{-x^2}$,
$y(2) = 3$

21. $y' - (\sin x) y = 2$,
$y = e^{-\cos x} \int_0^x e^{\cos s}\, 2\, ds + c e^{-\cos x}$, $\quad y(0) = 3$

22. $y' + (\cos x) y = 1$,
$y = e^{-\sin x} \int_\pi^x e^{\sin s}\, ds + c e^{-\sin x}$, $\quad y(\pi) = 2$

In Exercises 23 through 28 determine for what values, if any, of the constant r, that the function $y = e^{rx}$ is a solution of the given differential equation.

23. $y' + 2y = 0$
24. $y'' - 5y' + 6y = 0$
25. $y'' + 6y' + 9y = 0$
26. $y''' + 3y'' + 2y' = 0$
27. $y'' - y = 0$
28. $y'' + 2y' = 0$

In Exercises 29 through 32, determine for what values of the constant r, $y = x^r$ is a solution of the differential equation.

29. $x^2 y'' - 3xy' + 3y = 0$
30. $x^2 y'' + 5xy' + 4y = 0$
31. $x^2 y'' + 2xy' = 0$
32. $x^2 y'' + xy' = 0$

33. Determine the general solution of $d^3 y/dx^3 = \sin x$.

34. Verify that $y = x^2 + c_1 x + c_2$ is the general solution of $y'' = 2$.

The set of functions given by $y = x + c_1$, where c_1 is an arbitrary constant, is sometimes called a *family* of functions. Note that $y = x + 3c_2$ yields the same family since $3c_2$ is an arbitrary constant if c_2 is one. For example, the family member $y = x - 3$ is obtained by taking $c_1 = -3$ or $c_2 = -1$. In Exercises 35 through 40 show that the two families are the same set of functions.

35. $y = 3x + c_1$, $\quad y = 3x - 5c_2$
36. $y = x^2 + 2c_1 + 1$, $\quad y = x^2 - 3c_2 + 5$
37. $y = c_1 x + c_2$, $\quad y = 3c_3 x - 4c_4$
38. $y = c_1 x^2 + c_1 + c_2$, $\quad y = 2c_3 x^2 + c_4$
39. $y = c_1 x^2 + c_2 + 3$, $\quad y = c_3 x^2 + c_3 + c_4$
40. $y = c_1 e^{c_2 x}$, $\quad y = 3c_3 e^{4c_4 x}$

In each of Exercises 41 through 46, verify that the expression in the right column implicitly defines a solution of the differential equation; c is a constant.

41. $\dfrac{dy}{dx} = -\dfrac{x}{y}$, $\quad 2x^2 + 2y^2 = 8$

42. $\dfrac{dr}{d\sigma} = \dfrac{r+1}{\sigma+1}$, $\quad \dfrac{r+1}{\sigma+1} = c$

43. $\dfrac{dy}{dx} = \dfrac{y^2}{x^2+1}$, $\quad x = \tan(c - 1/y)$

44. $\dfrac{dy}{dx} = \dfrac{x^3}{y^3}$, $\quad y^4 - x^4 = 76$

45. $\dfrac{dy}{dx} = \dfrac{x^2+1}{y^3+y}$, $\quad y^4 + 2y^2 = \dfrac{4x^3}{3} + 4x - 13$

46. $\dfrac{dy}{dx} = \dfrac{2x}{x^2+y^2-2y}$, $\quad \ln(x^2 + y^2) = y$

The differential equation
$$\frac{dy}{dx} = x$$
is linear, while
$$\left(\frac{dy}{dx}\right)^2 = x^2$$
is not. In fact, $(dy/dx)^2 = x^2$ has solutions that are not solutions of either
$$\frac{dy}{dx} = x \quad \text{or} \quad \frac{dy}{dx} = -x.$$
This is illustrated in Exercises 47 through 49.

47. Let
$$g(x) = \begin{cases} -\dfrac{x^2}{2} & \text{if } x \le 0, \\ \dfrac{x^2}{2} & \text{if } x > 0. \end{cases}$$
Graph $g(x)$ and show that $g(x)$ is differentiable for all x and that
$$g'(x) = \begin{cases} -x & \text{if } x \le 0, \\ x & \text{if } x > 0 \end{cases} = |x|.$$

48. Show that $g(x)$ from Exercise 47 is a solution of $(y')^2 = x^2$ but is not a solution of either $y' = x$ or $y' = -x$ on $(-\infty, \infty)$.

49. a) Find the general solution $y' = x$ and graph several particular solutions.
b) Find the general solution of $y' = -x$ and graph several particular solutions.
c) Graph $g(x)$ from Exercise 47 and compare the result to the graphs in parts (a) and (b).

We will see later that $y(t) = c_1 \cos t + c_2 \sin t$ is the general solution of $d^2 y/dt^2 + y = 0$. Exercises 50 through 53 use this information.

50. Find all solutions, if any, of the initial-value problem

$$y'' + y = 0, \quad y(0) = 3, \quad y'(0) = 4.$$

51. Find all solutions, if any, of the boundary-value problem

$$y'' + y = 0, \quad y(0) = 3, \quad y'(\pi) = 4.$$

52. Find all solutions, if any, of the boundary-value problem

$$y'' + y = 0, \quad y(0) = 3, \quad y'(\tfrac{\pi}{2}) = 4.$$

53. Find all solutions, if any, of the boundary-value problem

$$y'' + y = 0, \quad y(0) = 0, \quad y(\pi) = 0.$$

In Exercises 54 through 57, express the given law as a differential equation. In each case, consider x to be the dependent variable (unknown quantity) and t to be the independent variable.

54. x is the displacement, at time t, of an object from its equilibrium position. Its velocity is proportional to the square of its displacement.

55. x is the displacement, at time t, of an object from its equilibrium position. Its acceleration is proportional to the difference between its displacement and $\cos t$.

56. x is the displacement, at time t, of an object from its equilibrium position. Its velocity is inversely proportional to the difference between its displacement and t^3.

57. x is the displacement, at time t, of an object from its equilibrium position. Its acceleration at time t is inversely proportional to the difference between the cube and the square of the displacement.

58. Verify that $y = -2/(x^2 + c)$ and $y = 0$ are solutions of $dy/dx = xy^2$. Show that the family $y = -2/(x^2 + c)$ does not include the solution $y = 0$ for any choice of the constant c; $y = 0$ is called a *singular solution*.

2

First-Order Equations

2.1 Existence and Uniqueness

This chapter is devoted to first-order ordinary differential equations. The first two sections will establish several basic facts about first-order differential equations. These facts are not just of theoretical interest but are helpful in analyzing real problems. They often tell us whether our problem has a solution and whether our solution is the only solution.

In many applications, for example, electrical circuits, the applications of the basic laws, such as the loop or node laws, may lead to an *implicit* first-order differential equation of the form

$$F\left(x, y, \frac{dy}{dx}\right) = 0. \tag{1}$$

The equation is called *implicit* since the derivative is not explicitly solved for in terms of y, x. For example,

$$\left(\frac{dy}{dx}\right)^3 + y\frac{dy}{dx} = \sin x \tag{2}$$

is an implicit first-order differential equation. The problem of writing an implicit differential equation in the *explicit* form (sometimes called state-variable form

Figure 2.1.1

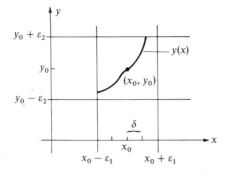

in engineering literature)

$$\frac{dy}{dx} = f(x, y) \tag{3}$$

can be a difficult problem mathematically, computationally, or both. In fact, there exist nonlinear circuits that cannot be written in the form (3). While important, such problems require sophisticated techniques and a thorough understanding of the explicit theory. Accordingly, we shall limit our discussion to equations that can be rewritten in the explicit form (3). The key result, which guarantees that there is even a solution to look for, is the following (here f_y denotes $\partial f/\partial y$):

Theorem 2.1.1   **Basic Existence and Uniqueness Theorem (technical version)** Let x_0, y_0 be real numbers such that (x_0, y_0) is in the domain of $f(x, y)$. Suppose that there exist positive numbers $\varepsilon_1 > 0$, $\varepsilon_2 > 0$ such that $f(x, y)$ and $f_y(x, y)$ are both continuous on the rectangle $\{(x, y): |x - x_0| < \varepsilon_1 \text{ and } |y - y_0| < \varepsilon_2\}$. Then there is a $\delta > 0$ such that, on the interval $|x - x_0| < \delta$, there is a unique solution to the initial-value problem:

$$\frac{dy}{dx} = f(x, y), \qquad y(x_0) = y_0. \tag{4}$$

(See Fig. 2.1.1.) A proof of Theorem 2.1.1 is given in Appendix C.

Theorem 2.1.1   **Basic Existence and Uniqueness Theorem (nontechnical version)** Let (x_0, y_0) be a point in the plane such that f and f_y are both continuous at and near (x_0, y_0). Then there is an interval containing x_0 and a unique solution $y(x)$ to $dy/dx = f(x, y)$, which is defined on this interval and whose graph passes through the point (x_0, y_0).

Note that, under the continuity conditions on f and f_y, we get two results:

1. There is a solution through (x_0, y_0).
2. The solution is unique on an interval containing x_0.

Figure 2.1.2
Some solutions of (5).

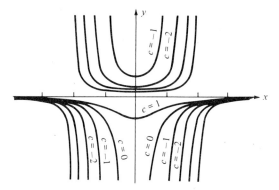

Example 2.1.1   Consider

$$\frac{dy}{dx} = xy^2. \tag{5}$$

Here $f(x, y) = xy^2$ and $f_y = 2xy$. Since these expressions are continuous everywhere, we see first of all that the solutions of (5) fill up the whole plane, that is, through every point (x_0, y_0) there is a solution. Second, the graphs of two different solutions can never cross, because if they were to cross, say at (x_0, y_0), then the uniqueness property would be violated at (x_0, y_0).

While the Basic Existence and Uniqueness Theorem guarantees a solution of (5) through any point (x_0, y_0), it does not say that the solution is defined for all x. For example, it is easy to verify that

$$y = \frac{2}{1 - x^2}$$

is a solution of (5) with $y(0) = 2$ ($x_0 = 0$, $y_0 = 2$).

However, this solution is not defined for all x. Using the techniques of Section 2.5, Separation of Variables, it is possible to show that $y = 0$ and $y = -2/(x^2 + c)$ gives all solutions of (5). Figure 2.1.2 shows several solutions of (5). Solutions have vertical asymptotes if $c \leq 0$. ∎

If the conditions of the Basic Existence and Uniqueness Theorem are not met at a point (x_0, y_0), then solutions may not exist at (x_0, y_0), there may be more than one solution passing through the same point (x_0, y_0), or the solution is not differentiable at x_0 (Exercise 12 at the end of this section). Such behavior has a physical interpretation in some boundary-value problems and will be discussed in Chapter 5, which deals with series solutions.

To illustrate this, we consider two examples.

Figure 2.1.3
Solutions of (6).

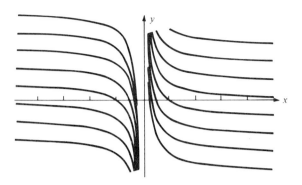

Example 2.1.2

The differential equation

$$\frac{dy}{dx} = -\frac{1}{x^2} \tag{6}$$

satisfies the Basic Existence and Uniqueness Theorem everywhere that $x_0 \neq 0$, since $f(x, y) = -1/x^2$ is continuous everywhere, $x \neq 0$, and $f_y = 0$ is continuous (except at $x = 0$ where the limit defining the partial is undefined). By anti-differentiation, the general solution of (6) is

$$y = \frac{1}{x} + c.$$

This family of solutions is shown in Fig. 2.1.3. Note that no solution ever touches the y-axis ($x = 0$). There is not a solution through any point (x_0, y_0) with $x_0 = 0$. ∎

Example 2.1.3

The functions $y = x$, and $y = -x$ both satisfy

$$y\frac{dy}{dx} = x, \qquad y(0) = 0.$$

The differential equation can be written as

$$\frac{dy}{dx} = \frac{x}{y}.$$

This is in the form (4) with $f(x, y) = x/y$, and $f(x, y)$ is not continuous at $y = 0$. ∎

We conclude this section with two additional examples of situations in which Theorem 2.1.1 does not hold.

Example 2.1.4

Determine the points (x_0, y_0) for which the differential equation

$$\frac{dy}{dx} = \frac{x^2 + y}{x - y}$$

satisfies the Basic Existence and Uniqueness Theorem.

Solution

In this problem

$$\frac{dy}{dx} = f(x, y),$$

where

$$f(x, y) = \frac{(x^2 + y)}{(x - y)}.$$

Both f and $f_y = (x + x^2)/(x - y)^2$ are discontinuous, in fact, undefined if $x = y$. Thus the Basic Existence and Uniqueness Theorem holds only if $x_0 \neq y_0$. If $x_0 \neq y_0$, then there is a (small) rectangle centered at (x_0, y_0) on which f and f_y are continuous. ∎

Example 2.1.5 Determine for what points (x_0, y_0) the differential equation

$$\frac{dy}{dx} = (2x - y)^{1/3}$$

satisfies the Basic Existence and Uniqueness Theorem.

Solution

In this problem $dy/dx = f(x, y)$, where $f(x, y) = (2x - y)^{1/3}$. Then

$$\frac{\partial f}{\partial y} = -\frac{1}{3}(2x - y)^{-2/3}.$$

The function f is continuous everywhere. However, f_y is discontinuous if $2x - y = 0$. Thus the Basic Existence and Uniqueness Theorem holds only if $2x_0 \neq y_0$. ∎

Exercises

For each of the differential equations in Exercises 1 through 10, give those points (x_0, y_0) in the xy-plane for which the Basic Existence and Uniqueness Theorem guarantees that a unique solution exists.

1. $\dfrac{dy}{dx} = \dfrac{y}{1 + x^2}$

2. $\dfrac{dy}{dx} = \dfrac{x - y}{3x - 7y}$

3. $\dfrac{dy}{dx} = (1 - x^2 - y^2)^{7/3}$

4. $\dfrac{dy}{dx} = (1 - x^2 - y^2)^{1/3}$

5. $\dfrac{dy}{dx} = (y + x)^{1/5}$

6. $x\dfrac{dy}{dx} = y^2 + 1$

7. $(y - 1)\dfrac{dy}{dx} = \cos x$

8. $\dfrac{dy}{dx} = \dfrac{x^3 + 1}{x(y + 1)}$

9. $\dfrac{dy}{dx} = (1 - x^2 - 2y^2)^{3/2}$

10. $\dfrac{dy}{dx} = y^{1/3}\sqrt{y - 1}$

Exercises 11 and 12 refer to the differential equation

$$\frac{dy}{dx} = x^{-1/3} \qquad (7)$$

11. Show that the Basic Existence and Uniqueness Theorem holds for (7) for all (x_0, y_0) such that $x_0 \neq 0$.

12. Show that, for any constant y_0, there exists a unique function $y(x)$ such that
   a) y satisfies (7) if $x \neq 0$,
   b) y is continuous for all x,
   c) $y(0) = y_0$,
   d) y is not differentiable at $x_0 = 0$

In Exercises 13 through 19, you are given a family of curves $g(x, y) = c$. For each exercise
   a) Verify that the curves implicitly define solutions of the given differential equation.
   b) Graph the curves on the same axis for the indicated values of c.
   c) Determine for which points (x_0, y_0) the assumptions of the Basic Existence and Uniqueness Theorem fail to hold, and graph these points on the same graph.

13. $x^2 - y^2 = c$, $\quad \dfrac{dy}{dx} = \dfrac{x}{y}$, $\quad c = 0, \pm 1, \pm 2$.

14. $x^2 + y^2 = c$, $\quad \dfrac{dy}{dx} = -\dfrac{x}{y}$, $\quad c = 1, 2, 3$.

15. $x + y^2 = c$, $\quad 2y\dfrac{dy}{dx} + 1 = 0$, $\quad c = 0, \pm 1$.

16. $y = cx$, $\quad \dfrac{dy}{dx} = \dfrac{y}{x}$, $\quad c = 0, \pm 1, \pm 2$.

17. $y = c \sin x$, $\quad \dfrac{dy}{dx} = y \cot x$, $\quad c = 0, \pm 1, \pm 2$.

18. $y = \tan(x + c)$, $\quad \dfrac{dy}{dx} = 1 + y^2$, $\quad c = 0, \pm 1$.

19. $y = \dfrac{1}{x + c}$, $\quad \dfrac{dy}{dx} = -y^2$, $\quad c = 0, \pm 1$.

20. a) Show that $y = 0$ and $y = (\tfrac{2}{3}x + c)^{3/2}$ are solutions of
$$y' = y^{1/3}. \tag{8}$$
   b) Show that there are at least two solutions of (8) through every point (x_0, y_0) with $y_0 = 0$.
   c) Sketch several solutions of (8), including $y = 0$, on the same graph.
   d) Note that $f(x, y) = y^{1/3}$ is continuous everywhere. Why aren't this fact and (b) a contradiction of the Basic Existence and Uniqueness Theorem?

21. a) Show that $y = (\tfrac{4}{5}x + c)^{5/4} + 1$ and $y = 1$ are solutions of
$$y' = (y - 1)^{1/5} \tag{9}$$
   b) Show that there are at least two solutions of (9) through every point (x_0, y_0) with $y_0 = 1$.
   c) Sketch several solutions of (9), including $y = 1$, on the same graph.
   d) Note that $f(x, y) = (y - 1)^{1/5}$ is continuous everywhere. Why aren't this fact and (b) a contradiction of the Basic Existence and Uniqueness Theorem?

2.2 Direction Fields

We can often get a good pictorial or graphic idea of what the solutions of a differential equation look like without actually solving the differential equation. Take as an example

$$\frac{dy}{dx} = x(y - 1). \tag{1}$$

We shall give several methods for solving this equation later. Right now we want to consider qualitative behavior based on tangent lines. This approach is easily implemented on a computer.

By the Basic Existence and Uniqueness Theorem (Theorem 2.1.1), there is a solution to $dy/dx = x(y - 1)$ through every point in the plane since

2.2 Direction Fields

$$f(x, y) = x(y - 1) \quad \text{and} \quad f_y(x, y) = x$$

are continuous everywhere. Take one such point, say (3, 7). There is a solution through this point. While we do not know the formula for this solution, we do know the slope of its tangent line at this point. It is given by the differential equation, since (1) tells us that

$$\text{at} \quad x = 3, \quad y = 7, \quad \frac{dy}{dx} = 3(7 - 1) = 18. \qquad (2)$$

We can now draw the tangent at (3, 7) to the solution through (3, 7) as a short line segment with slope 18. Continuing in this manner at other points, we can build up a picture of the solutions.

The above suggests a procedure that is fairly easy to implement on a computer with plot or graphic capabilities. In its simplest form it would be

1. Set up a grid (set of points) in part of the xy-plane. Call the grid points $(x_1, y_1), \ldots, (x_n, y_n)$.
2. Write the differential equation as $dy/dx = f(x, y)$.
3. At each grid point (x_i, y_i), draw a line segment with slope $f(x_i, y_i)$. This is the tangent line at (x_i, y_i) to the solution through (x_i, y_i).

Figure 2.2.1

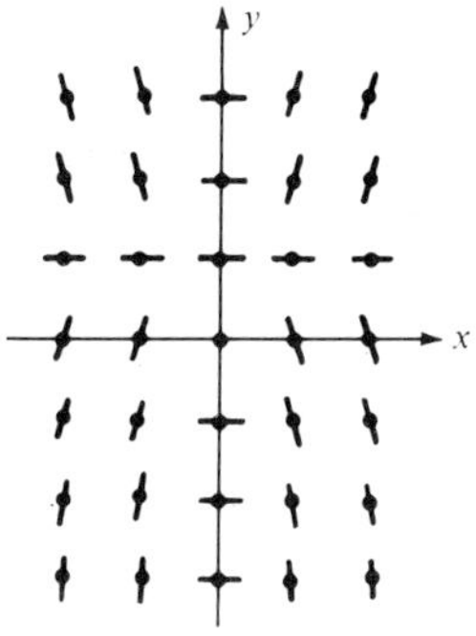

As a simple illustration, consider again $dy/dx = x(y - 1)$ and take the grid (x, y) with $x = 0, \pm 1, \pm 2, y = 0, \pm 1, \pm 2, \pm 3$. There are 35 points in this grid. The resulting plot, shown in Fig. 2.2.1, is sometimes referred to as a sketch of the *direction* or *vector field*. If a finer grid is chosen, say $x = 0, \pm 0.5, \pm 1, \pm 1.5, \pm 2, y = 0, \pm 0.5, \pm 1, \pm 1.5, \pm 2, \pm 2.5, \pm 3$, then there are 117 grid points. The resulting plot is given in Fig. 2.2.2(a), which suggests that the solutions may look like the graphs in Fig. 2.2.2(b).

From a plot such as that in Fig. 2.2.2(a), we can often get a good idea of what the solutions look like. This procedure, however, is not much different than that of determining the graph of a function by plotting a number of points. How do we know how fine a grid to choose? How do we know that the solutions

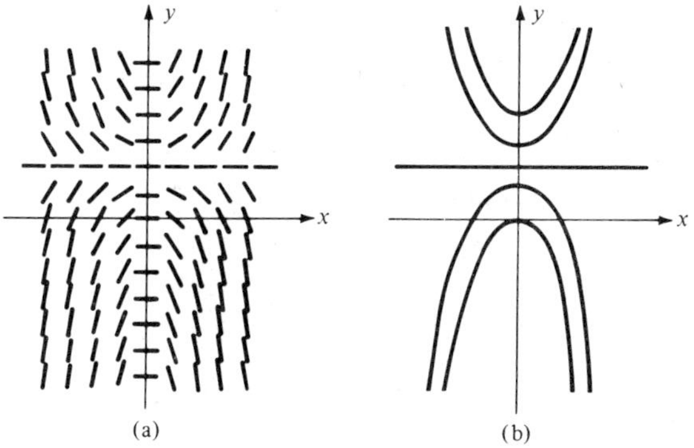

Figure 2.2.2

(a)   (b)

do not change their behavior at a distance from the grid? In practice, the simple procedure is usually combined with the following approach.

Procedure for Sketching the Direction Field

1. Determine at which points the Basic Existence and Uniqueness Theorem does not hold.

   Comment This is important numerically since near these points the slope may be large in absolute value, and a numerical method could run into overflow or underflow, that is, get numbers too large (or small) for the computer to handle. Analytically, the solutions may become discontinuous or cross at such points, so we need to be warned about their presence.

2. Determine whether there are any constant solutions, $y = $ constant.

   Comment Such constant or *equilibrium* solutions are often important in applications. When present, they also help in describing the behavior of other solutions, as will be seen shortly.

3. Determine the points where $dy/dx = 0$. At these points there will be a horizontal tangent.

4. The result of steps 1, 2, and 3 will be to split the xy-plane into regions where dy/dx does not change sign. If $dy/dx > 0$ in a region, all solutions are increasing while in that region. If $dy/dx < 0$ in a region, then all solutions are decreasing while in that region.

After step 4, we have several options. We may have enough information already, or several tangent lines may be graphed as discussed earlier. Possibly, we could perform a more detailed analysis checking such behavior as concavity.

Example 2.2.1

Sketch the solution curves of $dy/dx = x(y - 1)$, which is the same differential equation given in Eq. (1), using steps 1 through 4.

Solution

We shall go through the steps one at a time.

1. We have $f(x, y) = x(y - 1)$ and $f_y = x$. These are both continuous everywhere so there is a unique solution through every point.

2. Let $y = c$ and substitute into $y' = x(y - 1)$ to get $0 = x(c - 1)$. Since y is to be a constant function, $0 = x(c - 1)$ is to hold for all x. Thus $c = 1$ and $y = 1$ is a constant or equilibrium solution.

3. $dy/dx = x(y - 1) = 0$ if $x = 0$ or $y = 1$.

Figure 2.2.3

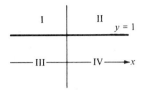

Figure 2.2.4

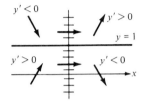

4. Drawing the lines $x = 0$, $y = 1$ in accordance with steps 2 and 3, we find that the plane is now split into four regions, as shown in Fig. 2.2.3. In each region, $y' = x(y - 1)$ is always positive or always negative.

By checking a value of y' in each region, we find that $y' < 0$ in regions I and IV, while $y' > 0$ in regions II and III. All of this information is summarized in Fig. 2.2.4, where some of the horizontal tangents along the $x = 0$ line have been drawn in.

Solutions $y = y(x)$ in region I decrease. By the Basic Existence and Uniqueness Theorem, they may not cross the solution $y = 1$. When they reach $x = 0$, the tangent line is horizontal. As x increases past zero, $y' > 0$ and the solutions increase. Similarly solutions starting in Region III first increase, then cross the y-axis, and then decrease. They may not cross the solution $y = 1$. We have then the qualitative picture given in Fig. 2.2.5. We could get a more detailed illustration by sketching in tangents as we did in Fig. 2.2.2. Figure 2.2.5 should be compared to Fig. 2.2.2(a) and (b). ∎

Concavity

As a quick check of Fig. 2.2.5, we shall compute the concavity of the solution of (1). Recall that if $y'' > 0$, then the curve is concave up. If $y'' < 0$, then the curve is concave down. (Assuming y'' exists), we can differentiate the original differential equation $y' = x(y - 1)$ with respect to x (remember y depends on x) to get

$$y'' = (y - 1) + x\, y'. \qquad (2)$$

Substituting $y' = x(y - 1)$ back into (2), we get

$$y'' = (y - 1) + x\, x(y - 1) = (1 + x^2)(y - 1).$$

Thus $y'' > 0$, and the solution is concave up at (x_0, y_0) if $y_0 > 1$, whereas $y'' < 0$ and the solution is concave down at (x_0, y_0) if $y_0 < 1$. This is consistent with Fig. 2.2.5.

Comment To utilize y'' we need to know it exists. If $y' = f(x, y)$ and f, f_x, f_y are all continuous, then the solution y will be twice differentiable (see Exercise 17 at the end of this section).

Example 2.2.2

Sketch the solutions of

$$y' = (y - 1)(y - 2). \qquad (3)$$

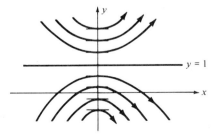

Figure 2.2.5

Solution

For Eq. (3), $y' = f(x, y) = (y - 1)(y - 2)$. Solutions go through every point and never cross since $f(x, y)$ and $f_y = 2y - 3$ are continuous everywhere. Next we look for constant solutions $y = c$. Substituting $y = c$ into (3) gives $0 = (c - 1)(c - 2)$ or $c = 1$ and $c = 2$ or $y = 1, y = 2$. If $y \neq 1$ and $y \neq 2$, then $y' \neq 0$, so there are no other horizontal tangents. The two equilibrium solutions $y = 1$ and $y = 2$ split the plane into three regions

$$\text{I} = \{(x_0, y_0): y_0 > 2\}, \quad \text{where } y' > 0,$$
$$\text{II} = \{(x_0, y_0): 1 < y_0 < 2\}, \quad \text{where } y' < 0,$$

and

$$\text{III} = \{(x_0, y_0): y_0 < 1\}, \quad \text{where } y' > 0.$$

This information is summarized in Fig. 2.2.6.

Note that the slope depends only on the y-coordinate. As an added piece of information, we compute y'' by differentiating (3) with respect to x, to get

$$y'' = y'(y - 2) + (y - 1)y'. \quad (4)$$

Using the original equation $y' = (y - 1)(y - 2)$, we substitute $(y - 1)(y - 2)$ for y' in (4) to obtain

$$(y - 1)(y - 2)(y - 2) + (y - 1)(y - 1)(y - 2) = (y - 1)(y - 2)(2y - 3).$$

The solutions thus are

concave up, $y'' > 0$ if $1 < y < \frac{3}{2}$ or $y > 2$;
concave down, $y'' < 0$ if $y < 1$ or $\frac{3}{2} < y < 2$.

Combining all this information, we have:

$y < 1$: solutions concave down, increasing;
$1 < y < \frac{3}{2}$: solutions concave up, decreasing;
$\frac{3}{2} < y < 2$: solutions concave down, decreasing;
$2 < y$: solutions concave up, increasing;
Solutions may not cross $y = 1$ or $y = 2$.

The final result is shown in Fig. 2.2.7(a). For comparison, the vector field is shown in Fig. 2.2.7(b). ∎

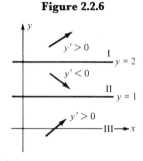

Figure 2.2.6

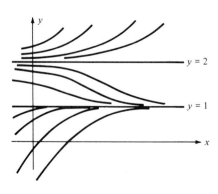

Figure 2.2.7(a)

Figure 2.2.7(b)

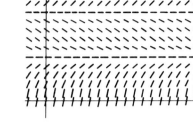

In Example 2.2.2, $y = 1$ is sometimes called a *stable equilibrium* or an *attractor*, since nearby solutions move toward it as x increases. On the other hand, $y = 2$ is called an *unstable equilibrium* since nearby solutions move away from it. If solutions on one side of an equilibrium move toward the equilibrium and solutions on the other side move away, then the equilibrium is called *semistable*.

Most physically observed equilibriums are stable for the following reason: every physical system is subject to occasional small perturbations which our equations did not take into account. The effect of these perturbations is to move our system slightly—onto a nearby solution. If this new solution moves the system back toward equilibrium, then it will seem to us that our physical system is "maintaining equilibrium."

Example 2.2.3

This example is somewhat more complicated. Sketch the solutions of

$$\frac{dy}{dx} = \frac{1 - x^2 - y^2}{y - x + 2}. \tag{5}$$

Solution

The function

$$f(x, y) = \frac{1 - x^2 - y^2}{y - x + 2}$$

is discontinuous if $y - x + 2 = 0$; f_y is also discontinuous along this same line, shown dashed in Fig. 2.2.8 to indicate that we do not know what happens there.

Figure 2.2.8

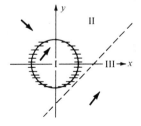

To check for equilibria, let us substitute $y = c$ into Eq. (5). Since $dc/dx = 0$, the substitution yields

$$0 = \frac{1 - x^2 - c^2}{c - x + 2} \quad \text{or} \quad x^2 = 1 - c^2.$$

Since x is a variable and c a constant, this result is impossible and there are no constant solutions. Next let us look for horizontal tangents. Setting $dy/dx = 0$ yields

Figure 2.2.9

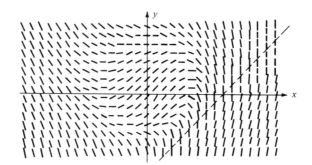

Figure 2.2.10

$$0 = \frac{1 - x^2 - y^2}{y - x + 2} \quad \text{or} \quad 0 = 1 - x^2 - y^2.$$

The points (x, y) for which $1 - x^2 - y^2 = 0$ form a circle of radius 1 centered at the origin. At each point on the circle, $dy/dx = 0$, and the tangent is horizontal. The circle $x^2 + y^2 = 1$ and the line $y - x + 2 = 0$ split the plane into the three regions shown in Fig. 2.2.8.

In each region, dy/dx is always either positive or negative, so it suffices to take one point in each region and check the slope there. In region I, say at $(0,0)$, we have

$$\text{I:} \quad \frac{dy}{dx} = \frac{1 - 0 - 0}{0 - 0 + 2} = \frac{1}{2} > 0.$$

In region II, say at $(0,2)$,

$$\text{II:} \quad \frac{dy}{dx} = \frac{1 - 0 - 4}{2 - 0 + 2} = -\frac{3}{4} < 0.$$

In region III, say at $(3,0)$,

$$\text{III:} \quad \frac{dy}{dx} = \frac{1 - 9 - 0}{0 - 3 + 2} = \frac{-8}{-1} > 0.$$

This information is summarized in Fig. 2.2.8. The direction field is shown in Fig. 2.2.9, which suggests the approximate sketch provided in Fig. 2.2.10. The

Figure 2.2.11

actual graph is given in Fig. 2.2.11. In order to obtain Fig. 2.2.11, we must examine what happens as the solutions approach the line $y - x + 2 = 0$. See Exercises 37 through 39 at the end of this section. ∎

Exercises

In Exercises 1 through 16, sketch several solution curves for each of the differential equations given below. Use the simple method illustrated in Figs. 2.2.1, 2.2.2, and 2.2.9 as well as the analytical method described in the other examples.

1. $y' = y(y - 1)$
2. $y' = y^2(y - 1)^2$
3. $y' = (y - 1)(y - 2)^2$
4. $y' = (y^2 - 1)(y^2 - 3)$
5. $y' = 3$
6. $y' = x^2 - y^2$
7. $y' = \dfrac{x - y}{2x + y}$
8. $y' = x^2 + y^2$
9. $y' = \dfrac{y - x^2}{1 + x^2}$
10. $y' = \dfrac{y - x^2}{x}$
11. $y' = \dfrac{-1 + x^2 + 4y^2}{y - 5x + 10}$
12. $y' = \dfrac{yx - y}{x}$
13. $y' = \sin y$
14. $y' = \tan y$
15. $y' = y(y - x)^2$
16. $y' = y(1 - x^2 - y^2)$

17. Show that if $f(x, y)$, $f_x(x, y)$, and $f_y(x, y)$ are all continuous on a rectangle containing (x_0, y_0), then the solution of $y' = f(x, y)$ is twice differentiable at x_0.

Exercises 18 through 22 refer to the differential equation

$$y' = g(y), \qquad (6)$$

where g is a twice differentiable function of y.

18. Show that $y = c$ is an equilibrium state of (6) if and only if $g(c) = 0$.

19. Show that if $g(c) = 0$ and $g'(c) < 0$, then the equilibrium $y = c$ is stable.

20. Show that if $g(c) = 0$ and $g'(c) > 0$, then the equilibrium $y = c$ is unstable.

21. Show that if $g(c) = 0$, $g'(c) = 0$, and $g''(c) > 0$, then solutions close to the equilibrium $y = c$ will approach the equilibrium if they are below it and move away if they are above it. (The equilibrium is semistable.)

22. Show that if $g(c) = 0$, $g'(c) = 0$, and $g''(c) < 0$, then solutions close to the equilibrium $y = c$ will approach the equilibrium if they are above it and move away if they are below it. (The equilibrium is semistable.)

In Exercises 23 through 28, use Exercises 18 through 22 to determine the stability, or lack of it, for the equilibria of the differential equation. Then compare your answer to the sketch found in the indicated exercise.

23. $y' = y(y - 1)$ (Exercise 1)
24. $y' = y^2(y - 1)^2$ (Exercise 2)
25. $y' = (y - 1)(y - 2)^2$ (Exercise 3)
26. $y' = (y^2 - 1)(y^2 - 3)$ (Exercise 4)
27. $y' = \sin y$ (Exercise 13)
28. $y' = \tan y$ (Exercise 14)

Another way to gain qualitative information about the solutions of a differential equation is through the use of isoclines. An *isocline* for $y' = f(x, y)$ is a curve along which the slope of the tangent of the solutions is constant, that is, $f(x, y) = c$. We have already found the horizontal tangents, $f(x, y) = 0$. This is the zero isocline. Similarly, $f(x, y) = 1$ and $f(x, y) = -1$ would be the points (x, y) at which the solutions have tangents with slopes 1 and -1, respectively.

29. Verify that the isoclines for $y' = x(y - 1)$ of $y' = 1$, $y' = -1$, and $y' = 0$ are the curves A, B, and C, respectively, shown in Fig. 2.2.12. The tangents to solutions of $y' = x(y - 1)$ at points along the isoclines are also drawn in Fig. 2.2.12. Compare Fig. 2.2.12 with Fig. 2.2.5. Note that the solutions of the differential equation move across the isoclines in the directions indicated by the tangents. An example is D in Fig. 2.2.12.

Figure 2.2.12

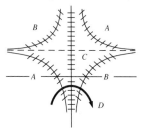

In Exercises 30 through 36, find the isoclines of the given differential equation and sketch them on the same axes. Try to sketch the solutions of the differential equation from this information.

30. $y' = x^2 - y^2$, $\quad y' = 0$, $\quad y' = 1$, $\quad y' = -1$
31. $y' = x^2 + y^2$, $\quad y' = 0$, $\quad y' = 1$, $\quad y' = 2$
32. $y' = x + y^2$, $\quad y' = 0$, $\quad y' = 1$, $\quad y' = -1$
33. $y' = xy$, $\quad y' = 0$, $\quad y' = 1$, $\quad y' = -1$
34. $y' = xy^2$, $\quad y' = 0$, $\quad y' = 1$, $\quad y' = -1$
35. $y' = x + y$, $\quad y' = 0$, $\quad y' = 1$, $\quad y' = -1$
36. $y' = x - y$, $\quad y' = 0$, $\quad y' = 1$, $\quad y' = -1$

Exercises 37 and 38 concern the differential equation

$$y' = \frac{g(x, y)}{1 - x + y}, \qquad (7)$$

where $g(x, y)$ is a differentiable function of x and y and $g(x, y) \neq 0$ for all (x, y) such that $1 - x + y = 0$.

37. Suppose that there is a $\delta > 0$ such that $g(x, y) < 0$ if $\delta > 1 - x + y > 0$. Show that if (x_0, y_0) is such that $\delta > 1 - x_0 + y_0 > 0$, then the solution to (7) starting at (x_0, y_0) intersects the $1 - x + y = 0$ line and has a vertical tangent at that point (Fig. 2.2.13).

Figure 2.2.13

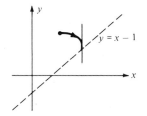

38. Suppose that there is a $\delta > 0$ such that $g(x, y) < 0$ for (x, y), such that $-\delta < 1 - x + y < 0$. Show that if (x_0, y_0) is such that $-\delta < 1 - x_0 + y_0 < 0$, then the solution to (7) starting at (x_0, y_0) intersects the $1 - x + y = 0$ line and has a vertical tangent there (Fig. 2.2.14).

Figure 2.2.14

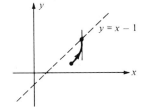

39. Use the ideas of Exercises 37 and 38 to show that Fig. 2.2.10 may be completed to give Fig. 2.2.11.

2.3 First-Order Linear Equations

The first-order linear equation

$$a_1(x)\frac{dy}{dx} + a_0(x)y = h(x) \qquad (1)$$

with *coefficients* a_1, a_0 is important in many applications. Most phenomena are not actually linear. However, if the solutions of interest do not differ too greatly from a given function (or operating point), the equations may often be approximated by a linear equation. This fundamental idea is investigated in Exercises 28 through 31 at the end of this section.

Equation (1) can be rewritten in the form

$$\frac{dy}{dx} = f(x, y), \quad \text{where } f(x, y) = \frac{h(x) - a_0(x)y}{a_1(x)}. \tag{2}$$

In order to apply the Basic Existence and Uniqueness Theorem 2.1 to (1) we compute that

$$\frac{\partial f}{\partial y} = \frac{-a_0(x)}{a_1(x)}.$$

The functions f, f_y will be continuous as long as $a_0(x)$, $a_1(x)$, $h(x)$ are continuous, and $a_1(x)$ is nonzero. The Basic Existence and Uniqueness Theorem of Section 2.1 thus provides the first half of the following useful fact.

Theorem 2.3.1   **Existence and Uniqueness Theorem for First-Order Linear Equations**
Suppose $a_0(x)$, $a_1(x)$, $h(x)$ are continuous on the interval I and $a_1(x)$ is never zero on I. Then, if $x_0 \in I$, and y_0 is any real number, there exists a unique solution to

$$a_1(x)\frac{dy}{dx} + a_0(x)y = h(x)$$

such that $y(x_0) = y_0$. Furthermore, this unique solution is defined on all of the interval I. (The interval I may be finite or infinite, and open, closed, or half-open.)

Comment That the unique solution is defined on all of the interval I is a special property of linear equations. For example, $y = (1 - x)^{-1}$ is a solution of the nonlinear initial-value problem $dy/dx = y^2$, $y(0) = 1$ and $f(x, y) = y^2$ and $f_y(x, y) = 2y$ are continuous everywhere, but the solution is not defined on all of $[0, 2]$.

Points x where $a_1(x) = 0$ are called *singular points* and will be studied more fully in Chapter 5. [Example 2.3.1 will have a singular point.]

We shall now develop a method of solution for first-order linear equations. This method will introduce several important ideas. It will also provide a proof of the entire Basic Existence and Uniqueness Theorem for first-order linear equations.

First, divide Eq. (1) by $a_1(x)$, to get

$$\frac{dy}{dx} + p(x)y = g(x), \tag{3}$$

where $p(x) = a_0(x)/a_1(x)$ and $g(x) = h(x)/a_1(x)$.

The only expressions we can antidifferentiate are those that we recognize as derivatives. Can we multiply (3) by a function of x, call it $u(x)$, so that the left-hand side turns into a derivative? The answer is yes. Such a function $u(x)$ is called an *integrating factor*.

Key Fact If $u(x) = e^{\int p(x)dx}$, then

$$u(x)\left(\frac{dy}{dx} + p(x)y\right) = \frac{d(uy)}{dx}. \tag{4}$$

Verification Let $u(x) = \exp[\int p(x)\,dx]$. Note that

$$\frac{du}{dx} = e^{\int p(x)dx}\frac{d\int p(x)\,dx}{dx} = e^{\int p(x)dx}p(x) = u(x)p(x). \tag{5}$$

Then

$$\frac{d(uy)}{dx} = \frac{du}{dx}y + u\frac{dy}{dx} \quad \text{(product rule)}$$

$$= up(x)y + u\frac{dy}{dx} = u\left(\frac{dy}{dx} + p(x)y\right) \quad \text{[by (5)]. ∎}$$

Technical Point The above derivation requires some assumption on $p(x)$. Continuity of $p(x)$ would suffice.

Using the key fact, we can now solve $dy/dx + p(x)y = g(x)$. Multiplying both sides of (3) by the integrating factor $u(x) = \exp[\int p(x)\,dx]$, we get

$$e^{\int p(x)dx}\left(\frac{dy}{dx} + p(x)y\right) = e^{\int p(x)dx}g(x). \tag{6}$$

Using the key fact, the left-hand side of (6) may be rewritten to give

$$\frac{d}{dx}(e^{\int p(x)dx}y) = e^{\int p(x)dx}g(x).$$

Antidifferentiate both sides with respect to x, to get

$$e^{\int p(x)dx}y = \int e^{\int p(x)dx}g(x)\,dx + C,$$

where C is an arbitrary constant. Then solve for y to get

$$y = e^{-\int p(x)dx}\int e^{\int p(x)dx}g(x)\,dx + Ce^{-\int p(x)dx}. \tag{7}$$

Formula (7) gives the *general solution* of the differential equation (3). That is, by appropriate choice of the constant C, Eq. (7) yields every solution of the differential equation. Also, (7) is a solution for every value of C.

While it is possible to memorize formula (7), it is usually easier to go through the solution procedure which may be summarized as follows.

Summary of First-Order Linear Procedure

a. Rewrite the differential equation as

$$\frac{dy}{dx} + p(x)y = g(x). \tag{8}$$

b. Compute the integrating factor

$$u(x) = e^{\int p(x)\,dx}.$$

c. Multiply both sides of (8) by $u(x)$ to get (by the Key Fact) that

$$\frac{d}{dx}(uy) = ug.$$

d. Antidifferentiate both sides with respect to x, being sure to introduce the arbitrary constant.

$$uy = \int ug\, dx + C.$$

e. If there are initial conditions, use them to find C.

f. Solve for y.

Comment In step (b) we need only find *an* integrating factor. Thus we do not need to introduce an arbitrary constant in evaluating $\int p(x)\, dx$. We leave it to an exercise to show that introducing an arbitrary constant in the integrating factor does not change the final solution.

Example 2.3.1 Find all solutions of

$$xy' - y = x^2, \qquad x > 0. \tag{9}$$

Solution First rewrite the equation as

$$\frac{dy}{dx} - \frac{1}{x}y = x \tag{10}$$

by dividing by x. Then $p(x) = -x^{-1}$. Thus

$$u = \exp\left[\int -x^{-1}\, dx\right] = \exp(-\ln x) = x^{-1}.$$

Upon multiplication by x^{-1}, the rewritten differential equation (10) becomes by step (c),

$$\frac{d}{dx}(x^{-1}y) = x^{-1}x = 1.$$

Antidifferentiating with respect to x gives

$$x^{-1}y = x + C$$

or

$$y = x^2 + Cx. \qquad (11)$$

Warning Note that in step (a) the equation is $y' + p(x)y = g(x)$. Equation (10), however, is $y' + [-1/x]y = x$. Thus we must include the minus sign in $p(x)$ so that $p(x) = -1/x$.

Before continuing, let us look at Example 2.3.1 more carefully. If we solve (9) for $x < 0$, then we again get (10) so that

$$u = e^{\int -x^{-1}dx} = e^{-\ln|x|} = |x|^{-1} = -x^{-1} \qquad \text{for } x < 0.$$

We could proceed as in Example 2.3.1, or note that if $u = -x^{-1}$ is an integrating factor, then $-u = x^{-1}$ is also an integrating factor (see Exercise 32 at the end of this section) and $y = x^2 + Cx$ is a solution for $x < 0$ also. Thus (11) gives differentiable solutions of (9) for all x, including $x = 0$, where the assumptions of the Existence and Uniqueness Theorem do not hold. However, $y(0) = 0$ for all solutions of (9), so that there does not exist a solution with $y(0) = y_0$, $y_0 \neq 0$. Also the solution for the initial condition $y(0) = 0$ is not unique.

In general, the solutions of (1) may not be differentiable or even defined at a singular point. Also, the arbitrary constant can sometimes take on different values on the two sides of the singular point (see Exercises 13, 14, 15, and 33).

Until now we have usually written our differential equations in terms of the variables y, x. It is important to be able to work with whatever variables are given. In later chapters we shall often use t as the independent variable.

Example 2.3.2 Find all solutions of

$$2\frac{dr}{d\theta} + r = 3.$$

Solution First rewrite the equation in the form (3),

$$\frac{dr}{d\theta} + \frac{1}{2}r = \frac{3}{2}, \qquad (12)$$

by dividing by 2. Then $p(\theta) = \frac{1}{2}$. The integrating factor is

$$u(\theta) = e^{\int 1/2 \, d\theta} = e^{\theta/2}.$$

On multiplication by $e^{\theta/2}$, the rewritten equation (12) becomes

$$\frac{d}{d\theta}(re^{\theta/2}) = \tfrac{3}{2}e^{\theta/2}.$$

Antidifferentiating with respect to θ gives

$$re^{\theta/2} = \frac{3}{2}\frac{e^{\theta/2}}{\frac{1}{2}} + C$$

or
$$r = 3 + Ce^{-\theta/2}. \blacksquare$$

In the applications of Sections 2.9 through 2.13, the functions $p(x)$, $g(x)$ in $dy/dx + p(x)y = g(x)$ will have physical interpretations as capacitances, flow rates, voltage sources, etc. In such problems the functions p, g and the solution y can be given by different formulas for different values of x.

From physical considerations, we often know that the solution y must be continuous. These ideas will be discussed more fully later. For now they emphasize the importance of learning how to do problems like the next example.

Example 2.3.3 Find the continuous solution for $x \geq 0$ of

$$\frac{dy}{dx} + 2y = g(x), \quad y(0) = 0, \tag{13}$$

where

$$g(x) = \begin{cases} 1, & 0 \leq x < 2, \\ 0, & 2 \leq x. \end{cases}$$

Solution We actually have a two-part problem here.

Part 1
For $0 \leq x < 2$, the differential equation is

$$\frac{dy}{dx} + 2y = 1, \quad y(0) = 0. \tag{14}$$

Part 2
For $2 \leq x$, the differential equation is

$$\frac{dy}{dx} + 2y = 0,$$

and

$$y(2) = \lim_{x \to 2^-} y(x).$$

That is, the solutions on $[0, 2)$ and $[2, \infty)$ should match up, or "meet," when $x = 2$, so that y can be continuous on $[0, \infty)$.

First we solve problem 1. The integrating factor is $u(x) = e^{\int 2\,dx} = e^{2x}$. Multiplying (14) by e^{2x} gives (using the Key Fact) that

$$(ye^{2x})' = e^{2x}.$$

Next we antidifferentiate, so that

$$ye^{2x} = \frac{e^{2x}}{2} + C. \tag{15}$$

To find C, let $x = 0$, and use the initial condition $y(0) = 0$ to get

$$0 \cdot 1 = \tfrac{1}{2} + C,$$

so that $C = -\tfrac{1}{2}$ and (15) is

$$ye^{2x} = \frac{e^{2x}}{2} - \frac{1}{2}$$

or

$$y = \frac{1 - e^{-2x}}{2}, \quad \text{for } 0 \le x < 2.$$

When $x \to 2^-$, this solution has the value

$$y(2) = \frac{1 - e^{-4}}{2}.$$

Thus problem 2 is

$$\frac{dy}{dx} + 2y = 0, \quad x \ge 2, \quad y(2) = \frac{1 - e^{-4}}{2}. \tag{16}$$

The integrating factor is again e^{2x}. Multiplying by e^{2x}, we see that Eq. (16) becomes

$$\frac{d}{dx}(ye^{2x}) = 0.$$

Antidifferentiation yields

$$ye^{2x} = C.$$

To find C, let $x = 2$, and use $y(2) = (1 - e^{-4})/2$, to find $C = (e^4 - 1)/2$. Thus $y = e^{-2x}(e^4 - 1)/2$.

Combining the solutions of parts 1 and 2, we obtain the final solution:

$$y = \begin{cases} (1 - e^{-2x})/2, & \text{for } 0 \le x < 2, \\ e^{-2x}(e^4 - 1)/2, & \text{for } 2 \le x. \end{cases} \tag{17}$$

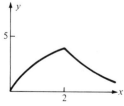

Figure 2.3.1 Graph of (17).

The graph of y is given in Fig. 2.3.1. Some students may recognize this graph as the charge-discharge curve of a capacitor in an RC-circuit. This will be discussed more carefully in Section 2.12. ∎

Technical Point Technically speaking, the function $y(x)$ given by (17) is not a solution of (13) for $x \ge 0$ since y is not differentiable at $x = 2$. On the other hand, since $g(x)$ is not continuous at $x = 2$, the Basic Existence and Uniqueness Theorem does not apply to (13) at $x = 2$, and we have no reason to even expect a solution there. To avoid this difficulty, we now adopt the concept of solution for linear equations usually applied in practice. We shall call $y(x)$ a *solution* of a linear differential equation on an x-interval I if the following hold:

1. $y(x)$ is differentiable and satisfies the differential equation for all x in the interval I at which the Basic Existence and Uniqueness Theorem holds.

2. The Basic Existence and Uniqueness Theorem fails to hold for at most a finite number of x-values in any finite subinterval $[a, b]$ of I.

That (13) has a unique continuous solution is shown in Exercise 34 at the end of this section.

Note also, that the consideration of solutions on closed intervals means that, technically speaking, we are using one-side derivatives at the endpoints. However, since many of our applications concern functions defined on closed intervals, and this technical point causes few problems in practice, we shall ignore it.

Exercises

In Exercises 1 through 12, solve the differential equation. If there is no initial condition, give the general solution.

1. $xy' + y = e^x$, $\quad y(1) = 1$
2. $y' + 2xy = x$
3. $y' = 3e^x$
4. $y' = \dfrac{y + x}{x}$
5. $(x^2 + 1)y' + 2xy = 1$
6. $\dfrac{du}{dt} = 3(u - 1)$
7. $y' + 4y = x$, $\quad y(0) = 0$
8. $(x + 1)y' + y = x$
9. $xy' = 2y$, $\quad y(1) = 4$
10. $xy' = -3y + \dfrac{\sin x}{x^2}$
11. $x^2 y' + xy = 1$
12. $t^3 \dfrac{dy}{dt} + 4t^4 = t^7$

For each of the linear differential equations in Exercises 13 through 15, the Existence and Uniqueness Theorem (Theorem 2.3.1) fails to hold at $x = 0$. Find the general solution of the differential equation and sketch several solutions ($C = 0$, $C = \pm 1$, $C = \pm 3$ would suffice) and describe the behavior of the solutions near $x = 0$.

13. $xy' + 3y = x$
14. $xy' + 2y = x^{-1}$
15. $xy' - y = x^2$

From calculus we know that

$$\frac{dx}{dy} = \frac{1}{dy/dx}.$$

That is, we can think of y as the independent variable and x the dependent variable provided $dy/dx \neq 0$. In Exercises 16 through 18 the differential equation is nonlinear in the form $dy/dx = f(x, y)$, but linear (after some algebraic manipulations) if written as $dx/dy = 1/f(x, y)$. Find the general solution of the differential equation.

16. $\dfrac{dy}{dx} = \dfrac{1}{y}$
17. $\dfrac{dy}{dx} = \dfrac{1}{y + x}$
18. $\dfrac{dy}{dx} = \dfrac{y}{x + y}$

In Exercises 19 through 24, find the continuous solution of

$$\frac{dy}{dx} + p(x)y = g(x), \qquad 0 \le x \le 2,$$

and sketch the solution.

19. $p(x) = \begin{cases} 2, & 0 \le x < 1, \\ 1, & 1 \le x \le 2, \end{cases}$ $g(x) = 0,$ $y(0) = 2$

20. $p(x) = 1,$ $g(x) = \begin{cases} 1, & 0 \le x < 1, \\ 0, & 1 \le x \le 2, \end{cases}$ $y(0) = 1$

21. $p(x) = 0,$ $g(x) = \begin{cases} 1, & 0 \le x < 1, \\ -1, & 1 \le x \le 2, \end{cases}$
    $y(0) = 0$

22. $p(x) = \begin{cases} 0, & 0 \le x < 1, \\ 1, & 1 \le x \le 2, \end{cases}$ $g(x) = 1,$ $y(0) = 0$

23. $p(x) = \begin{cases} 1, & 0 \le x < 1, \\ 0, & 1 \le x \le 2, \end{cases}$ $g(x) = \begin{cases} 0, & 0 \le x < 1, \\ 1, & 1 \le x \le 2, \end{cases}$
    $y(0) = 2$

24. $p(x) = \begin{cases} 1, & 0 \le x < 1, \\ -1, & 1 \le x \le 2, \end{cases}$ $g(x) = 2,$
    $y(0) = 1$

25. Verify that, if an arbitrary constant C_1 is introduced when $\int p(x)\,dx$ is computed, then the form of the general solution (7) is unchanged. (*Note*: If C is an arbitrary constant, then Ce^{-C_1} is still an arbitrary constant.)

26. The differential equation
$$\frac{dy}{dx} + p(x)y = 0$$
is sometimes called the homogeneous equation associated with
$$\frac{dy}{dx} + p(x)y = g(x).$$
Show that the solution (7) of
$$\frac{dy}{dx} + p(x)y = g(x)$$
is in the form $y = y_p + y_h$, where $y_p(x)$ is a particular solution of
$$\frac{dy}{dx} + p(x)y = g(x)$$
and $y_h(x)$ is the general solution of the *associated homogeneous equation* $dy/dx + p(x)y = 0$.

27. Show that if $u(x)$ is a function of x, and u satisfies the property
$$u\left(\frac{dy}{dx} + p(x)y\right) = \frac{d(uy)}{dx},$$
then $u = k\exp[\int p(x)\,dx]$, k a constant.

Taylor's Theorem for a function $f(y)$ which is n times differentiable at and near a number y_0 is
$$f(y) = f(y_0) + f'(y_0)(y - y_0)$$
$$+ \frac{f''(y_0)}{2}(y - y_0)^2 + \cdots + E_n(y, y_0),$$
where
$$\lim_{y \to y_0} \frac{E_n(y, y_0)}{(y - y_0)^n} = 0.$$

In particular the *first-order* or *linear approximation* to $f(y)$ centered at y_0 is $f(y_0) + f'(y_0)(y - y_0)$. As a simple example, the linear approximation to $f(y) = y^3$ centered at $y_0 = 2$ would be $8 + 3\cdot 2^2(y - 2) = 12y - 16$. Thus near $y_0 = 2$, the solutions of the differential equation $dy/dx = y^3$ "should be close" to those of $dy/dx = 12y - 16$ since $12y - 16$ is "close" to y^3.

In Exercises 28 through 31 give the differential equation which is the linear approximation centered at y_0 to the given differential equation.

28. $\dfrac{dy}{dx} = \sin y,$ $y_0 = 0$

29. $\dfrac{dy}{dx} = \sin y,$ $y_0 = \dfrac{\pi}{2}$

30. $\dfrac{dy}{dx} = y^2,$ $y_0 = 1$

31. $\dfrac{dy}{dx} = y^3 + y,$ $y_0 = -1.$

32. Show that if u is an integrating factor for (3), then so is ku for any constant k.

Exercise 33 illustrates a subtlety concerning solutions when there is a singular point.

33. Show that
$$y = \begin{cases} \frac{3}{2}x + C_1 x^{1/3} & \text{for } x > 0, \\ \frac{3}{2}x + C_2 x^{1/3} & \text{for } x \le 0, \end{cases}$$
where C_1, C_2 are two arbitrary constants, is con-

tinuous for all x and satisfies the first-order linear equation $xy' - y/3 = x$ for all $x \neq 0$.

34. Suppose that $p(x)$ is continuous on $[a, b]$ and $g(x)$ is continuous on $[a, b]$ except at a finite number of points where the left- and right-hand limits exist. Show that $y' + p(x)y = g(x)$, $y(a) = y_0$, has a unique continuous solution on $[a, b]$.

Sometimes the integrals in (7) cannot be evaluated in terms of elementary functions we are familiar with. In Exercises 35 through 40, solve the differential equations.

35. $y' + 2xy = 1$
36. $y' + (\sin x)y = x$
37. $y' + x^2 y = x$
38. $y' - 2xy = e^x$
39. $y' + e^x y = 3$
40. $y' - y = e^{x^2}$

2.4 Bernoulli Equations

A *Bernoulli equation* is a first-order differential equation in the form

$$\frac{dy}{dx} + p(x)y = q(x)y^n. \tag{1}$$

[handwritten annotation: makes it non linear — so divide through by y^n]

Bernoulli equations are often used for problems in which a linear model is not sufficient and it is necessary to take into account higher-order effects. Bernoulli equations also illustrate how a *substitution* can be used to help solve a differential equation.

If $n = 0$ or $n = 1$, then the Bernoulli equation is already linear and can be solved by the method of the previous section. If $n \neq 0$ and $n \neq 1$, then the substitution $v = y^{1-n}$ changes the Bernoulli equation (1) to a linear differential equation in v and x. To see this, let $v = y^{1-n}$. Then

$$v' = (1-n)y^{-n}y' \quad \text{or} \quad y' = \frac{y^n v'}{1-n}.$$

Substituting into (1) gives

$$\frac{y^n}{1-n}\frac{dv}{dx} + p(x)y = q(x)y^n.$$

We divide both sides by $y^n/(1-n)$ and use $y^{1-n} = v$, to get

$$\frac{dv}{dx} + (1-n)p(x)v = (1-n)q(x), \tag{2}$$

which is a linear first-order differential equation in v. If $n > 1$, the substitution $v = y^{1-n}$ assumes $y \neq 0$. Since $y = 0$ is a solution of (1) if $n > 0$, we must check whether this solution has been lost.

Example 2.4.1 Solve the differential equation

$$\frac{dy}{dx} = y + y^3. \tag{3}$$

Solution This is a Bernoulli equation:

$$\frac{dy}{dx} - y = y^3 \quad \text{with } n = 3, p = -1, \text{ and } q = 1.$$

Let $v = y^{1-3} = y^{-2}$. Then $y = v^{-1/2}$. We could go directly to (2). Alternatively substituting for y in the differential equation (3) yields

$$\frac{d(v^{-1/2})}{dx} = v^{-1/2} + (v^{-1/2})^3$$

or

$$-\frac{1}{2}v^{-3/2}\frac{dv}{dx} - v^{-1/2} = v^{-3/2}.$$

Dividing by $-\frac{1}{2}v^{-3/2}$ gives

$$\frac{dv}{dx} + 2v = -2,$$

which is a first-order linear differential equation in v. The integrating factor is $u = e^{\int 2\,dx} = e^{2x}$. Multiplication by the integrating factor gives

$$(ve^{2x})' = -2e^{2x}.$$

Antidifferentiating with respect to x gives

$$ve^{2x} = -e^{2x} + C,$$

and solving for v, we obtain

$$v = -1 + Ce^{-2x}. \qquad (4)$$

Since the original problem was in terms of x, y, we probably want the answer in terms of x, y. Since $v = y^{-2}$, the solution (4) is

$$y^{-2} = -1 + Ce^{-2x} \quad \text{and} \quad y = 0,$$

or

$$y = \pm[-1 + Ce^{-2x}]^{-1/2} \quad \text{and} \quad y = 0. \;\blacksquare$$

Exercises

In Exercises 1 through 16, solve the Bernoulli equation.

1. $y' = 3y - y^2$
2. $xy' = y + x^3y^3$
3. $y' = y^4$
4. $y' = y(y-1)$
5. $xy' + y = 3x^2y^2$
6. $y' = y^4 - y, \quad y(0) = 1$
7. $y' + xy = xy^3$
8. $y' = xy^5, \quad y(0) = 1$
9. $y' - y = y^{1/2}$
10. $y' + 3y = y^{1/3}$
11. $y' = y^{7/3}$
12. $xy' + y = x^2y^{1/5}$
13. $xy' + 2y = x^3y^{3/2}$
14. $xy' + y = -3x^2y^4$
15. $y' = y^5 + y$
16. $y' = e^xy^3$

In Exercises 28 through 31 of Section 2.3, we considered linear approximations of a nonlinear differential equation $dy/dx = f(y)$. If we are approximating $dy/dx = f(y)$ by using more than just the linear terms, then a Bernoulli equation may result. If $f(y_0) = 0$, then the quadratic approximation at y_0 to $dy/dx = f(y)$ is

$$\frac{dy}{dx} = f(y_0) + f'(y_0)(y - y_0) + \frac{f''(y_0)}{2}(y - y_0)^2$$

$$= f'(y_0)(y - y_0) + \frac{f''(y_0)}{2}(y - y_0)^2.$$

Letting $w = y - y_0$ yields

$$\frac{dw}{dx} = f'(y_0)w + \frac{f''(y_0)}{2}w^2,$$

which is a Bernoulli equation in w with $n = 2$. Approximate each of the following equations by a Bernoulli equation by taking the indicated number of terms in the Taylor series of $f(y)$ centered at y_0. (Review provided in Section 2 of Chapter 5.)

17. $\dfrac{dy}{dx} = \sin y$, $y_0 = 0$; use four terms. $\sin y \approx y - \dfrac{y^3}{3!} + \dfrac{y^5}{5!} - \dfrac{y^7}{7!}$

18. $\dfrac{dy}{dx} = \cos y - 1$, $y_0 = 0$; use quadratic approximation.

19. $\dfrac{dy}{dx} = y^3 - 1$, $y_0 = 1$; use quadratic approximation

20. $\dfrac{dy}{dx} = e^y - 1$, $y_0 = 0$; use quadratic approximation and change of variables $w = y - 1$.

21. Show that, if r is a real root of the polynomial $ay^2 + by + c$, then the change of variables, $y = z + r$, changes $y' = ay^2 + by + c$ into a Bernoulli equation in z.

Solve the differential equations in Exercises 22 through 25 by using Exercise 21 to change them to Bernoulli equations and then solving the Bernoulli equation.

22. $y' = y^2 - 1$
23. $y' = y^2 + 2y + 1$
24. $y' = 2y^2 - 12y + 18$
25. $y' = y^2 - 3y + 2$

26. Show that if the Basic Existence and Uniqueness Theorem (Theorem 2.1.1) is going to be used to ensure a solution to (1) at every initial condition $y(x_0) = y_0$, then $p(x)$ and $q(x)$ must be continuous everywhere and n must be greater than or equal to one.

27. Suppose that $n > 1$ and $p(x)$ and $q(x)$ are continuous on the interval $[a, b]$ with $a < b$. Suppose that $q(x)$ is never zero on $[a, b]$. Show that, for every x_0 in (a, b), there is a solution of (1) which has a vertical asymptote at x_0.

2.5 Separable Equations

A differential equation

$$\frac{dy}{dx} = f(x, y) \tag{1}$$

is called *separable* if it can be written as

$$\frac{dy}{dx} = h(x)g(y). \tag{2}$$

That is, $f(x, y)$ factors into a function of x times a function of y. Either $h(x)$ or $g(y)$ may be constant so that every differential equation of the form

$$\frac{dy}{dx} = h(x) \quad \text{or} \quad \frac{dy}{dx} = g(y)$$

is separable. Examples of functions $f(x, y)$ that factor are:

$$e^{x+y} = e^x \cdot e^y, \qquad x^2 y = x^2 \cdot y$$
$$xy + 2x + y + 2 = (x+1)(y+2)$$
$$3y^3 \quad (\text{here } h(x) = 1), \qquad \cos x \quad (\text{here } g(y) = 1)$$

Some functions do not factor. Examples are:

$$x + y, \qquad e^x + e^y, \qquad x^2 + x + y + y^2.$$

It is not always easy to tell whether a function will factor. If $f(x, y)$ has been factored so that the differential equation is written as in (2), then we divide by $g(y)$ to get

$$\frac{1}{g(y)} \frac{dy}{dx} = h(x).$$

Next we antidifferentiate both sides with respect to x:

$$\int \frac{1}{g(y)} \frac{dy}{dx} dx = \int h(x) \, dx.$$

By the chain rule

$$dy = \frac{dy}{dx} dx,$$

so that the integral on the left may be replaced by $\int g(y)^{-1} \, dy$ to yield

$$\int \frac{1}{g(y)} dy = \int h(x) \, dx.$$

The preceding derivation can be somewhat shortened by using differential notation. Before giving examples, we summarize the above comments.

Solution by Separation of Variables

To solve $dy/dx = f(x, y)$ by separation of variables, we proceed by the following steps.

a. Factor: $f(x, y) = h(x)g(y)$.
b. Rewrite $dy/dx = h(x)g(y)$ in differential form as

$$\frac{1}{g(y)} dy = h(x) \, dx.$$

c. The solution is

$$\int \frac{1}{g(y)}\, dy = \int h(x)\, dx.$$

d. If there is a constant a so that $g(a) = 0$, then $y = a$ is a solution of the differential equation. Check to see if it is included in the solutions found in part (c).

Example 2.5.1 Solve the differential equation $dy/dx = y^2/x$ for y.

Solution $y^2/x = y^2(1/x)$, so the differential equation is separable. Let us rewrite it in differential form as

$$\frac{1}{y^2}\, dy = \frac{1}{x}\, dx.$$

Thus

$$\int \frac{1}{y^2}\, dy = \int \frac{1}{x}\, dx$$

or

$$-\frac{1}{y} = \ln|x| + C.$$

Hence

$$y = -\frac{1}{\ln|x| + C}. \tag{3}$$

This completes parts (a), (b), (c). Since we divided by y^2, the calculations assumed that $y \neq 0$. The function $y = 0$ is a solution but it is not given by formula (3) for any choice of the constant C. The solutions of $dy/dx = y^2/x$ are thus

$$y = -\frac{1}{\ln|x| + C} \quad \text{or} \quad y = 0.$$

The solution $y = 0$ missing from (3) is called a *singular solution*. ∎

Example 2.5.2 Solve the differential equation $dy/dx = y^2 + 1$ for y.

Solution Rewrite the differential equation in differential form as

$$\frac{1}{y^2 + 1}\, dy = dx,$$

so that

$$\int \frac{1}{y^2 + 1}\, dy = \int dx.$$

Then
$$\tan^{-1} y = x + C. \tag{4}$$
Taking the tangent of both sides of (4) gives
$$y = \tan(x + C). \tag{5}$$
Since $y^2 + 1$ is never zero, (5) gives the general solution. ∎

Caution In going from (4) to (5) we take the tangent of both sides. A common error is to get $y = \tan x + \tan C$ instead of $\tan(x + C)$.

In Examples 2.5.1 and 2.5.2 we were able to solve for y explicitly in terms of x. This is not always possible. In order to clarify how far the reader is to go in doing the exercises, we introduce terminology which will be used throughout this chapter. *Solve the differential equation for y* means get an explicit formula for y in terms of x as in (3) or (5) if possible. *Solve the differential equation* means to obtain an implicit solution.

Example 2.5.3   Solve
$$\frac{dy}{dx} = \frac{1 + x + x^3}{2 + y^2 + y^6}.$$

Solution   Rewrite in differential form,
$$(2 + y^2 + y^6)\, dy = (1 + x + x^3)\, dx,$$
so that
$$\int (2 + y^2 + y^6)\, dy = \int (1 + x + x^3)\, dx$$
or, upon antidifferentiation,
$$2y + \frac{y^3}{3} + \frac{y^7}{7} = x + \frac{x^2}{2} + \frac{x^4}{4} + C. \tag{6}$$

Formula (6) gives all the solutions implicitly. An actual formula for y in terms of x would be difficult, if not impossible, to obtain. For a given value of x, C, values of y could be obtained by solving (6) numerically, for example, by Newton's method from calculus. ∎

Example 2.5.4   Solve $dy/dx = (y^2 - 1)/x$ for y.

Solution   Rewrite in differential notation as
$$\frac{1}{y^2 - 1}\, dy = \frac{1}{x}\, dx,$$
so that

$$\int \frac{1}{y^2 - 1}\, dy = \int \frac{1}{x}\, dx. \tag{7}$$

The integral on the left can be evaluated in several ways. Using partial fractions (Appendix B), for example, we get

$$\frac{1}{y^2 - 1} = \frac{1}{(y-1)(y+1)} = \frac{A}{y-1} + \frac{B}{y+1}.$$

Multiplying by $y^2 - 1$ gives $1 = A(y+1) + B(y-1)$. Since $y^2 - 1$ has simple linear factors $(y+1), (y-1)$, we may evaluate at the roots $-1, 1$ to get $A = \frac{1}{2}$, $B = -\frac{1}{2}$. Thus, (7) is

$$\int \frac{\frac{1}{2}}{y-1} + \frac{-\frac{1}{2}}{y+1}\, dy = \int \frac{1}{x}\, dx$$

so that $\frac{1}{2} \ln|y-1| - \frac{1}{2} \ln|y+1| = \ln|x| + C$ or

$$\ln \frac{|y-1|^{1/2}}{|y+1|^{1/2}} = \ln|x| + C. \tag{8}$$

Exponentiation of both sides of (8) yields

$$\frac{|y-1|^{1/2}}{|y+1|^{1/2}} = e^{\ln|x|+C} = e^C |x| = K|x| \tag{9}$$

where $K = e^C$ is a new constant. Equation (9) provides an implicit solution. In this example, we can compute an explicit solution as follows:

We square both sides to eliminate the square root:

$$\left| \frac{y-1}{y+1} \right| = K^2 x^2. \tag{10}$$

Eliminating the absolute value sign and writing $\tilde{C}$ for $\pm K^2$ we obtain

$$\frac{y-1}{y+1} = \tilde{C} x^2. \tag{11}$$

We then multiply by $y + 1$ and solve for y to get

$$y = \frac{1 + \tilde{C} x^2}{1 - \tilde{C} x^2}. \tag{12}$$

In obtaining (7), we assumed that $y^2 - 1 \neq 0$. However, $y = 1$, $y = -1$ are solutions of $y' = (y^2 - 1)/x$. The solution $y = 1$ is included in (12) when $\tilde{C} = 0$. However $y = -1$ is not included. All solutions are thus given by

$$y = \frac{1 + \tilde{C} x^2}{1 - \tilde{C} x^2} \quad \text{or} \quad y = -1. \tag{13}$$

Readers are advised to make sure they follow the calculations involved in going from (8) to (12), since they give many students difficulties. ∎

Initial Conditions When initial conditions are given it is sometimes simpler to find the value of the arbitary constant right after the integration. However, it is then necessary to keep track of how that constant is changed.

Example 2.5.5 Solve the initial-value problem
$$\frac{dy}{dx} = \frac{y^2 - 1}{x}, \quad y(1) = 2,$$
for y.

Solution This is the same differential equation as in the preceding example. Applying the initial condition to the solution (13) would involve solving for C. Instead we shall use (8), where the C is "out in the open." Letting $x = 1$, $y = 2$ in (8) gives
$$\tfrac{1}{2}\ln|2 - 1| - \tfrac{1}{2}\ln|2 + 1| = \ln 1 + C$$
or
$$C = -\tfrac{1}{2}\ln 3.$$
Now from Example 2.5.4, Eq. (11) shows $\tilde{C} > 0$ so that
$$\tilde{C} = K^2 = (e^C)^2 = e^{2C} = e^{-\ln 3} = 3^{-1}.$$
Thus from (13) the solution is
$$y = \frac{1 + x^2/3}{1 - x^2/3} = \frac{3 + x^2}{3 - x^2}. \tag{14}$$

Alternatively, we could enter $x = 1$, $y = 2$ into the solution (13) to get $2 = (1 + \tilde{C})/(1 - \tilde{C})$, and solve for $\tilde{C}$ to find $\tilde{C} = \tfrac{1}{3}$, in which case we again get (14). Which method of applying initial conditions is preferable depends on the individual problem and the solver's preferences. ∎

Exercises

In Exercises 1 through 10, solve the differential equation for y (or r in Ex. 2 and u in Ex. 6).

1. $\dfrac{dy}{dx} = \dfrac{y + 1}{x}$

2. $\dfrac{dr}{d\theta} = \dfrac{r^2 + r}{\theta}$

3. $\dfrac{dy}{dx} = e^x$

4. $\dfrac{dy}{dx} = e^{x+y}$

5. $\dfrac{dy}{dx} = xy + 4y + 3x + 12$

6. $\dfrac{du}{dt} = \dfrac{u^2 + 4}{t^2 + 4}$

7. $\dfrac{dy}{dx} = 3$

8. $\dfrac{dy}{dx} = x^2 y^2$

9. $\dfrac{dy}{dx} = y^5$, $\quad y(2) = 1$

10. $\dfrac{dy}{dx} = xy - 2y + x - 2$

In Exercises 11 through 14, solve the differential equation.

11. $\dfrac{du}{dt} = \dfrac{t^2 + 1}{u^2 + 4}$, $\quad u(0) = 1$

12. $\dfrac{dr}{d\theta} = \sin r$

13. $\dfrac{dy}{dx} = x^2 y^2 + y^2 + x^2 + 1$, $\quad y(0) = 2$

14. $\dfrac{du}{dt} = u^3 - u$

The differential equations in Exercises 15 through 17 were sketched in Exercises 1 through 3 of Section 2.2. Solve the differential equations and compare the solutions to your sketches. (This requires partial fractions.)

15. $y' = y(y - 1)$
16. $y' = y^2(y - 1)^2$
17. $y' = (y - 1)(y - 2)^2$

In Exercises 18 through 24, a first-order differential equation is written in differential form. Solve the differential equation.

18. $y^2\, dx + x^3\, dy = 0$
19. $(xy + y)\, dx + (xy + x)\, dy = 0$
20. $r \sin \theta\, dr + \cos \theta\, d\theta = 0$
21. $(t^2 - 4)\, dz + (z^2 - 9)\, dt = 0$
22. $(y^2 + y)\, dx + (x^3 + 4x^2)\, dy = 0$
23. $e^{x+y}\, dx + e^{2x-3y}\, dy = 0$
24. $xe^y\, dx + ye^{-x}\, dy = 0$
25. Show that if we make the substitution $z = ax + by + c$ and $b \neq 0$, then the differential equation

$$\dfrac{dy}{dx} = f(ax + by + c)$$

is changed to a differential equation in z and x that can be solved by separation of variables.

In Exercises 26 through 30, solve the differential equation using Exercise 25.

26. $\dfrac{dy}{dx} = (x + y)^2$

27. $\dfrac{dy}{dx} = (x + 4y - 1)^2$

28. $\dfrac{dy}{dx} = \tan(-x + y + 1) + 1$

29. $\dfrac{dy}{dx} = e^{x+y}(x + y)^{-1} - 1$

30. $\dfrac{dy}{dx} = \dfrac{x + y + 2}{x + y + 1}$

2.6 Exact Equations

Suppose that a first-order differential equation is written in differential form,

$$M(x, y)\, dx + N(x, y)\, dy = 0. \tag{1}$$

For example,

$$\dfrac{dy}{dx} = -\dfrac{(y + 2x)}{(x + 4y)}$$

could be written as

$$(y + 2x)\,dx + (x + 4y)\,dy = 0. \tag{2}$$

If there is a function $F(x, y)$ such that

$$dF = M(x, y)\,dx + N(x, y)\,dy,$$

then the differential equation $M\,dx + N\,dy = 0$ would become $dF = 0$, and the general solution would be $F = C$. Such differential equations are called *exact*, since $M\,dx + N\,dy$ is exactly the differential of a function F.

In (2), if $F(x, y) = x^2 + xy + 2y^2$, then

$$dF = \frac{\partial F}{\partial x}\,dx + \frac{\partial F}{\partial y}\,dy = (2x + y)\,dx + (x + 4y)\,dy,$$

so that $(y + 2x)\,dx + (x + 4y)\,dy = 0$ is the same as

$$d(x^2 + xy + 2y^2) = 0.$$

The general solution of (2) is then $F = C$ or

$$x^2 + xy + 2y^2 = C.$$

To develop a method based on this approach, we must address two issues.

1. How do we know when such an $F(x, y)$ exists?
2. If $F(x, y)$ exists, how can we find it?

Notation Just as we have used F_x to denote $\partial F/\partial x$, and F_y to denote $\partial F/\partial y$, we let

$$F_{xx} = \frac{\partial^2 F}{\partial x^2}, \quad F_{yy} = \frac{\partial^2 F}{\partial y^2}, \quad F_{xy} = \frac{\partial^2 F}{\partial y\,\partial x}, \quad F_{yx} = \frac{\partial^2 F}{\partial x\,\partial y}.$$

Suppose that $dF = M\,dx + N\,dy$. Since from calculus, $dF = F_x\,dx + F_y\,dy$, it follows that

$$F_x = M, \quad F_y = N. \tag{3}$$

If F has continuous second partials, then from calculus we also know that $F_{xy} = F_{yx}$. But then $(F_x)_y = M_y$ and $(F_y)_x = N_x$ from (3). Thus if an F having continuous second partials exists so that $dF = M\,dx + N\,dy$, then $M_y = N_x$. The somewhat surprising fact is that this condition is not only necessary but sufficient.

Theorem 2.6.1 Let $M(x, y)$, $N(x, y)$ and their first partials be continuous on a rectangle $a \leq x \leq b$, $c \leq y \leq d$. Then there is an $F(x, y)$ (defined on the same rectangle) such that $dF = M(x, y)\,dx + N(x, y)\,dy$ if and only if $M_y = N_x$.

In Eq. (2),

$$\frac{\partial M}{\partial y} = \frac{\partial(y + 2x)}{\partial y} = 1 \quad \text{and} \quad \frac{\partial N}{\partial x} = \frac{\partial(x + 4y)}{\partial x} = 1,$$

so that $M_y = N_x$. Theorem 2.6.1 then guarantees that an $F(x, y)$ exists such that $dF = M\, dx + N\, dy$.

Once we know that there is an $F(x, y)$, we may get it by solving (3). For (2), Eqs. (3) are

$$F_x = M = y + 2x, \tag{4}$$

$$F_y = N = x + 4y. \tag{5}$$

Antidifferentiating (4) with respect to x and thinking of x, y as independent variables give

$$F(x, y) = \int F_x\, dx = \int (y + 2x)\, dx = yx + x^2 + h(y), \tag{6}$$

where $h(y)$ is an unknown function of y. We need to introduce an unknown function of y because $F_x = G_x$ for two functions, $F(x, y)$ and $G(x, y)$, means that they differ by a function that depends only on y. [Similarly, $F_y = G_y$ would mean that $F(x, y)$ and $G(x, y)$ differ by a function of x only.] Substituting expression (6), that is, $F = yx + x^2 + h(y)$, back into the other equation (5), that is, $F_y = x + 4y$, in order to find $h(y)$, we get

$$\frac{\partial F}{\partial y} = \frac{\partial(yx + x^2 + h(y))}{\partial y} = x + 4y$$

or

$$x + h'(y) = x + 4y, \quad \text{so that} \quad h'(y) = 4y.$$

Hence $h(y) = 2y^2$ and $F = yx + x^2 + 2y^2$, as noted earlier. We do not need to add a constant C_1 when we find $h(y)$ from $h'(y)$ since the solution of (2) is obtained by setting $F = C$, and $F + C_1 = C$ would be an equivalent answer.

This procedure may be summarized as follows.

Summary of Method for Solution of Exact Equations

1. Write the differential equation in the form $M(x, y)\, dx + N(x, y)\, dy = 0$.
2. Compute M_y, N_x. If $M_y \ne N_x$, the equation is not exact and this technique will not work. If $M_y = N_x$, the equation is exact and this technique will work.
3. Either antidifferentiate $F_x = M$ with respect to x or $F_y = N$ with respect to y. Antidifferentiating will introduce an arbitrary function of the other variable.
4. Take the result for F from step 3 and substitute for F in the other equation from (3) to find the arbitrary function.
5. The solution is $F(x, y) = C$.

Example 2.6.1 Solve $(2x + 1 + 2xy)\,dx + (x^2 + 4y^3)\,dy = 0$.

Solution

In this problem, $M = 2x + 1 + 2xy$, $N = x^2 + 4y^3$. Since $M_y = N_x = 2x$, the equation is exact. The partial differential equations for $F(x, y)$ are

$$F_x = M = 2x + 1 + 2xy,$$
$$F_y = N = x^2 + 4y^3.$$

Either equation can be antidifferentiated. We shall antidifferentiate the second one:

$$F = \int F_y\,dy = \int (x^2 + 4y^3)\,dy = x^2 y + y^4 + k(x),$$

where $k(x)$ is an unknown function of x. We now substitute this expression for F in the other equation ($F_x = 2x + 1 + 2xy$) in order to find $k(x)$:

$$\frac{\partial(x^2 y + y^4 + k(x))}{\partial x} = 2x + 1 + 2xy.$$

Then

$$2xy + k'(x) = 2x + 1 + 2xy$$

or

$$k'(x) = 2x + 1 \quad \text{and} \quad k(x) = x^2 + x.$$

Thus $F(x, y) = x^2 y + y^4 + x^2 + x$ and the general solution is

$$F(x, y) = x^2 y + y^4 + x^2 + x = C. \blacksquare$$

A common mistake is to give $F + C$ as an answer instead of $F = C$. Note that $F + C$ does not give any relationship between x and y, and hence does not give us a function of x as an answer.

The above technique usually produces solutions in implicit form. It may not be possible to algebraically solve for y in terms of x.

In some problems the choice of which equation to antidifferentiate can greatly affect the ease of solution.

Example 2.6.2 Solve the initial-value problem

$$(3x^2 + y^3 e^y)\,dx + (3xy^2 e^y + xy^3 e^y + 3y^2)\,dy = 0, \qquad y(2) = 0.$$

Solution

In this example, $M = 3x^2 + y^3 e^y$ and $N = 3xy^2 e^y + xy^3 e^y + 3y^2$. Since $M_y = 3y^2 e^y + y^3 e^y = N_x$, the differential equation is exact. The partial differential equations for $F(x, y)$ are

$$F_x = M = 3x^2 + y^3 e^y,$$
$$F_y = N = 3xy^2 e^y + xy^3 e^y + 3y^2.$$

We have a choice now of evaluating

$$F = \int F_x\, dx = \int (3x^2 + y^3 e^y)\, dx$$

or

$$F = \int F_y\, dy = \int (3xy^2 e^y + xy^3 e^y + 3y^2)\, dy.$$

Note that $\int F_x\, dx$ is quite easy to carry out, whereas $\int F_y\, dy$ is considerably more complicated and requires either integration by parts or the use of tables. Therefore we shall compute the first integral,

$$F = \int F_x\, dx = \int (3x^2 + y^3 e^y)\, dx = x^3 + y^3 e^y x + h(y),$$

where $h(y)$ is an unknown function of y. Substituting this formula for F into $F_y = N$ in order to find $h(y)$, we have

$$\frac{\partial(x^3 + y^3 e^y x + h(y))}{\partial y} = N = 3xy^2 e^y + xy^3 e^y + 3y^2,$$

or

$$3y^2 e^y x + y^3 e^y x + h'(y) = 3xy^2 e^y + xy^3 e^y + 3y^2,$$

so that

$$h'(y) = 3y^2 \quad \text{and} \quad h(y) = y^3.$$

Thus $F(x, y) = x^3 + y^3 e^y x + y^3$ and the general solution is

$$F = x^3 + y^3 e^y x + y^3 = C.$$

Applying the initial condition $x = 2$, $y = 0$ gives $C = 8$. ∎

If we start with a first-order equation, for example,

$$\frac{dy}{dx} = \frac{-y^2}{x^2 + 1}, \tag{7}$$

there are many ways of rewriting the equation in the form $M\, dx + N\, dy = 0$. For (7), some possibilities are

$$(x^2 + 1)\, dy + y^2\, dx = 0, \tag{8}$$
$$y^{-2}\, dy + (x^2 + 1)^{-1}\, dx = 0, \tag{9}$$

and

$$dy + \frac{y^2}{x^2 + 1}\, dx = 0. \tag{10}$$

Exercises

In Exercises 1 through 20, determine whether the differential equation is exact. If it is exact, solve it.

1. $(2x + y)\, dx + (2y + x)\, dy = 0$
2. $(x + 3y)\, dx + (y + 2x)\, dy = 0$
3. $x^{-1}y\, dx + (\ln x + 3y^2)\, dy = 0$
4. $(1 + y^2)\, dx + (2xy + 4y)\, dy = 0$
5. $(1 + y^2)\, dx + (3xy + 4y)\, dy = 0$
6. $(3x^2 + 2xy + y^3)\, dx + (x^2 + 3xy^2 + \cos y)\, dy = 0$, $y(0) = 0$
7. $(-2x^{-3} + 2xe^{x^2}y)\, dx + (e^{x^2} + 2y)\, dy = 0$
8. $(2r\theta + 1)\, dr + (r^2 + 1)\, d\theta = 0$
9. $(2u + \theta + 2u\theta^2)\, du + (u + 2u^2\theta + 3\theta^2)\, d\theta = 0$
10. $(2x + y^5 e^{3y} \sin y)\, dx + (5xy^4 e^{3y} \sin y + 3xy^5 e^{3y} \sin y + xy^5 e^{3y} \cos y + 2y)\, dy = 0$, $y(0) = 2$
11. $x\, dx + (x + y)\, dy = 0$
12. $x^{-1}\, dx + y^{-1}\, dy = 0$
13. $(2x + 3x^2 y^2)\, dx + (2x^3 y - 5y^4)\, dy = 0$
14. $(\sin y + e^x)\, dx + (x \cos y - 2y)\, dy = 0$
15. $(2xe^{xy} + x^2 y e^{xy})\, dx + (x^3 e^{xy} + 3y^2)\, dy = 0$

16. $\dfrac{1}{(x + y)^2 + 1}\, dx + \left(1 + \dfrac{1}{(x + y)^2 + 1}\right) dy = 0$

17. $\dfrac{1}{x + y}\, dx + \left(\dfrac{1}{x + y} + 3y^2\right) dy = 0$

18. $\left(3x^2 + \dfrac{2x}{(x^2 + y^2 + 1)^2}\right) dx + \left(\dfrac{2y}{(x^2 + y^2 + 1)^2} - 4y^3\right) dy = 0$

19. $\cos(x + y)\, dx + (\cos(x + y) - 2y \sin(y^2))\, dy = 0$

20. $\dfrac{x + 2y}{x + y}\, dx + \left(\ln(x + y) + \dfrac{y}{x + y}\right) dy = 0$

21. Suppose that a differential equation can be solved by separation of variables so that

$$\frac{dy}{dx} = g(x)h(y).$$

This equation is equivalent to

$$\frac{1}{h(y)}\, dy - g(x)\, dx = 0.$$

Show that $[1/h(y)]\, dy - g(x)\, dx = 0$ is an exact differential equation and compare the solution obtained by separation of variables to that arrived at by the exact method.

Exercises 22 through 26 develop one application of exact differential equations. They may be assigned now or in conjunction with Section 2.11. Exercises 22 and 23 review some concepts from calculus.

Exact differentials also arise in calculus when we evaluate line integrals. We recall from calculus that if the differential $M\,dx + N\,dy$ is exact in a rectangular region R and (x_0, y_0) and (x_1, y_1) are two points in R, then

$$\int_{(x_0,y_0)}^{(x_1,y_1)} M(x,y)\,dx + N(x,y)\,dy = \int_{(x_0,y_0)}^{(x_1,y_1)} dF$$
$$= F(x_1, y_1) - F(x_0, y_0). \quad (11)$$

One application of line integrals occurs in the computation of work done. For example, let us suppose that at each point (x, y) in R, there is a force which in vector form is written as $(M(x, y), N(x, y))$. Suppose that $M\,dx + N\,dy$ is an exact differential and M, N have continuous first partials in R. We also assume that the points (x_0, y_0) and (x_1, y_1) are in R.

22. Show that the work in going from (x_0, y_0) to (x_1, y_1) is independent of the path taken as long as the path stays in R and that the work done is given by (11).

23. Show that we can move through R from (x_0, y_0) to (x_1, y_1) without any work being done during any part of the trip if and only if we move along a solution of $M\,dx + N\,dy = 0$.

24. The function $F(x, y)$ is sometimes called a *potential* and the curves $F(x, y) = C$ are called *level curves* or *equipotentials*. In order to have the maximum work done, our motion must be perpendicular to the curves $F(x, y) = C$. Show that at any point (x, y) in R where the solutions of

$$-F_y\,dx + F_x\,dy = 0 \quad (12)$$

exist, they are perpendicular (have perpendicular tangents) to the curves $F = C$. The solutions of (12) and $F = C$ are sometimes called *orthogonal families* (Section 2.11).

25. Show that (12) is exact if and only if F satisfies the partial different equation

$$F_{xx} + F_{yy} = 0. \quad (13)$$

This is *Laplace's equation*.

26. The solutions of (13) are called *harmonic functions*. Verify that $F(x, y) = x^2 - y^2$ is a harmonic function. Solve (12) for this F and sketch the solutions of (12) and $F = C$ on the same axes.

2.7 Integrating Factors

As we pointed out in Section 2.6, there are many ways of writing $dy/dx = f(x, y)$ in the form

$$M(x, y)\,dx + N(x, y)\,dy = 0,$$

most of which will not be exact. Since it is impossible to check all these various ways to see which are exact, we shall instead show how to make some nonexact equations exact.

In solving first-order linear equations, we multiplied the equation by a function that turned the left-hand side of $dy/dx + p(x)y = g(x)$ into a derivative. Perhaps the same idea would work here. A function $u(x, y)$ is an *integrating factor* for the differential equation

$$M(x, y)\,dx + N(x, y)\,dy = 0$$

if the equation obtained by multiplying by $u(x, y)$,

$$u(x, y)M(x, y)\,dx + u(x, y)N(x, y)\,dy = 0,$$

is exact. In general, finding an integrating factor may be difficult. However, in the two special cases when the integrating factor depends just on x or just on y, it is possible to compute it with relative ease.

To see how the integrating factor u can be computed in these special cases, let us suppose that

$$M(x, y)\,dx + N(x, y)\,dy = 0$$

has an integrating factor that depends just on x, say, $u(x)$. Then

$$u(x)M(x, y)\,dx + u(x)N(x, y)\,dy = 0 \tag{1}$$

is exact. But $u(x)M(x, y)\,dx + u(x)N(x, y)\,dy$ is exact if and only if

$$\frac{\partial(u(x)M(x, y))}{\partial y} = \frac{\partial(u(x)N(x, y))}{\partial x},$$

that is,

$$u(x)M_y(x, y) = u'(x)N(x, y) + u(x)N_x(x, y).$$

But then

$$\frac{u'(x)}{u(x)} = \frac{M_y(x, y) - N_x(x, y)}{N(x, y)}. \tag{2}$$

Since the left-hand side of (2) depends only on x, the right-hand side must also depend only on x. Let $Q(x)$ be the right-hand side of (2). Then $u'/u = Q$ so that

$$\ln|u(x)| = \int \frac{u'(x)}{u(x)}\,dx = \int Q(x)\,dx.$$

Thus the integrating factor is

$$u(x) = \exp\left[\int Q(x)\,dx\right].$$

(Since any constant multiple of an integrating factor is an integrating factor, we may omit arbitrary constants and drop the absolute value when finding u.) A similar argument works if there is an integrating factor that is a function of y only. These ideas may be summarized as follows:

Integrating Factor Method

For the differential equation $M(x, y)\,dx + N(x, y)\,dy = 0$, first compute M_y, N_x.

1a. If $(M_y - N_x)/N$ cannot be expressed as a function of x only, then we do not have an integrating factor that is a function of x only. If $(M_y - N_x)/N = Q(x)$ is a function of x, then

$$u(x) = \exp\left[\int Q(x)\,dx\right]$$

is an integrating factor.

1b. If $(N_x - M_y)/M$ cannot be expressed as a function of y only, then we do not have an integrating factor that is a function of y only. If $(N_x - M_y)/M = R(y)$ is a function of y, then

$$u(y) = \exp\left[\int R(y)\, dy\right]$$

is an integrating factor.
2. Multiply $M\, dx + N\, dy = 0$ by the integrating factor.
3. Solve the exact equation $uM\, dx + uN\, dy = 0$.

Example 2.7.1 Solve the differential equation

$$(3y^2 + 4x)\, dx + (2yx)\, dy = 0 \tag{3}$$

Solution In this example,

$$M = 3y^2 + 4x, \quad N = 2yx,$$

so that $M_y = 6y$, $N_x = 2y$, and the differential equation is not exact. However,

$$\frac{M_y - N_x}{N} = \frac{6y - 2y}{2yx} = \frac{2}{x}$$

is a function of x. Thus there is an integrating factor

$$u(x) = \exp\left[\int \frac{2}{x}\, dx\right] = \exp[2\ln|x|] = x^2.$$

Multiplying the differential equation by x^2 gives the new differential equation

$$(3x^2y^2 + 4x^3)\, dx + (2yx^3)\, dy = 0, \tag{4}$$

which is exact. For (4) we have $M = 3x^2y^2 + 4x^3$ and $N = 2yx^3$. Note $M_y = 6x^2y = N_x$ as expected. The partial differential equations for F are

$$F_x = M = 3x^2y^2 + 4x^3,$$
$$F_y = N = 2yx^3.$$

From the second equation,

$$F = \int F_y\, dy = \int 2yx^3\, dy = y^2x^3 + h(x),$$

where $h(x)$ is an unknown function of x. Substituting $F = y^2x^3 + h(x)$ for F in $F_x = 3x^2y^2 + 4x^3$, we get

$$\frac{\partial(y^2x^3 + h(x))}{\partial x} = 3x^2y^2 + 4x^3$$

or
$$3y^2x^2 + h'(x) = 3x^2y^2 + 4x^3,$$
so that $h'(x) = 4x^3$ and $h(x) = x^4$. Thus $F = y^2x^3 + x^4$ and the solution of the original differential equation is
$$F = y^2x^3 + x^4 = C. \blacksquare$$

Exercises

In Exercises 1 through 14, solve the differential equation.

1. $(2xy^3 + y^4)\,dx + (xy^3 - 2)\,dy = 0$
2. $(2y^3 + 2)\,dx + (3xy^2)\,dy = 0$
3. $(-2 + x^3y)\,dx + (x^4 + 6yx^3)\,dy = 0$
4. $(3x^2y^{-1} + y)\,dx + (2x + 4y^2)\,dy = 0$
5. $(y + e^{-x})\,dx + dy = 0$
6. $(y\cos(x+y)\sec y - 1)\,dx + \sec y(\sin(x+y) + y\cos(x+y) + x\sin y)\,dy = 0$
7. $2x\,dx + (6ye^y - x^2)\,dy = 0$
8. $(2y^2 + 1 + 2x)\,dx + 2y\,dy = 0$
9. $3x^2y\,dx + (3x^3 + 5y^2)\,dy = 0$
10. $(y + 2e^x)\,dx + (1 + e^{-x})\,dy = 0$
11. $4x^3y\,dx + (5x^4 + 6y)\,dy = 0$
12. $(e^y + e^{-x})\,dx + (e^y + 2ye^{-x})\,dy = 0$
13. $(3x^2y + 2xy^{-2})\,dx + (3x^3 + 2y^{-1})\,dy = 0$
14. $\left[\dfrac{x^2 + x}{x^2 + 2xy + y^2 + 1}\right] dx$
   $+ \left[\dfrac{x^2 + x}{(x^2 + 2xy + y^2 + 1)} + 3x^2y^2 + 3xy^2\right] dy = 0$
15. Verify that if the linear differential equation $dy/dx + p(x)y = q(x)$ is rewritten as
$$(p(x)y - q(x))\,dx + dy = 0,$$
then the method of this section and that of Section 2.3 produce the same function of x as an integrating factor. Compare the form of the solution obtained by the two methods.
16. Suppose $M\,dx + N\,dy$ is exact, and M, N have continuous first partials and are nonzero. Show that if u is a function of x or y only and $uM\,dx + uN\,dy$ is exact, then u is constant.
17. Show that
$$(y + 2x)\,dx + (x + 2y)\,dy = 0$$
is exact. Also show that if we multiply the given equation by
$$u(x, y) = xy + y^2 + x^2,$$
then the new differential equation is still exact. Compare this example to the conclusion of Exercise 16.
18. Derive the facts about the integrating factor in terms of y in statement 1b of the Integrating Factor Method.
19. Show that
$$(y^2 + 1)\,dx + 2y\,dy = 0$$
has integrating factors $u(x)$ and $\tilde{u}(y)$. Solve the differential equation both ways and show the answers are equivalent. (Note that this differential equation can also be solved by separation of variables.)

In Exercises 20 through 23, there is an integrating factor of the form $u(x, y) = x^r y^s$. Find r, s and solve the differential equation. [Use $(x^r y^s M)_y = (x^r y^s N)_x$ to find r, s.]

20. $6y^5\,dx + (7xy^4 + 3x^{-5})\,dy = 0$
21. $(2y + 3xy^2)\,dx + (3x + 4x^2y)\,dy = 0$
22. $(3 + 4xy^{-1})\,dx + (-2xy^{-1} - 3x^2y^{-2})\,dy = 0$
23. $(-3x^{-1} - 2y^4)\,dx + (-3y^{-1} + xy^3)\,dy = 0$

2.8 Homogeneous Equations

In this section we develop a substitution technique that can sometimes be used when other techniques fail. Suppose that we have the differential equation

$$\frac{dy}{dx} = f(x, y), \qquad (1)$$

and the value of $f(x, y)$ depends only on the ratio $v = y/x$, so that we can think of $f(x, y)$ as a function F of y/x,

$$f(x, y) = F(y/x) = F(v)$$

Examples of such functions are:

$$\frac{x + 3y}{2x + y} = \frac{1 + 3(y/x)}{2 + (y/x)}, \qquad F(v) = \frac{1 + 3v}{2 + v},$$

$$e^{y/x}, \qquad\qquad\qquad F(v) = e^v$$

and

$$\frac{x^2 + y^2}{3xy + y^2} = \frac{1 + (y/x)^2}{3(y/x) + (y/x)^2}, \qquad F(v) = \frac{1 + v^2}{3v + v^2}.$$

Using F, we can rewrite (1) as

$$\frac{dy}{dx} = F\left(\frac{y}{x}\right). \qquad (2)$$

A differential equation in the form (1) that may be written in the form (2) is sometimes called *homogeneous* (*not* the same as the homogeneous linear equations discussed in Chapter 3).

Let $y = xv$ in (2) so that (2) becomes

$$\frac{d(xv)}{dx} = F(v) \qquad \text{or} \qquad v + x\frac{dv}{dx} = F(v),$$

which may always be solved by separation of variables:

$$\int \frac{dv}{F(v) - v} = \int \frac{dx}{x}.$$

There is an alternative definition of "homogeneous" that is easier to verify:

$dy/dx = f(x, y)$ is a homogeneous equation if $f(tx, ty) = f(x, y)$ for all t such that (x, y) and (tx, ty) are in the domain of f. \qquad (3)

The above two definitions of homogeneous equation are equivalent (see Exercise 26).

Example 2.8.1

Solve the differential equation

$$\frac{dy}{dx} = \frac{-2x + 5y}{2x + y}. \tag{4}$$

Solution

First we use (3) to verify that (4) is homogeneous:

$$f(tx, ty) = \frac{-2tx + 5ty}{2tx + ty} = \frac{-2x + 5y}{2x + y} = f(x, y).$$

Let $y = xv$ in (4), so that (4) becomes

$$v + x\frac{dv}{dx} = \frac{-2x + 5xv}{2x + xv} = \frac{-2 + 5v}{2 + v}.$$

It is usually best to combine the v-terms first:

$$x\frac{dv}{dx} = \frac{-2 + 5v}{2 + v} - v = \frac{-2 + 3v - v^2}{2 + v}.$$

Now by separation of variables,

$$\int \frac{2 + v}{v^2 - 3v + 2} dv = \int -\frac{1}{x} dx = -\ln|x| + C. \tag{5}$$

The v-integral will be evaluated using partial fractions (Appendix B) so that

$$\frac{2 + v}{(v^2 - 3v + 2)} = \frac{2 + v}{(v - 2)(v - 1)} = \frac{A}{v - 2} + \frac{B}{v - 1}.$$

Multiplying by $v^2 - 3v + 2$ yields $2 + v = A(v - 1) + B(v - 2)$. Since we have distinct linear factors $v - 2, v - 1$, we may evaluate this expression at the roots 1, 2 of $(v - 2)(v - 1)$ to get $B = -3, A = 4$. Thus:

$$\int \frac{2 + v}{v^2 - 3v + 2} dv = \int \left(\frac{4}{v - 2} + \frac{-3}{v - 1}\right) dv$$
$$= 4\ln|v - 2| - 3\ln|v - 1|$$

and Eq. (5) is

$$4\ln|v - 2| - 3\ln|v - 1| = -\ln|x| + C. \tag{6}$$

We shall algebraically simplify (6). Taking the exponential of both sides of (6) yields

$$\frac{(v - 2)^4}{(v - 1)^3} = \frac{\tilde{C}}{x},$$

where the absolute values have been dropped by allowing $\tilde{C}$ to take on negative or positive values. Then let $v = y/x$ and multiply both sides by $x^4(y/x - 1)^3$, to get

$$(y - 2x)^4 = \tilde{C}(y - x)^3. \blacksquare$$

This technique may be summarized as follows:

Summary of Method for (Nonlinear) Homogeneous Equations

1. Verify that the differential equation is homogeneous, using Eq. (2) or (3).
2. Let $y = xv$ to get $x(dv/dx) + v = F(v)$.
3. Solve this differential equation by separation of variables.
4. Let $v = y/x$ to get the answer in terms of y and x.

Homogeneous equations are often readily recognizable because of the following common characteristics: They are expressions involving y/x such as $e^{y/x}$, $\cos(y/x)$, and constants, or they are made up of fractions of two polynomials such that all terms have the same *degree*. (By the degree of a term we mean the sum of the powers contained in it. For example, x^3, x^2y, xy^2, y^3 are all third-degree terms, whereas x^5, x^2y^3, x^4y, y^5 are fifth-degree terms. A polynomial all of whose terms have the same degree is sometimes called a *homogeneous polynomial*. Thus $x^2y + y^3$ and $x^4 + x^2y^2 + xy^3 + y^4$ are homogeneous polynomials of degrees 3 and 4, respectively.)

If $f(x, y) = p(x, y)/q(x, y)$ and $p(x, y)$ and $q(x, y)$ are homogeneous polynomials of the same degree, then $dy/dx = f(x, y)$ is a homogeneous differential equation.

As an example of this fact, consider

$$\frac{dy}{dx} = \frac{x^3 + x^2y + y^3}{xy^2 + x^2y + x^3} = \frac{p(x, y)}{q(x, y)}.$$

Both $p(x, y)$ and $q(x, y)$ are third-degree homogeneous polynomials. Dividing them both by x^3 yields

$$\frac{dy}{dx} = \frac{1 + (y/x) + (y/x)^3}{(y/x)^2 + (y/x) + 1} = F\left(\frac{y}{x}\right).$$

Other examples of homogeneous equations appear in the exercises.

Exercises

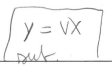

In Exercises 1 through 6, verify that the differential equation is homogeneous. Do not solve.

1. $\dfrac{dy}{dx} = \dfrac{y}{x} + \ln y - \ln x$

2. $\dfrac{dy}{dx} = \dfrac{(x^3 + y^3)^{1/3}}{x + y}$

3. $\dfrac{dy}{dx} = \ln(x^2 + y^2) - 2\ln y$

4. $\dfrac{dy}{dx} = 3\ln(x + y) - \ln(x^3 + y^3)$

5. $\dfrac{dy}{dx} = \dfrac{x\sqrt{x^3 + y^3}}{y^2\sqrt{x + y}}$

6. $\dfrac{dy}{dx} = \dfrac{4y\sqrt{x} - 5x\sqrt{y}}{\sqrt{x}(-x + 7y)}$

In Exercises 7 through 20, verify that the differential equation is homogeneous and solve it.

7. $\dfrac{dy}{dx} = e^{y/x} + \dfrac{y}{x}$

8. $\dfrac{dy}{dx} = \dfrac{-2x + 4y}{x + y}$

9. $\dfrac{dy}{dx} = \dfrac{-x + 3y}{x + y}$

10. $\dfrac{dy}{dx} = \dfrac{2x + y}{x + 2y}$

11. $\dfrac{dy}{dx} = \dfrac{-3x - y}{x + y}$

12. $\dfrac{dy}{dx} = \dfrac{2x + y}{y}$

13. $\dfrac{dy}{dx} = \dfrac{2xy + y^2}{x^2 + xy + y^2}$

14. $\dfrac{dy}{dx} = \dfrac{x + y}{x - y}$

15. $\dfrac{dy}{dx} = \dfrac{x^2 + 2y^2}{2xy + y^2}$

16. $\dfrac{dy}{dx} = \dfrac{x \tan(y/x) + y}{x}$

17. $\dfrac{dy}{dx} = \dfrac{y^4 + x^3 y}{x^4}$

18. $\dfrac{dy}{dx} = \dfrac{y^5 + x^4 y}{x^5}$

19. $\dfrac{dy}{dx} = \dfrac{y^2 + xy + x^2}{x^2}$

20. $\dfrac{dy}{dx} = \csc\left(\dfrac{y}{x}\right) + \dfrac{y}{x}$

21. Suppose that a, b, c, d, e, f are constants and at least one of d, e is nonzero.

   a) Show that $(ax + by)/(dx + ey)$ is constant if and only if $ae - bd = 0$.

   b) Suppose $ae - bd \neq 0$. Show that for any c and f, there exist constants k and l such that the change of variables $u = x + k$ and $w = y + l$ changes the nonhomogeneous equation

$$\dfrac{dy}{dx} = \dfrac{ax + by + c}{dx + ey + f}$$

into the homogeneous equation

$$\dfrac{du}{dw} = \dfrac{au + bw}{du + ew}.$$

In Exercises 22 through 25, use Exercise 21 to solve the differential equation given.

22. $\dfrac{dy}{dx} = \dfrac{2x + y - 13}{y + 1}$

23. $\dfrac{dy}{dx} = \dfrac{-8x + 3y + 17}{-3x + y + 6}$

24. $\dfrac{dy}{dx} = \dfrac{2x + y - 8}{x + 2y - 7}$

25. $\dfrac{dy}{dx} = \dfrac{-3x - y - 2}{x + y + 2}$

26. Verify that the definitions of "homogeneous" given in (2) and (3) are equivalent.

2.9 Heat and Decay Problems

The remainder of this chapter introduces several applications of first-order equations. In these applications the reader should pay close attention to how the equations are derived from the given physical problem and should not seek to merely memorize formulas. Problems are grouped into sections according to the similarity of the thought processes used to analyze them.

Let $x(t)$ be a quantity we are interested in at time t. If we know what dx/dt is in terms of t, say

$$\dfrac{dx}{dt} = 3t^2 + t, \tag{1}$$

then antidifferentiating (1) gives a formula for $x(t)$:

$$x(t) = t^3 + \dfrac{t^2}{2} + C.$$

In the applications of this chapter, instead of getting dx/dt directly in terms of t as in (1), dx/dt will usually be given in terms of x, that is,

> The rate of change of the quantity at time t will depend on how much of that quantity there is at time t.

2.9.1 Radioactive Decay

For a given atom of a radioactive material, not acted on by radiation or other particles, there is a fixed probability that it will decay in a given time period. For example, suppose the probability is 0.001 that the atom will decay in one year. Now suppose that we have $x(t)$ of these atoms at time t. We would expect at that time to have atoms decaying at a rate of $(0.001)x(t)$ atoms per year. This leads to the following law of radioactive decay:

> The rate of decay for a radioactive material is proportional to the number of atoms present. (2)

Note that the decay law is not really a law. However, if we have a reasonably large number of atoms and the material is not bombarded by other radiation or particles, then the decay law is often sufficiently accurate. We do not intend to belabor the point, but it is important to realize that the suitability of a law like (2) depends on the problem and the intended application of the answer.

Since $x(t)$ is the amount of radioactive material at time t, the law (2) may be written as

$$\frac{dx}{dt} = kx, \tag{3}$$

where (3) is based on the assumption that $x(t)$ is a differentiable function. Again this is an approximation that is accurate enough for many purposes. The constant k will be *negative* since there is a *loss* of material.

Equation (3) is a first-order linear differential equation. It may be solved by the integrating-factor technique (Section 2.3) or separation of variables (Section 2.5). Using separation of variables on Eq. (3), we find that

$$\int \frac{dx}{x} = \int k\, dt,$$

and hence

$$\ln|x| = kt + C.$$

We exponentiate both sides and use $x(0) > 0$ to get

$$x(t) = e^{kt}e^C = e^{kt}\tilde{c}.$$

Let $t = 0$ and note that $x(0) = \tilde{c}$ to arrive finally at

$$x(t) = e^{kt}x(0). \tag{4}$$

Thus radioactive decay is *exponential decay* since $k < 0$ implies that $x(t) \to 0$ as $t \to \infty$.

The *half-life* of a radioactive substance is the length of time it takes the material to decay to half its original amount. If T is the half-life, we have by definition

$$x(T) = \tfrac{1}{2}x(0). \tag{5}$$

The half-life and the constant of proportionality k are closely related. To see what this relationship is, substitute the formula (4) for $x(t)$ into (5) to get

$$e^{kT}x(0) = \tfrac{1}{2}x(0) \quad \text{or} \quad e^{kT} = \tfrac{1}{2}.$$

Taking the natural logarithm of both sides and solving for T gives the half-life as

$$T = -\frac{\ln 2}{k}. \tag{6}$$

Example 2.9.1 Three grams of a radioisotope decay in two years to 0.9 g. Determine both the half-life T and the constant k.

Solution Let $x(t)$ be the amount of the radioisotope at time t. Measure x in grams and t in years. The basic law is given by Eq. (3): $dx/dt = kx$. Solving this equation gives (4):

$$x(t) = e^{kt}x(0). \tag{7}$$

These equations hold for any radioisotope. For the isotope of this example, we know that

$$3 = x(0), \tag{8}$$
$$0.9 = x(2). \tag{9}$$

Substituting (8) and (9) into the general formula (7) with $t = 2$ yields

$$0.9 = e^{k2}3.$$

Thus

$$k = \frac{\ln(0.3)}{2} \approx -0.6.$$

Figure 2.9.1
Graph of solution of Example 2.9.1.

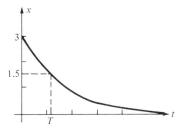

From (6), the half-life is

$$T = -\frac{2\ln 2}{\ln(0.3)} \approx 1.2 \text{ years.}$$

See Fig. 2.9.1. ∎

2.9.2 Cooling

Ignoring circulation and other effects, Newton's law of cooling states that:

> The rate of change of the surface temperature of an object is proportional to the difference between the temperature of the object and the temperature of its surroundings (also called the *ambient temperature*) at that time. (10)

If $T(t)$ is the surface temperature of the object at time t and T_0 is the ambient temperature at time t, (10) becomes

$$\frac{dT}{dt} = k(T - T_0). \qquad (11)$$

Note that (10) does not require T_0 to be constant. However, in our problems we shall assume that T_0 is constant unless stated otherwise. If $T(t) > T_0$, the body's surface temperature is hotter than the ambient temperature, and hence there is a loss of surface temperature, it follows that dT/dt must be negative and consequently k is negative. If $T(t) < T_0$, then $dT/dt > 0$ and the surface temperature increases.

The differential equation (11) is a first-order linear differential equation that can be solved by the methods of separation of variables (Section 2.5) if T_0 is constant, or integrating factors (Section 2.3) if T_0 depends on t. We shall use integrating factors.

Let us rewrite (11) as

$$\frac{dT}{dt} - kT = -kT_0.$$

The integrating factor is $e^{\int -k\,dt} = e^{-kt}$. Multiplying both sides by e^{-kt} gives

Figure 2.9.2
Graph of three solutions to Eq. (11).

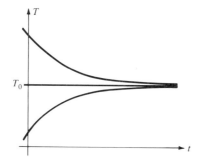

$$\frac{d}{dt}(Te^{-kt}) = -ke^{-kt}T_0.$$

Antidifferentiating (we now use the assumption that T_0 is constant for the first time), we get

$$Te^{-kt} = e^{-kt}T_0 + C,$$

and multiplying by e^{kt}, we obtain

$$T = T_0 + Ce^{kt}. \tag{12}$$

Letting $t = 0$ in (12) to find C, we have

$$T(0) = T_0 + C,$$

so that

$$T(0) - T_0 = C. \tag{13}$$

Finally, we substitute (13) into (12) to get

$$T(t) = T_0 + e^{kt}(T(0) - T_0). \tag{14}$$

Note that since k is negative

$$\lim_{t \to \infty} [T_0 + e^{kt}(T(0) - T_0)] = T_0 + 0(T(0) - T_0) = T_0.$$

That is, the surface temperature approaches the ambient temperature. Also note that T_0 is an *equilibrium* since $T(t) = T_0$ is a constant solution of the differential equation. (See Fig. 2.9.2.)

Example 2.9.2

The room temperature in your office is 70°F. Experience has taught you that the temperature of a cup of coffee brought to your office will drop from 120°F to 100°F in 10 minutes. What should be the temperature of your cup of coffee if you want it to take 20 minutes before it drops to 100°F?

Solution

First, we define our notation, set up the basic equations, and express the given data in terms of these. Let t be the time in minutes, $T(t)$ the temperature at time

t in degrees Fahrenheit. The cooling law (10) is

$$\frac{dT}{dt} = k(T - T_0). \tag{15}$$

From the problem description,

$$T_0 = 70 \tag{16}$$

and

$$T(10) = 100 \quad \text{when} \quad T(0) = 120. \tag{17}$$

We shall break this problem into two subproblems. The first is to determine k for our coffee cup. Solving (15) we get from (14) that

$$T(t) = T_0 + e^{kt}(T(0) - T_0). \tag{18}$$

To use condition (17), let $t = 10$ in (18),

$$T(10) = T_0 + e^{10k}(T(0) - T_0), \tag{19}$$

and use the given data from Eqs. (16) and (17) in (19),

$$100 = 70 + e^{10k}(120 - 70),$$

so that

$$e^{10k} = \frac{3}{5} \quad \text{and} \quad k = \frac{\ln(3/5)}{10} \approx -0.05.$$

Now we can solve the second subproblem, which is "Find $T(0)$ if $T(20) = 100$." From (18) with $t = 20$,

$$T(20) = T_0 + e^{k20}(T(0) - T_0).$$

Using $T(20) = 100$, $T_0 = 70$, and $k = \frac{1}{10}\ln\frac{3}{5}$, we have

$$100 = 70 + e^{2\ln(3/5)}(T(0) - 70)$$

or

$$30 = \frac{9}{25}(T(0) - 70).$$

Finally, solving this equation for $T(0)$, we obtain

$$T(0) \approx 153.33°F.$$

In this example the coffee-cup surface temperature dropped from 120° to 100° in 10 min, and from 153° to 100° in 20 min. This shows the impossibility of guessing the answer without using the differential equation and its solutions. ■

2.9.3 Evaporation

Let us suppose that a quantity of fluid is exposed to a gas. Then, assuming that the temperature is held constant, that the gas is well circulated, and that not enough fluid evaporates to significantly change the concentration in the gas, the following statement would, in many cases, be a sufficiently accurate model (law) of the described situation:

> The rate of evaporation at time t is proportional to the amount of surface area at time t. (20)

Example 2.9.3 You have a V-shaped tank full of water (Fig. 2.9.3). The tank is 10 m long, 3 m deep, and 6 m wide. Only the top surface is exposed to the air. After 1 hr the water level has dropped 0.1 m.

a) Set up an equation for the water depth, and solve for the depth as a function of time.

b) Determine how long it takes for the tank to empty.

c) Set up [without referring to part (a)] a differential equation for the volume, and solve it to obtain the volume as a function of t.

Solution Let $V(t)$ be the volume of water at time t in cubic meters, $h(t)$ be the depth at time t in meters, and $A(t)$ the surface area at time t in square meters. Measure time in hours. The evaporation law (20) is then

$$\frac{dV}{dt} = kA. \tag{21}$$

The data known to us are the shape of the tank and the fact that

$$h(0) = 3, \tag{22}$$
$$h(1) = 2.9. \tag{23}$$

Note that we cannot solve (21) as written since V and A are both unknown functions of t.

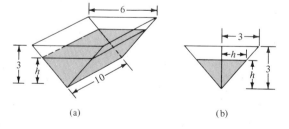

Figure 2.9.3   (a)   (b)

Parts (a) and (b)

We need to express V and A in terms of h. By similar triangles we see that the area of the shaded section in the end view of the tank (Fig. 2.9.3b) is $\frac{1}{2}$ base $\times$ height $= \frac{1}{2}(2h)h = h^2$. The length of the tank is 10, so that

$$V = 10h^2, \qquad V(0) = 10 \cdot 3^2 = 90. \tag{24}$$

The surface area of the water is

$$A = 10(2h) = 20h. \tag{25}$$

We now substitute these expressions for V and A into (21) to obtain a differential equation for h,

$$\frac{d}{dt}(10h^2) = k 20h.$$

Thus

$$20h\frac{dh}{dt} = k 20h.$$

and finally

$$\frac{dh}{dt} = k.$$

Antidifferentiating gives

$$h(t) = kt + C.$$

To find k and C apply the initial data (22) and (23), which yield $h(0) = C = 3$, and $k = -0.1$ so that

$$h(t) = -(0.1)t + 3.$$

The tank is empty when $h(t) = 0$ or $t = 30$ hr.

Part (c)

Again we work with Eqs. (21) through (25), but this time we wish to get a differential equation in terms of V and t. In order to write (21) in terms of V and t, we need to express A in terms of V. Using (24) and (25), we have

$$A = 20h = 20\left(\frac{V}{10}\right)^{1/2} = 2\sqrt{10}V^{1/2}.$$

The differential equation (21) now is

$$\frac{dV}{dt} = 2\sqrt{10}kV^{1/2}, \qquad V(0) = 90, \tag{26}$$

which is a nonlinear differential equation in V that can be solved by separation of variables. We rewrite (26) as

$$\frac{dV}{V^{1/2}} = 2\sqrt{10}k\, dt.$$

We then antidifferentiate to get

$$2V^{1/2} = 2\sqrt{10k}t + C$$

or

$$V^{1/2} = \sqrt{10k}t + \frac{C}{2}.$$

Let $t = 0$ in this equation to find

$$\frac{C}{2} = V(0)^{1/2} = \sqrt{90} = 3\sqrt{10}.$$

Thus

$$V^{1/2} = \sqrt{10k}t + 3\sqrt{10}$$

and

$$V(t) = (\sqrt{10k}t + 3\sqrt{10})^2 = 10(kt + 3)^2. \blacksquare$$

This example shows that our choice of variables and problem formulation determines whether or not a problem is linear and the techniques we must use to solve it.

2.9.4 Flow Due to Pressure

Figure 2.9.4

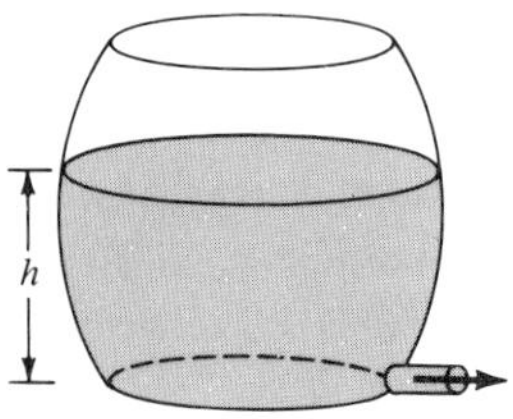

Suppose that we have a container of fluid, as in Fig. 2.9.4, which has a hole at the bottom. The hole has cross-sectional area B. Let us assume that the hole is small relative to the size of the tank and that the depth of the fluid at time t is $h(t)$. The loss of potential energy due to the drop in fluid level must equal the kinetic energy of the outflow (where we have ignored friction). Consider a short time period Δt for which there is a small change in fluid level. If m is the mass of the lost fluid and v is its exit velocity, then the potential energy is approximately mgh, where g is the gravitational constant in the units being used. The kinetic energy is approximately $\frac{1}{2}mv^2$. Thus

$$mgh = \tfrac{1}{2}mv^2 \quad \text{or} \quad v = \sqrt{2gh}.$$

The rate of water flow out of the hole is Bv. We can make this argument precise by taking the limit of estimates as $\Delta t \to 0$ or by using differentials, but it suffices to make the following "law" plausible.

The rate of flow of fluid out of a hole is proportional to the square root of the depth. (27)

Of course (27) is no longer valid once the water level becomes even with the hole. We shall ignore this point and assume that the amount of water at that time is "small." That is, either the hole is relatively small or it is on the bottom of the container.

Before giving a specific example we review the relationship between the volume and surface area for a general container (ignoring capillary effects). If $A(s)$ represents the surface area of the fluid when the depth is s, then the volume at depth h, $V(h)$, is

$$V(h) = \int_0^h A(s)\,ds,$$

so that, by the fundamental theorem of calculus,

$$\frac{dV}{dh} = A(h). \tag{28}$$

Using the chain rule, we can relate the rate of change of volume to the rate of change of depth:

$$\frac{dV}{dt} = \frac{dV}{dh}\frac{dh}{dt} = A(h)\frac{dh}{dt}. \tag{29}$$

Example 2.9.4

A cube-shaped open tank is 10 ft on a side (Fig. 2.9.5) and is initially full of water. There is a hole at the bottom from which water flows at a rate proportional to the square root of the depth. The constant of proportionality in ft²/sec is -5. Ignore evaporation.

a) Find the volume and depth as a function of time t.

b) Determine when the tank will be empty.

Solution

Let V be the volume of the tank. Then the given information is

$$\frac{dV}{dt} = -5\sqrt{h}, \quad V(0) = 10^3, \quad h(0) = 10. \tag{30}$$

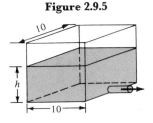

Figure 2.9.5

We need to express (30) in terms of a single dependent variable, say h. At time t, $V = h \cdot 10 \cdot 10 = 100h$. Substituting $V = 100h$, into (30) gives

$$100\frac{dh}{dt} = -5\sqrt{h} \tag{31}$$

or

$$\frac{dh}{dt} = -\frac{1}{20}\sqrt{h}. \tag{32}$$

Equation (32) is easily solved by separation of variables,

$$\int \frac{dh}{\sqrt{h}} = -\int \frac{1}{20}\,dt,$$

and integration,

$$2\sqrt{h} = -\frac{1}{20}t + C.$$

Letting $t = 0$ and using $h(0) = 10$ gives $C = 2\sqrt{10}$, so that

$$\sqrt{h} = -\frac{t}{40} + \sqrt{10}$$

and finally

$$h = \left(-\frac{t}{40} + \sqrt{10}\right)^2. \tag{33}$$

Also,

$$V = 100h = 100\left(-\frac{t}{40} + \sqrt{10}\right)^2.$$

The tank is empty when $h = 0$ or $t = 40\sqrt{10}$ sec (see Fig. 2.9.6). ∎

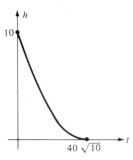

Figure 2.9.6
Graph of depth for Example 2.9.4.

In Example 2.9.4, $h = 0$ is an equilibrium and the solution $h(t)$ actually reaches the equilibrium in a finite amount of time ($40\sqrt{10}$ sec). Contrast this result with the behavior of the coffee-cup temperature (Fig. 2.9.2) or the law of radioactive decay (Fig. 2.9.1). In those examples the solution approached an equilibrium but never actually reached it (in theory) in a finite amount of time. To have a solution of a differential equation such as (33) reach an equilibrium (in this example $h(t) \equiv 0$) in finite time, the Basic Existence and Uniqueness Theorem (Theorem 2.1.1) must not hold at the equilibrium, since both the nonequilibrium solution and the equilibrium solution are satisfying the same initial condition, in this case $h(40\sqrt{10}) = 0$. One reason that the Basic Existence and Uniqueness Theorem fails to hold for Example 2.9.4 is that $\sqrt{h}$ is not differentiable with respect to h at $h = 0$.

Exercises

1. A radioactive isotope has a half-life of 16 days. You wish to have 30 g at the end of 30 days. How much radioisotope should you start with?

2. A radioisotope is going to be used in an experiment. At the end of 10 days only 5% is to be left. What should the half-life be?

3. A radioactive isotope sits unused in your laboratory for 10 yr, at which time it is found to contain only 80% of the original amount of radioactive material.

a) What is the half-life of this isotope?
b) How many *additional* years will it take until only 15% of the original amount is left?

4. At the time an item was produced, 0.01 of the carbon it contained was carbon-14, a radioisotope with a half-life of about 5745 yr.

a) You examine the item and discover that only 0.0001 of the carbon is carbon-14. How old is the object? (This process of determining the age of an object from the amount of carbon-14 it contains is known as carbon-14 dating.)

b) Derive a formula that gives the age A of the object in terms of the fraction of carbon that is carbon-14 at the present time, T.

5. The temperature of an engine at the time it is shut off is 200°C. The surrounding air temperature is 30°C. After 10 min have elapsed, the surface temperature of the engine is 180°C.

   a) How long will it take for the surface temperature of the engine to cool to 40°C?

   b) For a given temperature T between 200°C and 30°C, let $t(T)$ be the time it takes to cool the engine from 200°C to T. [For example, $t(200) = 0$ and $t(40)$ is the answer to part (a).] Find the formula for $t(T)$ in terms of T and graph the function. (The ambient temperature is still 30°C.)

6. Earlier experiments have shown that a certain component cools in air according to the cooling law (Eq. 15) with constant of proportionality -0.2. At the end of the first processing stage the temperature of the component is 120°C. The component remains for 10 min in a large room and then enters the next processing stage. At that time the surface temperature is supposed to be 60°C.

   a) What must the room temperature be for the desired cooling to take place?

   b) Suppose that the entrance and exit temperatures are still set at 120°C and 60°C, respectively, but the length of the wait in the room is w, a constant. Find the desired room temperature as a function of w and graph it.

7. An object at 100°C is to be placed in a 40°C room. What should the constant of proportionality be in Eq. (15) in order that the object be at 60°C after 10 min?

8. The air in a room is cooling. At time t (in hours) the air temperature is $T_0(t) = 70 + 20e^{-t/2}$. An object is placed in the room at time $t = 0$. The object is initially at 50°C and changes temperature according to Eq. (15) with $k = -\frac{1}{2}$.

   a) Find the temperature, $T(t)$, of the object for $0 \le t \le 5$.

   b) Graph both T_0 and T on the same axes.

9. An instrument at an initial temperature of 40°C is placed in a room whose temperature is 20°C. For the next 3 hr the room temperature $T_0(t)$ gradually rises and is given by $T_0(t) = 20 + 10t$, t in hours.

   a) Give the form the cooling law Eq. (15) takes for the instrument.

   b) From prior experience, you know that your instrument cools according to Eq. (15) with $k = -1$ if t is measured in hours. If $T(t)$ is the surface temperature at time t, solve the equation in part (a) for $T(t)$.

   c) Graph $T_0(t)$ and $T(t)$ on the same axes for $0 \le t \le 5$.

10. A conical tank (Fig. 2.9.7), open at the top, is 4 m high and has a top radius of 1 m. It is filled with water. After 2 hr the depth of the water has dropped 1 m due to evaporation.

Figure 2.9.7

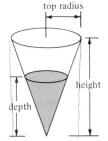

   a) Set up the equation for the depth of the water as a function of t and solve it.

   b) Set up the equation for the volume of water as a function of t and solve it. [Do not use the results of part (a).]

   c) When will half the water have evaporated?

11. A conical tank (Figure 2.9.7) is 6 m high and has a top radius of 2 m. Initially it is full of water. A valve is opened at the bottom and water is pumped out at a rate proportional to the depth of the water. After 1 min the depth has decreased to 3 m. Ignoring evaporation,

a) Set up the equation for the water's depth as a function of t and solve it.

b) Determine how long will it take to empty the tank.

12. A conical tank 6 m high with base radius 2 m is initially full of water. The tank sits on its base (Figure 2.9.8). A valve is opened at the bottom and water is pumped out at a rate proportional to the depth. After π min the depth of the water has decreased to 1 m. Ignoring evaporation,

Figure 2.9.8

a) Set up the equation for the depth of the water as a function of t and solve it.

b) Determine how long will it take to empty the tank.

13. A cylindrical tank has a radius of 3 m and is filled with water to a depth of 4 m. Water flows out of a hole at the bottom at a rate that is proportional to the square root of the depth of the water. The constant of proportionality in $m^2/\sec$ is -4. Determine the volume and depth as functions of t.

14. The tank in the shape of an inverted V shown in Fig. 2.9.9 is 10 ft high, 20 ft wide, and 20 ft long. The water is initially 4 ft deep and flows out of a hole in the bottom at a rate proportional to the square root of its depth. The constant of proportionality is -8 ft^2/min.

Figure 2.9.9

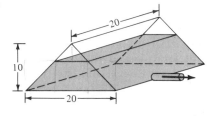

a) Find the (implicit) relationship between depth and time.

b) When will the tank be empty?

15. (*Interest*) Let us assume that an amount of money is invested at interest that is compounded daily. Since a good approximation to daily compounding is provided by *continuous compounding*, we can use, in this exercise, the rule for continuous compounding, assuming a constant interest rate. This rule is:

> The rate of change of the amount of money at time t is proportional to the amount of money. If t is measured in years, then the constant of proportionality is the annual interest rate. (34)

a) Express Statement (34) as a differential equation and solve it.

b) The *effective interest rate* is the percentage by which the money has increased in one year. Due to compounding, it is higher than the annual interest rate. If the annual interest rate is 6%, what is the effective interest rate?

c) Let r be the annual interest rate, and R the effective interest rate. Using the result from part (a), express R as a function r and graph the expression you obtained.

16. You have $10,000 and intend to invest it for 5 yr in a bank which offers continuous compounding. If you want to have $15,000 in your account at the end of these 5 yr, what annual interest rate do you have to get? Use (34).

17. You invest $2000 in an account paying 6% a year, compounded daily, on the amount in excess of $500.

a) Express this as a differential equation by assuming continuous compounding.

b) Solve the differential equation and determine the amount in the account after 10 yr.

c) How much more would the account have after 10 yr if the full amount earned interest?

Many animal populations can be assumed to grow (at least for a while) at a rate proportional to the population size if we ignore such factors as limitations on food supplies, competition for shelter, or the effects of predation. The *doubling time* for a population is how long it takes for the population to double. Exercises 18 and 19 assume that the population grows at a rate proportional to its size.

18. Initially you have 0.1 g of a bacteria in a large container. Two hr later you have 0.15 g. What is the doubling time for this bacteria?

19. The doubling time for a certain virus is 3 yr. How long will it take for the virus to increase to 10 times its current population level?

Note: Additional flow and money problems will be given in Section 2.10.

2.10 Mixing and Flow Problems

The problems in this section will all be analyzed in the following manner: A quantity $Q(t)$—say, the amount of water in a tank—varies with time. Further amounts of this quantity are being added. This addition will be called *inflow*. Simultaneously, some of this quantity is being lost. In the case of a water tank, the loss could be due to evaporation, overflow, an open valve, or all three. The quantity lost will be called *outflow*. The beginning point for analyzing these problems is

$$\begin{bmatrix} \text{Rate of change} \\ \text{of } Q \text{ at} \\ \text{time } t \end{bmatrix} = \frac{dQ}{dt} = \begin{bmatrix} \text{Inflow rate} \\ \text{of } Q \text{ at} \\ \text{time } t \end{bmatrix} - \begin{bmatrix} \text{Outflow rate} \\ \text{of } Q \text{ at} \\ \text{time } t \end{bmatrix}. \quad (1)$$

Procedure for Flow Problems

1. Label quantities and note given data.
2. Express "inflow rates" and "outflow rates" in terms of given variables and substitute into (1).
3. Does (1) now involve only one dependent variable? If yes, go to Step 4. If no, then use the relationships between the dependent variables to eliminate all but one dependent variable. (This procedure is like that illustrated in Example 2.9.3.)
4. Solve the differential equation.
5. Answer any questions, such as "how long"?

Example 2.10.1 A 100-m^3 tank is full of water. The water contains a pollutant z at a concentration of 0.5 g/m^3. Cleaner water, with a pollutant concentration of 0.1 g/m^3, is pumped into the well-mixed tank at a rate of 4 m^3/sec. Water flows out of the tank through an overflow valve at the same rate as it is pumped in.

a. Determine the amount and concentration of pollutant in the tank as a function of time. Graph the result.

b. At what time will the concentration be 0.3 g/m³? At what time will the concentration be 0.2 g/m³?

Solution

In order to illustrate the general procedure, we shall include a few more steps than necessary to do this particular problem.

1. Let $Q(t)$ be the amount in grams of pollutant in the tank, $c(t)$ the concentration in g/m³ of pollutant in the tank, and t the time in seconds. The information given is $c(0) = 0.5$, and

$$\begin{bmatrix} \text{Inflow rate} \\ \text{of water} \end{bmatrix} = \begin{bmatrix} \text{Outflow rate} \\ \text{of water} \end{bmatrix} = \frac{4 \text{ m}^3}{\text{sec}}.$$

2. The relationship (1) is

$$\frac{dQ}{dt} = \begin{bmatrix} \text{Inflow rate} \\ \text{of } z \end{bmatrix} - \begin{bmatrix} \text{Outflow rate} \\ \text{of } z \end{bmatrix},$$

which in turn is

$$\frac{dQ}{dt} = \begin{bmatrix} \text{Inflow rate} \\ \text{of water} \\ \text{at time } t \end{bmatrix} \begin{bmatrix} \text{Concentration of } z \\ \text{in inflow} \\ \text{water} \\ \text{at time } t \end{bmatrix}$$

$$- \begin{bmatrix} \text{Outflow rate} \\ \text{of water} \\ \text{at time } t \end{bmatrix} \begin{bmatrix} \text{Concentration of } z \\ \text{in outflow} \\ \text{water} \\ \text{at time } t \end{bmatrix}. \quad (2)$$

But

$$\begin{bmatrix} \text{Concentration of } z \\ \text{in outflow} \\ \text{water} \end{bmatrix} = \begin{bmatrix} \text{Concentration of } z \\ \text{in tank} \end{bmatrix}$$

$$= \frac{[\text{Amount of } z \text{ in tank}]}{[\text{Volume of tank}]} = \frac{Q}{100}.$$

Thus (2) is

$$\frac{dQ}{dt} = [4][0.1] - [4]\left[\frac{Q}{100}\right]$$

or

$$\frac{dQ}{dt} = 0.4 - \frac{Q}{25}. \quad (3)$$

Also

Figure 2.10.1
Graph of $10 + 40e^{-t/25}$.

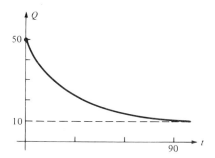

$$Q(0) = 100c(0) = 100(0.5) = 50. \tag{4}$$

3. The differential equation (3) is a first-order linear differential equation in terms of one dependent variable Q. We shall solve it by using integrating factors.

4. Rewrite the differential equation as

$$\frac{dQ}{dt} + \frac{Q}{25} = 0.4.$$

The integrating factor is $\exp(\int \frac{1}{25} \, dt) = \exp(t/25)$. Multiplying by the integrating factor gives

$$\frac{d}{dt}(e^{t/25}Q) = 0.4e^{t/25},$$

and, upon antidifferentiating, we obtain

$$e^{t/25}Q = 10e^{t/25} + C.$$

Letting $t = 0$ and using $Q(0) = 50$, we find $C = 40$. Thus

$$Q(t) = 10 + 40e^{-t/25}. \tag{5}$$

Figure 2.10.1 gives the graph of $Q(t)$.
Next we calculate the concentration $c(t)$:

$$c(t) = \frac{Q(t)}{\text{volume}} = \frac{10 + 40e^{-t/25}}{100} = 0.1 + 0.4e^{-t/25}. \tag{6}$$

To determine when $c(t) = 0.3$, use (6):

$$0.3 = 0.1 + 0.4e^{-t/25}.$$

Then $t = -25 \ln(0.5) \approx 17.32$ sec. Similarly, the concentration is 0.2 when $t = -25 \ln(0.25) \approx 34.66$ sec. ∎

For this particular problem, it would have been just as easy to set up the equation for concentration instead of amount of pollutant. However, if the volume varies, then it is usually easier to proceed as we did in this example.

Example 2.10.2 A 100-gal tank is initially half full of pure water. Then water containing 0.1 lb/gal of salt is added at a rate of 4 gal/min. The tank contents are well mixed. Water flows out of the tank at a rate of 2 gal/min until the tank is full. Once the tank is full it overflows. Find the amount and concentration of salt in the tank as functions of t for the first 2 hr. Graph the result.

Solution Let $Q(t)$ be the amount of salt in the tank and $c(t)$ the concentration of salt in the tank. Use pounds, gallons, and minutes as units. We have $Q(0) = 0$ since initially the tank is filled with pure water. The volume $V(t)$ of water in the tank varies. At the outset, the volume of water is 50 gal; it increases by 2 gal/min until the tank is full, and then the volume remains constant at 100 gal. Thus

$$V(t) = \begin{cases} 50 + 2t, & 0 \leq t \leq 25, \\ 100, & 25 \leq t \leq 120. \end{cases} \tag{7}$$

Again starting with (1), the basic relationship is

$$\frac{dQ}{dt} = \begin{bmatrix} \text{Inflow} \\ \text{rate of} \\ \text{salt} \end{bmatrix} - \begin{bmatrix} \text{Outflow} \\ \text{rate of} \\ \text{salt} \end{bmatrix}$$

$$= \begin{bmatrix} \text{Inflow} \\ \text{rate of} \\ \text{water} \end{bmatrix} \begin{bmatrix} \text{Concentration of} \\ \text{salt in} \\ \text{inflow water} \end{bmatrix}$$

$$- \begin{bmatrix} \text{Outflow} \\ \text{rate of} \\ \text{water} \end{bmatrix} \begin{bmatrix} \text{Concentration of} \\ \text{salt in} \\ \text{outflow water} \end{bmatrix}. \tag{8}$$

Both the outflow rate of water and the formula for the volume change when the tank is full ($t = 25$). Thus we break the problem into two parts.

Part I
$0 \leq t \leq 25$. Equation (8) for $0 \leq t \leq 25$ is

$$\frac{dQ}{dt} = [4][0.1] - [2]\frac{[\text{Salt in tank}]}{[\text{Water in tank}]}$$

$$= 0.4 - 2\frac{Q}{50 + 2t}, \tag{9}$$

or

$$\frac{dQ}{dt} + \frac{2}{50 + 2t}Q = 0.4. \tag{10}$$

We solve this linear differential equation by the integrating-factor method. The

2.10 Mixing and Flow Problems

integrating factor is

$$\exp\left(\int \frac{2}{50+2t}\, dt\right) = \exp[\ln(50+2t)] = 50+2t.$$

We next multiply (10) by this integrating factor to get

$$\frac{d}{dt}((50+2t)Q) = (50+2t)0.4 = 20+0.8t.$$

Antidifferentiating yields

$$(50+2t)Q = 20t + 0.4t^2 + C.$$

The initial condition $Q(0) = 0$ implies $C = 0$. Thus

$$Q(t) = \frac{20t + 0.4t^2}{50+2t} \tag{11}$$

and

$$c(t) = \frac{Q(t)}{V(t)} = \frac{20t + 0.4t^2}{(50+2t)^2}.$$

Part II

$25 \le t \le 120$. For t between 25 and 120 min, $V = 100$ gal, and the outflow rate of water is 4 gal/min. Thus (8) is now

$$\frac{dQ}{dt} = [4][0.1] - [4]\frac{Q}{V},$$

or

$$\frac{dQ}{dt} = 0.4 - 4\frac{Q}{100} = 0.4 - \frac{Q}{25} \tag{12}$$

instead of the result in (10). From (11), which is valid for $0 \le t \le 25$, we have

$$Q(25) = \frac{500 + 0.4(25)^2}{100} = 7.5. \tag{13}$$

Equation (12) is again linear. In Example 2.10.1, it was solved, to yield

$$e^{t/25}Q(t) = 10e^{t/25} + C. \tag{14}$$

To find C, let $t = 25$ in (14), and use $Q(25) = 7.5$ from (13), to get

$$e(7.5) = 10e + C,$$

so that $C = -2.5e$. Hence (14) is

$$e^{t/25}Q(t) = 10e^{t/25} - 2.5e$$

or

$$Q(t) = 10 - 2.5ee^{-t/25} = 10 - 2.5e^{(25-t)/25}. \tag{15}$$

Combining (11) and (15), we obtain the final result

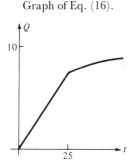

Figure 2.10.2
Graph of Eq. (16).

Figure 2.10.3
Graph of Eq. (17).

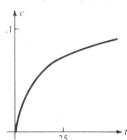

$$Q(t) = \begin{cases} \dfrac{20t + 0.4t^2}{50 + 2t} & \text{if } 0 \le t \le 25, \\ 10 - 2.5e^{(25-t)/25} & \text{if } 25 \le t, \end{cases} \quad (16)$$

and

$$c(t) = \dfrac{Q(t)}{V(t)} = \begin{cases} \dfrac{20t + 0.4t^2}{(50 + 2t)^2} & \text{if } 0 \le t \le 25, \\ 0.1 - (0.025)e^{(25-t)/25} & \text{if } 25 \le t, \end{cases} \quad (17)$$

which are graphed in Figs. 2.10.2 and 2.10.3. ∎

Mixing-flow problems can also be nonlinear.

Example 2.10.3

A conical tank is 12 m deep and its open top has a radius of 12 m. Initially the tank is empty. Water is added at a rate of π m³/hr. Water evaporates at a rate proportional to the surface area of the water. The constant of proportionality is 0.01 in m/hr (Fig. 2.10.4).

a. Does the tank ever fill up?
b. If the tank does not fill up, what is the maximum depth the water reaches?
c. Find the depth as a function of time. (Explicity if possible, otherwise implicitly.)

Solution

Let V be the volume of water in cubic meters, A the surface area, h the depth in meters, t the time in hours. Then

Figure 2.10.4

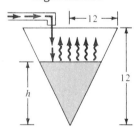

$$\dfrac{dV}{dt} = \begin{bmatrix} \text{Inflow} \\ \text{rate} \\ \text{of water} \end{bmatrix} - \begin{bmatrix} \text{Outflow} \\ \text{rate} \\ \text{of water} \end{bmatrix}$$

$$= \pi - k[\text{Surface area}] = \pi - (0.01)A. \quad (18)$$

The quantities V and A are both dependent on t. We shall express them in terms of the common variable h since parts (b) and (c) refer to depth: $V = \tfrac{1}{3}\pi h^3$ and $A = \pi h^2$. Thus (18) is

$$\dfrac{d(\tfrac{1}{3}\pi h^3)}{dt} = \pi - (0.01)\pi h^2$$

or

$$\pi h^2 \dfrac{dh}{dt} = \pi - (0.01)\pi h^2.$$

Dividing by πh^2, we have

$$\dfrac{dh}{dt} = \dfrac{1 - (0.01)h^2}{h^2}. \quad (19)$$

We can answer parts (a) and (b) without solving (19), using the ideas of Section 2.2. The tank will stop filling if inflow balances outflow. That is,

Figure 2.10.5
Graph of Eq. (23).

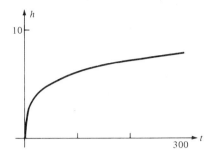

$$0 = \frac{dh}{dt} = \frac{1 - (0.01)h^2}{h^2}.$$

Solving for h, we find that $h = 10$ m. Also, $h' > 0$ if $h < 10$ and $h' < 0$ if $h > 10$. Thus in answer to part (a), the tank never fills completely. The answer to (b) is that the depth approaches the equilibrium value of 10 m.

To answer part (c) we must solve (19). Equation (19) is nonlinear and can be solved by separation of variables. Separating the variables h, t in (19) gives

$$\int \frac{h^2}{1 - (0.01)h^2} \, dh = \int dt. \tag{20}$$

In order to apply the method of partial fractions on the left, we first divide to get

$$\frac{h^2}{1 - (0.01)h^2} = -100 + \frac{100}{1 - (0.01)h^2}. \tag{21}$$

Using partial fractions (Appendix B), we get

$$\frac{h^2}{1 - (0.01)h^2} = -100 + \frac{A}{1 - 0.1h} + \frac{B}{1 + 0.1h},$$

where $100 = A(1 + 0.1h) + B(1 - 0.1h)$ so that $A = 50$, $B = 50$. Thus

$$\frac{h^2}{1 - (0.01)h^2} = -100 + \frac{50}{1 - 0.1h} + \frac{50}{1 + 0.1h}. \tag{22}$$

Returning to (20), use (22) and antidifferentiate both sides, to get

$$-100h - 500 \ln|1 - 0.1h| + 500 \ln|1 + 0.1h| = t + C.$$

Letting $t = 0$ and using $h(0) = 0$, we obtain $C = 0$, and the relationship between h and t is (for $h \neq 10$)

$$-100h + 500 \ln \left| \frac{1 + 0.1h}{1 - 0.1h} \right| = t. \tag{23}$$

It is easiest to graph this relationship by using a calculator or computer, and assigning values of h in the interval $0 \leq h < 10$ (see Fig. 2.10.5). ∎

That the rate of evaporation increases with surface area seems self-evident. However, we have examples of dams built in arid regions whose reservoirs have never been completely filled due to the large surface area of the water and the resulting high evaporation rate.

Many different kinds of problems can be analyzed as mixing-flow problems.

Example 2.10.4 You open a high-yield bank account with $1000 and deposit $10 a day. Interest is compounded daily at a rate of 10% per year.

a. Find a formula for the amount of money in the account as a function of time.

b. How much money is in the account after 10 years? after 20 years?

Solution Let $Q(t)$ be the amount of money in the account in dollars. Let t be time measured in years. Daily compounding will be approximated by continuous compounding. We look at the account as a "reservoir" of money with two sources of inflow, interest and deposits, and no outflow. Equation (1) takes the form

$$\begin{bmatrix} \text{Rate of} \\ \text{change of} \\ \text{money} \end{bmatrix} = \begin{bmatrix} \text{Rate of inflow} \\ \text{due to} \\ \text{deposits} \end{bmatrix} + \begin{bmatrix} \text{Rate of} \\ \text{inflow due} \\ \text{to interest} \end{bmatrix}. \tag{24}$$

All three terms in (24) must have the same units, which we take as dollars/year. The deposit rate is

$$10\frac{\text{dollars}}{\text{day}} = \frac{10 \text{ dollars}}{\text{day}} \frac{365 \text{ days}}{\text{year}} = 3650\frac{\text{dollars}}{\text{year}}.$$

At time t, $Q(t)$ dollars are earning interest at the yearly rate of $0.1Q$. (See Exercise 15 of Section 2.9.) Since, at the outset, we have $1000, Q(0) = $1000. Thus (24) is

$$\frac{dQ}{dt} = 3650 + 0.1Q, \qquad Q(0) = 1000. \tag{25}$$

This is a first-order linear differential equation. We will solve it by the integrating-factor method. We rewrite (25) as

$$\frac{dQ}{dt} - 0.1Q = 3650. \tag{26}$$

The integrating factor is

$$\exp\left(\int -0.1 \, dt\right) = \exp(-0.1t).$$

Multiplying both sides of (26) by the integrating factor, we get

$$\frac{d}{dt}(Qe^{-0.1t}) = 3650e^{-0.1t},$$

and antidifferentiating, we obtain

$$Qe^{-0.1t} = -\frac{3650}{0.1}e^{-0.1t} + C. \tag{27}$$

We now let $t = 0$ to find that $C = 37{,}500$. Substituting $C = 37{,}500$ into (27) and multiplying by $e^{0.1t}$ finally yields

$$Q(t) = -36{,}500 + 37{,}500e^{0.1t}. \tag{28}$$

This answers part (a). To answer part (b), we evaluate (28) at $t = 10$ and $t = 20$.

$$Q(10) = -36{,}500 + 37{,}500e \approx \$65{,}435,$$
$$Q(20) = -36{,}500 + 37{,}500e^2 \approx \$240{,}590. \quad \blacksquare$$

Exercises

1. A well-mixed tank contains 300 gal of water with a salt concentration of 0.2 lb/gal. Water containing salt at a concentration of 0.4 lb/gal enters at a rate of 2 gal/min. An open valve allows water to leave the tank at the same rate.

   a) Determine the amount and the concentration of salt in the tank as functions of time.

   b) How long will it take for the concentration to increase to 0.3 lb/gal?

2. A room has a volume of 800 ft^3. The air in the room contains chlorine at a concentration of 0.1 g/ft^3. Fresh air enters at a rate of 8 ft^3/min. The air in the room is well mixed and flows out of a door at the same rate as the fresh air comes in.

   a) Find the chlorine concentration in the room as a function of t.

   b) Suppose that the flow rate of fresh air is adjustable. Determine the flow rate required to reduce the chlorine concentration to 0.001 g/ft^3 within 20 min.

3. A well-mixed tank contains 100 liter of water with a salt concentration of 0.1 kg/liter. Water containing salt at a concentration of 0.2 kg/liter enters at a rate of 5 liter/hr. An open valve allows water to leave at 4 liter/hr. Water evaporates from the tank at 1 liter/hr.

   a) Determine the amount and concentration of salt as a function of time.

   b) Is the limiting concentration the same as that of the inflow?

4. A well-circulated pond contains 1 million liters of water which contains a pollutant at a concentration of 0.01 kg/liter. Pure water enters from a stream at 100 liter/hr. Water evaporates from the pond (leaving the pollutant behind) at 50 liter/hr and flows out an outlet pipe at 50 liter/hr. How many days will it take for the concentration of pollutant to drop to 0.001 kg/liter?

5. A 1000-gal tank is initially half full of water and contains 10 lb of iodine in solution. Pure water enters the tank at a rate of 6 gal/min. An open valve allows water to leave at a rate of 1 gal/min. When full, the tank overflows.

   a) Find the amount and concentration of iodine in the tank for the first 200 min.

   b) Graph both concentration and amount of iodine.

6. A 100-gal tank is initially full of water containing 10 lb of salt in solution. The tank will overflow whenever additional water is added. A pump is attached to a sensor. It pumps fresh water into the tank at a rate proportional to the concentration of salt in the tank. The constant of proportionality is 10(gal)^2/lb·min. Find the amount and concentration of salt in the tank as functions of t.

Exercises 7 and 8 are based on the following situation. A well-mixed rectangular tank with an open top is 10 m wide, 20 m deep, and 100 m long. It initially contains pure water to a depth of 2 m. Water containing salt at a concentration of 30 g/m³ is entering the tank at a rate of 20 m³/hr. Water evaporates from the tank at a rate proportional to the surface area. The constant of proportionality k is given in m/hr. An open valve permits water to flow from the tank at 10 m³/hr. When full, the tank overflows. *Hint:* For Exercises 7 and 8, first find $V(t)$.

7. Determine the amount and concentration of salt in the tank as functions of t if $k = 0.01$.

8. Suppose $k = 0.005$.
   a) When does overflow occur?
   b) Determine the amount and concentration of salt in the tank at the time at which the tank overflows.

9. A conical tank is 8 m deep and has a top radius of 8 m. The tank is initially empty. Water is added at a rate of 2π m³/min, a valve at the bottom releases water at a rate proportional to the water's depth (proportionality constant = π), and evaporation takes place at the surface at a rate proportional to the surface area of the water (proportionality constant = 1). (See Fig. 2.10.6.)

Figure 2.10.6

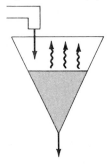

   a) Does the tank ever fill up? If it does, determine the time it takes for it to fill up. If it never fills up, give the limiting value of the volume.
   b) Find the (implicit) relationship between depth and time.

10. An amount of $10,000 is deposited in a bank that pays 9% annual interest compounded daily.
    a) If you withdraw $10 a day, how much money do you have after 3 yr?
    b) How much can you withdraw each day so that the account will be depleted in exactly 10 yr?

11. An amount of $1000 is deposited in a bank that pays 8% annual interest compounded daily. A deposit of B dollars is made daily.
    a) What should B be in order to have $10,000 after 5 yr?
    b) Determine the function $B(x)$ which gives the daily deposit needed to have $10,000 after x years. [$B(5)$ is computed in part (a).]

12. An amount of $10,000 is invested at 12% annual interest compounded daily. An additional investment of $$B$ is made daily. What should B be in order for the investment account to be $100,000 after 10 yr?

13. An amount of $100 is deposited in a foreign bank that pays 20% annual interest compounded daily. Each day you make a transaction of amount $f(t)$$/day. Part of the year you are able to deposit money [$f(t) > 0$] and part of the year you make withdrawals [$f(t) < 0$]. If t is in years, these transactions occur in the following cyclical yearly pattern:
$$f(t) = \frac{400}{365} \cos(2\pi t) \$/\text{day} = 400 \cos(2\pi t) \$/\text{year}.$$
    a) Find the amount of money in the account as a function of t.
    b) Graph your answer to (a) for 10 years (a sketch will do if no computer is available).

Exercises 14 and 15 are a continuation of Exercises 18 and 19 of Section 2.9.

14. A bacteria is reproducing in a large vat of nutrients according to an exponential growth law which would cause the population to double in .5 hr. However, bacteria are continuously siphoned off at a rate of 5 g/hr. Initially, there are 10 g of bacteria. How much bacteria is there in the vat after 2 hr?

15. An organism living in a pond reproduces at a rate proportional to the population size. Organisms also die off at a rate proportional to the population size. In addition, organisms are continuously added at a rate of k g/yr. Give the differential equation that models this situation.

16. Explain why *every* first-order differential equation can be viewed as a mixing-flow problem.

2.11 Orthogonal Trajectories

Many physical problems are described using a function $\phi(x, y)$ called a *potential* which represents a quantity such as height, temperature, or pressure. The curves $\phi(x, y) = c$ are called *equipotentials* or *level curves* because the potential is constant along these curves. In the case of height, the equipotentials are the familiar lines connecting points of equal altitude on a topographic map. For temperature, the equipotentials are usually called *isotherms*, and for pressure they are called *isobars*. (See Fig. 2.11.1.)

A potential often causes an action perpendicular to the equipotentials. These curves perpendicular to the equipotentials are usually called *flux* (or *stream*) *lines*. On a topographic map, the flux lines show the direction in which an object would roll downhill (at least initially). In the case of isotherms, the flux lines show the direction of heat flow.

Given the equipotentials, we can often find the flux lines. Since the flux lines are everywhere orthogonal (perpendicular) to the equipotentials, they are sometimes also called an *orthogonal family* of curves. (Recall Exercises 22 through 25 of Section 2.6.)

Procedure for Calculating Flux Lines from Equipotentials

1. First write the family of curves (equipotentials) in the form

$$\phi(x, y) = c.$$

2. Differentiate with respect to x to get

$$\phi_x(x, y) + \phi_y(x, y) \frac{dy}{dx} = 0.$$

3. Solve for dy/dx:

$$\frac{dy}{dx} = -\frac{\phi_x(x, y)}{\phi_y(x, y)}. \qquad (1)$$

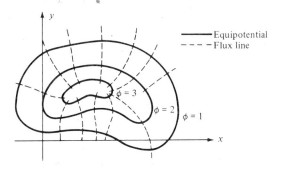

Figure 2.11.1 Equipotentials and flux lines.

Equation (1) gives the slope of the equipotentials at (x, y). Since the flux lines are to be orthogonal to the equipotentials, the flux slopes will be the negative reciprocals of the equipotential slopes. Thus the flux lines satisfy the differential equation

$$\frac{dy}{dx} = \frac{-1}{-\phi_x(x, y)/\phi_y(x, y)} = \frac{\phi_y(x, y)}{\phi_x(x, y)}. \tag{2}$$

4. Solve Eq. (2).

5. try and describe this family in words

Example 2.11.1 The expression $x^2 = c - 2y^2$ defines a family of curves (equipotentials). Find the orthogonal family (flux lines).

Solution First let us write the family of curves in the form $x^2 + 2y^2 = c$. We next differentiate $x^2 + 2y^2 = c$ with respect to x:

$$2x + 4y\frac{dy}{dx} = 0.$$

Solving for dy/dx yields

$$\frac{dy}{dx} = -\frac{x}{2y}.$$

The slope of the orthogonal family at (x, y) is thus given by

$$\frac{dy}{dx} = -\frac{1}{-x/2y} = \frac{2y}{x}.$$

Solving this differential equation by separation of variables, we get

$$\int \frac{1}{y}\,dy = \int \frac{2}{x}\,dx,$$

so that

$$\ln|y| = 2\ln|x| + c_1.$$

We now exponentiate both sides to arrive at the orthogonal family of curves

$$y = kx^2, \qquad k = \pm e^{c_1}.$$

Note that $k = 0$ also gives a solution. Figure 2.11.2 shows both families of curves. ∎

Sometimes the original family is not of the form $\phi(x, y) = c$ but rather $F(x, y, c) = 0$. That is, the parameter c is not solved for. We still need to rewrite

2.11 Orthogonal Trajectories

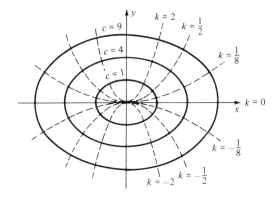

Figure 2.11.2
The families of curves
$x^2 + 2y^2 = c$ and
$y = kx^2$.

the family as $\phi(x, y) = c$. However, if ϕ is complicated to differentiate, we may differentiate $F(x, y, c) = 0$ with respect to x to get

$$F_x(x, y, c) + F_y(x, y, c)y' = 0$$

and then substitute ϕ for c.

Example 2.11.2 Find the differential equation for the orthogonal family to $x^3 + 3y^2 = x^2c$.

Solution Differentiating gives $3x^2 + 6yy' = 2xc$. But

$$c = \frac{x^3 + 3y^2}{x^2}$$

thus

$$y' = \frac{1}{6y}(2xc - 3x^2) = \frac{6y^2 - x^3}{6xy}$$

and the orthogonal family satisfies

$$y' = -\frac{6xy}{6y^2 - x^3}. \blacksquare$$

Exercises

For each family of curves (equipotentials) in Exercises 1–18, compute the orthogonal family (flux lines). Sketch both families for Exercises 1 through 3.

1. $y = x + c$
2. $y = cx$
3. $yx = c$
4. $y^2 = c(x + 3y)$
5. $y = x^c$
6. $y = (x + c)^3$
7. $y = x^2 + c$
8. $x^2 + y^2 = 4cx$
9. $x = (y - c)^2$
10. $y = e^{cx}$
11. $y = \tan(x + c)$
12. $x^{1/5} + y^{1/5} = c$
13. $y = c \cos x$
14. $y = \cos x + c$
15. $y = cx^2$
16. $e^y - e^x = c$
17. $y^3 + x^2 = c$
18. $\tan y + \tan x = c$

If they have not already been assigned, do Exercises 22 through 26 of Section 2.6.

In Exercises 19 through 21, suppose that a family of curves is the solution of a differential equation of the indicated type. Determine when the equation for the orthogonal family can be solved by the same method.

19. Linear **20.** Homogeneous **21.** Separable

22. Suppose that a family of curves is the solution of the exact differential equation $M\,dx + N\,dy = 0$.

a) Show that the orthogonal family satisfies $N\,dx - M\,dy = 0$.

b) Determine when $N\,dx - M\,dy = 0$ is also exact.

23. (Continuation of 22.) Suppose that $M\,dx + N\,dy = 0$ is exact. Determine when $N\,dx - M\,dy = 0$ has an integrating factor in terms of either x or y only.

2.12 Circuits

One of the applications of differential equations which will frequently recur throughout this book is the theory of electric circuits. There are several reasons for this, among them the importance of circuit theory and the pervasiveness of differential equations in circuit theory.

Circuits will be covered again in greater detail later. Right now we shall give some simple first-order applications and introduce the basic circuit concepts we shall use.

We consider only *lumped-parameter circuits*. In circuit theory, the word *circuit* means that quantities such as current are determined solely by position along the path. Thus field effects are ignored. *Lumped parameter* means that the effects of the various electrical components may be considered to be concentrated at one point. The converse of a lumped parameter circuit is a *distributed parameter* circuit. The analysis of distributed-parameter circuits often involves partial differential equations which we do not cover. Antennas are an example of distributed-parameter systems.

At each point in the circuit there are two quantities of interest: *voltage* (or potential) and *current* (or flow of charge). Current is, by convention, the net flow of positive charge. A *branch* is part of a circuit with two terminals to which connections can be made, a *node* is the point where two or more branches come together (a node is denoted ——•——), and a *loop* is a closed path formed by connecting branches. Our basic modeling laws are Kirchhoff's circuit laws: the current and voltage laws.

Current Law The algebraic sum of the currents entering a node at any instant is zero.

Voltage Law The algebraic sum of the voltage drops around a loop at any instant is zero.

To set up the circuit equations a current variable is assigned to each branch. One can talk either of the potentials (voltages) at the nodes or of the potential drops across the branches. Kirchhoff's current law may then be applied to each node, and the voltage law to each loop. This procedure exhibits a certain

Figure 2.12.1
A two-terminal device. The nodal voltages are denoted by v_0 and v_1, the voltage drop $v = v_0 - v_1$, and the current is denoted by i.

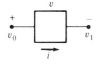

amount of arbitrariness. There is usually some redundancy among the equations and the determination of a minimal number of equations is generally computationally nontrivial.

In this section we discuss only single- or double-loop circuits. Let us consider the branch containing a two-terminal device shown in Fig. 2.12.1. For our purposes, the behavior of the device is completely determined if we know v and i at any time t. The relationship between v and i is called the *v-i characteristic* of the device.

We shall consider only five basic types of devices, as follows.

Resistor

If the voltage "drop" is uniquely determined by the time and the current,

$$v = f(i, t),$$

the device is called a *current-controlled resistor*. If

$$v = Ri$$

and R depends only on t, then the device is a *linear resistor* and R is called the *resistance*. A resistor that is not linear is called a *nonlinear resistor*. A resistor will be denoted by •—⋀⋀⋀—•; a nonlinear resistor by •—⎡⋀⋀⋀⎤—•.

For many devices, such as transistors, we can design models by considering them to be made up of linear capacitors, inductors, and current-controlled resistors.

We shall define only linear capacitors and inductors.

Capacitor

A capacitor stores energy in the form of a charge q. The charge q and voltage drop v across the capacitor are related by

$$q = Cv, \tag{1}$$

where C is the *capacitance*. If C is constant, we may differentiate (1) and use $i = dq/dt$ to get the v–i characteristic

$$i = C\frac{dv}{dt}. \tag{2}$$

The symbol for a capacitor is •—⊣⊢—•.

Table 2.12.1

Term	Symbol	Unit	Unit abbreviation
Charge	q	Coulomb	C
Current	i	Ampère	A
Resistance	R	Ohm	Ω
Capacitance	C	Farad = H·A/sec	F
Inductance	L	Henry	H
Voltage	v	Volt = Ω·A = C/F	V

Inductor

An inductor stores energy in a magnetic field. The voltage–current relationship for a (linear) inductor is

$$v = L\frac{di}{dt}, \tag{3}$$

where L is called the *inductance*. The symbol for an inductor is —⁀⁀⁀⁀⁀—.

Voltage Source

A voltage source is denoted by —⊖—.

Its voltage drop is $-e$ for any current. Batteries are an example of a voltage source.

Current Source

A current source is denoted by —⊖—.

In a current source, the current is i for any voltage. Some solar cells are current sources.

No device, of course, is only a resistor or an inductor. However, we can analyze many circuits by considering them to be made up of resistors, inductors, and capacitors. Also, no device is truly linear. However, if restrictions are put on the allowable current and voltage, we can often make the assumption that the device is linear.

Throughout this book the units and variables listed in Table 2.12.1 will be used in discussing circuits.

Applying Kirchhoff's Laws In applications of the loop and node laws, it is important to use the correct signs. When we apply the current law at a node, currents entering the node are given the opposite sign to that given the currents

Figure 2.12.2

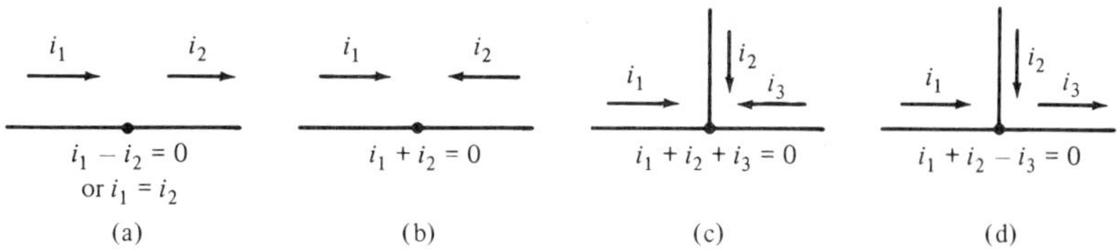

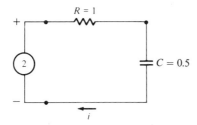

Figure 2.12.3

leaving the node. Figure 2.12.2 shows several nodes and the corresponding current equations. When applying the voltage law, we take a loop and designate a direction around it. The voltage drops for resistors, capacitors, and inductors are added if the current variable in that branch is in the same direction as we are moving around the loop. The voltage drops are subtracted if the current variable is in the direction opposite to that in which we are moving around the loop. The converse holds for a voltage source since in that case we have a voltage gain (negative drop).

Example 2.12.1

For the linear circuit shown in Fig. 2.12.3 find the charge $q(t)$ on the capacitor given that $q(0) = 0$.

Solution

Taking the current in each branch to move in a clockwise direction, we see, by the current law (Fig. 2.12.2a), that the current is the same in all three branches. Call it i. By the voltage law, the sum of the voltage drops, starting at the voltage source, is (note different sign for voltage gain at source)

$$-2 + 1i + 2q = 0, \tag{4}$$

or in terms of q,

$$-2 + \frac{dq}{dt} + 2q = 0.$$

This gives the linear differential equation

$$\frac{dq}{dt} + 2q = 2, \quad q(0) = 0, \tag{5}$$

which may be solved by the method of integrating factors or separation of variables to give

$$q(t) = 1 - e^{-2t}.$$

Letting $t \to \infty$, we see that the capacitor charge approaches 1 C. Note that if (5) is differentiated with respect to t, a first-order differential equation in i results,

$$\frac{di}{dt} + 2i = 0. \blacksquare$$

Example 2.12.2 Consider the circuit in Fig. 2.12.4, where the resistor and the inductor are linear. Find the current as a function of time given that its initial value is 6.

Solution The current is the same in each branch. From the loop law we have

Figure 2.12.4

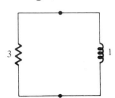

$$Ri + L\frac{di}{dt} = 0$$

or

$$3i + \frac{di}{dt} = 0.$$

The integrating factor of this first-order linear equation is e^{3t}, and the solution is $i(t) = 6e^{-3t}$. $\blacksquare$

Example 2.12.3 In the circuit shown in Fig. 2.12.5, the current is initially zero, there is a voltage source of 1 V, an inductor with inductance 2 H, and a nonlinear resistor whose v–i characteristic is $v = i^2$. Find the current as a function of time t.

Solution Let i be the current. The voltage law postulates that

(Voltage drop inductor) + (Voltage drop resistor)
+ (Voltage drop voltage source) = 0,

Figure 2.12.5 or

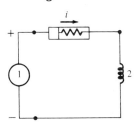

$$2\frac{di}{dt} + i^2 - 1 = 0.$$

This is a nonlinear equation that can be solved by separation of variables. We first separate the variables,

$$\int \frac{2}{1 - i^2} di = \int dt,$$

then expand the left-hand side by partial fractions,

$$\int \left[\frac{1}{1 - i} + \frac{1}{1 + i} \right] di = \int dt,$$

and integrate, so that

$$-\ln|1-i| + \ln|1+i| = t + C_1.$$

We next let $t = 0$ and use $i(0) = 0$ to get $0 = C_1$. Thus

$$\ln\frac{|1+i|}{|1-i|} = t.$$

Exponentiating both sides, we find

$$\left|\frac{1+i}{1-i}\right| = e^t.$$

We now use $i(0) = 0$ to eliminate the absolute-value signs (the quantity inside the bars is positive) and solve for i, to finally get

$$i(t) = \frac{e^t - 1}{e^t + 1}. \blacksquare$$

Example 2.12.4

As our last example, let us consider the circuit shown in Fig. 2.12.6. The figure shows a current source of 3 A, an inductor of 1 H, and a nonlinear resistor with v–i characteristic $v = i^3 - 4i$. Set up a differential equation for the current in the resistor. Sketch the solutions and find all equilibria, as done in Section 2.2. Then find all solutions.

Solution

Let i_R, v_R, i_L, v_L be the currents and voltages in the resistor and the inductor. Applying the current law at node ⓐ, we get

$$3 + i_R + i_L = 0 \tag{6}$$

Figure 2.12.6

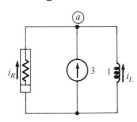

and, applying the voltage law to the outside loop consisting of just the inductor and resistor, we have

$$v_L - v_R = 0. \tag{7}$$

Since

$$v_L = L\frac{di_L}{dt} \quad \text{and} \quad v_R = i_R^3 - 4i_R,$$

Eq. (7) becomes

$$\frac{di_L}{dt} - (i_R^3 - 4i_R) = 0. \tag{8}$$

This equation has two dependent variables, i_R and i_L. We use (6) to eliminate i_L by letting $i_L = -3 - i_R$ so that finally (8) is

$$-\frac{di_R}{dt} - (i_R^3 - 4i_R) = 0$$

or

Figure 2.12.7
Solutions of Eq. (9).

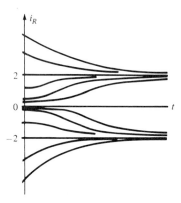

$$\frac{di_R}{dt} = 4i_R - i_R^3. \tag{9}$$

This is a nonlinear equation in i_R. To simplify what follows, let $i = i_R$. To find the equilibrium points (also known as steady-state operating points), set $di/dt = 0$, that is,

$$0 = \frac{di}{dt} = 4i - i^3 = i(4 - i^2).$$

The equilibrium points are $i = 0$ and $i = \pm 2$. If we analyze Eq. (9) as in Section 2.2, we see that

$$\frac{di}{dt} > 0 \quad \text{if} \quad 0 < i < 2, \quad \text{or} \quad i < -2,$$

$$\frac{di}{dt} < 0 \quad \text{if} \quad i > 2, \quad \text{or} \quad -2 < i < 0.$$

The solutions are graphically represented in Fig. 2.12.7.

Looking at Fig. 2.12.7, we see that the circuit shown in Fig. 2.12.6 always approaches an equilibrium—an equilibrium, however, which depends on the initial currents. Since the zero equilibrium is unstable, it would be difficult to physically observe it, so there would appear to be only the two equilibria, ± 2.

To actually solve (9), use separation of variables. Note that separation of variables works only if i is not an equilibrium solution. Separating variables in (9) gives

$$\int \frac{di}{4i - i^3} = \int dt. \tag{10}$$

Now $4i - i^3 = -i(i - 2)(i + 2)$, so that, by partial fractions (see Appendix B),

$$\frac{1}{4i - i^3} = \frac{-1}{i^3 - 4i} = \frac{A}{i} + \frac{B}{i - 2} + \frac{C}{i + 2}.$$

Thus

$$-1 = A(i^2 - 4) + Bi(i + 2) + Ci(i - 2).$$

Evaluating this expression at $i = 0, 2, -2$ yields $A = \frac{1}{4}$, $B = -\frac{1}{8}$, $C = -\frac{1}{8}$. It follows that integration of (10) gives

$$\tfrac{1}{4} \ln |i| - \tfrac{1}{8} \ln |i - 2| - \tfrac{1}{8} \ln |i + 2| = t + C$$

or

$$\ln \left| \frac{i^2}{(i-2)(i+2)} \right| = 8t + 8C.$$

Exponentiating to simplify, we obtain

$$\frac{i^2}{|i^2 - 4|} = e^{8t} K, \qquad K = e^{8C}. \tag{11}$$

We allow K to be negative in order to remove the absolute-value signs and solve for i^2. Then

$$i^2 = \frac{4e^{8t} K}{e^{8t} K - 1}$$

and finally the solutions are

$$i = \pm 2 \left[\frac{e^{8t} K}{e^{8t} K - 1} \right]^{1/2} \quad \text{and} \quad i = \pm 2. \ \blacksquare$$

The v–i characteristic of the resistor in Fig. 2.12.6 was chosen to simplify the calculations. However, similarly shaped v–i characteristics occur in certain diodes and transistors.

Exercises

Note: Not all of the following exercises can be done by linear techniques (Section 2.3) and separation of variables. Some will require other techniques. This is indicated in the solutions manual and in the solutions appendix. All quantities are in the units of Table 2.12.1.

Exercises 1 throught 8 refer to the circuit given in Fig. 2.12.8. The inductor is linear with inductance of L, the voltage source is $e(t)$.

Figure 2.12.8

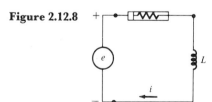

1. The voltage source is constant, $e = 1$, the resistor is linear with v–i characteristic $v = 2i$ and $L = 1$. Set up the differential equation for the current. Determine the current as a function of time for any initial current $i(0)$. Find any steady-state (equilibrium) solutions.

2. The voltage source is constant, $e = 4$, the resistor is linear with v–i characteristic $v = 6i$, and $L = 2$. Set up the differential equation for the current. Determine the current as a function of time for any initial current $i(0)$. Find any steady-state (equilibrium) solutions.

3. The voltage source is $e(t) = \sin t$, the resistor is linear with v–i characteristic $v = i$, and $L = 1$. Set up the differential equation for the current. Determine the current as a function of time for any initial current $i(0)$.

4. The voltage is constant, $e = 1$. The resistor is nonlinear with v–i characteristic $v = -i^2$, and $L = 1$. Set up the differential equation for the current. Solve the differential equation and sketch the solutions. (That the sign of the voltage in the resistor is opposite to that of the current merely means that we are dealing with the model of a device which puts power into the circuit rather than one that dissipates power.)

5. The voltage source is a constant e, the resistor is linear with v–i characteristic $v = iR$, $R > 0$, and the inductance is $L > 0$.
    a) Show that $\lim_{t \to \infty} i(t)$ exists and is the same for any initial current.
    b) If $e = 8.5$, for what values of R, L will $\lim_{t \to \infty} i(t) = 4.2$?

6. The resistor has v–i characteristic $v = i$ and $L = 1$. The voltage source is a 9-volt battery that is shorted out after 10 sec. That is,
$$e(t) = \begin{cases} 9 & 0 \le t < 10 \\ 0 & t \ge 10 \end{cases}$$
The current is initially zero. Find $i(t)$ for $0 \le t \le 20$ and graph the result.

7. The resistor has v–i characteristic $v = 2i$ and $L = 1$. The voltage source is a 1.5-volt battery that is shorted out for the first 10 sec and then unshorted (by a switch) so that
$$e(t) = \begin{cases} 0 & 0 \le t < 10 \\ 1.5 & t \ge 10 \end{cases}$$
The current is initially zero. Find $i(t)$ for $0 \le t \le 20$ and graph the result.

8. The resistor has v–i characteristic $v = i^3$, $e = 0$, and $L = 1$. The current is initially 4. Find the current as a function of time.

Exercises 9 through 11 refer to the circuit in Fig. 2.12.9. The capacitor is linear and has capacitance C. The voltage source is $e(t)$.

Figure 2.12.9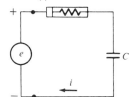

9. The capacitance is 0.5, the resistor is linear with v–i characteristic $v = 2i$, and the sinusoidal voltage source is given by $e(t) = 6 \sin t$. Find the charge on the capacitor as a function of t given that it was initially 1.

10. The capacitance is 1, the nonlinear resistor has v–i characteristic $v = i^2$, the voltage source is $e(t) = t$. Suppose that the current in the resistor is initially 2. Apply the voltage law, differentiate the resulting equation with respect to t, and use $dq/dt = i$ to get a first-order equation in the current. Solve this equation for the relationship (which may be implicit) between current and time.

11. The capacitance is 1, the resistor is nonlinear with v–i characteristic $v = i^2/2$. The voltage source is $e(t) = t^2$. The current in the resistor at $t = 1$ is $i(1) = 2$. Derive a differential equation for the current in the resistor and solve it (note Exercise 10).

12. Consider the circuit in Fig. 2.12.10. The capacitor has capacitance $C = 1$. The current source is $(t + 1)^{-1}$ and the resistor's v–i characteristic is $v = \ln i$ for $i > 0$. Proceeding as in Example 2.12.4, set up a differential equation for the current in the resistor, and solve the resulting differential equation, given that the current in the resistor is initially 2.

Figure 2.12.10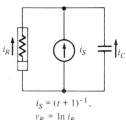

$i_S = (t + 1)^{-1}$,
$v_R = \ln i_R$

2.13 Mechanics and Chemical Reactions

This section deals with two applications of first-order equations to mechanics and chemical reactions. Many other mechanics problems which initially appear to require second-order equations can be reduced to first-order equations and solved as such. The "trick" is part of two useful variable changes to be discussed in Chapter 3. Additional examples of applications of differential equations to mechanics will be given after these variable changes have been discussed.

2.13.1 Mechanics

The mechanics problem to be considered is that of the straight-line (linear) motion of a mass m. Let $v(t)$ be the velocity at time t. The *momentum* of the object is mv. If the velocity is not too large, we may use Newton's second law of motion:

> The time rate of change of momentum of a body is proportional to the net force acting on the body and has the same direction (sign) as the force. (1)

We will assume that our body is rigid and does not rotate so that angular momentum and angular velocity need not be considered. Newton's law (1) may be written mathematically as

$$\frac{d(mv)}{dt} = kF_T, \qquad (2)$$

where $F_T(t)$ is the total or net force acting on the body at time t. If the mass is constant, this law is

$$m\frac{dv}{dt} = kF_T. \qquad (3)$$

Let $a(t)$ be the acceleration at time t. Recall that $a = dv/dt$. Then (3) is $ma = kF$. Units are often chosen so that $k = 1$ and the familiar relationship

$$\text{Force} = \text{Mass} \times \text{Acceleration}$$

results. Table 2.13.1 lists two sets of units which give $k = 1$. We shall use these units throughout this text.

The force due to gravity is called the *weight* w of the body. In this section it will be assumed that the object is near the surface of a much larger body and

Table 2.13.1 Systems of Units

Quantity	Centimeter/gram/seconds (cgs)		Foot/pound/second (fps)	
	Unit	Abbreviation	Unit	Abbreviation
Distance	centimeters	cm	feet	ft
Time	seconds	sec	seconds	sec
Mass	grams	g	slug (lb-sec^2/ft)	
Force	dynes (g-cm/sec^2)		pounds	lb

that the assumption of constant weight is sufficiently accurate. Then

$$w = mg,$$

where g is the acceleration due to gravity. Near the surface of the planet earth g is approximately

$$g = 980 \text{ cm/sec}^2 \quad \text{(cgs)}, \qquad g = 32 \text{ ft/sec}^2 \quad \text{(fps)}.$$

In the examples to follow we shall initially describe the problem in terms of velocity so that Eq. (2) leads to a first-order equation in v. Second-order equations dealing with problems in terms of position will be examined in Chapter 3.

If the object is moving through a fluid or gas, the fluid or gas exerts a force called *resistance* on the body. If the fluid or gas has uniform density and the velocity is not too large, resistance may often be described by the following law:

Resistance is proportional to the magnitude of the velocity and acts in a direction opposite to that of the velocity. (4)

Thus resistive force is $-kv$ with $k \geq 0$. The constant of proportionality depends on the shape of the body and the nature of the medium the object travels through.

Example 2.13.1

A 4-g mass is dropped from a height of 6 m. Air resistance acts on the mass with constant of proportionality of 12 g/sec.

a) Find the velocity as a function of time.

b) If the gravitational force is assumed constant, then what is the limiting velocity?

Solution

Since the mass is constant, we may use (3),

$$m\frac{dv}{dt} = F_T. \tag{5}$$

Let us take our starting position at $t = 0$ and measure the distance the object has dropped so that $v > 0$ is a downward motion. Then

$$F_T = \text{Force due to gravity} + \text{Resistive force}$$
$$= mg - kv$$
$$= 4 \cdot 980 - 12v \tag{6}$$

in cgs units. At the instant the mass is released its velocity is zero. Thus $v(0) = 0$.

Substituting the formula for the total force, Eq. (6), into (5) gives the linear differential equation

$$4\frac{dv}{dt} = 4 \cdot 980 - 12v, \quad v(0) = 0.$$

or

$$\frac{dv}{dt} + 3v = 980, \quad v(0) = 0, \tag{7}$$

which may be solved by the integrating-factor method. Multiplying (7) by the integrating factor e^{3t} changes (7) to

$$(e^{3t}v)' = 980e^{3t}.$$

Next we antidifferentiate,

$$e^{3t}v = \frac{980}{3}e^{3t} + C,$$

and solve for v,

$$v = \frac{980}{3} + Ce^{-3t}.$$

But $v(0) = 0$; hence $C = -980/3$, and finally

$$v(t) = \frac{980}{3}(1 - e^{-3t}) \text{ cm/sec.} \tag{8}$$

This answers part (a). To arrive at an answer to part (b), note that $e^{-3t} \to 0$ as t increases so that $v(t)$ approaches a velocity of $(980/3)$ cm/sec. ∎

2.13.2 Chemical Reactions

Consider a chemical reaction occurring in a well-mixed solution. Such reactions can be quite complicated. We shall present a model that is sometimes sufficiently accurate.

It will be assumed that the reaction is irreversible and no other processes take place to affect the amount of each reactant. Temperature and volume are assumed constant. It will be convenient to measure amounts in moles and concentrations in moles/liter, since a mole of a given chemical always contains the same number of molecules.

Suppose that at the start, our reaction involves two reactants, A and B, and a product E. Furthermore, one molecule of A and one molecule of B produce one molecule of E. This could be written as

$$A + B \to E. \tag{9}$$

Let a and b be the concentrations at time t of reactants A and B. The probability of a collision between a molecule of A and a molecule of B should be proportional to the product of the concentrations. Similarly a certain portion of the collisions between molecules of A and B would be expected to react to form a molecule of E. This leads us to the *mass action law*.

> The instantaneous rate of production of product per unit volume is proportional to the product of the concentrations of the reactants.

Let $x(t)$ be the amount of product E per unit volume produced by time t. Then the mass action law for (9) would be

$$\frac{dx}{dt} = kab, \tag{10}$$

k is always positive and is called the *rate constant*. Since, according to (9), one mole of both A and B are needed to produce one mole of E, we have $a = a(0) - x$, $b = b(0) - x$, and (9) may be rewritten in terms of one dependent variable as

$$\frac{dx}{dt} = k(a(0) - x)(b(0) - x), \qquad x(0) = 0, \tag{11}$$

which is a first-order nonlinear equation.

Example 2.13.2

Two reactants, A and B, are in solution. The initial concentrations are $a(0) = 3$ moles/liter, $b(0) = 1$ mole/liter, and the rate constant $k = 1$ liter/mole·sec. Assume that the mass action law is applicable.

a. Find the concentration of A at time t.
b. What is the limiting concentration of the product E?
c. How long does it take to produce one-half of this limiting amount? three-quarters of this limiting amount?

Solution

Our model (11) gives the differential equation

$$\frac{dx}{dt} = (3 - x)(1 - x), \qquad x(0) = 0. \tag{12}$$

This equation may be solved by separation of variables;

$$\int \frac{1}{(3-x)(1-x)} dx = \int dt. \tag{13}$$

Use partial fractions (Appendix B) and expand the integrand on the left as

$$\frac{1}{(3-x)(1-x)} = \frac{G}{x-3} + \frac{H}{x-1}.$$

Solve for G, H to get $G = \frac{1}{2}, H = -\frac{1}{2}$. Thus (13) is

$$\tfrac{1}{2}\ln|x-3| - \tfrac{1}{2}\ln|x-1| = t + C.$$

To find C, let $t = 0$, $x(0) = 0$, so that $C = (\ln 3)/2$. Hence

$$\tfrac{1}{2}\ln|x-3| - \tfrac{1}{2}\ln|x-1| = t + \frac{\ln 3}{2} \tag{14}$$

or

$$\ln\left|\frac{x-3}{x-1}\right| = 2t + \ln 3.$$

Exponentiate both sides to arrive at

$$\left|\frac{x-3}{x-1}\right| = 3e^{2t}.$$

Since $x(0) = 0$, we are interested only in the solution for which $0 < x < 1$ so that

$$\frac{3-x}{1-x} = 3e^{2t}. \tag{15}$$

Multiply both sides of this equation by $(1-x)$ and solve for x to finally get

$$x = \frac{3 - 3e^{2t}}{1 - 3e^{2t}} = \frac{3 - 3e^{-2t}}{3 - e^{-2t}}. \tag{16}$$

We can now answer all three questions:

a. $a = 3 - x = 3 - \dfrac{3 - 3e^{-2t}}{3 - e^{-2t}} = \dfrac{6}{3 - e^{-2t}}.$

b. $\lim\limits_{t \to \infty} x = \dfrac{3}{3} = 1.$

c. It is easiest to use (14) to solve for t, given $x(t)$. If $x(t) = \frac{1}{2}$, then $t_1 = 0.26$ sec. If $x(t_2) = \frac{3}{4}$, then $t_2 = 0.55$ sec. ∎

Exercises

1. A mass of 20 g is dropped from an airplane flying horizontally. Air resistance acts according to (4) with a constant of proportionality of 10 g/sec. Considering only vertical motion:
   a) Find the velocity as a function of time.
   b) Find the velocity after 10 sec, assuming the body has not hit the ground (or a bird).
   c) If the gravitational force is assumed constant, what is the limiting velocity?

2. A weight of 64 lb is flung vertically up into the air from the earth's surface. At the instant it leaves the launcher it has a velocity of 192 ft/sec.
   a) Ignoring air resistance, determine how long it takes for the object to reach its maximum altitude.
   b) If air resistance acts according to (4) with a constant of proportionality of 4 slug/sec, how long does it take for the object to reach its maximum altitude?

3. A mass of 70 g is to be ejected downward from a stationary helicopter. Air resistance acts according to (4) with the constant of proportionality of 7 g/sec. At what velocity should the mass be ejected if it is to have a velocity of 12,600 cm/sec after 5 sec?

4. Suppose that an object of mass m is ejected vertically downward with velocity v_0 over the surface of a planet whose gravitational constant is G. The atmosphere exerts resistance on the body according to (4) with constant of proportionality r. (All parameters are in cgs units.)
   a) Find the formula for $v(t)$.
   b) Find the formula for the limiting velocity.

   Assume that the object consists of a payload and a parachute. The payload is 80% of the total mass, and the parachute takes up the balance of the total mass. Assume also that the parachute accounts for essentially all the resistive force.
   c) How much would the payload have to be reduced to cut the limiting velocity in half?

5. A 32-lb weight is dropping through a gas near the earth's surface. The resistance is proportional to the square of the velocity, with constant of proportionality 1. At time $t = 0$ the velocity is 1000 ft/sec.
   a) Find the velocity as a function of time.
   b) Is there a limiting velocity? If there is, find it.

6. The chemical reaction $A + B \to E$ obeys the mass-action law with rate constant $k = 2$(liter/mole·sec). Suppose that the initial concentrations are $a(0) = 5$, $b(0) = 2$ moles/liter.
   a) Find the concentration of A at time t.
   b) What are the limiting concentrations of the product E and reactant B?
   c) How long does it take for 95% of B to be used up?

7. The chemical reaction $A + B \to E$ obeys the mass-action law with rate constant k. Suppose the initial concentrations are $a(0) = \alpha$, $b(0) = \beta$, and assume $\alpha < \beta$.
   a) Find the concentration of A at time t.
   b) Give the limiting concentrations of the product E and reactants A and B.

8. The chemical reaction $A + B \to E$ obeys the mass-action law with rate constant k. Suppose the initial concentrations are the same: $a(0) = b(0) = \alpha$. Find the concentration of the product E as a function of time.

9. The chemical reaction $A + B + C \to E$ obeys the mass-action law with rate constant k. If x is the concentration of E produced, the resulting differential equation is

$$\frac{dx}{dt} = k(a(0) - x)(b(0) - x)(c(0) - x), \quad (17)$$

$$x(0) = 0.$$

Suppose that the initial concentrations are $a(0) = 1$, $b(0) = 2$, $c(0) = 3$, and $k = 1$(liter/mole·sec).
   a) Find the (implicit) relationship between the concentration of product E and time t.
   b) Using the analysis of Section 2.2, determine the limiting value of x. (Note that the limiting value is an equilibrium point.)
   c) Determine the time it will take for 90% of the limiting value found in part (b) to be produced.

10. The chemical reaction $A + B + C \to E$ again obeys the mass action law, and the differential equation (17) is a valid model. Suppose that $k = 1$ and $a(0) = b(0) = c(0) = 2$. Find the

concentration of the product as a function of time t.

11. Suppose that resistance depends on velocity according to a formula $R(v)$. The presence of sound barriers and similar phenomena shows that the relationship is not really linear as v increases. Explain why $R(0) = 0$ is a reasonable assumption. Assume that R is a differentiable function of v. Using calculus, explain why law (4) is equivalent to taking the linear approximation (tangent line) to $R(v)$ for v near zero.

(*Logistic Equation*) Exercises 18 and 19 of Section 2.9 and Exercises 14 and 15 of Section 2.10 modeled population growth using an exponential growth law $dQ/dt = rQ$. For any species, factors such as limited food supply slow growth as the population gets large and can even cause a population decrease. These ideas are discussed more carefully in Section 9.6. For now, we consider the nonlinear model

$$\frac{dQ}{dt} = rQ - kQ^2, \quad r > 0, k > 0 \quad (18)$$

which is often called the *logistic equation*. Note that if $k = 0$, then (18) is just the exponential growth model.

12. Use the solution-sketching techniques of Section 2.2 to do a), b), c).

   a) Show (18) has two equilibrium solutions which are $Q = 0, Q = r/k$.

   b) The quantity r/k is sometimes called the *carrying capacity*. Show that if $Q(0) > 0$, then $\lim_{t \to \infty} Q(t) = r/k$.

   c) Sketch the solutions of (18) for $Q(0) \geq 0$.

13. Suppose that $r = 1, k = 1$. Find the solutions $Q(t)$ of (18) for $Q(0) > 0$.

14. Suppose that $r = 4, k = 1$. Find the solutions $Q(t)$ of (18) for $Q(0) > 0$.

15. Find the solutions of (18) for any $r > 0, k > 0$, $Q(0) > 0$.

16. Let $Q(t)$ be the population of a particular species at time t. Suppose that the rate of population growth depends only on the population size so that

$$\frac{dQ}{dt} = f(Q). \quad (19)$$

Explain, in biological terms, why each of the following three statements could be reasonable assumptions.

   a) $f(0) = 0$

   b) $f'(Q) > 0$ for small positive Q

   c) $f'(Q) < 0$ for large Q

17. Suppose that $f(Q) = a + bQ + cQ^2$ is a second-degree polynomial in Q. Show that if f satisfies (a), (b), and (c) of Exercise 16, then $a = 0, b > 0$, and $c < 0$ so that (19) becomes (18).

3

Higher-Order Differential Equations

3.1 Introduction and Basic Theory

Chapter 2 discussed first-order equations and gave several examples of their properties. Many applications, however, require second- or higher-order derivatives. Equations involving Newton's law, as discussed in Section 2.13, may invoke position x, velocity dx/dt, and acceleration d^2x/dt^2. Similarly, as discussed in Section 2.12, the equations describing an *RLC*-circuit may involve charge q, current dq/dt, and the current's rate of change d^2q/dt^2.

This section discusses some of the basic theory for second-order ordinary differential equations. The following two sections will deal with two special techniques that can be used to solve certain second-order nonlinear equations and apply these procedures to nonlinear mechanics problems. The rest of the chapter will be devoted to second-order (and nth-order if Sections 3.6, 3.9, 3.10, 3.13, and 3.15 are included) linear differential equations and their applications.

Both $y = x + 1$ and $y = 1$ are solutions of the second-order differential equation with one initial condition.

$$\frac{d^2y}{dx^2} = 0, \qquad y(0) = 1. \tag{1}$$

Thus, in contrast to first-order equations, solutions for second-order differential

Figure 3.1.1

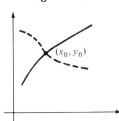

equations are not uniquely determined by their value at a point, and the graphs of distinct solutions may intersect (Fig. 3.1.1).

Consider, for a moment, an arrow shot vertically upward. To determine the path of the arrow, we need to know not only the initial position (where the bow is) but also the initial velocity. The next theorem states that if we have decided on the differential equation (friction, effect of gravity, etc.) to be used and it is a "nice" second-order equation, then initial position and velocity will uniquely determine the solutions. Let

$$y' = \frac{dy}{dx}, \qquad y'' = \frac{d^2y}{dx^2}.$$

Let a *box* about a point (x_0, y_0, z_0) in 3-space be

$$\{(x, y, z): \ |x - x_0| \leq \varepsilon_1, \ |y - y_0| \leq \varepsilon_2, \ |z - z_0| \leq \varepsilon_3\}$$

for some constants $\varepsilon_1 > 0, \varepsilon_2 > 0, \varepsilon_3 > 0$.

Theorem 3.1.1

Basic Existence and Uniqueness Theorem for Second-Order Ordinary Differential Equations Let $f(x, y, z)$ be a function of three variables such that f, f_y, f_z are continuous in a box around the point (x_0, y_0, z_0). Then there is a nontrivial interval I containing x_0 such that the differential equation

$$y'' = f(x, y, y') \tag{2a}$$

has a unique solution $y(x)$ defined on I that satisfies the initial conditions

$$y(x_0) = y_0, \qquad y'(x_0) = z_0. \tag{2b}$$

Example 3.1.1

For the initial-value problem

$$y'' = xy \sin(yy'), \qquad y(0) = 1, \qquad y'(0) = 2, \tag{3}$$

we have $f(x, y, z) = xy \sin(yz)$. Since f, f_y, f_z are continuous at and near $(0, 1, 2)$, there is a solution to (3) defined on an interval containing $x_0 = 0$ and the solution is unique. ∎

Figure 3.1.2

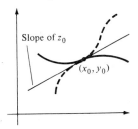

Slope of z_0

(x_0, y_0)

Graphically, the Existence and Uniqueness Theorem says that if the assumptions hold at (x_0, y_0, z_0), then two solutions can cross at (x_0, y_0) as in Fig. 3.1.1 but they cannot also be tangent at (x_0, y_0) as in Fig. 3.1.2. The theorem says even more. It says that there will be an infinite number of solutions through (x_0, y_0), one with each possible slope z_0, for which f, f_y, f_z are continuous on a box containing (x_0, y_0, z_0). For example, the solutions of Eq. (1) are $y = cx + 1$, and the slope $y' = c$ is arbitrary.

Since the solution of a second-order equation can be found to satisfy two initial conditions (2b), we expect to see two arbitrary constants in solving second-order differential equations.

Exercises

In Exercises 1 through 12, determine for which points (x_0, y_0) and slopes z_0, the Existence and Uniqueness Theorem (Theorem 3.1.1) guarantees a unique solution through (x_0, y_0) with $y'(x_0) = z_0$.

1. $y'' = (y')^2 + 1$
2. $y'' = (x + y + y')^{1/3}$
3. $y'' = x^{1/3}(y + y')^2$
4. $y'' = y'/y^2$
5. $y'' = \dfrac{y}{(y')^2 + y^2}$
6. $(y' + 1)y'' = 1 + y^2$
7. $y'' = \sin y$
8. $y'' = \tan(y' + xy)$
9. $xy'' = y(y' + 1)^{1/5}$
10. $y'y'' = \sin x$
11. $y(y' + 1)y'' = 1$
12. $y'' = xyy'$

In Exercises 13 through 20, find all solutions of the differential equation and sketch several solutions on the same axes.

13. $y'' = 2$, $y(0) = 0$
14. $y'' = x/6$, $y'(0) = 0$
15. $y'' = 2$, $y'(0) = 0$
16. $y'' = x/6$, $y(0) = 0$
17. $y'' = x + 1$, $y(0) = 1$
18. $y'' = -x^2$, $y(0) = 1$
19. $y'' = x + 1$, $y'(0) = -1$
20. $y'' = -x^2$, $y'(0) = 1$

21. Verify that $y = 0$ and $y = x^2$ are both solutions of the initial-value problem
$$yy'' = xy', \quad y(0) = 0, \quad y'(0) = 0.$$
Explain why this is not a contradiction of Theorem 3.1.1.

In Chapter 2 we saw that, if f and f_y were continuous in the xy-plane, then the graphs of solutions of $y' = f(x, y)$ filled up the xy-plane. Exercises 22 and 23 develop the second-order analog of this fact.

22. Suppose that f, f_y, f_z are continuous everywhere in three-dimensional xyz-space. Each solution $y(x)$ of $y'' = f(x, y, y')$ defines a curve $(x, y(x), y'(x))$ in three-dimensional space. The curve is parametrized by x. Show that no two of these curves ever intersect.

23. (Continuation of Exercise 22.) Show that these curves fill up three-dimensional space.

3.2 Two Helpful Substitutions

In general, second-order (nonlinear) differential equations are hard to solve. In principle, since the second-order differential equation involves second derivatives, it should be possible to solve it by antidifferentiating twice, that is, by treating the second-order equation as two first-order equations. This section will present two substitutions that allow us to solve several important equations of the form

$$\frac{d^2y}{dx^2} = f\left(x, y, \frac{dy}{dx}\right), \tag{1}$$

by solving two first-order equations. Applications of this technique are provided in Section 3.3.

Case 1: Dependent Variable Missing

Suppose the differential equation involves only the independent variable x and derivatives of the dependent variable y:

$$\frac{d^2y}{dx^2} = f\left(x, \frac{dy}{dx}\right). \tag{2}$$

Let $v = dy/dx$. Then (2) becomes a first-order equation $dv/dx = f(x, v)$ in v, which can often be solved using the techniques of Chapter 2.

Example 3.2.1 Solve the initial-value problem

$$y'' = 2x(y')^2, \qquad y(0) = 2, \qquad y'(0) = 1. \tag{3}$$

Solution Note that the dependent variable y does not appear explicitly in (3). Let $v = y'$. The initial condition $y'(0) = 1$ is then

$$v(0) = 1, \tag{4}$$

and the differential equation is

$$\frac{dv}{dx} = 2xv^2,$$

which can be solved by separation of variables:

$$\int \frac{dv}{v^2} = \int 2x\, dx$$

or

$$-\frac{1}{v} = x^2 + C_1. \tag{5}$$

[Note that (5) does not include the singular solution $v = 0$, that is, y constant.] The initial condition (4) applied to (5) implies that $C_1 = -1$. Solving (5) for v gives

$$v = \frac{1}{1-x^2} \qquad \text{or} \qquad \frac{dy}{dx} = \frac{1}{1-x^2};$$

hence

$$y = \int \frac{dy}{dx}\, dx = \int \frac{1}{1-x^2}\, dx$$
$$= \int \frac{1/2}{1-x} + \frac{1/2}{1+x}\, dx \qquad \text{(by partial fractions)}$$
$$= -\tfrac{1}{2}\ln|1-x| + \tfrac{1}{2}\ln|1+x| + C_2,$$

where C_2 is a new arbitrary constant. The initial condition $y(0) = 2$ from (3) implies that $C_2 = 2$, and the solution of (3) is

$$y = -\tfrac{1}{2}\ln|1-x| + \tfrac{1}{2}\ln|1+x| + 2. \blacksquare$$

Example 3.2.2 Solve the differential equation

$$y'' - \frac{y'}{x} = 0, \qquad x > 0. \tag{6}$$

Solution Again the dependent variable y is missing from (6). Let $v = y'$. Then (6) is a first-order linear equation (separation of variables could also be used)

$$\frac{dv}{dx} - \frac{v}{x} = 0 \tag{7}$$

with integrating factor

$$\exp\left[\int -\frac{1}{x}\,dx\right] = \exp[-\ln x] = x^{-1}.$$

Multiply (7) by the integrating factor to get

$$\left(\frac{1}{x}v\right)' = 0.$$

Antidifferentiate and solve for v to obtain

$$v = C_1 x \qquad \text{or} \qquad y' = C_1 x.$$

Antidifferentiate again to find y:

$$y = C_1 \frac{x^2}{2} + C_2 = \tilde{C}_1 x^2 + C_2, \tag{8}$$

where $\tilde{C}_1 = C_1/2$ is a different arbitrary constant. ∎

Note that (6) does not satisfy the Existence and Uniqueness Theorem of Section 3.1 at (x_0, y_0, z_0) if $x_0 = 0$. From (8) we see that, at $x_0 = 0$, the solutions may take on any value y_0 but the slope $C_1 x$ is zero and cannot be specified arbitrarily.

We summarize this technique as follows.

Technique if Dependent Variable Missing in Eq. (1)

1. Let $v = dy/dx$.
2. Solve the resulting first-order differential equation $dv/dx = f(x, v)$ for v. (*Note*: $y'(x_0) = z_0$ is now $v(x_0) = z_0$.) Check whether a singular solution has been lost. Use the initial condition if present.
3. Express v as a function of x.
4. Antidifferentiate v with respect to x to get y. Use the initial condition if there is one.

Case 2: Independent Variable Missing

In this case Eq. (1) involves only the dependent variable and its derivatives so that it is in the form

$$\frac{d^2y}{dx^2} = f\left(y, \frac{dy}{dx}\right).$$

Again let $v = dy/dx$ to get

$$\frac{dv}{dx} = f(y, v). \tag{9}$$

In order to reduce (9) to an equation in just y and v, observe that, by the chain rule,

$$\frac{dv}{dx} = \frac{dv}{dy}\frac{dy}{dx} = \frac{dv}{dy}v.$$

Thus (9) can be written as

$$v\frac{dv}{dy} = f(y, v),$$

which is a first-order equation in v and y with initial condition $v(y_0) = z_0$.

Example 3.2.3   Solve the initial-value problem

$$y'' = (y')^3 y, \qquad y(0) = 1, \qquad y'(0) = -2. \tag{10}$$

Solution   Note that the independent variable is missing in (10). Let

$$v = \frac{dy}{dx}, \qquad v\frac{dv}{dy} = \frac{d^2y}{dx^2}.$$

When $x = 0$, then $y = 1$, $v = -2$, so $v(1) = -2$. The initial-value problem (10) is now

$$v\frac{dv}{dy} = v^3 y, \qquad v(1) = -2. \tag{11}$$

Proceed by separation of variables, assuming $v \neq 0$. [Note: $v = 0$ is a solution of the differential equation in (11).] You obtain

$$\frac{1}{v^2}dv = y\,dy,$$

and antidifferentiate, getting

$$-\frac{1}{v} = \frac{y^2}{2} + C_1. \tag{12}$$

The initial condition in (11) implies $C_1 = 0$, so that solving (12) for v gives

$$v = -2/y^2.$$

Thus

$$\frac{dy}{dx} = -\frac{2}{y^2}.$$

This again is a nonlinear first-order equation which can be solved by separation of variables,

$$\int y^2 \, dy = -2 \int dx$$

and antidifferentiation,

$$\frac{y^3}{3} = -2x + C_2.$$

The initial condition $y(0) = 1$ implies $C_2 = \frac{1}{3}$, and the final result is

$$\frac{y^3}{3} = -2x + \frac{1}{3} \quad \text{or} \quad y = (1 - 6x)^{1/3}. \blacksquare$$

Example 3.2.4 Solve the initial-value problem

$$y'' = (y')^3 y, \qquad y(0) = 1, \qquad y'(0) = 0. \tag{13}$$

Solution This is Eq. (10) with a different initial condition. However, (12) does not satisfy the initial condition $y'(0) = v(0) = 0$ for any C_1. Thus we use the singular solution $v = 0$. But $y' = v$, so $y' = 0$ and $y = C_2$. The initial condition gives $C_2 = 1$, and $y = 1$ is the solution of (13). $\blacksquare$

To summarize:

Technique if Independent Variable Missing in Eq. (1)

1. Let $v = dy/dx$, $\quad v \, dv/dy = d^2y/dx^2$.
2. Solve the resulting first-order equation in v and y. The initial conditions $y(x_0) = y_0$, $y'(x_0) = z_0$ are now $v(y_0) = z_0$.
3. Replace v by dy/dx. Solve the resulting first-order differential equation in y and x.
4. In steps 2 and 3, check whether any singular solutions are present.

If initial conditions are given, it is sometimes easier to use them to eliminate the arbitrary constant before step 4 of Case 1 or before Step 3 in Case 2. In either case, different values of the arbitrary constant may require using different techniques. Note Exercises 18, 19.

Exercises

In Exercises 1 through 17, solve the given second-order differential equation.

1. $\dfrac{d^2y}{dx^2} + 3\dfrac{dy}{dx} = 2$
2. $y'' - y' = e^x$
3. $y'' + y = 0, \quad y(0) = 1, \quad y'(0) = 0$
4. $y'' + y = 0, \quad y(0) = 0, \quad y'(0) = 1$
5. $y'' = (y')^3 - (y')^2, \quad y(0) = 3, \quad y'(0) = 1$
6. $y'' = (y')^3 + y'$
7. $(1 + x^2)\dfrac{d^2y}{dx^2} + 2x\dfrac{dy}{dx} = 0$
8. $y'' = 2yy', \quad y(0) = 0, \quad y'(0) = -1$
9. $y^2 y'' = y'$
10. $-x + y'y'' = 0, \quad y(1) = 0, \quad y'(1) = 1$
11. $y'' = 2x(y')^2, \quad y(0) = 0, \quad y'(0) = -1$
12. $xy'' + 4y' = x^3$
13. $y'' = (y')^2 - y'$
14. $y'' = (y')^{-2}$
15. $y'' = (y')^2$
16. $y'' = y'e^y, \quad y(0) = 0, \quad y'(0) = 1$
17. $y'' = \exp(-y')$
18. In Example 3.2.3 we found that $\tilde{y} = (1 - 6x)^{1/3}$ was the solution of $y'' = (y')^3 y, \; y(0) = 1$, $y'(0) = -2$.
    a) Show that $\tilde{y}$ is not a solution of $y'' = (y')^3 y$, $y(\tfrac{1}{6}) = 0$.
    b) Find the solution of $y'' = (y')^3 y, \; y(\tfrac{1}{6}) = 0$, $y'(\tfrac{1}{6}) = 2$.
19. Show that in solving $y'' = 2x(y')^2$, three different integration formulas need to be used depending on whether $K = -y'(x_0)^{-1} - (x_0)^2$ is positive, negative, or zero.
20. Solve $y'' = (y')^3 y$.
21. Show that $y'' = y'f(y)$ can always be solved, in principle, by two applications of separation of variables.

Additional problems are given in the next exercise set.

3.3 Applications to Mechanics

This section is a continuation of Section 2.13 and will use the substitutions of Section 3.2 to study the linear motion of a rigid body governed by Newton's law

$$\frac{d(mv)}{dt} = F_T \qquad (1)$$

where m is the mass, v is the velocity, mv is the momentum, and F_T is the total force. Pick a reference point on the line of motion and let $x(t)$ be the distance at time t from that reference point. Then $v = dx/dt$.

Example 3.3.1 If F_T is composed of resistance (as discussed in Section 2.13) and gravity, m is constant, the motion is vertical near the earth's surface, and the downward x-direction is positive, then (1) takes the form

$$m\frac{dv}{dt} = -[\text{resistance}] + [\text{gravity}].$$

that is,
$$m\frac{dv}{dt} = -kv + mg \tag{2}$$
or
$$m\frac{d^2x}{dt^2} = -k\frac{dx}{dt} + mg. \tag{3}$$

Note that (3) is a second-order linear differential equation in x and t but letting $v = dx/dt$ yields a first-order equation, Eq. (2), for the velocity. This fact was made use of in Section 2.13, where several examples of (2) were solved for v. ∎

Variable Mass

In all our applications so far, the mass has been constant. Two variable-mass problems will now be given.

Example 3.3.2

A flexible cable or cord 10 cm long is coiled neatly on the ground. A steady upward force of 9800 dynes is applied to one end of the cable. Assume that the cable has uniform density 3 g/cm and that friction and any nonvertical motion are negligible. Find the equation for the position of the end of the cable as a function of time while the cable is in touch with the ground.

Solution

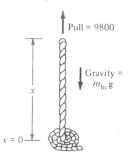

Figure 3.3.1

Let x be the height (in centimeters) of the end above the coil at time t (in seconds), as shown in Fig. 3.3.1. The mass of the hanging cable m_{hc} will be $3x$ since the density is 3. Then Newton's law (1) is

$$\frac{d(m_{\text{hc}}v)}{dt} = [\text{Total force}] = [\text{Pull}] - [\text{Gravity}]$$
$$= 9800 - m_{\text{hc}}g$$

or

$$\frac{d(3xx')}{dt} = 9800 - 3x \cdot 980, \tag{4}$$

which is the nonlinear second-order differential equation

$$3x'x' + 3xx'' = 9800 - 2940x. \tag{5}$$

The independent variable t is missing in (5) so that the second technique of Section 3.2 is applicable. Let

$$v = \frac{dx}{dt}, \quad v\frac{dv}{dx} = \frac{d^2x}{dt^2}$$

and (5) becomes

$$3v^2 + 3xv\frac{dv}{dx} = 9800 - 2940x.$$

Variables do not separate. However, if we rewrite the equation in the form $M(x, v)\, dv + N(x, v)\, dx = 0$,

$$3xv\, dv + (3v^2 + 2940x - 9800)\, dx = 0, \tag{6}$$

where

$$M(x, v) = 3xv, \qquad N(x, v) = 3v^2 + 2940x - 9800,$$

we have

$$\frac{N_v - M_x}{M} = \frac{6v - 3v}{3xv} = \frac{1}{x}.$$

Thus, by Section 2.7, $\exp[\int (1/x)\, dx] = x$ is an integrating factor. We multiply (6) by x to get the exact equation

$$3x^2 v\, dv + (3v^2 x + 2940x^2 - 9800x)\, dx = 0. \tag{7}$$

The solution of (7) (see Section 2.6) is $G(x, v) = C_1$, where $G(x, v)$ is the solution of

$$G_v = 3x^2 v, \tag{8a}$$
$$G_x = 3v^2 x + 2940x^2 - 9800x. \tag{8b}$$

Antidifferentiate Eq. (8a) with respect to v to get

$$G(x, v) = \tfrac{3}{2} x^2 v^2 + h(x).$$

To find $h(x)$, substitute this expression for G into Eq. (8b),

$$3xv^2 + h'(x) = 3v^2 x + 2940x^2 - 9800x$$

or

$$h'(x) = 2940x^2 - 9800x.$$

Hence

$$h(x) = 980x^3 - 4900x^2$$

and the solution of (6) is $G(x, v) = C_1$ or

$$\tfrac{3}{2} x^2 v^2 + 980x^3 - 4900x^2 = C_1. \tag{9}$$

From the problem description, we want a solution with zero initial position and zero initial momentum. Thus our initial conditions will be

$$x(0) = 0, \qquad (xv)(0) = 0. \tag{10}$$

Note that the existence and uniqueness result does *not* hold for (5) if $x = 0$, so we do not have any guarantee that a solution satisfying (10) will exist. However, (10) applied to (9) at $t = 0$ gives $C_1 = 0$. Solving (9) for v and using the fact that $v(t) > 0$, at least initially, gives

$$v = 14\sqrt{\frac{10}{3}} \sqrt{5 - x} = \alpha \sqrt{5 - x},$$

where $\alpha = 14\sqrt{10/3}$. Using α will simplify the remaining calculations. Thus

$$\frac{dx}{dt} = \alpha\sqrt{5-x}.$$

This nonlinear equation can be solved by separation of variables

$$\frac{1}{\sqrt{5-x}}\,dx = \alpha\,dt$$

and antidifferentiation,

$$-2(5-x)^{1/2} = \alpha t + C_2.$$

Let $t = 0$, use $x(0) = 0$, and find $C_2 = -2\sqrt{5}$ so that

$$(5-x)^{1/2} = -\frac{\alpha}{2}t + \sqrt{5}.$$

Squaring, using the value of α, and solving for x finally yields

$$x = 5 - \left(-7\sqrt{\frac{10}{3}}t + \sqrt{5}\right)^2$$

or

$$x = 5\left[1 - \left(1 - 7\sqrt{\frac{2}{3}}t\right)^2\right]. \tag{11}$$

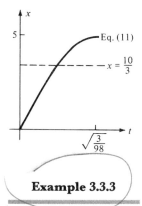

Figure 3.3.2
Solutions (11) and $x = 10/3$ of (5).

Given (11), we can readily verify that it satisfies the original differential equation not only for all $t > 0$ but also for $t = 0$. However, (11) is physically meaningful only if $10 \geq x \geq 0$, $v \geq 0$. ∎

Observe that the original differential equation (5) has an equilibrium solution. To find it, set $x = k$, a constant, in (5). Then $x = 10/3$. The two solutions, $x = 10/3$ and (11) are graphed in Fig. 3.3.2.

The solutions cross but have different slopes at their intersection, as predicted by the Existence and Uniqueness Theorem 3.1.1. What happens physically is that the cable accelerates until $x = 10/3$. After that it decelerates but the built-up momentum carries it past the equilibrium until $x = 5$. Then it falls back under the pull of gravity.

Example 3.3.3 A flexible cable 10 cm long is pulled vertically past a frictionless pulley of negligible mass by a steady upward force of 9800 dynes applied to the free end of the cable. Assume that the cable has a uniform density of 2 g/cm and friction is negligible. Find the equation of motion that applies to the end of the cable so long as the cable is in contact with the pulley (Fig. 3.3.3).

3.3 Applications to Mechanics

Solution

Let m_c be the mass of the whole cable and m_{hc} the mass of the vertical portion, which has length x. Gravity acts only on the vertical section of cable whereas the whole cable has momentum. Thus, Newton's law (1) is

$$\frac{d(m_c v)}{dt} = [\text{Pull}] - [\text{Gravity}] = 9800 - m_{hc}g$$

Figure 3.3.3

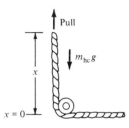

or

$$\frac{d(20v)}{dt} = 9800 - 2x980,$$

which is the second-order linear differential equation with constant coefficients

$$20\frac{d^2 x}{dt^2} + 1960x = 9800. \qquad (12)$$

The initial conditions are that the original position and momentum are zero. Since the mass of the moving portion of cable is constant, we have

$$x(0) = 0, \qquad v(0) = 0. \qquad (13)$$

Again, the independent variable t is missing. Setting

$$\frac{dx}{dt} = v, \qquad \frac{d^2 x}{dt^2} = v\frac{dv}{dx},$$

as described in Section 3.2, we see that (12) becomes

$$20v\frac{dv}{dx} = -1960x + 9800,$$

which can be solved by separation of variables,

$$2v\,dv = (-196x + 980)\,dx,$$

and by integration,

$$v^2 = -98x^2 + 980x + C_1.$$

Letting $t = 0$ and using the initial conditions (13) gives $C_1 = 0$. Since $v \geq 0$, we take the positive square root and obtain

$$v = \frac{dx}{dt} = 7\sqrt{2}\sqrt{10x - x^2}. \qquad (14)$$

This differential equation can also be solved by separation of variables,

$$\int \frac{dx}{\sqrt{10x - x^2}} = \int 7\sqrt{2}\,dt$$

or

$$\int \frac{dx}{\sqrt{25 - (x-5)^2}} = 7\sqrt{2}\,t + C_2.$$

Evaluating the integral, we get

$$\sin^{-1}\left(\frac{x-5}{5}\right) = 7\sqrt{2}t + C_2. \tag{15}$$

At $t = 0$, (15) becomes $\sin^{-1}(-1) = C_2$. Taking the principal branch of $\sin^{-1}$ to ensure that $x \geq 0$, we get $C_2 = -\pi/2$. Thus, from (15),

$$\frac{x-5}{5} = \sin(7\sqrt{2}t - \pi/2) = -\cos(7\sqrt{2}t)$$

and

$$x = -5\cos(7\sqrt{2}t) + 5. \tag{16}$$

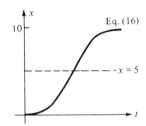

Figure 3.3.4
Graph of Eq. (16) and $x = 5$.

While (16) is a solution of (12) for all t, Eq. (12) is a valid model only when $x \geq 0$, $v \geq 0$.

Equation (12) also has an equilibrium $x = 5$ (Fig. 3.3.4). The solution (16) will be valid for all t only if we assume the cable to be rigid enough that, as it falls back down, it rewinds under the pulley in exactly the same way as it was pulled out. ∎

An alternative method of solving second-order linear differential equations with constant coefficients will be developed later in this chapter.

Exercises

1. A flexible cable 20 cm long is pulled vertically past a frictionless pulley as illustrated in Fig. 3.3.3 by a steady upward force of 9800 dynes. The cable has a uniform density of 1 g/cm and friction is negligible. Find the equation of motion that applies to the end of the cable so long as the cable is in contact with the pulley and moving upward.

*2. At the surface of a planet with gravitational constant G, a flexible cable ℓ cm long is pulled vertically past a frictionless pulley (see Fig. 3.3.3) by a steady upward force of F dynes. The cable has a uniform density of δ g/cm and friction is negligible. Find the equation that describes the motion of the end of the cable so long as the cable is in contact with the pulley and moving upward.

3. A flexible cable 20 cm long is being pulled over the edge of a frictionless table by a steady downward force of 9800 dynes. The cable has a uniform density of 1 g/cm and friction is negligible. The table is 80 cm high. The cable is initially at rest, with the end being pulled over the edge of the table (Fig. 3.3.5). Find the equation for the position of the end of the cable hanging down as a function of time. (The second integration may require integral tables.)

Figure 3.3.5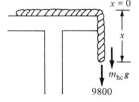

4. A flexible cable 30 cm long is coiled neatly on the ground. A steady upward force of $(980)^2$ dynes is

applied to one end of the cable. Assume that the cable has a uniform density of 9 g/cm, and friction and nonvertical motion are negligible (Fig. 3.3.1). Find the equation for the position of the end of the cable while the cable is still in contact with the ground.

*5. Suppose that a flexible cable ℓ cm long is coiled neatly on the ground. A steady upward force of F dynes is applied to one end of the cable. Assume the cable has uniform density δ g/cm, and friction and nonvertical motion are negligible (Fig. 3.3.1). Find the equation for the position of the end of the cable as a function of time while the cable is still in contact with the ground.

6. A 20-cm long thin flexible cable of density δ is suspended near the earth's surface by one end, with the other end just touching the ground. At time zero, the cable is released and falls under its own weight. Assume that the height of the pile it forms is negligible (Fig. 3.3.6). Find the equation which describes the motion of the top end until it hits the ground. (*Note:* The rope that hits the ground loses its momentum. Thus the ground must exert an upward force on the rope equal to the rate at which momentum is lost. This upward force is $vd(m_{hc})/dt$ or $\delta x'v$.)

Figure 3.3.6

Exercises 7 and 8 involve second-order equations solvable by the method of Section 3.2. Each is derived from an application but will not be discussed in detail.

7. The equation $1 + (y')^2 = yy''$ arises both in descriptions of the shape of a cable hanging suspended at both ends and in minimum-time problems. Solve this equation under the initial conditions $y(0) = 4$ and $y'(0) = 0$. The resulting graph is called a *catenary* (Fig. 3.3.7).

Figure 3.3.7

*8. Let x be the distance of an object of mass m from the center of the earth. The force of gravitational attraction on the mass (assuming x is greater than the radius of the earth) is $F = mgR^2/x^2$, where R is the radius of the earth. Suppose that the mass is ejected radially outward from a satellite 500 mi above the earth's surface with an initial velocity of v_0 mph. Take $R = 4000$ mi, $g = 32$ ft/(sec)^2. Ignore air resistance and the gravitational effect of other planets and moons.

a) Using Newton's law (1), set up the differential equation for x as a function of time.

b) Using the techniques of Section 3.2, reduce the differential equation obtained in (a) to a first-order differential equation in x and $v = x'$.

c) If the mass is to keep moving outward indefinitely, v will remain positive and $x \to \infty$. Find the minimum v_0 such that $dx/dt > 0$ and $x \to \infty$ as $t \to \infty$. This v_0 is the *escape velocity*. Does the mass affect the escape velocity?

*9. Show that, for the solution (11), we have $v(0) \neq 0$. Thus the initial velocity is nonzero but the initial momentum is zero. Is this reasonable or not? Try to explain what is happening in physical terms.

3.4 Second-Order Linear Equations

The preceding section showed that many physical problems are modeled by second-order differential equations. Some of these equations were linear, but many were not. For many applications involving nonlinear equations, if there is a "moderate" change in position and velocity (or charge and current), the nonlinear equation may be approximated by a linear differential equation. The remainder of this chapter considers only linear differential equations.

In this and the next section, we develop some of the basic theory for second-order linear differential equations. This theory will be extensively used in the solution techniques developed later in this chapter. This section uses an intuitive definition of an important concept called linear independence. This concept is examined in greater depth in Section 3.5.

Recall from Section 1.1 that the general second-order linear differential equation is of the form

$$a_2(x)\frac{d^2y}{dx^2} + a_1(x)\frac{dy}{dx} + a_0(x)y = h(x), \tag{1}$$

where a_0, a_1, a_2, h are functions of x, and $a_2(x)$ is not identically zero. The functions a_2, a_1, a_0 are called *coefficients*. The function h, for reasons to become clear when we introduce the applications at the end of this chapter, will be called the *forcing function* or *input function*. It is more convenient to state some of the results if both sides of (1) are divided by $a_2(x)$, to give

$$\frac{d^2y}{dx^2} + p(x)\frac{dy}{dx} + q(x)y = f(x), \tag{2}$$

where

$$p(x) = \frac{a_1(x)}{a_2(x)}, \qquad q(x) = \frac{a_0(x)}{a_2(x)}, \qquad f(x) = \frac{h(x)}{a_2(x)}.$$

Values of x for which $a_2(x) = 0$ are called *singular points* of the differential equation. If $f(x)$ is identically zero, so that (2) is

$$\frac{d^2y}{dx^2} + p(x)\frac{dy}{dx} + q(x)y = 0, \tag{3}$$

the differential equation is called *homogeneous*. [This is not the same type of homogeneous equation that appeared in Section 2.8.] Throughout the rest of this book, "homogeneous" will mean "homogeneous linear" as in (3) and not "homogeneous" as in Section 2.8. If f is not identically zero, then (3) is called the *associated homogeneous equation* of (2).

The following theorem summarizes several key facts about second-order linear equations.

Theorem 3.4.1

Key Facts about Second-Order Linear Differential Equations Suppose that $p(x)$, $q(x)$, and $f(x)$ are continuous functions on the nontrivial interval I. Let x_0 be any number in the interval I and let y_0, v_0 be any two real numbers. Then

1. There exists a unique solution to the initial-value problem

$$y'' + py' + qy = f, \qquad y(x_0) = y_0, \qquad y'(x_0) = v_0, \tag{4}$$

and it is defined on all of the interval I.

Let y_p be a particular solution of $y'' + py' + qy = f$.

2. If y is a solution of $y'' + py' + qy = f$, then $y = y_p + y_h$, where y_h is a solution of the associated homogeneous equation

$$y'' + py' + qy = 0.$$

3. If y_1, y_2 are solutions of the homogeneous equation $y'' + py' + qy = 0$, then

$$c_1 y_1 + c_2 y_2, \quad \text{(what about } c_3 y_3\text{)} \tag{5}$$

where c_1 and c_2 are arbitrary constants, is a solution of $y'' + py' + qy = 0$ also. If neither y_1 nor y_2 is a constant multiple of the other, then y_1 and y_2 form a *fundamental set of solutions* for $y'' + py' + qy = 0$ and all solutions of $y'' + py' + qy = 0$ are of the form (5). The expression (5) is called a *linear combination* of y_1 and y_2.

4. If y_1 and y_2 is a fundamental set of solutions for the associated homogeneous equation $y'' + py' + qy = 0$, and y_p is a particular solution of $y'' + py' + qy = f$, then

$$y = y_p + c_1 y_1 + c_2 y_2 \tag{6}$$

is the general solution of $y'' + py' + qy = f$.

5. *Superposition Principle.* If y_p is a solution of $y'' + py' + qy = f_1$, and $\tilde{y}_p$ is a solution of $y'' + py' + qy = f_2$, then $y_p + \tilde{y}_p$ is a solution of $y'' + py' + qy = f_1 + f_2$.

Since these facts are so important in both the theory and techniques to follow, we shall carefully discuss them one at a time.

Fact 1

That there exists a unique solution near x_0 follows from the Existence and Uniqueness Theorem (Theorem 3.1.1) since (2) may be written as

$$y'' = F(x, y, y'),$$

where

$$F(x, y, v) = f(x) - p(x)v - q(x)y,$$

and $F_y = -q(x)$ and $F_v = -p(x)$ are continuous as long as p and q are continuous. However, Fact 1 gives us even more information. The solutions exist not only near x_0 as promised in Section 3.1 but over the entire interval I on which p, q, f are continuous.

Example 3.4.1 The coefficients 1, x^2, x, and the forcing function e^x of

$$y'' + x^2 y' + xy = e^x$$

are continuous for all x so that we take $I = (-\infty, \infty)$. Therefore the solutions will be defined and continuously differentiable for all x. ∎

Example 3.4.2 Consider
$$(x - 2)y'' + xy' + 6y = (x - 2)^3. \tag{7}$$
Dividing by $(x - 2)$, we rewrite (7) in the form (2):
$$y'' + \frac{x}{x - 2} y' + \frac{6}{x - 2} y = (x - 2)^2. \tag{8}$$

The coefficients 1, $x/(x - 2)$ and $6/(x - 2)$ are continuous everywhere except at $x = 2$. The forcing function $(x - 2)^2$ is continuous everywhere. Thus we are guaranteed that solutions will exist on $(-\infty, 2)$ and $(2, \infty)$. What happens at $x = 2$ depends on the differential equation and will be discussed in the chapter on series solutions. ∎

Fact 2
The associated homogeneous equation (3) is often easier to solve than the non-homogeneous problem (2). Fact 2 says that if we can get *one* particular solution of the nonhomogeneous equation $y'' + py' + qy = f$, then to find all solutions, we need find only the general solution of the associated homogeneous equation $y'' + py' + qy = 0$.

Facts 3 and 4
These make solving the homogeneous equation easier. We do not need to worry about where the arbitrary constants go. If we can find y_1 and y_2, which are solutions of $y'' + py' + qy = 0$ and neither y_1 nor y_2 is a constant multiple of the other, then the general linear combination $c_1 y_1 + c_2 y_2$ gives *all* solutions of $y'' + py' + qy = 0$ and $y_p + c_1 y_1 + c_2 y_2$ gives the general solution of $y'' + py' + qy = f$.

Procedures for determining the y_1, y_2, and y_p will be presented later in this chapter.

Example 3.4.3
a. Verify that $\sin x$, $\cos x$ are solutions of $y'' + y = 0$ and that e^{-3x} is a solution of $y'' + y = 10e^{-3x}$.
b. Give the general solution of
$$y'' + y = 10e^{-3x}. \tag{9}$$
c. Solve the initial-value problem
$$y'' + y = 10e^{-3x}, \quad y(0) = 0, \quad y'(0) = 1. \tag{10}$$

Solution

a. The verification is straightforward but provides good practice.

b. Note that sin x is not a *constant* multiple of cos x since sin x/cos x is not constant. Also $y'' + y = 0$ is of second order. Thus by Fact 3, $\{\sin x, \cos x\}$ is a fundamental set of solutions and $c_1 \sin x + c_2 \cos x$ gives all solutions of $y'' + y = 0$. Our particular solution of the nonhomogeneous equation is $y_p = e^{-3x}$. Thus, by Fact 4,

$$y = e^{-3x} + c_1 \sin x + c_2 \cos x \qquad (11)$$

is the general solution of $y'' + y = 10e^{-3x}$.

c. To solve (10) we apply the initial conditions in (10) to the general solution (11). Evaluating (11) at zero and using the initial condition in (10) gives

$$0 = y(0) = 1 + c_2 \qquad \text{so that } c_2 = -1.$$

Differentiate (11) to get

$$y' = -3e^{-3x} + c_1 \cos x - c_2 \sin x.$$

Evaluate y' at $x = 0$ and use the initial condition $y'(0) = 1$ to get

$$1 = y'(0) = -3 + c_1$$

so that $c_1 = 4$. Thus the solution of (10) is

$$y = e^{-3x} + 4 \sin x - \cos x. \quad \blacksquare$$

Example 3.4.4

The functions $\{\sin 2x, \cos 2x\}$ form a fundamental set of solutions of the homogeneous linear differential equation

$$y'' + 4y = 0. \qquad (12)$$

The function $\tilde{y}(x) = \sin\left(2x + \frac{\pi}{4}\right)$ is also a solution of (12). Find c_1, c_2 so that

$$\sin\left(2x + \frac{\pi}{4}\right) = c_1 \sin 2x + c_2 \cos 2x.$$

Solution

Since $\{\sin 2x, \cos 2x\}$ is a fundamental set of solutions for the second-order linear homogeneous differential equation (12), by Fact 3 every solution of (12) is of the form

$$y_h = c_1 \sin 2x + c_2 \cos 2x.$$

Thus $\tilde{y}(x) = \sin\left(2x + \frac{\pi}{4}\right) = c_1 \sin 2x + c_2 \cos 2x$ for some c_1, c_2. By Fact 1, solutions are uniquely determined once $y(x_0)$, $y'(x_0)$ are specified. For convenience, take $x_0 = 0$ and let

$$y_0 = \tilde{y}(0) = \sin\frac{\pi}{4} = \frac{1}{\sqrt{2}},$$

$$v_0 = \tilde{y}'(0) = 2\cos\frac{\pi}{4} = \frac{2}{\sqrt{2}}. \quad (13)$$

To find c_1, c_2 we apply the initial conditions (13) that $\sin\left(2x + \frac{\pi}{4}\right)$ satisfies to the general solution of (12), $y_h = c_1 \sin 2x + c_2 \cos 2x$:

$$\frac{1}{\sqrt{2}} = y_0 = y_h(0) = c_2,$$

$$\frac{2}{\sqrt{2}} = v_0 = y_h'(0) = 2c_1.$$

Thus $c_1 = c_2 = 1/\sqrt{2}$. Hence

$$y_h = \frac{1}{\sqrt{2}}\sin 2x + \frac{1}{\sqrt{2}}\cos 2x \quad \text{and} \quad \tilde{y} = \sin\left(2x + \frac{\pi}{4}\right)$$

satisfy the same second-order linear differential equation and initial conditions at $x = 0$. By uniqueness of solutions (Fact 1),

$$\frac{1}{\sqrt{2}}\sin 2x + \frac{1}{\sqrt{2}}\cos 2x = \sin\left(2x + \frac{\pi}{4}\right). \quad \blacksquare$$

Depending on our method of solution, the same differential equation may have different-appearing general solutions.

Example 3.4.5 a. Verify that both

$$y = c_1 e^{2x} + c_2 e^{3x} - 1 \quad (14)$$

and

$$\tilde{y} = \tilde{c}_1 e^{2x} + \tilde{c}_2 e^{3x} + 2e^x \sinh x \quad (15)$$

are general solutions of

$$y'' - 5y' + 6y = -6. \quad (16)$$

b. Verify that, if the initial conditions $y(0) = 0$, $y'(0) = 1$ are applied to (14) and (15), both give the same solution.

Solution a. To verify that (14) and (15) are general solutions of (16), it suffices to show that $\{e^{2x}, e^{3x}\}$ is a fundamental set of solutions for $y'' - 5y' + 6y = 0$ and -1, $2e^x \sinh x$ are particular solutions of $y'' - 5y' + 6y = -6$. But

$$(e^{2x})'' - 5(e^{2x})' + 6e^{2x} = 4e^{2x} - 10e^{2x} + 6e^{2x} = 0,$$

$$(e^{3x})'' - 5(e^{3x})' + 6e^{3x} = 9e^{3x} - 15e^{3x} + 6e^{3x} = 0,$$

so that $\{e^{2x}, e^{3x}\}$ are solutions of $y'' - 5y' + 6y = 0$. Since e^{3x} is not a nonzero constant multiple of e^{2x}, $\{e^{2x}, e^{3x}\}$ is a fundamental set of solutions. Similarly, one can verify directly that -1, $2e^x \sinh x$ are particular solutions.

Alternatively, note that -1 is a solution of (16) and

$$2e^x \sinh x = 2e^x \left(\frac{e^x - e^{-x}}{2}\right) = e^{2x} - 1,$$

which is a solution from (14) (take $c_1 = 1, c_2 = 0$).

b. Applying the initial conditions to (14) gives

$$\left.\begin{array}{l}0 = y(0) = c_1 + c_2 - 1 \\ 1 = y'(0) = 2c_1 + 3c_2\end{array}\right\} \Rightarrow c_1 = 2, \quad c_2 = -1$$

or

$$y = 2e^{2x} - e^{3x} - 1, \tag{17}$$

whereas applying the initial conditions to (15) gives

$$\left.\begin{array}{l}0 = \tilde{y}(0) = \tilde{c}_1 + \tilde{c}_2 \\ 1 = \tilde{y}'(0) = 2\tilde{c}_1 + 3\tilde{c}_2 + 2\end{array}\right\} \Rightarrow \tilde{c}_1 = 1, \quad \tilde{c}_2 = -1$$

or

$$\tilde{y} = e^{2x} - e^{3x} + 2e^x \sinh x = e^{2x} - e^{3x} + e^{2x} - 1$$
$$= 2e^{2x} - e^{3x} - 1,$$

which is the same as (17). ∎

Fact 5
This tells us that, if the forcing function f is a sum of several terms, we may find a particular solution for each term and then add these particular solutions to get a particular solution for the whole forcing function. This fact plays an important role in several of our techniques. Fact 5 may also be expressed as the response (output) to a sum of two inputs is the sum of the responses (output) to each input. This is an important property of linear devices and systems.

Example 3.4.6

Suppose you wish to solve

$$y'' - y = e^{2x} - e^{3x}$$

and you determine

$y_p(x) = \frac{1}{3}e^{2x}$ is a solution of $y'' - y = e^{2x}$,

$\tilde{y}_p(x) = -\frac{1}{8}e^{3x}$ is a solution of $y'' - y = -e^{3x}$.

Then Fact 5 says that

$$y_p(x) + \tilde{y}_p(x) = \tfrac{1}{3}e^{2x} - \tfrac{1}{8}e^{3x}$$

is a solution of $y'' - y = e^{2x} - e^{3x}$. ∎

Verification of Facts 2, 3, 5

Proofs of some of the key facts will be omitted. However, it is easy and instructive to verify Facts 2, 5, and part of 3.

Fact 2

Suppose that y_p is a particular solution of $y'' + py' + qy = f$ and y is any other solution. Let $y_h = y - y_p$. Then $y = y_p + y_h$. To verify (2) we need only show that y_h is a solution of the associated homogeneous equation. But

$$\begin{aligned} y_h'' + py_h' + qy_h &= (y - y_p)'' + p(y - y_p)' + q(y - y_p) \\ &= y'' - y_p'' + py' - py_p' + qy - qy_p \\ &= (y'' + py' + qy) - (y_p'' + py_p' + qy_p) \\ &= f - f = 0 \quad \text{(since } y, y_p \text{ are solutions)} \end{aligned}$$

so that y_h is a solution of (3), as desired.

Fact 3

Suppose that y_1, y_2 are solutions of the homogeneous equation $y'' + py' + qy = 0$ and c_1, c_2 are constants. We shall verify the first half of Fact 3, namely that $c_1 y_1 + c_2 y_2$ is also a solution of $y'' + py' + qy = 0$. Substitute $c_1 y_1 + c_2 y_2$ for y in $y'' + py' + qy = 0$;

$$\begin{aligned} (c_1 y_1 + c_2 y_2)'' &+ p(c_1 y_1 + c_2 y_2)' + q(c_1 y_1 + c_2 y_2) \\ &= c_1 y_1'' + c_2 y_2'' + pc_1 y_1' + pc_2 y_2' + qc_1 y_1 + qc_2 y_2 \\ &= c_1(y_1'' + py_1' + qy_1) + c_2(y_2'' + py_2' + qy_2) \\ &= c_1 0 + c_2 0 = 0 \end{aligned}$$

(since y_1, y_2 are solutions of $y'' + py' + qy = 0$). Thus $c_1 y_1 + c_2 y_2$ is a solution of $y'' + py' + qy = 0$, as desired.

Fact 5

Suppose y_p is a solution of $y'' + py' + qy = f_1$ and $\tilde{y}_p$ is a solution of $y'' + py' + qy = f_2$. We shall show that $y_p + \tilde{y}_p$ is a solution of $y'' + py' + qy = f_1 + f_2$. Substituting $y_p + \tilde{y}_p$ for y in $y'' + py' + qy$, we get

$$\begin{aligned} (y_p + \tilde{y}_p)'' &+ p(y_p + \tilde{y}_p)' + q(y_p + \tilde{y}_p) \\ &= y_p'' + \tilde{y}_p'' + py_p' + p\tilde{y}_p' + qy_p + q\tilde{y}_p \\ &= (y_p'' + py_p' + qy_p) + (\tilde{y}_p'' + p\tilde{y}_p' + q\tilde{y}_p) \\ &= f_1 + f_2, \end{aligned}$$

as desired.

Exercises

For Exercises 1 through 8, verify that the given set is a fundamental set of solutions for the associated homogeneous equation and y_p is a particular solution. Then give the general solution of the differential equation, and solve the initial-value problem.

1. $y'' + y = 1$, $\quad y(0) = 0$, $\quad y'(0) = 0$,
   $\{\sin x, \cos x\}$, $\quad y_p = 1$

2. $y'' - y = e^{3x}$, $\quad y(0) = 0$, $\quad y'(0) = 1$,
   $\{e^x, e^{-x}\}$, $\quad y_p = \frac{1}{8}e^{3x}$

3. $y'' - 3y' + 2y = 2x$, $\quad y(0) = 1$, $\quad y'(0) = 0$,
   $\{e^x, e^{2x}\}$, $\quad y_p = x + 3/2$

4. $y'' - 2y' + y = 4e^{2x}$, $\quad y(0) = 0$, $\quad y'(0) = 0$,
   $\{e^x, xe^x\}$, $\quad y_p = 4e^{2x}$

5. $y'' + 2y' + 2y = 6$, $\quad y(0) = 1$, $\quad y'(0) = 1$,
   $\{e^{-x}\cos x, e^{-x}\sin x\}$, $\quad y_p = 3$

6. $x^2 y'' - 2xy' + 2y = 2x^3$, $\quad y(1) = 2$,
   $y'(1) = 3$, $\quad \{x, x^2\}$, $\quad y_p = x^3$, $\quad x > 0$

7. $y'' + y = 2\cos x$, $\quad y(0) = 1$, $\quad y'(0) = -1$,
   $\{\sin x, \cos x\}$, $\quad y_p = x \sin x$

8. $x^2 y'' + 4xy' + 2y = 2 \ln x + 3$, $\quad y(1) = 0$,
   $y'(1) = 2$, $\quad \{x^{-1}, x^{-2}\}$, $\quad y_p = \ln x$, $\quad x > 0$

9. a) Verify that both
   $$y = c_1 e^x + c_2 e^{2x} + 2 \cosh x \quad (18)$$
   and
   $$\tilde{y} = \tilde{c}_1 e^x + \tilde{c}_2 e^{2x} + e^{-x} \quad (19)$$
   are the general solution of $y'' - 3y' + 2y = 6e^{-x}$.

   b) Verify that, if the initial conditions $y(0) = 4$, $y'(0) = 3$, are applied to (18) and (19), both give the same solution.

10. a) Verify that both
    $$y = c_1 + c_2 x + x^2 \quad (20)$$
    and

$$\tilde{y} = \tilde{c}_1(1 + x) + \tilde{c}_2(1 - x) + x^2 + 3x + 1 \quad (21)$$
are the general solution of $y'' = 2$.

b) Verify that, if the initial conditions $y(0) = 0$, $y'(0) = 1$, are applied to (20) and (21), then both give the same solution.

11. Verify that $-\frac{1}{3}\sin 2x$ is a solution of $y'' + y = \sin 2x$, $-\frac{1}{3}\cos 2x$ is a solution of $y'' + y = \cos 2x$, and $\{\sin x, \cos x\}$ is a fundamental set of solutions for $y'' + y = 0$. Find the general solution of $y'' + y = \sin 2x + \cos 2x$.

12. Verify that e^{2x} is a solution of $y'' - 2y' + y = e^{2x}$, 1 is a solution of $y'' - 2y' + y = 1$, and $\{e^x, xe^x\}$ is a fundamental set of solutions for $y'' - 2y' + y = 0$. Find the general solution of $y'' - 2y' + y = 1 + e^{2x}$.

13. Suppose that $a_2(x)$, $a_1(x)$, $a_0(x)$, $h(x)$ are continuous on the interval $[a, b]$. Show that Theorem 3.1.1 guarantees a unique solution to (1) such that $y(x_0) = y_0$, $y'(x_0) = v_0$ only if $a \le x_0 \le b$ and $a_2(x_0) \ne 0$.

14. Suppose that y_1, y_2 are differentiable functions. Show that if $y_1 = cy_2$ for a constant c, then $y_1' y_2 - y_1 y_2' = 0$ for all x.

15. Suppose that y_1, y_2 are differentiable functions and neither y_1 nor y_2 are ever zero. Show that if $y_1' y_2 - y_1 y_2' = 0$ for all x, then $y_1 = cy_2$ for some constant c.

16. Verify that $y = c_1 x^2 + c_2 x^3$ is the general solution of
    $$x^2 y'' - 4xy' + 6y = 0 \quad (22)$$
    on either $(-\infty, 0)$ or $(0, \infty)$.

17. Verify that $y = c_1 x^2 + c_2 x^3$ is actually a solution of (22) on $(-\infty, \infty)$. Show that $\tilde{y} = |x|^3$ is a solution of (22) on $(-\infty, \infty)$. Show that there do not exist constants c_1, c_2 such that $|x|^3 = c_1 x^2 + c_2 x^3$ for $-\infty < x < \infty$.

3.5 Linear Independence and the Wronskian

In this section we will explore more carefully the concept of a fundamental set of solutions introduced in Section 3.4.

Two functions $y_1(x)$, $y_2(x)$ defined on an interval I of nonzero length are said to be *linearly independent* if the equation

$$c_1 y_1(x) + c_2 y_2(x) = 0, \tag{1}$$

for all x in I, implies that $c_1 = c_2 = 0$. For two functions y_1, y_2, this is equivalent to saying that neither function is a nonzero constant multiple of the other. (See Exercise 9 at the end of this section.) However, the definition of linear independence extends readily to more than two functions. If functions are not linearly independent, they are called *linearly dependent*.

Example 3.5.1 Show that the functions $y_1 = e^x$ and $y_2 = \sin x$ are linearly independent.

Solution Suppose that

$$c_1 e^x + c_2 \sin x = 0 \tag{2}$$

for all x. In particular, (2) must hold for $x = 0$ and $x = 1$. Thus (2), evaluated at $x = 0$, $x = 1$, gives

$$x = 0: \quad c_1 e^0 + c_2 \sin 0 = 0 \quad \text{or} \quad c_1 = 0,$$
$$x = 1: \quad c_1 e + c_2 \sin 1 = 0.$$

But $\sin 1 \neq 0$; thus $c_1 = c_2 = 0$ and $\{e^x, \sin x\}$ are linearly independent. ∎

To understand best what follows we need to review a few facts from algebra. The exposition is self-contained except that some proofs are deleted. More detail is found in Section 6.8.

For a 2×2 matrix of numbers

$$\mathbf{A} = \begin{bmatrix} a & b \\ c & d \end{bmatrix},$$

recall that the *determinant of* $\mathbf{A}$ is the number $ad - bc$. The determinant, denoted $\det(\mathbf{A})$, is closely related to properties of systems of algebraic equations.

Suppose a, b, c, d, e, and f are constants and z, w are variables. Consider the following system of two equations in the two unknowns z, w;

$$az + bw = e,$$
$$cz + dw = f. \tag{3}$$

In elementary algebra the following facts are proved:

Equations (3) have a solution for every e, f if and only if

$$\det \begin{bmatrix} a & b \\ c & d \end{bmatrix} \neq 0. \tag{4}$$

Solutions of (3) are unique for a given e, f if and only if

$$\det \begin{bmatrix} a & b \\ c & d \end{bmatrix} \neq 0. \tag{5}$$

Example 3.5.2 Consider the system of equations

$$\begin{aligned} z + 2w &= e, \\ 2z + 4w &= f. \end{aligned} \tag{6}$$

Since

$$\det \begin{bmatrix} 1 & 2 \\ 2 & 4 \end{bmatrix} = 4 - 4 = 0,$$

the equations (6) do not have a solution for some e, f such as $e = 1, f = 7$. When solutions (z, w) do exist, they are not unique. For example, both $(1, 3)$ and $(3, 2)$ are solutions of

$$\begin{aligned} z + 2w &= 7, \\ 2z + 4w &= 14. \end{aligned}$$ ∎

Example 3.5.3 Now consider

$$\begin{aligned} z + 2w &= e, \\ 3z + 4w &= f. \end{aligned}$$

Since

$$\det \begin{bmatrix} 1 & 2 \\ 3 & 4 \end{bmatrix} = 4 - 6 = -2 \neq 0,$$

there is a unique solution for any e, f. ∎

For two functions $y_1(x), y_2(x)$, the *Wronskian* of y_1, y_2, denoted $W[y_1, y_2]$, is given by

$$W[y_1, y_2] = \det \begin{bmatrix} y_1 & y_2 \\ y_1' & y_2' \end{bmatrix} = y_1(x)y_2'(x) - y_2(x)y_1'(x). \tag{7}$$

The Wronskian is a function of x that plays an important role in the theory of linear differential equations. The value of $W[y_1, y_2]$ at x will sometimes be denoted $W[y_1, y_2](x)$.

The next theorem summarizes the key relationships between the Wronskian and the concepts of linear independence and fundamental set of solutions.

Theorem 3.5.1

Suppose that p, q are continuous functions on the interval I and y_1, y_2 are solutions of the linear homogeneous differential equation $y'' + py' + qy = 0$. Then the following are equivalent.

i. $\{y_1, y_2\}$ are linearly independent.
ii. $W[y_1, y_2] \neq 0$ for some x in I.
iii. $W[y_1, y_2] \neq 0$ for all x in I.
iv. If y_3 is any solution of

$$y'' + py' + qy = 0, \tag{8}$$

then there exist constants c_1, c_2 such that $y_3 = c_1 y_1 + c_2 y_2$ and the c_1, c_2 are uniquely determined by y_3. That is, $\{y_1, y_2\}$ is a *fundamental set of solutions* for (8) and any other solution y_3 may be written as a *linear combination* of y_1, y_2.

Note that parts (i), (ii), (iii) of Theorem 3.5.1 are not equivalent for an arbitrary pair of functions $\{y_1, y_2\}$. The functions must be solutions of the same second-order homogeneous linear differential equation. (See Exercises 10, 11.)

Verification

It is worthwhile to take the time to understand why Theorem 3.5.1 is true. If y_1, y_2 are solutions of $y'' + py' + qy = 0$, then they are differentiable. Let $W(x) = W[y_1, y_2]$.

The equivalence of (ii) and (iii) says that the Wronskian is either always zero or never zero. To show this, we first show that $W(x)$ satisfies the differential equation

$$W'(x) + p(x)W(x) = 0, \tag{9}$$

where $p(x)$ is the same p that appears in (8). To verify (9), note that

$W' + pW$

$= [y_1 y_2' - y_1' y_2]' + p[y_1 y_2' - y_1' y_2]$
$= y_1' y_2' + y_1 y_2'' - y_1' y_2' - y_1'' y_2 + p y_1 y_2' - p y_1' y_2$ (by differentiation)
$= y_1 y_2'' - y_1'' y_2 + p y_1 y_2' - p y_1' y_2$ (cancellation of two terms)
$= y_1 [y_2'' + p y_2'] - [y_1'' + p y_1'] y_2$ (by rearrangement)
$= y_1 [-q y_2] - [-q y_1] y_2 = 0$ (y_1, y_2 are solutions of (8).)

But from Section 2.3, the solution of (9) is

$$W(x) = \exp\left(\int_{x_0}^{x} p(t)\, dt\right) W(x_0). \tag{10}$$

The exponential in (10) is never zero and $W(x_0)$ is a constant. Thus $W(x)$ is either always zero or never zero, and (ii) and (iii) are equivalent.

If $\{y_1, y_2\}$ are linearly independent, then neither $y_1 - c y_2$ nor $y_2 - c y_1$ is identically zero. Thus, from Theorem 3.4.1, $\{y_1, y_2\}$ is a fundamental set of solutions for (8) and (i) implies (iv). Let y_3 be another solution of (8). Then from Theorem 3.4.1,

$$y_3 = c_1 y_1 + c_2 y_2, \tag{11}$$

so that
$$y_3(x_0) = c_1 y_1(x_0) + c_2 y_2(x_0) \tag{12}$$
and
$$y_3'(x_0) = c_1 y_1'(x_0) + c_2 y_2'(x_0). \tag{13}$$

On the other hand, if we can find c_1, c_2 so that (12) and (13) hold, then since y_3 and $c_1 y_1 + c_2 y_2$ satisfy the same initial conditions (12) and (13) at x_0, the equations (12) and (13) will imply, by the uniqueness part of Theorem 3.4.1, that $y_3 = c_1 y_1 + c_2 y_2$. Thus being able to write (11) for some c_1, c_2 is equivalent to being able to solve (12) and (13) for some c_1, c_2. But, by (4), we can always solve (12) and (13) for c_1, c_2 if and only if

$$\det \begin{bmatrix} y_1 & y_2 \\ y_1' & y_2' \end{bmatrix} \neq 0,$$

which implies that having nonzero Wronskian (iii) and being a fundamental set of solutions (iv) is equivalent. By (5), we get the uniqueness of c_1, c_2 in (iv).

There remains only to show that (iv) (or (iii) or (ii)) implies (i). Suppose that $W(x) \neq 0$ and that

$$c_1 y_1(x) + c_2 y_2(x) = 0 \tag{14a}$$

for constants c_1, c_2. Then differentiating gives

$$c_1 y_1'(x) + c_2 y_2'(x) = 0. \tag{14b}$$

Since $W(x) \neq 0$ for all x, we have $c_1 = c_2 = 0$ by (5), and y_1, y_2 are linearly independent. ∎

Example 3.5.4

Determine two solutions of $y'' - 3y' + 2y = 0$ that are of the form e^{rx}, and decide whether they form a fundamental set of solutions.

Solution

Substituting e^{rx} into $y'' - 3y' + 2y = 0$ gives
$$(e^{rx})'' - 3(e^{rx})' + 2e^{rx} = 0,$$
$$r^2 e^{rx} - 3r e^{rx} + 2e^{rx} = 0$$
$$e^{rx}(r^2 - 3r + 2) = 0$$
or
$$r^2 - 3r + 2 = 0;$$

that is, $(r - 2)(r - 1) = 0$. The roots are $r = 2, 1$, so e^{2x}, e^x are solutions. To determine whether they are a fundamental set of solutions, we compute the Wronskian;

$$W[e^{2x}, e^x] = \det\begin{bmatrix} e^{2x} & e^x \\ 2e^{2x} & e^x \end{bmatrix} = e^{3x} - 2e^{3x} = -e^{3x} \neq 0 \quad \text{for all } x.$$

Since the Wronskian is nonzero, $\{e^{2x}, e^x\}$ is a linearly independent set of solutions, and every solution of $y'' - 3y' + 2y = 0$ is of the form $c_1 e^{2x} + c_2 e^x$ by Theorem 3.5.1. ∎

Example 3.5.5 Let y_1 be the solution of

$$y'' + x^2 y' + x^3 y = 0, \quad y(1) = 1, \quad y'(1) = 3, \tag{15}$$

and let y_2 be the solution of

$$y'' + x^2 y' + x^3 y = 0, \quad y(1) = -1, \quad y'(1) = 2. \tag{16}$$

a. Verify $\{y_1, y_2\}$ is a fundamental set of solutions of $y'' + x^2 y' + x^3 y = 0$.
b. Find constants c_1, c_2 such that $y_3 = c_1 y_1 + c_2 y_2$ is the solution of

$$y'' + x^2 y' + x^3 y = 0, \quad y(1) = 2, \quad y'(1) = 0. \tag{17}$$

Solution a. By Theorem 3.5.1, to show that $\{y_1, y_2\}$ is a fundamental set of solutions, it suffices to verify that $W[y_1, y_2] \neq 0$ at any point. We shall verify that $W[y_1, y_2] \neq 0$ at $x = 1$. Now, at $x = 1$,

$$W[y_1, y_2] = \begin{bmatrix} y_1(1) & y_2(1) \\ y_1'(1) & y_2'(1) \end{bmatrix}$$

$$= \det\begin{bmatrix} 1 & -1 \\ 3 & 2 \end{bmatrix} \quad \text{(By (15) and (16))}$$

$$= 5 \neq 0,$$

and hence y_1, y_2 are linearly independent.

b. We shall use the fact that (11) is equivalent to (12) and (13). Evaluating (12) and (13) at $x_0 = 1$ gives

$$\begin{aligned} y_3(1) &= c_1 y_1(1) + c_2 y_2(1), \\ y_3'(1) &= c_1 y_1'(1) + c_2 y_2'(1). \end{aligned} \tag{18}$$

The initial conditions in (15), (16), and (17) applied to (18) yield two equations in c_1, c_2:

$$2 = c_1 1 + c_2(-1)$$
$$0 = c_1 3 + c_2 2.$$

Solving for c_1, c_2 yields $c_2 = -\frac{6}{5}$, $c_1 = \frac{4}{5}$. Thus $y_3 = \frac{4}{5} y_1 - \frac{6}{5} y_2$. ∎

Exercises

1. Verify that $\sin x$, $\cos x$ are solutions of $y'' + y = 0$. Determine whether they are a fundamental set of solutions.

2. Find all solutions of the form x^r for $x^2 y'' - 6y = 0$ on $(0, \infty)$ and determine whether they form a fundamental set of solutions.

3. Find all solutions of the form x^r for $x^2 y'' - xy' + y = 0$ on $(0, \infty)$ and determine whether they form a fundamental set of solutions.

4. Find all solutions of the form e^{rx} of $y'' - 4y' + 4y = 0$ on $(-\infty, \infty)$ and determine whether they form a fundamental set of solutions.

5. Let y_1 be the solution on $(0, \infty)$ of
$$x^2 y'' + y' + xy = 0, \quad y(1) = 1, \quad y'(1) = 1,$$
and y_2 the solution on $(0, \infty)$ of
$$x^2 y'' + y' + xy = 0, \quad y(1) = 0, \quad y'(1) = -1.$$
   a) Verify that $\{y_1, y_2\}$ is a fundamental set of solutions of $x^2 y'' + y' + xy = 0$.
   b) Let y_3 be the solution of
$$x^2 y'' + y' + xy = 0, \quad y(1) = 2, \quad y'(1) = 0.$$
   Find constants c_1, c_2 such that
$$y_3 = c_1 y_1 + c_2 y_2.$$

6. Let y_1 be the solution of
$$y'' + xy' + y = 0, \quad y(0) = 1, \quad y'(0) = 2,$$
and y_2 the solution of
$$y'' + xy' + y = 0, \quad y(0) = 1, \quad y'(0) = -1.$$
   a) Verify that $\{y_1, y_2\}$ is a fundamental set of solutions of $y'' + xy' + y = 0$.
   b) Let y_3 be the solution of
$$y'' + xy' + y = 0, \quad y(0) = 2, \quad y'(0) = 2.$$
   Find constants c_1, c_2 such that
$$y_3 = c_1 y_1 + c_2 y_2.$$

7. Let y_1 be the solution of
$$y'' + py' + qy = 0, \quad y(x_0) = 1, \quad y'(x_0) = 0,$$
and y_2 the solution of
$$y'' + py' + qy = 0, \quad y(x_0) = 0, \quad y'(x_0) = 1.$$
   Verify that y_1, y_2 is a fundamental set of solutions and that the solution y_3 of
$$y'' + py' + qy = 0, \quad y(x_0) = \alpha, \quad y'(x_0) = \beta$$
is $y_3 = \alpha y_1 + \beta y_2$.

8. Let $y_1 = x^3 - x$, $y_2 = x^2 - 1$.
   a) Verify that y_1, y_2 are linearly independent functions.
   b) Can y_1, y_2 be solutions of a differential equation $y'' + py' + qy = 0$ where p, q are continuous on $(-2, \infty)$?
   c) Suppose that y_1, y_2 are solutions of a differential equation $y'' + py' + qy = 0$. Where must p, q have discontinuities?

9. Verify directly that y_1, y_2 are linearly dependent if and only if one of them is a constant multiple of the other.

10. Consider the differentiable functions y_1, y_2 on $[0, 2]$ with graphs in Fig. 3.5.1.

Figure 3.5.1

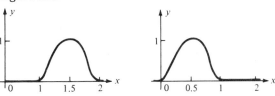

   a) Verify that y_1, y_2 are linearly independent.
   b) Verify that $W[y_1, y_2] = 0$ for all x in $[0, 2]$.
   c) Explain why this does not contradict Theorem 3.5.1.

11. In Example 3.5.1, we showed that $y_1 = e^x$, $y_2 = \sin x$ are linearly independent. Find the values of x such that $W[e^x, \sin x] = 0$. Can y_1, y_2 be solutions of a linear homogeneous second-order differential equation with continuous coefficients at these values of x?

12. Suppose that y_1, y_2 are linearly independent solutions of $y'' + py' + qy = 0$ on $(-\infty, \infty)$. Let $\phi(t)$ be a differentiable function on $(-\infty, \infty)$ such that ϕ' is not identically zero. Show that $y_1(\phi(t)), y_2(\phi(t))$ are linearly independent.

Exercises 3.4.14 and 3.4.15 may be worked now if they were not assigned earlier.

3.6 Operator Notation

Well-designed notation not only helps in the solving of a problem but sometimes suggests new approaches and relationships. The notation of this section is not essential for solving simple differential equations, although it can add insight. For more complicated problems, it becomes progressively more important.

The expression

$$L(y) = a_n(x)\frac{d^n y}{dx^n} + a_{n-1}(x)\frac{d^{n-1} y}{dx^{n-1}} + \cdots + a_1(x)\frac{dy}{dx} + a_0(x)y \qquad (1)$$

is called an *nth-order linear differential operator*.

Example 3.6.1 $L_1(y) = 3y'' + 2y' + y$ is a second-order differential operator, while $L_2(y) = y' + xy$ is a first-order differential operator. ∎

An operator is a function of functions. Given a function y, $L(y)$ is another function. We already have two examples of operators; differentiation and anti-differentiation. In Chapter 4 the Laplace transform operator is discussed in detail. In this chapter, "operator" will mean linear differential operator.

Example 3.6.2 If $L(y) = xy'' + y$, compute $L(\sin x)$.

Solution $L(\sin x) = x(\sin x)'' + \sin x = -x \sin x + \sin x.$ ∎

Just as we sometimes write f for the function whose value at x is $f(x)$, L denotes the operator whose value at y is $L(y)$.

If $L(y)$ is given by (1), then

$$L = a_n(x)\frac{d^n}{dx^n} + a_{n-1}(x)\frac{d^{n-1}}{dx^{n-1}} + \cdots + a_1(x)\frac{d}{dx} + a_0(x). \qquad (2)$$

It is important to note that the $a_0(x)$ in (2) stands for multiplication by $a_0(x)$.

Example 3.6.3 Suppose

$$L = x^2 \frac{d^2}{dx^2} + x\frac{d}{dx} + x^3.$$

Compute $L(x^4)$.

Solution By (1) with $y = x^4$,

$$L(x^4) = \left(x^2 \frac{d^2}{dx^2} + x\frac{d}{dx} + x^3\right)x^4$$

$$= x^2 \frac{d^2 x^4}{dx^2} + x\frac{dx^4}{dx} + x^3 x^4$$

$$= x^2 12x^2 + x4x^3 + x^3 x^4 = 16x^4 + x^7. \blacksquare$$

If it is obvious what the independent variable is, D^n is often written for d^n/dx^n, the operation of differentiating n times with respect to the independent variable. For example,

$$x^2 \frac{d^2}{dx^2} + x\frac{d}{dx} + x^3 = x^2 D^2 + xD + x^3.$$

In order to consider a differential operator as a function of functions, we need to specify its domain. We shall always take as the domain those functions that are n-times differentiable on a fixed interval.

The theory for linear differential equations is closely intertwined with that of differential operators. We shall now show some of that relationship. The arguments here should be compared with those at the end of Section 3.4, which proved Facts 2, 3, and 5 of that section.

Proposition 3.6.1

Differential operators defined by (1) are *linear*. That is, if L is a differential operator, c_1, c_2 are constants and y_1, y_2 are functions, then

$$L(c_1 y_1 + c_2 y_2) = c_1 L(y_1) + c_2 L(y_2). \tag{3}$$

(There are nonlinear differential operators, but we shall not discuss them.)

Verification

Suppose that c_1, c_2 are constants and y_1, y_2 are functions. We shall prove (3) for the second-order operator

$$L = a_2 D^2 + a_1 D + a_0.$$

The proof for higher-order differential operators is essentially the same. Now

$$\begin{aligned}
L(c_1 y_1 &+ c_2 y_2) \\
&= (a_2 D^2 + a_1 D + a_0)(c_1 y_1 + c_2 y_2) \\
&= a_2 D^2(c_1 y_1 + c_2 y_2) + a_1 D(c_1 y_1 + c_2 y_2) + a_0(c_1 y_1 + c_2 y_2) \\
&= a_2(c_1 y_1 + c_2 y_2)'' + a_1(c_1 y_1 + c_2 y_2)' + a_0(c_1 y_1 + c_2 y_2) \\
&= a_2(c_1 y_1'' + c_2 y_2'') + a_1(c_1 y_1' + c_2 y_2') + a_0(c_1 y_1 + c_2 y_2) \\
&= c_1(a_2 y_1'' + a_1 y_1' + a_0 y_1) + c_2(a_2 y_2'' + a_1 y_2' + a_0 y_2) \\
&= c_1(a_2 D^2 y_1 + a_1 D y_1 + a_0 y_1) + c_2(a_2 D^2 y_2 + a_1 D y_2 + a_0 y_2) \\
&= c_1(a_2 D^2 + a_1 D + a_0) y_1 + c_2(a_2 D^2 + a_1 D + a_0) y_2 \\
&= c_1 L(y_1) + c_2 L(y_2). \blacksquare
\end{aligned}$$

Note that the linear differential equation

$$a_n \frac{d^n y}{dx^n} + a_{n-1} \frac{d^{n-1} y}{dx^{n-1}} + \cdots + a_1 \frac{dy}{dx} + a_0 y = f$$

may be written as

$$L(y) = f,$$

where L is given by (2). The next theorem is the operator analogue of the key facts of Theorem 3.4.1. The arguments should be compared with those in Section 3.4.

Theorem 3.6.1 Let y_p be a particular solution of the linear differential equation $L(y) = f$.

1. If $\tilde{y}$ is any other solution of $L(y) = f$, then

$$\tilde{y} = y_p + y_h,$$

where y_h is a solution of the associated homogeneous equation $L(y) = 0$.

2. If y_1, y_2 are solutions of the homogeneous equation $L(y) = 0$, then $c_1 y_1 + c_2 y_2$ is also a solution $L(y) = 0$ for any constants c_1, c_2.

Verification The key to the verification of parts (1) and (2) of Theorem 3.6.1 is the linearity of L, as shown in (3). To verify (1): Suppose that y_p, $\tilde{y}$ are solutions of $L(y) = f$. Let $y_h = \tilde{y} - y_p$. Clearly $\tilde{y} = y_p + y_h$. We need only show that y_h is a solution of $L(y) = 0$. But

$$L(y_h) = L(\tilde{y} - y_p) = L(\tilde{y}) - L(y_p) = f - f = 0 \quad \text{(by linearity),}$$

so that y_h is a solution of $L(y) = 0$, as desired.

To verify (2), suppose that y_1, y_2 are solutions of $L(y) = 0$ and c_1, c_2 are constants. Then

$$L(c_1 y_1 + c_2 y_2) = c_1 L(y_1) + c_2 L(y_2) = c_1 \cdot 0 + c_2 \cdot 0 = 0 \quad \text{(by linearity)}$$

and $c_1 y_1 + c_2 y_2$ is also a solution of $L(y) = 0$. ∎

The product of two differential operators is defined to be their composition as functions:

If L_1, L_2 are differential operators, then $L_1 L_2$ is another differential operator, and it is defined by

$$(L_1 L_2)(y) = L_1(L_2(y)). \tag{4}$$

Note that this means, for example, that $D^2 D^3 = D^5$, since

$$D^2 D^3 y = \frac{d^2}{dx^2}\left(\frac{d^3 y}{dx^3}\right) = \frac{d^5 y}{dx^5} = D^5 y.$$

If the differential operators have constant coefficients, then they may be easily multiplied.

Theorem 3.6.2   If L_1, L_2 have *constant coefficients*, then

1. $L_1 L_2 = L_2 L_1$ (commutativity).
2. $L_1 L_2$ can be computed by multiplying out the expressions for L_1, L_2, as if they were ordinary polynomials.

The verification of Theorem 3.6.2 is left to the exercises.

Example 3.6.4   Let $L_1 = D + 2$ and $L_2 = D + 1$. Then
$$L_1 L_2 = (D + 2)(D + 1) = D^2 + 3D + 2,$$
and
$$L_2 L_1 = (D + 1)(D + 2) = D^2 + 3D + 2.$$
Similarly
$$L_2^3 = (D + 1)^3 = D^3 + 3D^2 + 3D + 1. \blacksquare$$

Division by differential operators is more subtle then multiplication and is probably best avoided at first. The next example illustrates why. (Note also Exercises 20 and 21.)

Example 3.6.5   Note that
$$(D - 1)(e^t + t) = (D - 1)t = 1 - t,$$
but that $e^t + t \neq t$. $\blacksquare$

Other than in proving Theorem 3.6.1, this book will consider only differential operators with constant coefficients. Greater care must be taken if nonconstant-coefficient differential operators are used, since

1. In general, $L_1 L_2 \neq L_2 L_1$.
2. In general, $L_1 L_2$ cannot be computed by multiplying out the expressions for L_1, L_2.

Example 3.6.6   Let $L_1 = xD + 1$, $L_2 = D$. Compute $L_1 L_2$ and $L_2 L_1$.

Solution   We compute $L_1 L_2$ by evaluating $L_1(L_2(y))$ and similarly, for $L_2 L_1$, we evaluate $L_2(L_1(y))$:
$$\begin{aligned}(L_1 L_2)(y) &= (xD + 1)(Dy) \\ &= (xD + 1)y' \\ &= xDy' + y' = xy'' + y' = (xD^2 + D)y.\end{aligned}$$
Thus
$$L_1 L_2 = (xD + 1)D = xD^2 + D. \tag{5}$$

But

$$(L_2 L_1)(y) = D(xD + 1)y = D(xDy + y)$$
$$= D(xy' + y) = (xy')' + y'$$
$$= y' + xy'' + y' = xy'' + 2y'$$
$$= xD^2 y + 2Dy = (xD^2 + 2D)y,$$

so that $L_2 L_1 = D(xD + 1) = xD^2 + 2D$, which is not equal to $L_1 L_2$ in (5). ∎

Exercises

In all of the exercises, L denotes a differential operator.

1. If $L = 2D + x$, compute $L(x^3)$.
2. If $L = xD^2 + 1$, compute $L(\sin x)$.
3. If $L = D^2 + xD$, compute $L(3e^{-x})$.
4. If $L = D^2 + 1$, compute $L(\sin x + 3)$.
5. If $L = D^3 - D + 1$, compute $L(e^{-x})$.
6. If $L = xD^2 + x^2 D - x + 1$, compute $L(x^4 + 1)$.

For Exercises 7 through 13, compute $L_1 L_2$ and $L_2 L_1$.

7. $L_1 = D$, $L_2 = xD$
8. $L_1 = D + x$, $L_2 = D - x$
9. $L_1 = D + 3$, $L_2 = D^2 + 1$
10. $L_1 = xD$, $L_2 = D^2$
11. $L_1 = D^2 + D$, $L_2 = D - 1$
12. $L_1 = D - 1$, $L_2 = e^x D$
13. $L_1 = D^2 + 1$, $L_2 = \sin x \, D$

14. Verify that, if y_1 is a solution of $L(y) = e^x$ and y_2 is a solution of $L(y) = \sin x$, then $y_1 + y_2$ is a solution of $L(y) = e^x + \sin x$. (This is sometimes referred to as the *superposition principle*.)

15. Verify that, if y_1 is a solution of $L_1(y) = f$ and f is a solution of $L_2(y) = 0$, then y_1 is a solution of the homogeneous equation $L_2 L_1(y) = 0$. (This is the idea behind the method of undetermined coefficients, to be discussed later in this chapter.)

16. Suppose that $L = L_1 L_2$. Verify that, if y_1 is a solution of $L_1(y_1) = f$ and y_2 is a solution of $L_2(y_2) = y_1$, then y_2 is a solution of $L(y) = f$.

17. Note that $L = D^2 - 1 = (D + 1)(D - 1) = L_1 L_2$. By finding y_1, y_2 as in Exercise 16, solve the second-order differential equation $y'' - y = 3$.

18. Note that $L = D^2 - 3D + 2 = (D - 1)(D - 2) = L_1 L_2$. By finding y_1, y_2 as in Exercise 16, solve the second-order linear differential equation $y'' - 3y' + 2y = 2$.

19. Verify Theorem 3.6.2 if L_1, L_2 are second-order constant-coefficient differential operators.

20. Suppose that f, g are differentiable functions of x. Show that, if $(D - 1)f = (D - 1)g$, then $f = g + ce^x$ for some constant c.

21. Suppose that f, g are twice-differentiable functions and L is a second-order differential operator. Show that if $L(f) = L(g)$ then $f = g + h$, where h is a solution of $L(y) = 0$.

Exercises 22 through 25 introduce some facts about differential operators that can be used to reduce the amount of computation when using the method of undetermined coefficients (Sections 3.12 and 3.13). In these exercises, $p(\lambda) = a_n \lambda^n + a_{n-1} \lambda^{n-1} + \cdots + a_1 \lambda + a_0$ is an nth degree polynomial with constant coefficients and $p(D) = a_n D^n + a_{n-1} D^{n-1} + \cdots + a_1 D + a_0$.

22. a) Show that $p(D)e^{\alpha x} = p(\alpha)e^{\alpha x}$ for any scalar α.
    b) Compute $(D^3 - 2D^2 + D - 1)e^{3x}$ using part (a).

23. Suppose that $p(\lambda)$, $q(\lambda)$ are polynomials with degree $(p(\lambda)) >$ degree $(q(\lambda))$. Let $r(\lambda)$ be the remainder after dividing $q(\lambda)$ into $p(\lambda)$ using polynomial division. That is, $p(\lambda) = m(\lambda)q(\lambda) + r(\lambda)$

with m, r polynomials and degree $(r(\lambda)) <$ degree $(q(\lambda))$. Suppose that $f(x)$ is a function such that $q(D)f = 0$. Show that $p(D)f = r(D)f$.

24. **a)** Verify that $(D^2 + 2D + 2)e^{-x} \sin x = 0$.
   b) Compute $(D^4 + 3D^3 - D)e^{-x} \sin x$ by letting $p(D) = D^4 + 3D^3 - D$, $q(D) = D^2 + 2D + 2$, $f = e^{-x} \sin x$ and using Exercise 23.

25. **a)** Verify that $(D^2 + 2D + 1)xe^{-x} = 0$.
   b) Compute $(D^5 + 3D^2 + 1)xe^{-x}$ by letting $p(D) = D^5 + 3D^2 + 1$, $q(D) = D^2 + 2D + 1$, $f = xe^{-x}$ and using Exercise 23.

3.7 Reduction of Order

Several techniques for solving linear differential equations, including some in this chapter, are based on knowing the *form* for *some* of the solutions. Suppose we wish to solve the second-order linear differential equation

$$a_2(x)y''(x) + a_1(x)y'(x) + a_0(x)y(x) = f(x), \tag{1}$$

and we have determined one solution $y_1(x)$ of the associated homogeneous equation

$$a_2(x)y''(x) + a_1(x)y'(x) + a_0(x)y(x) = 0. \tag{2}$$

We know from Sections 3.4 and 3.5 that it would take two functions to form a fundamental set of solutions to (2). Since we need only *one* more, it might be possible to reduce (1) to a *first*-order problem. The method of *reduction of order* does this.

We shall discuss the method, work two examples, and then summarize it. Suppose that y_1 is a solution of (2). We now show how to reduce (1) to a first-order equation. The key is to look for solutions of the form

$$y = vy_1, \tag{3}$$

where v is an unknown function of x and y_1 is our known solution of the associated homogeneous equation (2). Substituting $y = vy_1$ into (1) gives

$$a_2(vy_1)'' + a_1(vy_1)' + a_0(vy_1) = f.$$

Perform the differentiations,

$$a_2(v''y_1 + 2v'y_1' + vy_1'') + a_1(v'y_1 + vy_1') + a_0vy_1 = f,$$

and regroup by derivatives of v:

$$[a_2y_1]v'' + [2a_2y_1' + a_1y_1]v' + [a_2y_1'' + a_1y_1' + a_0y_1]v = f. \tag{4}$$

But by assumption, $a_2y_1'' + a_1y_1' + a_0y_1 = 0$, so that (4) is

$$[a_2y_1]v'' + [2a_2y_1' + a_1y_1]v' = f.$$

Now $[a_2y_1]$ and $[2a_2y_1' + a_1y_1]$ are known functions of x, so that letting

$w = v'$ gives us a first-order linear equation in w (as in Section 3.2):
$$[a_2 y_1]w' + [2a_2 y_1' + a_1 y_1]w = f.$$
This differential equation may be solved by the integrating-factor method of Section 2.3.

Example 3.7.1 You are given $y_1 = x$ is a solution of
$$x^2 y'' - xy' + y = 0, \qquad x > 0. \tag{5}$$
Find all solutions of
$$x^2 y'' - xy' + y = x^4, \qquad x > 0. \tag{6}$$

Solution Let $y = vy_1 = vx$ and substitute into (6):
$$x^2(vx)'' - x(vx)' + vx = x^4$$
$$x^2(v''x + 2v') - x(v'x + v) + vx = x^4,$$
or
$$x^3 v'' + x^2 v' = x^4.$$
Let $w = v'$ and divide by x^3:
$$\frac{dw}{dx} + \frac{1}{x} w = x.$$
This is a first-order linear equation. The integrating factor is $\exp\left[\int \frac{1}{x} dx\right] = \exp[\ln x] = x$. Multiply by the integrating factor to get:
$$(wx)' = x^2.$$
Antidifferentiate:
$$wx = \frac{x^3}{3} + c_1,$$
and solve for w:
$$w = \frac{x^2}{3} + \frac{c_1}{x},$$
so that
$$v' = \frac{x^2}{3} + \frac{c_1}{x}.$$
Now antidifferentiate to find v:
$$v = \frac{x^3}{9} + c_1 \ln x + c_2.$$
Finally,

$$y = vy_1 = \frac{x^4}{9} + c_1 x \ln x + c_2 x \tag{7}$$

is the general solution of (6). ∎

Note that $x^4/9$ is a particular solution of the equation (6) and $\{x \ln x, x\}$ is a fundamental set of solutions for the associated homogeneous equation (5), since (5) is second-order and $W[x \ln x, x] = -x \neq 0$. Thus, (7) is in the form $y_p + y_h$ discussed in Sections 3.4 and 3.5.

Reduction of order can also be used to find a fundamental set of solutions for (2).

Example 3.7.2

i. Find a solution of $y'' - 2y' + y = 0$ of the form e^{rx}.

ii. Use the solution from part (i) to find a fundamental set of solutions for $y'' - 2y' + y = 0$, using reduction of order.

Solution

i. Let $y = e^{rx}$ and substitute into

$$y'' - 2y' + y = 0, \tag{8}$$

to get

$$r^2 e^{rx} - 2re^{rx} + e^{rx} = 0$$

or

$$r^2 - 2r + 1 = 0.$$

Thus $(r - 1)^2 = 0$ and $r = 1$, so e^x is a solution of (8).

ii. We have one solution $y_1 = e^x$ of (8). We shall use reduction of order to find the rest. Let $y = vy_1 = ve^x$ so that (8) becomes:

$$(ve^x)'' - 2(ve^x)' + ve^x = 0.$$

Differentiate

$$(v''e^x + 2v'e^x + ve^x) - 2(v'e^x + ve^x) + ve^x = 0$$

and simplify, to get

$$v''e^x = 0, \quad \text{or} \quad v'' = 0.$$

Antidifferentiate twice:

$$v' = c_1, \tag{9}$$

$$v = c_1 x + c_2. \tag{10}$$

Thus,

$$y = vy_1 = (c_1 x + c_2)e^x = c_1 xe^x + c_2 e^x$$

is the general solution of (8). A fundamental set of solutions would be $\{e^x, xe^x\}$. The second solution xe^x is obtained by taking $c_1 = 1$ in (9) and $c_2 = 0$ in (10). ∎

Summary of Reduction of Order

Reduction of order can be used to solve $a_2 y'' + a_1 y' + a_0 y = f$, given a solution y_1 of $a_2 y'' + a_1 y' + a_0 y = 0$, as follows:

1. Let $y = vy_1$ and substitute into $a_2 y'' + a_1 y' + a_0 y = f$.
2. Differentiation and simplification leads to a second-order linear equation in v with the v term absent. Let $w = v'$ to get the first-order equation

$$(a_2 y_1) w' + (2 a_2 y_1' + a_1 y_1) w = f. \qquad (11)$$

3. Solve this differential equation for w.
4. Antidifferentiate w to get v.
5. The general solution of $a_2 y'' + a_1 y' + a_0 y = f$ is given by $y = vy_1$.

Reduction of order can also be used to find a fundamental set of solutions for $a_2 y'' + a_1 y' + a_0 y = 0$, given a nonzero solution y_1.

1. Let $y = vy_1$, and substitute into $a_2 y'' + a_1 y' + a_0 y = 0$.
2. This leads to a second-order linear equation in v with the v term absent. Let $w = v'$ to get the first-order equation (11) with $f = 0$.
3. Find a nonzero solution w.
4. Let v be an antiderivative of w.
5. Let $y_2 = vy_1$. Then $\{y_1, y_2\}$ is a fundamental set of solutions.

Exercises

For Exercises 1 through 8, find one solution of the associated homogeneous equation of the form e^{rx}. Then find the general solution by reduction of order.

1. $y'' - 4y' + 4y = e^x$
2. $y'' - 6y' + 9y = 0$
3. $y'' - 3y' + 2y = e^x$
4. $y'' + 2y' + y = e^{-x}$
5. $y'' - y = e^x$
6. $y'' + 3y' + 2y = e^{-x} \sin x$
7. $y'' - 4y = e^{2x}$
8. $y'' - y' - 2y = x$
9. Verify that $y_1 = x^{-1}$ is a solution of $x^2 y'' + 3xy' + y = 0$. Then find a fundamental set of solutions of $x^2 y'' + 3xy' + y = 0$ and give the general solution of $x^2 y'' + 3xy' + y = 0$ for $x > 0$.
10. Find a solution of $x^2 y'' - 3xy' + 4y = 0$ of the form x^r. Then find the general solution of $x^2 y'' - 3xy' + 4y = x^5$ for $x > 0$.
11. Find a solution of $x^2 y'' + 5xy' + 4y = 0$ of the form x^r. Then find the general solution of $x^2 y'' + 5xy' + 4y = x^6$ for $x > 0$.

12. Find a solution of $x^2y'' + 7xy' + 9y = 0$ of the form x^r. Then find the general solution of $x^2y'' + 7xy' + 9y = 1$ for $x > 0$.

13. Find a solution of $x^2y'' - 5xy' + 9y = 0$ of the form x^r. Then find the general solution of $x^2y'' - 5xy' + 9y = x^3$ for $x > 0$.

14. Find a solution of $y'' + 4y = 0$ of the form $\sin rx$. Then find the general solution of $y'' + 4y = 1$.

15. Verify $y_1(x) = e^x$ is a solution of $xy''(x) - (1+x)y'(x) + y(x) = 0$, and then find the general solution for $x > 0$.

16. Verify $y_1(x) = e^{-x}$ is a solution of $xy''(x) + (x-1)y'(x) - y(x) = 0$ and then find the general solution for $x > 0$.

17. Verify that $y_1(x) = x$ is a solution of (Legendre equation of order one) $(1 - x^2)y'' - 2xy' + 2y = 0$ on the interval $(-1, 1)$. Find the general solution on $(-1, 1)$. (Note: The integrations are a little more difficult, but can be worked out, using our techniques).

18. Verify statement (v) in the summary of reduction of order applied to $a_2 y'' + a_1 y' + a_0 y = 0$. That is, verify that $\{y_1, y_2\}$ is a fundamental set of solutions.

3.8 Homogeneous Linear Constant-Coefficient Equations (Second-Order)

Linear constant-coefficient differential equations form an important class of differential equations that appear both in physical models and as approximations for more complicated equations. Applications to electrical circuits and mechanical systems will be given in Sections 3.16, 3.17, and 3.18.

This section will consider the general linear, second-order homogeneous, constant-coefficient differential equation,

$$ay'' + by' + cy = 0,$$

where the coefficients a, b, and c are real constants and $a \neq 0$. From Sections 3.4 and 3.5, we know that the general solution of $ay'' + by' + cy = 0$ will be

$$c_1 y_1 + c_2 y_2,$$

where $\{y_1, y_2\}$ is a fundamental set of solutions for $ay'' + by' + cy = 0$.

The key to finding $\{y_1, y_2\}$ is to look for a solution of the form $y = e^{rx}$, where r is a constant. Substituting $y = e^{rx}$ into $ay'' + by' + cy = 0$ gives

$$a(e^{rx})'' + b(e^{rx})' + ce^{rx} = 0,$$

and, upon differentiation,

$$ar^2 e^{rx} + bre^{rx} + ce^{rx} = 0.$$

Finally, divide by e^{rx}, which is always nonzero:

$$ar^2 + br + c = 0.$$

The polynomial $ar^2 + br + c$ is called the *characteristic polynomial* of
$$ay'' + by' + cy = 0.$$
The equation $ar^2 + br + c = 0$ is the *characteristic equation* of
$$ay'' + by' + cy = 0.$$

We have shown that

If r is a root of the characteristic polynomial $ar^2 + br + c$, then e^{rx} is a solution of $ay'' + by' + cy = 0$. (1)

Every second-degree polynomial has two roots. For $ar^2 + br + c$, the roots are given by $(-b \pm \sqrt{b^2 - 4ac})/2a$. There are three cases, depending on whether $b^2 - 4ac > 0$, $b^2 - 4ac = 0$, or $b^2 - 4ac < 0$.

Case 1: Characteristic Polynomial Has Distinct Real Roots ($b^2 - 4ac > 0$)

Suppose the characteristic polynomial $ar^2 + br + c$ has two distinct real roots r_1, r_2. Then by (1), $e^{r_1 x}, e^{r_2 x}$ are solutions of $ay'' + by' + cy = 0$. These solutions are linearly independent, since their Wronskian

$$W[e^{r_1 x}, e^{r_2 x}] = \det \begin{bmatrix} e^{r_1 x} & e^{r_2 x} \\ r_1 e^{r_1 x} & r_2 e^{r_2 x} \end{bmatrix} = (r_2 - r_1) e^{(r_1 + r_2) x}$$

is never zero if $r_1 \neq r_2$.
Thus
$$c_1 e^{r_1 x} + c_2 e^{r_2 x}$$
would be the general solution of $ay'' + by' + cy = 0$.

Example 3.8.1 Find the general solution of
$$y'' + 4y' = 0.$$

Solution The characteristic equation is $r^2 + 4r = r(r + 4) = 0$. There are two distinct real roots $r = 0$, $r = -4$. A fundamental set of solutions would be $\{1, e^{-4x}\}$ since $e^{0x} = 1$. The general solution is
$$y = c_1 \cdot 1 + c_2 e^{-4x} = c_1 + c_2 e^{-4x}. \blacksquare$$

Case 2: Characteristic Polynomial Has a Repeated Real Root ($b^2 - 4ac = 0$)

Suppose that the characteristic polynomial has a single repeated root r_1. This can happen only if the characteristic equation is of the form

$$ar^2 + br + c = a(r - r_1)^2 = 0$$

or

$$ar^2 - 2ar_1 r + ar_1^2 = 0,$$

so that the differential equation must have been

$$ay'' - 2ar_1 y' + ar_1^2 y = 0 \tag{2}$$

The root r_1 provides one solution $e^{r_1 x}$. To get the second solution we will use reduction of order. (See Section 3.7.)

Let $y = ve^{r_1 x}$ and substitute into (2):

$$a(ve^{r_1 x})'' - 2ar_1(ve^{r_1 x})' + ar_1^2(ve^{r_1 x}) = 0,$$

or

$$a(v''e^{r_1 x} + 2v'r_1 e^{r_1 x} + vr_1^2 e^{r_1 x}) - 2ar_1(v'e^{r_1 x} + vr_1 e^{r_1 x}) + a^2 r_1 ve^{r_1 x} = 0,$$

which is

$$av''e^{r_1 x} = 0 \quad \text{or} \quad v'' = 0.$$

Antidifferentiating twice yields

$$v = c_1 x + c_2.$$

Thus,

$$y = ve^{r_1 x} = c_1 x e^{r_1 x} + c_2 e^{r_1 x} \tag{3}$$

is the general solution of $ay'' + by' + cy = 0$ if r_1 is a repeated real root of the characteristic equation.

Example 3.8.2 Find the general solution of $y'' + 2y' + y = 0$.

Solution The characteristic equation is $r^2 + 2r + 1 = (r + 1)^2 = 0$. Thus -1 is a repeated root. The general solution according to (3) is

$$y = c_1 e^{-x} + c_2 x e^{-x}. \blacksquare$$

Case 3: The Characteristic Polynomial Has Complex Roots $(b^2 - 4ac < 0)$.

Suppose that

$$r_1 = \alpha + i\beta \quad (i^2 = -1),$$

with α, β real numbers, is a complex root of $ar^2 + br + c = 0$. Since a, b, c are real, the other root r_2 must be the *complex conjugate* of r_1. That is,

$$r_2 = \alpha - i\beta.$$

(For a review of complex numbers see Appendix A.) Both $e^{r_1 x}$, $e^{r_2 x}$ are still solutions but they involve complex numbers. We shall replace them by a different fundamental set of solutions. From Appendix A we have that

$$e^{r_1 x} = e^{\alpha x + i\beta x} = e^{\alpha x} \cos \beta x + i e^{\alpha x} \sin \beta x$$

and

$$e^{r_2 x} = e^{\alpha x - i\beta x} = e^{\alpha x} \cos \beta x - i e^{\alpha x} \sin \beta x.$$

Since $ay'' + by' + cy = 0$ is a linear homogeneous differential equation, any linear combination, $c_1 e^{r_1 x} + c_2 e^{r_2 x}$, will also be a solution of $ay'' + by' + cy = 0$ (Theorem 3.4.1). This is true even if c_1, c_2 are complex. In particular

$$\frac{1}{2} e^{r_1 x} + \frac{1}{2} e^{r_2 x} = e^{\alpha x} \cos \beta x$$

and

$$\frac{1}{2i} e^{r_1 x} - \frac{1}{2i} e^{r_2 x} = e^{\alpha x} \sin \beta x$$

are also solutions of $ay'' + by' + cy = 0$. They are also linearly independent (see Exercise 21 at the end of this section). Thus,

In the case of complex roots $\alpha \pm \beta i$, the general solution of $ay'' + by' + cy = 0$ is $c_1 e^{\alpha x} \cos \beta x + c_2 e^{\alpha x} \sin \beta x$. (4)

Example 3.8.3 Find the general solution of

$$y'' + y' + y = 0.$$

Solution The characteristic polynomial is $r^2 + r + 1 = 0$. By the quadratic formula, the roots are

$$r = \frac{-1 \pm \sqrt{1-4}}{2} = -\frac{1}{2} \pm i \frac{\sqrt{3}}{2},$$

so that $\alpha = -\frac{1}{2}$, $\beta = \sqrt{3}/2$. Thus,

$$y = c_1 e^{-x/2} \cos\left(\frac{\sqrt{3}}{2} x\right) + c_2 e^{-x/2} \sin\left(\frac{\sqrt{3}}{2} x\right)$$

is the general solution of $y'' + y' + y = 0$.

Note that the root $-\frac{1}{2} + i\frac{\sqrt{3}}{2}$ does not give the solution $e^{-x/2}\cos\left(\frac{\sqrt{3}}{2}x\right)$. Rather the *pair* of solutions $e^{-x/2}\cos\left(\frac{\sqrt{3}}{2}x\right), e^{-x/2}\sin\left(\frac{\sqrt{3}}{2}x\right)$ comes from the *pair* of roots

$$-\frac{1}{2} + i\frac{\sqrt{3}}{2}, \quad -\frac{1}{2} - i\frac{\sqrt{3}}{2}. \quad \blacksquare$$

Example 3.8.4 Find the general solution of $y'' + 4y = 0$.

Solution The characteristic polynomial is $r^2 + 4 = 0$ so that the roots are $r = \pm 2i$. Thus $\alpha = 0$, $\beta = 2$, and a fundamental set of solution is

$$e^{0x}\cos 2x = \cos 2x \quad \text{and} \quad e^{0x}\sin 2x = \sin 2x.$$

The general solution is

$$y = c_1 \cos 2x + c_2 \sin 2x. \quad \blacksquare$$

For convenience, we summarize the three cases.

Solution of $ay'' + by' + cy = 0$ with a, b, c Real Constants

First find the roots r_1, r_2 of the characteristic equation $ar^2 + br + c = 0$. There are three cases

1. If r_1, r_2 are distinct real roots, then the general solution is $y = c_1 e^{r_1 x} + c_2 e^{r_2 x}$.
2. If $r_1 = r_2$ is a repeated real root, the general solution is $y = c_1 e^{r_1 x} + c_2 x e^{r_1 x}$.
3. If r_1, r_2 are complex roots, they are a conjugate pair:

$$r_1 = \alpha + i\beta, \quad r_2 = \alpha - i\beta.$$

The general solution is

$$y = c_1 e^{\alpha x}\cos \beta x + c_2 e^{\alpha x}\sin \beta x.$$

As will be shown in Sections 3.16 and 3.18 there is a close relationship between electrical circuits and mechanical systems and linear differential equations with constant coefficients. In these problems one often starts knowing the desired response (solution) and wants to design the device (make up the differential equation).

Example 3.8.5 Find a second-order linear homogeneous constant-coefficient differential equation that has $c_1 e^{-3x} + c_2 e^{-2x}$ as its general solution.

Solution $c_1 e^{-3x} + c_2 e^{-2x}$ will be the general solution if -3 and -2 are roots of the characteristic equation. One such characteristic polynomial would be

$$(r - (-3))(r - (-2)) = (r + 3)(r + 2) = r^2 + 5r + 6.$$

A corresponding differential equation is

$$y'' + 5y' + 6y = 0.$$

Note that $2y'' + 10y' + 12y = 0$ would be another correct answer, since the roots determine the polynomial only up to a constant factor. ∎

Exercises

In Exercises 1 through 25, solve the differential equation. Determine the general solution if no initial conditions are given.

1. $y'' + y' - 6y = 0$
2. $y'' - y = 0$
3. $y'' + y = 0$
4. $y'' + 4y' + 4y = 0$
5. $y'' + 4y' + 5y = 0$
6. $y'' - 2y' + y = 0$
7. $y'' - 3y' + 2y = 0$
8. $2y'' - 2y' + y = 0$
9. $y'' - y' = 0$
10. $4y'' + 8y' + 3y = 0$
11. $3y'' = 0$
12. $y'' - 2y' + 2y = 0$
13. $3y'' + 2y' - y = 0$
14. $y'' + 9y = 0$, $\quad y(0) = 1$, $\quad y'(0) = 1$
15. $y'' + y' - 2y = 0$, $\quad y(0) = 0$, $\quad y'(0) = 1$
16. $2y'' + 12y' + 18y = 0$, $\quad y(0) = 1$, $\quad y'(0) = 0$
17. $2y'' + 4y = 0$
18. $3y'' - 24y' + 45y = 0$
19. $2y'' + 8y' + 6y = 0$, $\quad y(0) = 2$, $\quad y'(0) = 0$
20. $y'' - 16y = 0$
21. $y'' + 10y' + 25y = 0$
22. $2y'' + 3y' = 0$
23. $y'' - 14y' + 49y = 0$
24. $y'' + 4y' + 20y = 0$
25. $y'' - 6y' + 25y = 0$
26. Suppose that r is a real number. Verify e^{rx}, xe^{rx} are linearly independent.
27. Suppose α, β are real numbers and $\beta \neq 0$. Verify $e^{\alpha x} \cos \beta x$, $e^{\alpha x} \sin \beta x$ are linearly independent.

In Exercises 28 through 39, determine a homogeneous second-order linear constant-coefficient differential equation with the given expression as its general solution.

28. $y = c_1 e^{3x} + c_2 e^{-4x}$
29. $y = c_1 e^{-x} + c_2 e^{-2x}$
30. $y = c_1 e^{2x} + c_2 x e^{2x}$
31. $y = c_1 e^{3x} + c_2 x e^{3x}$
32. $y = c_1 + c_2 e^{-5x}$
33. $y = c_1 \sin 4x + c_2 \cos 4x$
34. $y = c_1 e^{-x} \sin 2x + c_2 e^{-x} \cos 2x$
35. $y = c_1 \sin 3x + c_2 \cos 3x$
36. $y = c_1 e^{2x} \sin 3x + c_2 e^{2x} \cos 3x$
37. $y = c_1 + c_2 x$
38. $y = c_1 \sin 2x + c_2 \cos 2x$
39. $y = c_1 e^x \sin x + c_2 e^x \cos x$

3.9 nth-Order Linear Differential Equations

This section will present the basic theory for nth-order linear differential equations

$$a_n(x)\frac{d^n y}{dx^n} + a_{n-1}(x)\frac{d^{n-1} y}{dx^{n-1}} + \cdots + a_1(x)\frac{dy}{dx} + a_0(x)y = f(x), \qquad (1)$$

which is very similar to the theory for second-order equations developed in Sections 3, 4, and 5. Rather than writing (1) repeatedly, we shall utilize the operator notation of Section 3.6 and write

$$L(y) = f \qquad (2)$$

to represent (1). The *associated homogeneous equation* for (1), (2) is then

$$L(y) = 0.$$

Thus $y''' + y = 0$ is the associated homogeneous equation for $y''' + y = \sin x$.

For first-order linear differential equations, we had one arbitrary constant in the general solution, and for second-order equations there were two. Thus we would expect to have n arbitrary constants for an nth-order differential equation.

Theorem 3.9.1 Suppose in (1) that $a_0, \ldots, a_n, f$ are continuous functions on the interval I and $a_n(x)$ is never zero on I. Then for any x_0 in I and any n real numbers $v_0, \ldots, v_{n-1}$, there exists a unique solution to

$$L(y) = f, \quad y(x_0) = v_0, \quad y'(x_0) = v_1, \quad \ldots, \quad y^{(n-1)}(x_0) = v_{n-1}. \qquad (3)$$

and this unique solution is defined on all of the interval I.

Places where $a_n(x) = 0$ are called *singular points* of (1). The solutions of (1) may be broken into a particular solution and a homogeneous solution, just as for second-order linear differential equations. From Theorem 3.6.1 we have

Theorem 3.9.2 Suppose y_p is a solution of (1), $L(y) = f$.

Then any solution y of $L(y) = f$ may be written as $y = y_p + y_h$, where y_h is a solution of the associated homogeneous equation $L(y) = 0$. $\qquad (4)$

Also $y_p + y_h$ is a solution of $L(y) = f$ for every y_h that is a solution of $L(y) = 0$.

If y_1, y_2 are solutions of $L(y) = 0$, then $c_1 y_1 + c_2 y_2$ is a solution of $L(y) = 0$ for any constants c_1, c_2. $\qquad (5)$

A set of functions $\{y_1, \ldots, y_n\}$ is *linearly independent* if

$$c_1 y_1 + c_2 y_2 + \cdots + c_n y_n = 0 \qquad (6)$$

for constants $c_1, \ldots, c_n$ implies that $c_1 = c_2 = \cdots = c_n = 0$. This is equivalent to saying that none of the y_i can be written as a sum of constant multiples of the other y_i's. That is, no y_i is a *linear combination* of the other y_i. If a set of functions is not linearly independent, it is *linearly dependent*.

Example 3.9.1 Show that the set of functions $\{1, x, x^2\}$ is linearly independent.

Solution We shall verify (6). Suppose, then, that

$$c_1 1 + c_2 x + c_3 x^2 = 0 \quad \text{for all } x. \tag{7}$$

In general, we could evaluate (7) at several x-values, to get equations for the c_i. In this example, we may differentiate (7) with respect to x

$$c_2 + 2c_3 x = 0, \tag{8}$$

and then differentiate again

$$2c_3 = 0. \tag{9}$$

From (7), (8), and (9) we have $c_3 = 0$, $c_2 = 0$, $c_1 = 0$, and thus $\{1, x, x^2\}$ is linearly independent, by the definition (6). ∎

Example 3.9.2 Show that $\{e^x, e^{-x}, \cosh x\}$ is a linearly dependent set of functions.

Solution Recall that

$$\cosh x = \tfrac{1}{2}e^x + \tfrac{1}{2}e^{-x},$$

so that $\cosh x$ is a linear combination of e^x, e^{-x}. Alternatively, by taking $c_1 = \tfrac{1}{2}$, $c_2 = \tfrac{1}{2}, c_3 = -1$, we get

$$c_1 e^x + c_2 e^{-x} + c_3 \cosh x = 0,$$

and c_1, c_2, c_3 are not all zero, so, by definition (6), $\{e^x, e^{-x}, \cosh x\}$ is not linearly independent. ∎

A set of solutions $\{y_1, \ldots, y_n\}$ of the nth-order linear homogeneous equation $L(y) = 0$ is a *fundamental set of solutions* on the interval I if

- Every solution $\tilde{y}_h$ of $L(y) = 0$ may be written as a linear combination of $y_1, \ldots, y_n$:

$$\tilde{y}_h = c_1 y_1 + c_2 y_2 + \cdots + c_n y_n, \tag{10}$$

- The coefficients in (10) are unique. This is equivalent to $\{y_1, \ldots, y_n\}$ being linearly independent.

Theorem 3.9.3 Suppose $a_0, \ldots, a_n$ in (1) are continuous on the interval I and $a_n(x)$ is never zero on I. Then the following are equivalent for a set of solutions $\{y_1, \ldots, y_n\}$ of $L(y) = 0$:

$\{y_1, \ldots, y_n\}$ is a linearly independent set of functions. (11)

Every solution of $L(y) = 0$ can be written as a linear combination of $y_1, \ldots, y_n$. (12)

$\{y_1, \ldots, y_n\}$ is a fundamental set of solutions. (13)

Example 3.9.3

Consider (1) with $a_3 = 1$ and $n = 3$. Then,

$$y''' + a_2(x)y'' + a_1(x)y' + a_0(x)y = 0, \qquad (14)$$

or $L(y) = 0$, and assume a_0, a_1, a_2, are continuous on $[-1, 1]$. Let

y_1 be the solution of $L(y) = 0$, $\quad y(0) = 1, \quad y'(0) = y''(0) = 0;$
y_2 be the solution of $L(y) = 0$, $\quad y'(0) = 1, \quad y(0) = y''(0) = 0;\quad (15)$
y_3 be the solution of $L(y) = 0$, $\quad y''(0) = 1, \quad y(0) = y'(0) = 0.$

Verify that $\{y_1, y_2, y_3\}$ is a fundamental set of solutions of $L(y) = 0$.

Solution

We shall show that $\{y_1, y_2, y_3\}$ is linearly independent by verifying (6). The y_i are differentiable since they are solutions of (14). Assume that

$$c_1 y_1(x) + c_2 y_2(x) + c_3 y_3(x) = 0. \qquad (16)$$

Differentiate (16) two times:

$$\begin{aligned} c_1 y_1'(x) + c_2 y_2'(x) + c_3 y_3'(x) &= 0, \\ c_1 y_1''(x) + c_2 y_2''(x) + c_3 y_3''(x) &= 0. \end{aligned} \qquad (17)$$

Evaluating (16) and (17) at $x = 0$ and using (15) yields:

$$c_1 \cdot 1 + c_2 \cdot 0 + c_3 \cdot 0 = 0,$$
$$c_1 \cdot 0 + c_2 \cdot 1 + c_3 \cdot 0 = 0,$$
$$c_1 \cdot 0 + c_2 \cdot 0 + c_3 \cdot 1 = 0,$$

so that $c_1 = c_2 = c_3 = 0$ and $\{y_1, y_2, y_3\}$ is linearly independent by (6). ∎

Example 3.9.4

Given that $\{\sin x, \cos x, e^x, e^{-x}\}$ is a fundamental set of solutions of $y'''' - y = 0$ and $e^{3x}/10$ is a solution of $y'''' - y = 8e^{3x}$, find the general solution of $y'''' - y = 8e^{3x}$.

Solution

Since $\{\sin x, \cos x, e^x, e^{-x}\}$ is a fundamental set of solutions of the homogeneous equation $y'''' - y = 0$,

$$y_h = c_1 \sin x + c_2 \cos x + c_3 e^x + c_4 e^{-x}$$

is the general solution of $y'''' - y = 0$. We are given that

$$y_p = \frac{e^{3x}}{10}$$

is a particular solution of $y'''' - y = 8e^{3x}$. Thus, by Theorem 3.9.2,

$$y = y_p + y_h = \frac{e^{3x}}{10} + c_1 \sin x + c_2 \cos x + c_3 e^x + c_4 e^{-x}$$

is the general solution of $y'''' - y = 8e^{3x}$. ∎

As with second-order equations, the *Wronskian* may also be utilized. For n functions $\{y_1, \ldots, y_n\}$, the Wronskian is defined as

$$W[y_1, \ldots, y_n] = \det \begin{bmatrix} y_1(x) & y_2(x) & \cdots & y_n(x) \\ y_1'(x) & y_2'(x) & \cdots & y_n'(x) \\ \vdots & \vdots & \cdots & \vdots \\ y_1^{(n-1)}(x) & y_2^{(n-1)}(x) & \cdots & y_n^{(n-1)}(x) \end{bmatrix}, \quad (18)$$

where det again denotes the determinant. In particular, from Section 3.5,

$$W[y_1, y_2] = \det \begin{bmatrix} y_1(x) & y_2(x) \\ y_1'(x) & y_2'(x) \end{bmatrix}$$

and

$$W[y_1, y_2, y_3] = \det \begin{bmatrix} y_1(x) & y_2(x) & y_3(x) \\ y_1'(x) & y_2'(x) & y_3'(x) \\ y_1''(x) & y_2''(x) & y_3''(x) \end{bmatrix}.$$

Section 3.5 showed how to evaluate $W[y_1, y_2]$. The evaluation of the determinant (18) is covered in Section 6.8. All of the examples and exercises of this section may be done using the technique of Example 3.9.3, instead of using the Wronskian.

Theorem 3.9.4

Suppose the nth-order homogeneous linear differential equations (1) and (2) have continuous coefficients and $a_n(x)$ is nonzero on the interval I. Let $\{y_1, \ldots y_n\}$ be a set of n solutions to $L(y) = 0$. Then

a. $W[y_1, \ldots, y_n]$ is either always zero or never zero on I.
b. $\{y_1, \ldots, y_n\}$ is a fundamental set of solutions if and only if $W[y_1, \ldots, y_n] \neq 0$.

Example 3.9.5

a. Verify that $\{\sin 2x, \cos 2x, 1\}$ is a fundamental set of solutions for

$$y''' + 4y' = 0. \quad (19)$$

b. Find the solution of

$$y''' + 4y' = 0, \quad y(0) = 0, \quad y'(0) = 0, \quad y''(0) = 2. \quad (20)$$

Solution

a) First we must verify that $\sin 2x, \cos 2x, 1$ are solutions of (19). This is left to the reader. Since there are *three* solutions and (19) is *third*-order, by Theorem 3.9.3, Eq. (11), it suffices to show that $1, \sin 2x, \cos 2x$ are linearly independent. We shall use (6). Suppose

$$c_1 \cdot 1 + c_2 \sin 2x + c_3 \cos 2x = 0, \qquad (21)$$

and differentiate this equation twice, to also get

$$\begin{aligned} 2c_2 \cos 2x - 2c_3 \sin 2x &= 0, \\ -4c_2 \sin 2x - 4c_3 \cos 2x &= 0. \end{aligned} \qquad (22)$$

Evaluate these equations [(21) and (22)] at $x = 0$, to yield:

$$\begin{aligned} c_1 + c_2 \cdot 0 + c_3 \cdot 1 &= 0, \\ 2c_2 \cdot 1 - 2c_3 \cdot 0 &= 0, \\ -4c_2 \cdot 0 - 4c_3 \cdot 1 &= 0. \end{aligned} \qquad (23)$$

The last two equations imply $c_2 = c_3 = 0$, and the first then gives $c_1 = 0$. Thus $\sin 2x, \cos 2x, 1$ are linearly independent.

Alternatively (for those using determinants),

$$W[\sin 2x, \cos 2x, 1](0) = \det \begin{bmatrix} 0 & 1 & 1 \\ 2 & 0 & 0 \\ 0 & -4 & 0 \end{bmatrix} = -8 \neq 0,$$

so $\sin 2x, \cos 2x, 1$ are linearly independent by Theorem 3.9.4.

b) Since $\{\sin 2x, \cos 2x, 1\}$ is a fundamental set of solutions for $y''' + 4y' = 0$, the general solution of $y''' + 4y' = 0$ must be a linear combination of $\sin 2x$, $\cos 2x$, 1:

$$y = c_1 \sin 2x + c_2 \cos 2x + c_3 \cdot 1. \qquad (24)$$

Applying the initial conditions in (20) to (24), we get:

$$\begin{aligned} 0 &= y(0) = c_1 \cdot 0 + c_2 \cdot 1 + c_3, \\ 0 &= y'(0) = c_1 \cdot 2 - c_2 \cdot 0, \\ 2 &= y''(0) = c_1 \cdot 0 - c_2 \cdot 4. \end{aligned}$$

Starting with the last equation and working up, we find $c_2 = -\frac{1}{2}$, $c_1 = 0$, $c_3 = \frac{1}{2}$, and the solution of the initial-value problem (20) is

$$y = -\tfrac{1}{2} \cos 2x + \tfrac{1}{2}. \blacksquare$$

Exercises

1. Is $\{x, x^{-1}, 1\}$ a linearly independent set of functions on $(0, 1]$?

2. Is $\{x + 1, x^2 + x, x^2 - 1\}$ a linearly independent set of functions on $[0, 1]$?

3. For the differential equation $L(y) = y''' - x^2 y'' + x^3 y = 0$, let:

y_1 be the solution of

$L(y) = 0, \qquad y(0) = y'(0) = 1, \qquad y''(0) = 0;$

y_2 be the solution of

$L(y) = 0$, $\quad y(0) = y''(0) = 1$, $\quad y'(0) = 0$;

y_3 be the solution of

$L(y) = 0$, $\quad y''(0) = 3$, $\quad y(0) = y'(0) = 0$.

a) Verify that $\{y_1, y_2, y_3\}$ is a fundamental set of solutions.

b) Find constants c_1, c_2, c_3 such that $\tilde{y} = c_1 y_1 + c_2 y_2 + c_3 y_3$ is the solution of $L(y) = 0$, $y(0) = y''(0) = y'(0) = 1$.

4. For the differential equation $L(y) = y''' + \cos xy' + y = 0$, let

y_1 be the solution of

$L(y) = 0$, $\quad y(0) = y'(0) = 1$, $\quad y''(0) = 0$;

y_2 be the solution of

$L(y) = 0$, $\quad y(0) = y''(0) = 1$, $\quad y'(0) = 0$;

y_3 be the solution of

$L(y) = 0$, $\quad y'(0) = y''(0) = 1$, $\quad y(0) = 0$.

a) Verify that $\{y_1, y_2, y_3\}$ is a fundamental set of solutions.

b) Find constants c_1, c_2, c_3 such that $\tilde{y} = c_1 y_1 + c_2 y_2 + c_3 y_3$ is the solution of $L(y) = 0$, $y(0) = 0$, $y'(0) = 1$, $y''(0) = 2$.

5. a) Verify that $\{1, e^x, xe^x\}$ is a fundamental set of solutions of $y''' - 2y'' + y' = 0$ and that $y_p = x$ is a solution of $y''' - 2y'' + y' = 1$.

b) Give the general solution of $y''' - 2y'' + y' = 1$.

6. a) Verify that $\{\sin x, \cos x, \sin 2x, \cos 2x\}$ is a fundamental set of solutions of $y'''' + 5y'' + 4y = 0$ and that $y_p = 1$ is a solution of $y'''' + 5y'' + 4y = 4$.

b) Give the general solution of $y'''' + 5y'' + 4y = 4$.

7. a) Verify that $\{e^x, xe^x, x^2 e^x\}$ is a fundamental set of solutions of $y''' - 3y'' + 3y' - y = 0$ and that $y_p = -e^{-x}$ is a solution of $y''' - 3y'' + 3y' - y = 8e^{-x}$.

b) Give the general solution of $y''' - 3y'' + 3y' - y = 8e^{-x}$.

8. a) Verify that $\{x, x^2, x^3\}$ is a fundamental set of solutions of $x^3 y''' - 3x^2 y'' + 6xy' - 6y = 0$ on $(0, \infty)$ and that $y_p = 3$ is a solution of $x^3 y''' - 3x^2 y'' + 6xy' - 6y = -18$.

b) Find the solution of

$x^3 y''' - 3x^2 y'' + 6xy' - 6y = -18$,

$y(1) = 0$, $\quad y'(1) = 0$, $\quad y''(1) = 0$.

9. Determine on which of the following intervals $\{x, |x|\}$ is a linearly independent set of functions.

a) $[-1, 0]$ $\quad$ **b)** $[0, 1]$ $\quad$ **c)** $[-1, 1]$.

10. Show that a set of functions $\{y_1, y_2, \ldots, y_n\}$ is linearly independent if and only if no one of the functions is a linear combination of the rest.

11. Suppose that $\{y_1, \ldots, y_n\}$ are solutions of $L(y) = 0$ and L is nth-order. Suppose also that the coefficients of L are continuous on $[a, b]$ and that each y_i is n-times differentiable on $[a, b]$. Suppose that $W(x) = W[y_1, \ldots, y_n]$ is nonzero except at a finite set of points $x_1, \ldots, x_r$. Explain why the x_i are singular points of the differential equation $L(y) = 0$.

12. Make the same assumptions on $y_1, \ldots, y_n$ and L, but not $W(x)$, as in Exercise 11. Suppose that x_0 is a singular point of the differential equation $L(y) = 0$ but not all coefficients are zero at x_0. Explain why $W(x_0) \stackrel{?}{=} 0$. (Note: This utilizes the fact (from matrix theory) that a square matrix has zero determinant if and only if its columns are linearly dependent. This is discussed in Chapter 6.)

3.10 Homogeneous Linear Constant-Coefficient Equations (nth-Order)

The solution of the nth-order homogeneous linear constant-coefficient differential equation

$$a_n y^{(n)} + a_{n-1} y^{(n-1)} + \cdots + a_1 y' + a_0 y = 0 \tag{1}$$

(here $y^{(m)} = d^m y/dx^m$), where the coefficients $a_n, \ldots, a_0$ are constants and $a_n \neq 0$ proceeds almost exactly as for the second-order case ($n = 2$) in Section 3.8. Again, e^{rx} will be a solution if r is a root of the characteristic equation

$$a_n r^n + a_{n-1} r^{n-1} + \cdots + a_1 r + a_0 = 0. \tag{2}$$

The nth-degree polynomial (2) has n roots, but they need not be distinct. The *multiplicity* of a root is the number of times it is repeated. The sum of the multiplicities of all the distinct roots equals the degree of the polynomial.

Example 3.10.1 $r^3 - 3r^2 + 3r - 1 = (r-1)^3$ has 1 as a root of multiplicity 3. ∎

Example 3.10.2 $(r-4)^2(r-3)$ has two distinct roots; 4 is a root of multiplicity 2, and 3 is a root of multiplicity 1. ∎

Example 3.10.3 $r^4 + 2r^2 + 1 = (r^2+1)^2 = (r-i)^2(r+i)^2$ has two complex roots $\pm i$, each of multiplicity 2. ∎

The procedure for the solution of (1) is summarized in the following algorithm.

Procedure for Solution of $a_n y^{(n)} + a_{n-1} y^{(n-1)} + \cdots + a_1 y' + a_0 y = 0$ when $a_n, a_{n-1}, \ldots, a_0$ Are Constants

1. Form the characteristic equation $a_n r^n + a_{n-1} r^{n-1} + \cdots + a_1 r + a_0 = 0$ and determine its roots and their multiplicities.
2. A fundamental set of solutions for (1) is determined as follows:
a. If r_1 is a real root of multiplicity m, then include the m functions

$$\{e^{r_1 x}, xe^{r_1 x}, \ldots, x^{m-1} e^{r_1 x}\} \tag{3}$$

in the fundamental set of solutions.
b. If $\alpha \pm \beta i$ is a pair of complex conjugate roots and they each have multiplicity m, then include the $2m$ functions

$$\{e^{\alpha x} \cos \beta x, e^{\alpha x} \sin \beta x, xe^{\alpha x} \cos \beta x, xe^{\alpha x} \sin \beta x, \ldots,$$
$$x^{m-1} e^{\alpha x} \cos \beta x, x^{m-1} e^{\alpha x} \sin \beta x\} \tag{4}$$

in the fundamental set of solutions. (If $\alpha + \beta i$ is a root of multiplicity m, then $\alpha - \beta i$ will be a root of multiplicity m also, since $a_n, \ldots, a_0$ are real.)

Example 3.10.4   Find the general solution of $y'''' + 2y'' + y = 0$.

Solution   The characteristic equation is $r^4 + 2r^2 + 1 = (r^2 + 1)^2 = 0$. Thus $\pm i$ are complex conjugate roots of multiplicity 2. In the notation of (4):

$$i = 0 + 1 \cdot i, \quad \text{so that } \alpha = 0 \quad \text{and} \quad \beta = 1.$$

Thus, a fundamental set of solutions is:

$$\{\cos x, \sin x, x \cos x, x \sin x\}.$$

The general solution is

$$y = c_1 \cos x + c_2 \sin x + c_3 x \cos x + c_4 x \sin x. \quad \blacksquare$$

Example 3.10.5   Find the general solution of

$$y^{(5)} - 3y^{(4)} + 3y^{(3)} - y^{(2)} = 0.$$

Solution   The characteristic equation is

$$r^5 - 3r^4 + 3r^3 - r^2 = r^2(r^3 - 3r^2 + 3r - 1) = r^2(r - 1)^3 = 0.$$

There are two distinct roots 0, 1, of multiplicities 2 and 3. Using (3), the root 0 of multiplicity 2 means that we include

$$\{e^{0x}, xe^{0x}\} = \{1, x\}$$

in the fundamental set of solutions. The root 1 of multiplicity 3 means that we include $\{e^x, xe^x, x^2 e^x\}$ in the fundamental set of solutions. The general solution is thus

$$y = c_1 + c_2 x + c_3 e^x + c_4 x e^x + c_5 x^2 e^x. \quad \blacksquare$$

In actual applications the roots of the characteristic equation will usually not be integers. They are often found (estimated) using a numerical procedure. In some cases this could be Newton's method from calculus. In this case, care must be taken when determining whether values such as say 1.123, 1.124 represent distinct real roots that are close together, or one root of multiplicity 2 whose computation has been influenced by roundoff error or the numerical method.

Exercises

For Exercises 1 through 23, find the general solution.

1. $y''' - 6y'' + 12y' - 8y = 0$
2. $y'''' + 5y'' + 4y = 0$
3. $y'''' - 5y'' + 4y = 0$
4. $y'''' + 8y'' + 16y = 0$
5. $y''' + y'' - 2y' = 0$
6. $y'''' - 2y''' = 0$
7. $y'''' + 4y''' + 6y'' + 4y' + y = 0$

8. $y''' - 2y'' - y' + 2y = 0$
9. $y' - 3y = 0$
10. $y' + 4y = 0$
11. $y''' + y'' - 2y = 0$
12. $y'''' + 4y'' + 4y = 0$
*13. $y'''' + 4y''' + 8y'' + 8y' + 4y = 0$
14. $y''' + 3y'' + 3y' + y = 0$
15. $y^{(4)} - 4y^{(3)} + 6y^{(2)} - 4y^{(1)} + y = 0$
16. $y^{(6)} + 3y^{(4)} + 3y^{(2)} + y = 0$
17. $y^{(6)} - 3y^{(4)} + 3y^{(2)} - y = 0$
18. $y^{(4)} - 16y = 0$
19. $y^{(4)} - y = 0$
20. $y''' + 2y''' + 2y'' = 0$
21. $y'''' + 50y'' + 625y = 0$
22. $3y' + 4y = 0$
23. $y'' + 3y'' + y' - 5y = 0$

For Exercises 24 through 33, you are given the general solution. Write down a homogeneous linear constant-coefficient differential equation that has that general solution.

24. $y = c_1 + c_2 x + c_3 x^2 + c_4 x^3$
25. $y = c_1 e^{2x} + c_2 x e^{2x} + c_3$
26. $y = c_1 e^x \sin 2x + c_2 e^x \cos 2x + c_3 e^{-x} + c_4 x e^{-x}$
27. $y = c_1 \sin 5x + c_2 \cos 5x + c_3 x \sin 5x + c_4 x \cos 5x$
28. $y = c_1 e^x + c_2 e^{2x} + c_3 e^{-x}$
29. $y = c_1 e^{-2x} + c_2 x e^{-2x} + c_3 x^2 e^{-2x}$
30. $y = c_1 e^x + c_2 x e^x + c_3 e^{-x} + c_4 x e^{-x}$
31. $y = c_1 e^x \sin x + c_2 e^x \cos x + c_3 x e^x \sin x + c_4 x e^x \cos x$
32. $y = c_1 e^x + c_2 e^{2x} + c_3 e^{3x} + c_4 e^{4x}$
33. $y = c_1 \sin x + c_2 \cos x + c_3 x \sin x + c_4 x \cos x$
34. Find all solutions of $y''' - 3y'' + 3y' - y = 0$ of the form e^{rx}. Then find the general solution by reduction of order (Section 3.7). Compare your answer to that obtained by using (3).

3.11 Euler's Equation

Linear differential equations with variable coefficients are frequently solved by the method of power series (Chapter 5) or numerically (Chapter 8). However, they can occasionally be solved by combining our earlier techniques with a change of variable. One such equation is *Euler's equation*. The second-order Euler's equation is

$$ax^2 \frac{d^2y}{dx^2} + bx \frac{dy}{dx} + cy = 0, \qquad (1)$$

where a, b, c are constants and $a \neq 0$. The third-order Euler's equation is

$$a_3 x^3 \frac{d^3y}{dx^3} + a_2 x^2 \frac{d^2y}{dx^2} + a_1 x \frac{dy}{dx} + a_0 y = 0, \qquad (2)$$

with a_3, a_2, a_1, a_0 constants and $a_3 \neq 0$, and in general the nth-order Euler's equation is

$$a_n x^n \frac{d^n y}{dx^n} + a_{n-1} x^{n-1} \frac{d^{n-1} y}{dx^{n-1}} + \cdots + a_1 x \frac{dy}{dx} + a_0 y = 0. \qquad (3)$$

Euler's equation is also known as the Cauchy-Euler or the equidimensional equation. (See Exercise 9 at the end of this section.) It arises, for example, in some drag problems in uniformly viscous flows.

Note that, at $x = 0$, the leading coefficient vanishes, so that the basic Existence and Uniqueness Theorem does not hold on an interval including zero. Thus it is necessary to consider Euler's equation separately for $x < 0$ and $x > 0$. We shall consider it for $x > 0$ only.

There are two ways generally used to solve Euler's equation. We shall give the one that closely parallels the technique for a series solution near a regular singular point (Chapter 5), and develop it for the second-order equation (1). The alternative method is developed in the exercises for both second- and nth-order equations.

Since the coefficients of $ax^2 y'' + bxy' + cy = 0$ are powers of x, it is possible that there is a solution of the form x^r for some constant r. (This is actually a special case of a general technique to be explained later.)

Substituting $y = x^r$ into $ax^2 y'' + bxy' + cy = 0$ gives:

$$ax^2 r(r-1) x^{r-2} + bxr x^{r-1} + cx^r = 0,$$
$$(ar(r-1) + br + c) x^r = 0,$$

so that $ar(r-1) + br + c = 0$. Thus,

$$ar^2 + (b-a)r + c = 0. \tag{4}$$

Case 1: Distinct Real Roots

If the polynomial $ar^2 + (b-a)r + c = 0$ has two distinct real roots, r_1, r_2, then x^{r_1}, x^{r_2} provide a fundamental set of solutions.

Example 3.11.1 Find the general solution of

$$2x^2 \frac{d^2 y}{dx^2} + 7x \frac{dy}{dx} - 3y = 0, \quad x > 0. \tag{5}$$

Solution The student is encouraged to rederive the polynomial (4) the first few times. Letting $y = x^r$ in (5) and then dividing by x^r yields

$$2r(r-1) + 7r - 3 = 0$$

or

$$2r^2 + 5r - 3 = (2r - 1)(r + 3) = 0.$$

The roots are $r = \frac{1}{2}, -3$ and the general solution is

$$y = c_1 x^{1/2} + c_2 x^{-3}. \blacksquare$$

3.11 Euler's Equation

Case 2: Complex Conjugate Roots

If $ar^2 + (b - a)r + c = 0$ has a complex conjugate pair of roots $\alpha \pm \beta i$, then $x^{\alpha+\beta i}$, $x^{\alpha-\beta i}$ will be (complex) solutions. In order to get real solutions, recall (Appendix A) that

$$a^b = e^{b \ln a}, \quad a > 0.$$

Thus for $x > 0$,

$$x^{\alpha+\beta i} = e^{(\alpha+\beta i)\ln x} = e^{\alpha \ln x + i\beta \ln x}$$
$$= e^{\alpha \ln x}(\cos(\beta \ln x) + i \sin(\beta \ln x))$$
$$= x^{\alpha}(\cos(\beta \ln x) + i \sin(\beta \ln x)).$$

Similarly

$$x^{\alpha-\beta i} = x^{\alpha}(\cos(\beta \ln x) - i \sin(\beta \ln x)).$$

Since $x^{\alpha-\beta i}$, $x^{\alpha+\beta i}$ are solutions of a linear homogeneous differential equation, so is any linear combination of $x^{\alpha-\beta i}$ and $x^{\alpha+\beta i}$. Thus

$$\frac{1}{2}x^{\alpha+\beta i} + \frac{1}{2}x^{\alpha-\beta i} = x^{\alpha} \cos(\beta \ln x),$$

$$\frac{1}{2i}x^{\alpha+\beta i} - \frac{1}{2i}x^{\alpha-\beta i} = x^{\alpha} \sin(\beta \ln x),$$

are also solutions.

Example 3.11.2 Find the general solution of

$$9x^2 y'' + 15xy' + 5y = 0, \quad x > 0.$$

Solution The polynomial $ar^2 + (b - a)r + c$ is $9r^2 + 6r + 5 = (3r + 1)^2 + 4$, so the roots are

$$r = -\frac{1}{3} \pm i\frac{2}{3}.$$

The general solution is thus

$$y = c_1 x^{-1/3} \cos\left(\frac{2}{3} \ln x\right) + c_2 x^{-1/3} \sin\left(\frac{2}{3} \ln x\right). \blacksquare$$

Case 3: Repeated Real Roots

Suppose $ar^2 + (b - a)r + c = 0$ has a repeated real root of α. This happens if and only if $(b - a)^2 - 4ac = 0$. Then the root $r_1 = (a - b)/2a$.

Since we have one solution x^{r_1}, the other can be found by reduction of order (see Section 3.7). Let $y = vx^{r_1}$. Then (1) becomes

$$ax^2(v''x^{r_1} + 2v'r_1 x^{r_1-1} + vr_1(r_1 - 1)x^{r_1-2})$$
$$+ bx(v'x^{r_1} + vr_1 x^{r_1-1}) + cvx^{r_1} = 0.$$

Using the fact that x^{r_1} is a solution reduces this equation to:

$$av''x^{r_1+2} + v'(2r_1 a + b)x^{r_1+1} = 0.$$

Divide by x^{r_1+1}; then let $r_1 = (a - b)/2a$, to get

$$axv'' + av' = 0.$$

Divide by a and let $w = v'$. We thus obtain

$$xw' + w = 0.$$

This equation is first-order linear (also, variables separate). Its solution is

$$w = \frac{c_1}{x}.$$

Thus

$$v' = \frac{c_1}{x}$$

and

$$v = c_1 \ln x + c_2,$$

so that

$$y = x^{r_1} v = c_1 x^{r_1} \ln x + c_2 x^{r_1}.$$

Example 3.11.3 Find the general solution of

$$x^2 y'' - xy' + y = 0, \quad x > 0.$$

Solution The polynomial $ar^2 + (b - a)r + c = r^2 - 2r + 1$ has repeated roots 1, 1. Thus

$$y = c_1 x \ln x + c_2 x, \quad x > 0. \blacksquare$$

We summarize these three cases in the following theorem.

Theorem 3.11.1 The general solution of the second-order Euler's equation $ax^2 y'' + bxy' + cy = 0$, $x > 0$, is given as follows: Find the roots r_1, r_2 of the polynomial $ar^2 + (b - a)r + c = 0$.

1. If $r_1 \neq r_2$ and r_1, r_2 are real, then

$$y = c_1 x^{r_1} + c_2 x^{r_2}. \tag{6}$$

3.11 Euler's Equation

2. If $r_1 = \alpha + \beta i$, $r_2 = \alpha - \beta i$, $\beta \neq 0$, then
$$y = c_1 x^\alpha \cos(\beta \ln x) + c_2 x^\alpha \sin(\beta \ln x). \tag{7}$$

3. If $r_1 = r_2$, then
$$y = c_1 x^{r_1} + c_2 x^{r_1} \ln x. \tag{8}$$

Exercises

In Exercises 1 through 8, solve the second-order Euler's equation for $x > 0$. If no initial conditions are given, find the general solution.

1. $x^2 y'' + xy' - y = 0$
2. $x^2 y'' - 4xy' + 6y = 0$
3. $x^2 y'' + 3xy' + y = 0$
4. $x^2 y'' - xy' + 2y = 0$, $y(1) = 0$, $y'(1) = 2$
5. $4x^2 y'' + 8xy' + y = 0$
6. $x^2 y'' + xy' + y = 0$
7. $x^2 y'' + 4xy' + 2y = 0$, $y(1) = 1$, $y'(1) = 0$
8. $9x^2 y'' + 15xy' + 2y = 0$

9. (*Equidimensionality*) Let k be a constant, and perform the change of variables $x = ks$.

   i) Show that the Euler equation $ax^2(d^2y/dx^2) + bx(dy/dx) + cy = 0$ becomes
   $$as^2 \frac{d^2y}{ds^2} + bs \frac{dy}{ds} + cy = 0.$$
   That is, Euler's equation is unaltered by a change of scale in the independent variable.

   ii) Contrast this with what happens when the same change of scale is performed on the constant-coefficient equation $ay'' + by' + cy = 0$.

   iii) Verify that, if $k > 0$ is a real constant and r_1, r_2 are distinct real constants, then $x = ks$ changes $c_1 x^{r_1} + c_2 x^{r_2}$ into $\tilde{c}_1 s^{r_1} + \tilde{c}_2 s^{r_2}$.

   iv) Verify that if $k > 0$ and r_1 is a real constant, then $x = ks$ changes $c_1 x^{r_1} + c_2 x^{r_1} \ln x$ into $\tilde{c}_1 s^{r_1} + \tilde{c}_2 s^{r_1} \ln s$.

   v) Verify that if $k > 0$, and α, β are real constants, $\beta \neq 0$, then $x = ks$ changes $c_1 x^\alpha \cos(\beta \ln x) + c_2 x^\alpha \sin(\beta \ln x)$ into $\tilde{c}_1 s^\alpha \cos(\beta \ln s) + \tilde{c}_2 s^\alpha \sin(\beta \ln s)$.

10. (Alternative Method) Let $t = \ln x$. ($x = e^t$)

    i) Verify that
    $$x \frac{dy}{dx} = \frac{dy}{dt}, \tag{9}$$
    $$x^2 \frac{d^2y}{dx^2} = \frac{d^2y}{dt^2} - \frac{dy}{dt}. \tag{10}$$

    ii) Show that the change of variables $t = \ln x$ changes
    $$ax^2 \frac{d^2y}{dx^2} + bx \frac{dy}{dx} + cy = 0$$
    into the constant-coefficient equation
    $$a \frac{d^2y}{dt^2} + (b-a) \frac{dy}{dt} + cy = 0. \tag{11}$$

Note that the polynomial $ar^2 + (b-a)r + c$ in Theorem 3.11.1 is just the characteristic polynomial of (11). Solve Exercises 11 through 16 by changing to a constant-coefficient equation, as shown in Exercise 10, solving the constant-coefficient equation, and then changing back to the x-variable.

11. $x^2 y'' + 5xy' + 4y = 0$
12. $4x^2 y'' + 8xy' + 2y = 0$
13. $x^2 y'' + 5xy' + 3y = 0$
14. $x^2 y'' = 0$
15. $x^2 y'' + xy' = 0$
16. $x^2 y'' + xy' + 9y = 0$

17. (Uses operator notation of Section 3.6.) If D denotes the operation of differentiation by the variable t, where $t = \ln x$, then formulas (9) and (10) take the form
$$x \frac{dy}{dx} = Dy, \tag{12}$$
$$x^2 \frac{d^2y}{dx^2} = (D^2 - D)y = D(D-1)y. \tag{13}$$

i) Show that

$$x^3 \frac{d^3y}{dx^3} = D(D-1)(D-2)y \quad (14)$$

and

$$x^4 \frac{d^4y}{dx^4} = D(D-1)(D-2)(D-3)y. \quad (15)$$

(Exercises 18 through 21 require Section 3.10.) Use formulas (12)–(15) above to change the following Euler equations to constant-coefficient equations, solve by the method of Section 3.10, and write your solution in terms of x.

18. $x^4 y'''' + 6x^3 y''' + 7x^2 y'' + xy' - y = 0$
19. $x^4 y'''' + 6x^3 y''' + 9x^2 y'' + 3xy' + y = 0$
20. $x^3 y''' + xy' - y = 0$
21. $x^3 y''' + 4x^2 y'' = 0$

3.12 Undetermined Coefficients (Second-Order)

As pointed out in Section 3.4, the general solution of

$$ay''(x) + by'(x) + cy(x) = f(x), \quad (1)$$

where a, b, c are constants and $a \neq 0$, may be written in the form $y = y_p + y_h$, where y_h is the general solution of the associated homogeneous equation $ay'' + by' + cy = 0$ and y_p is a particular solution of $ay'' + by' + cy = f$. The homogeneous equation was solved in Section 3.8. Two methods for finding the particular solution y_p will be given in this chapter. The method of undetermined coefficients will be developed in this section. The method of variation of parameters will be discussed in Sections 3.14 and 3.15.

The method of undetermined coefficients can be presented in (at least) two ways. While mathematically equivalent, the approaches appear different. In this section we present an approach that utilizes only our second-order results. In Section 3.13.2 an alternative approach is presented that requires operator notation (Section 3.6) and nth-order homogeneous equations (Section 3.10).

If we think of the differential equation (1) as modeling a physical system, say a circuit, then f often stands for the input or outside influence, and the solution y is the response. The response is often similar to the input function.

Example 3.12.1 Find a particular solution of

$$y'' + y = 3e^{-x}. \quad (2)$$

Here $f(x) = 3e^{-x}$ is an exponential. Let us see whether there is a solution of (2) of the form $y_p = Ae^{-x}$ for some constant A. Substituting this *form* for y_p into (2) for y gives

$$(Ae^{-x})'' + Ae^{-x} = 3e^{-x}$$

or $2A = 3$. Thus $y_p = \tfrac{3}{2} e^{-x}$ is a particular solution of (2). ∎

Example 3.12.2   Let us now try to find a particular solution of
$$y'' - y = 3e^{-x}. \tag{3}$$
Like Example 3.12.1, we have $f(x) = 3e^{-x}$, so that we might try $y_p = Ae^{-x}$ again. Substituting this form for y_p into (3), we get
$$(Ae^{-x})'' - (Ae^{-x}) = 3e^{-x},$$
or, after simplification, $0 = 3$, which is impossible. Thus there is no particular solution of the form Ae^{-x}. Note that e^{-x} is also a solution of the associated homogeneous equation $y'' - y = 0$. Letting $y_1 = e^{-x}$ and using reduction of order (Section 3.7), we find, after some calculation, that
$$y = c_1 e^{-x} + c_2 e^x - \tfrac{3}{2} x e^{-x}$$
is the general solution of (3) and $-\tfrac{3}{2} x e^{-x}$ is a particular solution. ∎

Examples 1 and 2 are typical of how the method of undetermined coefficients works. The f in (1) determines the form of a particular solution up to certain powers of x. The powers of x are determined by the solutions of the associated homogeneous equation, or equivalently, by the roots of the characteristic equation. Once the form for a particular solution is determined, the actual particular solution is determined by substituting the form into the original differential equation and solving for the constants.

We shall give several important special cases of the method of undetermined coefficients and work several examples. These special cases are summarized in Table 3.12.1. There is a pattern to these special cases which the reader should look for. Finally, the general procedure will be given.

Special Case 1

If $f(x)$ in (1) is an mth-degree polynomial, then y_p is of the form $x^k(A_0 + A_1 x + \cdots + A_m x^m)$, where k is the multiplicity of 0 as a root of the characteristic polynomial.

Example 3.12.3

Solution

Find the general solution of
$$y'' - 3y' = 2x^2 + 1. \tag{4}$$
The characteristic polynomial $r^2 - 3r$ has roots $r_1 = 0$, $r_2 = 3$. Thus $y_h = c_1 e^{0x} + c_2 e^{3x} = c_1 + c_2 e^{3x}$. Here $f(x) = 2x^2 + 1$ is a second-degree polynomial and 0 is a root of multiplicity 1, so that $k = 1$. Thus, by Special Case 1, y_p has the form $x(A_0 + A_1 x + A_2 x^2)$. Substituting this form into (4) gives

Table 3.12.1 Special Cases

f includes summands of form	y_p then includes	k is the multiplicity of the root
1. $p(x)$, an mth-degree polynomial	$x^k(A_0 + A_1 x + \cdots + A_m x^m)$	0
2. $E e^{\alpha x}$	$x^k A e^{\alpha x}$	α
3. $p(x) e^{\alpha x}$, $p(x)$ an mth-degree polynomial	$x^k(A_0 + A_1 x + \cdots + A_m x^m) e^{\alpha x}$	α
4. $E_1 \cos \beta x + E_2 \sin \beta x$	$x^k(A_0 \cos \beta x + B_0 \sin \beta x)$	βi
5. $p(x) \cos \beta x + q(x) \sin \beta x$ where $p(x)$ is an mth-degree polynomial and $q(x)$ is an nth-degree polynomial	$x^k(A_0 + A_1 x + \cdots + A_s x^s) \cos \beta x +$ $x^k(B_0 + B_1 x + \cdots + B_s x^s) \sin \beta x$; $s = $ larger of m, n	βi
6. $E_1 e^{\alpha x} \cos \beta x + E_2 e^{\alpha x} \sin \beta x$	$x^k e^{\alpha x}(A_0 \cos \beta x + B_0 \sin \beta x)$	$\alpha + \beta i$

$$(A_0 x + A_1 x^2 + A_2 x^3)'' - 3(A_0 x + A_1 x^2 + A_2 x^3)' = 2x^2 + 1$$

or

$$2A_1 + 6A_2 x - 3A_0 - 6A_1 x - 9A_2 x^2 = 2x^2 + 1.$$

Equating coefficients of like powers of x,

$$1: \quad 2A_1 - 3A_0 = 1,$$
$$x: \quad 6A_2 - 6A_1 = 0,$$
$$x^2: \quad -9A_2 = 2,$$

we get $A_2 = -\frac{2}{9}$, $A_1 = -\frac{2}{9}$, and $A_0 = -\frac{13}{27}$. Thus the particular solution is $y_p = -\frac{13}{27}x - \frac{2}{9}x^2 - \frac{2}{9}x^3$ and the general solution is

$$y = y_p + y_h = -\frac{13}{27}x - \frac{2}{9}x^2 - \frac{2}{9}x^3 + c_1 + c_2 e^{3x}. \blacksquare$$

Special Case 2

If $f(x)$ is a constant times $e^{\alpha x}$, then y_p has the form $A x^k e^{\alpha x}$, where k is the multiplicity of α as a root of the characteristic polynomial.

Example 3.12.4

Solution

Find the general solution of

$$y'' - 5y' + 6y = 4e^{2x}. \tag{5}$$

First we solve the homogeneous equation $y'' - 5y' + 6y = 0$. The characteristic equation is $r^2 - 5r + 6 = (r-2)(r-3) = 0$ and it has roots $r = 2, 3$. Thus $y_h = c_1 e^{2x} + c_2 e^{3x}$. According to Special Case 2, y_p is of the form $y = x^k A e^{2x}$ (here $\alpha = 2$) and $k = 1$, since 2 is a root of the characteristic equation of multiplicity 1. Thus $y_p = Axe^{2x}$ for some constant A. Substituting this form for y_p into the original equation (5), we obtain

$$(Axe^{2x})'' - 5(Axe^{2x})' + 6(Axe^{2x}) = 4e^{2x}$$

or

$$4Axe^{2x} + 4Ae^{2x} - 5(Ae^{2x} + 2Axe^{2x}) + 6Axe^{2x} = 4e^{2x}.$$

Then $-Ae^{2x} = 4e^{2x}$ and $A = -4$. Thus $y_p = -4xe^{2x}$ is a particular solution of (5) and

$$y = y_p + y_h = -4xe^{2x} + c_1 e^{2x} + c_2 e^{3x}$$

is the general solution of (5). ∎

Example 3.12.5

Find the general solution of

$$y'' - 5y' - 6y = 4e^{2x}. \tag{6}$$

Solution

First we solve the associated homogeneous equation $y'' - 5y' - 6y = 0$. The characteristic polynomial is $r^2 - 5r - 6 = (r - 6)(r + 1)$ and has roots $6, -1$. Thus $y_h = c_1 e^{6x} + c_2 e^{-x}$. According to Special Case 2, y_p is of the form $x^k A e^{2x}$ (here $\alpha = 2$) and $k = 0$ since 2 is *not a root* of the characteristic equation. Thus $y_p = Ae^{2x}$ for some constant A. Substitute this form for y_p into the original equation (6)

$$(Ae^{2x})'' - 5(Ae^{2x})' - 6(Ae^{2x}) = 4e^{2x},$$

and simplify

$$4Ae^{2x} - 10Ae^{2x} - 6Ae^{2x} = 4e^{2x},$$

so that $-12A = 4$ or $A = -\frac{1}{3}$. Thus $y_p = -\frac{1}{3}e^{2x}$ and the general solution of (4) is

$$y = y_p + y_h = -\frac{1}{3}e^{2x} + c_1 e^{6x} + c_2 e^{-x}. \blacksquare$$

Note that differential equations (5) and (6) had exactly the same forcing function $4e^{2x}$ and the coefficients differed in only one place (-6 vs. 6), yet the form for y_p was different. This points out an important fact.

In general, the form for y_p cannot be determined by looking only at the forcing function f. The solution of the homogeneous equation (roots of the characteristic polynomial) must also be considered.

In the examples to follow we shall make frequent use of the fact that

If $f(x) = f_1(x) + \cdots + f_m(x)$, then there is a particular solution of the form obtained by adding up the forms of the particular solutions for each $f_i(x)$. (7)

Also, in several of the examples we shall merely determine the form for y_p and not actually find the constants in the form.

Special Case 3

If $f(x)$ is $p(x)e^{\alpha x}$, where $p(x)$ is an mth-degree polynomial, then there is a particular solution of the form

$$x^k[A_0 + A_1 x + \cdots + A_m x^m]e^{\alpha x},$$

where k is the multiplicity of α as a root of the characteristic polynomial. Note that Special Case 3 includes Special Case 1 and 2 by taking $\alpha = 0$ and $m = 0$, respectively.

Example 3.12.6 Give the form for y_p if

$$y'' - y' = x^3 + x + e^x - 2xe^x$$

is to be solved by the method of undetermined coefficients.

Solution The characteristic polynomial is $r^2 - r = r(r-1)$, which has roots $r = 0, 1$ so that $y_h = c_1 + c_2 e^x$. The forcing term is

$$f = \underbrace{(x^3 + x)} + \underbrace{(1 - 2x)e^x}.$$

The first term is a third-degree polynomial. Since 0 is a root of multiplicity 1 of the characteristic equation (Special Case 1), y_p must include a term of the form $x^k(A_0 + A_1 x + A_2 x^2 + A_3 x^3)$ with $k = 1$. The second term is of the form $p(x)e^{\alpha x}$, where $p(x) = 1 - 2x$ is a first-degree polynomial and $\alpha = 1$. Since 1 is a root of the characteristic equation by Special Case 3, y_p must include a term of the form $x^k(A_4 + A_5 x)e^x$ with $k = 1$. Thus y_p has the form

$$y_p = x(A_0 + A_1 x + A_2 x^2 + A_3 x^3) + x(A_4 + A_5 x)e^x. \blacksquare$$

Example 3.12.7 Give the form for y_p if

$$y'' - 2y' + y = 7xe^x$$

is to be solved by the method of undetermined coefficients.

Solution The characteristic polynomial $r^2 - 2r + 1 = (r-1)^2$ has a root 1 of multiplicity 2. Thus $y_h = c_1 e^x + c_2 x e^x$. The forcing term is of the form $p(x)e^{\alpha x}$, where $p(x) = 7x$ is a first-degree polynomial and $\alpha = 1$. Since $\alpha = 1$ is a root of multiplicity 2, by Special Case 3 with $k = 2$, $m = 1$, the form for y_p is $x^2(A_0 + A_1 x)e^x. \blacksquare$

Special Case 4

If $f(x) = E_1 \cos \beta x + E_2 \sin \beta x$, where at least one of the constants E_1, E_2 is nonzero, then y_p has the form $x^k(A_0 \cos \beta x + B_0 \sin \beta x)$, where k is the multiplicity of βi as a root of the characteristic polynomial.

Example 3.12.8 Find the general solution of

$$y'' + 2y' + 2y = 3e^{-x} + 4 \cos x. \tag{8}$$

Solution First we solve the associated homogeneous equation. The characteristic equation is $r^2 + 2r + 2 = 0$. Its roots are

$$r = -1 \pm i.$$

Thus $y_h = c_1 e^{-x} \cos x + c_2 e^{-x} \sin x$. Now we will determine y_p. Note that f is a sum of two terms, $3e^{-x}$ and $4 \cos x$. Consider first $3e^{-x}$. Since -1 is not a root of the characteristic equation, Special Case 2 says that y_p includes a term of the form $A_1 e^{-x}$. Now consider $4 \cos x$. Since i is not a root of the characteristic equation, Special Case 4 with $\beta = 1$ says that y_p includes terms of the form $A_2 \cos x + A_3 \sin x$. Thus $y_p = A_1 e^{-x} + A_2 \cos x + A_3 \sin x$ for some constants A_1, A_2, A_3. Substituting this expression into (8) gives

$$[A_1 e^{-x} + A_2 \cos x + A_3 \sin x]'' + 2[A_1 e^{-x} + A_2 \cos x + A_3 \sin x]'$$
$$+ 2[A_1 e^{-x} + A_2 \cos x + A_3 \sin x] = 3e^{-x} + 4 \cos x.$$

That is,

$$A_1 e^{-x} - A_2 \cos x - A_3 \sin x - 2A_1 e^{-x} - 2A_2 \sin x + 2A_3 \cos x$$
$$+ 2A_1 e^{-x} + 2A_2 \cos x + 2A_3 \sin x = 3e^{-x} + 4 \cos x.$$

There are three functions e^{-x}, $\cos x$, $\sin x$ that appear in this equation. Equating the coefficients of like terms gives

$$\begin{aligned} e^{-x}: &\quad A_1 = 3, \\ \cos x: &\quad A_2 + 2A_3 = 4, \\ \sin x: &\quad A_3 - 2A_2 = 0. \end{aligned} \tag{9}$$

This system of three equations in the three unknowns A_1, A_2, A_3 has the solution

$$A_1 = 3, \qquad A_2 = \tfrac{4}{5}, \qquad A_3 = \tfrac{8}{5}.$$

Thus $y_p = 3e^{-x} + \tfrac{4}{5} \cos x + \tfrac{8}{5} \sin x$, and

$$y = y_p + y_h = 3e^{-x} + \tfrac{4}{5} \cos x + \tfrac{8}{5} \sin x + c_1 e^{-x} \cos x + c_2 e^{-x} \sin x. \quad \blacksquare$$

This example emphasizes another common source of errors:

Even though only $\cos \beta x$ appeared in the forcing term f, the form for y_p may require both $x^k A \cos \beta x$ and $x^k B \sin \beta x$.

The system of equations (9) consisted of three equations in three unknowns and had a unique solution. It can be proved, using properties of linear independence and the Wronskian, that

> If a, b, c are constants and f is the type of function described in the Method of Undetermined Coefficients, then the method always works (perhaps messily). In particular, if the equations for the undetermined constants are not consistent (don't have a solution), then an error has been made.

For example, if one arrives at

$$A_1 \sin x + A_2 e^{-x} = \cos x + 3 \sin x + 5e^{-x},$$

equating coefficients of *all* the functions that appear gives

$$\begin{aligned} \sin x: &\quad A_1 = 3, \\ \cos x: &\quad 0 = 1, \\ e^{-x}: &\quad A_2 = 5, \end{aligned}$$

which is impossible. Since the method always works for appropriate f, we know that we have made an error. Frequently the error is in finding the form for y_p.

Example 3.12.9 Give the form for y_p if

$$y'' + 4y = \sin 2x$$

is to be solved by the method of undetermined coefficients.

Solution The roots of the characteristic polynomial $r^2 + 4$ are $\pm 2i$ and $y_h = c_1 \cos 2x + c_2 \sin 2x$. The forcing term $\sin 2x$ is $\sin \beta x$, where $\beta = 2$. Since βi is a root of the characteristic polynomial of multiplicity 1, we have $k = 1$ and by Special Case 4, y_p has the form $x[A_0 \cos 2x + B_0 \sin 2x]$. ∎

Special Case 5

If $f(x) = p(x) \sin \beta x + q(x) \cos \beta x$, where $p(x)$ is an mth-degree polynomial in x and $q(x)$ is an nth-degree polynomial in x, then there is a particular solution of the form

$$x^k[(A_0 + A_1 x + \cdots + A_s x^s) \cos \beta x + (B_0 + B_1 x + \cdots + B_s x^s) \sin \beta x],$$

3.12 Undetermined Coefficients (Second-Order)

where k is the multiplicity of βi as a root of the characteristic polynomial and s is the larger of m, n.

Special Case 5 includes Special Cases 3 ($m = 0$, $n = 0$), and 1 ($\beta = 0$, $n = 0$).

Example 3.12.10 Give the form for y_p if

$$y'' + 4y = x^2 \cos 2x - x \sin 2x + \sin 2x = x^2 \cos 2x + (1-x) \sin 2x$$

is to be solved by the method of undetermined coefficients.

Solution The roots of the characteristic polynomial $r^2 + 4$ are $\pm 2i$ and $y_h = c_1 \cos 2x + c_2 \sin 2x$. The forcing term is of the form

$$p(x) \cos \beta x + q(x) \sin \beta x,$$

where $p(x) = x^2$ is a second-degree polynomial, $q(x) = 1 - x$ is a first-degree polynomial, and $\beta = 2$. Since $\beta i = 2i$ is a root of the characteristic equation of multiplicity 1 by Special Case 5, with $k = 1$, we have

$$y_p = x[(A_1 + A_2 x + A_3 x^2) \cos 2x + (A_4 + A_5 x + A_6 x^2) \sin 2x]. \blacksquare$$

Our final special case is:

Special Case 6

If $f(x) = E_1 e^{\alpha x} \cos \beta x + E_2 e^{\alpha x} \sin \beta x$, where E_1, E_2 are constants at least one of which is nonzero, then there is a particular solution of the form

$$x^k [A_0 e^{\alpha x} \cos \beta x + B_0 e^{\alpha x} \sin \beta x],$$

where k is the multiplicity of $\alpha + \beta i$ as a root of the characteristic polynomial.

Example 3.12.11 Give the form for y_p if

$$y'' + 2y' + 2y = 5e^{-x} \cos x$$

is to be solved by the method of undetermined coefficients.

Solution The roots of the characteristic polynomial $r^2 + 2r + 2$ are $-1 \pm i$, so that

$$y_h = c_1 e^{-x} \cos x + c_2 e^{-x} \sin x.$$

The forcing term is of the form $e^{\alpha x} \cos \beta x$, where $\alpha = -1$, $\beta = 1$. Since $-1 + i$ is a root of the characteristic equation of multiplicity 1, by Special Case 6, with $k = 1$,

$$y_p = x(A_0 e^{-x} \cos x + B_0 e^{-x} \sin x). \blacksquare$$

Example 3.12.12 Give the form for y_p if

$$y'' + 2y' + 2y = e^{-x}\cos 2x + e^{-x}\sin 2x + e^{-x} - 3\cos x \qquad (10)$$

is to be solved by undetermined coefficients.

Solution Equation (10) has the same characteristic polynomial as Example 3.12.11, so the roots are $-1 \pm i$. The forcing term in (10) is the sum of three groups of terms:

$e^{-x}\cos 2x + e^{-x}\sin 2x$: Since $-1 + 2i$ is not a root, we include $A_0 e^{-x}\cos 2x + B_0 e^{-x}\sin 2x$, by Special Case 6.

e^{-x}: Since -1 is not a root, we include $A_2 e^{-x}$, by Special Case 2.

$-3\cos x$: Since i is not a root, we include $A_1 \cos x + B_1 \sin x$, by Special Case 4.

Thus the form for y_p is

$$A_0 e^{-x}\cos 2x + B_0 e^{-x}\sin 2x + A_1 \cos x + B_1 \sin x + A_2 e^{-x}. \blacksquare$$

When initial conditions are present and the forcing function is nonzero, it is important to remember to apply the initial conditions to the general solution $y_p + y_h$ and not just to y_h.

Example 3.12.13 Solve

$$y'' - 5y' + 6y = 4e^{2x}, \qquad y(0) = 0, \qquad y'(0) = 1. \qquad (11)$$

Solution First find the general solution of $y'' - 5y' + 6y = 4e^{2x}$. This was done in Example 3.12.4 and is

$$y = -4xe^{2x} + c_1 e^{2x} + c_2 e^{3x}. \qquad (12)$$

In order to apply the initial conditions, we must compute y'. Differentiate (12), to get

$$y' = -4e^{2x} - 8xe^{2x} + 2c_1 e^{2x} + 3c_2 e^{3x}.$$

The initial conditions then give

$$0 = y(0) = c_1 + c_2,$$
$$1 = y'(0) = -4 + 2c_1 + 3c_2.$$

Solving for c_1, c_2 yields $c_1 = -5$, $c_2 = 5$, and the solution of (11) is

$$y = -4xe^{2x} - 5e^{2x} + 5e^{3x}. \blacksquare$$

3.12.1 The Superposition Principle

The Superposition Principle (Section 3.4) for linear differential equations says that

> If y_{p1} is a particular solution of $ay'' + by' + cy = f_1$ and y_{p2} is a particular solution of $ay'' + by' + cy = f_2$, then $y_{p1} + y_{p2}$ is a particular solution of $ay'' + by' + cy = f_1 + f_2$.

Intuitively, the superposition principle means that the response (output) that results from the sum (superposition) of two forcing terms (inputs) is the sum of the response from each forcing term. In applications, knowing whether our device or physical problem acts in this manner is a key factor in deciding whether linear equations can be used to analyze the problem. Fact (7) is one version of the superposition principle.

The superposition principle can also be used as follows: Suppose we wish to solve

$$y'' + 3y' + 2y = x^2 e^{-x} + \cos 2x - 3x^2 \sin x. \tag{13}$$

If the method of undetermined coefficients is applied to (13), the form for y_p will have 11 constants to determine. Alternatively, the superposition principle says if

y_{p1} is a particular solution of $y'' + 3y' + 2y = x^2 e^{-x}$,
y_{p2} is a particular solution of $y'' + 3y' + 2y = \cos 2x$.
y_{p3} is a particular solution of $y'' + 3y' + 2y = -3x^2 \sin x$,

then $y_p = y_{p1} + y_{p2} + y_{p3}$ would be a particular solution of (13).

Using the superposition principle in this manner will never reduce the amount of calculation. It will, however, often reduce the length of the expressions being worked with and thus reduce human error. If undetermined coefficients are being implemented symbolically on a computer, there is no advantage.

The superposition principle, combined with the theory of *Fourier series*, enables the method of undetermined coefficients to be used on many additional kinds of forcing functions f. The idea is to write $f(x)$ as an (infinite) linear combination of functions $\cos \beta x$, $\sin \beta x$ for different β, and then use an infinite-series version of the superposition principle. This important idea is somewhat beyond the immediate aims of this book and will not be pursued further.

Undetermined coefficients can also be applied to first-order linear constant-coefficient differential equations if f is in the right form and is often quicker than using integrating factors.

Method of Undetermined Coefficients

In summary, the *method of undetermined coefficients* can be used on $ay'' + by' + cy = f$ if a, b, c are constants and f is a linear combination of functions of the form

$$x^m e^{\alpha x} \cos \beta x, \quad x^m e^{\alpha x} \sin \beta x,$$

where m is a nonnegative integer and α, β are real numbers. Special cases are:

$$x^m, \quad x^m e^{\alpha x}, \quad e^{\alpha x}, \quad e^{\alpha x} \cos \beta x, \quad e^{\alpha x} \sin \beta x, \quad x^m \cos \beta x, \quad x^m \sin \beta x.$$

The method is as follows.

1. First solve the associated homogeneous equation

$$ay'' + by' + cy = 0.$$

2. Determine y_p as a linear combination of functions with unknown coefficients using the following rules.

   R1. If f includes a sum of terms of the form $p(x)e^{\alpha x}$, where $p(x)$ is an mth-degree polynomial, then the form for y_p should include

   $$x^k[A_0 + A_1 x + \cdots + A_m x^m]e^{\alpha x},$$

   where k is the multiplicity of α as a root of the characteristic polynomial $ar^2 + br + c$.

   R2. If f includes a sum of terms of the form

   $$p(x)e^{\alpha x} \cos \beta x + q(x)e^{\alpha x} \sin \beta x,$$

   where $p(x)$ is an mth-degree polynomial and $q(x)$ is an nth-degree polynomial, then the form for y_p should include

   $$x^k[A_0 + A_1 x + \cdots + A_s x^s]e^{\alpha x} \cos \beta x$$
   $$+ x^k[B_0 + B_1 x + \cdots + B_s x^s]e^{\alpha x} \sin \beta x,$$

   where s is the larger of m and n and k is the multiplicity of $\alpha + \beta i$ as a root of the characteristic polynomial $ar^2 + br + c$.

3. Substitute the expression for y_p into the differential equation $ay'' + by' + cy = f$ to determine the unknown coefficients A_i, B_i.
4. The general solution of $ay'' + by' + cy = f$ is $y = y_p + y_h$.
5. Apply any initial conditions to $y_p + y_h$ in order to determine arbitrary constants.

Note that if $f(x)$ includes terms like $\ln x$, $x^{1/3} \sin x$, $\tan x$, then undetermined coefficients will not generally work.

Exercises

In Exercises 1 through 12, state whether undetermined coefficients can be applied to the differential equation. If it cannot, explain why not.

1. $y'' + y = x \sin x$
2. $y'' + 3y = x^{1/2} \sin x$
3. $y'' + y = x^2 + x + \ln|x|$
4. $y'' + y = e^{x+1}$
5. $y'' + y = \dfrac{\sin x}{\cos x}$

6. $y'' + y' + y = \cosh x$
7. $y'' + xy' = 3e^{2x}$
8. $y'' + y' = x \sinh 2x$
9. $y'' + y = x^{-1} e^x$
10. $y'' + yy' = e^{2x}$
11. $y' + 3y = e^{-2x} \cos 3x + \sinh 3x$
12. $y'' + 3y' + 4y = \sin^2 x$

In Exercises 13 through 36, solve the differential equation using the method of undetermined coefficients. If no initial conditions are given, give the general solution.

13. $y'' - 3y' + 2y = 2e^x$
14. $y'' - 3y' + 2y = 2e^{-x}$
15. $y'' + 2y' + 5y = 3 \sin x$
16. $y'' + 4y' + 8y = x^2 + 1$; $y(0) = 0$, $y'(0) = 0$
17. $y'' + y = \cos 2x$; $y(0) = 0$, $y'(0) = 1$
18. $y'' - 4y' + 3y = xe^{2x}$
19. $y'' - 4y' + 3y = xe^x$
20. $y'' + y = \sin x$
21. $y'' - 2y' + 5y = e^x \cos 3x$
22. $y'' + 2y' + y = 3e^x$; $y(0) = 0$, $y'(0) = 2$
23. $y'' + 4y' = 12x^2 + e^x$; $y(0) = 1$, $y'(0) = 1$
24. $y'' + y' + y = \cos 2x$
25. $y'' + 2y' + y = 3e^{-x}$
26. $y'' + 2y' + y = 3xe^{-x} + 2e^{-x}$
27. $y'' - 4y' + 4y = xe^x - e^x + 2e^{3x}$
28. $y'' - 5y' + 4y = 17 \sin x + 3e^{2x}$
29. $y'' + 5y' + 4y = 8x^2 + 3 + 2 \cos 2x$
30. $y' + y = 2e^{-x}$
31. $y' + 3y = x^2 + 1$
32. $y' - y = \sin x$
33. $y'' + 4y = \sin 2x$
34. $2y' + 4y = x$
35. $3y' - 2y = xe^x$
36. $y' - 3y = e^x \sin x$

In Exercises 37 through 49, give the form for y_p if the method of undetermined coefficients were to be used. You need not actually compute y_p.

37. $y'' - 2y' + 5y = 3e^x \sin 2x$
38. $y'' + 2y' - 3y = x^3 e^x - e^x + e^{-2x} + e^{-3x}$
39. $y'' - 6y' + 9y = x^4 e^{3x}$
40. $y'' + 9y = x^2 \sin 3x + \cos 2x$
41. $y'' + 9y = xe^{-x} \sin 2x$
42. $y'' + 2y' + 2y = 3x^2 e^{-x} \cos 2x + xe^{-x} \sin 2x$
43. $y'' + 2y' + 2y = e^{-x} \cos x + e^x \sin x$
44. $y'' + 3y' - 10y = x^2 \cos x$
45. $y'' + 3y' - 10y = x^2 e^{2x} + e^{5x}$
46. $y'' + y = e^{-x} - e^x + e^x \cos x$
47. $y'' + 16y = x \cos 4x + e^{-x} \sin 4x + 3e^{-4x}$
48. $y'' + 2y' + 2y = x^2 e^{-x} \cos x$
49. $y'' - 2y' + 2y = x^3 e^{-x} \sin x + e^x \cos x$

3.13 Undetermined Coefficients (*n*th-Order)

3.13.1 The Procedure

The method of undetermined coefficients described in Section 3.12 works with only slight modification on *n*th-order linear differential equations with real constant coefficients:

$$a_n y^{(n)}(x) + a_{n-1} y^{(n-1)}(x) + \cdots + a_1 y'(x) + a_0 y(x) = f(x). \tag{1}$$

The forcing or input function $f(x)$ still needs to be a linear combination of functions of the form $x^m e^{\alpha x} \cos \beta x$, $x^m e^{\alpha x} \sin \beta x$ for integers $m \geq 0$ and real numbers α, β. There is only one change in the method of undetermined coefficients from the second-order case.

The characteristic polynomial is now nth-degree, so that its roots may have multiplicity greater than two.

We still have

R1'. If $f(x)$ includes a sum of terms of the form $p(x)e^{\alpha x}$, where $p(x)$ is an mth-degree polynomial, then the form for y_p should include

$$x^k [A_0 + A_1 x + \cdots + A_m x^m] e^{\alpha x},$$

where k is the multiplicity of α as a root of the characteristic polynomial.

R2'. If $f(x)$ includes a sum of terms of the form

$$p(x)e^{\alpha x} \cos \beta x + q(x)e^{\alpha x} \sin \beta x,$$

where $p(x)$ is an mth-degree polynomial and $q(x)$ is an nth-degree polynomial, then the form for y_p should include

$$x^k [A_0 + A_1 x + \cdots + A_s x^s] e^{\alpha x} \cos \beta x$$
$$+ x^k [B_0 + B_1 x + \cdots + B_s x^s] e^{\alpha x} \sin \beta x,$$

where s is the larger of m and n and k is the multiplicity of $\alpha + \beta i$ as a root of the characteristic polynomial.

Example 3.13.1 Find the general solution of

$$y''' - 6y'' + 12y' - 8y = xe^{2x}. \tag{2}$$

Solution First we need to solve the associated homogeneous equation

$$y''' - 6y'' + 12y' - 8y = 0.$$

The characteristic polynomial is $r^3 - 6r^2 + 12r - 8 = (r-2)^3$, so that 2 is a root of multiplicity 3. A fundamental set of solutions for the associated homogeneous equation is

$$\{e^{2x}, xe^{2x}, x^2 e^{2x}\}.$$

The forcing term is $f(x) = xe^{2x}$. By Rule R1' with $\alpha = 2$, $m = 1$, y_p must include $x^k [A_0 + A_1 x] e^{2x}$, where k is the multiplicity of 2 as a root of the characteristic equation. Thus $k = 3$ and y_p is in the form

$$y_p = x^3 [A_0 + A_1 x] e^{2x} = A_0 x^3 e^{2x} + A_1 x^4 e^{2x}.$$

To find A_0, A_1 substitute the form for y_p into the original differential equation (2):

$$(A_0 x^3 e^{2x} + A_1 x^4 e^{2x})''' - 6(A_0 x^3 e^{2x} + A_1 x^4 e^{2x})''$$
$$+ 12(A_0 x^3 e^{2x} + A_1 x^4 e^{2x})' - 8(A_0 x^3 e^{2x} + A_1 x^4 e^{2x}) = xe^{2x}.$$

After differentiating and combining like terms, we get

$$6A_0 e^{2x} + (30A_0 + 24A_1)xe^{2x} = xe^{2x}$$

Thus equating coefficients of like terms:

$$e^{2x}: \quad 6A_0 = 0,$$
$$xe^{2x}: \quad 30A_0 + 24A_1 = 1$$

and $A_0 = 0$, $A_1 = \frac{1}{24}$. Thus $y_p = \frac{1}{24} x^4 e^{2x}$ and the general solution of (2) is

$$y = \frac{1}{24} x^4 e^{2x} + c_1 e^{2x} + c_2 xe^{2x} + c_3 x^2 e^{2x}. \quad \blacksquare$$

Example 3.13.2 Give the form for y_p if

$$y'''' + 8y'' + 16y = x \sin x + x^2 \cos 2x \tag{3}$$

is to be solved by the method of undetermined coefficients.

Solution First we solve the associated homogeneous equation

$$y'''' + 8y'' + 16y = 0.$$

The characteristic equation is $r^4 + 8r^2 + 16 = (r^2 + 4)^2 = 0$ and has repeated complex roots $\pm 2i$, $\pm 2i$. A fundamental set of solutions for the associated homogeneous equation is thus

$$\{\sin 2x, \quad \cos 2x, \quad x \sin 2x, \quad x \cos 2x\}.$$

Now we find the form for y_p. Consider the forcing function

$$f = x \sin x + x^2 \cos 2x.$$

By R2', the $x \sin x$ term implies that y_p includes

$$x^{k_1}[(A_0 + A_1 x) \sin x + (B_0 + B_1 x) \cos x] \tag{4}$$

with $k_1 = 0$ since i is not a root of the characteristic polynomial. The $x^2 \cos 2x$ term implies that y_p includes

$$x^{k_2}[(A_2 + A_3 x + A_4 x^2) \sin 2x + (B_2 + B_3 x + B_4 x^2) \cos 2x] \tag{5}$$

with $k_2 = 2$, since $2i$ has multiplicity two as a root of the characteristic equation. In actually using (4) and (5), one may add (4) and (5) to get the form for y_p, substitute into the original differential equation (3), and solve for $A_0, \ldots, A_4$, $B_0, \ldots, B_4$. Alternatively, one could use (4) to find a particular solution of

$$y'''' + 8y'' + 16y = x \sin x,$$

and then use (5) to find a particular solution of

$$y'''' + 8y'' + 16y = x^2 \cos 2x.$$

Adding these two particular solutions gives, by the superposition principle, a solution of $y'''' + 8y'' + 16y = x \sin x + x^2 \cos 2x$ as desired. ∎

Exercises

In Exercises 1 through 10, solve the differential equation by the method of undetermined coefficients. If no initial conditions are given, give the general solution.

1. $y''' - 3y'' + 3y' - y = e^{2x}$
2. $y''' - 3y'' + 3y' - y = e^x$
3. $y''' - y' = \sin x$
4. $y'''' - 25y'' + 144y = x^2 - 1$
5. $y''' + y' = 3 + 2 \cos x, \quad y(0) = y'(0) = y''(0) = 0$
6. $y'''' - y = 2e^{3x} - e^x$
7. $y'''' - 16y = 5xe^x$
8. $y'''' + 4y'' + 4y = \cos 2x$
9. $y'''' - 5y'' + 4y = e^{2x} - e^{3x}$
10. $y'''' + y''' - 6y'' = 72x + 24, \quad y(0) = 0, \\ y'(0) = 0, \quad y''(0) = -6, \quad y'''(0) = -57$

In Exercises 11 through 18, give the form for y_p that you would use to find a particular solution by the method of undetermined coefficients. Do not actually solve for y_p.

11. $y''' - 3y'' + 3y' - y = x^2 e^x - 3e^x$
12. $y'''' + 2y'' + y = x \sin x$
13. $y'''' - 4y''' + 6y'' - 4y' + y = x^3 e^x + x^2 e^{-x}$
14. $y'''' + 5y'' + 4y = \sin x + \cos 2x + \sin 3x$
15. $y''' + 2y'' + 2y' = 3e^{-x} \cos x$
16. $y''' + 2y'' + 2y' = x^2 e^{-x} \cos x - xe^{-x} \sin x$
17. $y'''' + 4y''' + 8y'' + 8y' + 4y = 7e^{-x} \cos x$
18. $y'''' - 2y'' + y = xe^x + x^2 e^{-x} + e^{2x}$

3.13.2 Undetermined Coefficients Using Annihilators (Requires Section 3.6)

As noted in Sections 3.12 and 3.13, the method of undetermined coefficients always works if we have a linear constant-coefficient differential equation and the forcing function f is of the right kind. This section will provide one explanation of why that is true. If can also be used as an alternative approach for carrying out the method of undetermined coefficients.

Suppose that we have the linear differential equation with constant coefficients

$$a_n y^{(n)} + a_{n-1} y^{(n-1)} + \cdots + a_1 y' + a_0 y = f$$

or

$$L_1 y = f,$$

where $L_1 = a_n D^n + a_{n-1} D^{n-1} + \cdots + a_1 D + a_0$ (Section 3.6) and to simplify our notation we now write $L_1 y$ instead of $L_1(y)$.

If f is a linear combination of terms of the form $x^m e^{\alpha x} \cos \beta x$, $x^m e^{\alpha x} \sin \beta x$, with nonnegative integer m and real α, β, then f is the *solution* of another linear constant-coefficient homogeneous differential equation, $L_2 y = 0$. That is, $L_2 f = 0$. Thus

If y_p is a particular solution of $L_1 y = f$, then y_p is a solution

$$L_2 L_1 y = L_2 f = 0.$$

But $L_2 L_1 y = 0$ is a homogeneous equation with constant coefficients that can be solved to give a form for y_p. We may delete the solution of the associated homogeneous equation $L_1 y = 0$ from the form for y_p.

Rather than give a detailed proof we shall work several examples.

Example 3.13.3

Consider the differential equation

$$y'' - 3y' + 2y = 3xe^x$$

or

$$(D^2 - 3D + 2)y = 3xe^x. \tag{6}$$

We need to find a homogeneous differential equation for which xe^x is a solution. This amounts to applying the rules of Sections 3.8 and 3.10 backwards. Thus xe^x would correspond to a root 1 of the characteristic polynomial of multiplicity two. Hence $(r - 1)^2 = r^2 - 2r + 1$ would be the characteristic polynomial of a linear homogeneous differential equation for which xe^x is a solution. That is,

$$(D^2 - 2D + 1)xe^x = 0.$$

Multiply both sides of the original equation (6) by $(D^2 - 2D + 1)$ to get

$$(D^2 - 2D + 1)(D^2 - 3D + 2)y = 0. \tag{7}$$

The characteristic polynomial of this linear homogeneous equation is

$$(r^2 - 2r + 1)(r^2 - 3r + 2) = (r - 1)^3(r - 2).$$

Thus the solution of (7) is

$$y = c_1 e^x + c_2 x e^x + c_3 x^2 e^x + c_4 e^{2x}. \tag{8}$$

Now every solution of the original equation (6) is a solution of (7), so (8) gives a *form* for all solutions of our original equation (6). The terms $c_1 e^x + c_4 e^{2x}$ in (8) are the general solution of the associated homogeneous equation for our original problem (6). Thus the remaining terms give a form for a particular solution:

$$y_p = c_2 x e^x + c_3 x^2 e^x. \tag{9}$$

The constants c_2, c_3 can be found by substituting into the original equation (6). Note that the form (9) for y_p is precisely that provided by the rules of Sections 3.12 and 3.13. ∎

Example 3.13.4 Consider the differential equation

$$y'' + 2y' + 2y = 3 \sin x$$

or

$$(D^2 + 2D + 2)y = 3 \sin x. \tag{10}$$

The term $\sin x$ occurs as a solution of a homogeneous differential equation if the roots of the characteristic polynomial are $\pm i$, or the characteristic polynomial is $r^2 + 1$. Thus $(D^2 + 1) \sin x = 0$. Applying $D^2 + 1$ to both sides of (10) gives

$$(D^2 + 1)(D^2 + 2D + 2)y = 3(D^2 + 1) \sin x = 0.$$

The characteristic polynomial for this differential equation is $(r^2 + 1) \times (r^2 + 2r + 2)$, and it has roots $\pm i, -1 \pm i$, so that its solution is

$$y = c_1 \sin x + c_2 \cos x + c_3 e^{-x} \sin x + c_4 e^{-x} \cos x.$$

The terms $c_3 e^{-x} \sin x + c_4 e^{-x} \cos x$ are the general solution of the associated homogeneous equation for our original problem (10). The remaining terms give a form for y_p,

$$y_p = c_1 \sin x + c_2 \cos x. \ ∎$$

Example 3.13.5 Consider the differential equation

$$y'' - y' = 2e^x + 3e^{2x} + 5 \sin 2x$$

or

$$(D^2 - D)y = 2e^x + 3e^{2x} + 5 \sin 2x. \tag{11}$$

If $2e^x + 3e^{2x} + 5 \sin 2x$ is to be the solution of a linear homogeneous differential equation with constant coefficients, the roots of the characteristic polynomial would have to be 1 (for the e^x), 2 (for the e^{2x}) and $\pm 2i$ (for the $\sin 2x$). Thus, the characteristic polynomial would be $(r - 1)(r - 2)(r^2 + 4)$ and

$$(D - 1)(D - 2)(D^2 + 4)(2e^x + 3e^{2x} + 5 \sin 2x) = 0.$$

Now multiply both sides of the original equation (11) by $(D - 1)(D - 2) \times (D^2 + 4)$ to get

$$(D - 1)(D - 2)(D^2 + 4)(D^2 - D)y = 0.$$

The characteristic polynomial of this equation is

$$(r - 1)(r - 2)(r^2 + 4)(r^2 - r) = (r - 1)^2 (r - 2)(r^2 + 4)r$$

with roots $1, 1, 2, \pm 2i, 0$. Thus

$$y = c_1 e^x + c_2 xe^x + c_3 e^{2x} + c_4 \cos 2x + c_5 \sin 2x + c_6.$$

Since $c_1 e^x + c_6$ is the general solution of the associated homogeneous equation of the original problem (11), the remaining terms give a form for y_p:

$$y_p = c_2 xe^x + c_3 e^{2x} + c_4 \cos 2x + c_5 \sin 2x. \blacksquare$$

Exercises

Each of these differential equations is in the form of $L_1 y = f$, where L_1 has constant coefficients.

i) Find a constant-coefficient differential operator L_2 so that $L_2 f = 0$.

ii) Then find the form for a particular solution of $L_1 y = f$ by removing terms from the solution of $L_1 y = 0$ from the solution of $L_2 L_1 y = 0$.

19. $y'' + y = \sin x$
20. $y'' + 4y' + 3y = 2e^{-x} + 3e^{3x}$
21. $y'' - y = x \sin x + 3 \cos x$
22. $y'' - 2y' + y = x^2 e^x$
23. $y'' + 2y' + 5y = 3e^{-x} \cos 2x$
24. $y'' + 4y = \sin x + \cos 2x$
25. $y'' + y = x^2 + 3e^{-x}$
26. $y'' - y = xe^x - e^{-x} \cos x$

Exercises 13 through 49 of Section 3.12 and Exercises 1 through 18 of Section 3.13 may also be worked with this approach.

3.14 Variation of Parameters (Second-Order)

In Sections 3.12 and 3.13 the method of undetermined coefficients was used to find a particular solution of

$$a(x)y''(x) + b(x)y'(x) + c(x)y(x) = f(x). \tag{1}$$

There were two major restrictions on the method of undetermined coefficients. First, a, b, c had to be constants. Secondly, f had to be in a special form. This section will present a method for finding a particular solution of $ay'' + by' + cy = f$ provided we have first solved the associated homogeneous equation:

$$ay'' + by' + cy = 0.$$

This new method will not require a, b, c to be constants nor f to be in a special form. (One method for solving $ay'' + by' + cy = 0$ with nonconstant a, b, c was given in Section 3.11.)

Suppose that we have solved $ay'' + by' + cy = 0$ and have a fundamental set of solutions $\{y_1, y_2\}$. In Section 3.7, we showed how to use one solution y_1

of the associated homogeneous equation $ay'' + by' + cy = 0$ to reduce $ay'' + by' + cy = f$ to a first-order equation by the substitution

$$y = vy_1.$$

This was the method of reduction of order. Since we now have two solutions y_1, y_2 of $ay'' + by' + cy = 0$, it should be possible to reduce the equation $ay'' + by' + cy = f$ twice, in which case we would have the solution. Fortunately, this double reduction of order takes a simple form so that we never have to actually perform the substitution and differentiation.

We shall first derive the method and then work several examples. We begin by looking for functions v_1, v_2 so that

$$y = v_1 y_1 + v_2 y_2 \tag{2}$$

is a particular solution of $ay'' + by' + cy = f$. Substituting (2) into $ay'' + by' + cy = f$, gives

$$a(v_1 y_1 + v_2 y_2)'' + b(v_1 y_1 + v_2 y_2)' + c(v_1 y_1 + v_2 y_2) = f$$

or

$$a(v_1'' y_1 + 2v_1' y_1' + v_1 y_1'' + v_2'' y_2 + 2v_2' y_2' + v_2 y_2'')$$
$$+ b(v_1' y_1 + v_1 y_1' + v_2' y_2 + v_2 y_2') + c(\underline{v_1 y_1} + \underline{v_2 y_2}) = f.$$

Since y_1, y_2 are solutions of $ay'' + by' + cy = 0$, the underlined terms cancel, leaving

$$a(v_1'' y_1 + 2v_1' y_1' + v_2'' y_2 + 2v_2' y_2') + b(v_1' y_1 + v_2' y_2) = f. \tag{3}$$

If v_1, v_2 are chosen so that

$$v_1' y_1 + v_2' y_2 = 0, \tag{4}$$

then we also have $0 = (v_1' y_1 + v_2' y_2)' = v_1'' y_1 + v_1' y_1' + v_2'' y_2 + v_2' y_2'$. Thus, assuming $v_1' y_1 + v_2' y_2 = 0$, (3) reduces to

$$a(v_1' y_1' + v_2' y_2') = f. \tag{5}$$

If v_1', v_2' satisfy the algebraic equations (4) and (5), then $y = v_1 y_1 + v_2 y_2$ is a particular solution of the original differential equation (1). The preceding discussion is summarized in the following procedure:

Summary of Variation of Parameters

Variation of Parameters (Second-Order) is a method of calculating a particular solution of $ay'' + by' + cy = f$ given a fundamental set of solutions $\{y_1, y_2\}$ of the associated homogeneous equation $ay'' + by' + cy = 0$. The method is as follows:

1. Find a fundamental set of solutions $\{y_1, y_2\}$ of $ay'' + by' + cy = 0$.
2. Solve the algebraic system of equations

$$v_1' y_1 + v_2' y_2 = 0,$$
$$v_1' y_1' + v_2' y_2' = \frac{f}{a}, \tag{6}$$

for the functions v_1', v_2'.

3. Antidifferentiate to find v_1, v_2.
4. Then $y = v_1 y_1 + v_2 y_2$ is a particular solution of $ay'' + by' + cy = f$.
5. $y = v_1 y_1 + v_2 y_2 + c_1 y_1 + c_2 y_2$ is the general solution of $ay'' + by' + cy = f$.

Note If arbitrary constants are introduced in step 3 when finding v_1, v_2, then $y = v_1 y_1 + v_2 y_2$ will be the general solution.

Note The coefficients of v_1', v_2' in (6) are the entries of the matrix

$$\begin{bmatrix} y_1 & y_2 \\ y_1' & y_2' \end{bmatrix},$$

whose determinant is the Wronskian (see Section 3.5). Since the Wronskian of a fundamental set of solutions is never zero, it follows from matrix theory (Cramer's rule) that (6) can always be solved for v_1', v_2'. Thus the only difficulty is in Steps 1 and 3.

Example 3.14.1 Find the general solution of

$$2y'' - 4y' + 2y = x^{-1} e^x, \qquad x > 0. \tag{7}$$

Solution Note that $x^{-1} e^x$ is not the kind of forcing term to which the method of undetermined coefficients can be applied. However, the differential equation has constant coefficients so that we know how to solve the associated homogeneous equation. We shall solve (7) by the method of variation of parameters.

Step 1
We must solve $2y'' - 4y' + 2y = 0$. The characteristic polynomial is $2r^2 - 4r + 2 = 2(r-1)^2$, which has roots 1, 1. Thus $\{e^x, xe^x\}$ is a fundamental set of solutions. Let $y_1 = e^x$, $y_2 = xe^x$.

Step 2
The equations (6) are

$$v_1' e^x + v_2' xe^x = 0,$$
$$v_1' (e^x)' + v_2' (xe^x)' = \frac{x^{-1} e^x}{2}. \tag{8}$$

(*Note:* $a = 2$) That is,

$$v_1' e^x + v_2' x e^x = 0,$$
$$v_1' e^x + v_2'(xe^x + e^x) = \frac{x^{-1}e^x}{2}.$$

For students familiar with solving systems of equations, there are several options available, such as using augmented matrices (Sections 6.8, 6.9) or Cramer's rule (see formula (a) at the end of this section). We shall just solve for v_1', v_2', using basic algebra.

Divide both equations by e^x to simplify:

$$v_1' + v_2' x = 0$$
$$v_1' + v_2'(x + 1) = \frac{x^{-1}}{2}.$$

Solve the first equation for v_1',

$$v_1' = -v_2' x,$$

and substitute into the second equation,

$$-v_2' x + v_2'(x + 1) = \frac{x^{-1}}{2},$$

so that

$$v_2' = \frac{1}{2x},$$

and

$$v_1' = -v_2' x = -\frac{1}{2}.$$

Step 3

$$v_2 = \int v_2' \, dx = \int \frac{1}{2x} dx = \frac{\ln x}{2},$$
$$v_1 = \int v_1' \, dx = \int -\frac{1}{2} dx = -\frac{x}{2}.$$

Step 4

$$y_p = v_1 y_1 + v_2 y_2 = -\frac{x}{2} e^x + \frac{\ln x}{2} x e^x$$

is a particular solution of (7).

Step 5

$$y = v_1 y_1 + v_2 y_2 + c_1 y_1 + c_2 y_2$$
$$= \frac{-x}{2} e^x + \frac{\ln x}{2} xe^x + c_1 e^x + c_2 xe^x$$
$$= \frac{xe^x \ln x}{2} + c_1 e^x + \tilde{c}_2 xe^x \quad \left(\tilde{c}_2 = c_2 - \frac{1}{2} \right)$$

is the general solution of $2y'' - 4y' + 2y = x^{-1}e^x$. ∎

Example 3.14.2 $\{x, x^3\}$ is a fundamental set of solutions of $x^2 y'' - 3xy' + 3y = 0$. Find the general solution of

$$x^2 y'' - 3xy' + 3y = 4x^7.$$

Solution Step 1 has been performed for us. (The technique is in Section 3.11.) Thus we must solve (6)

$$v_1' x + v_2' x^3 = 0,$$
$$v_1'(x)' + v_2'(x^3)' = \frac{4x^7}{x^2},$$

or

$$v_1' x + v_2' x^3 = 0,$$
$$v_1' + v_2' 3x^2 = 4x^5.$$

Solve the first equation for v_1'; $v_1' = -v_2' x^2$, and substitute into the second equation:

$$-v_2' x^2 + v_2' 3x^2 = 4x^5,$$

and solve for v_2',

$$v_2' = 2x^3.$$

Then

$$v_1' = -v_2' x^2 = -2x^5.$$

Thus antidifferentiating v_1', v_2' gives

$$v_2 = \frac{x^4}{2}, \qquad v_1 = -\frac{x^6}{3}.$$

A particular solution is

$$y_p = v_1 y_1 + v_2 y_2 = \left(-\frac{x^6}{3} \right) x + \left(\frac{x^4}{2} \right) x^3 = \frac{x^7}{6},$$

and the general solution is

$$y = \frac{x^7}{6} + c_1 x + c_2 x^3. \quad \blacksquare$$

Two comments are in order. First, a fact from algebra called Cramer's rule can be applied to the system of equations (6) to give formulas for v_1', v_2'. They are

$$v_1' = \frac{\det\begin{bmatrix} 0 & y_2 \\ f/a & y_2' \end{bmatrix}}{\det\begin{bmatrix} y_1 & y_2 \\ y_1' & y_2' \end{bmatrix}} \quad \text{and} \quad v_2' = \frac{\det\begin{bmatrix} y_1 & 0 \\ y_1' & f/a \end{bmatrix}}{\det\begin{bmatrix} y_1 & y_2 \\ y_1' & y_2' \end{bmatrix}}.$$

Evaluating the determinants on top gives

$$v_1' = -\frac{fy_2}{aW[y_1, y_2]}, \quad v_2' = \frac{fy_1}{aW[y_1, y_2]}, \tag{9}$$

where $W[y_1, y_2]$ is the Wronskian of y_1, y_2. In practice it is probably quicker to use (9) to find v_1', v_2'. For example, if (9) is applied to Example 3.14.1 we have $y_1 = e^x$, $y_2 = xe^x$,

$$W[e^x, xe^x] = \det\begin{bmatrix} e^x & xe^x \\ e^x & e^x + xe^x \end{bmatrix} = e^{2x},$$

and $f = x^{-1}e^x$, $a = 2$. Thus (9) gives

$$v_1' = -\frac{[x^{-1}e^x] \cdot [xe^x]}{2 \cdot [e^{2x}]} = -\frac{1}{2}$$

and

$$v_2' = \frac{[x^{-1}e^x]e^x}{2[e^{2x}]} = \frac{x^{-1}}{2}.$$

Secondly, even if we cannot antidifferentiate v_1', v_2' explicitly, the antidifferentiation can usually be done numerically and an approximate solution obtained.

Exercises

In Exercises 1 through 14, find the general solution by the method of variation of parameters. Decide whether the method of undetermined coefficients could also have been used.

1. $y'' - y = e^{2x}$
2. $y'' - 4y' + 4y = e^{2x}$
3. $y'' + y = \dfrac{1}{\sin x}$
4. $y'' + 4y = \dfrac{4}{\cos 2x}$
5. $y'' + y = \tan x$
6. $y'' + 2y' + y = \dfrac{e^{-x}}{1 + x^2}$
7. $y'' + 3y' + 2y = \dfrac{1}{1 + e^{2x}}$

8. $4y'' - y = x$
9. $y'' - 6y' + 9y = e^{3x}x^{3/2}$, $x > 0$
10. $y'' - 4y' + 3y = e^{-x}$
11. $4y'' + 4y' + y = x^{-2}e^{-x/2}$
12. $y'' + 5y' + 4y = e^x$
13. $y'' - y' - 6y = e^{-2x}$
14. $y'' - 3y' + 2y = e^{3x}\cos(e^x)$

In Exercises 15 through 18, a fundamental set of solutions is given for the associated homogeneous equation for $x > 0$. Solve the differential equation, using variation of parameters, and give the general solution. In each case the fundamental set of solutions could be found by the method in Section 3.11 for Euler's equation.

15. $x^2 y'' - 2xy' + 2y = x^3$, $\{x, x^2\}$
16. $x^2 y'' + xy' - y = x^{1/2}$, $\{x, x^{-1}\}$
17. $x^2 y'' + 2xy' = x^{-1}$, $\{1, x^{-1}\}$
18. $x^2 y'' - 3xy' + 3y = x$, $\{x, x^3\}$

The remaining exercises develop some additional aspects of variation of parameters. They are all based on the following assumptions: Assume that $\{y_1, y_2\}$ is a fundamental set of solutions of $ay'' + by' + cy = 0$ on the interval $I = [0, T]$. We also assume that a, b, c, f are continuous on I and a is never zero on I.

19. Show that a particular solution of $ay'' + by' + cy = f$ given by variation of parameters may be written as:

$$y_p(x) = \int_0^x \frac{y_2(x)y_1(s) - y_1(x)y_2(s)}{a(s)W[y_1, y_2](s)} f(s)\, ds. \quad (10)$$

20. Show that (10) gives the unique solution of the initial-value problem $ay'' + by' + cy = f$, $y(0) = 0$, $y'(0) = 0$.

In Exercises 21 through 24, use (10) to derive an integral formula for a particular solution of the differential equation for $x \geq 0$.

21. $y'' - y = e^{-x^2}$
22. $y'' - 4y = \sin(x^{1/3})$
23. $y'' + 5y' + 6y = \dfrac{1}{x+1}$
24. $y'' - 3y' + 2y = x^{1/5}$

If $G(x, s)$ is a continuous function of x and s for $0 \leq x \leq T$, $0 \leq s \leq T$, then G can be used to define an integral operator K by

$$(Kf)(x) = \int_0^x G(x, s)f(s)\, ds. \quad (11)$$

25. Show that if $G(x, s) = 1$, then Kf is an antiderivative of f. Let D stand for differentiation with respect to x.

  i) Show that if f is continuous on I, then $D(Kf) = f$.

  ii) show that if f is continuously differentiable, then $K(Df) = f - f(0)$.

The preceding example can be viewed as saying that the operator of antidifferentiation defined by (11) with $G = 1$ acts like an inverse to the operator of differentiation. The next two examples show that this is in fact a general property of variation of parameters. The remaining two exercises require Section 3.6. Let $L = aD^2 + bD + c$ and

$$G(x, s) = \frac{y_2(x)y_1(s) - y_1(x)y_2(s)}{a(s)W[y_1, y_2](s)} \quad (12)$$

where $\{y_1, y_2\}$ is a fundamental set of solutions of $L(y) = 0$. Let K be defined by (11) with G given by (12).

26. Show that if f is continuous on I, then $L(Kf) = f$.
27. Show that if f is twice continuously differentiable, then

$$K(Lf) = f - c_1 y_1 - c_2 y_2,$$

where

$$c_1 = \frac{W[f, y_2](0)}{W[y_1, y_2](0)},$$

$$c_2 = \frac{W[y_1, f](0)}{W[y_1, y_2](0)}.$$

3.15 Variation of Parameters (nth-Order)

The method of variation of parameters given in Section 3.14 may also be used for nth-order linear differential equations.

Theorem 3.15.1 Suppose that $\{y_1, \ldots, y_n\}$ is a fundamental set of solutions of the associated homogeneous equation for

$$a_n(x)y^{(n)}(x) + a_{n-1}(x)y^{(n-1)}(x) + \cdots + a_1(x)y'(x) + a_0(x)y(x) = f(x). \quad (1)$$

If $v_1'(x), \ldots, v_n'(x)$ satisfy the system of equations

$$\begin{aligned} v_1'(x)y_1(x) + v_2'(x)y_2(x) + \cdots + v_n'(x)y_n(x) &= 0 \\ v_1'(x)y_1'(x) + v_2'(x)y_2'(x) + \cdots + v_n'(x)y_n'(x) &= 0 \\ \vdots \qquad \vdots \qquad \vdots \qquad \vdots \qquad &\vdots \\ v_1'(x)y_1^{(n-1)}(x) + v_2'(x)y_2^{(n-1)}(x) + \cdots + v_n'(x)y_n^{(n-1)}(x) &= \frac{f(x)}{a_n(x)}, \end{aligned} \quad (2)$$

then $y_p = v_1 y_1 + v_2 y_2 + \cdots + v_n y_n$ is a particular solution of (1).

In general, solving (2) for $n > 2$ can become quite complicated. For small n, say 3 or 4, the system (2) can be solved by Cramer's rule. Cramer's rule and ways to evaluate determinants are covered in Section 6.8. For larger n, computer programs doing symbolic manipulations can be used.

Cramer's rule applied to (2) takes the form

$$v_1' = \frac{\det \begin{bmatrix} 0 & y_2 & \cdots & y_n \\ \vdots & \vdots & & \vdots \\ 0 & y_2^{(n-2)} & & y_n^{(n-2)} \\ f/a_n & y_2^{(n-1)} & & y_n^{(n-1)} \end{bmatrix}}{W[y_1, \ldots, y_n]}$$

$$= (-1)^{n+1} f \frac{\det \begin{bmatrix} y_2 & \cdots & y_n \\ \vdots & & \vdots \\ y_2^{(n-2)} & & y_n^{(n-2)} \end{bmatrix}}{a_n W[y_1, \ldots, y_n]}.$$

and

$$\begin{aligned} v_i' &= (-1)^{n+i} \frac{f}{a_n} \frac{\det \begin{bmatrix} y_1 & \cdots & y_{i-1} & y_{i+1} & \cdots & y_n \\ \vdots & & \vdots & \vdots & & \vdots \\ y_1^{(n-2)} & & y_{i-1}^{(n-2)} & y_{i+1}^{(n-2)} & \cdots & y_n^{(n-2)} \end{bmatrix}}{W[y_1, \ldots, y_n]} \\ &= (-1)^{n+i} \frac{f}{a_n} \frac{W_i}{W[y_1, \ldots, y_n]}, \end{aligned} \quad (3)$$

where W_i is the Wronskian of the $(n-1)$ functions obtained by deleting y_i from the set $\{y_1, \ldots, y_n\}$.

Example 3.15.1   Solve

$$y''' - y'' = e^x, \tag{4}$$

using the method of variation of parameters.

Solution   The differential equation (4) has constant coefficients and has characteristic polynomial $r^3 - r^2 = r^2(r-1)$ with roots of 0, 0, 1. Thus $\{1, x, e^x\}$ is a fundamental set of solutions of the associated homogeneous equation $y''' - y'' = 0$. Let $y_1 = 1, y_2 = x, y_3 = e^x$. Then $a_3 = 1, n = 3$, and $f = e^x$. The equations (2) are

$$\begin{aligned} v_1' \cdot 1 + v_2' x + v_3' e^x &= 0, \\ v_1' \cdot 0 + v_2' \cdot 1 + v_3' e^x &= 0, \\ v_1' \cdot 0 + v_2' \cdot 0 + v_3' e^x &= e^x. \end{aligned} \tag{5}$$

This particular example may be easily solved to yield

$$v_3' = 1, \qquad v_2' = -e^x, \qquad v_1' = xe^x - e^x \tag{6}$$

or, upon antidifferentiation,

$$v_3 = x, \qquad v_2 = -e^x, \qquad v_1 = xe^x - 2e^x.$$

Thus

$$y_p = v_1 y_1 + v_2 y_2 + v_3 y_3 = (xe^x - 2e^x)1 + (-e^x)x + x(e^x) = xe^x - 2e^x.$$

The general solution is

$$y = y_p + y_h = xe^x - 2e^x + c_1 + c_2 x + c_3 e^x = xe^x + c_1 + c_2 x + \tilde{c}_3 e^x.$$

Suppose, however, instead of solving (5) directly we had used Cramer's rule (3). Then

$$W[1, x, e^x] = \det \begin{bmatrix} 1 & x & e^x \\ 0 & 1 & e^x \\ 0 & 0 & e^x \end{bmatrix} = e^x$$

and

$$v_1' = (-1)^{3+1} \frac{e^x W[x, e^x]}{W[1, x, e^x]} = e^x \frac{\det \begin{bmatrix} x & e^x \\ 1 & e^x \end{bmatrix}}{e^x} = xe^x - e^x,$$

$$v_2' = (-1)^{3+2} \frac{e^x W[1, e^x]}{W[1, x, e^x]} = -e^x \frac{\det \begin{bmatrix} 1 & e^x \\ 0 & e^x \end{bmatrix}}{e^x} = -e^x,$$

$$v_3' = (-1)^{3+3} \frac{e^x W[1, x]}{W[1, x, e^x]} = e^x \frac{\begin{bmatrix} 1 & x \\ 0 & 1 \end{bmatrix}}{e^x} = 1,$$

which agrees with (6). ∎

Exercises

In Exercises 1 through 4, solve the differential equation by the method of variation of parameters and give the general solution.

1. $y''' - y' = e^{2x}$
2. $y''' + y' = \dfrac{1}{\sin x}$
3. $y'''' - y''' = x$
4. $y''' - 6y'' + 11y' - 6y = e^x$
5. Verify that
$$W[e^{ax}, e^{bx}, e^{cx}] = e^{(a+b+c)x}(b-a)(c-a)(c-b).$$

In Exercises 6 through 12, use Exercise 5 and variation of parameters to find the general solution.

6. $y''' - y'' - 4y' + 4y = e^{-x}$
7. $y''' - 2y'' - y' + 2y = e^x$
8. $y''' + y'' - 4y' - 4y = x$
9. $y''' + 2y'' - y' - 2y = 1$
10. $y''' - 3y'' - y' + 3y = \sin x$
11. $y''' + 3y'' - y' - 3y = e^x$
12. $y''' - 3y'' + 2y' = e^{-x}$
13. Verify that
$$W[e^{ax}, xe^{ax}, x^2 e^{ax}] = 2e^{3ax}.$$

In Exercises 14 through 17, use Exercise 13 and variation of parameters to find the general solution.

14. $y''' - 3y'' + 3y' - y = x^{1/2} e^x$
15. $y''' + 3y'' + 3y' + y = x^{-3} e^{-x}$
16. $y''' + 6y'' + 12y' + 8y = x^{-1} e^{-2x}$
17. $y''' - 6y'' + 12y' - 8y = x^{7/2} e^{2x}$

In Exercises 18 through 21, you are given a fundamental set of solutions for the associated homogeneous equation and f and a_n. Solve the differential equation using variation of parameters.

18. $\{1, x, x^2, x^3\}$, $f = x$, $a_4 = x$
19. $\{e^x, xe^x, x^2 e^x\}$, $f = e^x$, $a_3 = 2$
20. $\{1, x^2, x^3\}$, $f = x^{1/2}$, $a_3 = 1$
21. $\{1, x, x^{1/2}\}$, $f = x^3$, $a_3 = x^2$.

3.16 Mechanical Vibrations

3.16.1 Formulation of Equations

This section, like Sections 3.3 and 2.13, will study a problem involving the linear motion of a rigid body governed by *Newton's law*,

$$\frac{d(mv)}{dt} = F_T, \tag{1}$$

where m is mass, v is velocity, mv is momentum, t is time, and F_T is the total force acting on the body. The resulting differential equations will be analyzed using the preceding sections on linear constant-coefficient differential equations. We shall again use the centimeter-gram-second (cgs) and foot-pound-second (fps) systems of measurement given in Table 2.13.1.

The problem to be studied is that of a constant mass suspended on a spring moving vertically under the influence of external forces. The problem is actually quite general since many mechanical systems and structures may be considered to be made up of point masses connected with "springs."

Suppose, then, we have a mass m hanging from a spring. We shall assume that the total force F_T is made up of four forces.

$F_g = $ The force of gravity,

$F_r = $ A friction or resistive force acting on the mass (air resistance, for example)

$F_s = $ Force exerted by the spring,

$F_e = $ Any other external forces (such as wind, magnetism, etc.)

Then Newton's law (1) takes the form:

$$\frac{d(mv)}{dt} = F_T = F_g + F_s + F_r + F_e. \tag{2}$$

Figure 3.16.1

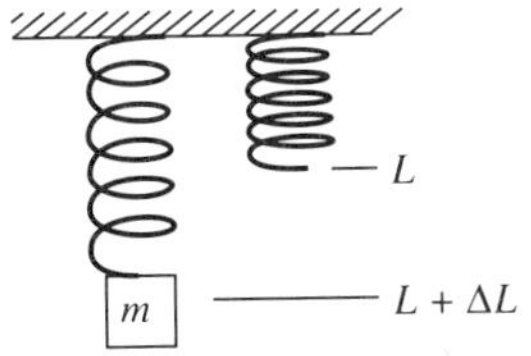

Let L be the length of the spring with no mass attached, and $L + \Delta L$ the length of the spring with the mass attached but at rest. (See Figure 3.16.1.) This rest position is the *equilibrium* position of the spring-mass system.

Let $x(t)$ be the distance of the mass from the equilibrium position at time t, with the downward direction being positive. Thus we have Fig. 3.16.2(a) if $x > 0$ and Fig. 3.16.2(b) if $x < 0$.

We can now express the different forces in (2) in terms of x. To begin with, assume that the spring satisfies *Hooke's law*,

The force exerted by the spring is proportional to the distance the spring is stretched (or compressed).

Figure 3.16.2

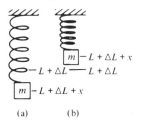

This is an assumption, not only on the type of spring but also on the mass, external forces, and initial conditions. In particular, we are assuming that the resulting motion is not of too great an amplitude, so that linear equations may be used. Since the spring length is changed by $\Delta L + x$, we have

$$F_s = -k(\Delta L + x), \tag{3}$$

where k is a positive constant; k is positive since the spring force acts opposite to the sign of $x + \Delta L$. That is, the spring is extended if $x + \Delta L > 0$ and is compressed if $x + \Delta L < 0$. We also assume that we may neglect the mass of the spring itself. The force of gravity acting on the mass (the weight) is

$$F_g = mg, \qquad (4)$$

where $g = 32$ ft/sec^2 (fps) or 980 cm/sec^2 (cgs).

If the mass is at rest ($x = 0$), then the force of gravity must balance out the spring force, so that $mg = k\,\Delta L$, or

$$mg - k\,\Delta L = 0. \qquad (5)$$

Many types of resistive forces at low velocities are (approximately) proportional to the velocity. That is,

$$F_r = -\delta v = -\delta \frac{dx}{dt} \quad \text{with } \delta > 0. \qquad (6)$$

The constant δ is called the *damping constant*. The constant of proportionality, $-\delta$, is negative since the force of friction acts in the opposite direction to the motion. Note that (6), like Hooke's law, represents restriction on the motion to be studied and not just an assumption on the type of spring and the type of resistance. Finally, we assume that the external forces F_e depend only on time t and not on x or x'.

Then substituting (3), (4), and (6) into (2), we get the linear differential equation

$$m\frac{d^2x}{dt^2} = mg - k(\Delta L + x) - \delta \frac{dx}{dt} + f(t),$$

where $f(t) = F_e(t)$. But $mg - k\,\Delta L = 0$ from (5), and we have finally

$$m\frac{d^2x}{dt^2} + \delta \frac{dx}{dt} + kx = f(t). \qquad (7)$$

Note that m, δ, and k are all nonnegative constants. The initial conditions usually used with (7) are the initial position $x(0)$ and the initial velocity $v(0) = x'(0)$.

Sometimes it is convenient to think of the resistive force as being due to an attached device such as a piston moving in a fluid. One such device is called a *dashpot*. Conceptually, this approach is similar to modeling the resistance in a wire by the inclusion of a small resistor in the circuit equations instead of thinking of the resistance as distributed throughout the wire.

3.16.2 Free Response

Suppose that there is no external force other than gravity, so that the external force is $f(t) = 0$. The solutions of the resulting homogeneous differential equation

$$m\frac{d^2x}{dt^2} + \delta \frac{dx}{dt} + kx = 0 \qquad (8)$$

are called the *free* (or *natural*) *response* of the spring-mass system. This is how the system reacts for given initial conditions if it is allowed to proceed without external interference. Note that the free response is the same thing as the solution of the associated homogeneous equation. The resulting behavior is quite different, depending on whether or not there is any damping present.

I. Simple Harmonic Motion (No damping, $\delta = 0$)

Suppose that there is no damping ($\delta = 0$), so that the free response is given by the differential equation

$$m\frac{d^2x}{dt^2} + kx = 0. \tag{9}$$

We can solve (9) using the techniques of Section 3.8. The characteristic equation is $mr^2 + k = 0$, with roots $\pm i\sqrt{k/m}$, since $m > 0, k > 0$. Thus

$$x = A \cos \sqrt{\frac{k}{m}} t + B \sin \sqrt{\frac{k}{m}} t. \tag{10}$$

Using the trigonometric identity $\sin(u + v) = \sin u \cos v + \cos u \sin v$, it is possible to show that

$$A \cos \omega t + B \sin \omega t = \sqrt{A^2 + B^2} \sin(\omega t + \phi), \tag{11a}$$

where

$$\sin \phi = \frac{A}{\sqrt{A^2 + B^2}}, \qquad \cos \phi = \frac{B}{\sqrt{A^2 + B^2}}. \tag{11b}$$

The variable ϕ is sometimes called the *phase angle*, and $\sqrt{A^2 + B^2}$ the *amplitude*. Formula (11) shows that the resulting free response, in the absence of friction, is simple harmonic motion (given by a sine function). The *period* of (11a) is $2\pi/\omega$ and the *frequency* is $\omega/2\pi$.

Example 3.16.1

The spring constant of a 24-inch steel spring is measured by hanging a 1-slug mass from the spring and observing that the spring stretches 3 inches. Now a $\frac{1}{2}$-slug mass is attached to the spring. The mass is pulled down 3 inches and released with a velocity of 1 ft/sec downward. Determine the resulting motion.

Solution

First we need to determine the spring constant. From (5),

$$k \Delta L = mg.$$

In fps units this is

$$k\left(\frac{1}{4}\text{ft}\right) = 1\frac{\text{lb sec}^2}{\text{ft}} \cdot \left(32\frac{\text{ft}}{\text{sec}^2}\right),$$

or $k = 128$ lb/ft. The equation of motion is (9),

$$\frac{1}{2}\frac{d^2x}{dt^2} + 128x = 0,$$

or $d^2x/dt^2 + 256x = 0$. The characteristic equation is $r^2 + 256 = 0$, which has roots $r = \pm 16i$. Thus the general solution of the differential equation is

$$x = A\cos 16t + B\sin 16t. \tag{12}$$

The initial conditions are

$$x(0) = \frac{1}{4}\text{ft}, \qquad x'(0) = 1\frac{\text{ft}}{\text{sec}}.$$

Thus

$$\frac{1}{4} = x(0) = A, \qquad 1 = x'(0) = 16B,$$

and the resulting motion is

$$x = \frac{1}{4}\cos 16t + \frac{1}{16}\sin 16t. \tag{13}$$

In order to more easily visualize this motion, we shall rewrite the solution in the form (11). The amplitude is

$$\text{Amplitude} = \sqrt{A^2 + B^2} = \sqrt{\left(\frac{1}{4}\right)^2 + \left(\frac{1}{16}\right)^2} = \frac{\sqrt{17}}{16}$$

and

$$\sin\phi = \frac{1/4}{\sqrt{17/16}} = \frac{4}{\sqrt{17}},$$

so that

$$\phi = \sin^{-1}\left(\frac{4}{\sqrt{17}}\right) \approx 1.326 \text{ radians}.$$

Thus the motion (13) may be rewritten as

$$x = \frac{\sqrt{17}}{16}\sin(16t + 1.326).$$

The period is $2\pi/16$ sec and the frequency is $(16/2\pi)/\text{sec}$ (see Fig. 3.16.3). ■

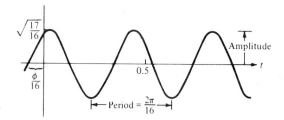

Figure 3.16.3
Graph of
$x = \frac{\sqrt{17}}{16}\sin(16t + 1.326)$.

Note When finding ϕ, both equations in (11b) must hold. If A, B are not both positive, then ϕ may not be given by the principal branch of the inverse sine. One can determine ϕ from the fact that ϕ is the angle the vector (B, A) makes with the positive real axis. Sometimes it is convenient to use $\tan \phi = A/B$.

Exercises

Free Response

Exercises 1 through 14 we assume that Hooke's law is applicable, resistance is negligible, and the mass of the spring is negligible.

1. A mass of 30 g is attached to a spring. At equilibrium the spring has stretched 20 cm. The spring is pulled down another 10 cm and released. Set up the differential equation for the motion, and solve it to determine the resulting motion, ignoring friction.

2. A mass of 400 g is attached to a spring. At equilibrium the spring has stretched 245 cm. The spring is pulled down and released. At noon we observe that the mass is 10 cm below equilibrium and traveling upward at $-\sqrt{84}$ cm/sec. Set up the differential equation for the motion. Solve the differential equation and express the solution in the phase-amplitude form (11).

3. A mass of 8 slugs is attached to a long spring. The spring stretches two feet before coming to rest. The 8-slug mass is removed and a 2-slug mass is attached and placed at equilibrium. The mass is pushed down and released. At the time it is released, the mass is 2 ft below equilibrium and traveling downward at 1 ft/sec. Derive the differential equation for the motion of the spring-mass system. Solve the differential equation.

4. A spring-mass system has a spring contant of $k = 5$ g/sec^2. What mass should be attached to make the resulting motion have a frequency of 30 cycles/sec?

5. A spring is to be attached to a mass of 10 slugs. What should the spring constant be to make the resulting motion have a frequency of 5 cycles/sec?

6. A mass of 6 g is attached to a spring-mass system with spring constant 30 g/sec^2. What should the initial conditions be to give a response with amplitude 3 and phase angle $\pi/4$?

7. A mass of 16 g is attached to a spring-mass system with spring constant 64 g/sec^2. What should the initial conditions be to give a response with amplitude 2 and phase angle $\pi/3$?

8. Suppose you have a spring-mass system as in Fig. 3.16.2.
   a) What is the effect on the period and frequency of doubling the mass?
   b) What is the effect on the period and frequency of doubling the spring constant?
   c) What is the effect on the period and frequency of doubling both the spring constant and the mass?

9. A spring-mass system with mass m and spring constant k is subjected to a sudden impulse. The result is that, at time $t = 0$, the mass is at the equilibrium position but has a velocity of 10 cm/sec downward.
   a) Determine the subsequent motion.
   b) Determine the amplitude of the resulting motion as a function of m and k.
   c) What is the effect on the amplitude of increasing k?
   d) What is the effect on the amplitude of increasing m?

10. (Same situation as Exercise 9). If the mass is 50 g, what should the spring constant be to give an amplitude of 20 cm?

11. At time $t = 0$ the mass in a spring-mass system with mass m and spring constant k is observed to be 1 ft below equilibrium and traveling downward at 1 ft/sec.
   a) Determine the subsequent motion.
   b) Determine the amplitude as a function of m and k.
   c) What is the effect on the amplitude of increasing m or k?
   d) Determine the phase angle ϕ as a function of m and k.
   e) What is the effect on the phase angle of increasing m or k?

12. (Same situation as Exercise 11.) If the spring constant is $k = 8$ ft/sec^2, what should the mass be to give an amplitude of 4 ft? At what time will this amplitude be first attained?

13. (*Conservation of mechanical energy*) Suppose that a spring-mass system is modeled by $mx'' + kx = 0$. The quantities $\frac{1}{2}mv^2$ and $\frac{1}{2}kx^2$ are the *kinetic energy* and the *elastic potential energy*. Their sum is the total mechanical energy. Multiply $mx'' + kx = 0$ by x' and integrate to show that $\frac{1}{2}mv^2 + \frac{1}{2}kx^2 = $ Constant.

14. Let $E = \frac{1}{2}mv^2 + \frac{1}{2}kx^2 = \frac{1}{2}m(x')^2 + \frac{1}{2}kx^2$. Show that if x is a solution of $mx'' + \delta x' + kx = 0$ with $\delta > 0$, then $dE/dt < 0$. Thus E is monotonically decreasing so that mechanical energy decreases in the presence of resistance.

II. Free Response with Friction $(\delta > 0)$

Suppose now that friction is not negligible so that the dynamics of the spring-mass system in Fig. 3.16.1 are described by the differential equation

$$m\frac{d^2x}{dt^2} + \delta\frac{dx}{dt} + kx = 0. \tag{14}$$

The characteristic equation of (14) is $mr^2 + \delta r + k = 0$ and the roots of this equation are

$$r = -\frac{\delta}{2m} \pm \frac{\sqrt{\delta^2 - 4mk}}{2m}. \tag{15}$$

There are three cases to consider, depending on whether $\delta^2 - 4mk$ is negative, zero, or positive. Since we have just discussed the $\delta = 0$ (no friction) case, it is convenient to think of the mass m and spring constant k as fixed, and explain what happens as the friction coefficient δ increases.

Case 1: Moderate (Under) Damping $[0 < \delta < \sqrt{4mk}\,]$

In this case, $\delta^2 - 4mk < 0$, and there is a pair of complex conjugate roots

$$r = -\alpha \pm \beta i, \qquad \alpha = \frac{\delta}{2m}, \qquad \beta = \frac{\sqrt{4mk - \delta^2}}{2m}, \tag{16}$$

so that the solutions of the differential equation $mx'' + \delta x' + kx = 0$ are

$$x = e^{-\alpha t}(A\cos\beta t + B\sin\beta t), \tag{17}$$

or, using (11),

$$x = e^{-\alpha t}\sqrt{A^2 + B^2}\sin(\beta t + \phi). \tag{18}$$

This is a *damped oscillation*. For a given mass m and spring constant k, increas-

Figure 3.16.4
Graph of (18) with $\phi > 0$.

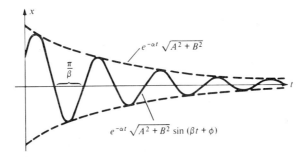

ing the resistance δ (but keeping $\delta < \sqrt{4mk}$) has two effects. First, the oscillation is damped faster; that is, the *time-varying amplitude* $e^{-\alpha t}\sqrt{A^2 + B^2}$ decreases faster. Second, the period of the oscillation being damped, $2\pi/\beta$, increases (Fig. 3.16.4).

Example 3.16.2   A mass of 0.5 slugs is suspended on a spring with spring constant of $k = 2$ lb/ft. The damping is δ lb-sec/ft. The mass is pulled down one foot and released. Describe and graph the motion for $\delta = 0.25$, 1, and 1.75.

Solution   The differential equation describing the motion is

$$\frac{1}{2}\frac{d^2 x}{dt^2} + \delta\frac{dx}{dt} + 2x = 0, \qquad x(0) = 1, \qquad x'(0) = 0. \tag{19}$$

The characteristic polynomial of (19) is $\frac{1}{2}r^2 + \delta r + 2 = 0$, which has roots $r = -\delta \pm i\sqrt{4 - \delta^2}$, so that the solution of the differential equation (19) is

$$x = e^{-\delta t}(A\cos(\sqrt{4 - \delta^2}\, t) + B\sin(\sqrt{4 - \delta^2}\, t)). \tag{20}$$

The initial conditions imply that

$$1 = x(0) = A,$$
$$0 = x'(0) = -\delta A + \sqrt{4 - \delta^2}\, B.$$

Thus $A = 1$, $B = \delta(4 - \delta^2)^{-1/2}$, and

$$x = e^{-\delta t}\left(\cos(\sqrt{4 - \delta^2}\, t) + \frac{\delta}{(4 - \delta^2)^{1/2}}\sin(\sqrt{4 - \delta^2}\, t)\right). \tag{21}$$

We shall rewrite (21) using (11). Now

$$1 + \left(\frac{\delta}{(4 - \delta^2)^{1/2}}\right)^2 = \frac{4}{4 - \delta^2}.$$

Thus (21) is

$$x = e^{-\delta t}\frac{2}{\sqrt{4 - \delta^2}}\sin(\sqrt{4 - \delta^2}\, t + \phi), \tag{22}$$

Figure 3.16.5
Graph of (22) for $\delta = 0.25$, 1, and 1.75.

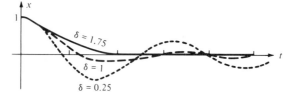

where $\phi = \sin^{-1}(\sqrt{4-\delta^2}/2)$. The graphs of (22) for $\delta = 0.25$, 1, and 1.75 are shown in Fig. 3.16.5.

The solution for $\delta = 1.75$ differs by no more than 0.061 from the solution of (19) when $\delta = 2$, which is $x = (1 + 2t)e^{-2t}$. When $\delta = 2$ there is no damped oscillation. This situation, known as *critical damping*, is studied next. ∎

Case 2: Critical Damping ($\delta = \sqrt{4mk}$)

In this case, $\delta^2 - 4mk = 0$ and (15) shows there is a repeated real root $r = -\delta/(2m)$. The general solution of the differential equation $mx'' + \delta x' + kx = 0$ is then

$$x = Ae^{-(\delta/2m)t} + Bte^{-(\delta/2m)t}. \tag{23}$$

Depending on the initial conditions, this solution for x takes one of the three forms shown in Fig. 3.16.6, or their negative (see Exercises 23, 24). Figure 3.16.6(a) results if the mass is initially moving away from equilibrium ($x = 0$). The mass slows down and then returns toward the equilibrium. If the mass is initially moving toward the equilibrium, it either approaches equilibrium (Fig. 3.16.6(b)) or, if the initial velocity is great enough, overshoots the equilibrium and then returns toward the equilibrium position as in Fig. 3.16.6(c).

Example 3.16.3

A device is being designed that can be modeled as a spring-mass system. The spring constant is $k = 10$ g/sec^2 and the damping constant is $\delta = 20$ g/sec.

a. Determine the mass so that the resulting spring-mass system will be critically damped.

b. The mass is pulled down 5 cm from the rest position and released with a downward velocity of 10 cm/sec. Determine and solve the equations of motion. Graph the resulting motion.

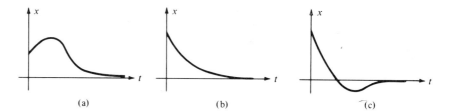

Figure 3.16.6
$x = Ae^{-t} + Bte^{-t}$

Solution

a. Critical damping occurs if $\delta^2 - 4mk = 0$. That is, $400 - 40m = 0$. Thus the desired mass is 10 g.

b. The equation of motion is

$$10x'' + 20x' + 10x = 0, \quad x(0) = 5, \quad x'(0) = 10. \tag{24}$$

The characteristic equation for (24) is $10r^2 + 20r + 10 = 0$, or $10(r+1)^2 = 0$, which has a repeated real root of -1. The general solution of $10x'' + 20x' + 10x = 0$ is thus

$$x = Ae^{-t} + Bte^{-t}.$$

Applying the initial conditions in (24) to this general solution gives

$$5 = x(0) = A,$$
$$10 = x'(0) = -A + B.$$

Solve for A, B to get $A = 5$, $B = 15$. Thus

$$x = 5e^{-t} + 15te^{-t},$$

which has a graph like that in Fig. 3.16.6(a). ∎

Case 3: Overdamping ($\delta > \sqrt{4mk}$)

In this case the characteristic equation $mr^2 + \delta r + k = 0$ for the differential equation $mx'' + \delta x' + kx = 0$ has two distinct negative roots

$$-r_1 = -\frac{\delta}{2m} + \frac{\sqrt{\delta^2 - 4mk}}{2m}, \tag{25a}$$

$$-r_2 = -\frac{\delta}{2m} - \frac{\sqrt{\delta^2 - 4mk}}{2m}, \tag{25b}$$

[Note: $-r_1 < 0$ since $\sqrt{\delta^2 - 4mk} < \sqrt{\delta^2} < \delta$.] Thus the solution is

$$x = Ae^{-r_1 t} + Be^{-r_2 t}. \tag{26}$$

The solution (26) of the overdamped case is different from that of the critically damped case (23). However, depending on the values of A, B, the overdamped solution (26) has a graph that has the same general shape as one of those given in Fig. 3.16.6 for the critically damped case.

Exercises

Free Response with Friction

In Exercises 15 through 22, set up the differential equation that describes the motion under the assumptions of this section. Solve the differential equation. State whether the motion of the spring-mass system is harmonic, damped oscillation, critically damped, or overdamped. If the motion is a damped oscillation, rewrite it in the form (11).

15. The spring-mass system has an attached mass of 10 g. The spring constant is 30 g/sec². A dashpot mechanism is attached, which has a damping coefficient of 40 g/sec. The mass is pulled down and released. At time $t = 0$, the mass is 3 cm below the rest postion and moving upward at 5 cm/sec.

16. A long spring has a mass of 1 slug attached to it. The spring stretches 16/13 ft and comes to rest. The damping coefficient is 2 slug/sec. The mass is subjected to an impulsive force at time $t = 0$, which imparts a velocity of 5 ft/sec downward.

17. A mass of 1 g is attached to a spring-mass system for which friction is negligible. The spring stretches 20 cm and comes to rest. The mass is pulled down 1 cm from rest and released with a velocity of 7 cm/sec downward.

18. A spring-system has an attached mass of 1 g, a spring constant of 5 g/sec², and a damping coefficient of 4 g/sec. The mass is pushed upward 1 cm from rest position and released with a velocity of 3 cm/sec downward.

19. A spring with spring constant $k = 12$ slug/sec² has a mass attached that stretches the spring $2\frac{2}{3}$ ft. The damping coefficient is 7 slug/sec. The mass is pushed 1 ft above the rest position and then released with a velocity of 1 ft/sec downward.

20. A long spring with spring constant $k = 8$ g/sec² has a mass attached that stretches the spring 245 cm. The damping coefficient is $\delta = 8$ g/sec. At time $t = 0$, the mass is at the equilibrium position and has a velocity of 3 cm/sec downward.

21. A spring-mass system has a spring with spring constant $k = 5$ g/sec², attached mass of $m = 1$ g, and a friction coefficient of $\delta = 4$ g/sec. The mass is pulled 2 cm below the equilibrium position and released.

22. A spring-mass system has a spring constant of $k = 1$ lb/ft, a damping coefficient of 2 slug/sec, and an attached mass of 1 slug. The mass is pushed upward and released 1 ft above the equilibrium postion with a velocity of 1 ft/sec upward.

23. Show that, in the case of critical damping, $\delta = \sqrt{4mk}$, the mass may change direction at most once (the general solution (23) has at most one horizontal tangent).

24. Show that, in the case of overdamping, the mass can change direction at most once.

25. Suppose $\delta > 0$, $k > 0$ are fixed. Describe how varying the mass affects whether the motion is a damped oscillation, overdamped, or critically damped.

Exercises 26 through 29 illustrate a general fact but are given only for $m = 1$, $k = 1$. For Exercises 26 through 27, let x_δ be the solution of

$$x'' + \delta x' + x = 0, \qquad x(0) = 1, \qquad x'(0) = -1. \quad (27)$$

for a given δ. Note that $\sqrt{4mk} = \sqrt{4} = 2$.

26. Solve (27) for $0 < \delta < 2$ and $\delta = 2$. Show that

$$\lim_{\delta \to 2^-} x_\delta(t) = x_2(t) \qquad \text{for all } t \geq 0.$$

This shows that, as the damping coefficient approaches critical damping, the damped oscillation approaches the critical damping solution.

27. Solve (27) for $2 < \delta$ and $\delta = 2$. Show that

$$\lim_{\delta \to 2^+} x_\delta(t) = x_2(t) \qquad \text{for all } t \geq 0.$$

This shows that the overdamped solution approaches that of critical damping.

For Exercises 28 and 29 let x_δ be the solution of

$$x'' + \delta x' + x = 0, \qquad x(0) = 0, \qquad x'(0) = 1. \quad (28)$$

28. Solve (28) for $0 < \delta < 2$, and $\delta = 2$. Show that

$$\lim_{\delta \to 2^-} x_\delta(t) = x_2(t) \qquad \text{for all } t.$$

29. Solve (28) for $\delta > 2$ and $\delta = 2$. Show that

$$\lim_{\delta \to 2^+} x_\delta(t) = x_2(t) \qquad \text{for all } t.$$

30. (Requires computer with fairly good graphic capabilities.) Illustrate the limits in Exercises 26 through 29 by graphing x_δ for several values of δ on the same graph. For example, in Exercise 26 you might take $\delta = 2 - (1/n)$ or $\delta = 2 - (1/n^2)$ for $n = 1, 2, 3, \ldots$, and plot on an interval of, say, $0 \leq t \leq 5$. (See Fig. 3.16.5.)

3.16.3 Forced Response

Now that we have carefully examined the free response of the spring-mass system, it is time to investigate what happens when an additional external vertical force $f(t)$, which depends only on the time t, is applied to the mass. The equation of motion is then (7), that is,

$$mx'' + \delta x' + kx = f(t). \tag{29}$$

The external force $f(t)$ is often referred to as the *forcing function* or the *input*. The solution x is then called the *response* or the *output* of the system. From our theory developed earlier, we know that the solution x of (29) has the form $x = x_p + x_h$, where x_p is a particular solution of (29) and x_h is the general solution of the associated homogeneous equation

$$mx'' + \delta x' + kx = 0. \tag{30}$$

But then x_h is just what we have been calling the free response. Note that the arbitrary constants appear only in x_h. We have then

> The possible responses to the input $f(t)$ consist of a particular response to $f(t)$ added to the possible free responses of the system.

In order to simplify our examples, we shall consider only forcing functions $f(t)$ for which (29) can be solved by the method of undetermined coefficients. This is not a major restriction, since the use of series permits many other kinds of functions to be written in terms of either powers of t, or sines and cosines (if Fourier series are used). The study of forced responses will be broken into two cases, depending on whether or not friction is present.

I. Friction Is Present $(\delta > 0)$

If friction is present, then the solution of $mx'' + \delta x' + kx = f(t)$ takes the form

$$x = x_p + x_h = x_p + c_1 x_1 + c_2 x_2.$$

As noted earlier there are three possibilities for the fundamental set of solutions $\{x_1, x_2\}$;

$$\{e^{-\alpha t} \cos \beta t, e^{-\alpha t} \sin \beta t\}: \quad \text{underdamped,}$$
$$\{e^{-\alpha t}, te^{-\alpha t}\}: \quad \text{critically damped,}$$
$$\{e^{-\alpha_1 t}, e^{-\alpha_2 t}\}; \quad \text{overdamped.}$$

The initial conditions determine the arbitrary constants c_1, c_2, but $\lim_{t \to \infty} x_1 = 0$, $\lim_{t \to \infty} x_2 = 0$ in all three cases. Thus

> If friction is present $(\delta > 0)$, the free response x_h is always *transient*. That is, $\lim_{t \to \infty} x_h(t) = 0$.

Another way to say the same thing is that, if friction is present, the effect of the initial conditions dies out (is transient), and the response is eventually determined almost completely by the forcing function $f(t)$. In many situations, particularly the circuits discussed in the next section, this is a desirable phenomenon.

Example 3.16.4 A spring-mass system with an attached mass of $m = 1$ g, has a spring constant of 10 g/sec² and a damping coefficient of 2 g/sec. The mass is pushed up 1 cm and released. A constant force of 20 dynes acts on the mass in the downward direction. Determine the resulting motion.

Solution The equation of motion is

$$x'' + 2x' + 10x = 20, \quad x(0) = -1, \quad x'(0) = 0. \tag{31}$$

First, we find the general free response (solution of the associated homogeneous equation $x'' + 2x' + 10x = 0$). The characteristic equation is $r^2 + 2r + 10 = 0$, which has roots $r = -1 \pm 3i$. Thus

$$x_h = c_1 e^{-t} \cos 3t + c_2 e^{-t} \sin 3t.$$

Using the method of undetermined coefficients (Section 3.12), the form for the particular solution x_p is A, with A a constant. Substituting this into the differential equations (31) gives $10A = 20$, or $A = 2$. Thus

$$x = 2 + c_1 e^{-t} \cos 3t + c_2 e^{-t} \sin 3t$$

gives the possible (or general) forced response. Applying the initial conditions in (31), we obtain

$$-1 = x(0) = 2 + c_1,$$
$$0 = x'(0) = -c_1 + 3c_2,$$

so that $c_1 = -3, c_2 = -1$. Thus

$$x = 2 - 3e^{-t} \cos 3t - e^{-t} \sin 3t,$$

which can be written, using (11), as

$$x = 2 - e^{-t}\sqrt{10} \sin(3t + \phi),$$

where $\phi \approx 1.25$ radians is the phase angle (Fig. 3.16.7).

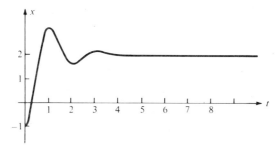

Figure 3.16.7 Graph of $x = 2 - e^{-t}\sqrt{10} \sin(3t + 1.25)$.

In this example, the long range or asymptotic response is $x = 2$ while $-e^{-t}\sqrt{10} \sin(3t + \phi)$ is the transient free response. ∎

Example 3.16.5 A spring-mass system, with an attached mass of $m = 1$ g, has a spring constant of 50 g/sec^2 and a damping coefficient of 2 g/sec. At time $t = 0$, the mass is pushed down $\frac{23}{26}$ cm and released with a velocity of $28\frac{2}{13}$ cm/sec downward. A force of $41 \cos 2t$ dynes acts downward on the mass for $t \geq 0$. Determine the resulting motion and graph it.

Solution The differential equation governing the motion is

$$x'' + 2x' + 50x = 41 \cos 2t, \quad x(0) = \tfrac{23}{26}, \quad x'(0) = 28\tfrac{2}{13}. \tag{32}$$

The characteristic equation $r^2 + 2r + 50 = 0$ has roots $r = -1 \pm 7i$. Thus the solution of the associated homogeneous equation is

$$x_h = c_1 e^{-t} \cos 7t + c_2 e^{-t} \sin 7t.$$

From Section 3.12, the method of undetermined coefficients tells us that there will be a particular solution of the form

$$x_p = A \cos 2t + B \sin 2t.$$

Substituting this form into the differential equation (32) yields

$$(A \cos 2t + B \sin 2t)'' + 2(A \cos 2t + B \sin 2t)' + 50(A \cos 2t + B \sin 2t)$$
$$= 41 \cos 2t$$

or

$$-4A \cos 2t - 4B \sin 2t - 4A \sin 2t + 4B \cos 2t + 50A \cos 2t + 50B \sin 2t$$
$$= 41 \cos 2t.$$

Equating coefficients of $\cos 2t$, $\sin 2t$ gives us equations in A, B;

$$\cos 2t: \quad 46A + 4B = 41,$$
$$\sin 2t: \quad -4A + 46B = 0,$$

so that $B = \tfrac{1}{13}$, $A = \tfrac{23}{26}$, and

$$x = x_p + x_h = \tfrac{23}{26} \cos 2t + \tfrac{1}{13} \sin 2t + c_1 e^{-t} \cos 7t + c_2 e^{-t} \sin 7t.$$

Applying the initial conditions in (32) in order to determine c_1, c_2, we find

$$\tfrac{23}{26} = x(0) = \tfrac{23}{26} + c_1,$$
$$28\tfrac{2}{13} = x'(0) = \tfrac{2}{13} - c_1 + 7c_2.$$

Thus $c_1 = 0$, $c_2 = 4$, and

$$x = \tfrac{23}{26} \cos 2t + \tfrac{1}{13} \sin 2t + 4e^{-t} \sin 7t.$$

Figure 3.16.8
Graph of Eq. (33).

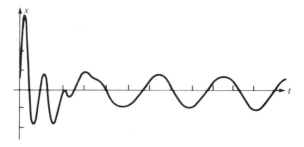

This solution may be simplified, using the form (11), to give

$$x = \sqrt{\tfrac{41}{52}} \sin(2t + \phi) + 4e^{-t} \sin 7t, \qquad (33)$$

where $\phi \approx 1.484$. Note that the long term or steady response is periodic, with the same period as the input. It is out of phase, however (achieves its maxima at a different time determined by ϕ). As Fig. 3.16.8 shows, this motion may be viewed as a damped (transient) oscillation superimposed (added) onto the forced (steady-state) harmonic motion. ∎

II. Friction Is Absent ($\delta = 0$)

Suppose, now, that friction is negligible ($\delta = 0$) so that the model for the spring-mass system is

$$mx'' + kx = f(t). \qquad (34)$$

The solution of (34) has the form

$$x = x_p + x_h,$$

where x_p is a particular solution of (34) and

$$x_h = c_1 \cos\left(\sqrt{\tfrac{k}{m}}\, t\right) + c_2 \sin\left(\sqrt{\tfrac{k}{m}}\, t\right) \qquad (35)$$

is the general solution of the associated homogeneous equation $mx'' + kx = 0$. Note that the free response x_h given in (35) is no longer transient.

Now consider a particular type of forcing term

$$f(t) = \sin \omega t, \qquad (36)$$

where ω is a constant. This is an important case to consider for two reasons. First, many forcing functions, such as AC currents in circuits, are approximately in this form. Also, the theory of Fourier series, which we will not develop, may be used to express many forcing terms as a sum (series) of terms of the form $\sin \alpha nt$, $\cos \alpha nt$, α a constant.

Substituting the forcing term $\sin \omega t$ into the differential equation $mx'' + kx = f$ gives

$$mx'' + kx = \sin \omega t. \qquad (37)$$

We shall now solve (37) by the method of undetermined coefficients (Section 3.12).

Case 1: $\omega \neq \sqrt{k/m}$

(Frequency of input unequal to frequency of free response) In this case, the form for x_p given by the method of undetermined coefficients, is

$$x_p = A \cos \omega t + B \sin \omega t. \tag{38}$$

Substituting this form (38) into the differential equation (37) gives

$$m(A \cos \omega t + B \sin \omega t)'' + k(A \cos \omega t + B \sin \omega t) = \sin \omega t$$

or

$$-mA\omega^2 \cos \omega t - mB\omega^2 \sin \omega t + kA \cos \omega t + kB \sin \omega t = \sin \omega t.$$

Equating coefficients of like terms in order to find A, B:

$$\cos \omega t: \quad -mA\omega^2 + kA = 0 \Rightarrow A = 0,$$

$$\sin \omega t: \quad -mB\omega^2 + kB = 1 \Rightarrow B = \frac{1}{k - m\omega^2},$$

so that

$$x = \frac{1}{k - m\omega^2} \sin \omega t + c_1 \cos\left(\sqrt{\frac{k}{m}} t\right) + c_2 \sin\left(\sqrt{\frac{k}{m}} t\right). \tag{39}$$

This is the superposition of two harmonic motions. One is the free response of the system and the other is a periodic forced response with the same period as the forcing function. Figure 3.16.9 gives the graph of (39) for one choice of the parameters m, k, ω, c_1, c_2.

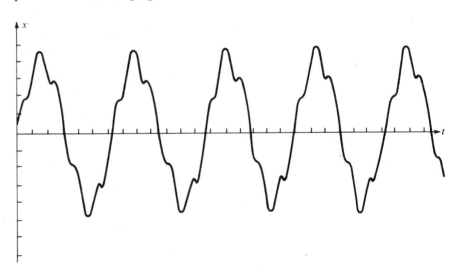

Figure 3.16.9
Graph of (39) with $k = 0.25, m = 0.01, c_1 = c_2 = 0.5, \omega = 1$.

Case 2: $\omega = \sqrt{k/m}$

(Frequency of input equals frequency of free response) If $\omega = \sqrt{k/m}$, then using the method of undermined coefficients gives the form of the particular solution of (37) as

$$x_p = At \cos \omega t + Bt \sin \omega t. \tag{40}$$

Substituting this form into the differential equation (37) gives

$$m(At \cos \omega t + Bt \sin \omega t)'' + k(At \cos \omega t + Bt \sin \omega t) = \sin \omega t$$

or

$$m(-2A\omega \sin \omega t - At\omega^2 \cos \omega t + 2B\omega \cos \omega t - B\omega^2 t \sin \omega t)$$
$$+ kAt \cos \omega t + kBt \sin \omega t = \sin \omega t.$$

Equating coefficients of like terms gives four equations in A, B:

$$\sin \omega t: \quad -2Am\omega = 1,$$
$$\cos \omega t: \quad 2mB\omega = 0,$$
$$t \sin \omega t: \quad -Bm\omega^2 + kB = 0,$$
$$t \cos \omega t: \quad -Am\omega^2 + kA = 0.$$

Since $\omega^2 = k/m$ by assumption, the last two equations are trivially true and thus $B = 0$, $A = -(2m\omega)^{-1}$, from the first two equations. Thus,

$$x = -\frac{1}{2m\omega} t \cos \omega t + c_1 \sin \omega t + c_2 \cos \omega t \tag{41}$$

is the general solution. The particular solution

$$x_p = -\frac{1}{2m\omega} t \cos \omega t \tag{42}$$

is of special interest since it illustrates the phenomenon of *resonance*. The graph of (42) is given in Fig. 3.16.10.

The forced response is now an unbounded oscillation. Intuitively, the phenomena of resonance may be summarized as follows.

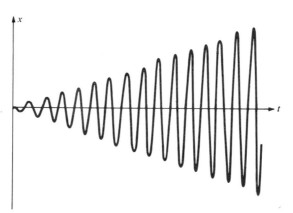

Figure 3.16.10
Graph of
$x = -(2m\omega)^{-1} t \cos \omega t.$

3.16 Mechanical Vibrations

If the period (frequency) of the forcing function is the same as the period (frequency) of the free response of the system, then large-amplitude oscillations may result from forcing terms of small amplitude.

In structures such as bridges or aircraft wings, resonance is to be avoided. Bridges and aircraft wings have both failed on occasion due to resonance. On the other hand, resonance is sometimes desirable in the manipulation of sound waves in musical instruments and detectors of various kinds. Of course, in practice the oscillations do not become arbitrarily large. Rather, either the system is changed (the spring breaks, for example) or the original linear model $mx'' + \delta x' + kx = f$ is no longer a valid model.

Also, "pure resonance" is never observed. In every system there is some friction, so δ, while perhaps very small, is greater than zero. Also, it is almost impossible to make ω exactly $\sqrt{k/m}$. However, if δ is close to zero and ω close to $\sqrt{k/m}$, then the free response, even if transient, may exhibit a large response. The next example illustrates this.

Example 3.16.6

Suppose that the model for a spring-mass system with forcing term $\sin \omega t$ is

$$x'' + x = \sin \omega t, \quad x(0) = 0, \quad x'(0) = 0. \tag{43}$$

Determine the solution for all ω, and graph the solution for $\omega = 1$ and for several values of ω close to 1.

Solution

Using the method of undetermined coefficients (or (39) and (41)), we find that the general solution of $x'' + x = \sin \omega t$ is

$$\omega \neq 1: \quad x = \frac{1}{1 - \omega^2} \sin \omega t + c_1 \cos t + c_2 \sin t, \tag{44}$$

$$\omega = 1: \quad x = -\tfrac{1}{2} t \cos t + c_1 \cos t + c_2 \sin t. \tag{45}$$

Applying the initial conditions to (44) gives

$$0 = x(0) = c_1,$$

$$0 = x'(0) = \frac{\omega}{1 - \omega^2} + c_2,$$

so $c_1 = 0, c_2 = -\omega/(1 - \omega^2)$. Thus $x = (1 - \omega^2)^{-1} \sin \omega t - \omega(1 - \omega^2)^{-1} \sin t$, while applying the initial conditions to (45) gives

$$0 = x(0) = c_1,$$

$$0 = x'(0) = -\tfrac{1}{2} + c_2.$$

Thus the solution of (43) is

Figure 3.16.11
Graph of (46) for $\omega = 0.7$, 0.8, 0.9, and 1.0.

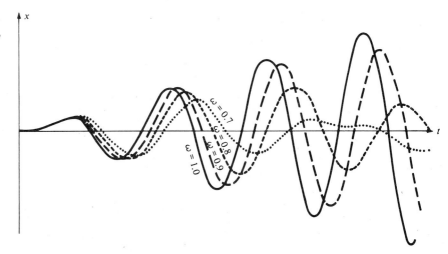

$$x(t) = \begin{cases} \dfrac{1}{1-\omega^2}(\sin \omega t - \omega \sin t) & \text{if } \omega \neq 1, \\ -\frac{1}{2}t \cos t + \frac{1}{2}\sin t & \text{if } \omega = 1. \end{cases} \quad (46)$$

The graph of (46) for several values of ω is given in Fig. 3.16.11. Notice that for ω close to 1, the solutions exhibit large-amplitude oscillations. ∎

Similarly, one can show that if δ is small and ω is close to $\sqrt{k/m}$, then there will be a response similar to resonance (see Exercise 43).

Exercises

Forced Response

31. You are building a detector that is to be sensitive (exhibit resonance) to harmonic vibrations at a frequency of 30 cycles/sec. The detector will be a spring-mass system with spring constant $k = 15$ g/sec^2. Assuming friction is negligible, what should the attached mass be?

32. A device is being built that can be modeled as a simple spring-mass system with negligible friction. The attached mass is 20 g. The spring constant is adjustable. What values of the spring constant are to be avoided if the device is subjected to external harmonic forces in the *range* of 10 to 50 cycles/sec, and your design goal is to avoid resonance?

33. (*Beats*) Using the trigonometric identities $\cos(\theta \pm \phi) = \cos \theta \cos \phi \mp \sin \phi \sin \theta$ verify that

$$\cos \omega t - \cos \beta t = 2 \sin\left(\dfrac{(\beta - \omega)}{2}t\right) \sin\left(\dfrac{(\beta + \omega)}{2}t\right). \quad (47)$$

34. A spring-mass system has an attached mass of 4 g, a spring constant of 16 g/sec^2 and negligible friction. It is subjected to a force of $4\cos(2.2\,t)$ downward and at time $t = 0$ is initially at rest. Determine the subsequent motion. Using (47) from Exercise 33, rewrite the solution as the product of two sine functions, and graph the result. The resulting function has a periodic variation in amplitude, or a *beat*.

35. A spring-mass system has mass $m = 4$ g, friction coefficient $\delta = 24$ g/sec, and spring constant $k = 52$ g/sec^2. The mass is pulled down 3 cm and released. Throughout the subsequent motion, the mass is subjected to a constant external force of 4 dynes downward. Find the subsequent motion. Determine which terms are transient and which are steady-state.

36. Suppose that a spring-mass system has mass m, damping $\delta > 0$, and spring constant k. The system is subjected to a constant force of F dynes downward.
   a) Show that the system has an equilibrium and that the equilibrium depends only on k, F and not m, δ.
   b) Show that the solution has the equilibrium as a limit as $t \to \infty$, regardless of the initial condition. Thus, this equilibrium is an attractor.

37. A spring-mass system has an attached mass of 1 g, friction coefficient $\delta = 3$ g/sec, and spring constant $k = 2$ g/sec^2. The mass is initially at rest. There is an external force of $\sin t$ dynes downward.
   a) Determine the subsequent motion.
   b) State which terms are transient.

38. A spring-mass system has an attached mass of 2 g, friction coefficient $\delta = 4$ g/sec, and spring constant $k = 2$ g/sec^2. The mass is initially at rest. There is an external force of $1 - e^{-t}$ dynes downward.
   a) Determine the subsequent motion.
   b) State which terms are transient.

39. A spring-mass system has an attached mass of 1 g, friction coefficient $\delta = 6$ g/sec, and spring constant $k = 25$ g/sec^2. The mass is initially 1 cm below the rest position with a velocity of 1 cm/sec downward. There is an external force of $3 - e^{-2t}$ dynes downward.
   a) Determine the subsequent motion.
   b) State which terms are transient.

40. A spring-mass system has an attached mass of 4 g, spring constant $k = 16$ g/sec^2 and friction is negligible. There is an external force of $3 \sin 4t$ dynes downward. The mass is initially 0.2 cm below the rest position with a velocity of 1 cm/sec downward.
   a) Determine the subsequent motion.
   b) State which terms are transient.

41. A spring-mass system has an attached mass of 1 g, spring constant $k = 1$ g/sec^2 and friction is negligible. The mass is initially at rest. There is an external force of $t \sin t$. (The forcing function already "looks like" resonance.)
   a) Determine the subsequent motion.
   b) State which terms are transient.
   c) Graph the nontransient terms for $0 \le t \le 5$.

42. Suppose that friction is negligible for a spring-mass system with external force $f(t) = \sin \omega t$. The spring-mass system is exhibiting resonance.
   a) Show that there is a constant M such that the forced response (42) satisifes $|x_p(t)| \le Mt$ for $t \ge 0$.
   b) Find all values of m, k such that $|x_p(t)| \le 10 t$ for $t \ge 0$.

43. Let x_δ be the solution of $x'' + \delta x' + x = \sin t$, $x(0) = 0$, $x'(0) = 1$. Find the formula for x_δ when $\delta = 0$ and $0 < \delta < 2$. Show that, for all $t > 0$, $\lim_{\delta \to 0^+} x_\delta(t) = x_0(t)$. This illustrates the point made at the end of this section that if $\omega = \sqrt{k/m}$ and δ is small, there will be a response similar to resonance.

44. Let $x_\delta(t)$ be the function of t found in Exercise 43. Graph $x_\delta(t)$ for $0 \le t \le 30$ for $\delta = 0.5, 0.2, 0.1, 0.05$ on the same set of coordinates. (Requires computer with graphics capability. Illustrates the limit of Exercise 43.)

45. Suppose that the top of our spring-mass system is solidly attached to a mechanism that causes the point of attachment to also move in a vertical direction. Let $h(t)$ measure the vertical displacement of the attachment point, with a downward displacement being positive. Explain why a reasonable model for small movements of the mass in the spring-mass system would be

$mx'' + \delta x' + kx = kh(t)$, which is in the form of (29).

46. A rectangular closed box is floating in the ocean with its top always parallel to the water's surface. The buoyancy force is proportional to the volume of water displaced, which is also proportional to the depth of the bottom of the box. Let m be the mass of the box and let d be the depth of the bottom of the box when the box is at rest. Let $x(t)$ measure how much deeper than this rest depth the bottom of the box is at time t. Ignore resistance and assume small displacements and velocities. Explain why $mx'' + kx = 0$ is a reasonable model for the vertical movement of the box.

3.17 Linear Electrical Circuits

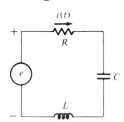

Figure 3.17.1

We have already considered nonlinear circuits in Section 2.12, and will use the units from that section. In this section we shall consider single-loop linear RLC circuits, as shown in Fig. 3.17.1. The figure shows a linear resistor of resistance R ohms, capacitor of capitance C farads, inductor of inductance L henries, and voltage source with voltage $e(t)$ volts. Furthermore, we shall assume that R, L, C are nonnegative, which is often the case. (A negative resistance means that we have a device that puts power into the loop rather than dissipating it, as a positive resistor does.)

Recall from Section 2.12 that the *voltage law* says that

> The algebraic sum of the voltage drops around a loop at any instant is zero.

Since there is a single loop, the *current law* says the current is the same in the resistor, inductor, and capacitor. Let this current at time t be $i(t)$ amps. Let $q(t)$ be the charge in coulombs in the capacitor at time t. From Section 2.12 we know the voltage drops for each device are:

$$\text{Resistor:} \quad V_R = iR,$$

$$\text{Capacitor:} \quad V_C = \frac{1}{C} q,$$

$$\text{Inductor:} \quad V_I = L \frac{di}{dt},$$

where we assume R, L, C are constants. Thus the voltage law may be written as

$$L \frac{di}{dt} + Ri + \frac{1}{C} q = e.$$

However, current is the time rate of change of charge so that $i = dq/dt$, and we have the second-order linear differential equation in q

$$L \frac{d^2 q}{dt^2} + R \frac{dq}{dt} + \frac{1}{C} q = e(t), \qquad (1)$$

or, upon differentiation,

$$L\frac{d^2i}{dt^2} + R\frac{di}{dt} + \frac{1}{C}i = e'(t), \tag{2}$$

which is a second-order linear differential equation in i. Note that this differential equation has exactly the same form as the spring-mass system and L, R, C are also nonnegative by assumption. Thus all of the analysis of the preceding section is still valid, but now mass has become inductance, friction is resistance, and the spring constant is the reciprocal of capacitance. In particular, the discussion of resonance, damping, amplitude, and phase angle is still appropriate.

That many mechanical and nonelectrical problems have the same differential equation (1) as a model is the idea behind the analogue computer. To solve the differential equation (1) for a given L, R, C, e, with an analogue computer, one would build the circuit and then measure the resulting charge (or current) to determine the values of the solution.

Most modern circuits, of course, involve many loops and hence are modeled by systems of differential equations. As circuits have become more and more complex, it has become increasingly expensive and cumbersome to design them by building numerous prototypes. Increasingly, preliminary design work is done by computer simulations, which often involve the solution of systems of differential equations.

Exercises

1. An RLC circuit, given by Fig. 3.17.1, has a voltage source of $e(t) = 3 \cos t$ volts. The values for the components are $R = 3$ ohms, $L = 0.5$ henries, and $C = 0.4$ farads. Initially, the charge on the capacitor is zero and the current in the resistor is 1 A. Find the charge on the capacitor and the current as functions of time.

2. An RLC circuit, given by Fig. 3.17.1, has a voltage source of $e(t) = 5 \cos 2t$ volts. The values for the components are $R = 2$ ohms, $L = 1$ henries, and $C = \frac{1}{17}$ farads. Initially, the charge on the capacitor and the current in the resistor are zero. Find the charge on the capacitor and the current as functions of time.

3. An RLC circuit, given by Fig. 3.17.1, has a 1.5-volt battery as a voltage source ($e(t) = 1.5$). The values for the components are $R = 1.5$ ohms, $L = 1$ henries, and $C = 2$ farads. Initially, the charge on the capacitor is 2 coloumbs, and the current in the resistor is 4 A. Find the charge on the capacitor and the current as functions of time.

4. An RLC circuit, given by Fig. 3.17.1, has a 9-volt battery as a voltage source ($e(t) = 9$). The values for the components are $R = 5$ ohms, $L = 6$ henries, and $C = 1$ farad. Initially, the charge on the capacitor is 1 coloumb, and the current in the resistor is zero. Find the charge on the capacitor and the current as functions of time.

5. An LC circuit, ($R = 0$) given by Fig. 3.17.1, has $C = 0.1$ farad and $e(t) = \sin \omega t$. Suppose that ω is a constant such that $e(t)$ has a frequency between 20 and 30 cycles/sec. What values of L will not lead to resonance for any such ω?

6. The RLC circuit in Fig. 3.17.1 has $R = 20$ ohms, $L = 1$ henries, $C = 0.005$ farads. The voltage

source is shorted out ($e(t) \equiv 0$). At time $t = 0$, there is a charge of 10 coulombs on the capacitor and no current. Solve the differential equation for the charge, and put in phase amplitude form [(11) of Section 3.16]. How many seconds will it take the variable amplitude to be reduced 99%?

7. The *RLC* circuit in Fig. 3.17.1 has $R = 2$ ohms, $L = 1$ henry, $C = 0.5$ farad. The initial charge on the capacitor is zero, and the initial current is zero. The voltage source $e(t)$ is a 1-volt battery which is shorted out after π seconds. Find the charge on the capacitor and graph for $0 \le t \le 2\pi$ (sec). (Assume that charge and current are continuous at $t = \pi$.)

8. The *LC* circuit in Fig. 3.17.1 has ($R = 0$), $L = 4$ henries, $C = 0.25$ farad. The voltage source is a 2-volt battery which is shorted out ($e = 0$) after 4π seconds. The initial charge is zero and the initial current is 1 A. Find the charge on the capacitor and current in the inductor for $0 \le t \le 8\pi$. (Assume that charge and current are continuous at $t = 4\pi$.)

9. Let $R = 0$ and $e(t) = 0$ in Fig. 3.17.1 so that we have an *LC* circuit with no voltage source. Let $E(i, q) = \frac{L}{2}i^2 + \frac{1}{2C}q^2$. Show that E is constant by verifying that $dE/dt = 0$. (This is the electrical equivalent of conservation of mechanical energy.)

10. An *LC* circuit ($R = 0$) given by Fig. 3.17.1 has $L = 8$ henries, $C = 2$ farads. For what value of ω will the voltage source $e(t) = 2 \cos \omega t$ volts create resonance?

11. An *LC* circuit ($R = 0$) given by Fig. 3.17.1 has $L = 9$ henries, $C = 1$ farad. The voltage source is $e(t) = 4 \cos 2t$. Since $R = 0$, the free response will not be transient. However, there is one choice for $q(0)$, $i(0)$, for which the free response will be absent. What are the values of $q(0)$, $i(0)$?

For the next five exercises, use the correspondence

$$\text{inductance} \leftrightarrow \text{mass},$$
$$\text{resistance} \leftrightarrow \text{friction},$$

and

$$\frac{1}{\text{capacitance}} \leftrightarrow \text{spring constant}$$

between the circuit given by Fig. 3.17.1 and the spring-mass system in Section 3.16. (A mass of m in slugs or grams becomes an inductance of m henries, etc.)

12. Rewrite Exercise 3.16.8 in terms of this circuit and answer parts (a)–(c).

13. Rewrite Exercise 3.16.9 in terms of this circuit and answer parts (a)–(c).

14. Rewrite Exercise 3.16.10 in terms of this circuit and solve.

15. Rewrite Exercise 3.16.11 in terms of this circuit and solve.

16. Rewrite Exercise 3.16.12 in terms of this circuit and solve.*

(*Input-Output Voltages*) Frequently it is helpful to consider circuits as input-output devices. (This is done more carefully in the next section.) Suppose for the circuit in Fig. 3.17.1 that $R > 0$ so that the free response is transient. We assume that these transient terms are not important and consider the solution $q(t)$ to be given by the terms which are not transient, that is, the forced response. The *input voltage* will be taken as $e(t)$, the *output voltage* as the voltage across the capacitor which is $\frac{1}{C}q$.

17. Suppose that the input voltage is $e(t) = E \sin \omega t$, with E, ω constants. Show that the output voltage is $M \sin(\omega t + \phi)$ for constants M, ϕ. Thus an *RLC* circuit can only change the amplitude M and the phase angle ϕ.

18. Suppose that $E = 1$, $R = 6$, $L = 1$, $C = \frac{1}{13}$, and $\omega = 3$. Find M, ϕ from Exercise 17.

19. Suppose that $E = 1$, $R = 7$, $L = 12$, $C = 1$, and $\omega = 2$. Find M, ϕ from Exercise 17.

* The next section also discusses the relationship between differential equations and electrical circuits. Additional exercises on circuits appear at the end of that section.

3.18 Phase Angles, Complex Arithmetic, and Phasors

In working particular types of problems, it is often possible to develop short cuts and symbol manipulations which, when presented to the student, may appear to have little connection with the theory of this chapter. This section will present one example from electrical engineering. The discussion requires some familiarity with the polar form $re^{i\theta}$ of a complex number (see Appendix A).

The situation is as follows: We have a device modeled by a second-order linear differential equation with constant coefficients,

$$ay'' + by' + cy = f(t), \qquad (1)$$

where we assume that $a > 0, b > 0, c > 0$. In this section we shall refer to (1) as a *system* even though there is only one equation. System is being used for "model of a physical system" rather than "system of equations." This use of "system" is common in parts of the engineering literature. Also in this section the function $f(t)$ will be considered the *input* and the resulting solution y the *output*. For convenience, we may consider (1) an *RLC* circuit with $f(t)$ a voltage source and y the charge on the capacitor, as in Section 3.17. Since $b > 0$, we know, from Section 3.16, that the free response is transient. We assume that the free response then is unimportant, and consider only the forced response. Using the theory of Fourier series (which we have not discussed) and the superposition principle for (1), one may usually reduce the problem of analyzing (1) to finding the outputs for inputs of the form

$$f(t) = \sin \omega t, \qquad (2)$$

where ω is a constant. (One could consider $\cos \omega t$ also; see the exercises at the end of this section.) The method of undetermined coefficients of Section 3.12 says that a particular solution of (1), (2) will be of the form

$$y_p = A \sin \omega t + B \cos \omega t,$$

which may be written (Section 3.16) in phase amplitude form,

$$\begin{aligned} y_p &= \sqrt{A^2 + B^2} \sin(\omega t + \phi) \\ &= R \sin(\omega t + \phi). \end{aligned}$$

Note R is the amplitude and not the resistance of Section 3.17.

Thus the output of the system (1) in response to the input (2) is obtained by

Multiplying the amplitude of (2) by R \qquad (3)

and

Shifting the input to be ϕ radians out of phase with the original input (2). (4)

We wish to develop a formula for R, ϕ in terms of the system and input parameters a, b, c, ω, in two different ways. We will first obtain the formula directly in terms of the approach of Section 3.16. The second method involves a complex version of undetermined coefficients and leads to a quick method of finding R and ϕ, using complex arithmetic.

Method 1

(Outlined.) Using the method of undetermined coefficients we let $y_p = A \sin \omega t + B \cos \omega t$, substitute into (1), and equate coefficients of $\sin \omega t$, $\cos \omega t$ to get

$$-aA\omega^2 - \omega bB + cA = 1,$$
$$-aB\omega^2 + b\omega A + cB = 0. \qquad (5)$$

Solving (5) for A, B will yield

$$y_p = \frac{-(a\omega^2 - c)}{(a\omega^2 - c)^2 + \omega^2 b^2} \sin \omega t - \frac{b\omega}{(a\omega^2 - c)^2 + \omega^2 b^2} \cos \omega t. \qquad (6)$$

By Section 3.16, (6) may be written as

$$= R \sin(\omega t + \phi), \qquad (7)$$

where

$$R = [(a\omega^2 - c)^2 + \omega^2 b^2]^{-1/2} \qquad (8)$$

and

$$\sin \phi = \frac{-b\omega}{[(a\omega^2 - c)^2 + \omega^2 b^2]^{1/2}}, \quad \cos \phi = \frac{-(a\omega^2 - c)}{[(a\omega^2 - c)^2 + \omega^2 b^2]^{1/2}}. \qquad (9)$$

There is an elegant relationship between (1), (7), (8), (9) but it is not obvious unless we use complex arithmetic.

Method II

For solving (1), (2). Note that $\sin \omega t$ is the imaginary part of $e^{i\omega t}$ (Appendix A). That is,

$$\sin \omega t = \text{Im}(e^{i\omega t}) = \text{Im}(\cos \omega t + i \sin \omega t).$$

Since a, b, c are real constants, we may left $f(t) = e^{i\omega t}$ and take the imaginary part of the resulting particular solution y_p to get the forced response for $\sin \omega t$.

If $f(t) = e^{i\omega t}$, then using a complex version of undetermined coefficients, y_p is of the form

$$y_p = Ae^{i\omega t}. \qquad (10)$$

Substituting (10) into (1) yields

$$a(Ae^{i\omega t})'' + b(Ae^{i\omega t})' + c(Ae^{i\omega t}) = e^{i\omega t}$$

or

$$aA(i\omega)^2 e^{i\omega t} + bAi\omega e^{i\omega t} + cAe^{i\omega t} = e^{i\omega t}.$$

That is,

$$[a(i\omega)^2 + b(i\omega) + c]A = 1 \tag{11}$$

or

$$p(i\omega)A = 1,$$

where

$$p(\lambda) = a\lambda^2 + b\lambda + c \tag{12}$$

is the characteristic polynomial of (1). Thus $A = 1/(p(i\omega))$ so that (10) is

$$y_p = \frac{1}{p(i\omega)} e^{i\omega t}. \tag{13}$$

Let $Re^{i\phi}$ be the polar form of $1/(p(i\omega))$. Then, from (13),

$$y_p = Re^{i\phi} e^{i\omega t} = Re^{i(\omega t + \phi)}$$
$$= R\cos(\omega t + \phi) + iR\sin(\omega t + \phi).$$

Thus the solution of (1), (2) is

$$\text{Im}(y_p) = R\sin(\omega t + \phi).$$

We have then the following nice algebraic technique.

Algorithm 3.18.1   If $a \neq 0$, $b \neq 0$, the forced response y_p for

$$ay'' + by' + cy = \sin \omega t \tag{14}$$

is given by

$$y_p = R\sin(\omega t + \phi).$$

$Re^{i\phi}$ is the polar form of $1/(p(i\omega))$, where $p(\lambda) = a\lambda^2 + b\lambda + c$ is the characteristic polynomial of (14).

Note   Since $Re^{i\phi} = 1/(p(i\omega))$ implies that $p(i\omega) = (1/R)e^{-i\phi}$, it follows that R, ϕ may also be computed from the polar form of $p(i\omega)$.

Example 3.18.1   Compute the forced response of

$$y'' + y' + y = \sin 2t.$$

Solution   The characteristic polynomial is $p(\lambda) = \lambda^2 + \lambda + 1$ and $\omega = 2$. Thus

$$p(2i) = (2i)^2 + 2i + 1 = 2i - 3.$$

Figure 3.18.1

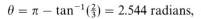

We shall compute the polar decomposition $\tilde{R}e^{i\theta}$ of $-3 + 2i$. First,
$$\tilde{R} = |2i - 3| = \sqrt{4 + 9} = \sqrt{13}.$$
Also (note Fig. 3.18.1),
$$\theta = \pi - \tan^{-1}(\tfrac{2}{3}) = 2.544 \text{ radians,}$$
so that
$$p(2i) = \sqrt{13}e^{i(2.544)}.$$
Thus,
$$\frac{1}{p(2i)} = \frac{1}{\sqrt{13}}e^{-i(2.544)}.$$
Hence,
$$R = \frac{1}{\sqrt{13}}, \qquad \phi = -2.544 \text{ radians,}$$
and the forced response is
$$y_p = \frac{1}{\sqrt{13}}\sin(2t - 2.544). \blacksquare$$

Example 3.18.2 For a fixed value of ω, design a system
$$ay'' + by' + cy = \sin \omega t,$$
such that $R = \sqrt{2}$, $\phi = -\frac{\pi}{4}$. That is, design a system so that the output y is $\sqrt{2}\sin\left(\omega t - \frac{\pi}{4}\right)$.

Solution The problem is to find a, b, c such that $p(\lambda) = a\lambda^2 + b\lambda + c$ has the property that
$$\frac{1}{p(i\omega)} = \sqrt{2}e^{-i(\pi/4)} = \sqrt{2}\left(\cos\frac{\pi}{4} - i\sin\frac{\pi}{4}\right) = 1 - i,$$
that is,
$$p(i\omega) = \frac{1}{\sqrt{2}}e^{i(\pi/4)} = \frac{1}{2} + \frac{i}{2}.$$
But
$$p(i\omega) = (c - a\omega^2) + ib\omega.$$
Thus we need only choose a, b, c, so that
$$b\omega = \tfrac{1}{2},$$
$$c - a\omega^2 = \tfrac{1}{2}. \blacksquare$$

If Example 2 is interpreted in terms of an *RLC* circuit, it is interesting to note that the resistance b is uniquely determined by ω, whereas there is some freedom in the choice of the inductor and capacitor.

In the context of a particular multiloop circuit, there may be several voltages of the form $v = E \sin(\omega t + \phi)$ with the same ω but different E, ϕ, since, as we have seen, the effect of a single *RLC* loop is often to change the phase and amplitude but not the frequency. Thus the amplitude E and phase angle ϕ completely determine the voltage v if they are known. Since

$$v = E \sin(\omega t + \phi) = \text{Im}(E e^{i\phi} e^{i\omega t}),$$

the expression $E e^{i\phi}$, which is known as a *phasor*, contains a complete description of the voltage $v(t)$, if ω is fixed, and is sometimes called a *transform* of the voltage. Using phasors, it is possible to algebraically analyze many circuits and devices.

Note Which variables are considered inputs and which are considered outputs depends on the application or design problem and is not an intrinsic property of a given circuit.

Exercises

In Exercises 1 through 5, use Algorithm 3.18.1 to compute the forced response.

1. $y'' + 2y' + 2y = \sin t$
2. $y'' + 2y' + 2y = \sin 3t$
3. $y'' + 3y' + 2y = \sin t$
4. $y'' + 2y' + y = \sin 2t$
5. $y'' + 2y' + y = \sin t$
6. Design a system $ay'' + by' + cy = \sin t$ with output $4 \sin(t - \pi/3)$.
7. Design a system $ay'' + by' + cy = \sin 2t$ with output $2 \sin(2t - \pi/4)$.
8. Show that, if $a\omega^2 - c \neq 0$, then Method II may still be used to get the forced response of $ay'' + cy = \sin \omega t$.
9. Let $p(\lambda) = a\lambda^2 + b\lambda + c$. Express $1/(p(i\omega))$ in the form $\alpha + i\beta$. Then verify that
$$y_p = \text{Im}\left(\frac{1}{p(i\omega)} e^{i\omega t}\right)$$
$$= \text{Im}[(\alpha + i\beta)(\cos \omega t + i \sin \omega t)],$$

given by Method II, is the same as the solution (6) found by Method I.

Exercises 10 through 12 should be done consecutively.

10. Suppose $p(\lambda) = a\lambda^2 + b\lambda + c$ and $1/(p(i\omega)) = Re^{i\phi}$. Show that the forced response of
$$ay'' + by' + cy = \cos \omega t$$
is $y_p = R \cos(\omega t + \phi)$.

11. Show that, if $f(t) = \sin(\omega t + \psi)$, $p(\lambda) = a\lambda^2 + b\lambda + c$ and $1/(p(i\omega)) = Re^{i\phi}$, then the forced response of
$$ay'' + by' + cy = \sin(\omega t + \psi)$$
is $R \sin(\omega t + \psi + \phi)$.

12. Two systems are *cascaded* if the output of the first is the input of the second. That is,
$$a_1 y_1'' + b_1 y_1' + c_1 y_1 = f, \quad (15)$$
$$a_2 y_2'' + b_2 y_2' + c_2 y_2 = y_1. \quad (16)$$
Here the input f in system (15) results in the output y_1, which is inputted into (16), producing the output y_2. An *RLC* realization is Fig. 3.18.2,

where we have adopted the traditional port notation. Show that if

$$\frac{1}{p_1(i\omega)} = R_1 e^{i\phi_1} \quad \text{where } p_1(\lambda) = a_1\lambda^2 + b_1\lambda + c_1,$$

$$\frac{1}{p_2(i\omega)} = R_2 e^{i\phi_2} \quad \text{where } p_2(\lambda) = a_2\lambda^2 + b_2\lambda + c_2,$$

then the forced response y_2 corresponding to an input $f(t) = \sin \omega t$ is

$$y_2 = \text{Im}(R_1 R_2 e^{i(\phi_1 + \phi_2 + \omega t)})$$
$$= R_1 R_2 \sin(\omega t + \phi_1 + \phi_2).$$

Figure 3.18.2

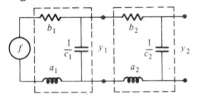

13. Suppose that we have two systems, as in Exercise 12,

$$y_1'' + 2y_1' + 2y_1 = \sin t,$$
$$y_2'' + 2y_2' + 2y_2 = y_1.$$

Find the output y_2 using Exercise 12.

14. Suppose that we have two systems, as in Exercise 12,

$$y_1'' + 2y_1' + y_1 = \sin 2t,$$
$$y_2'' + 3y_2' + 2y_2 = y_1.$$

Find the output y_2 using Exercise 12.

15. Explain why the system (1) can never shift the input 180° out of phase if the resistance is greater than zero.

16. Suppose $\omega > 0$ is fixed and $a, b, c > 0$. Explain why we always have $-\pi < \phi < 0$ (or, equivalently, $\pi < \phi < 2\pi$). That is, the effect is always to translate the voltage curve $\sin \omega t$ to the right.

17. (Uses Exercises 12, 16, 17.) Explain how, given an input $\sin \omega t$, it is possible to produce an output $R \sin(\omega t + \phi)$ with $0 \leq \phi \leq 2\pi$ by cascading three RLC circuits.

4

The Laplace Transform

4.1 Definition and Basic Properties

We have developed several methods for solving differential equations. In this chapter we shall introduce a different type of approach that is very important in many areas of applied mathematics. The idea is to use a *transformation* that changes one set of objects and operations into a different set of objects and operations. Our transformation will be the *Laplace transform*, and it will change a linear differential equation with constant coefficients into a problem in algebra. Many design procedures in such areas as circuit and control theory are based on the algebraic form of the problem provided by the Laplace transform (see Section 4.7). The Laplace transform is also especially well suited for handling discontinuous forcing functions and impulses. But first we need to develop the basic properties of the Laplace transform. In this chapter we will use t instead of x as the independent variable.

4.1.1 Definition of Laplace Transform

Let $f(t)$ be a function defined on the interval $[0, \infty)$. The Laplace transform of $f(t)$ is a new function denoted

$$F(s) \quad \text{or} \quad \mathscr{L}[f(t)] \quad \text{or} \quad \mathscr{L}[f](s),$$

of the variable s, and it is given by

$$F(s) = \mathscr{L}[f](s) = \int_0^\infty e^{-st} f(t)\, dt,$$

provided the improper integral exists. The notation $\mathscr{L}[f]$ will sometimes be used instead of $\mathscr{L}[f](s)$.

Throughout this chapter lower-case letters will denote the function of t and capital letters will denote its Laplace transform. Thus,

$$\mathscr{L}[g(t)] = G(s) \quad \text{and} \quad \mathscr{L}[y(t)] = Y(s).$$

The one exception is H for the Heaviside (unit-step) function in Section 4.3.

Example 4.1.1 Calculate the Laplace transform of e^t.

Solution Recall that, by definition, for any function $h(t)$ defined on $[0, \infty)$,

$$\int_0^\infty h(t)\, dt = \lim_{b \to \infty} \int_0^b h(t)\, dt,$$

and the integral is said to *converge* if this limit exists. If the limit does not exist, the integral is said to *diverge*. Thus, if $f(t) = e^t$, then,

$$\mathscr{L}[e^t](s) = \int_0^\infty e^{-st} e^t\, dt = \lim_{b \to \infty} \int_0^b e^{t(1-s)}\, dt$$

$$= \lim_{b \to \infty} \begin{cases} b & \text{if } s = 1, \\ \dfrac{e^{b(1-s)}}{1-s} - \dfrac{1}{1-s}, & \text{if } s \ne 1. \end{cases}$$

If $s < 1$, then $1 - s > 0$ and $\lim_{b \to \infty} e^{b(1-s)} = \infty$, so that the integral diverges. If $s = 1$, then $\lim_{b \to \infty} b = \infty$ and the integral diverges. On the other hand, if $s > 1$, then $1 - s < 0$, and

$$\lim_{b \to \infty} \frac{e^{b(1-s)}}{1-s} - \frac{1}{1-s} = \frac{0}{1-s} - \frac{1}{1-s} = \frac{1}{s-1}.$$

Thus $F(s) = \mathscr{L}[e^t] = 1/(s-1)$ and the domain of the Laplace transform of e^t is $1 < s < \infty$. ∎

Example 4.1.2 Let

$$f(t) = \begin{cases} 0, & \text{if } 0 \le t \le 2, \\ 3, & \text{if } 2 < t \le 4, \\ 0, & \text{if } 4 < t. \end{cases}$$

Calculate $\mathscr{L}[f(t)]$.

Figure 4.1.1

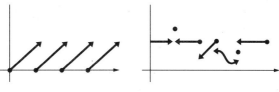

Solution

$$\mathscr{L}[f(t)] = \int_0^\infty e^{-st} f(t)\, dt$$

$$= \int_0^2 e^{-st} f(t)\, dt + \int_2^4 e^{-st} f(t)\, dt + \int_4^\infty e^{-st} f(t)\, dt$$

$$= \int_0^2 e^{-st} 0\, dt + \int_2^4 e^{-st} 3\, dt + \int_4^\infty e^{-st} 0\, dt$$

$$= 3\int_2^4 e^{-st}\, dt = \left.\frac{3e^{-st}}{-s}\right|_2^4 = \frac{3e^{-4s}}{-s} + \frac{3e^{-2s}}{s}.\ \blacksquare$$

Obviously, we do not want to have to compute every Laplace transform from the definition. A table of Laplace transforms will be developed, and is given in Section 4.2, but first we need to establish when the Laplace transform exists.

A function $f(t)$ defined on $[0, \infty)$ has a *jump discontinuity* at $a \in [0, \infty)$ if the one-sided limits

$$\lim_{t \to a^+} f(t) = l_+ \quad \text{and} \quad \lim_{t \to a^-} f(t) = l_-$$

exist but either

$$l_+ \neq l_- \quad \text{or} \quad l_+ \neq f(a) \quad \text{or} \quad l_- \neq f(a).$$

A function $f(t)$ is *piecewise continuous* on $[0, \infty)$ if, for every number $B > 0$, $f(t)$ is continuous on $[0, B]$ except possibly for a finite number of jump discontinuities.

If $f(t)$ is continuous on $[0, \infty)$, it is piecewise continuous. Other examples of piecewise continuous functions are shown in Fig. 4.1.1.

A function $f(t)$ on $[0, \infty)$ is said to be of *exponential order* if there exists constants α, M so that $|f(t)| \leq Me^{\alpha t}$ for $t \geq 0$. That is, as $t \to \infty$, $f(t)$ grows more slowly than a multiple of some exponential. All of the examples in Fig. 4.1.1 are bounded functions and hence of exponential order (take $\alpha = 0$). Note that e^{t^2} is not of exponential order since t^2 eventually grows faster than αt for any constant α (see Exercise 52 at the end of this section).

Theorem 4.1.1

Existence of the Laplace Transform If $f(t)$ is piecewise continuous and of exponential order on $[0, \infty)$, so that $|f(t)| \leq Me^{\alpha t}$, then $\mathscr{L}[f] = F(s)$ exists and is defined at least for $s > \alpha$.

This theorem takes care of the existence of the Laplace transform. In the applications to be developed, we shall often have to take an $F(s)$ and then find a function $f(t)$, so that $\mathscr{L}[f] = F(s)$. That is, we will have to invert the Laplace transform. In this situation, it is important to know that $f(t)$ is unique. Since the Laplace transform is given in terms of an integral, changing a few values of $f(t)$ will not change the Laplace transform (see Exercise 51).

Theorem 4.1.2

Uniqueness of the Laplace Transform If $f(t)$ and $g(t)$ are piecewise continuous functions of exponential order and $\mathscr{L}[f] = \mathscr{L}[g]$, then $f(t) = g(t)$ on any interval $[0, B]$ except possibly at a finite number of points.

If $F(s)$ is given and there is a continuous function $f(t)$ so that $\mathscr{L}[f] = F(s)$, then $f(t)$ is the *only* continuous function for which $\mathscr{L}[f] = F(s)$. Since, for all practical purposes (at least ours), $F(s)$ uniquely determines $f(t)$, we may denote this by $f(t) = \mathscr{L}^{-1}[F(s)]$ and call $\mathscr{L}^{-1}$ the *inverse Laplace transform*.

The Laplace transform has several important properties. One is that it is a *linear transformation*. That is; if $f(t)$, $g(t)$ are piecewise continuous functions of exponential order on $[0, \infty)$ and a, b are constants, then

$$\mathscr{L}[af + bg] = a\mathscr{L}[f] + b\mathscr{L}[g], \tag{1}$$

and as a corollary

$$\mathscr{L}^{-1}[aF(s) + bG(s)] = a\mathscr{L}^{-1}[F(s)] + b\mathscr{L}^{-1}[G(s)].$$

Verification

$$\mathscr{L}[af + bg] = \int_0^\infty e^{-st}(af(t) + bg(t))\, dt$$

$$= a\int_0^\infty e^{-st}f(t)\, dt + b\int_0^\infty e^{-st}g(t)\, dt$$

$$= a\mathscr{L}[f] + b\mathscr{L}[g]. \quad \blacksquare$$

There is a slight technical problem with the formula (1) in that the domain of $\mathscr{L}[af + bg]$ might not be quite the same as that of $\mathscr{L}[f]$ and $\mathscr{L}[g]$. However, there is an interval of the form (α, ∞) on which (1) holds, and that turns out to be sufficient. Before illustrating the linearity property, we need one formula.

Example 4.1.3

Compute $\mathscr{L}[e^{at}]$ where a is a constant.

Solution

$$\mathscr{L}[e^{at}] = \int_0^\infty e^{-st}e^{at}\, dt = \lim_{b\to\infty}\int_0^b e^{t(a-s)}\, dt$$

$$= \lim_{b\to\infty} \frac{e^{t(a-s)}}{a-s}\bigg|_{t=0}^{t=b} \quad (a \neq s)$$

4.1 Definition and Basic Properties

$$= \lim_{b \to \infty} \frac{e^{b(a-s)}}{a-s} - \frac{1}{a-s}$$

$$= \begin{cases} -\dfrac{1}{a-s} & \text{if } a - s < 0, \\ \text{diverges} & \text{if } a - s > 0, \end{cases}$$

so that

$$\mathscr{L}[e^{at}] = \frac{1}{s-a}, \qquad s > a. \blacksquare \tag{2}$$

An important special case of (2) is when $a = 0$. In this case,

$$\mathscr{L}[1] = \frac{1}{s}, \qquad s > 0. \tag{3}$$

Example 4.1.4 Compute $\mathscr{L}[3e^t + 5e^{-2t} + 6]$, using (1), (2), and (3).

Solution Using formulas (1), (2), and (3),

$$\mathscr{L}[3e^t + 5e^{-2t} + 6] = 3\mathscr{L}[e^t] + 5\mathscr{L}[e^{-2t}] + 6\mathscr{L}[1]$$

$$= \frac{3}{s-1} + \frac{5}{s+2} + \frac{6}{s}. \blacksquare$$

Example 4.1.5 Compute $\mathscr{L}[\cosh at]$.

Solution
$$\mathscr{L}[\cosh at] = \mathscr{L}\left[\frac{1}{2}(e^{at} + e^{-at})\right] = \mathscr{L}\left[\frac{1}{2}e^{at} + \frac{1}{2}e^{-at}\right]$$

$$= \frac{1}{2}\mathscr{L}[e^{at}] + \frac{1}{2}\mathscr{L}[e^{-at}] = \frac{1}{2}\frac{1}{(s-a)} + \frac{1}{2}\frac{1}{(s+a)}$$

$$= \frac{s}{s^2 - a^2},$$

so that

$$\mathscr{L}[\cosh at] = \frac{s}{s^2 - a^2}. \blacksquare \tag{4}$$

An important step in applying the Laplace transform to differential equations will be computing the inverse Laplace transform.

Example 4.1.6 Given $F(s) = \dfrac{3}{s} + \dfrac{6}{s-3}$, find $f(t) = \mathscr{L}^{-1}[F(s)]$.

Solution From (2) and (3), we have

$$e^{at} = \mathscr{L}^{-1}\left[\dfrac{1}{s-a}\right] \quad \text{and} \quad 1 = \mathscr{L}^{-1}\left[\dfrac{1}{s}\right].$$

Thus

$$\mathscr{L}^{-1}\left[\dfrac{3}{s} + \dfrac{6}{s-3}\right] = 3\mathscr{L}^{-1}\left[\dfrac{1}{s}\right] + 6\mathscr{L}^{-1}\left[\dfrac{1}{s-3}\right]$$
$$= 3 \cdot 1 + 6e^{3t} = 3 + 6e^{3t}. \blacksquare$$

In Exercises 9 and 10 at the end of this section, we develop the important formulas

$$\mathscr{L}[\sin at] = \dfrac{a}{s^2 + a^2}, \quad \mathscr{L}[\cos at] = \dfrac{s}{s^2 + a^2}. \tag{5}$$

When using formula (5) and several of the formulas from the next section to compute inverse Laplace transforms, we will often have to adjust the constants so that the expression corresponds to one in the table. This is similar to what is done in introductory calculus when we compute

$$\int \dfrac{dx}{1 + 4x} = \dfrac{1}{4}\int \dfrac{d(4x)}{1 + (4x)} = \dfrac{1}{4}\ln|1 + 4x| + C.$$

Example 4.1.7 Given $F(s) = \dfrac{11}{s^2 + 17}$, find $f(t) = \mathscr{L}^{-1}[F(s)]$.

Solution From (5) we have $\mathscr{L}^{-1}\left[\dfrac{a}{s^2 + a^2}\right] = \sin at$. Then

$$\dfrac{11}{s^2 + 17} = \dfrac{11}{s^2 + (\sqrt{17})^2} = \dfrac{11}{\sqrt{17}}\dfrac{\sqrt{17}}{s^2 + (\sqrt{17})^2},$$

which is (5) with $a = \sqrt{17}$. Hence,

$$\mathscr{L}^{-1}\left[\dfrac{11}{s^2 + 17}\right] = \dfrac{11}{\sqrt{17}} \sin \sqrt{17}t. \blacksquare$$

Frequently, $F(s)$ will have to be expressed as a sum of several terms that can each be evaluated from known formulas.

Example 4.1.8

Given $F(s) = \dfrac{2s+5}{s^2+9}$, find $f(t) = \mathscr{L}^{-1}[F(s)]$.

Solution

$$\mathscr{L}^{-1}\left[\frac{2s+5}{s^2+9}\right] = \mathscr{L}^{-1}\left[2\frac{s}{s^2+9} + \frac{5}{3}\cdot\frac{3}{s^2+9}\right]$$

$$= 2\mathscr{L}^{-1}\left[\frac{s}{s^2+9}\right] + \frac{5}{3}\mathscr{L}^{-1}\left[\frac{3}{s^2+9}\right]$$

$$= 2\cos 3t + \frac{5}{3}\sin 3t. \quad \text{(by (5))} \blacksquare$$

More complicated expressions for $F(s)$ may require other techniques, such as partial fractions (Appendix B), to express $F(s)$ as a sum of simpler terms. This important idea is discussed more fully in Section 4.2. For now, we consider two examples that can be solved using (1), (2), (3), (4), (5).

Example 4.1.9

Given $F(s) = \dfrac{3}{s^2-4}$, find $f(t) = \mathscr{L}^{-1}[F(s)]$.

Solution

Using partial fractions,

$$\frac{3}{s^2-4} = \frac{3}{(s-2)(s+2)} = \frac{A}{s-2} + \frac{B}{s+2}.$$

Letting $s = 2, -2$ in $3 = A(s+2) + B(s-2)$ gives $A = \tfrac{3}{4}$, $B = -\tfrac{3}{4}$, and

$$\mathscr{L}^{-1}\left[\frac{3}{s^2-4}\right] = \frac{3}{4}\mathscr{L}^{-1}\left[\frac{1}{s-2}\right] - \frac{3}{4}\mathscr{L}^{-1}\left[\frac{1}{s+2}\right]$$

$$= \frac{3}{4}e^{2t} - \frac{3}{4}e^{-2t}.$$

(An alternative approach would be to use the formula from Exercise 11 for $\sinh at$.) $\blacksquare$

Example 4.1.10

Given $F(s) = \dfrac{2s+1}{s^3 - 3s^2 + 2s}$, find $f(t) = \mathscr{L}^{-1}[F(s)]$.

Solution

Using partial fractions,

$$\frac{2s+1}{s^3-3s^2+2s} = \frac{2s+1}{s(s-1)(s-2)} = \frac{A}{s} + \frac{B}{s-1} + \frac{C}{s-2},$$

so that $2s + 1 = A(s-1)(s-2) + Bs(s-2) + Cs(s-1)$. Letting $s = 0, 1, 2$ gives $A = \tfrac{1}{2}$, $B = -3$, $C = \tfrac{5}{2}$. Thus

$$\mathscr{L}^{-1}\left[\frac{2s+1}{s^3-3s^2+2s}\right] = \frac{1}{2}\mathscr{L}^{-1}\left[\frac{1}{s}\right] - 3\mathscr{L}^{-1}\left[\frac{1}{s-1}\right] + \frac{5}{2}\mathscr{L}^{-1}\left[\frac{1}{s-2}\right]$$

$$= \frac{1}{2} - 3e^t + \frac{5}{2}e^{2t}. \blacksquare$$

Exercises

1. Sketch the following function and explain why it is not piecewise continuous:
$$f(t) = \begin{cases} t, & 0 \le t < 2, \\ 3, & 2 = t, \\ \dfrac{1}{t-2}, & 2 < t. \end{cases}$$

In Exercises 2 through 4, sketch the function and explain why it is piecewise continuous.

2. $f(t) = \begin{cases} 1, & 0 \le t \le 1, \\ -1, & 1 < t \le 2, \\ 1, & 2 < t. \end{cases}$

3. $f(t) = \begin{cases} t, & 0 \le t < 1, \\ \frac{1}{2}, & 1 = t \\ t-1, & 1 < t. \end{cases}$

4. $f(t) = \begin{cases} e^{-t}, & 0 \le t < 1, \\ 3, & 1 = t, \\ e^{-t}, & 1 < t. \end{cases}$

In Exercises 5 through 7 compute $\mathcal{L}[f(t)]$, using the formula $\mathcal{L}[f] = \int_0^\infty e^{-st} f(t)\, dt$.

5. $f(t) = \begin{cases} 1 & \text{if } 0 \le t \le 1, \\ 0 & \text{if } 1 < t. \end{cases}$

6. $f(t) = \begin{cases} 1 & \text{if } 0 \le t < 1, \\ -1 & \text{if } 1 \le t < 2, \\ 0 & \text{if } 2 \le t. \end{cases}$

7. $f(t) = \begin{cases} t, & \text{if } 0 \le t < 1, \\ 2-t, & \text{if } 1 \le t \le 2, \\ 0, & \text{if } 2 < t. \end{cases}$

8. Verify that $\mathcal{L}[t] = 1/s^2$, using the formula $\mathcal{L}[f] = \int_0^\infty e^{-st} f(t)\, dt$.

9. Verify that $\mathcal{L}[\sin at] = a/(s^2 + a^2)$, using the formula $\mathcal{L}[f] = \int_0^\infty e^{-st} f(t)\, dt$.

10. Verify that $\mathcal{L}[\cos at] = s/(s^2 + a^2)$, using the formula $\mathcal{L}[f] = \int_0^\infty e^{-st} f(t)\, dt$.

11. Verify that $\mathcal{L}[\sinh at] = a/(s^2 - a^2)$.

In Exercises 12 through 26, use Exercises 8 through 11 and formulas (1), (2), (3), and (4) to compute $\mathcal{L}[f]$.

12. $f(t) = 3t + 2$
13. $f(t) = 3 \cosh 2t$
14. $f(t) = 4e^{3t} + 6e^{-t}$
15. $f(t) = 5 \sin 6t$
16. $f(t) = 4 \sin 3t + 5 \cos 7t$
17. $f(t) = -t + 3$
18. $f(t) = e^t - e^{-t} + e^{2t}$
19. $f(t) = 2 + \cos 5t$
20. $f(t) = 2 \sin 3t + 4 \sin 5t$
21. $f(t) = \sinh 3t$
22. $f(t) = t + 3 - e^t$
23. $f(t) = 2e^{-t} + 6e^{3t}$
24. $f(t) = -4 \sin 2t$
25. $f(t) = 3t - 1 + \cosh 2t$
26. $f(t) = 7e^{-5t} - 9e^{3t} - 6$

In Exercises 27 through 50, $F(s)$ is given. Use Exercises 8 through 11 and formulas (1) through (5) to compute $f(t) = \mathcal{L}^{-1}[F(s)]$.

27. $\dfrac{1}{s^2} - \dfrac{1}{s}$

28. $\dfrac{3}{s-3} + \dfrac{4}{s+3}$

29. $\dfrac{1}{s^2 + 9}$

30. $\dfrac{7s+1}{s^2+4}$

31. $\dfrac{1+s}{s^2}$

32. $\dfrac{2}{s^2 - 16}$

33. $\dfrac{3}{s} - \dfrac{7}{s^2} + \dfrac{19}{s^2+1}$

34. $\dfrac{2}{s} + \dfrac{3}{s+1} - \dfrac{7}{s-8}$

35. $\dfrac{3s+7}{s^2+16}$

36. $\dfrac{11}{s^2} - \dfrac{2}{s}$

37. $\dfrac{s+2}{s^2-1}$

38. $\dfrac{3}{s^2 - s}$

39. $\dfrac{2s-1}{s^2 - s}$

40. $\dfrac{5}{s^2 + 5s + 6}$

41. $\dfrac{-s-1}{s^2+s-2}$

42. $\dfrac{2s-13}{s^2+8s+15}$

43. $\dfrac{4}{s^2+s-6}$

44. $\dfrac{s-1}{s^2+6s+5}$

45. $\dfrac{1}{s^3 - 3s^2 + 2s}$

46. $\dfrac{2s+1}{s^3+3s^2+2s}$

47. $\dfrac{s^2-2}{s^3+8s^2+7s}$

48. $\dfrac{s^2+s}{s^3+4s^2-5s}$

49. $\dfrac{s+3}{s^3-s}$

50. $\dfrac{3}{s^3-4s}$

51. Let $f(t) = \begin{cases} 2 & \text{if } t = 1, \\ t & \text{if } t \neq 1. \end{cases}$

Show that $\mathscr{L}[f(t)] = \mathscr{L}[t]$.

52. Show that $\lim\limits_{t \to \infty} e^{-\alpha t} e^{t^2} = \infty$ for any α. Use this to show that e^{t^2} is not of exponential order.

53. Determine if $e^{\sqrt{t}}$ is of exponential order.

54. Show that e^{t^β} is of exponential order if $0 \leq \beta \leq 1$ and is not of exponential order if $\beta > 1$.

55. Prove Theorem 4.1.1, using the following two facts:

i) If $\int_0^\infty |h(t)|\, dt$ converges, then $\int_0^\infty h(t)\, dt$ converges.

ii) If $0 < g(t) < r(t)$ and $\int_0^\infty r(t)\, dt$ converges, then $\int_0^\infty g(t)\, dt$ converges.

4.2 Initial-Value Problems

The formulas developed in Section 4.1.1 told how to take the Laplace transform of particular functions. This section will develop two important operational formulas and then begin to show how they may be used in solving differential equations. A table (4.2.1) of all the basic formulas appears later in this section. Formulas from that table are referenced throughout this book by a T. We developed (T1) through (T4), (T7) and (T8) in Section 4.1.1.

Theorem 4.2.1

Shifting Theorem If $\mathscr{L}[f(t)] = F(s)$, then

$$\mathscr{L}[e^{ct}f(t)] = F(s - c). \tag{1}$$

Verification

$$\mathscr{L}[e^{ct}f(t)] = \int_0^\infty e^{-st} e^{ct} f(t)\, dt$$

$$= \int_0^\infty e^{-(s-c)t} f(t)\, dt$$

$$= \mathscr{L}[f](s - c) = F(s - c). \blacksquare$$

Combining Theorem 4.2.1 with (T3) and (T4) of the last section gives the two important formulas

$$\mathscr{L}[e^{bt} \sin at] = \frac{a}{(s - b)^2 + a^2} \tag{T5}$$

and

$$\mathscr{L}[e^{bt} \cos at] = \frac{s - b}{(s - b)^2 + a^2}. \tag{T6}$$

The next result is the key to solving differential equations using the Laplace transform. For convenience we shall give two special versions and the general version.

Theorem 4.2.2 (**First Derivative Case**) Suppose that $f(t)$ is continuous and of exponential order on $[0, \infty)$ and $f'(t)$ is piecewise continuous on $[0, \infty)$. Then $\mathscr{L}[f'(t)]$ exists and

$$\mathscr{L}[f'(t)] = s\mathscr{L}[f(t)] - f(0) = sF(s) - f(0).$$

(**Second Derivative Case**) Suppose that $f(t)$ and $f'(t)$ are continuous and of exponential order on $[0, \infty)$ and $f''(t)$ is piecewise continuous on $[0, \infty)$. Then $\mathscr{L}[f''(t)]$ exists and

$$\mathscr{L}[f''(t)] = s^2\mathscr{L}[f(t)] - sf(0) - f'(0) = s^2F(s) - sf(0) - f'(0).$$

(**General Case**) If $f(t), f'(t), \ldots, f^{(n-1)}(t)$ are continuous and of exponential order on $[0, \infty)$ and $f^{(n)}(t)$ is piecewise continuous on $[0, \infty)$, then $\mathscr{L}[f^{(n)}(t)]$ exists and

$$\begin{aligned}\mathscr{L}[f^{(n)}(t)] &= s^n\mathscr{L}[f(t)] - s^{n-1}f(0) - s^{n-2}f'(0)\cdots - f^{(n-1)}(0) \\ &= s^nF(s) - s^{n-1}f(0) - s^{n-2}f'(0)\cdots - f^{(n-1)}(0).\end{aligned}$$

Verification of First Derivative Case

Under the assumption of the theorem, we have

$$\mathscr{L}[f'(t)] = \lim_{b\to\infty}\int_0^b e^{-st}f'(t)\,dt \quad \text{(by definition)}$$

(if it exists)

$$= \lim_{b\to\infty}\int_0^b e^{-st}\,df(t)$$

$$= \lim_{b\to\infty}\left[e^{-st}f(t)\Big|_{t=0}^{t=b} - \int_0^b f(t)\,de^{-st}\right] \quad \text{(integration by parts)}$$

$$= \lim_{b\to\infty}\left[e^{-sb}f(b) - f(0) + s\int_0^b f(t)e^{-st}\,dt\right] \tag{2}$$

But $\lim\limits_{b\to\infty}\int_0^b f(t)e^{-st}\,dt = \mathscr{L}[f(s)]$, which exists since $f(t)$ is continuous and of exponential order, and $\lim\limits_{b\to\infty} e^{-sb}f(b) = 0$ for s large enough, since $f(t)$ is of exponential order. Thus (2) is $\mathscr{L}[f'(t)] = 0 - f(0) + s\mathscr{L}[f(t)]$, as desired. ∎

The general case may now be done by induction. For example, to show the second derivative case, compute as follows:

$$\mathscr{L}[f''(t)] = \mathscr{L}[(f'(t))'] = s\mathscr{L}[f'(t)] - f'(0), \quad \text{(by Case 1)}$$
$$= s(s\mathscr{L}[f(t)] - f(0)) - f'(0), \quad \text{(by Case 1 again)}$$
$$= s^2\mathscr{L}[f(t)] - sf(0) - f'(0), \quad \text{as desired.}$$

For convenience in working problems, we will now give Table 4.2.1, which includes the basic formulas developed and used in this chapter. Many of the formulas are special cases of each other. The first group of formulas give the Laplace transforms of particular functions. The second group are general operational formulas. Some of these will be developed in later sections or in the exercises at the end of this section. Table 4.2.2 gives additional formulas developed in the exercises and Section 4.4.

In this table $H(t)$ is a function to be discussed in the next section. The symbol $\delta(t)$ will be explained in Section 5 on impulses. $\Gamma(p)$ is the *Gamma function* discussed in Section 5.7. If n is a positive integer, then $\Gamma(n) = (n-1)!$. Thus, for example, $\Gamma(4) = 3! = 6$.

We are now ready to solve linear constant-coefficient differential equations using the Laplace transform. The general procedure is as follows.

Procedure for Solving Initial-Value Problems Using Laplace Transforms

Given the differential equation

$$a_n y^{(n)} + a_{n-1} y^{(n-1)} + \cdots + a_0 y = f(t),$$

with $a_n, \ldots, a_0$ constants; and f a function that has a Laplace transform:

1. Take the Laplace transform of both sides.
2. Use the initial conditions on the solution $y(t)$ when applying Theorem 4.2.2 (formulas (T18), (T19), (T20)).
3. Solve for $Y(s) = \mathscr{L}[y(t)]$.
4. Express $Y(s)$ in terms of functions of s appearing in the second column of Table 4.2.1 (and Table 4.2.2 if needed).
5. Compute $y(t)$ from $Y(s)$, using these formulas.

Note that, in many ways, finding $y(t)$ from $Y(s)$ is much like antidifferentiating. We have a table to work from and must often adjust constants in order to complete the problem. Before illustrating the entire procedure, we shall work two additional examples of Step 4.

Example 4.2.1

If $Y(s) = \dfrac{3}{s^5}$, find $y(t)$.

Solution

Using (T9) with $n = 4$,

$$y(t) = \mathscr{L}^{-1}[Y(s)] = \mathscr{L}^{-1}\left[\frac{3}{s^5}\right] = \frac{3}{4!}\mathscr{L}^{-1}\left[\frac{4!}{s^5}\right] = \frac{3}{4!}t^4. \quad \blacksquare$$

Table 4.2.1 Summary of Formulas with n, p, a, b, c Constants

	$g(t) = \mathscr{L}^{-1}[G(s)]$	$G(s) = \mathscr{L}[g(t)]$	Comments
(T1)	1	$\dfrac{1}{s}, \quad s > 0$	Special case of (T10)
(T2)	e^{at}	$\dfrac{1}{s-a}, \quad s > a$	
(T3)	$\sin at$	$\dfrac{a}{s^2 + a^2}, \quad s > 0$	
(T4)	$\cos at$	$\dfrac{s}{s^2 + a^2}, \quad s > 0$	
(T5)	$e^{bt} \sin at$	$\dfrac{a}{(s-b)^2 + a^2}, \quad s > |a|$	From (T3) and (T16)
(T6)	$e^{bt} \cos at$	$\dfrac{s-b}{(s-b)^2 + a^2}, \quad s > |a|$	From (T4) and (T16)
(T7)	$\sinh at$	$\dfrac{a}{s^2 - a^2}, \quad s > 0$	
(T8)	$\cosh at$	$\dfrac{s}{s^2 - a^2}, \quad s > 0$	
(T9)	t^n, n a positive integer	$\dfrac{n!}{s^{n+1}}, \quad s > 0$	Special case of (T10)
(T10)	$t^p, \; p > 0$	$\dfrac{\Gamma(p+1)}{s^{p+1}}, \quad s > 0$	
(T11)	$t^n e^{at}$	$\dfrac{n!}{(s-a)^{n+1}}, \quad s > a$	From (T9) and (T16)
(T12)	$\delta(t)$	1	Delta function (Section 4.5)
(T13)	$\delta(t - c)$	e^{-cs}	
(T14)	$H(t - c)$	$\dfrac{e^{-cs}}{s}$	Heaviside or unit-step function (Section 4.3)
(T15)	$f(t-c)H(t-c)$	$e^{-cs}F(s)$	
(T16)	$e^{ct}f(t)$	$F(s - c)$	Theorem 4.2.1
(T17)	$f(ct)$	$\dfrac{1}{c}F\!\left(\dfrac{s}{c}\right)$	
(T18)	$f'(t)$	$sF(s) - f(0)$	Theorem 4.2.2
(T19)	$f''(t)$	$s^2 F(s) - sf(0) - f'(0)$	Theorem 4.2.2
(T20)	$f^{(n)}(t)$	$s^n F(s) - s^{n-1}f(0) \cdots - f^{(n-1)}(0)$	Theorem 4.2.2
(T21)	$(-t)^n f(t)$, n a positive integer	$F^{(n)}(s)$	
(T22)	$\displaystyle\int_0^t f(t-\tau)g(\tau)\,d\tau$	$F(s)G(s)$	Convolution formula (Section 4.6)

Table 4.2.2 Other Formulas Involving the Laplace Transform

(T23) $\quad \mathscr{L}[t \sin at] = \dfrac{2as}{(s^2 + a^2)^2}$

(T24) $\quad \mathscr{L}[t \cos at] = \dfrac{s^2 - a^2}{(s^2 + a^2)^2}$

(T25) $\quad \mathscr{L}\left[\dfrac{1}{a}\sin at - t \cos at\right] = \dfrac{2a^2}{(s^2 + a^2)^2}$

(T26) $\quad \mathscr{L}[g] = \dfrac{\int_0^T e^{-st}g(t)\,dt}{1 - e^{-sT}} = \dfrac{\mathscr{L}[g(t)[1 - H(t - T)]]}{1 - e^{-sT}} \quad$ for g periodic with period T

(T27) $\quad \mathscr{L}^{-1}\left[\dfrac{F(s)}{1 - e^{-sT}}\right] = \mathscr{L}^{-1}\left[\sum_{n=0}^{\infty} e^{-nsT}F(s)\right] = \sum_{n=0}^{\infty} f(t - nT)H(t - nT)$

(T28) $\quad \mathscr{L}^{-1}\left[\dfrac{F(s)}{1 + e^{-sT}}\right] = \mathscr{L}^{-1}\left[\sum_{n=0}^{\infty} (-1)^n e^{-nsT}F(s)\right] = \sum_{n=0}^{\infty} (-1)^n f(t - nT)H(t - nT)$

Example 4.2.2

If $Y(s) = \dfrac{s}{s^2 + 2s + 5}$, find $y(t)$.

Solution

$$Y(s) = \dfrac{s}{(s + 1)^2 + 4} = \dfrac{s}{(s + 1)^2 + 2^2}.$$

The two formulas we have available with $(s + c)^2 + a^2$, with c, a constants in the denominator, are (T5) and (T6) with $c = -b$

$$\mathscr{L}^{-1}\left[\dfrac{s + 1}{(s + 1)^2 + 2^2}\right] = e^{-t}\cos 2t$$

and

$$\mathscr{L}^{-1}\left[\dfrac{2}{(s + 1)^2 + 2^2}\right] = e^{-t}\sin 2t.$$

We need to write $\dfrac{s}{(s + 1)^2 + 4}$ in terms of

$$\dfrac{s + 1}{(s + 1)^2 + 4} \quad \text{and} \quad \dfrac{2}{(s + 1)^2 + 4}.$$

Note that this is *not* partial fractions. If

$$\dfrac{s}{(s + 1)^2 + 4} = A\dfrac{(s + 1)}{(s + 1)^2 + 4} + B\dfrac{2}{(s + 1)^2 + 4},$$

then $s = A(s + 1) + 2B$, so that $A = 1$, $B = -\tfrac{1}{2}$. Thus,

$$Y(s) = \frac{s+1}{(s+1)^2 + 4} - \left(\frac{1}{2}\right)\frac{2}{(s+1)^2 + 4},$$

and we conclude, using (T5) and (T6), that

$$y(t) = e^{-t}\cos 2t - \tfrac{1}{2}e^{-t}\sin 2t. \quad \blacksquare$$

We can now solve many linear constant-coefficient differential equations using the Laplace transform.

Example 4.2.3 Solve the initial-value problem

$$y'' + 3y' + 2y = 1, \quad y(0) = 0, \quad y'(0) = 2.$$

Solution Take the Laplace transform of both sides of the differential equation, and obtain (using the linearity of $\mathscr{L}$),

$$\mathscr{L}[y''] + 3\mathscr{L}[y'] + 2\mathscr{L}[y] = \mathscr{L}[1].$$

From Theorem 4.2.2 (or T19),

$$\mathscr{L}[y''] = s^2 Y(s) - sy(0) - y'(0), \qquad \mathscr{L}[y'] = sY(s) - y(0).$$

Thus, using the initial conditions $y(0) = 0$, $y'(0) = 2$, we have

$$s^2 Y(s) - 2 + 3sY(s) + 2Y(s) = \frac{1}{s}.$$

Now solve for $Y(s)$:

$$(s^2 + 3s + 2)Y(s) = 2 + \frac{1}{s} = \frac{2s+1}{s}, \qquad Y(s) = \frac{2s+1}{s(s^2 + 3s + 2)}.$$

To find $y(t)$, expand $Y(s)$ using partial fractions (Appendix B),

$$Y(s) = \frac{2s+1}{s(s+1)(s+2)} = \frac{A}{s} + \frac{B}{s+1} + \frac{C}{s+2}.$$

To find A, B, C, multiply both sides by $s(s+1)(s+2)$ and get

$$2s + 1 = A(s+1)(s+2) + Bs(s+2) + Cs(s+1).$$

Evaluating this expression at the roots $s = 0, -1, -2$ of the denominator yields:

$$s = 0: \qquad 1 = A2, \text{ so } A = \tfrac{1}{2};$$
$$s = -1: \qquad -1 = B(-1)(1), \text{ so } B = 1;$$
$$s = -2: \qquad -3 = C(-2)(-1), \text{ so } C = -\tfrac{3}{2},$$

and hence

$$Y(s) = \frac{1}{2}\cdot\frac{1}{s} + \frac{1}{s+1} + \left(-\frac{3}{2}\right)\frac{1}{s+2}. \tag{3}$$

From (T2) and (T1)

$$1 = \mathscr{L}^{-1}\left[\frac{1}{s}\right], \quad e^{-t} = \mathscr{L}^{-1}\left[\frac{1}{s+1}\right], \quad e^{-2t} = \mathscr{L}^{-1}\left[\frac{1}{s+2}\right],$$

so that from (3),

$$y(t) = \tfrac{1}{2} + e^{-t} - \tfrac{3}{2}e^{-2t}. \blacksquare$$

This problem, of course, could also be done by the methods of undetermined coefficients or variation of parameters discussed in Chapter 3. In the next section we shall consider problems for which the Laplace transform works more easily than either of these techniques.

In concluding this section, two additional examples of solving differential equations using the Laplace transform will be given.

Example 4.2.4

Solve the initial-value problem

$$y'' - 2y' + y = 3e^t, \quad y(0) = 1, \quad y'(0) = 1.$$

Solution

Taking the Laplace transform of both sides gives

$$\mathscr{L}[y''] - 2\mathscr{L}[y'] + \mathscr{L}[y] = 3\mathscr{L}[e^t].$$

Using (T18), (T19) and the initial conditions on the left, and (T2) on the right, we obtain

$$s^2 Y(s) - s - 1 - 2(sY(s) - 1) + Y(s) = \frac{3}{s-1}.$$

Solving for $Y(s)$,

$$(s^2 - 2s + 1)Y(s) = \frac{3}{s-1} + s - 1 = \frac{s^2 - 2s + 4}{s-1},$$

$$Y(s) = \frac{s^2 - 2s + 4}{(s^2 - 2s + 1)(s-1)} = \frac{s^2 - 2s + 4}{(s-1)^3}.$$

Expanding $Y(s)$ by partial fractions, we obtain:

$$Y(s) = \frac{s^2 - 2s + 4}{(s-1)^3} = \frac{A}{s-1} + \frac{B}{(s-1)^2} + \frac{C}{(s-1)^3}.$$

Multiply by $(s-1)^3$, obtaining:

$$s^2 - 2s + 4 = A(s-1)^2 + B(s-1) + C$$
$$= A(s^2 - 2s + 1) + B(s-1) + C.$$

Equating the coefficients of like powers of s,

$$\begin{aligned} 1: \quad & 4 = A - B + C, \\ s: \quad & -2 = -2A + B, \\ s^2: \quad & 1 = A, \end{aligned}$$

and, solving from the last equation up, we get $A = 1$, $B = 0$, $C = 3$. Thus,

$$Y(s) = \frac{1}{s-1} + \frac{3}{(s-1)^3}.$$

Now, from (T2) and (T11),

$$e^t = \mathscr{L}^{-1}\left[\frac{1}{s-1}\right] \quad \text{and} \quad t^2 e^t = \mathscr{L}^{-1}\left[\frac{2}{(s-1)^3}\right].$$

Since $\dfrac{2}{(s-1)^3}$ appears in the table, but not $\dfrac{3}{(s-1)^3}$, we need to rewrite $Y(s)$ as

$$Y(s) = \frac{1}{s-1} + \frac{3}{2} \cdot \frac{2}{(s-1)^3},$$

in order to get

$$y(t) = e^t + \tfrac{3}{2} t^2 e^t. \quad \blacksquare$$

Example 4.2.5 Solve $y'' + 3y = \sin 5t$, $y(0) = 0$, $y'(0) = 0$.

Solution Taking the Laplace transform of both sides yields

$$\mathscr{L}[y''] + 3\mathscr{L}[y] = \mathscr{L}[\sin 5t]$$

or, by (T3), (T18), and (T19),

$$s^2 Y(s) + 3Y(s) = \frac{5}{s^2 + 25}.$$

Thus

$$Y(s) = \frac{5}{(s^2 + 3)(s^2 + 25)}.$$

By partial fractions,

$$Y(s) = \frac{5}{(s^2 + 3)(s^2 + 25)} = \frac{As + B}{s^2 + 3} + \frac{Cs + D}{s^2 + 25}.$$

Multiplying both sides by $(s^2 + 25)(s^2 + 3)$, we find that

$$5 = (As + B)(s^2 + 25) + (Cs + D)(s^2 + 3).$$

Equating like powers of s we get

$$
\begin{aligned}
1: & \quad 5 = 25B + 3D, \\
s: & \quad 0 = 25A + 3C, \\
s^2: & \quad 0 = B + D, \\
s^3: & \quad 0 = A + C.
\end{aligned}
$$

The second and fourth equations yield $A = C = 0$. The first and third equations give, after a little algebra,

$$B = 5/22, \qquad D = -5/22,$$

so that

$$Y(s) = \frac{5/22}{s^2 + 3} - \frac{5/22}{s^2 + 25}.$$

Formula (T3) is that

$$\mathcal{L}^{-1}\left[\frac{\sqrt{3}}{s^2 + 3}\right] = \sin\sqrt{3}t, \qquad \mathcal{L}^{-1}\left[\frac{5}{s^2 + 25}\right] = \sin 5t.$$

Thus, we need to write $Y(s)$ as

$$Y(s) = \frac{5}{22\sqrt{3}} \cdot \frac{\sqrt{3}}{s^2 + 3} - \left(\frac{1}{22}\right)\frac{5}{s^2 + 25},$$

to conclude that

$$y(t) = \frac{5}{22\sqrt{3}} \sin\sqrt{3}t - \frac{1}{22} \sin 5t. \ \blacksquare$$

Computer programs have been written that do symbolic manipulations (they appear to manipulate formulas rather than numbers). This means that Laplace-transform techniques and their accompanying algebra involving functions of s may soon be handled as simply as ordinary arithmetic on a pocket calculator.

Exercises

In Exercises 1 through 24 find $y(t) = \mathcal{L}^{-1}[Y(s)]$.

1. $Y(s) = \dfrac{2}{s^2 + 9}$

2. $Y(s) = \dfrac{2}{(s-3)^5}$

3. $Y(s) = \dfrac{s}{s^2 - 2s + 26}$

4. $Y(s) = \dfrac{2s}{s^2 + 4s + 13}$

5. $Y(s) = \dfrac{4}{(s+3)^6}$

6. $Y(s) = \dfrac{3s - 1}{s^2 + 2s + 3}$

7. $Y(s) = \dfrac{1}{s^2 - 9}$

8. $Y(s) = \dfrac{1}{s^4 + 5s^2 + 4}$

9. $Y(s) = \dfrac{2s + 5}{s^2 - 6s + 18}$

10. $Y(s) = \dfrac{17 - 3s}{s^2 + 2s + 26}$

11. $Y(s) = \dfrac{3s - 2}{s^2 + 10s + 26}$

12. $Y(s) = \dfrac{2}{s^2 - 13s + 42}$

13. $Y(s) = \dfrac{3}{s^{10}}$

14. $Y(s) = \dfrac{s^2 - 3}{s^4 - 1}$

15. $Y(s) = \dfrac{s}{s^4 - 16}$

16. $Y(s) = \dfrac{2}{s^5}$

17. $Y(s) = \dfrac{s^3 - 1}{s^4 + 10s^2 + 9}$

18. $Y(s) = \dfrac{s - 3}{(s + 1)^2 (s - 1)^2}$

19. $Y(s) = \dfrac{s^3}{s^4 - 5s^2 + 4}$

20. $Y(s) = \dfrac{s^2}{s^4 - 10s^2 + 9}$

21. $Y(s) = \dfrac{3}{2s^2 + 7}$

22. $Y(s) = \dfrac{-s + 1}{3s^2 + 11}$

23. $Y(s) = \dfrac{4}{(3s + 1)^5}$

24. $Y(s) = \dfrac{s - 2}{(5 - 7s)^9}$

In Exercises 25 through 44, use the Laplace transform to solve the initial-value problems.

25. $y'' - 4y = 1$, $y(0) = 0, y'(0) = 1$
26. $y'' - 6y' + 5y = 2$, $y(0) = 0, y'(0) = -1$
27. $y'' + 3y' - 4y = 0$, $y(0) = 1, y'(0) = 1$
28. $y'' - 3y' - 4y = t$, $y(0) = 0, y'(0) = 0$
29. $y'' + 5y' + 6y = 1$, $y(0) = 1, y'(0) = 0$
30. $y'' + y = 1$, $y(0) = y'(0) = 2$
31. $y'' - 3y' + 2y = 1$, $y(0) = 0, y'(0) = 1$
32. $y'' - y = e^t$, $y(0) = 1, y'(0) = 0$
33. $y'' + 9y = e^{-t}$, $y(0) = 1, y'(0) = 2$
34. $y'' + 4y = 3 \cos 5t$, $y(0) = 0, y'(0) = 3$
35. $y'' + 4y' + 13y = 2$, $y(0) = 1, y'(0) = 0$
36. $y'' - 2y' + 5y = 0$, $y(0) = 2, y'(0) = 1$
37. $y^{(4)} - y = 1$, $y(0) = 3, y'(0) = 5, y''(0) = 0, y'''(0) = 0$
38. $y' + y = e^t \sin t$, $y(0) = 1$
39. $y' - y = 1 - t$, $y(0) = -1$
40. $y'' + 2y' + 2y = e^{-t} \sin 2t$, $y(0) = 0, y'(0) = 0$
41. $y'' + 2y' + y = e^{-t}$, $y(0) = 0, y'(0) = 1$
42. $y''' + y'' - y' - y = 0$, $y(0) = 0, y'(0) = 0, y''(0) = 1$
43. $y' + 2y = e^{-t} \cos t$, $y(0) = 0$
44. $y' - 3y = t^2 e^t$, $y(0) = 1$

45. Show that, if $\mathscr{L}[f(t)] = F(s)$, then $\mathscr{L}[f(ct)] = \dfrac{1}{c} F\left(\dfrac{s}{c}\right)$, where c is a positive constant.

46. Verify that $\mathscr{L}[tf(t)] = -F'(s)$, assuming that $\int_0^\infty e^{-st} f(t)\,dt$ may be differentiated with respect to s by differentiating inside the integral sign.

47. Using Exercise 46, verify by induction that $\mathscr{L}[t^n f(t)] = (-1)^n F^{(n)}(s)$. This is (T21).

48. Using Exercise 47 and (T1), show that $\mathscr{L}[t^n] = n!/s^{n+1}$. This is (T9).

49. Using Exercise 47 and (T2) show that
$$\mathscr{L}[t^n e^{at}] = \dfrac{n!}{(s - a)^{n+1}}.$$
This is (T11).

In Sections 3.16, 3.17 we saw that resonance involved terms like $t \cos \omega t$ and $t \sin \omega t$. The remaining exercises develop the Laplace transform formulas needed to handle these problems.

50. Using (T3) and (T21) show that
$$\mathscr{L}[t \sin at] = \dfrac{2as}{(s^2 + a^2)^2} \qquad (4)(\text{T23})$$

51. Using (T4) and (T21) show that
$$\mathscr{L}[t \cos at] = \dfrac{s^2 - a^2}{(s^2 + a^2)^2} \qquad (5)(\text{T24})$$

Partial fractions results in terms like $(As + B)(s^2 + a^2)^{-2}$, which is not readily related to (4), (5). Thus (6) is helpful.

52. Using (4), (5), and (T3) show that

$$\mathscr{L}^{-1}\left[\frac{2a^2}{(s^2 + a^2)^2}\right] = \frac{1}{a}\sin at - t\cos at \quad (6)(T25)$$

In Exercises 53 through 58, use (4), (5), (6) to find $y(t) = \mathscr{L}^{-1}[Y(s)]$.

53. $Y(s) = \dfrac{2s + 1}{(s^2 + 1)^2}$

54. $Y(s) = \dfrac{3s - 5}{(s^2 + 4)^2}$

55. $Y(s) = \dfrac{s^2 - s}{(s^2 + 4)^2}$

56. $Y(s) = \dfrac{2s^3 - 1}{(s^2 + 1)^2}$

57. $Y(s) = \dfrac{s^3 - 1}{(s^2 + 9)^2}$

58. $Y(s) = \dfrac{s^2 + 3s}{(s^2 + 16)^2}$

In Exercises 59 through 62, solve the initial value problem using the Laplace transform and (T3), (T4), (4), (5), (6).

59. $y'' + y = \sin t, \quad y(0) = y'(0) = 0$
60. $y'' + 4y = 3\cos 2t, \quad y(0) = y'(0) = 0$
61. $y'' + 16y = 5\cos 4t, \quad y(0) = y'(0) = 1$
62. $y'' + 9y = \sin 3t, \quad y(0) = y'(0) = 1$

4.3 Discontinuous Forcing Functions

In the previous section we saw how the Laplace transform could be used to solve linear differential equations with constant coefficients. All of the examples given, however, could easily have been solved by either the method of undetermined coefficients or variation of parameters. In this and the next two sections, we shall examine problems where the Laplace transform works better than either of these two methods.

The key is a notational one. We need a way to write a piecewise-continuous function as a simple formula so that it may be handled the way the simpler functions $\cos t$, $\sin 5t$, and $e^{3t} \sin 2t$ were treated.

The needed notation involves the unit-step or Heaviside function. Since manipulating the unit-step function often gives students difficulty, we shall discuss it very carefully.

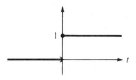

Figure 4.3.1 Graph of Heaviside function.

Definition 4.3.1

The *unit step*, or *Heaviside function* $H(t)$ is defined as

$$H(t) = \begin{cases} 1 & \text{if } t \geq 0, \\ 0 & \text{if } t < 0. \end{cases}$$

Its graph is given in Fig. 4.3.1.

The notation $u(t)$, $u_0(t)$, or $1(t)$, are sometimes also used to denote the Heaviside function. Other notations are also used.

The graph of $H(t - a)$, $a > 0$, (Fig. 4.3.2) is just a translate of that of $H(t)$. Thus

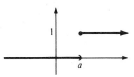

Figure 4.3.2 Graph of $H(t - a)$.

$$H(t-a) = \begin{cases} 0, & \text{if } t < a, \\ 1, & \text{if } t \geq a. \end{cases}$$

Example 4.3.1 Graph $f(t) = H(t-1) - H(t-3)$.

Solution Since the formula for $H(t-1)$ changes at $t = 1$, and the formula for $H(t-3)$ changes at $t = 3$, the formula for $f(t)$ will undergo changes at $t = 1$ and $t = 3$. Accordingly, we divide the t-interval into the intervals $(-\infty, 1)$, $[1, 3)$, and $[3, \infty)$.

If $t < 1$, then $f(t) = H(t-1) - H(t-3) = 0 - 0$, since $H(t-1) = 0$ if $t < 1$ and $H(t-3) = 0$ if $t < 3$.

If $1 \leq t < 3$, then $f(t) = H(t-1) - H(t-3) = 1 - 0 = 1$, since $H(t-1) = 1$ if $t \geq 1$ and $H(t-3) = 0$ if $t < 3$.

If $3 \leq t$, then $f(t) = H(t-1) - H(t-3) = 1 - 1 = 0$, since $H(t-1) = 1$ if $t \geq 1$ and $H(t-3) = 1$ if $t \geq 3$.

Figure 4.3.3 gives the graph of $f(t) = H(t-1) - H(t-3)$. ∎

Figure 4.3.3 Graph of $H(t-1) - H(t-3)$.

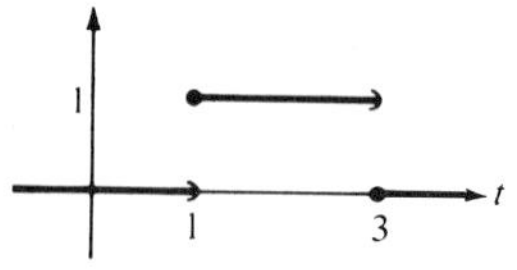

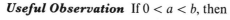

From this example we make the following deduction.

Useful Observation If $0 < a < b$, then

$$H(t-a) - H(t-b) = \begin{cases} 0 & \text{if } t < a, \\ 1 & \text{if } a \leq t < b, \\ 0 & \text{if } b \leq t, \end{cases}$$

and $H(t-a) - H(t-b)$ has the graph in Fig. 4.3.4.

Figure 4.3.4 Graph of $H(t-a) - H(t-b)$.

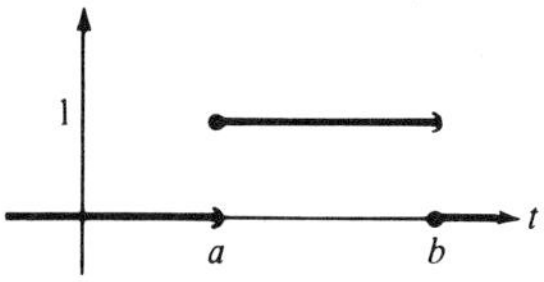

Now let $g(t)$ be some function of t. Then if $a < b$,

$$g(t)[H(t-a) - H(t-b)] = \begin{cases} 0 & \text{if } t < a, \\ g(t) & \text{if } a \leq t < b, \\ 0 & \text{if } b \leq t. \end{cases}$$

Example 4.3.2 Let $f(t) = t^2[H(t-1) - H(t-2)]$. Then

$$t^2[H(t-1) - H(t-2)] = \begin{cases} 0 & \text{if } t < 1, \\ t^2 & \text{if } 1 \leq t < 2, \\ 0 & \text{if } 2 \leq t, \end{cases}$$

and the graph of $t^2[H(t-1) - H(t-2)]$ is given in Fig. 4.3.5. The formula $t^2[H(t-1) - H(t-2)]$ picks out that part of the graph of t^2 between 1 and 2, and sets the rest equal to zero. ∎

Figure 4.3.5 Graph of $t^2[H(t-1) - H(t-2)]$.

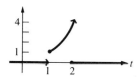

We are now ready to write a piecewise-continuous function in terms of unit-step functions.

Table 4.3.1

t interval	$f(t) = 2H(t) + tH(t-1) + (3-t)H(t-2) - 3H(t-4)$
$0 \le t < 1$	$f(t) = 2 \cdot 1 + t \cdot 0 + (3-t) \cdot 0 - 3 \cdot 0 = 2$
$1 \le t < 2$	$f(t) = 2 \cdot 1 + t \cdot 1 + (3-t) \cdot 0 - 3 \cdot 0 = 2 + t$
$2 \le t < 4$	$f(t) = 2 \cdot 1 + t \cdot 1 + (3-t) \cdot 1 - 3 \cdot 0 = 5$
$4 \le t$	$f(t) = 2 \cdot 1 + t \cdot 1 + (3-t) \cdot 1 - 3 \cdot 1 = 2.$

Example 4.3.3 Write in terms of unit-step functions the function $f(t)$ given by the graph in Fig. 4.3.6.

Solution The function $f(t)$ is

Figure 4.3.6

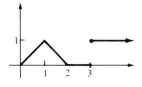

$$f(t) = \begin{cases} t & \text{if } 0 \le t < 1, \\ 2-t & \text{if } 1 \le t < 2, \\ 0 & \text{if } 2 \le t < 3, \\ 1 & \text{if } 3 \le t. \end{cases}$$

Thus

$$f(t) = \underbrace{t[H(t) - H(t-1)]}_{} + \underbrace{(2-t)[H(t-1) - H(t-2)]}_{}$$
$$+ \underbrace{1H(t-3)}_{}. \qquad (1)$$

Note that the first term in (1) is zero except when $0 \le t < 1$ and there it takes on the value t. The second term in (1) is zero except when $1 \le t < 2$ and then it takes on the value $2 - t$. The third term in (1) is zero except when $3 \le t$ and then it takes on the value 1. ∎

Note Since we consider only $t \ge 0$, the first term in (1) could have been written $t[1 - H(t-1)]$.

Example 4.3.4 Graph $f(t) = 2H(t) + tH(t-1) + (3-t)H(t-2) - 3H(t-4)$ for $t \ge 0$.

Solution Looking at the Heaviside functions, we see that the formula for $f(t)$ undergoes changes at $t = 1$, $t = 2$, and $t = 4$. Thus we break up the interval $[0, \infty)$ into four subintervals $[0, 1)$, $[1, 2)$, $[2, 4)$, $[4, \infty)$ and see what $f(t)$ looks like on each subinterval. Using the definition of the Heaviside function, we form Table 4.3.1. The graph of this function is given in Fig. 4.3.7. ∎

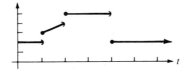

Figure 4.3.7

Now we shall consider how to use the Heaviside function in conjunction with the Laplace transform. From formula (T14) of the table in Section 4.2, if $c \geq 0$,

$$\mathscr{L}[H(t-c)] = \frac{e^{-cs}}{s}, \qquad s > 0. \qquad (2)(T14)$$

Technical Point If $f(t)$ is a piecewise-continuous function of exponential order, then $\mathscr{L}[f]$ is the same whether left- or right-hand limits are used for the values of f at the jump discontinuities. This is not really a problem, since physically all measurements take place over a nonzero time period. Thus, asking whether the value of f at a jump discontinuity is a right- or left-hand limit is not a physically meaningful question. We defined $H(0) = 1$. We could just as well have defined $H(0) = 0$. Since, for our purposes, the value of the function at the jump discontinuity is not important, we will sometimes omit this information from the graphs.

Verification of (T14)

$$\mathscr{L}[H(t-c)] = \int_0^\infty e^{-st} H(t-c)\, dt$$

$$= \int_0^c e^{-st} H(t-c)\, dt + \int_c^\infty e^{-st} H(t-c)\, dt$$

$$= \int_0^c e^{-st} 0\, dt + \int_c^\infty e^{-st} 1\, dt = \int_c^\infty e^{-st}\, dt$$

$$= \lim_{m \to \infty} \int_c^m e^{-st}\, dt = \lim_{m \to \infty} \frac{e^{-st}}{-s}\bigg|_{t=c}^{t=m}$$

$$= \lim_{m \to \infty} \frac{e^{-cs}}{s} - \frac{e^{-sm}}{s}$$

$$= \frac{e^{-cs}}{s} \qquad \text{if } s > 0,\, c > 0. \quad \blacksquare$$

(T14) is a special case of formula (T15),

$$\mathscr{L}[f(t-c)H(t-c)] = e^{-cs} F(s), \qquad (3)(T15)$$

with $f(t-c) = 1$, $F(s) = 1/s$. This formula may be verified much as (2) was. Formula (T15) is tricky to use at first, so several examples will be given.

Example 4.3.5 Compute $\mathscr{L}[t^2 H(t-1)]$.

Solution In order to use (3)(T15), we must take

$$t^2 H(t-1) = f(t-c) H(t-c).$$

Thus $c = 1$ and $t^2 = f(t-1)$. To compute $e^{-cs} F(s)$ in (3), we need $F(s) = \mathscr{L}[f(t)]$. But we have $f(t-1)$, which is not $f(t)$. To compute $f(t)$, introduce a new variable, say τ, and let $\tau = t - 1$, or $t = \tau + 1$. Then,

$$t^2 = f(t-1),$$
$$(\tau + 1)^2 = f(\tau),$$

or
$$\tau^2 + 2\tau + 1 = f(\tau). \tag{4}$$

Now in the formula (4), τ plays the role of a dummy (independent) variable, so that, from (T9),

$$F(s) = \mathscr{L}[f(\tau)] = \mathscr{L}[\tau^2 + 2\tau + 1]$$
$$= \frac{2}{s^3} + \frac{2}{s^2} + \frac{1}{s},$$

and (3) gives that

$$\mathscr{L}[t^2 H(t-1)] = e^{-s}\left[\frac{2}{s^3} + \frac{2}{s^2} + \frac{1}{s}\right]. \blacksquare$$

Example 4.3.6 Compute $\mathscr{L}[(e^t + 1)H(t-2)]$ using (3).

Solution Taking
$$(e^t + 1)H(t-2) = f(t-2)H(t-2)$$

gives $c = 2$ and $f(t-2) = e^t + 1$. Let $\tau = t - 2$ or $\tau + 2 = t$. Then $f(\tau) = e^{\tau+2} + 1 = e^2 e^\tau + 1$. Thus

$$F(s) = \mathscr{L}[e^2 e^\tau + 1] = e^2 \frac{1}{s-1} + \frac{1}{s}$$

and
$$\mathscr{L}[(e^t + 1)H(t-2)] = e^{-2s}\left[\frac{e^2}{s-1} + \frac{1}{s}\right]. \blacksquare$$

Example 4.3.7 If $Y(s) = e^{-3s}/s^2$, find $y(t) = \mathscr{L}^{-1}[Y(s)]$.

Solution The presence of e^{-3s} in $Y(s)$ is an indication that we should use (3)(T15). Taking

$$\frac{e^{-3s}}{s^2} = e^{-cs} F(s),$$

we see that $c = 3$ and $F(s) = 1/s^2$. Thus $f(t) = \mathscr{L}^{-1}[F(s)] = t$. The result for $y(t)$ is then

$$y(t) = f(t-3)H(t-3) = (t-3)H(t-3). \blacksquare$$

Example 4.3.8

If $Y(s) = e^{-2s}\left[\dfrac{3}{s} + \dfrac{s}{s^2+4}\right]$, find $y(t) = \mathcal{L}^{-1}[Y(s)]$.

Solution

Again the e^{-2s} suggests that we should use formula (T15). Taking

$$e^{-2s}\left[\dfrac{3}{s} + \dfrac{s}{s^2+4}\right] = e^{-cs}F(s),$$

we see that $c = 2$ and $F(s) = \dfrac{3}{s} + \dfrac{s}{s^2+4}$. Thus $f(t) = 3 + \cos 2t$. Hence,

$$y(t) = f(t-2)H(t-2) = [3 + \cos(2(t-2))]H(t-2)$$
$$= [3 + \cos(2t-4)]H(t-2). \blacksquare$$

4.3.1 Solution of Differential Equations

We can now use the Laplace transform to solve linear constant-coefficient differential equations with piecewise continuous forcing functions of exponential order.

Example 4.3.9

Solve the differential equation

$$y'' + 3y' + 2y = g(t), \qquad y(0) = 0, \qquad y'(0) = 1, \qquad (5)$$

using the Laplace transform where $g(t)$ has the graph in Fig. 4.3.8.

Solution

The function $g(t)$ is given by

$$g(t) = \begin{cases} 1 & \text{if } 0 \le t < 1, \\ 0 & \text{if } 1 \le t. \end{cases} \qquad (6)$$

Figure 4.3.8

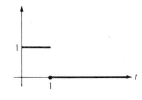

Thus $g(t) = [1 - H(t-1)]$ (or equivalently, $H(t) - H(t-1)$), and the differential equation is

$$y'' + 3y' + 2y = 1 - H(t-1), \qquad y(0) = 0, \qquad y'(0) = 1.$$

Taking the Laplace transform of both sides of this differential equation gives

$$s^2 Y(s) - sy(0) - y'(0) + 3(sY(s) - y(0)) + 2Y(s) = \dfrac{1}{s} - \dfrac{e^{-s}}{s},$$

so that

$$Y(s) = \left[\dfrac{s+1}{s(s^2+3s+2)}\right] - e^{-s}\left[\dfrac{1}{s(s^2+3s+2)}\right]. \qquad (7)$$

First, we shall find the inverse Laplace transform of the first term in $Y(s)$. Note that

$$\dfrac{s+1}{s(s^2+3s+2)} = \dfrac{s+1}{s(s+1)(s+2)} = \dfrac{1}{s(s+2)}.$$

By partial fractions,
$$\frac{1}{s(s+2)} = \frac{A}{s} + \frac{B}{s+2}$$
so that
$$1 = A(s+2) + Bs.$$
Evaluating at $s = 0$ and $s = -2$ yields $A = \frac{1}{2}$, $B = -\frac{1}{2}$. Thus
$$\mathscr{L}^{-1}\left[\frac{1}{s(s+2)}\right] = \mathscr{L}^{-1}\left[\frac{1/2}{s} + \frac{-1/2}{s+2}\right] = \frac{1}{2} - \frac{1}{2}e^{-2t}. \tag{8}$$
Now the second term of (7) is in the form $e^{-cs}F(s)$, where $c = 1$ and
$$F(s) = \frac{1}{s(s^2 + 3s + 2)} = \frac{1}{s(s+2)(s+1)},$$
so that (3), or (T15), should be used. Again using partial fractions,
$$F(s) = \frac{1}{s(s+2)(s+1)} = \frac{A}{s} + \frac{B}{s+2} + \frac{C}{s+1},$$
so that
$$1 = A(s+2)(s+1) + Bs(s+1) + Cs(s+2).$$
Since the denominator has distinct linear factors, evaluating at their roots $s = 0$, $s = -1$, $s = -2$ yields $A = \frac{1}{2}$, $C = -1$, and $B = \frac{1}{2}$. Hence,
$$F(s) = \frac{1/2}{s} + \frac{1/2}{s+2} + \frac{-1}{s+1}$$
and
$$f(t) = \frac{1}{2} + \frac{1}{2}e^{-2t} - e^{-t}.$$
Thus,
$$\mathscr{L}^{-1}[e^{-s}F(s)] = f(t-1)H(t-1)$$
$$= [\tfrac{1}{2} + \tfrac{1}{2}e^{-2(t-1)} - e^{-(t-1)}]H(t-1). \tag{9}$$
Finally, combining the two terms (8), (9),
$$y(t) = [\tfrac{1}{2} - \tfrac{1}{2}e^{-2t}] - [\tfrac{1}{2} + \tfrac{1}{2}e^{-2(t-1)} - e^{-(t-1)}]H(t-1). \quad \blacksquare \tag{10}$$

The approach of this section should be contrasted with the examples in Section 2.3, where we had to match initial conditions with terminal values at every discontinuity of the "forcing function" $g(t)$.

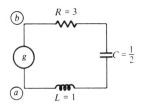

Figure 4.3.9

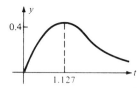

Figure 4.3.10
Graph of the solution (10) of (5).

A Physical Interpretation

As explained in Section 3.17, under certain operating conditions, equation (5) of the preceding exercise is a mathematical model for the *RLC* circuit in Fig. 4.3.9.

In Fig. 4.3.9, g is a voltage source that varies with time, $y(0) = 0$ is the initial charge in the capacitor, and $y'(0) = 1$ is the initial current in the capacitor. There are several ways to interpret $g(t)$ as given by (6). One is that it is the voltage across nodes ⓐ and ⓑ in Fig. 4.3.9, and these nodes are connected to other circuit elements not shown. Alternatively, g can be thought of as a one-volt battery that is shorted out for $t > 1$. Intuitively, one expects that, because of the relatively large resistance (the circuit is overdamped, in the terminology of Section 3.16), the capacitor will charge for $0 \le t < 1$ and discharge for $t \ge 1$. Figure 4.3.10 is the graph of $y(t)$. Note that, in fact, the capacitor keeps charging for 1.127 time units before beginning to discharge. That is because the inductor gives the current charging the capacitor the electrical equivalent of momentum. It takes a while for the current charging the capacitor to stop and reverse direction.

Example 4.3.10

Solve the differential equation

$$y'' + 3y' + 2y = g(t), \qquad y(0) = 0, \qquad y'(0) = 1, \qquad (11)$$

using the Laplace transform, where $g(t)$ has the graph shown in Fig. 4.3.11.

This is the same as the previous example, except that $g(t)$ is now a periodic function. In the next section, we shall discuss periodic functions in more detail, since they are very important.

Solution

First note that

$$g(t) = [1 - H(t-1)] + [H(t-2) - H(t-3)] + [H(t-4) - H(t-5)] + \cdots$$
$$= 1 - H(t-1) + H(t-2) - H(t-3) + H(t-4) - H(t-5) + \cdots$$
$$= \sum_{n=0}^{\infty} (-1)^n H(t-n).$$

While $g(t)$ is given by an infinite series, for any specific value of t only a finite number of terms in the series are nonzero. For example, if $t = 13.7$, then $H(13.7 - n)$ is zero if $n > 13.7$. The differential equation is now written as

$$y'' + 3y' + 2y = \sum_{n=0}^{\infty} (-1)^n H(t-n). \qquad (12)$$

It can be shown that it makes sense to take the Laplace transform of the series in (12), and this Laplace transform can be done term by term. Taking the Laplace transform of both sides of (12) gives

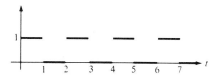

Figure 4.3.11

Figure 4.3.12
Graph of (14).

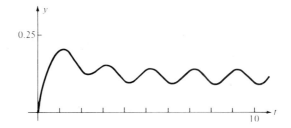

$$s^2 Y(s) - 1 + 3s Y(s) + 2Y(s) = \sum_{n=0}^{\infty} (-1)^n \frac{e^{-ns}}{s}.$$

Solving for $Y(s)$,

$$Y(s) = \frac{1}{s^2 + 3s + 2} + \sum_{n=0}^{\infty} (-1)^n e^{-ns} \left[\frac{1}{s(s^2 + 3s + 2)} \right]$$

$$= \frac{1}{s+1} + \frac{-1}{s+2} + \sum_{n=0}^{\infty} (-1)^n e^{-ns} \left[\frac{1}{s(s^2 + 3s + 2)} \right]. \quad (13)$$

From the preceding example, we know that

$$\mathcal{L}^{-1} \left[e^{-ns} \frac{1}{s(s^2 + 3s + 2)} \right] = f(t-n) H(t-n)$$

and

$$f(t) = \mathcal{L}^{-1} \left[\frac{1}{s(s^2 + 3s + 2)} \right] = \frac{1}{2} + \frac{1}{2} e^{-2t} - e^{-t}.$$

Thus (13) gives that

$$y(t) = e^{-t} - e^{-2t} + \sum_{n=0}^{\infty} (-1)^n [\tfrac{1}{2} + \tfrac{1}{2} e^{-2(t-n)} - e^{-(t-n)}] H(t-n). \quad (14)$$

Again, for any specific value of t, only a finite number of terms in the series in (14) are nonzero. Figure 4.3.12 is the graph of (14). ∎

Exercises

In Exercises 1 through 3, write the function defined on $[0, \infty)$ in terms of unit-step functions and graph.

1. $f(t) = \begin{cases} 0 & 0 \le t < 2, \\ 3 & 2 \le t < 5, \\ t & 5 \le t. \end{cases}$

2. $f(t) = \begin{cases} 1 & 0 \le t < 1, \\ -1 & 1 \le t < 2, \\ 1 & 2 \le t < 3, \\ -1 & 3 \le t. \end{cases}$

3. $f(t) = \begin{cases} \sin t & 0 \leq t < \pi, \\ 0 & \pi \leq t < 2\pi, \\ \sin t & 2\pi \leq t < 3\pi, \\ 0 & 3\pi \leq t. \end{cases}$

In Exercises 4 through 9, write the function defined on $[0, \infty)$ in terms of unit-step functions.

4.
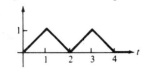

5.
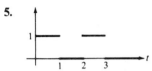

6.
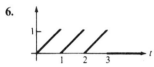

7.
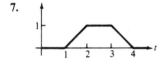

8.
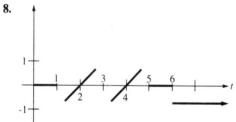

9.
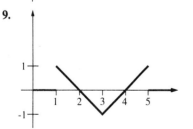

In Exercises 10 through 21, compute the Laplace transform $Y(s)$ of $y(t)$.

10. $y(t) = 2 - 5H(t-1) + 6H(t-3)$
11. $y(t) = tH(t-2)$
12. $y(t) = (t-2)H(t-2)$
13. $y(t) = (t^3 + 1)H(t-1)$
14. $y(t) = e^{3t}H(t-2) + 6H(t-3)$
15. $y(t) = \sin t H(t-\pi)$
16. $y(t) = \sin t H(t-\pi/2)$
17. $y(t) = \cos t H(t-2\pi)$
18. $y(t) = (t^3 + t)H(t-2)$
19. $y(t) = e^{2t}H(t-3)$
20. $y(t) = t^2 e^{-3t}H(t-1)$
21. $y(t) = te^{5t}H(t-2)$

In Exercises 22 through 29, compute the inverse Laplace transform $y(t)$ of $Y(s)$.

22. $Y(s) = \dfrac{e^{-3s}}{s^3}$

23. $Y(s) = e^{-2s}\dfrac{1}{s^2+s} + e^{-3s}\dfrac{1}{s^2+s}$

24. $Y(s) = \dfrac{1}{s^2} + \dfrac{e^{-s}}{s^2+4} + e^{-2s}\dfrac{s}{s^2+9}$

25. $Y(s) = \dfrac{e^{-3s}}{s^2+2s+2}$

26. $Y(s) = \dfrac{e^{-2s}}{(s+4)^5}$

27. $Y(s) = \dfrac{e^{-s}}{s^2+1} - \dfrac{e^{-2s}}{s^2+4}$

28. $Y(s) = e^{-3s}\left[\dfrac{1}{s} + \dfrac{1}{s-2}\right]$

29. $Y(s) = e^{-s}\left[\dfrac{4}{s} + \dfrac{6s}{s^2+9}\right]$

30. Using only the definitions of the Laplace transform and the Heaviside function, verify that
$$\mathscr{L}[f(t-c)H(t-c)] = e^{-cs}F(s)$$
if $f(t)$ is piecewise-continuous of exponential order and $\mathscr{L}[f(t)] = F(s)$.

In Exercises 31 through 38, solve the differential equation using the Laplace transform.

31. $y'' + y = g(t)$, $\quad y(0) = y'(0) = 0$,
where $g(t) = \begin{cases} 0, & 0 \leq t < 2, \\ 1, & 2 \leq t < 5, \\ 0, & 5 \leq t. \end{cases}$

32. $y'' - y = g(t)$, $\quad y(0) = y'(0) = 0$
where $g(t) = \begin{cases} 0, & 0 \leq t < 1, \\ 1, & 0 \leq 1 \leq t < 3, \\ 2, & 3 \leq t. \end{cases}$

33. $y' - 3y = g(t)$, $y(0) = 1$, $g(t)$ given by

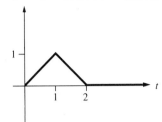

34. $y' + y = g(t)$, $y(0) = 0$, $g(t)$ given by

35. $y'' + 2y' + 10y = g(t)$, $y(0) = 1$, $y'(0) = 0$,

where $g(t) = \begin{cases} 1, & 0 \le t < 3, \\ 0, & 3 \le t. \end{cases}$

36. $y'' + 5y' + 6y = g(t)$, $y(0) = y'(0) = 0$,

$g(t) = \begin{cases} 0, & 0 \le t < \pi/2, \\ \cos t, & \pi/2 \le t < 3\pi/2, \\ 0, & 3\pi/2 \le t. \end{cases}$

37. $y'' + 4y = g(t)$, $y(0) = 0$, $y'(0) = 0$, g given by the graph in Exercise 4.3.5.

38. $y'' - 2y' + y = g(t)$, $y(0) = 1$, $y'(0) = 1$,

where $g(t) = \begin{cases} t^2, & 0 \le t < 1, \\ 0, & 1 \le t. \end{cases}$

In Exercises 39 and 40, use the method given in Example 4.3.10 to solve the differential equation using Laplace transforms.

39. $y'' + 4y = g(t)$, $y(0) = 0$, $y'(0) = 0$, where $g(t)$ is periodic and given by:

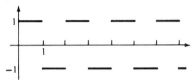

40. $y' + 2y = g(t)$, $y(0) = 1$, where $g(t)$ is periodic and given by:

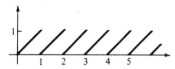

4.4 Periodic Functions

A function $g(t)$ defined on $[0, \infty)$ is *periodic with period T* if $g(t + T) = g(t)$ for all $t > 0$. Periodic functions appear in many applications. We have already seen examples of periodic functions and the Laplace transform. This section will consider them in more detail.

The basic formula is

If $g(t)$ is periodic on $[0, \infty)$ with period T and has a Laplace transform, then:

$$\mathscr{L}[g(t)] = \frac{\int_0^T e^{-st} g(t)\, dt}{1 - e^{-sT}}. \tag{1}$$

Verification Suppose $g(t)$ is periodic with period T and has a Laplace transform. Then

$$\mathscr{L}[g(t)] = \int_0^\infty e^{-st}g(t)\,dt = \int_0^T e^{-st}g(t)\,dt + \int_T^\infty e^{-st}g(t)\,dt$$

(Let $t = T + \tau$ in the second integral.)

$$= \int_0^T e^{-st}g(t)\,dt + \int_0^\infty e^{-s(T+\tau)}g(T+\tau)\,d\tau$$

$$= \int_0^T e^{-st}g(t)\,dt + e^{-sT}\int_0^\infty e^{-s\tau}g(\tau)\,d\tau,$$

(since $g(t)$ has period T.) Thus

$$\mathscr{L}[g(t)] = \int_0^T e^{-st}g(t)\,dt + e^{-sT}\mathscr{L}[g(t)].$$

Solving this equation for $\mathscr{L}[g(t)]$ gives (1). ∎

Example 4.4.1 Find $\mathscr{L}[g(t)]$, where

$$g(t) = \begin{cases} 1 & \text{if } 0 \le t < 1, \\ 0 & \text{if } 1 \le t < 2; \end{cases} \quad \text{period is 2.}$$

Figure 4.4.1 gives the graph of $g(t)$.

Solution Since the period is 2, formula (1) gives

$$\mathscr{L}[g(t)] = \frac{1}{1-e^{-2s}}\int_0^2 e^{-st}g(t)\,dt$$

$$= \frac{1}{1-e^{-2s}}\int_0^1 e^{-st}\,dt$$

$$= \frac{1}{1-e^{-2s}}\left(\frac{e^{-st}}{-s}\bigg|_{t=0}^{t=1}\right)$$

$$= \frac{1}{1-e^{-2s}}\left(\frac{e^{-s}}{-s} + \frac{1}{s}\right)$$

which can be simplified to

$$\frac{1}{s(1+e^{-s})}. \quad \blacksquare$$

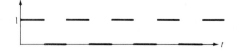

Figure 4.4.1

The most difficult part of (1) is computing the integral $\int_0^T e^{-st}g(t)\, dt$. This can be done using Laplace transform tables, as follows. If

$$\hat{g}(t) = \begin{cases} g(t) & \text{for } 0 \leq t \leq T, \\ 0 & \text{for } t > T, \end{cases}$$

then

$$\int_0^T e^{-st}g(t)\, dt = \mathscr{L}[\hat{g}(t)], \qquad (2)$$

since

$$\int_0^T e^{-st}g(t)\, dt = \int_0^T e^{-st}\hat{g}(t)\, dt$$

$$= \int_0^\infty e^{-st}\hat{g}(t)\, dt = \mathscr{L}[\hat{g}(t)].$$

Note that $\hat{g}(t) = g(t)[1 - H(t - T)]$.

Example 4.4.2 Let $g(t)$ have period 2, where

$$g(t) = \begin{cases} t & \text{for } 0 \leq t \leq 1, \\ 2 - t & \text{for } 1 \leq t \leq 2. \end{cases}$$

Find $G(s)$.

Solution Let

$$\hat{g}(t) = t[1 - H(t-1)] + (2-t)[H(t-1) - H(t-2)]$$
$$= t + 2(1-t)H(t-1) + (t-2)H(t-2).$$

Then $\hat{g}(t) = g(t)$ for $0 \leq t \leq 2$ and $\hat{g}(t) = 0$ for $t > 2$. Graphs of g and $\hat{g}$ are given in Fig. 4.4.2. Thus by (1), (2),

$$\mathscr{L}[g(t)] = \frac{1}{1-e^{-2s}} \int_0^2 e^{-st}g(t)\, dt$$

$$= \frac{1}{1-e^{-2s}} \mathscr{L}[\hat{g}(t)]$$

$$= \frac{1}{1-e^{-2s}} \mathscr{L}[t + 2(1-t)H(t-1) + (t-2)H(t-2)]$$

$$= \frac{1}{1-e^{-2s}}\left[\frac{1}{s^2} - 2\frac{e^{-s}}{s^2} + \frac{e^{-2s}}{s^2}\right]. \blacksquare$$

Figure 4.4.2

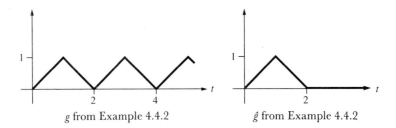

g from Example 4.4.2   ĝ from Example 4.4.2

Inverse Laplace transforms are not usually computed using (1). One method to compute inverse Laplace transformations is by the use of series. Recall that, for $|x| < 1$,

$$\frac{1}{1-x} = 1 + x + x^2 + \cdots = \sum_{n=0}^{\infty} x^n. \tag{3}$$

If $s > 0$, $T > 0$, then $e^{-sT} < 1$. Letting $x = e^{-sT}$, (3) becomes

$$\frac{1}{1 - e^{-sT}} = 1 + e^{-sT} + e^{-2sT} + \cdots = \sum_{n=0}^{\infty} e^{-nsT}. \tag{4}$$

Thus, if $\mathscr{L}[f(t)] = F(s)$

$$\frac{1}{1 - e^{-sT}} F(s) = \sum_{n=0}^{\infty} e^{-nsT} F(s).$$

Taking the inverse Laplace transform term by term using (T15) with $c = nT$ gives

$$\mathscr{L}^{-1}\left[\frac{1}{1 - e^{-sT}} F(s)\right] = \sum_{n=0}^{\infty} f(t - nT) H(t - nT). \tag{5}$$

Similarly, letting $x = -e^{-sT}$ in (3) gives

$$\frac{1}{1 + e^{-sT}} = \sum_{n=0}^{\infty} (-1)^n e^{-nsT},$$

and

$$\mathscr{L}^{-1}\left[\frac{1}{1 + e^{-sT}} F(s)\right] = \sum_{n=0}^{\infty} (-1)^n f(t - nT) H(t - nT). \tag{6}$$

Example 4.4.3

Let $G(s) = \dfrac{e^{-2s} - 1}{s^2(1 - e^{-3s})}$. Find $g(t) = \mathscr{L}^{-1}[G(s)]$.

Solution

Using (4) we get

$$G(s) = \frac{1}{s^2}(e^{-2s} - 1) \sum_{n=0}^{\infty} e^{-3ns}$$

$$= \sum_{n=0}^{\infty} \frac{1}{s^2} e^{-(2+3n)s} - \sum_{n=0}^{\infty} \frac{1}{s^2} e^{-3ns}.$$

Now, by (T15), $\mathscr{L}^{-1}\left[\dfrac{1}{s^2}e^{-cs}\right] = (t-c)H(t-c)$. Thus,

$$g(t) = \sum_{n=0}^{\infty}(t-2-3n)H(t-2-3n) - \sum_{n=0}^{\infty}(t-3n)H(t-3n).$$

This example also illustrates that, if $G(s)$ has a $1/[1-e^{-sT}]$-factor, it is not necessarily the case that $g(t)$ is periodic. ∎

Alternative Solution   We could also do Example 4.4.3 using formula (5) with $T=3$ and $F(s) = -\dfrac{1}{s^2} + \dfrac{1}{s^2}e^{-2s}$. Then $f(t) = -t + (t-2)H(t-2)$, and (5) is

$$\sum_{n=0}^{\infty} f(t-3n)H(t-3n)$$

$$= \sum_{n=0}^{\infty}[-(t-3n) + (t-3n-2)H(t-3n-2)]H(t-3n)$$

$$= \sum_{n=0}^{\infty} -(t-3n)H(t-3n) + (t-3n-2)H(t-3n-2)H(t-3n)$$

$$= \sum_{n=0}^{\infty} -(t-3n)H(t-3n) + (t-3n-2)H(t-3n-2).$$

The last equality follows, since

$$H(t-a)H(t-b) = H(t-c),$$

where c is the larger of a, b.

Exercises

In Exercises 1 through 8, sketch $g(t)$ and find $G(s)$ using formula (1), or (1) and (2).

1. $g(t) = |\sin t|$ (rectified sine)
2. $g(t) = |\cos t|$ (rectified cosine)
3. $g(t) = \begin{cases} t^2 & \text{for } 0 \leq t < 1, \\ 2-t^2 & \text{for } 1 \leq t < 2, \end{cases}$ period 2.
4. $g(t) = t$ for $0 \leq t < 1$, period 1.
5. $g(t) = e^t$ for $0 \leq t < 1$, period 1.
6. $g(t) = t^3$ for $0 \leq t < 1$, period 1.
7. $g(t) = \begin{cases} 1 & \text{for } 0 \leq t < 1, \\ -1 & 1 \leq t < 2, \end{cases}$ period 2
8. $g(t) = \begin{cases} t & \text{for } 0 \leq t < 1, \\ 1 & 1 \leq t < 2, \\ -1 & \text{for } 2 \leq t < 3, \end{cases}$ period 3

In Exercises 9 through 20, find $g(t)$, given $G(s)$.

9. $G(s) = \dfrac{1}{1-e^{-s}}\left[\dfrac{1}{s^2}+\dfrac{1}{s^3}\right]$

10. $G(s) = \dfrac{1}{1+e^{-3s}}\left[\dfrac{2s}{s^2+6s+13}\right]$

11. $G(s) = \dfrac{1}{1-e^{-s}}\left[\dfrac{s}{s^2+4}\right]$

12. $G(s) = \dfrac{1}{s}\tanh\dfrac{s}{2}$

13. $G(s) = \dfrac{1}{1-e^{-2s}}\left[\dfrac{1}{s^2}+\dfrac{e^{-s}}{s^3}\right]$

14. $G(s) = \dfrac{1}{s}\operatorname{sech} s$

15. $G(s) = \dfrac{1}{1-e^{-\pi s}}\left[\dfrac{1}{s}+\dfrac{e^{-\pi s/2}}{s^2+1}\right]$

16. $G(s) = \dfrac{1}{1-e^{-3s}}\left[\dfrac{1}{s}+\dfrac{e^{-s}}{s^2}+\dfrac{e^{-2s}}{s^3}\right]$

17. $G(s) = \dfrac{1}{1+e^{-5s}}\left[\dfrac{1}{s^3}+\dfrac{e^{-2s}}{s^4}\right]$

18. $G(s) = \dfrac{1}{1+e^{-s}}\left[\dfrac{1}{s^2-1}+\dfrac{e^{-2s}}{s^2-4}\right]$

19. $G(s) = \dfrac{1}{s^2}\operatorname{csch} s$

20. $G(s) = \dfrac{1}{1+e^{-3s}}\left[\dfrac{1}{s+2}+\dfrac{e^{-2s}}{s-2}\right]$

21. Verify that $H(t-a)H(t-b) = H(t-c)$, where c is the larger of a, b.

In Exercises 22 through 28, solve the initial-value problem, using the Laplace transform.

22. $y' + y = g(t)$, $\quad y(0) = 0$, $g(t)$ from Example 1.

23. $y'' + y = |\sin 2t|$, $\quad y(0) = 0$, $y'(0) = 0$

24. $y'' - y = \sin t - |\sin t|$, $\quad y(0) = 0$, $y'(0) = 1$

25. $y' + y = g(t)$, $\quad y(0) = 1$, $g(t)$ as in Exercise 5.

26. $y' - 2y = g(t)$, $\quad y(0) = 0$, $g(t)$ as in Exercise 4.

27. $y'' - 4y = g(t)$, $\quad y(0) = y'(0) = 0$, $g(t)$ as in Exercise 7.

28. $y'' + 2y' + y = g(t)$, $\quad y(0) = y'(0) = -1$, $g(t)$ as in Exercise 8.

29. Suppose that the series $\sum_{n=0}^{\infty} t^n/n!$ for e^t can be Laplace-transformed term by term. Show that the resulting series gives $\mathscr{L}[e^t]$, which is $1/(s-1)$.

4.5 Impulses and Distributions

Suppose that a capacitor having charge Q_0 for time $t \leq t_0$ is suddenly partially discharged (think of the spark in a spark plug) to a lower charge Q_1 at time t_1. The capacitor then maintains the charge Q_1 for $t \geq t_1$. The amount of charge lost is $Q_1 - Q_0$. If this discharge happens quickly, then, for a very brief period of time, the current $i(t)$, which is the rate of change of the charge $q(t)$, must be very large. The faster the discharge, the higher the current gets. The relationship between change in charge and current is

$$\Delta Q = Q_1 - Q_0 = \int_{t_0}^{t_1} \frac{dq}{dt}\, dt = \int_{t_0}^{t_1} i(t)\, dt. \tag{1}$$

Typical graphs of $q(t)$ and $i(t)$ are given in Fig. 4.5.1. If the time interval is shortened, we have the graphs shown in Fig. 4.5.2.

Since $i(t)$ is zero outside of $[t_0, t_1]$, (1) may be replaced by

$$\Delta Q = \int_{-\infty}^{\infty} i(t)\, dt.$$

Figure 4.5.1

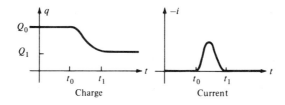

Figure 4.5.2

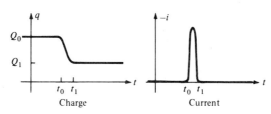

Figure 4.5.3

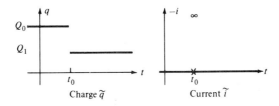

Similarly, one may think of a mass m being acted on by a large force for a brief period of time, leading to a change in momentum. In this case the momentum mv plays the role of the charge, and the instantaneous force $(mv)'$ plays the role of the current.

Suppose that we take the limit as the length of the time interval $[t_0, t_1]$ goes to zero by letting $t_1 \to t_0^+$. The charge $q(t)$ approaches the function (graphed in Fig. 4.5.3):

$$\tilde{q}(t) = Q_0 - (Q_0 - Q_1)H(t - t_0).$$

Let $\tilde{i}(t)$ be the limit of $i(t)$ as $t_1 \to t_0^+$. But $i(t)$ goes to zero for $t \neq t_0$ and $i(t_0) \to \infty$. On the other hand, we still want (1) to hold:

$$\Delta Q = Q_1 - Q_0 = \int_{-\infty}^{\infty} \tilde{i}(t)\, dt,$$

which means $\tilde{i}(t)$ should have a finite area. Clearly, this limiting current $\tilde{i}(t)$ is not a function in the sense we are familiar with. It is an example of a *distribution* or *impulse function*. These generalized functions are a very convenient mathematical notation.

Definition 4.5.1 $\delta(t)$ is a mathematical object known as the *delta function*. It is an example of a *distribution* or *generalized function*. It has the following properties:

i) $\delta(t) = 0$ if $t \neq 0$.

ii) $\delta(0)$ is not defined.

iii) $\int_{-\infty}^{\infty} \delta(t)\, dt = 1$.

iv) If $g(t)$ is a continuous function on $(-\infty, \infty)$, then
$$\int_{-\infty}^{\infty} g(t)\delta(t)\, dt = g(0).$$

It is possible to build a logical, rigorous definition of $\delta(t)$, but we shall not do so. Intuitively, we may think of $\delta(t)$ as an approximation of a physical impulse of magnitude 1 at time $t = 0$. For example, it could be the rapid transfer of one unit of charge at time zero. It can be shown that if a is a constant, then

v) $\delta(t - a) = 0$ if $t \neq a$.

vi) $\int_{-\infty}^{\infty} \delta(t - a)\, dt = 1$.

vii) If $g(t)$ is a continuous function on $(-\infty, \infty)$, then
$$\int_{-\infty}^{\infty} g(t)\delta(t - a)\, dt = g(a).$$

From these formulas, we get that
$$\int_{-\infty}^{t} \delta(t)\, dt = \begin{cases} 1 & \text{if } t > 0, \\ 0 & \text{if } t < 0. \end{cases}$$

Thus, formally,
$$\int_{-\infty}^{t} \delta(t)\, dt = H(t),$$

and $\delta(t)$ may be considered, in some sense, to be the derivative of the Heaviside function. One nice property of the Laplace transform is that it works almost as easily for distributions and impulses as it does for ordinary functions.

Proceeding formally,
$$\mathscr{L}[\delta(t - a)] = \int_0^{\infty} e^{-st}\delta(t - a)\, dt,$$
$$= [\text{by (vii) above}] e^{-sa} \qquad (\text{this is (T13)}).$$

In particular,
$$\mathscr{L}[\delta(t)] = 1,$$

which is (T12). There are many other distributions, for example, "derivatives" of $\delta(t)$, but they will not be considered here (note Exercise 11).

Example 4.5.1

Solve $y' + y = \delta(t - 1)$, $y(0) = 1$.

Solution   Taking the Laplace transform of both sides gives

$$sY(s) - y(0) + Y(s) = e^{-s}$$

Figure 4.5.4

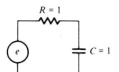

or

$$Y(s) = \frac{1}{s+1} + \frac{e^{-s}}{s+1},$$

so that, by (T2) and (T15),

$$y(t) = \mathcal{L}^{-1}\left[\frac{1}{s+1}\right] + \mathcal{L}^{-1}\left[e^{-s}\frac{1}{s+1}\right] = e^{-t} + e^{-(t-1)}H(t-1). \blacksquare \quad (2)$$

Figure 4.5.5
Graph of (2).

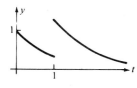

Physically, this example could be viewed as the simple linear *RC* circuit in Fig. 4.5.4, where y is the charge on the capacitor at time t and there is an initial charge of one on the capacitor. For $0 \leq t < 1$, the voltage e is zero and the capacitor is discharging. At $t = 1$, there is a voltage impulse; that is, a very large voltage is applied for a brief period, which recharges the capacitor. Then the voltage is again zero, and the capacitor resumes discharging.

Figure 4.5.6

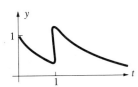

The graph of (2) is given in Fig. 4.5.5. This graph should be interpreted as meaning that, in a real problem, $y(t)$ would be given by a function like that in Fig. 4.5.6.

Needless to say, in real problems involving impulses, some care should be taken to ensure that the equations being used still provide accurate models in the presence of the large, but brief, values given by the impulse.

Exercises

In Exercises 1 through 8, solve the differential equation.

1. $y' + 8y = \delta(t - 1) + \delta(t - 2)$, $y(0) = 0$
2. $y'' + 2y' + 2y = \delta(t - 5)$, $y(0) = 1, y'(0) = 0$
3. $y'' + 6y' + 109y = \delta(t - 1) - \delta(t - 7)$,
   $y(0) = 0, y'(0) = 0$
4. $y'' + 3y' + 2y = -2\delta(t - 1)$, $y(0) = 1$,
   $y'(0) = 0$
5. $y'' + 4y' + 3y = 1 + \delta(t - 3)$, $y(0) = 0$,
   $y'(0) = 1$
6. $y' + 4y = 1 - \delta(t - 4)$, $y(0) = 1$
7. $y'' + y = 1 + \delta(t - 2\pi)$, $y(0) = 1, y'(0) = 0$
8. $y'' + 4y = t + 4\delta(t - 4\pi)$, $y(0) = y'(0) = 1$

*9. (Requires personal computer or access to computer facilities for Exercises 2 through 4.) For Exercises 1 through 4, above:
   a) Sketch the solution of the differential equation.
   b) Obtain the graph of the solution, and compare it to your sketch.

*10. For Exercises 5 through 8, above:
   a) Sketch the solution of the differential equation.
   b) Obtain the graph of the solution, and compare to your sketch.

11. Let $\delta^{(n)}(t)$, $n \geq 0$, n an integer, have the properties that

i) $\delta^{(n)}(t) = 0$ if $t \neq 0$,

ii) $\int_{-\infty}^{\infty} \delta^{(n)}(t-a)g(t)\,dt = (-1)^n g^{(n)}(a)$ for continuous g.

Show formally that $\mathscr{L}[\delta^{(n)}(t-a)] = s^n e^{-as}$.

12. This exercise considers one sequence of approximations to $\delta(t)$. Let $f_n(t)$ for each positive integer n be defined as

$$f_n(t) = \begin{cases} 0, & \text{if } t \leq 0, \\ n^2 t, & \text{if } 0 \leq t \leq 1/n, \\ 2n - n^2 t, & \text{if } 1/n \leq t \leq 2/n, \\ 0, & \text{if } 2/n < t. \end{cases}$$

a) Graph $f_n(t)$ for $n = 1, 2, 3, 4$ on the same graph.

b) Verify $\lim_{n \to \infty} f_n(t) = 0$ for each value of t.

c) Verify that $\int_{-\infty}^{\infty} f_n(t)\,dt = 1$ for all n.

13. This exercise considers a slightly different sequence of approximations to $\delta(t)$. Let $f_n(t)$, for each positive integer n, be defined by:

$$f_n(t) = \begin{cases} 0, & \text{if } t < 0, \\ n, & \text{if } 0 \leq t < 1/n, \\ 0, & \text{if } 1/n \leq t. \end{cases}$$

a) Graph $f_n(t)$, on the same graph, for $n = 1, 2, 4, 6$.

b) Verify $\lim_{n \to \infty} f_n(t) = 0$ if $t \neq 0$ and $\lim_{n \to \infty} f_n(0) = \infty$.

c) Verify $\int_{-\infty}^{\infty} f_n(t)\,dt = 1$ for all n.

d) Verify

$$\lim_{n \to \infty} \int_{-\infty}^{\infty} (t^3 + 2) f_n(t)\,dt = 2.$$

e) Verify that, if g is a differentiable function at zero and continuous on $(-\infty, \infty)$, then

$$\lim_{n \to \infty} \int_{-\infty}^{\infty} g(t) f_n(t)\,dt = g(0).$$

[*Hint*: Use mean value theorem for integrals.]

Note Parts (d) and (e) of this exercise should be compared with property (iv), which says that $\int_{-\infty}^{\infty} g(t)\delta(t)\,dt = g(0)$.

4.6 Convolution Integrals

If $f(t)$, $g(t)$ are two piecewise-continuous functions on $[0, \infty)$, then the *convolution* of f and g is a new function, denoted $f * g$, and it is given by

$$(f * g)(t) = \int_0^t f(\tau) g(t - \tau)\,d\tau. \tag{1}$$

Example 4.6.1 Suppose $f(t) = t^3$, and $g(t) = t^2$; compute $f * g$.

Solution Using (1),

$$(t^3 * t^2) = \int_0^t \tau^3 (t - \tau)^2 \,d\tau$$

$$= \int_0^t \tau^5 - 2\tau^4 t + t^2 \tau^3 \,d\tau$$

$$= \left. \frac{\tau^6}{6} - \frac{2\tau^5}{5} t + t^2 \frac{\tau^4}{4} \right|_{\tau=0}^{\tau=t} = \frac{1}{60} t^6. \blacksquare$$

Convolution integrals have at least two important applications. They can be used to evaluate some inverse Laplace transforms. They also appear in certain problems expressed in terms of integral equations. Both applications will be considered in this section.

Convolution is like a product. Two useful algebraic identities are:

If f, g, h are piecewise-continuous functions on $[0, \infty)$, then

$$f * g = g * f. \qquad (2)$$

If, in addition, α, β are constants, then

$$(\alpha f + \beta g) * h = \alpha (f * h) + \beta (g * h). \qquad (3)$$

That is, the commutative and distributive laws hold. We shall verify (2). The verification of (3) is left to an exercise.

Verification of (2)
By definition (1),

$$(f * g)(t) = \int_0^t f(\tau) g(t - \tau) \, d\tau. \qquad (4)$$

Let $s = t - \tau$, then $ds = -d\tau$ and (4) is

$$= \int_t^0 f(t - s) g(s)(-ds) = -\int_t^0 f(t - s) g(s) \, ds$$

$$= \int_0^t g(s) f(t - s) \, ds = (g * f)(t). \quad \blacksquare$$

In general, if $\mathscr{L}[f(t)] = F(s)$, $\mathscr{L}[g(t)] = G(s)$, then $\mathscr{L}[f(t)g(t)] \neq F(s)G(s)$. For example,

$$\mathscr{L}[t] = \frac{1}{s^2}, \qquad \mathscr{L}[t^2] = \frac{2}{s^3}$$

but

$$\mathscr{L}[tt^2] = \mathscr{L}[t^3] = \frac{6}{s^4}$$

and

$$\frac{1}{s^2} \cdot \frac{2}{s^3} \neq \frac{6}{s^4}.$$

What then is $\mathscr{L}^{-1}[F(s)G(s)]$? It turns out that the convolution integral provides the answer.

Relationship of Convolution to the Laplace Transform Suppose that $f(t)$, $g(t)$ are piecewise-continuous functions of exponential order. Let $\mathscr{L}[f(t)] = F(s)$, $\mathscr{L}[g(t)] = G(s)$. Then

$$\mathscr{L}[(f * g)(t)] = F(s)G(s) \tag{5}$$

and

$$\mathscr{L}^{-1}[F(s)G(s)] = (f * g)(t) = \int_0^t f(\tau)g(t-\tau)\,d\tau$$

$$= \int_0^t f(t-\tau)g(\tau)\,d\tau. \tag{6}$$

Example 4.6.2 Use (6) to evaluate

$$\mathscr{L}^{-1}\left[\frac{1}{(s+1)(s+2)}\right]$$

Solution Let $F(s) = 1/(s+1)$, $G(s) = 1/(s+2)$. Then

$$f(t) = \mathscr{L}^{-1}[F(s)] = \mathscr{L}^{-1}\left[\frac{1}{s+1}\right] = e^{-t},$$

$$g(t) = \mathscr{L}^{-1}\left[\frac{1}{s+2}\right] = e^{-2t}.$$

Formula (6) now gives

$$\mathscr{L}^{-1}\left[\frac{1}{(s+1)}\cdot\frac{1}{(s+2)}\right] = e^{-t} * e^{-2t} = \int_0^t e^{-\tau}e^{-2(t-\tau)}\,d\tau$$

$$= e^{-2t}\int_0^t e^{\tau}\,d\tau = e^{-2t}[e^t - 1] = e^{-t} - e^{-2t}.$$

Of course, this example could also have been done using a partial-fractions expansion. ■

Example 4.6.3 Using (6), find

$$\mathscr{L}^{-1}\left[\frac{1}{(s^2+1)^2}\right].$$

Solution Note that $1/(s^2+1)^2$ is already reduced as far as possible by partial fractions. This inverse Laplace transform is given by (T25). However, we shall use (5). Observe that

$$\frac{1}{(s^2+1)^2} = \frac{1}{(s^2+1)}\cdot\frac{1}{(s^2+1)}.$$

Let $F(s) = 1/(s^2+1)$, so that $f(t) = \sin t$. Then, by (6),

$$\mathscr{L}^{-1}\left[\frac{1}{s^2+1}\cdot\frac{1}{s^2+1}\right] = \mathscr{L}^{-1}\left[\frac{1}{s^2+1}\right] * \mathscr{L}^{-1}\left[\frac{1}{s^2+1}\right]$$

$$= \sin t * \sin t$$

$$= \int_0^t \sin\tau \sin(t-\tau)\, d\tau$$

$$= \int_0^t \sin\tau(\sin t \cos\tau - \cos t \sin\tau)\, d\tau \qquad \text{(trig identity)}$$

$$= \sin t \int_0^t \sin\tau \cos\tau\, d\tau - \cos t \int_0^t \sin^2\tau\, d\tau$$

$$= \sin t \left(\frac{\sin^2\tau}{2}\right)\bigg|_{\tau=0}^{\tau=t} - \cos t \left(-\frac{1}{2}\sin\tau\cos\tau + \frac{\tau}{2}\bigg|_{\tau=0}^{\tau=t}\right)$$

$$= \frac{\sin^3 t}{2} - \cos t \left(-\frac{1}{2}\sin t \cos t + \frac{t}{2}\right)$$

$$= \frac{\sin t}{2}(\sin^2 t + \cos^2 t) - \frac{t\cos t}{2}$$

$$= \frac{\sin t}{2} - \frac{t\cos t}{2}. \blacksquare$$

4.6.1 Integral Equations

Another application of the Laplace transform and formulas (5) and (6) is to certain types of integral equations.

A differential equation, such as those we have been studying,

$$\frac{dx}{dt} = f(x, t)$$

relates the rate of change of x at the instant t to the time t and the value of x at time t. Other physical laws, however, relate the current value of x at time t to the cumulative effect of previous values of x.

Example 4.6.4 Consider the "physical law":

The value of $x(t)$ at time t is three times the average of x over the time interval $[0, t]$. (7)

The *average* of a function $f(t)$ over an interval $[0, T]$ is

$$\frac{1}{T}\int_0^T f(s)\,ds.$$

Thus (7) is the integral equation,

$$x(t) = \frac{3\int_0^t x(s)\,ds}{t}. \quad \blacksquare \qquad (8)$$

Sometimes an integral equation may be rewritten as a differential equation.
For example, we may rewrite (8) as

$$tx(t) = 3\int_0^t x(s)\,ds.$$

Differentiating both sides with respect to t gives the differential equation $tx'(t) + x(t) = 3x(t)$, or

$$tx'(t) = 2x(t). \qquad (9)$$

Integral equations have many important uses. They are sometimes more numerically stable than differential equations. They can be used to study problems involving "noise" and random effects. Many times measuring devices (instruments) actually measure not an instantaneous value x but an "average" value over a time period. This also occurs in signal and image processing.

In these problems, $x(t)$ is often a weighted average of previous values and we get

$$x(t) = \int_0^t k(t, \tau)x(\tau)\,d\tau + g(t), \qquad (10)$$

where $k(t, \tau)$, $g(t)$ are known functions. One important special case of (10) is

$$x(t) = \int_0^t h(t - \tau)x(\tau)\,d\tau + g(t)$$
$$= (h * x)(t) + g(t), \qquad (11)$$

where h, g are known functions. Equation (11) can be solved using Laplace transforms.

Procedure for Solving (11) by Laplace Transforms

1. Take Laplace transforms of both sides of (11), using (5).
2. Solve for $X(s)$.
3. Compute $\mathscr{L}^{-1}[X(s)]$.

Example 4.6.5 Solve the integral equation

$$x(t) = \int_0^t \sin(t - \tau)x(\tau)\,d\tau + 1. \tag{12}$$

Solution Taking the Laplace transform of (12), we have

$$\mathscr{L}[x(t)] = \mathscr{L}\left[\int_0^t \sin(t - \tau)x(\tau)\,d\tau\right] + \mathscr{L}[1],$$

and using (5) obtain

$$X(s) = \frac{1}{s^2 + 1}X(s) + \frac{1}{s}.$$

Solving for $X(s)$, we get

$$X(s) = \frac{s^2 + 1}{s^3} = \frac{1}{s} + \frac{1}{s^3}.$$

Thus,

$$x(t) = \mathscr{L}^{-1}[X(s)] = 1 + \frac{t^2}{2}. \quad \blacksquare$$

Exercises

1. If $f(t) = t$, $g(t) = e^t$, find $f * g$ from the definition (1).

2. If $f(t) = e^t$, $g(t) = \cos t$, find $f * g$ from the definition (1).

3. Compute $1 * 1$.

4. Compute $1 * t$.

5. Using formula (6), compute

$$\mathscr{L}^{-1}\left[\frac{1}{s^2} \cdot \frac{1}{(s^2 + 1)}\right].$$

6. Using formula (6), compute

$$\mathscr{L}^{-1}\left[\frac{s}{(s^2 + 1)^2}\right].$$

7. Verify that the differential equation

$$\frac{dx}{dt} = p(t)x(t) + q(t), \quad x(0) = a,$$

is equivalent to the integral equation

$$x(t) = \int_0^t p(\tau)x(\tau)\,d\tau + g(t) + a,$$

where $g(t) = \int_0^t q(\tau)\,d\tau$.

In Exercises 8 through 15, solve the given integral equation using the Laplace transform.

8. $x(t) = \dfrac{1}{2}\displaystyle\int_0^t x(t)\,ds + 1$

9. $x(t) = \displaystyle\int_0^t \cos(t - \tau)x(\tau)\,d\tau + \sin t$

10. $x(t) = \displaystyle\int_0^t (t - \tau)x(\tau)\,d\tau + 1$

11. $x(t) = \displaystyle\int_0^t e^{-t+\tau}x(\tau)\,d\tau + 2$

12. $6x(t) = \displaystyle\int_0^t (t - \tau)^3 x(\tau)\,d\tau + t$

13. $x(t) = \displaystyle\int_0^t 2\sin(2t - 2\tau)x(\tau)\,d\tau + \sin t$

14. $x(t) = \displaystyle\int_0^t \cos(2t - 2\tau)x(\tau)\,d\tau + e^{3t}$

15. $x(t) = -\displaystyle\int_0^t \sinh(t - \tau)x(\tau)\,d\tau + 3$

16. Suppose that f, g, h are piecewise-continuous

functions on $[0, \infty)$ and α, β are constants. Using the definition (1) of convolution, verify that

a) $(\alpha f + \beta g) * h = \alpha(f * h) + \beta(g * h)$.

b) $f * (\alpha g + \beta h) = \alpha(f * g) + \beta(f * h)$.

Using (5) verify that

c) $f * (g * h) = (f * g) * h$.

17. Suppose that f is piecewise-continuous, of exponential order ($|f(t)| \leq Me^{\alpha t}$). Show that, for $s > \alpha$,

$$|F(s)| \leq \frac{M}{s - \alpha}.$$

18. Using Exercise 17 and the Laplace transform, show that there do not exist f, g both piecewise-continuous, of exponential order, such that $f * g = 1$.

19. The following example illustrates one way that integral equations are "nicer" than differential equations. Let

$$g(t) = \begin{cases} e^t & \text{for } 0 \leq t < 1, \\ -1 + e^t + e^{t-1} & \text{for } 1 \leq t. \end{cases}$$

i) Show that $g(t)$ is continuous for $t \geq 0$, but $g'(1)$ is not defined. Also, show that $g(t)$ is a solution of $y' - y = H(t - 1)$, $y(0) = 1$ for $t \geq 0$, $t \neq 1$.

ii) Show that $g(t)$ satisfies the integral equation $y(t) = \int_0^t (y(\tau) + H(\tau - 1)) \, d\tau + 1$ for all $t \geq 0$, including $t = 1$.

20. Suppose that $x(t)$ is a piecewise-continuous function for $-\infty < t < \infty$. Think of t as being the location along the real axis. That is, t measures a position rather than time. Let $x(t)$ be a quantity that varies along the t-axis, such as light intensity or frequency. A measurement of $x(t)$ would actually be a (possibly weighted) average of x over an interval, say $[t - a, t + a]$, which contains t. The *unweighted average* over $[t - a, t + a]$ would be

$$\frac{1}{2a} \int_{-a}^{a} x(t + \tau) \, d\tau. \tag{13}$$

Show that (13) can be written as

$$\frac{1}{2a}[(1 * x)(t + a) - (1 * x)(t - a)].$$

4.7 Transfer Functions

The Laplace transform plays a fundamental role in many of the design and modeling procedures used today, particularly in electrical engineering. However, it is often not used in the way we applied it in Section 4.2 to solve initial-value problems. Rather, it is used to turn the whole problem into an algebra problem. The entire analysis is then performed in terms of the algebra problem. In this situation the original differential equation is called the *state variable model*. The algebraic problem is said to be in the *frequency domain*. This section will briefly discuss these ideas.

To begin, consider a simple *RLC*-circuit (1) with linear elements [Sections 2.12 and 3.17]. Written as a two-port, it is shown below.

(1)

Here we have drawn $v_1 = e$ as a voltage source, but it could be a voltage coming from another circuit; v_2 is the voltage across the capacitor. From Sections 2.12 and 3.17, we have the state variable model.

$$L\frac{d^2q}{dt^2} + R\frac{dq}{dt} + \frac{1}{C}q = e \tag{2a}$$

$$\frac{1}{C}q = v_2, \qquad e = v_1, \tag{2b}$$

where q is the charge in the capacitor. Suppose that we are concerned only with the possible frequencies and rates of decay of the solutions. This is always determined by the denominator when taking the inverse Laplace transform, so we ignore initial conditions and take $q(0) = 0$, $q'(0) = 0$. (Recall also that, if $R > 0$, then $q(0)$, $q'(0)$ affect only the free-response part of the solution, and that is transient anyway.)

"Laplacing" (2) gives

$$\left[Ls^2 + Rs + \frac{1}{C}\right]Q(s) = E(s) \tag{3a}$$

$$\frac{1}{C}Q(s) = V_2(s), \qquad E(s) = V_1(s). \tag{3b}$$

Using (3b) to eliminate Q, E in (3a), and then solving for V_2 yields

$$V_2(s) = \left[\frac{1}{CLs^2 + RCs + 1}\right]V_1(s). \tag{4}$$

The function

$$T(s) = \frac{1}{CLs^2 + RCs + 1}$$

is called the *transfer function* of the *two-port* (1). The transfer function $T(s)$ gives the relationship between the Laplace transforms of the input ($V_1(s)$) and output ($V_2(s)$) voltages. The circuit (1) is now represented by

$$V_1(s) \rightarrow \boxed{T(s)} \rightarrow V_2(s), \tag{5}$$

where V_1, V_2, T satisfy (4). That is, $V_2(s) = T(s)V_1(s)$. Variables other than voltages can be used, but we shall consider only voltages. We can now construct circuit diagrams with components such as (5) and voltage dividers

$$V_1 \longrightarrow \begin{matrix} V_1 \\ \\ V_1 \end{matrix}$$

and voltage adders (summers)

$$\underset{V_2}{\overset{V_1}{\downarrow}} \Sigma \longrightarrow V_1 + V_2.$$

For example, the circuit shown in (6)

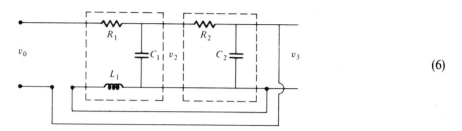

(6)

could be written [under certain loading assumptions, which we ignore] as

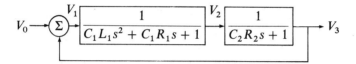

or, more simply, as shown in (7).

$$V_0 \longrightarrow \Sigma \overset{V_1}{\longrightarrow} \boxed{T_1(s)} \overset{V_2}{\longrightarrow} \boxed{T_2(s)} \longrightarrow V_3 \qquad (7)$$

The circuit in (6) is a simple example of a *feedback* circuit. The transfer function of (6) and (7) may be computed as follows:

$$V_3 = T_2 V_2 = T_2[T_1 V_1] = T_2 T_1[V_0 + V_3]$$

(since $V_1 = V_0 + V_3$). Thus

$$[1 - T_2 T_1]V_3 = T_2 T_1 V_0,$$

and, finally,

$$V_3 = \frac{T_2 T_1}{[1 - T_2 T_1]} V_0.$$

Thus the transfer function of (6) and (7) is

$$\frac{T_2(s)T_1(s)}{1-T_2(s)T_1(s)}.$$

Feedback circuits are often used to stabilize or control electrical and mechanical (and chemical) systems.

Exercises

In Exercises 1 through 7, calculate the transfer functions for each of the indicated circuits; that is, find $T(s)$ so that:

$$\frac{V_1(s)}{V_0(s)} = T(s).$$

1.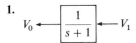

2.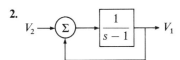

3.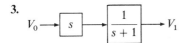

4.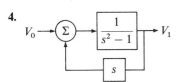

5.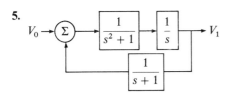

6.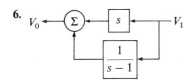

7.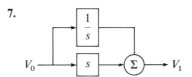

8. A device with transfer function $T(s) = 1/s$ is sometimes called an *integrator*. Using Laplace transforms, explain why.

9. A device with transfer function $T(s) = s$ is sometimes called a *differentiator*. Using Laplace transforms, explain why.

The remaining exercises refer to the following situation: A device has transfer function $T(s) = p(s)/q(s)$, where $p(s)$, $q(s)$ are real polynomials. We assume that $p(s)$ is an mth-degree polynomial whose roots $s_1, \ldots, s_m$ are called *zeros* of $T(s)$. The denominator $q(s)$ is an nth-degree polynomial, whose roots $\lambda_1, \ldots, \lambda_n$ are called the *poles* of $T(s)$. We assume that $s_i \neq \lambda_j$ for all i, j. That is, p, q are *relatively prime*. Suppose that

$$V_0(s) \rightarrow \boxed{T(s)} \rightarrow V_1(s)$$

and

$$v_0(t) = A\cos(\omega t + \phi).$$

10. Explain why $v_1(t)$ will have no terms of frequency $\omega/2\pi$ if ωi is a zero of $T(s)$.

11. Explain why resonance might occur if ωi is a pole of $T(s)$.

12. Let $v_0(t)$ be a unit-step voltage for one second; $v_0(t) = [1 - H(t-1)]$. Explain why we are assured that the response is transient if and only if all poles λ of $T(s)$ have negative real part. Such a circuit is called *stable*.

13. Using Exercises 10 and 11, give a transfer function that will filter out $v_0(t) = \cos(\omega t + \phi)$ at frequencies of 2 and 4 cycles/second (remove these frequencies) but will exhibit resonance at 36 cycles/second.

14. Show that, if the denominator of $T(s)$ has higher degree than the numerator ($T(s)$ is *proper*), then a piecewise-continuous input $v_0(t)$, of exponential order, will result in a continuous output $v_1(t)$.

15. Show that, if the numerator of $T(s)$ has higher degree than the denominator, then the step voltage $v_0(t)$ of Exercise 12 will result in an impulse (Section 4.5).

16. Give a transfer function that satisfies the requirements of Exercise 13 but has its denominator of higher degree than its numerator. Make sure any new poles have negative real part.

5

Series Solution of Linear Equations

5.1 Introduction

In Section 3.12 we saw how the method of undetermined coefficients could be used to solve linear constant-coefficient differential equations of the form

$$ay'' + by' + cy = f, \tag{1}$$

provided f was a linear combination of the "right sort" of functions. However, in many applications the coefficients a, b, c can also depend on x. Frequently the best, and sometimes the only, way to handle such problems is by using series.

There are many types of series solutions to consider. The choice depends on the particular application. Two examples are

$$\sum_{n=0}^{\infty} a_n x^n, \quad \text{a power series,}$$

and

$$\sum_{n=0}^{\infty} a_n \cos n\pi x, \quad \text{a Fourier series.}$$

This chapter will show how power series can be used to solve a variety of differential equations, some of which cannot be solved analytically any other way. In the process we shall introduce an important group of functions in applied mathematics, the Bessel functions, and discuss very briefly their application to the solution of partial differential equations.

We begin with a review of series and power series.

5.2 Review of Power Series

This section will review from calculus some of the basic facts about series and power series.

If $\{a_0, a_1, \ldots\} = \{a_n\}_{n=0}^{\infty}$ is a *sequence* of numbers, then the expression

$$\sum_{n=0}^{\infty} a_n = a_0 + a_1 + a_2 + \cdots$$

is called an *infinite series*. The mth *partial sum* s_m is given by

$$s_m = \sum_{n=0}^{m} a_n = a_0 + a_1 + \cdots + a_m.$$

The series $\sum_{n=0}^{\infty} a_n$ is said to *converge* if the partial sums have a limit. If L is that limit, that is

$$\lim_{m \to \infty} s_m = L,$$

then L is called the *sum* of the series, and we write

$$\sum_{n=0}^{\infty} a_n = L.$$

Thus, by definition,

$$\sum_{n=0}^{\infty} a_n = \lim_{m \to \infty} \sum_{n=0}^{m} a_n = \lim_{m \to \infty} s_m. \tag{1}$$

If the series does not converge, it is called *divergent*. Note that there are two sequences here: the sequence being added up, $\{a_n\}$, and the sequence of partial sums, $\{s_m\}$.

$$\text{If the series } \sum_{n=0}^{\infty} a_n \text{ converges,} \quad \text{then } \lim_{n \to \infty} a_n = 0. \tag{2}$$

Essentially, (2) says that if we can add up an infinite number of terms $\{a_n\}$ and get a finite sum L, then the terms a_n must become smaller and smaller, eventually approaching zero.

Example 5.2.1   Verify that $\sum_{n=0}^{\infty} \frac{1}{2^n} = 2$.

Solution   In this example, $a_n = 1/2^n$, and

$$s_m = \sum_{n=0}^{m} \frac{1}{2^n} = 1 + \frac{1}{2} + \cdots + \frac{1}{2^m}. \tag{3}$$

The series is a *geometric series*, and we use the algebraic technique of multiplying both sides of (3) by $(1 - \frac{1}{2})$ to get

$$\left(1 - \frac{1}{2}\right) s_m = 1 + \frac{1}{2} + \cdots + \frac{1}{2^m} - \frac{1}{2} - \frac{1}{2^2} - \cdots - \frac{1}{2^{m+1}} = 1 - \frac{1}{2^{m+1}}.$$

Solving for s_m:

$$s_m = \frac{1 - \frac{1}{2^{m+1}}}{\frac{1}{2}} = 2 - \frac{1}{2^m}.$$

Since $\lim_{m \to \infty} s_m = 2 - 0 = 2$, we have finished. Note also that $\lim_{n \to \infty} a_n = 0$. ∎

This example is to illustrate the concept of convergence. We will not actually need to sum series in this manner. Section 7.4 provides a method of summing some series.

Note the similarity between the series $\sum_{n=0}^{\infty} a_n$ and the improper integral $\int_0^{\infty} f(x)\, dx$. Whether they converge depends on $\{a_n\}$ or $f(x)$. Both are defined as the limit of a sum (integral) over an infinite interval:

$$\sum_{n=0}^{\infty} a_n = \lim_{m \to \infty} \sum_{n=0}^{m} a_n, \qquad \int_0^{\infty} f(x)\, dx = \lim_{m \to \infty} \int_0^{m} f(x)\, dx.$$

While $\lim_{n \to \infty} a_n = 0$, if $\sum_{n=0}^{\infty} a_n$ converges, the converse is not true. It is possible to have $\lim_{n \to \infty} a_n = 0$, but $\sum_{n=0}^{\infty} a_n$ does not converge.

Example 5.2.2   Consider the series $\sum_{n=0}^{\infty} (n+1)^{-1} = 1 + \frac{1}{2} + \frac{1}{3} + \cdots$. We have $a_n = 1/(n+1)$, so that $\lim_{n \to \infty} a_n = 0$. However, the series diverges, since $\lim_{m \to \infty} s_m = \infty$. This may be proved; however, it is easy to convince yourself of this fact by noting $s_{m+1} > s_m$ and that (good computer or calculator exercise)

$$s_1 = 1.50,$$
$$s_4 = 2.28,$$
$$s_{16} = 3.44,$$
$$s_{64} = 4.76,$$

and, in general, $s_{4^r} \geq r$. ∎

A series is called *absolutely convergent* if $\sum_{n=0}^{\infty} a_n$ and $\sum_{n=0}^{\infty} |a_n|$ both converge. Absolutely convergent series have the nice property that the terms can be arranged in any order and the series still has the same sum. For example, we could add up all the odd-numbered terms and then add up all the even-numbered terms.

A series $\sum_{n=0}^{\infty} a_n$ is *conditionally convergent* if $\sum_{n=0}^{\infty} a_n$ converges but $\sum_{n=0}^{\infty} |a_n|$ is divergent. As noted earlier, $\lim_{n \to \infty} a_n = 0$ is not usually enough to guarantee convergence of $\sum_{n=0}^{\infty} a_n$. There is an important exception.

If $a_n > 0$, $a_n \geq a_{n+1}$ for all $n > 0$ and, $\lim_{n \to \infty} a_n = 0$, then $\sum_{n=0}^{\infty} (-1)^n a_n$ is a convergent (*alternating*) series. (4)

Example 5.2.3 Consider the series $\sum_{n=0}^{\infty} \frac{(-1)^n}{n+1} = 1 - \frac{1}{2} + \frac{1}{3} - \cdots$. Since

$$\sum_{n=0}^{\infty} \left|(-1)^n \frac{1}{n+1}\right| = \sum_{n=0}^{\infty} \frac{1}{n+1},$$

we know from Example 5.2.2 that the series does not converge absolutely. However,

$$\frac{1}{n+1} > 0, \qquad \frac{1}{n+1} \geq \frac{1}{n+2}, \qquad \text{and} \qquad \lim_{n \to \infty} \frac{1}{n+1} = 0,$$

so that, by (4), the series does converge. Since the convergence of the series is not absolute, it is conditional. ∎

In general, testing for conditional convergence can be difficult. Fortunately, the following test for absolute convergence will be sufficient for most of our examples.

Theorem 5.2.1   **Ratio Test** Suppose $a_n \neq 0$ for all n.

1. If $\lim_{n \to \infty} \frac{|a_{n+1}|}{|a_n|} = L < 1$, then $\sum_{n=0}^{\infty} a_n$ converges absolutely.

2. If $\lim_{n \to \infty} \frac{|a_{n+1}|}{|a_n|} = L > 1$, then $\sum_{n=0}^{\infty} a_n$ diverges.

3. If $\lim_{n \to \infty} \frac{|a_{n+1}|}{|a_n|} = 1$, then this test does not tell whether the series converges or diverges.

The expression

$$\sum_{n=0}^{\infty} a_n (x - b)^n \tag{5}$$

is called a *power series centered at b*. It is also sometimes called a *Taylor series centered at b* or *Maclaurin's series* if $b = 0$. The numbers x for which (5) converges are called the *convergence set* of the power series (5). A power series defines a function of x on its convergence set, and we often write

$$f(x) = \sum_{n=0}^{\infty} a_n(x-b)^n. \tag{6}$$

The *convergence set* is the domain of the function defined by the power series. The convergence set is relatively simple.

> The convergence set of a power series is always an interval of some radius r centered at b. This includes the two special cases where $r = 0$ and the convergence set is just the point $\{b\}$, and where $r = \infty$ and the convergence set is $(-\infty, \infty)$.

The number r is called the *radius of convergence*. We shall give two formulas for the radius of convergence.

Theorem 5.2.2

Root-Test Formula The radius of convergence r of the power series $\sum_{n=0}^{\infty} a_n(x-b)^n$ is given by:

$$\frac{1}{r} = \lim_{n \to \infty} \sqrt[n]{|a_n|}, \tag{7}$$

provided the limit in (7) exists [$r = \infty$ if the limit in (7) is zero, and $r = 0$ if the limit is ∞].

This formula can be difficult to apply if a_n involves factorials, which is often the case. The ratio test (Theorem 5.2.1) provides a simpler way of computing r, which we shall also call the ratio test (see Exercise 35 at the end of this section).

Theorem 5.2.3

Ratio Test The radius of convergence r of the power series $\sum_{n=0}^{\infty} a_n(x-b)^n$ is given by

$$\frac{1}{r} = \lim_{n \to \infty} \frac{|a_{n+1}|}{|a_n|}, \tag{8}$$

if the limit in (8) exists and the $0/\infty$ convention of Theorem 5.2.2 is used.

In the later sections we will not usually be finding the radius of convergence, but we will use the concept.

Example 5.2.4 Determine the convergence set of

$$\sum_{n=0}^{\infty} n! x^n.$$

Solution By the ratio test (Theorem 5.2.3),

$$\lim_{n \to \infty} \frac{|a_{n+1}|}{|a_n|} = \lim_{n \to \infty} \frac{(n+1)!}{n!} = \lim_{n \to \infty} (n+1) = +\infty.$$

Thus $1/r = \infty$ or (by convention) $r = 0$ and the convergence set is just the center $b = 0$. Alternatively we could use Theorem 1:

$$\lim_{n \to \infty} \frac{(n+1)!|x|^{n+1}}{n!|x|^n} = \lim_{n \to \infty} (n+1)|x| = \begin{cases} 0 & \text{if } x = 0, \\ \infty & \text{if } x \neq 0. \end{cases}$$

Thus the series converges only for $x = 0$. ∎

Example 5.2.5 Determine the convergence set of

$$\sum_{n=0}^{\infty} \frac{1}{n+1} (x-2)^n.$$

Solution Using the power-series version of the ratio test (Theorem 5.2.3), we obtain

$$\frac{1}{r} = \lim_{n \to \infty} \frac{1/(n+2)}{1/(n+1)} = \lim_{n \to \infty} \frac{n+1}{n+2} = 1,$$

so that $r = 1$. Thus the series converges for $|x - 2| < 1$ and diverges for $|x - 2| > 1$. The endpoints of this interval are

$$|x - 2| = 1 \quad \text{or} \quad x = 1, 3.$$

To determine whether they are included in the convergence set, we must examine them separately. If $x = 1$, then the series is

$$\sum_{n=0}^{\infty} \frac{1}{n+1} (1-2)^n = \sum_{n=0}^{\infty} \frac{(-1)^n}{n+1},$$

which converges, from Example 5.2.3. On the other hand, if $x = 3$, then the series is

$$\sum_{n=0}^{\infty} \frac{1}{n+1} (3-2)^n = \sum_{n=0}^{\infty} \frac{1}{n+1},$$

which diverges, from Example 5.2.2. Thus the convergence set is the interval $[1, 3)$. ∎

Example 5.2.6 Find the convergence set of $\sum_{n=0}^{\infty} \frac{x^n}{n!}$ (*Note*: $0! = 1$).

Solution

Again using the ratio test (Theorem 5.2.3),

$$\frac{1}{r} = \lim_{n \to \infty} \frac{1/(n+1)!}{1/(n!)} = \lim_{n \to \infty} \frac{n!}{(n+1)!} = \lim_{n \to \infty} \frac{1}{n+1} = 0.$$

Thus, by convention, $r = \infty$ and the series converges for all x. That is, the convergence set is the real line $(-\infty, \infty)$. ∎

The properties of power series that we shall need in order to solve differential equations derive from the following fundamental fact, which is not necessarily true for other types of series.

Theorem 5.2.4

Suppose $\sum_{n=0}^{\infty} a_n(x-b)^n$ is a power series with radius of convergence $r > 0$. Then the function defined on the convergence set by

$$f(x) = \sum_{n=0}^{\infty} a_n(x-b)^n \tag{9}$$

is a differentiable function everywhere inside the convergence set ($|x - b| < r$). Furthermore, the derivative $f'(x)$ can also be written as a power series centered at b, and that series is

$$f'(x) = \sum_{n=1}^{\infty} na_n(x-b)^{n-1}, \tag{10}$$

and the radius of convergence of this series is also r.

Note that the series (10) for $f'(x)$ is obtained by differentiating the series (9) for $f(x)$, term by term.

Since $f'(x)$ is given by a power series, we can apply Theorem 5.2.4 to $f'(x)$, and conclude that $f'(x)$ is also differentiable and

$$f''(x) = \sum_{n=2}^{\infty} (n-1)na_n(x-b)^{n-2}. \tag{11}$$

This series also has radius of convergence r. This process may be continued indefinitely. Thus,

If a function is given by a power series with radius of convergence $r > 0$, the function is infinitely differentiable. $\tag{12}$

If we set $x = b$ in (9), we find

$$f(b) = a_0.$$

Setting $x = b$ in (10) and (11) gives

$$f'(b) = 1 \cdot a_1,$$
$$f''(b) = 2 \cdot 1 \cdot a_2.$$

In general,

$$f^{(n)}(b) = n!a_n$$

or

$$\frac{f^{(n)}(b)}{n!} = a_n. \qquad (13)$$

Thus we have the following important fact.

Uniqueness

If $f(x)$ can be written as a power series centered at b, there is only one way to do it, and that is

$$f(x) = \sum_{n=0}^{\infty} \frac{f^{(n)}(b)}{n!}(x-b)^n. \qquad (14)$$

If a function f can be written as a power series centered at b, then f is sometimes said to be *analytic at b* or f is representable by a power series or f has a power-series *expansion*. Not all infinitely differentiable functions can be written as power series (see Example 5.2.8). However, except for Example 5.2.8 and Exercise 42, we shall avoid this difficulty and consider only functions that can be represented by a power series at most points in their domain. This will include most of our standard functions, such as polynomials, exponentials, and trigonometric functions, and fractions, compositions, and inverses of these functions.

Example 5.2.7 Verify that, for $|x| < 1$,

$$\frac{1}{1-x} = 1 + x + x^2 + \cdots = \sum_{n=0}^{\infty} x^n. \qquad (15)$$

Solution We could verify this as in Example 5.2.1. We shall use (14) instead. Let

$$f(x) = (1-x)^{-1}.$$

Then

$$f'(x) = (1-x)^{-2},$$
$$f''(x) = 2(1-x)^{-3},$$

and in general,

$$f^{(n)}(x) = n!(1-x)^{-n-1}. \qquad (16)$$

Thus, $f(0) = 1$, $f'(0) = 1$, and, in general, $f^{(n)}(0) = n!$, so that, by (14),

$$\frac{1}{1-x} = \sum_{n=0}^{\infty} \frac{f^{(n)}(0)}{n!} x^n = \sum_{n=0}^{\infty} \frac{n!}{n!} x^n = \sum_{n=0}^{\infty} x^n.$$

The ratio test can be used to show that $\sum_{n=0}^{\infty} x^n$ converges for $|x| < 1$. However, $1/(1-x)$ makes sense for all $x \neq 1$. To write $1/(1-x)$ as a power series for $x > 1$ or for $x < -1$, we must pick a different center. For example, if we take $b = 3$, then from (16), $f^{(n)}(3) = n!(1 - 3)^{-n-1}$, and

$$\frac{1}{1-x} = \sum_{n=0}^{\infty} \frac{f^{(n)}(3)}{n!}(x-3)^n = \sum_{n=0}^{\infty} \frac{(-1)^{n+1}}{2^{n+1}}(x-3)^n,$$

which converges for $|x - 3| < 2$, which is the interval $(1, 5)$. ∎

Technical Point What we have just shown is that if $1/(1 - x)$ can be written as a power series centered at zero, then it is given by (15). We have not shown that $1/(1 - x)$ can be written as a power series centered at zero! Example 5.2.8 gives an infinitely differentiable function that cannot be written as a power series centered at zero. However, we will have a theorem in the next section that ensures that solutions of the differential equations we consider can be written as power series.

Example 5.2.8 Define the function $f(x)$ by

$$f(x) = \begin{cases} e^{-1/x^2} & \text{if } x \neq 0, \\ 0 & \text{if } x = 0. \end{cases}$$

This function has the property that it is infinitely differentiable everywhere, including zero, and that $f^{(n)}(0) = 0$ for all n. Thus

$$\sum_{n=0}^{\infty} f^{(n)}(0)\frac{x^n}{n!} = 0 \qquad \text{for all } x, \tag{17}$$

which does not agree with f anywhere except at $x = 0$.

The series (17), which does not converge to the function except at $x = 0$, still carries information about the function. Such "nonconvergent" series are important in the theory of asymptotic expansions, which is used to analyze the boundary layer (edge effects) in fluid-flow problems, a topic well beyond the scope of this book. ∎

5.2.1 Taylor Polynomials

If $f(x)$ is l times continuously differentiable in an open interval containing b, then, for $m \leq l$,

$$p_m(x) = \sum_{n=0}^{m} \frac{f^{(n)}(b)}{n!}(x-b)^n \tag{18}$$

is called the mth-degree *Taylor polynomial* for $f(x)$, (centered at b). If $f(x)$ has a power-series expansion (14), then $p_m(x)$ is just the mth partial sum of the Taylor series for $f(x)$ and

$$\lim_{m \to \infty} p_m(x) = f(x)$$

for all x in the convergence set of the power series. Thus the Taylor polynomials can be used to approximate f. Even if f does not have a power series, we have, from Taylor's Theorem in calculus, that, if $f(x)$ is l times continuously differentiable in the interval $(b - x, b + x)$ and $m < l$, then

$$f(x) = f(b) + f'(b)(x - b) + \cdots + \frac{f^{(m)}(b)}{m!}(x - b)^m + R_m(x) \qquad (19)$$

or

$$f(x) = p_m(x) + R_m(x),$$

where $R_m(x)$ is the remainder. We may estimate $R_m(x)$ by using the fact that

$$R_m(x) = \frac{f^{(m+1)}(\xi)}{m!}(x - b)(x - \xi)^m, \qquad (20)$$

where ξ is some number between b and x.

Example 5.2.9 From Example 5.2.7, we have that if $f(x) = 1/(1 - x)$, then the mth-degree Taylor polynomial for $f(x)$ centered at zero is

$$p_m(x) = 1 + x + \cdots + x^m. \blacksquare$$

Exercises

For Exercises 1 through 13, determine which of the following expressions are power series.

1. $\sum_{n=0}^{\infty} x^n$

2. $\sum_{n=0}^{\infty} n^x$

3. $\sum_{n=0}^{\infty} \sin(n) x^n$

4. $\sum_{n=0}^{\infty} \sin(nx)$

5. 3

6. $x^{1/2}$

7. $\sum_{n=-2}^{\infty} 3x^n$

8. $\sum_{n=1}^{\infty} 2^n x^{2n}$

9. $\sum_{n=0}^{\infty} nx^{3n+1}$

10. $8 - x^2$

11. $\frac{1}{x}$

12. $x - 3^{1/2}$

13. e^{-3x}

For Exercises 14 through 16, use the ratio test (Theorem 5.2.1) to determine whether or not the series converges. If the ratio test fails, so state.

14. $\sum_{n=1}^{\infty} \frac{n}{2^n}$

15. $\sum_{n=0}^{\infty} \frac{1}{(n+1)^2}$

16. $\sum_{n=1}^{\infty} \frac{1}{3n}$

For Exercises 17 through 19, determine the radius of convergence and convergence set for the power series. Determine, if possible, whether the endpoints of the interval are in the convergence set.

17. $\sum_{n=0}^{\infty} \frac{x^n}{\sqrt{n!}}$

18. $\sum_{n=0}^{\infty} 3^n(x - 2)^n$

19. $\sum_{n=1}^{\infty} n! \frac{x^n}{2^n}$

For Exercises 20 through 28, write the indicated function as a Taylor series with the given center b.

20. $f(x) = e^x$, $\qquad b = 0$

21. $f(x) = e^{2x}$, $\quad b = 1$
22. $f(x) = \sin x$, $\quad b = 0$
23. $f(x) = \cos x$, $\quad b = 0$
24. $f(x) = \sin x$, $\quad b = \pi/2$
25. $f(x) = x^3$, $\quad b = 0$
26. $f(x) = x^3$, $\quad b = 1$
27. $f(x) = x$, $\quad b = 0$
28. $f(x) = x$, $\quad b = -3$

Since the power series for a function at a given center is unique, it is often possible to obtain the series by manipulating simpler series.

29. Take the series for $1/(1 - x)$ in Example 5.2.7. Make the substitution $-x^2$ for x to obtain the series for $1/(1 + x^2)$ centered at zero.

30. Antidifferentiate the result of Exercise 29 termwise, to get a series for $\tan^{-1} x$.

31. Use the answer for Exercise 20 to obtain a Taylor series for $f(x) = e^{x^3}$. Note that obtaining this series, using (13), becomes progressively more complicated.

*32. In Example 5.2.3, it was noted that $\sum_{n=0}^{\infty}(-1)^n/(n + 1)$ converges. It can be shown that

$$\sum_{n=0}^{\infty} (-1)^n \frac{1}{n + 1} = \ln 2. \quad (21)$$

Starting with $s_0 = 1$, compute

$$s_m = \sum_{n=0}^{m} (-1)^n \frac{1}{n + 1}$$

for larger and larger m, until s_m gives the correct answer to three decimal places. (This requires access to a computer or programmable calculator and a simple loop program. It illus- trates how convergence can be slow.) Now repeat the experiment until s_m is accurate to five places.

33. Find the power series for $\ln(1 + x)$ centered at zero. Show, using Examples 5.2.2 and 5.2.3 and the ratio test, that this series has the convergence set $(-1, 1]$. Assuming that the function defined by a power series is continuous on the entire convergence set and not just inside it, verify (21).

34. Can the interval $[0, \infty)$ be the convergence set for a power series? Explain your answer.

35. Using the series version of the ratio test (Theorem 5.2.1), prove the radius-of-convergence test given in Theorem 5.2.3.

In Exercises 36 through 41, find the mth-degree Taylor polynomial of $f(x)$ centered at b.

36. $m = 3$ and $f(x)$, b as in Exercise 20.
37. $m = 2$ and $f(x)$, b as in Exercise 21.
38. $m = 5$ and $f(x)$, b as in Exercise 22.
39. $m = 5$ and $f(x)$, b as in Exercise 24.
40. $m = 2$ and $f(x)$, b as in Exercise 25.
41. $m = 1$ and $f(x)$, b as in Exercise 27.

*42. Define the function $f(x)$ by

$$f(x) = \begin{cases} e^{-1/x^2} & \text{if } x \neq 0, \\ 0 & \text{if } x = 0. \end{cases}$$

Show that $f(x)$ is infinitely differentiable everywhere, including zero, and that $f^{(n)}(0) = 0$ for all n. Thus

$$\sum_{n=0}^{\infty} f^{(n)}(0) \frac{x^n}{n!} = 0 \quad \text{for all } x,$$

which does not agree with f anywhere except at $x = 0$.

5.3 Solution at an Ordinary Point (Theory)

While power series can be used to solve some nonlinear differential equations, we shall consider only first- and second-order linear differential equations.

As noted earlier, not all functions can be written as a power series at a given point. The next fundamental result gives conditions under which the solution of a differential equation may be written as a power series.

Theorem 5.3.1 Suppose that the coefficients $a(x)$, $b(x)$, $c(x)$, and the forcing function $f(x)$ of the differential equation

$$a(x)y''(x) + b(x)y'(x) + c(x)y(x) = f(x) \tag{1}$$

can all be written as power series centered at x_0 and their positive radii of convergence are r_a, r_b, r_c, r_f, respectively. If, in addition, $a(x_0) \neq 0$, then x_0 is called an *ordinary* point of the differential equation and every solution of (1) can be written as a power series centered at x_0. The radius of convergence for this power series solution is at least the smallest of the five numbers

$\{r_a, r_b, r_c, r_f,$ and the distance from x_0 to the first number (real or complex) such that $a(x)$ is zero$\}$.

If a, b, c, and f can be written as power series centered at x_0, but $a(x_0) = 0$, then x_0 is called a *singular* point of (1) and will be discussed in Section 5.7.

There are several important special cases of (1). One is

$$y''(x) + b(x)y'(x) + c(x)y(x) = f(x), \tag{2}$$

where the radius of convergence will be at least the smallest of $\{r_b, r_c, r_f\}$. A similar result holds for the first-order equation

$$a(x)y'(x) + b(x)y(x) = f(x)$$

and its special case

$$y'(x) + b(x)y(x) = f(x).$$

Example 5.3.1 Take $x_0 = 0$ and consider the differential equation

$$y''(x) + \frac{1}{x-1}y'(x) + xy(x) = e^x. \tag{3}$$

Both x, e^x have power series centered at $x_0 = 0$ with infinite radius of convergence while $1/(x-1)$ has a power series centered at zero with a radius of convergence 1 (Example 5.2.7). Thus, by Theorem 5.3.1, all solutions of (3) can be written as a power series centered at zero with radius of convergence at least 1. ∎

In "solving" a differential equation by power series, there are three levels of solution. Suppose the solution is

$$y(x) = \sum_{n=0}^{\infty} y_n(x-b)^n. \tag{4}$$

where the y_n are constants.

Level I
Given a number, say 5, we can determine the first five coefficients $\{y_0, y_1, y_2, y_3, y_4\}$ of (4). The Taylor polynomial $\sum_{n=0}^{4} y_n(x-b)^n$ then produces an approximation to the solution. The method of Section 4 produces this level of solution.

Level II
We have a recursion relationship (also known as a difference equation; see Chapter 7) for the coefficients. For example, it could take the form

$$y_{n+1} = n y_n + y_{n-1} + n^2 y_{n-2}. \tag{5}$$

With this relationship we can automatically generate as many coefficients as needed since, given $\{y_0, \ldots, y_n\}$, formula (5) gives y_{n+1}. Then it will give y_{n+2}, etc. The method of Section 5 produces this level of solution.

Level III
We have a formula for the coefficients themselves, for example, $y_n = n^2 + 1$. Occasionally, Level III can be reached by inspection of a Level I or Level II solution. Sometimes the expression for y_n can be obtained by solving (see Chapter 7) the recursion relationship found in a Level II solution. We will not emphasize Level III solutions further in this text, but their importance should not be overlooked.

Exercises

For Exercises 1 through 6, state at which points Theorem 5.3.1 will guarantee the existence of a power-series solution centered at that point.

1. $y'' + xy = e^x$
2. $y'' + xy = \tan x$
3. $xy'' + e^x y = \dfrac{1}{x-1}$
4. $(x^2 - 2x)y'' + xy' + y = \cos x$
5. $(\sin x)y'' + \dfrac{1}{x-2}y' + y = |x-3|$
6. $e^x y'' + (\cosh x)y' - y = \sinh x$

For Exercises 7 through 14 give the radius of convergence of the power series solution guaranteed by Theorem 5.3.1.

7. $y'' + \dfrac{1}{x-1}y = \sin x$, $\quad x_0 = 0$
8. $(x^2 + 4)y'' + y = e^x$, $\quad x_0 = 0$
9. $(x+1)y' + e^x y = \cos x$, $\quad x_0 = 2$
10. $e^{3x}y' + (\sinh x)y = \cosh x$, $\quad x_0 = 6$
11. $(x^2 + 1)y'' + xy' = e^x - \cos x$, $\quad x_0 = 0$
12. $y'' + \dfrac{1}{x+1}y' + \dfrac{1}{x-1}y = x$, $\quad x_0 = \tfrac{1}{2}$
13. $y'' + \dfrac{1}{x-2}y' + \dfrac{3}{x-5}y = e^x$, $\quad x_0 = 3.3$
14. $xy'' + \dfrac{1}{x-1}y' = e^x$, $\quad x_0 = 0.2$

5.4 Solution at an Ordinary Point (Taylor-Series Method)

The first method we present for solving a differential equation using series has certain advantages and disadvantages. They will be discussed at the end of this section.

The Taylor-series method is based on two ideas. In order to get a series for $y(x)$ centered at x_0, we need only compute the derivatives of y at x_0, $y^{(n)}(x_0)$. Also, the differential equation of which y is a solution can be used to express the higher derivatives in terms of lower derivatives.

Example 5.4.1 Find the first five terms of the Taylor-series expansion for the solution of

$$y'' + 3y' + xy = \sin x, \quad y(0) = 0, \quad y'(0) = 2, \tag{1}$$

centered at $x_0 = 0$, using the Taylor-series method.

Solution Note that, by Theorem 5.3.1, (1) has a series solution. For convenience, rewrite the differential equation (1) as

$$y'' = \sin x - 3y' - xy. \tag{2}$$

The initial conditions (1) give

$$y(0) = 0, \quad y'(0) = 2. \tag{3}$$

Evaluate (2) at $x = 0$, to get

$$y''(0) = \sin(0) - 3y'(0) - 0y(0) = -3 \cdot 2 = -6. \tag{4}$$

Now differentiate (2):

$$y''' = \cos x - 3y'' - y - xy'. \tag{5}$$

Evaluate (5) at $x = 0$ to find $y'''(0)$, using (3) and (4),

$$y'''(0) = 1 - 3y''(0) - y(0) = 1 - 3(-6) - 0 = 19. \tag{6}$$

Now differentiate (5):

$$y^{(4)} = -\sin x - 3y''' - y' - xy'' - y'. \tag{7}$$

Evaluate (7) at $x = 0$ to find $y^{(4)}(0)$ using (3), (4), and (6):

$$y^{(4)}(0) = -0 - 3y'''(0) - 2y'(0) = -3(19) - 2(2) = -61. \tag{8}$$

Thus the solution (to five terms) is

$$y(x) = y(0) + y'(0)x + \frac{y''(0)}{2}x^2 + \frac{y'''(0)}{3!}x^3 + \frac{y^{(4)}(0)}{4!}x^4 + \cdots$$

$$= 2x - 3x^2 + \frac{19}{6}x^3 - \frac{61}{24}x^4 + \cdots$$

If our purpose is to get an estimate for the solution $y(x)$, then

$$2x - 3x^2 + \frac{19}{6}x^3 - \frac{61}{24}x^4$$

may be sufficient. We will not address the important problem of determining the number of needed terms. ∎

Summary of Taylor-Series Method

To solve

$$a(x)y''(x) + b(x)y'(x) + c(x)y(x) = f(x), \qquad (9)$$

1. Choose the center x_0 and be sure a, b, c, f can be written as power series centered at x_0 with radii of convergence greater than zero, and $a(x_0) \neq 0$. (In our problems, x_0 will be given.)
2. Determine how many terms of the expansion

$$y(x) = \sum_{n=0}^{\infty} y_n(x - x_0)^n = \sum_{n=0}^{\infty} \frac{y^{(n)}(x_0)}{n!}(x - x_0)^n$$

   are needed. Note that the constants $\{y_n\}$ are the coefficients of the Taylor-series expansion of y and are given by

$$y_n = \frac{y^{(n)}(x_0)}{n!}.$$

3. If initial conditions for y are given at x_0, use them to determine y_0 and y_1. If no initial conditions are given for (9), take y_0, y_1 as arbitrary.
4. Evaluate (9) at x_0. Then solve to find $y''(x_0)$. Set $y_2 = y''(x_0)/2$.
5. Differentiate (9) to get

$$a'(x)y''(x) + a(x)y'''(x) + b'(x)y'(x) + b(x)y''(x) + c'(x)y(x) + c(x)y'(x)$$
$$= f'(x). \qquad (10)$$

6. Evaluate (10) at x_0 and use the known $y(x_0), y'(x_0), y''(x_0)$ to find $y'''(x_0)$. Let $y_3 = y'''(x_0)/3!$.
7. Continue in this manner by repeatedly differentiating (10) and evaluating at x_0, to find $y^{(n)}(x_0)$ for as many n as needed.

Modifications for First-Order Equation

For

$$a(x)y'(x) + b(x)y(x) = f(x),$$

the procedure is essentially the same, except that if no initial condition is given, only y_0 is taken to be arbitrary.

There are several advantages to this method.

Advantages of Method

i. Done just by evaluation and differentiation.
ii. Does not require knowing series for $a(x)$, $b(x)$, $c(x)$, $f(x)$.
iii. Does not require manipulation of series or summation notation.
iv. Can be used, in principle, to find any needed number of terms.

There are also disadvantages.

Disadvantages of Method

i. If $a(x)$, $b(x)$, $c(x)$, $f(x)$ are at all complicated, the repeated differentiations may become quite messy.
ii. If the expression for a general term is required, then it is more difficult to find this way, and requires some "insight."
iii. May sometimes be difficult to determine a recursion relationship for coefficients that would enable one to easily generate additionally needed terms from known terms.

If one has access to a symbolic differentiation program, then the first disadvantage is easily surmounted, since the computer does all the differentiation. The second problem still remains. The method of the next section will address these problems.

In concluding this section an example of computing the general solution of a linear second-order differential equation will be given.

Example 5.4.2   Find the first five terms of the Taylor-series expansion for the general solution of

$$y'' + xy' + x^2 y = 3 + x, \qquad (11)$$

centered at $x_0 = 0$, using the Taylor-series method.

Solution   Since no initial conditions are given and (11) is a second-order linear differential equation, we know that there are two arbitrary constants, and

$$y(0) = y_0, \qquad y'(0) = y_1,$$

may be taken as arbitrary. Solve (11) for y'':

$$y'' = 3 + x - xy' - x^2 y. \qquad (12)$$

Evaluate (12) at $x = 0$ to find $y''(0)$:

$$y''(0) = 3 + 0 - 0 - 0 = 3. \qquad (13)$$

Differentiate (12):

$$y''' = 1 - y' - xy'' - 2xy - x^2 y'. \qquad (14)$$

Evaluate (14) at $x = 0$:

$$y'''(0) = 1 - y'(0) = 1 - y_1. \tag{15}$$

Differentiate (14):
$$y^{(4)} = -2y'' - xy''' - 2y - 2xy' - 2xy' - x^2y''. \tag{16}$$

Evaluate (16) at $x = 0$:
$$y^{(4)}(0) = -2y''(0) - 2y(0) = -6 - 2y_0. \tag{17}$$

Thus,
$$y(x) = y_0 + y_1 x + \frac{3}{2!}x^2 + \frac{(1-y_1)}{3!}x^3 + \frac{(-6-2y_0)}{4!}x^4 + \cdots \tag{18}$$

Recall that, from the theory of linear differential equations (Section 3.4), the general solution $y(x)$ of (11) may be written as:
$$y(x) = y_p(x) + c_1 h_1(x) + c_2 h_2(x),$$

where $y_p(x)$ is a particular solution of (11), $h_1(x)$, $h_2(x)$ are solutions of the associated homogeneous equation, and c_1, c_2 are arbitrary constants. Note that the general solution (18) can be rewritten as

$$y(x) = \left[\frac{3}{2}x^2 + \frac{x^3}{3!} - \frac{x^4}{4} + \cdots \right] + y_0 \left[1 - \frac{x^4}{12} + \cdots \right] + y_1 \left[x - \frac{x^3}{3!} + \cdots \right].$$

That is,
$$\frac{3}{2}x^2 + \frac{x^3}{3!} - \frac{x^4}{4} + \cdots \quad \text{is the first five terms}$$

(the first two are zero) of a power series for the particular solution y_p, and

$$1 - \frac{x^4}{12} + \cdots$$
$$x - \frac{x^3}{3!} + \cdots \quad \text{are the first five terms}$$

of a power series for h_1, h_2, respectively, which are solutions of the associated homogeneous equation $y'' + xy' + x^2 y = 0$. ∎

Note In Examples 5.4.1 and 5.4.2, we first rewrote the differential equation by solving for y''. If $a(x)$ is not constant, it may be simpler to omit this step and follow the general procedure of repeatedly differentiating the differential equation.

Exercises

In Exercises 1 through 6, use the method of Section 5.4 to find the indicated number of terms (m) of the Taylor-series expansion of the solution of the given differential equation. The center x_0 is where the initial conditions are applied.

1. $y'' + y = x$, $\quad y(0) = 0$, $\quad y'(0) = 2$, $\quad m = 5$
2. $y'' + x^2 y' + y = \sin x$, $\quad y(0) = 1$, $\quad y'(0) = 1$, $\quad m = 5$
3. $y' + x^2 y = 1$, $\quad y(0) = 2$, $\quad m = 4$
4. $y' + x^2 y = 1$, $\quad y(1) = 2$, $\quad m = 4$
5. $xy' + y = x^2$, $\quad y(1) = 0$, $\quad m = 4$
6. $y'' + xy = e^x$, $\quad y(-1) = 0$, $\quad y'(-1) = 1$, $\quad m = 5$

In Exercises 7 through 16, use the method of Section 5.4 to find the first five terms of the general solution of the indicated differential equation. Center the series at x_0 and express your answer in the form $y_p + y_h$ (see Example 5.4.2).

7. $y'' + y = 0$, $\quad x_0 = 0$
8. $y' + xy = \sin x$, $\quad x_0 = 0$
9. $y'' + x^2 y = e^x$, $\quad x_0 = 0$
10. $xy' + y = \ln x$, $\quad x_0 = 1$
11. $y'' - (\sin x)y = \cos x$, $\quad x_0 = \pi/2$
12. $y'' + y' + y = x^{-1}$, $\quad x_0 = 1$
13. $xy' + 2y = x$, $\quad x_0 = 1$
14. $y'' - xy = 0$, $\quad x_0 = 1$
15. $y'' + \dfrac{1}{x-1} y = x^2$, $\quad x_0 = 0$
16. $y'' + xy' + xy = \sin x$, $\quad x_0 = 0$

An alternative method to find the series solution at a center $x_0 \neq 0$ is to do the change of variables $z = x - x_0$, and find the series centered at $z_0 = 0$.

17. Verify that the change of variables $z = x - 1$ changes

$$\frac{dy}{dx} + x^2 y = 1, \quad y(1) = 2,$$

to

$$\frac{dy}{dz} + (z+1)^2 y = 1, \quad y(0) = 2.$$

Find the first five terms of the power-series solution for $y(z)$ centered at $z_0 = 0$, and compare your answer to that of Exercise 4.

18. Verify that the change of variables $z = x + 1$ changes $d^2 y/dx^2 + xy = e^x$, $y(-1) = 0$, $y'(-1) = 1$, to:

$$\frac{d^2 y}{dz^2} + (z-1)y = e^{z-1}, \quad y(0) = 0, \quad y'(0) = 1. \quad (19)$$

Find the first five terms of the power-series solution to (19), centered at $z = 0$, and compare to your answer for Exercise 6.

19. Let $z = x - 1$, and rewrite the differential equation in Exercise 10 as a differential equation in y and z. Find the first five terms of a power-series solution of y, centered at $z_0 = 0$, and compare to your answer for Exercise 10.

20. Let $z = x - \pi/2$, and rewrite the differential equation in Exercise 11 as a differential equation in y and z. Find the first five terms of a power-series solution of y, centered at $z_0 = 0$, and compare your answer to that of Exercise 11.

21. Let $z = x - 1$, and rewrite the differential equation in Exercise 12 as a differential equation in y and z. Find the first five terms of a power-series solution of y, centered at $z_0 = 0$, and compare your answer to that of Exercise 12.

5.5 Solution at an Ordinary Point (Undetermined Coefficients)

This section will present an alternative method to that of Section 5.4 for the determination of a series solution of

$$a(x)y''(x) + b(x)y'(x) + c(x)y(x) = f(x), \quad (1)$$

at a point x_0, where a, b, c, f all have power-series expansions and $a \neq 0$. That is, at an *ordinary point* of the differential equation. The idea behind the method is much like the method of undetermined coefficients in Chapter 3. Given a form for the solution, in this case a Taylor series, we substitute the form into the differential equation. An algebraic problem in the unknown constant coefficients then must be solved.

Method of Undetermined Series Coefficients

1. Select the center x_0 and make sure that $a(x), b(x), c(x), f(x)$ all have power-series expansions centered at x_0 and $a(x_0) \neq 0$. (In our problems the center will be given.)
2. Let

$$y = \sum_{n=0}^{\infty} c_n (x - x_0)^n, \quad (2)$$

where the c_n are the unknown coefficients of the solution.

3. Write $a(x), b(x), c(x), f(x)$ as Taylor series centered at x_0.
4. Substitute the unknown expansion (2) for y and the series for a, b, c, f into the differential equation (1).
5. Algebraically manipulate the result so as to obtain a single series on each side of the equal sign. This procedure usually takes the following order:
    i. Multiply out any products between series for the coefficients and the series for y.
    ii. By shifting summation variables, get all series in terms of $(x - x_0)^n$.
    iii. By possibly removing the first few terms of some of the series and adding them as separate terms, get all series summed over the same values of n.
    iv. Combine all the series on each side of the equal sign into a single series.
6. Equate coefficients of corresponding powers of x to obtain equations for the coefficients.

This method works equally well with first-order equations. Our first example is a fairly simple first-order equation.

Example 5.5.1 Find the first five terms of the series solution, centered at $x_0 = 0$, of the differential equation

$$y' + xy = e^x, \qquad y(0) = 1, \tag{3}$$

using the method of this section.

Solution (Steps 1, 2.) Let

$$y = \sum_{n=0}^{\infty} c_n x^n. \tag{4}$$

(Step 3.) The coefficients 1, x of y', y are already written as power series centered at zero, while

$$e^x = \sum_{n=0}^{\infty} \frac{x^n}{n!}. \tag{5}$$

(Step 4.) Substitute these series into (3), to get

$$\left(\sum_{n=0}^{\infty} c_n x^n \right)' + x \left(\sum_{n=0}^{\infty} c_n x^n \right) = \sum_{n=0}^{\infty} \frac{x^n}{n!} \tag{6}$$

or

$$\sum_{n=1}^{\infty} n c_n x^{n-1} + \sum_{n=0}^{\infty} c_n x^{n+1} = \sum_{n=0}^{\infty} \frac{x^n}{n!}. \tag{7}$$

Step 5i was performed in going from (6) to (7).

(Step 5ii.) We now wish to get all three series in terms of the same x powers. There are several possibilities; we shall use x^n. In order to do this, we need to change the way the summation is written. This is quite similar to a change of variables in an integral. Since writing $n = n + 1$ can be confusing, we will introduce a dummy variable.

For the first series in (7), let $m = n - 1$ (so that $m + 1 = n$) and note that

$$\sum_{n=1}^{\infty} n c_n x^{n-1} = \sum_{m=0}^{\infty} (m+1) c_{m+1} x^m, \tag{8}$$

since $m = 0$ when $n = 1$. Now let $m = n$ in the second series in (8), and we have

$$\sum_{n=1}^{\infty} n c_n x^{n-1} = \sum_{n=0}^{\infty} (n+1) c_{n+1} x^n. \tag{9}$$

The student who is confused by n appearing in both series in (9) is encouraged to write out the first few terms of each series and see that they both give

$$c_1 + 2c_2 x + 3c_3 x^2 + 4c_4 x^3 + \cdots$$

Similarly, for the second series in (7),

$$\sum_{n=0}^{\infty} c_n x^{n+1} \xrightarrow{(m=n+1)} \sum_{m=1}^{\infty} c_{m-1} x^m \xrightarrow{(m=n)} \sum_{n=1}^{\infty} c_{n-1} x^n. \tag{10}$$

Substituting (9), (10) into (7) gives

$$\underbrace{\sum_{n=0}^{\infty} (n+1) c_{n+1} x^n + \sum_{n=1}^{\infty} c_{n-1} x^n = \sum_{n=0}^{\infty} \frac{x^n}{n!}}. \tag{11}$$

(Step 5iii.) We cannot combine the series yet, since they do not all begin with the same value of n. This is taken care of by writing the extra terms ($n = 0$) separately.

$$c_1 + \sum_{n=1}^{\infty} (n+1)c_{n+1}x^n + \sum_{n=1}^{\infty} c_{n-1}x^n = 1 + \sum_{n=1}^{\infty} \frac{x^n}{n!}.$$

(Step 5iv.) Now combine the two series on the left, to get

$$c_1 + \sum_{n=1}^{\infty} [(n+1)c_{n+1} + c_{n-1}]x^n = 1 + \sum_{n=1}^{\infty} \frac{x^n}{n!}. \tag{12}$$

(Step 6.) On each side of the equal sign in (12) we have a power series. Since a power-series expansion at a given center is unique, the corresponding coefficients must be equal. Thus equating coefficients of like powers of x^n gives

$$x^0: \quad c_1 = 1$$

and

$$x^n \text{ for } n \geq 1: \quad (n+1)c_{n+1} + c_{n-1} = \frac{1}{n!}. \tag{13}$$

The initial condition in the original differential equation (3) implies

$$1 = y(0) = c_0. \tag{14}$$

In comparison to the method of the last section, we have done more work to get the first two terms. However, we are now in a position to generate as many terms as needed very rapidly. Solving (13) for c_{n+1}:

$$c_{n+1} = -\frac{1}{n+1}c_{n-1} + \frac{1}{(n+1)n!}$$

$$= -\frac{1}{n+1}c_{n-1} + \frac{1}{(n+1)!} \quad \text{for } n \geq 1. \tag{15}$$

This formula enables us to find the c_n recursively, since we know $c_0 = 1, c_1 = 1$. At $n = 1$, the recursion (15) is

$$c_2 = -\frac{1}{2}c_0 + \frac{1}{2} = -\frac{1}{2}(1) + \frac{1}{2} = 0.$$

At $n = 2$, the recursion (15) is

$$c_3 = -\frac{1}{3}c_1 + \frac{1}{6} = -\frac{1}{3}(1) + \frac{1}{6} = -\frac{1}{6}.$$

At $n = 3$, the recursion (15) is

$$c_4 = -\frac{1}{4}c_2 + \frac{1}{4!} = -\frac{1}{4}(0) + \frac{1}{24} = \frac{1}{24}.$$

and

$$y = 1 + x - \frac{1}{6}x^3 + \frac{1}{24}x^4 + \cdots$$

Using (15), a programmable pocket calculator can generate a large number of terms with a simple loop program. ∎

Since manipulating summations can be confusing at first, we will work several additional examples.

Example 5.5.2 Find two linearly independent power-series solutions of the linear homogeneous differential equation

$$y'' + x^2 y = 0, \qquad (16)$$

centered at $x_0 = 0$.

Solution Let $y = \sum_{n=0}^{\infty} c_n x^n$ and substitute into (16),

$$\left(\sum_{n=0}^{\infty} c_n x^n\right)'' + x^2 \left(\sum_{n=0}^{\infty} c_n x^n\right) = 0$$

or

$$\sum_{n=2}^{\infty} n(n-1) c_n x^{n-2} + \sum_{n=0}^{\infty} c_n x^{n+2} = 0. \qquad (17)$$

In order to combine these two series, we need to shift the summation index. Both will be expressed in terms of x^n. (Other choices could be x^{n+2} or x^{n-2}.) The first series in (17) is

$$\sum_{n=2}^{\infty} n(n-1) c_n x^{n-2} \xrightarrow{(m=n-2)} \sum_{m=0}^{\infty} (m+2)(m+1) c_{m+2} x^m$$

$$\xrightarrow{(m=n)} \sum_{n=0}^{\infty} (n+2)(n+1) c_{n+2} x^n,$$

and the second series in (17) is

$$\sum_{n=0}^{\infty} c_n x^{n+2} \xrightarrow{(m=n+2)} \sum_{m=2}^{\infty} c_{m-2} x^m \xrightarrow{(m=n)} \sum_{n=2}^{\infty} c_{n-2} x^n.$$

Thus (17) can be written as

$$\sum_{n=0}^{\infty} (n+2)(n+1) c_{n+2} x^n + \sum_{n=2}^{\infty} c_{n-2} x^n = 0. \qquad (18)$$

Removing the first two terms from the first series in (18):

$$2c_2 + 3 \cdot 2 c_3 x + \sum_{n=2}^{\infty} (n+2)(n+1) c_{n+2} x^n + \sum_{n=2}^{\infty} c_{n-2} x^n = 0$$

enables the two series to be combined,

$$2c_2 + 6c_3 x + \sum_{n=2}^{\infty} [(n+2)(n+1)c_{n+2} + c_{n-2}]x^n = 0.$$

Equating coefficients,

$$x^0: \quad 2c_2 = 0, \tag{19}$$
$$x^1: \quad 6c_3 = 0, \tag{20}$$
$$x^n, n \geq 2: \quad (n+2)(n+1)c_{n+2} + c_{n-2} = 0. \tag{21}$$

The differential equation (16) is second-order so that $c_0 = y(0)$, $c_1 = y'(0)$, can be taken as arbitrary, and, from (19) and (20), $c_2 = c_3 = 0$. Solving (21) for c_{n+2}, we get:

$$c_{n+2} = -\frac{1}{(n+2)(n+1)} c_{n-2} \quad \text{for } n \geq 2,$$

is the recursion relationship. Thus for

$$n = 2: \quad c_4 = -\frac{1}{4 \cdot 3} c_0,$$

$$n = 3: \quad c_5 = -\frac{1}{5 \cdot 4} c_1,$$

$$n = 4: \quad c_6 = -\frac{1}{6 \cdot 5} c_2 = 0,$$

$$n = 5: \quad c_7 = -\frac{1}{7 \cdot 6} c_3 = 0,$$

$$n = 6: \quad c_8 = -\frac{1}{8 \cdot 7} c_4 = \frac{1}{8 \cdot 7 \cdot 4 \cdot 3} c_0,$$

and

$$y = c_0 + c_1 x - \frac{1}{4 \cdot 3} c_0 x^4 - \frac{1}{5 \cdot 4} c_1 x^5 + \frac{1}{8 \cdot 7 \cdot 4 \cdot 3} c_0 x^8 + \frac{1}{9 \cdot 8 \cdot 5 \cdot 4} c_1 x^9 + \cdots$$

$$= c_0 \left[1 - \frac{x^4}{12} + \frac{x^8}{672} + \cdots \right] + c_1 \left[x - \frac{x^5}{20} + \frac{x^9}{1440} + \cdots \right].$$

From the theory of linear differential equations developed in Chapter 3, we know the functions given by the series

$$g(x) = 1 - \frac{x^4}{12} + \frac{x^8}{672} + \cdots$$

and

$$h(x) = x - \frac{x^5}{20} + \frac{x^9}{1440} + \cdots$$

will give two linearly independent solutions of the original differential equation (16), since using the Wronskian gives

$$W[g, h](0) = \det \begin{bmatrix} 1 & 0 \\ 0 & 1 \end{bmatrix} = 1 \neq 0. \quad \blacksquare$$

Sometimes it is possible to solve a recursion relationship for the c_n directly in terms of n, either by inspection or by using the theory of difference equations (Chapter 7).

Example 5.5.3 Find the series solution, centered at $x_0 = 0$, of the differential equation

$$y' + xy = 0. \tag{22}$$

Solution Again let

$$y = \sum_{n=0}^{\infty} c_n x^n.$$

Note that (22) is the associated homogeneous equation for (3), so that all of the series manipulations of Example 5.5.1 are still valid, except that $f(x) = 0$ instead of e^x. The recursion relationship is now

$$c_0 = y(0) \text{ is arbitrary,}$$
$$c_1 = 0,$$

and

$$c_{n+1} = -\frac{1}{n+1} c_{n-1} \quad \text{for } n \geq 1. \tag{23}$$

Evaluating (23) at successive n, we see that

$$n = 1: \quad c_2 = -\frac{1}{2} c_0,$$

$$n = 2: \quad c_3 = -\frac{1}{3} c_1 = 0,$$

$$n = 3: \quad c_4 = -\frac{1}{4} c_2 = -\frac{1}{4}\left(-\frac{1}{2}\right) c_0 = \frac{1}{4 \cdot 2} c_0$$

$$n = 4: \quad c_5 = -\frac{1}{5} c_3 = -\frac{1}{5} 0 = 0,$$

$$n = 5: \quad c_6 = -\frac{1}{6} c_4 = -\frac{1}{6 \cdot 4 \cdot 2} c_0,$$

$$\vdots \qquad \vdots$$

Note that the odd terms c_1, c_3, c_5 are all zero, and the even terms have the pattern

$$c_8 = \frac{1}{8\cdot 6\cdot 4\cdot 2}c_0 = \frac{1}{2^4\cdot 4\cdot 3\cdot 2\cdot 1}c_0,$$

$$c_{10} = \frac{-1}{10\cdot 8\cdot 6\cdot 4\cdot 2}c_0 = \frac{-1}{2^5\cdot 5\cdot 4\cdot 3\cdot 2\cdot 1}c_0.$$

That is,

$$c_{2n} = \frac{(-1)^n}{2^n n!}c_0. \tag{24}$$

The solution to (22) is thus

$$y = c_0 \sum_{n=0}^{\infty} \frac{(-1)^n}{2^n n!} x^{2n}. \quad \blacksquare \tag{25}$$

Most students will not be able to arrive at a formula such as (24) without practice, and then only for simpler problems. Thus the answers (in the back of the book) for this section are given in three forms: recursion relationship, first few terms, and sometimes as (25).

Incidentally, the example just completed may be summed (written without summation notation) by recalling that

$$e^z = \sum_{n=0}^{\infty} \frac{z^n}{n!}.$$

Thus,

$$y = c_0 \sum_{n=0}^{\infty} \frac{(-1)^n}{2^n n!} x^{2n} = c_0 \sum_{n=0}^{\infty} \frac{\left(-\frac{x^2}{2}\right)^n}{n!} = c_0 e^{-x^2/2}.$$

Our last example has a power series centered at $x_0 \neq 0$.

Example 5.5.4 For the differential equation

$$y'' + xy' + 2y = x + 3, \quad y(1) = 0, \quad y'(1) = 1, \tag{26}$$

find the series solution centered at $x_0 = 1$, using the method of this section. Find the first five terms of the solution and the recursion relationship for finding additional terms.

Solution Let

$$y = \sum_{n=0}^{\infty} c_n(x-1)^n. \tag{27}$$

The coefficient x, and the forcing term $x + 3$, need to be written as power series centered at $x_0 = 1$, that is, in terms of powers of $(x - 1)$. This can be done either using the formulas of Section 5.2 or by inspection:

$$x = 1 + (x-1), \quad x + 3 = 4 + (x-1). \tag{28}$$

Alternatively, we may let $z = x - 1$, and consider the problem (26) centered at

$z = 0$. We shall use the latter approach since it reduces the amount of writing and simplifies the notation. If we let $z = x - 1$, the differential equation (26) becomes

$$\frac{d^2y}{dz^2} + (1 + z)\frac{dy}{dz} + 2y = 4 + z, \qquad y(0) = 0, \qquad y'(0) = 1, \qquad (29)$$

and the series (27) for y is

$$y = \sum_{n=0}^{\infty} c_n z^n. \qquad (30)$$

Substituting (30) into (29),

$$\left(\sum_{n=0}^{\infty} c_n z^n\right)'' + (1 + z)\left(\sum_{n=0}^{\infty} c_n z^n\right)' + 2\left(\sum_{n=0}^{\infty} c_n z^n\right) = 4 + z$$

or, upon differentiating,

$$\sum_{n=2}^{\infty} n(n-1)c_n z^{n-2} + (1 + z)\left(\sum_{n=1}^{\infty} nc_n z^{n-1}\right) + 2\sum_{n=0}^{\infty} c_n z^n = 4 + z. \qquad (31)$$

Multiplying out $(1 + z)$ times the second series yields:

$$\sum_{n=2}^{\infty} n(n-1)c_n z^{n-2} + \sum_{n=1}^{\infty} nc_n z^{n-1} + \sum_{n=1}^{\infty} nc_n z^n + 2\sum_{n=0}^{\infty} c_n z^n = 4 + z. \qquad (32)$$

We now rewrite the first two series of (32) in terms of z^n by shifting the summation index:

$$\sum_{n=2}^{\infty} n(n-1)c_n z^{n-2} \xrightarrow{(m=n-2)} \sum_{m=0}^{\infty} (m+2)(m+1)c_{m+2} z^m$$

$$\xrightarrow{(m=n)} \sum_{n=0}^{\infty} (n+2)(n+1)c_{n+2} z^n$$

and

$$\sum_{n=1}^{\infty} nc_n z^{n-1} \xrightarrow{(m=n-1)} \sum_{m=0}^{\infty} (m+1)c_{m+1} z^m \xrightarrow{(m=n)} \sum_{n=0}^{\infty} (n+1)c_{n+1} z^n.$$

Substituting these series back into (32) gives:

$$\sum_{n=0}^{\infty} (n+2)(n+1)c_{n+2} z^n + \sum_{n=0}^{\infty} (n+1)c_{n+1} z^n + \sum_{n=1}^{\infty} nc_n z^n + 2\sum_{n=0}^{\infty} c_n z^n$$
$$= 4 + z. \qquad (33)$$

In order to get all four series in (33) summed over the same n values, remove the first term from the first, second, and fourth series, and then combine the series:

$$2c_2 + c_1 + 2c_0 + \sum_{n=1}^{\infty} [(n+2)(n+1)c_{n+2} + (n+1)c_{n+1} + (n+2)c_n]z^n$$
$$= 4 + z.$$

Now equate coefficients of powers of z on both sides of the equal sign:

$$n = 0: \quad 2c_2 + c_1 + 2c_0 = 4, \tag{34}$$
$$n = 1: \quad 6c_3 + 2c_2 + 3c_1 = 1, \tag{35}$$

and

$$n \geq 2: \quad (n+2)(n+1)c_{n+2} + (n+1)c_{n+1} + (n+2)c_n = 0$$

or

$$c_{n+2} = -\frac{1}{n+2}c_{n+1} - \frac{1}{n+1}c_n \quad \text{for } n \geq 2. \tag{36}$$

Equations (34)–(36) give our recursion relationship. From the initial conditions,

$$c_0 = y(0) = 0, \quad c_1 = y'(0) = 1.$$

Using (34), we then find that

$$c_2 = 2 - \frac{1}{2}c_1 - c_0 = 2 - \frac{1}{2}(1) - 0 = \frac{3}{2},$$

and, from (35),

$$c_3 = \frac{1}{6} - \frac{1}{3}c_2 - \frac{1}{2}c_1 = \frac{1}{6} - \frac{1}{2} - \frac{1}{2} = -\frac{5}{6},$$

and, by (36), with $n = 2$,

$$c_4 = -\frac{1}{4}c_3 - \frac{1}{3}c_2 = -\frac{7}{24}.$$

Thus,

$$y = \sum_{n=0}^{\infty} c_n z^n = \sum_{n=0}^{\infty} c_n (x-1)^n$$

$$= (x-1) + \frac{3}{2}(x-1)^2 - \frac{5}{6}(x-1)^3 - \frac{7}{24}(x-1)^4 + \cdots \blacksquare$$

Exercises

In Exercises 1 through 14, find the series solution, centered at x_0, of the differential equation, using the method of this section. Give the recursion relationship for the coefficients and also find the first five coefficients.

1. $y' + 3y = 0$, $\quad y(0) = 2$, $\quad x_0 = 0$
2. $y'' + y = 1$, $\quad y(0) = 0$, $\quad y'(0) = 1$, $\quad x_0 = 0$
3. $xy' + y = 0$, $\quad y(1) = 2$, $\quad x_0 = 1$
4. $y'' + xy = 0$, $\quad y(0) = 1$, $\quad y'(0) = 1$, $\quad x_0 = 0$

5. $y'' + xy = \dfrac{1}{1-x}$, $y(0) = 0$, $y'(0) = 0$,
$x_0 = 0$

6. $y'' - 2xy' + 4y = 0$, $y(0) = 1$, $y'(0) = 0$,
$x_0 = 0$

7. $y'' - 2xy' + 6y = 0$, $y(0) = 0$, $y'(0) = 1$,
$x_0 = 0$

8. $x^2 y' - 2y = 1$, $y(1) = 0$, $x_0 = 1$

9. $y' + 3x^2 y = e^x$, $y(0) = 0$, $x_0 = 0$

10. $y' - 2x^2 y = xe^x$, $y(0) = 1$, $x_0 = 0$

11. $y'' + x^2 y = \sin x$, $y(0) = 0$, $y'(0) = 0$,
$x_0 = 0$

12. $y'' + xy' + x^3 y = \cos x$, $y(0) = 1$,
$y'(0) = 1$, $x_0 = 0$

13. $y' + (x^2 + 1)y = x^2$, $y(0) = 2$, $x_0 = 0$

14. $y' + (x^3 - x)y = e^x$, $y(0) = 1$, $x_0 = 0$

In Exercises 15 through 22, express the fundamental solution as a linear combination of a linearly independent set of solutions, using the method of this section. Take all power series centered at $x_0 = 0$.

15. $y'' - y = 0$

16. $y'' - x^3 y = 0$

17. $(1 + x^2) y'' + y = 0$

18. $(1 - x^2) y'' + xy' + 3y = 0$

19. $y'' + 2xy = 0$

20. $y' - 2xy = 0$

21. $y' + 3x^5 y = 0$

22. $y'' + 5x^2 y' + (1 + x)y = 0$

23. Suppose that $a(x)$, $b(x)$, $c(x)$ in (1) have power series expansions centered at $x_0 = 0$ and that $f \equiv 0$ and $a(0) \neq 0$.

   a) Show that (1) has solutions with power series expansions,
   $$g(x) = x + \sum_{n=2}^{\infty} \alpha_n x^n, \quad h(x) = 1 + \sum_{n=2}^{\infty} \beta_n x^n$$
   for unique constants α_n, β_n.

   b) Show that g, h are a fundamental set of solutions for (1).

   c) Show that if $y(x) = \sum_{n=0}^{\infty} y_n x^n$ is the solution of (1) with $y(0) = y_0$, $y'(0) = y_1$, then $y = y_0 h + y_1 g$ and $y_n = \beta_n y_0 + \alpha_n y_1$ for all $n \geq 2$.

5.6 Legendre Polynomials

Many different functions other than just powers of x are used to give series expansions. In Section 5.8 we shall discuss the important Bessel functions. In this section we will briefly discuss the Legendre polynomials.

The differential equation

$$(1 - x^2) \frac{d^2 y}{dx^2} - 2x \frac{dy}{dx} + v(v+1)y = 0, \tag{1}$$

with parameter v is called *Legendre's equation of order* v. It is assumed that $v \geq -\tfrac{1}{2}$, since $v^2 + v$ has the same values for $v < -\tfrac{1}{2}$. The values $x = \pm 1$ are singular points, and all other values of x are ordinary points. Thus, by Theorem 5.3.1, there are solutions on $-1 < x < 1$ of the form

$$y = \sum_{n=0}^{\infty} c_n x^n. \tag{2}$$

Substituting (2) into (1) yields

5.6 Legendre Polynomials

$$(1-x^2)\sum_{n=0}^{\infty} c_n n(n-1)x^{n-2} - 2x\sum_{n=0}^{\infty} c_n n x^{n-1} + v(v+1)\sum_{n=0}^{\infty} c_n x^n = 0,$$

or, after multiplying out and regrouping,

$$\sum_{n=0}^{\infty} c_n n(n-1)x^{n-2} - \sum_{n=0}^{\infty}[n(n-1)+2n-v(v+1)]c_n x^n = 0$$

or

$$\sum_{n=0}^{\infty} c_{n+2}(n+2)(n+1)x^n - \sum_{n=0}^{\infty}[n(n-1)+2n-v(v+1)]c_n x^n = 0.$$

Equating the x^n coefficients to zero for $n \geq 0$ gives

$$c_{n+2}(n+2)(n+1) - [n^2+n-v(v+1)]c_n = 0.$$

But $[n^2+n-v(v+1)] = [n+(v+1)][n-v]$, so that

$$c_{n+2} = \frac{(n+v+1)(n-v)}{(n+2)(n+1)}c_n, \qquad n \geq 0. \tag{3}$$

We may get a linearly independent set of solutions by taking (2), (3) and having

$$y_1(x) \text{ defined by } c_0 = 1, \quad c_1 = 0, \tag{4}$$
$$y_2(x) \text{ defined by } c_0 = 0, \quad c_1 = 1. \tag{5}$$

Note that, from (3), (4) and (5), $y_1(x)$ will involve only even powers of x and $y_2(x)$ will involve only odd powers of x.

If v is an integer m, then (3) implies that $c_{m+2} = 0$. Thus $c_{m+2r} = 0$ for all integers $r \geq 1$, and there is an mth-degree polynomial $p(x)$ for a solution (p is a multiple of y_1 if m is even, while p is a multiple of y_2 if m is odd). It can be shown that $p(1) \neq 0$. Thus, by taking a multiple of $p(x)$, we get a polynomial $P_m(x)$ such that $P_m(1) = 1$. This function $P_m(x)$ is known as the *Legendre polynomial of degree m*.

For example, if $m = 3$, then, for y_2, we have $c_2 = c_4 = \cdots = 0$ from (3), and $c_0 = 0$. Also, (3) implies that $c_5 = 0$, and hence, $c_7 = c_9 = c_{11} = \cdots = 0$. Finally, (3), for $n = 1$, gives

$$c_3 = \frac{(1+3+1)(1-3)}{(1+2)(1+1)}c_1 = -\frac{5}{3}c_1$$

and hence

$$p(x) = c_1 x - \frac{5}{3}c_1 x^3.$$

If $p(1) = 1$, then $c_1 - \frac{5}{3}c_1 = 1$, so that $c_1 = -\frac{3}{2}$. Thus,

Figure 5.6.1
Graphs of $P_0(x)$, $P_2(x)$, $P_4(x)$.

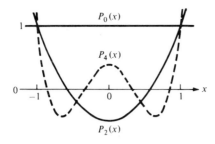

Figure 5.6.2
Graphs of $P_1(x)$, $P_3(x)$, $P_5(x)$.

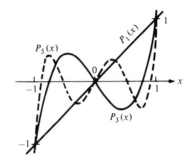

$$P_3(x) = \frac{-3x + 5x^3}{2}.$$

Similarly, one can show that $P_0(x) = 1$, $P_1(x) = x$, and

$$P_2(x) = \frac{-1 + 3x^2}{2}.$$

The graphs of the first six Legendre polynomials are shown in Figs. 5.6.1 and 5.6.2.

Laplace's equation $u_{xx} + u_{yy} + u_{zz} = 0$ is one of the more important partial differential equations in applied mathematics. One use is to describe potentials. If we are considering, for example, a hollow sphere whose surface is maintained at a certain electrical potential, then Laplace's equation in spherical coordinates (ρ, θ, ϕ) is used. The functions $u_n(\rho, \phi) = \rho^n P_n(\cos \phi)$, P_n the nth Legendre polynomial, are solutions of Laplace's equation in spherical coordinates. If the potential on the surface does not depend on θ, then the potential inside the sphere can be expressed as a series in the $u_n(\rho, \phi)$.

Exercises

1. Compute $P_4(x)$.
2. Compute $P_5(x)$.
3. It can be shown (Rodrigues's formula) that

$$P_n(x) = \frac{1}{2^n n!} \frac{d^n}{dx^n}[(x^2 - 1)^n]; \quad (6)$$

verify that (6) holds for $n = 1, 2, 3$.

4. The Binomial Theorem says that
$$(a+b)^n = \sum_{m=0}^{n} \frac{n!}{m!(n-m)!} a^m b^{n-m}. \quad (7)$$
Use (7) to express $(x^2-1)^n$ as a sum. Then insert this sum into (6), to show that
$$P_n(x) = \frac{1}{2^n} \sum_{k=0}^{[n/2]} \frac{(-1)^k (2n-2k)!}{k!(n-k)!(n-2k)!} x^{n-2k},$$
where $[n/2]$ is the greatest integer less than or equal to $n/2$.

5. Verify that
$$\int_{-1}^{1} P_m(x) P_n(x)\, dx = 0 \quad \text{if } m \neq n \quad (8)$$
for $0 \le m \le 3$, $0 \le n \le 3$. Property (8) actually holds for all nonnegative integers m, n.

6. The differential equation
$$(1-x^2)y'' - xy' + \alpha^2 y = 0 \quad (9)$$
is called Tchebycheff's differential equation. Show that if $\alpha = n$ is a nonnegative integer, then (9) has an nth-degree polynomial as a solution. When normalized so that $p(1) = 1$, this polynomial is called an nth-*degree Tchebycheff polynomial*, and is denoted $T_n(x)$.

7. (Continuation of Exercise 6.) Find $T_0(x)$, $T_1(x)$, $T_2(x)$, $T_3(x)$.

8. The differential equation
$$y'' - 2xy' + \lambda y = 0 \quad (10)$$
is known as the *Hermite equation*. Show that, if $\lambda = 2n$, n a nonnegative integer, then (10) has a solution that is an nth-degree polynomial. If the polynomial is chosen so that the coefficient of x^n is 2^n, it is called the *Hermite polynomial of degree n*, and denoted $H_n(x)$.

9. (Continuation of Exercise 8.) Find $H_0(x)$, $H_1(x)$, $H_2(x)$, $H_3(x)$.

10. On the basis of Figs. 5.6.1 and 5.6.2, sketch what you think $P_6(x)$, $P_7(x)$ should look like for $-1 \le x \le 1$.

11. Suppose that $p(x)$ is a polynomial solution of (1) with $v = m$ an integer, $p \neq 0$. Show that $p(1) \neq 0$.

5.7 Regular Singular Points (Frobenius Method)

The preceding sections have discussed how to solve
$$a(x)y'' + b(x)y' + c(x)y = f(x) \quad (1)$$
and $a(x)y' + b(x)y = f$, using power series at an ordinary point x_0. In many applications, however, it is necessary to consider the solution of (1) near a singular point. In this situation, power series are frequently inadequate.

Example 5.7.1 Consider
$$xy' + y = 0, \quad y(1) = 1, \quad (2)$$
and suppose we are interested in the solution of (2) for $0 < x < \infty$. The first difficulty is that we cannot get a power series that will converge on $(0, \infty)$, since the interval $(0, \infty)$ has no center and $x = 0$ is a singular point for (1). Let us settle, then, for a solution of (2) for $0 < x < 2$. If we use the technique of

Section 5.5, we find that the power series for the solution of (2) centered at $x_0 = 1$ is

$$y = \sum_{n=0}^{\infty} (-1)^n (x-1)^n. \tag{3}$$

However, (3) converges slowly for x near 0 (or 2) and does not readily tell us what the function looks like near zero.

Since the variables separate (also, it is linear), it is easy to show that the solution of (2) is

$$y = x^{-1}. \tag{4}$$

Formula (4) is much easier to work with than (3). ∎

We have also seen equations like (1) in Section 3.11, which considered *Euler's equation*,

$$ax^2 y'' + bxy' + cy = 0, \tag{5}$$

with a, b, c constant. In Section 3.11, it was shown that (5) always had at least one solution of the form

$$y = x^r, \tag{6}$$

where r could be any number. The exponent r was found as the root of a quadratic equation.

These comments suggest that perhaps a way to solve (1) at a singular point is to use series with fractional or negative powers. This, in fact, is often the case. This section will develop this method. The next section applies this method to an important equation that arises in many applications that have a circular geometry.

First, we need to specify the types of equations our method will work on. We begin by assuming that $x_0 = 0$ is the singular point. If the singular point x_0 is not equal to zero, then the change of independent variables $z = x - x_0$ gives an equation for which $z_0 = 0$ is a singular point. A function $g(x)$ is *analytic at zero* if it has a power-series expansion centered at zero.

Definition 5.7.1 The point $x_0 = 0$ is a *regular singular point* of $a(x)y'' + b(x)y' + c(x)y = 0$ if

$$a(x), b(x), c(x) \quad \text{are analytic at zero}, \tag{7}$$

$$\lim_{x \to 0} \frac{a(x)}{x^2} = p_0 \text{ exists} \quad \text{and} \quad p_0 \neq 0, \tag{8}$$

and

$$\lim_{x \to 0} \frac{b(x)}{x} = q_0 \text{ exists}. \tag{9}$$

For example, Euler's equation (5) is regular singular at 0, since

$$\lim_{x\to 0}\frac{a(x)}{x^2} = \lim_{x\to 0}\frac{ax^2}{x^2} = a, \qquad \lim_{x\to 0}\frac{b(x)}{x} = \lim_{x\to 0}\frac{bx}{x} = b.$$

An alternative definition of *regular singular* is:

Definition 5.7.2   The point $x_0 = 0$ is a *regular singular point* of (1) if (1) can be written as

$$x^2 P(x) y'' + x Q(x) y' + R(x) y = 0, \tag{10}$$

where

$$P(x), Q(x), R(x) \quad \text{are analytic at zero}, \tag{11}$$

and

$$P(0) \neq 0. \tag{12}$$

To relate these definitions, note that $p_0 = P(0)$ and $Q(0) = q_0$. Also, if $a(x)$, $b(x)$ have a power series centered at zero (are analytic at zero) and the limits (8) and (9) exist, then the functions P, Q defined by

$$P(x) = \begin{cases} a(x) x^{-2} & \text{if } x \neq 0, \\ p_0 & \text{if } x = 0, \end{cases} \qquad Q(x) = \begin{cases} b(x) x^{-1} & \text{if } x \neq 0, \\ q_0 & \text{if } x = 0, \end{cases}$$

can be shown to be analytic, and thus (1) can be written in the form (10). Thus, for example,

$$x^2 (\cos x) y'' + x^2 y' + 7 y = 0$$

has zero as a regular singular point, since we can take $P(x) = \cos x$, $Q(x) = x$, $R(x) = 7$. All these functions have power series centered at zero and $\cos 0 = 1 \neq 0$. On the other hand,

$$(x^3 + x^5) y'' + x y' - x y = 0$$

does not have zero as a regular singular point, since $x^2 P(x) = x^3 + x^5$ implies that $P(x) = x + x^3$ and $P(0) = 0$. If we divide by x to get:

$$x^2 (1 + x^2) y'' + y' - y = 0,$$

zero is still not a regular singular point, since $xQ(x) = 1$ implies that $Q(x) = x^{-1}$, which is not analytic at zero.

Some equations, such as

$$x^3 y'' + x^2 y' + x^2 y = 0,$$

which does not have zero as a regular singular point, can be converted to one that does. In this example, dividing by x gives

$$x^2 y'' + x y' + x y = 0,$$

which does have zero as a regular singular point. If "can be written" is interpreted to include such algebraic operations, Definition 5.7.2 is slightly more general than Definition 5.7.1.

We shall now show that if (1) has a regular singular point at $x_0 = 0$, then it has at least one solution of the form

$$y = x^r \sum_{n=0}^{\infty} C_n x^n = \sum_{n=0}^{\infty} C_n x^{n+r}, \tag{13}$$

where r is a number. By changing r if necessary, we may assume that $C_0 \neq 0$. As with the method for solving Euler's equation (5), r will be determined by a quadratic equation. Assume the differential equation is in the form of (10). Let

$$P(x) = \sum_{n=0}^{\infty} P_n x^n, \quad Q(x) = \sum_{n=0}^{\infty} Q_n x^n, \quad R(x) = \sum_{n=0}^{\infty} R_n x^n, \tag{14}$$

be power series for P, Q, R centered at zero. Substituting (13) and (14) into (10) gives

$$x^2 \left(\sum_{n=0}^{\infty} P_n x^n \right) \left(\sum_{n=0}^{\infty} (r+n)(r+n-1) C_n x^{r+n-2} \right)$$

$$+ x \left(\sum_{n=0}^{\infty} Q_n x^n \right) \left(\sum_{n=0}^{\infty} (r+n) C_n x^{r+n-1} \right) + \left(\sum_{n=0}^{\infty} R_n x^n \right) \left(\sum_{n=0}^{\infty} C_n x^{r+n} \right) = 0.$$

Since $x = 0$ is a singular point of (1), we only expect this equation to hold for $x \neq 0$. Multiply through by x^{-r} and simplify:

$$\left(\sum_{n=0}^{\infty} P_n x^n \right) \left(\sum_{n=0}^{\infty} (r+n)(r+n-1) C_n x^n \right) + \left(\sum_{n=0}^{\infty} Q_n x^n \right) \left(\sum_{n=0}^{\infty} (r+n) C_n x^n \right)$$

$$+ \left(\sum_{n=0}^{\infty} R_n x^n \right) \left(\sum_{n=0}^{\infty} C_n x^n \right) = 0. \tag{15}$$

Equating the constant terms in (15) to 0 gives

$$[P_0 r(r-1) + rQ_0 + R_0] C_0 = 0.$$

Since $C_0 \neq 0$, by assumption, we have

$$\rho(r) = P_0 r(r-1) + rQ_0 + R_0 = 0. \tag{16}$$

This is known as the *indicial equation* and $\rho(r)$ is the *indicial polynomial*. Note that, for Euler's equation (5), the indicial equation is the same as the one derived in Section 3.11.

Since $\rho(r)$ is a second-degree polynomial, there are at most two values of r, denoted r_1, r_2, such that $\rho(r) = 0$. If they are real, we take r_1 so that $r_1 \geq r_2$. Having determined r, we need to know whether it is possible to find the C_n. We omit the details, since they are fairly complicated. However, it can be shown that the coefficient of x^n in (15) is

$$\rho(r+n) C_n = \sum_{i=0}^{n-1} \alpha_i C_i, \tag{17}$$

where ρ is the polynomial in (16) and the α_i are numbers that can be expressed in terms of r, P_j, Q_j, R_j.

If $r = r_1$, then $\rho(r_1 + n) \neq 0$ for $n \geq 1$ and (17) shows that we can compute the C_n recursively. If also $r_1 - r_2$ is not a positive integer, then $\rho(r_2 + n) \neq 0$ for $n \geq 1$ (since $r_2 + n$ is never a root), and we can find the C_n recursively from (17). In each of these cases, it can be shown that the series converges for $x \neq 0$ and we get a solution in the form (13). However, if $r_1 = r_2$ or $r_1 - r_2$ is a positive integer, then this method may provide only one of the two solutions needed for a fundamental set of solutions. In this case we could use reduction of order. However, it is possible to show that the reduction of order always results in a certain form [given in (22)], and we are led to the following result, which summarizes all of the preceding discussion. If we only consider $x > 0$, the absolute values may be dropped.

Theorem 5.7.1

Method of Frobenius Suppose that zero is a regular singular point of (1) and that (1) is written as

$$x^2 P(x) y'' + x Q(x) y' + R(x) y = 0, \tag{18}$$

where P, Q, R have the expansions (14). Let

$$\rho(r) = P_0 r(r-1) + Q_0 r + R_0 \tag{19}$$

have roots r_1, r_2, and take $r_1 \geq r_2$ if the roots are real.

a) If $r_1 \neq r_2$, and $r_1 - r_2$ is not an integer, then (18) has a fundamental set of solutions of the form

$$y_1 = |x|^{r_1} \sum_{n=0}^{\infty} C_n x^n, \quad y_2 = |x|^{r_2} \sum_{n=0}^{\infty} \tilde{C}_n x^n, \tag{20}$$

where C_n, $\tilde{C}_n$ are constants. If case (a) does not hold, then r_1, r_2 must be real.

b) If $r_1 - r_2 = N > 0$, N an integer, then a fundamental set of solutions is

$$y_1(x) = |x|^{r_1} \sum_{n=0}^{\infty} C_n x^n, \tag{21}$$

$$y_2(x) = a y_1(x) \ln|x| + |x|^{r_2} \sum_{n=0}^{\infty} \tilde{C}_n x^n, \tag{22}$$

where a is a constant that may be zero.

c) If $r_1 = r_2$, then a fundamental set of solutions has the form (21), (22) except that $a = 1$.

Comment In all cases we may take $C_0 = 1$. In Case (b), if $a \neq 0$, then we may take $a = 1$ but not necessarily $\tilde{C}_0 = 1$. Also, r_1, r_2 in Case (a) may be complex. Then the real and imaginary parts of $y_1(x)$ will give two real solutions. However, we shall not work any complex examples. Our first, and simplest, example could also be done using the substitution $v = y'$; however, we shall use the Frobenius method.

Example 5.7.2 Find a fundamental set of solutions for the second-order linear differential equation

$$2x^2 y'' + (x + x^2) y' = 0. \tag{23}$$

Solution Equation (23) is in the form (18) with

$$x^2 P(x) = 2x^2, \qquad xQ(x) = x + x^2, \qquad R(x) = 0.$$

Thus $P(x) = 2$ and $Q(x) = 1 + x$. Hence, $P(0) = 2$, $Q(0) = 1$, $R(0) = 0$. The indicial equation (19) is

$$2r(r - 1) + r = 2r^2 - r = 0,$$

which has roots $r_1 = \frac{1}{2}$, $r_2 = 0$. Since the roots are distinct and do not differ by an integer, there will be two solutions of the form

$$y = x^r \sum_{n=0}^{\infty} C_n x^n = \sum_{n=0}^{\infty} C_n x^{n+r}, \tag{24}$$

where we consider $x > 0$ and drop the absolute values. It is quicker to initially consider both values of r simultaneously. Substituting (24) into (23) gives

$$2x^2 \sum_{n=0}^{\infty} (n + r)(n + r - 1) C_n x^{n+r-2} + (x + x^2) \sum_{n=0}^{\infty} (n + r) C_n x^{n+r-1} = 0.$$

Divide through by x^r and then multiply the coefficients times the series to get

$$\sum_{n=0}^{\infty} 2(n + r)(n + r - 1) C_n x^n + \sum_{n=0}^{\infty} (n + r) C_n x^n + \sum_{n=0}^{\infty} (n + r) C_n x^{n+1} = 0. \tag{25}$$

But

$$\sum_{n=0}^{\infty} (n + r) C_n x^{n+1} = \sum_{n=1}^{\infty} (n + r - 1) C_{n-1} x^n,$$

so that (25) becomes

$$\sum_{n=0}^{\infty} 2(n + r)(n + r - 1) C_n x^n + \sum_{n=0}^{\infty} (n + r) C_n x^n + \sum_{n=1}^{\infty} (n + r - 1) C_{n-1} x^n = 0.$$

Combining coefficients of like powers of x^n and setting them to zero yields

$$n = 0: \qquad 2r(r - 1) C_0 + r C_0 = 0, \tag{26}$$

$$n \geq 1: \qquad 2(n + r)(n + r - 1) C_n + (n + r) C_n + (n + r - 1) C_{n-1} = 0. \tag{27}$$

Equation (26) holds, since $2r(r - 1) + r = 0$ is the indicial equation. Then (27) can be rewritten as

$$(n + r)(2n + 2r - 1) C_n + (n + r - 1) C_{n-1} = 0, \qquad n \geq 1. \tag{28}$$

We must now consider the recursion formula (28) for each root separately.

Root $r_2 = 0$

Substituting $r = 0$ into (28) gives

$$n(2n - 1)C_n + (n - 1)C_{n-1} = 0, \qquad n \geq 1, \qquad (29)$$

so that

$$C_n = -\frac{(n - 1)}{n(2n - 1)}C_{n-1}, \qquad n \geq 1,$$

for $n = 1$, we have $C_1 = 0$. Thus $C_n = 0$ for all $n \geq 1$ and our solution with $C_0 = 1$ for $r_2 = 0$, is the constant solution

$$y_2 = x^0 \cdot 1 = 1.$$

Root $r_1 = \frac{1}{2}$

Substituting $r_1 = \frac{1}{2}$ into (28) and solving for C_n gives

$$C_n = -\frac{(2n - 1)}{2n(2n + 1)}C_{n-1}, \qquad n \geq 1. \qquad (30)$$

In general, we may be able to find the C_n only by recursively solving (30). In this example, there is a recognizable pattern. Take

$$C_0 = 1.$$

Then, from (30),

$$C_1 = -\frac{1}{2 \cdot 3}C_0 = -\frac{1}{2 \cdot 3},$$

$$C_2 = -\frac{3}{4 \cdot 5}C_1 = \frac{1}{5 \cdot 4 \cdot 2},$$

$$C_3 = -\frac{5}{6 \cdot 7}C_2 = -\frac{1}{7 \cdot 6 \cdot 4 \cdot 2},$$

$$C_4 = -\frac{7}{8 \cdot 9}C_3 = \frac{1}{9 \cdot 8 \cdot 6 \cdot 4 \cdot 2}.$$

Now

$$8 \cdot 6 \cdot 4 \cdot 2 = 2^4 \cdot 4 \cdot 3 \cdot 2 \cdot 1 = 2^4 4!$$

and

$$6 \cdot 4 \cdot 2 = 2^3 \cdot 3 \cdot 2 \cdot 1 = 2^3 \cdot 3!$$

It can be verified that, in fact,

$$C_n = \frac{(-1)^n}{(2n + 1)2^n n!}.$$

Our second solution is thus

$$y_1(x) = |x|^{1/2} \sum_{n=0}^{\infty} (-1)^n \frac{x^n}{(2n+1)2^n n!}. \tag{31}$$

The general solution of (23) on $(0, \infty)$ is

$$y = Ay_1 + By_2$$
$$= Ax^{1/2} \sum_{n=0}^{\infty} \frac{(-1)^n x^n}{(2n+1)2^n n!} + B,$$

where A, B are arbitrary constants. ∎

The next example is discussed more fully in the next section.

Example 5.7.3 Find a fundamental set of solutions of

$$x^2 y'' + xy' + x^2 y = 0. \tag{32}$$

Solution Note that (32) is in the form (10) with $x^2 P(x) = x^2$, $xQ(x) = x$, $R(x) = x^2$. Thus

$$P(x) = 1, \quad Q(x) = 1, \quad R(x) = x^2$$

and the indicial equation is

$$\rho(r) = r(r-1) + r = r^2 = 0,$$

which has $r_1 = r_2 = 0$ as a repeated root. From Theorem 5.7.1 we have that (32) has one solution of the form

$$y = \sum_{n=0}^{\infty} C_n x^n. \tag{33}$$

Substituting (33) into (32) gives

$$x^2 \sum_{n=0}^{\infty} n(n-1) C_n x^{n-2} + x \sum_{n=0}^{\infty} n C_n x^{n-1} + x^2 \sum_{n=0}^{\infty} C_n x^n = 0.$$

Multiplying out and rewriting the third term;

$$\sum_{n=0}^{\infty} n(n-1) C_n x^n + \sum_{n=0}^{\infty} n C_n x^n + \sum_{n=2}^{\infty} C_{n-2} x^n = 0.$$

Collecting coefficients of x^n and setting them equal to zero then yields

$$\begin{aligned} n = 0: & \quad 0 = 0, \\ n = 1: & \quad C_1 = 0, \\ n \geq 2: & \quad n^2 C_n + C_{n-2} = 0. \end{aligned} \tag{34}$$
$$\tag{35}$$

Set $C_0 = 1$. Note that (34) and (35) imply that

$$0 = C_1 = C_3 = C_5 = \cdots$$

and

$$C_0 = 1,$$

$$C_2 = -\frac{1}{2^2}C_0 = -\frac{1}{2^2},$$

$$C_4 = -\frac{1}{4^2}C_2 = \frac{1}{2^2 4^2} = \frac{1}{2^4 \cdot (1 \cdot 2)^2},$$

$$C_6 = -\frac{1}{6^2}C_4 = -\frac{1}{2^2 4^2 6^2} = -\frac{1}{2^6(1 \cdot 2 \cdot 3)^2},$$

$$C_8 = -\frac{1}{8^2}C_6 = \frac{1}{2^2 4^2 6^2 8^2} = \frac{1}{2^8(1 \cdot 2 \cdot 3 \cdot 4)^2}.$$

Thus $C_{2n} = \dfrac{(-1)^n}{2^{2n}(n!)^2}$ and

$$y_1(x) = \sum_{n=0}^{\infty} (-1)^n \frac{x^{2n}}{2^{2n}(n!)^2}. \tag{36}$$

There are several ways to find the second solution y_2, none of which are this simple. We shall just substitute the form (22) from Theorem 5.7.1,

$$y_2(x) = y_1(x) \ln x + \sum_{n=0}^{\infty} \tilde{C}_n x^n, \tag{37}$$

where the $\tilde{C}_n$ are new constants to be determined, and $a = 1$ by (c) of Theorem 5.7.1, into the differential equation (32), to get

$$x^2 \left(y_1'' \ln x + 2y_1' \frac{1}{x} - y_1 \frac{1}{x^2} + \sum_{n=0}^{\infty} n(n-1)\tilde{C}_n x^{n-2} \right)$$
$$+ x \left(y_1' \ln x + y_1 \frac{1}{x} + \sum_{n=0}^{\infty} n\tilde{C}_n x^{n-1} \right) + x^2 \left(y_1 \ln x + \sum_{n=0}^{\infty} \tilde{C}_n x^n \right) = 0. \tag{38}$$

Now

$$(x^2 y_1'' + xy_1' + x^2 y_1) \ln x = 0,$$

since y_1 is a solution of (32). Thus, all the terms involving $\ln x$ cancel. Multiplying out (38), putting the remaining y_1, y_1' terms on the right, and substituting the series (36) in for y_1 gives

$$\sum_{n=0}^{\infty} n(n-1)\tilde{C}_n x^n + \sum_{n=0}^{\infty} n\tilde{C}_n x^n + \sum_{n=2}^{\infty} \tilde{C}_{n-2} x^n = -2xy_1' = -\sum_{n=0}^{\infty} \frac{4n(-1)^n x^{2n}}{2^{2n}(n!)^2}.$$

Equating coefficients of x^n yields

$$\begin{aligned} n = 0: & \quad 0 = 0, \\ n = 1: & \quad \tilde{C}_1 = 0, \end{aligned} \tag{39}$$

and, for $n \geq 2$, there are two cases,

n even $[n = 2m]$: $\quad n^2 \tilde{C}_n + \tilde{C}_{n-2} = -\dfrac{2n(-1)^m}{2^n(m!)^2}$

or $\quad \tilde{C}_n = -\dfrac{1}{n^2}\left[\tilde{C}_{n-2} + \dfrac{n(-1)^m}{2^{n-1}(m!)^2}\right]$; (40)

n odd: $\quad [n = 2m+1]$: $\quad n^2 \tilde{C}_n + \tilde{C}_{n-2} = 0 \quad$ or $\quad \tilde{C}_n = -\dfrac{1}{n^2}\tilde{C}_{n-2}$. (41)

From (39) and (41), we have

$$0 = \tilde{C}_1 = \tilde{C}_3 = \tilde{C}_5 = \tilde{C}_7 = \cdots \qquad (42)$$

On the other hand, if we take $\tilde{C}_0 = 1$, which (40) allows, we can solve (40) recursively for $\tilde{C}_n$, n even. For example,

$$\tilde{C}_2 = -\dfrac{1}{2^2}[\tilde{C}_0 - 1] = 0,$$

$$\tilde{C}_4 = -\dfrac{1}{4^2}\left[\tilde{C}_2 + \dfrac{1}{8}\right] = -\dfrac{1}{128}.$$

Thus, (37) is

$$y_2(x) = (\ln x) \sum_{n=0}^{\infty} (-1)^n \dfrac{x^{2n}}{2^{2n}(n!)^2} + 1 - \dfrac{x^4}{128} + \cdots$$

In some applications, however, one wants to use only those solutions that are bounded, or even continuous at the singularity. In these cases only y_1 would be used if it can be shown that the other solution has an $\ln x$ term. ∎

Before giving an example where the roots of the indicial equation differ by an integer, we wish to briefly discuss the gamma function.

5.7.1 Gamma Function

In working with series, one often encounters products of the form

$$\alpha(\alpha + 1)(\alpha + 2) \cdots (\alpha + n).$$

These may be expressed in terms of the *Gamma Function* $\Gamma(x)$. The gamma function is defined for $x > 0$ by

$$\Gamma(x) = \int_0^{\infty} t^{x-1} e^{-t}\, dt,$$

and has the properties that

$$\Gamma(n+1) = n! \qquad \text{for any integer } n \geq 0, \tag{43}$$
$$\Gamma(x+1) = x\Gamma(x) \qquad \text{for any } x > 0. \tag{44}$$

Using (44) we can compute, for example, that, if $x > 0$, then

$$\Gamma(x+3) = (x+2)\Gamma(x+2) = (x+2)(x+1)\Gamma(x+1)$$
$$= (x+2)(x+1)x\Gamma(x).$$

A similar calculation shows that, if $\alpha > 0$ and n is a nonnegative integer, then

$$\alpha(\alpha+1)(\alpha+2)\cdots(\alpha+n) = \frac{\Gamma(\alpha+n+1)}{\Gamma(\alpha)}. \tag{45}$$

Using this, we can also rewrite products of terms that increase by multiples of n. Suppose $\alpha > 0$, $m > 0$, and n is a nonnegative integer. Then (45) shows that

$$\alpha(\alpha+m)(\alpha+2m)(\alpha+3m)\cdots(\alpha+nm)$$
$$= m^{n+1}\frac{\alpha}{m}\left(\frac{\alpha}{m}+1\right)\left(\frac{\alpha}{m}+2\right)\left(\frac{\alpha}{m}+3\right)\cdots\left(\frac{\alpha}{m}+n\right)$$
$$= m^{n+1}\frac{\Gamma\left(\frac{\alpha}{m}+n+1\right)}{\Gamma\left(\frac{\alpha}{m}\right)}. \tag{46}$$

Example 5.7.4

Using (45) or (46), rewrite

$$1 \cdot 6 \cdot 11 \cdot 16 \cdot 21 \cdot 26,$$

using the Gamma function.

Solution

This is a product in which each term, beginning with $\alpha = 1$, increases by 5, so that we have (46) with $m = 5$, and also $n = 5$. Thus,

$$1 \cdot 6 \cdot 11 \cdot 16 \cdot 21 \cdot 26 = 1 \cdot (1 + 5 \cdot 1)(1 + 5 \cdot 2)(1 + 5 \cdot 3)(1 + 5 \cdot 4)(1 + 5 \cdot 5)$$
$$= 5^6 \frac{\Gamma(\frac{1}{5} + 5 + 1)}{\Gamma(\frac{1}{5})} = 5^6 \frac{\Gamma(6\frac{1}{5})}{\Gamma(\frac{1}{5})}. \blacksquare$$

Example 5.7.5

Rewrite

$$2 \cdot 5 \cdot 8 \cdot 11 \cdots (2 + 3n),$$

in terms of the Gamma function.

Solution   The terms of the product start with $\alpha = 2$ and increase by 3, so that $m = 3$ in (46),

$$2 \cdot 5 \cdot 8 \cdot 11 \cdots (2 + 3n) = 2(2 + 3)(2 + 3 \cdot 2)(2 + 3 \cdot 3) \cdots (2 + 3n)$$
$$= 3^{n+1} \frac{\Gamma(\frac{2}{3} + n + 1)}{\Gamma(\frac{2}{3})} = 3^{n+1} \frac{\Gamma(\frac{5}{3} + n)}{\Gamma(\frac{2}{3})}. \blacksquare$$

Familiarity with the Gamma function is not necessary to find the first few terms of a Frobenius expansion. However, it is difficult in many cases to give an explicit expression for the coefficients without using the Gamma function.

Example 5.7.6   Find a fundamental set of solutions on $x > 0$ for

$$xy'' - 2y' + y = 0. \tag{47}$$

Solution   Multiply (47) by x, to make $x = 0$ a regular singular point.

$$x^2 y'' - 2xy' + xy = 0, \tag{48}$$

Then $P(x) = 1$, $Q(x) = -2$, $R(x) = x$. The indicial polynomial

$$r(r - 1) - 2r = r^2 - 3r$$

has roots $r_1 = 3$, $r_2 = 0$. By Theorem 5.7.1, (48) has solutions of the form

$$y_1 = x^{r_1} \sum_{n=0}^{\infty} C_n x^n, \tag{49}$$

$$y_2 = a y_1 \ln x + x^{r_2} \sum_{n=0}^{\infty} \tilde{C}_n x^n. \tag{50}$$

If $a = 0$, then both series have the same form. Thus we begin by substituting

$$y = x^r \sum_{n=0}^{\infty} C_n x^n = \sum_{n=0}^{\infty} C_n x^{n+r} \tag{51}$$

into (48). If we can get linearly independent solutions for $r = r_1$, $r = r_2$, then $a = 0$. If, however, both roots give the same solution, then we must use $a \neq 0$ in (50).

Substituting (51) into (48) yields

$$x^2 \sum_{n=0}^{\infty} (n + r)(n + r - 1) C_n x^{n+r-2} - 2x \sum_{n=0}^{\infty} (n + r) C_n x^{n+r-1} + x \sum_{n=0}^{\infty} C_n x^{n+r}$$
$$= 0.$$

Multiplying the series by the coefficients, dividing by x^r, and shifting summation in the third series gives

$$\sum_{n=0}^{\infty} (n + r)(n + r - 1) C_n x^n - 2 \sum_{n=0}^{\infty} (n + r) C_n x^n + \sum_{n=1}^{\infty} C_{n-1} x^n = 0. \tag{52}$$

Equating coefficients then yields

$$n = 0: \quad [r(r-1) - 2r]C_0 = 0, \tag{53}$$

$$n \geq 1: \quad [(n+r)(n+r-1) - 2(n+r)]C_n + C_{n-1} = 0. \tag{54}$$

Equation (53) holds if r is either root of the indicial equation. Then (54) is

$$(n+r)(n+r-3)C_n + C_{n-1} = 0, \quad n \geq 1. \tag{55}$$

Now we consider the two roots r_1, r_2. First, consider r_1 since there is always a solution for the larger root.

Root r_1

Let $r = r_1 = 3$. Then (55) is

$$(n+3)nC_n + C_{n-1} = 0, \quad n \geq 1,$$

and

$$C_n = (-1)\frac{C_{n-1}}{n(n+3)}.$$

Thus, taking $C_0 = 1$,

$$C_n = (-1)^n \frac{1}{\underbrace{n(n-1)\cdots 1}\underbrace{(n+3)((n-1)+3)\cdots(1+3)}},$$

which is

$$C_n = (-1)^n \frac{3!}{n!(n+3)!}$$

and

$$y_1 = x^3 \sum_{n=0}^{\infty} \frac{(-1)^n 6}{n!(n+3)!} x^n = \sum_{n=0}^{\infty} \frac{(-1)^n 6}{n!(n+3)!} x^{n+3}. \tag{56}$$

Root r_2

Let $r = r_2 = 0$. Then (55) becomes

$$n(n-3)\tilde{C}_n + \tilde{C}_{n-1} = 0, \quad n \geq 1. \tag{57}$$

If $n = 3$, then (57) implies that $\tilde{C}_2 = 0$. Letting $n = 2$ and $n = 1$, in (57) we find that $\tilde{C}_1 = 0$, $\tilde{C}_2 = 0$. But then we have

$$\sum_{n=3}^{\infty} \tilde{C}_n x^n = x^3 \sum_{n=0}^{\infty} \tilde{C}_{n+3} x^n,$$

which is the same as (51) with $\tilde{C}_{n+3} = C_n$. Since taking $a = 0$ does not give a second solution, we take $a = 1$ in (50), and substitute (50) into (48);

$$x^2\left[y_1'' \ln x + 2y_1'\frac{1}{x} - y_1\frac{1}{x^2} + \sum_{n=0}^{\infty}(n+r)(n+r-1)\tilde{C}_n x^{n+r-2}\right]$$
$$- 2x\left[y_1' \ln x + y_1\frac{1}{x} + \sum_{n=0}^{\infty}(n+r)\tilde{C}_n x^{n+r-1}\right]$$
$$+ x\left[y_1 \ln x + \sum_{n=0}^{\infty}\tilde{C}_n x^{n+r}\right] = 0.$$

The $\ln x$ terms cancel, since y_1 is a solution of (48). Thus we have, after regrouping and dividing by x^r, ($r = 0$, but we are illustrating a general calculation)

$$\sum_{n=0}^{\infty}(n+r)(n+r-1)\tilde{C}_n x^n - 2\sum_{n=0}^{\infty}(n+r)\tilde{C}_n x^n + \sum_{n=1}^{\infty}\tilde{C}_{n-1} x^n$$
$$= x^{-r}[-2xy_1' + 3y_1]. \tag{58}$$

Letting $r = r_2 = 0$ (this could have been done before substitution) gives

$$\sum_{n=0}^{\infty}[n(n-1) - 2n]\tilde{C}_n x^n + \sum_{n=1}^{\infty}\tilde{C}_{n-1} x^n = [-2xy_1' + 3y_1]. \tag{59}$$

Equating coefficients, using the series (56) for y_1, yields

$n = 0$: $\qquad\qquad 0 = 0,$ \hfill (60)

$n = 1$: $\qquad\qquad -2\tilde{C}_1 + \tilde{C}_0 = 0,$ \hfill (61)

$n = 2$: $\qquad\qquad -2\tilde{C}_2 + \tilde{C}_1 = 0,$ \hfill (62)

$n \geq 3$: $\qquad (n-3)n\tilde{C}_n + \tilde{C}_{n-1} = -2\dfrac{(-1)^{n-3}6n}{(n-3)!n!} + 3\dfrac{(-1)^{n-3}6}{(n-3)!n!}$

$$= \frac{(-1)^{n-3}6}{(n-3)!n!}(-2n + 3). \tag{63}$$

For $n = 3$ (63) implies that

$$\tilde{C}_2 = \frac{6}{3!}(-6 + 3) = -3.$$

Then, from (62),

$$\tilde{C}_1 = 2\tilde{C}_2 = -6,$$

and, from (61),

$$\tilde{C}_0 = 2\tilde{C}_1 = -12.$$

The remaining $\tilde{C}_n$ may be found recursively from (63) for $n \geq 4$. However, $\tilde{C}_3$ is arbitrary. (This happens since we can add a multiple of y_1 to y_2 and still have a linearly independent solution.) The simplest choice for $\tilde{C}_3$ is

$$\tilde{C}_3 = 0;$$

and then

$$\tilde{C}_n = \frac{1}{(n-3)n}\left\{-\tilde{C}_{n-1} + \frac{(-1)^{n-3}6(3-2n)}{(n-3)!n!}\right\}, \qquad n \geq 4. \tag{64}$$

Using the terms we have found so far, we have

$$y_2 = \ln x \sum_{n=0}^{\infty} \frac{(-1)^n 6}{n!(n+3)!} x^{n+3} - 12 - 6x - 3x^2 + \cdots \blacksquare$$

There are two facts that make using these techniques somewhat simpler. First, in using (22), the ln x terms will always cancel out. Secondly, if (52) and (58) are examined, we see that the homogeneous part of the difference equations for C_n and $\tilde{C}_n$ are identical before the value of r is substituted. The $\tilde{C}_n$, however, satisfy a nonhomogeneous difference equation.

To actually complete Example 5.7.6, we would need to show that the series converges for $0 < x < \infty$ and not on some smaller interval. This may be done using the ratio test and properties of factorials. However, we shall not do so.

Exercises

In Exercises 1 through 8, determine whether the differential equation has a regular singular point at $x = 0$.

1. $x^2 y'' + y = 0$
2. $x^2 y'' + y' = 0$
3. $x^3 y'' + x^2 y' = 0$
4. $x^3 y'' + 5xy' - xy = 0$
5. $x^2 y'' + x^{5/3} y' + 6y = 0$
6. $(x^2 + x^3) y'' + 3xy' + x^{-1} y = 0$
7. $x^2 y'' + (\sin x) y' - 3y = 0$
8. $(e^x - 1 - x) y'' + xy = 0$

In Exercises 9 through 15, express the product in terms of the Gamma or factorial functions.

9. $2 \cdot 4 \cdot 6 \cdot 8 \cdots (2n)$
10. $1 \cdot 3 \cdot 5 \cdot 7 \cdot 9 \cdot 11 \cdots (2n+1)$
11. $1 \cdot 4 \cdot 7 \cdot 10 \cdot 13 \cdot 14 \cdots (3n+1)$
12. $1 \cdot 5 \cdot 9 \cdot 13 \cdot 17 \cdots (4n+1)$
13. $2 \cdot 6 \cdot 10 \cdot 14 \cdot 18 \cdots (4n+2)$
14. $3 \left(3 + \frac{1}{2}\right)(3+1)\left(3 + \frac{3}{2}\right) \cdots \left(3 + \frac{n}{2}\right)$
15. $\left(1 + \frac{1}{5}\right)\left(1 + \frac{2}{5}\right)\left(1 + \frac{3}{5}\right) \cdots \left(1 + \frac{n}{5}\right)$

16. Show that $x^3 y'' + y = 0$ has no nonzero solutions of the form

$$y = x^r \sum_{n=0}^{\infty} C_n x^n.$$

17. Show that, if m is an integer and $m > 2$, then $x^m y'' + y = 0$ has no nonzero solutions of the form $y = x^r \sum_{n=0}^{\infty} C_n x^n$.

18. Show that zero is not a regular singular point of

$$\frac{2}{3} x^{3/2} y'' + \left(-\frac{1}{3} x^{1/2} + \frac{2}{3} x\right) y' - y = 0$$

even if we multiply through by a power of x, but $y = x^{3/2}$ is a solution of the form x^r. Thus the assumptions of Theorem 5.7.1 are only sufficient, and not necessary, to insure there is a solution of the form (13) on $(0, \infty)$.

In Exercises 19 through 28, find a fundamental set of solutions using Theorem 5.7.1. If possible, find a formula for the general term. In every case, find at least the first four terms.

19. $4x^2 y'' + 2xy' + xy = 0$

20. $36x^2y'' + x^2y' + 5y = 0$
21. $3xy'' + 2y' + y = 0$
22. $xy'' - y' - y = 0$
23. $2x(1-x)y'' + (1-5x)y' - y = 0$
24. $6x(1-x)y'' + (2-15x)y' - 3y = 0$
25. $x(1-x)y'' + (1-2x)y' = 0$
26. $x(1-x)y'' + (1-3x)y' - y = 0$
27. $x(1-x)y'' + (2-3x)y' - y = 0$
28. $x(1-x)y'' + (3-2x)y' = 0$
29. The differential equation
$$x(1-x)y'' + [\gamma - (\alpha + \beta + 1)x]y' - \alpha\beta y = 0 \quad (65)$$
is called the *hypergeometric equation*. Show that
   a) Equation (65) has regular singular points at $x_0 = 0, x_0 = 1$.
   b) For $x_0 = 0$, the indicial equation has roots 0, $1 - \gamma$.
   c) The solution for $r = 0$ with $\gamma > 1, \alpha > 0, \beta > 0$, is
$$\sum_{n=0}^{\infty} \frac{\Gamma(\alpha + n)}{\Gamma(\alpha)} \frac{\Gamma(\beta + n)}{\Gamma(\beta)} \frac{\Gamma(\gamma)}{n!\Gamma(\gamma + n)} x^n.$$
(Exercises 23 through 28 are hypergeometric equations.)

30. The differential equation
$$xy'' + (1-x)y' + \lambda y = 0 \quad (66)$$
has a regular singular point at $x = 0$ and is called *Laguerre's equation*. Show that if λ is a positive integer, then (66) has a polynomial for a solution. This polynomial, when normalized, is known as a *Laguerre polynomial* and denoted $L_m(x)$. Like the Legendre polynomials of Section 5.6, Laguerre polynomials can be used to give series expansions.

31. Find a polynomial solution of (66) for $\lambda = 0, 1, 2$.
32. Verify (17).

5.8 Bessel Functions

Many vibration problems are analyzed using the *wave equation*, which is a partial differential equation. The solutions for vibration problems involving rectangular membranes in cartesian coordinates take the form of series involving terms like sin αnx, cos αnx. Such *Fourier* series are interpreted as the sum of the basic harmonic motions. For example, for a violin string of length π with fixed ends, the first four basic harmonic motions are constant multiples of the functions shown in Fig. 5.8.1.

However, if we have a round membrane, fixed at the edges, such as a drum, the shape of the fundamental harmonic motions is different. There is still

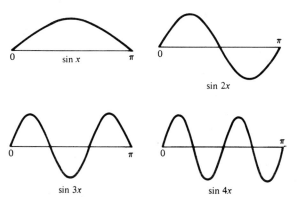

Figure 5.8.1

a function, call it $J(x)$, which has an infinite number of zeros, and the basic harmonic motions take the form $J(\alpha_n x)$ for constants α_n. However, this function is no longer the sine or cosine. The function $J(x)$ will be a *Bessel function*. Bessel functions appear in many problems involving partial differential equations. Since these functions play such an important role in applied mathematics and because our analysis of the preceding section can be used to study them, this section will briefly examine Bessel functions.

The differential equation

$$x^2 \frac{d^2 y}{dx^2} + x \frac{dy}{dx} + (x^2 - v^2)y = 0 \qquad (1)$$

with $v \geq 0$ a parameter is called *Bessel's equation*. Note that $x_0 = 0$ is a regular singular point for (1), so that (1) can be solved by the Frobenius method of Section 5.7. The indicial equation for (1) is

$$r(r-1) + r - v^2 = r^2 - v^2 = 0, \qquad (2)$$

which has the roots, $r_1 = v, r_2 = -v$. We have already considered (1) with $v = 0$ in Example 5.7.3. We now wish to consider it in general. Following the last section, we look for solutions of the form

$$y = x^r \sum_{n=0}^{\infty} C_n x^n. \qquad (3)$$

Substituting (3) into (1), we obtain:

$$x^2 \sum_{n=0}^{\infty} (n+r)(n+r-1) C_n x^{n+r-2} + x \sum_{n=0}^{\infty} (n+r) C_n x^{n+r-1}$$
$$+ (x^2 - v^2) \sum_{n=0}^{\infty} C_n x^{n+r} = 0.$$

Divide by x^r and multiply the series by the coefficients:

$$\sum_{n=0}^{\infty} (n+r)(n+r-1) C_n x^n + \sum_{n=0}^{\infty} (n+r) C_n x^n - v^2 \sum_{n=0}^{\infty} C_n x^n + \sum_{n=0}^{\infty} C_n x^{n+2} = 0.$$

The last term $\sum_{n=0}^{\infty} C_n x^{n+2}$ equals $\sum_{n=2}^{\infty} C_{n-2} x^n$.

Equating coefficients of powers of x to zero gives the equations:

$n = 0$: $\qquad [r(r-1) + r - v^2] C_0 = 0,$ (4)

$n = 1$: $\qquad [(1+r)r + (r+1) - v^2] C_1 = 0,$ (5)

$n \geq 2$: $\quad [(n+r)(n+r-1) + (n+r) - v^2] C_n + C_{n-2} = 0,$ (6)

and, after simplification,

$n = 0$: $\qquad [r^2 - v^2] C_0 = 0,$ (7)

$n = 1$: $\qquad [r + 1 - v][r + 1 + v] C_1 = 0,$ (8)

$n \geq 2$: $\quad [n + r - v][n + r + v] C_n + C_{n-2} = 0.$ (9)

Thus, for $r = v$, we have
$$[1 + 2v]C_1 = 0, \tag{10}$$
$$n[n + 2v]C_n + C_{n-2} = 0. \tag{11}$$

Since $v \geq 0$, (10) implies that $C_1 = 0$, and hence, by (11),
$$C_3 = C_5 = \cdots = 0.$$

To find the C_n with n even, note that, from (11),
$$C_n = (-1)\frac{1}{n(n + 2v)}C_{n-2}. \tag{12}$$

Let $n = 2m$. Then (12) is
$$C_{2m} = (-1)\frac{1}{2m(2m + 2v)}C_{2(m-1)} = \frac{(-1)}{2^2 m(m + v)}C_{2(m-1)}$$
$$= \frac{(-1)}{2^2 m(m + v)} \frac{(-1)}{2^2(m - 1)(m - 1 + v)}C_{2(m-2)}$$
$$\vdots$$
$$= \frac{(-1)^m}{2^{2m} m(m - 1)(m - 2)\cdots(m + v)(m - 1 + v)\cdots(1 + v)}C_0$$
$$= \frac{(-1)^m}{2^{2m} m!(m + v)(m - 1 + v)\cdots(1 + v)}C_0$$
$$= (-1)^m \frac{\Gamma(v + 1)}{2^{2m} m!\Gamma(v + m + 1)}C_0. \tag{13}$$

By convention, we take $C_0 = 1/(2^v \Gamma(v + 1))$ and the resulting solution is denoted by $J_v(x)$;
$$J_v(x) = x^v \sum_{m=0}^{\infty} \frac{(-1)^m x^{2m}}{2^{2m+v} m!\Gamma(v + m + 1)}. \tag{14}$$

The function $J_v(x)$ is called the *Bessel function of the first kind of order v*.

If $\tau < 0$ is a negative number that is not a negative integer, then there exists an integer k such that $\tau + k + 1 > 0$. Then we can define $\Gamma(\tau)$ by
$$\Gamma(\tau) = \frac{\Gamma(\tau + k + 1)}{\tau(\tau + 1)\cdots(\tau + k)}. \tag{15}$$

With this convention, we can try to solve (7), (8), and (9) with $r = -v$. The equations (7), (8), and (9) are then
$$n = 0: \qquad 0 = 0, \tag{16}$$
$$n = 1: \qquad [1 - 2v]\tilde{C}_1 = 0, \tag{17}$$
$$n \geq 2: \qquad [n - 2v]n\tilde{C}_n + \tilde{C}_{n-2} = 0. \tag{18}$$

If $2v$ is not an integer, then (17) and (18) again imply $0 = \tilde{C}_1 = \tilde{C}_3 = \tilde{C}_5 = \cdots$ and

$$\tilde{C}_{2m} = (-1)\frac{1}{2m(2m-2v)}\tilde{C}_{2(m-1)} = \frac{-1}{2^2 m(m-v)}\tilde{C}_{2(m-1)}. \tag{19}$$

Thus,

$$\tilde{C}_{2m} = \frac{(-1)}{2^2 m(m-v)}\frac{(-1)}{2^2(m-1)(m-1-v)}\tilde{C}_{2(m-2)}$$

$$= \frac{(-1)}{2^2 m(m-v)}\frac{(-1)}{2^2(m-1)(m-1-v)}\frac{(-1)}{2^2(m-2)(m-2-v)}\tilde{C}_{2(m-3)},$$

so that

$$\tilde{C}_{2m} = \frac{(-1)^m}{2^{2m} m(m-1)\cdots 1(m-v)(m-1-v)\cdots(1-v)}\tilde{C}_0$$

$$= (-1)^m \frac{\Gamma(-v+1)}{2^{2m} m!\,\Gamma(-v+m+1)}\tilde{C}_0.$$

Letting $C_0 = 1/(2^{-v}\Gamma(-v+1))$ gives the Bessel function

$$J_{-v}(x) = x^{-v} \sum_{m=0}^{\infty} \frac{(-1)^m x^{2m}}{2^{2m-v} m!\,\Gamma(-v+m+1)}. \tag{20}$$

Note that the series (20) for J_{-v} comes by replacing v by $-v$ in the series (14) for J_v.

If $v = (2p+1)/2$ for an integer p, then $2v$ is an *odd* integer. Thus we can again solve (7), (8), and (9) by taking $0 = \tilde{C}_1 = \tilde{C}_3 = \tilde{C}_5 = \cdots$ and $\tilde{C}_{2m}$ given by (13), so that (20) also holds if $2v$ is an odd integer. The only remaining possibility is that v is an integer and hence $2v$ is an even integer. Then (17) and (18) imply that $\tilde{C}_1 = \tilde{C}_3 = \tilde{C}_5 = \cdots = 0$. For $n = 2v$, in (18), we get $\tilde{C}_{n-2} = 0$, and hence, from (18), that

$$\tilde{C}_{n-2} = \tilde{C}_{n-4} = \cdots = \tilde{C}_0 = 0.$$

Thus we do not get a second solution in the form (3) and, according to Theorem 5.7.1, must look for a solution in the form

$$y_2(x) = J_v(x)\ln x + x^{-v}\sum_{n=0}^{\infty}\tilde{C}_n x^n. \tag{21}$$

Theorem 5.8.1   If v is not an integer, then $\{J_v(x), J_{-v}(x)\}$, given by (14), (20), is a fundamental set of solutions of Bessel's equation (1). If $v = n$ is an integer, then a second solution $Y_n(x)$ can be found that involves a $\ln x$ term, and $\{J_n(x), Y_n(x)\}$ will be a fundamental set of solutions.

There are several ways to define Y_n. We omit the details but, if $n = 0$, then it can be shown that the Frobenius method gives

$$y_2 = J_0(x) \ln x + \sum_{n=1}^{\infty} \frac{(-1)^{n-1} h_n}{2^{2n}(n!)^2} x^{2n},$$

where $h_n = 1 + \frac{1}{2} + \cdots + \frac{1}{n}$.

Rather than use y_2, we define

$$Y_0(x) = \frac{2}{\pi}(\gamma - \ln 2) J_0(x) + \frac{2}{\pi} y_2(x), \tag{22}$$

where γ is the *Euler constant* ($\gamma \approx 0.5772\ldots$). Then, since $Y_0(x)$ is a linear combination of the solutions J_0, y_2, it will also be a solution of (1) for $v = 0$. Y_0 is called the *Bessel function of the second kind of order 0*. For integers $n > 0$, one can derive a similar expansion. (See Exercise 9 at the end of this section.) Finally, for v not an integer, define

$$Y_v(x) = \frac{1}{\sin v\pi} [J_v(x) \cos v\pi - J_{-v}(x)], \tag{23}$$

which is a linear combination of J_v, J_{-v}. Then Y_v is called the *Bessel function of the second kind of order v*. It can be shown that, for an integer $n \geq 0$,

$$\lim_{v \to n} Y_v(x) = Y_n(x).$$

Theorem 5.8.2 For any $v \geq 0$, $\{J_v(x), Y_v(x)\}$ is a fundamental set of solutions of Bessel's equation of order v. Furthermore, the only bounded solutions of Bessel's equation are multiples of $J_v(x)$.

We now wish to consider what the graphs of the Bessel functions look like. We begin with the Bessel functions of the first kind $J_v(x)$. The important points are covered in the next theorem.

Theorem 5.8.3 The function $J_v(x)$ is continuous and bounded on the interval $0 \leq x < \infty$. The series in (14) converges for all x. Furthermore, $J_v(x)$ has an infinite number of real zeros denoted λ_{vs}; $s = 0, 1, 2, \ldots$ and $\lim_{s \to \infty} \lambda_{vs} = \infty$.

This theorem makes the Bessel functions sound somewhat like the sine and cosine. However, there are important differences. To illustrate, we consider $v = \frac{1}{2}$.

If $v = \frac{1}{2}$, then it turns out that $J_v(x)$, $J_{-v}(x)$ can be expressed in *closed form*. Returning to the derivation of $J_v(x)$, we see from (12) that if $v = \frac{1}{2}$, then

$$C_{2m} = (-1)^m ((2m+1)!)^{-1} C_0.$$

Thus,

Figure 5.8.2

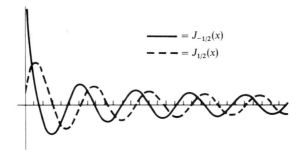

$$J_{1/2}(x) = x^{1/2} \sum_{m=0}^{\infty} \frac{(-1)^m x^{2m}}{(2m+1)!} \frac{1}{2^{1/2}\Gamma(3/2)}. \tag{24}$$

However,

$$\Gamma(\tfrac{1}{2}) = \sqrt{\pi},$$

so that (24) becomes

$$J_{1/2}(x) = \sqrt{\frac{2}{\pi x}} \sin x, \tag{25}$$

and, similarly,

$$J_{-1/2}(x) = \sqrt{\frac{2}{\pi x}} \cos x. \tag{26}$$

$J_{1/2}$ and $J_{-1/2}$ are graphed in Fig. 5.8.2.

The formulas for $J_{1/2}(x)$, $J_{-1/2}(x)$ illustrate the important fact that, if $y(x)$ is any solution of Bessel's equation (for any $v \geq 0$), then

$$\lim_{x \to \infty} y(x) = 0. \tag{27}$$

The formulas (26) and (27), however, are somewhat misleading, in that the zeros of $J_{1/2}(x)$, $\lambda_{1/2 s} = \pi s$ are evenly spaced. This is not true in general. (Note also Exercise 8.) For example, the first four zeros of $J_0(x)$ are:

$$\lambda_{01} = 2.40482,$$
$$\lambda_{02} = 5.52008,$$
$$\lambda_{03} = 8.65373,$$
$$\lambda_{04} = 11.7915.$$

Other than the fact that $J_0(0) = 1$ and $J_v(0) = 0$ if $v > 0$, all of the Bessel functions have graphs that are similar in general appearance to those in Fig. 5.8.2, the major differences being the rate at which $\lim_{x \to 0^+} J_v(x) = 0$ for $v > 0$, and the rate at which $\lim_{x \to 0^+} Y_v(x) = -\infty$.

Additional properties of the Bessel functions appear in the exercises.

Exercises

1. Verify that $J_{-1/2}(x) = \sqrt{2/(\pi x)} \cos x$.

2. Using the series (14), verify that, for $v \geq 1$,
$$J_{v-1}(x) + J_{v+1}(x) = \frac{2v}{x} J_v(x). \tag{28}$$

3. Using the series (14), verify that, for $v \geq 1$,
$$J_{v-1}(x) - J_{v+1}(x) = 2J_v'(x), \tag{29}$$
where $J_v'(x)$ is the derivative with respect to x of $J_v(x)$.

4. Using (28) and (29), verify that, for $v \geq 1$,
$$J_{v+1}(x) = \frac{v}{x} J_v(x) - J_v'(x). \tag{30}$$

5. Assuming that (30) holds for $v \geq 0$ (which is true), use formula (25) for $J_{1/2}(x)$ to derive a formula for $J_{3/2}(x)$.

6. Assuming that (30) holds for $v \geq 0$, use Exercise 5 and (30) to derive a formula for $J_{5/2}(x)$.

7. Assuming that (30) holds for $v \geq 0$, use (30) to conclude that $J_1(x) = -J_0'(x)$. Assume that J_v has the property that the only places where $J_v'(x) = 0$ are at a max or min, and that all maxima are positive and all minima are negative. Conclude that
$$\lambda_{00} < \lambda_{10} < \lambda_{01} < \lambda_{11} < \lambda_{02} < \lambda_{12} < \cdots$$
That is, the zeros of $J_0(x), J_1(x)$ have the *interlacing property*.

8. Using the formula for $J_{3/2}(x)$ found in Exercise 5, show that the zeros of $J_{3/2}(x)$ are the solutions of $\tan \lambda = \lambda$. By graphing $y = \tan \lambda$, $y = \lambda$, conclude that
$$\lim_{s \to \infty} \left| \lambda_{(3/2)s} - \frac{\pi}{2} - \pi s \right| = 0,$$
so that, for large s, the zeros of $J_{3/2}(x)$ are about π units apart.

9. Verify that, for $n > 0$, n an integer, and $h_m = 1 + \frac{1}{2} + \cdots + \frac{1}{m}$,
$$\pi Y_n(x) = 2J_n(x)\left(\ln \frac{x}{2} + \gamma\right)$$
$$+ x^n \sum_{m=0}^{\infty} \frac{(-1)^{m-1}(h_m + h_{m+n})}{2^{2m+n} m!(m+n)!} x^{2m}$$
$$- x^{-n} \sum_{m=0}^{n-1} \frac{(n-m-1)!}{2^{2m-n} m!} x^{2m}$$
is a solution of Bessel's equation (1) with $v = n$.

10. Graph the functions
$$f_n(x) = J_{1/2}(nx)$$
on the interval $0 \leq x \leq \pi$ for $n = 1, 2, 3, 4$ and compare them to the curves in Fig. 5.8.1.

11. Euler's constant γ can be defined as
$$\gamma = \lim_{n \to \infty} h_n - \ln n$$
where $h_n = 1 + (1/2) + \cdots (1/n)$. Compute the value of $q_n = h_n - \ln n$ for several values of n, and observe the apparent convergence.

6

Linear Systems of Differential Equations

6.1 Introduction

6.1.1 General Linear Systems

Up to this point we have considered only single differential equations with one dependent variable. In many applications, however, one is led to simultaneously consider several ordinary differential equations with several dependent variables and one independent variable. Such *systems of differential equations* may be linear or nonlinear. Nonlinear systems are studied in Chapter 9. If all of the differential equations are linear in the dependent variable, the resulting linear systems of differential equations are most naturally studied using vector notation and matrix theory, which we will cover beginning in Section 6.7. The remainder of this section introduces some basic concepts for linear systems. Sections 6.2 and 6.3 discuss methods that are helpful with linear systems consisting of a few equations. Sections 6.4, 6.5, and 6.6 cover three applications in which systems arise. The applications may also be covered after Section 6.11 if Section 6.2 and 6.3 are omitted.

A simple example of a linear system of differential equations is

$$x'(t) = 2x(t) + 3y(t) + \sin t,$$
$$y'(t) = 4x(t) - 5y(t) + \cos(t^2) + t. \tag{1}$$

In the system (1), there are two unknown functions $x(t)$, $y(t)$. A pair of functions $x(t)$, $y(t)$ would be a *solution* of (1) if they satisfied both equations.

Example 6.1.1 Verify that $x = e^t$, $y = e^{2t}$ is a solution of the system

$$x' = x + 3y - 3e^{2t}, \tag{2a}$$
$$y' = 4x + 2y - 4e^t. \tag{2b}$$

Solution We must show that $x = e^t$, $y = e^{2t}$ satisfy both (2a) and (2b). Substituting $x = e^t$, $y = e^{2t}$ into (2a) gives

$$(e^t)' = e^t + 3e^{2t} - 3e^{2t} \quad \text{or} \quad e^t = e^t,$$

while substituting $x = e^t$, $y = e^{2t}$ into (2b) gives

$$(e^{2t})' = 4e^t + 2e^{2t} - 4e^t \quad \text{or} \quad 2e^{2t} = 2e^{2t}.$$

Thus $x = e^t$, $y = e^{2t}$ is a solution of the system (2). ∎

Systems arise naturally in many applications. Three of these applications will be discussed in more detail in Sections 6.4, 6.5, and 6.6 of this chapter, but we shall briefly mention them here.

One application is to mechanical systems. For example, the linear motion of several interconnected masses as in Fig. 6.1.1 (a flexible structure) would have a dependent variable for each mass. If other than straight-line motion is considered, each mass would have the x, y, z coordinates of its position as dependent variables.

A second application involves multiloop circuits. Consider, for example, the circuit in Fig. 6.1.2. The voltage law applied to the top and bottom loop of this circuit yields:

$$\frac{di_1}{dt} + 3i_1 - i_2 = e_1(t), \tag{3}$$

$$2\frac{di_3}{dt} + 5i_3 + i_2 = e_2(t), \tag{4}$$

respectively. The current law at node (a) is

$$i_1 + i_2 - i_3 = 0.$$

Using this equation to eliminate i_2 from (3) and (4), gives the system

$$\frac{di_1}{dt} + 4i_1 - i_3 = e_1(t),$$
$$2\frac{di_3}{dt} - i_1 + 6i_3 = e_2(t). \tag{5}$$

The final application we shall consider will be mixing problems, in which there will be several interrelated concentrations and volumes of interest.

Figure 6.1.1

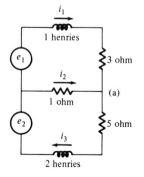

Figure 6.1.2

In this chapter we will consider only systems of linear differential equations, especially those with constant coefficients. Nonlinear systems, with a few simple exceptions, are either solved numerically (Chapter 8) or analyzed qualitatively (Chapter 9).

The existence and uniqueness properties of general systems of differential equations is not as straightforward as for single equations. In particular, determining the expected number of initial conditions or arbitrary constants is a little trickier. One cannot tell how many constants there will be merely by looking at the system.

Example 6.1.2

Consider the second-order linear system of differential equations with constant coefficients,

$$x''(t) + y'(t) + x(t) + 3y(t) = t, \tag{6a}$$

$$2x''(t) + 2y'(t) + 2x(t) + 7y(t) = 3. \tag{6b}$$

Since this system involves second derivatives of x and first derivatives of y, one might expect the answer to involve three arbitrary constants. Note, however, that subtracting twice equation (6a) from (6b) gives the new *equivalent system* (has same solutions)

$$x''(t) + y'(t) + x(t) + 3y(t) = t, \tag{7a}$$

$$y(t) = 3 - 2t; \tag{7b}$$

or, using (7b) to eliminate $y(t)$ in (7a),

$$x''(t) + x(t) = 7t - 7, \tag{8a}$$

$$y(t) = 3 - 2t. \tag{8b}$$

Using the method of undetermined coefficients, we can solve (8a) for x. The result is

$$x(t) = 7t - 7 + c_1 \cos t + c_2 \sin t, \tag{9a}$$

$$y(t) = 3 - 2t. \tag{9b}$$

Note that there are only two arbitrary constants in this solution, and there are no arbitrary constants in the y portion of the solution. ∎

There is a general procedure for determining how many arbitrary constants there will be in the general solution of a linear system of differential equations with constant coefficients. Its use is mostly restricted to small-sized systems, since it uses determinants, and determinants are "expensive" to compute for systems with many dependent variables. We shall illustrate each step by applying it to system (6) of Example 6.1.2.

Procedure for Determining the Number of Arbitrary Constants

Procedure for determining the number of arbitrary constants in a system of linear differential equations with constant coefficients:

1. *First rewrite the system in operator notation* (see Section 3.6). Let D stand for the operation of differentiation with respect to the independent variable, and define $D^n = d^n/dt^n$. Thus, for example, system (6) of Example 6.1.2 would be written as

$$(D^2 + 1)x + (D + 3)y = t,$$
$$(2D^2 + 2)x + (2D + 7)y = 3. \tag{10}$$

2. *Take the determinant of the coefficient matrix*, thinking of D as a variable. For (10) this would be:

$$\det \begin{bmatrix} D^2 + 1 & D + 3 \\ 2D^2 + 2 & 2D + 7 \end{bmatrix} = (D^2 + 1)(2D + 7) - (D + 3)(2D^2 + 2)$$
$$= D^2 + 1.$$

3. *The degree of the resulting polynomial in D will be the number of arbitrary constants.* In our example, $D^2 + 1$ is a second-degree polynomial in D so that the system (6) will have two arbitrary constants in its general solutions, which is what was observed in (9).

If the determinant is zero, the system is called *degenerate*. As the next two examples show, degenerate systems can have properties quite different from those we would expect from our study (in Chapter 3) of linear differential equations with constant coefficients.

Example 6.1.3   Consider the system of differential equations

$$x' + y' + x + y = e^t, \tag{11a}$$
$$x + y = e^t, \tag{11b}$$

or, equivalently,

$$(D + 1)x + (D + 1)y = e^t,$$
$$x + y = e^t.$$

Since

$$\det \begin{bmatrix} D + 1 & D + 1 \\ 1 & 1 \end{bmatrix} = 0,$$

this system is degenerate. It has no solution. If equation (11b) is substituted into (11a), we get the contradiction that $(x + y)' = 0$, so that $x + y$ is a constant, but, from (11b), $(x + y) = e^t$. A system with no solutions is called *inconsistent*. ∎

Example 6.1.4 Consider the system

$$x' + y = 0,$$
$$2x' + 2y = 0,$$

or

$$Dx + y = 0,$$
$$2Dx + 2y = 0.$$

Since

$$\det \begin{bmatrix} D & 1 \\ 2D & 2 \end{bmatrix} = 0,$$

this system is also degenerate. Note that the second equation, $2x' + 2y = 0$, is just a constant multiple of the first equation, $x' + y = 0$. In this example, one could take $x(t)$ to be an *arbitrary differentiable function* and then $y(t)$ would be $-x'(t)$. Having an arbitrary function in the solution is quite different from having an arbitrary constant. One difference is that no finite set of initial conditions of the types we have considered can determine an arbitrary function. ∎

The problem of determining what initial conditions are possible for a system of differential equations, and what equations can be simplified or eliminated is an important problem. In many fields, such as electrical engineering, procedures have been developed that can be implemented on a computer and are based on the area of mathematics known as graph theory.

6.1.2 First-Order Systems

The situation is simpler for some *first-order* linear systems. A system is first order if it involves only first derivatives.

A system of first-order linear differential equations is said to be *explicit* if it is in the form:

$$\begin{aligned}
x_1'(t) &= a_{11}(t)x_1(t) + a_{12}(t)x_2(t) + \cdots + a_{1n}(t)x_n(t) + f_1(t), \\
x_2'(t) &= a_{21}(t)x_1(t) + a_{22}(t)x_2(t) + \cdots + a_{2n}(t)x_n(t) + f_2(t), \\
&\vdots \qquad \qquad \vdots \qquad \qquad \qquad \qquad \vdots \\
x_n'(t) &= a_{n1}(t)x_1(t) + a_{n2}(t)x_2(t) + \cdots + a_{nn}(t)x_n(t) + f_n(t).
\end{aligned} \qquad (12)$$

Here the dependent variables are $x_1, \ldots, x_n$, the functions a_{ij}, with $1 \leq i \leq n$, $1 \leq j \leq n$, are the *coefficients*, and $f_1, \ldots, f_n$ are the *input* or *forcing functions*.

We have seen several examples of such systems already, for example, systems (1) and (2), as well as the circuit example (5). Another example is:

$$\begin{aligned} x_1'(t) &= 3x_1(t) - 5x_2(t) + \cos t, \\ x_2'(t) &= tx_1(t) + 3x_2(t) + 6x_3(t) + \sin t, \\ x_3'(t) &= x_1(t) - e^t x_3(t). \end{aligned} \tag{13}$$

In this example, $a_{11} = 3$, $a_{12} = -5$, $a_{13} = 0$, $f_1(t) = \cos t$; $a_{21} = t$, $a_{22} = 3$, $a_{23} = 6$, $f_2(t) = \sin t$; $a_{31} = 1$, $a_{32} = 0$, $a_{33} = -e^t$, and $f_3(t) = 0$.

While we will not go into matrix theory until Section 6.7, note that (12) may be simplified by using matrix notation. Let

$$\mathbf{x}(t) = \begin{bmatrix} x_1(t) \\ x_2(t) \\ \vdots \\ x_n(t) \end{bmatrix}, \quad \mathbf{A}(t) = \begin{bmatrix} a_{11}(t) & \cdots & a_{1n}(t) \\ a_{21}(t) & \cdots & a_{2n}(t) \\ \vdots & & \vdots \\ a_{n1}(t) & \cdots & a_{nn}(t) \end{bmatrix}, \quad \mathbf{f}(t) = \begin{bmatrix} f_1(t) \\ f_2(t) \\ \vdots \\ f_n(t) \end{bmatrix},$$

and recall from calculus that

$$\mathbf{x}'(t) = \begin{bmatrix} x_1'(t) \\ x_2'(t) \\ \vdots \\ x_n'(t) \end{bmatrix}.$$

Then (12) can be written as

$$\mathbf{x}'(t) = \mathbf{A}(t)\mathbf{x}(t) + \mathbf{f}(t), \tag{14}$$

or simply

$$\mathbf{x}' = \mathbf{A}\mathbf{x} + \mathbf{f}. \tag{15}$$

In particular, the example (13) would be

$$\begin{bmatrix} x_1'(t) \\ x_2'(t) \\ x_3'(t) \end{bmatrix} = \begin{bmatrix} 3 & -5 & 0 \\ t & 3 & 6 \\ 1 & 0 & -e^t \end{bmatrix} \begin{bmatrix} x_1(t) \\ x_2(t) \\ x_3(t) \end{bmatrix} + \begin{bmatrix} \cos t \\ \sin t \\ 0 \end{bmatrix}. \tag{16}$$

For a first-order linear system, the existence and uniqueness result is what we would expect.

Theorem 6.1.1 **Existence and Uniqueness for First-Order Linear Systems** Suppose that, in the explicit system (12), $f_1(t), \ldots, f_n(t)$ and the $a_{ij}(t)$ are continuous functions on the interval I. Let t_0 be in the interval I and let $x_{10}, \ldots, x_{n0}$ be n real numbers. Then there exists a unique solution $x_1(t), \ldots, x_n(t)$ to (12), defined for all t in the interval I, such that

$$x_1(t_0) = x_{10}, \quad x_2(t_0) = x_{20}, \quad \ldots, \quad x_n(t_0) = x_{n0}.$$

This theorem will be proved for the constant-coefficient case in Section 6.9.

Example 6.1.5

By Theorem 6.1.1, the system
$$x' = 3x - 2y + e^t,$$
$$y' = -2x + 5y - e^{3t}, \quad x(0) = 4, \quad y(0) = 5,$$

has a unique solution $x(t)$, $y(t)$ defined for all t. ∎

The explicit first-order form (12) is actually fairly general, as the next two facts show. When we talk of rewriting a system, it is understood that the new system is equivalent to the old one (has the same solutions).

Theorem 6.1.2

If a system of n linear, constant-coefficient, first-order differential equations with n dependent variables $x_1, \ldots, x_n$ has n arbitrary initial conditions in its general solution, then the system can be rewritten in the explicit form (12) by adding and subtracting constant multiples of the equations. In particular, $x_1(t_0), \ldots, x_n(t_0)$ may be specified arbitrarily at any fixed time t_0. Conversely, if a system of n linear, constant-coefficient differential equations has less than n arbitrary constants in its general solution, it cannot be written in the explicit form (12).

Example 6.1.6

Consider the system of two first-order linear, differential equations
$$x'(t) + 3y'(t) = x(t) - 5y(t) + t, \quad (17a)$$
$$x'(t) - 2y'(t) = 3x(t) + y(t) + e^t. \quad (17b)$$

Since
$$\det \begin{bmatrix} D-1 & 3D+5 \\ D-3 & -2D-1 \end{bmatrix} = -5D^2 + 5D + 16,$$

we have, from Theorem 6.1.1, that the system (17) will have two arbitrary constants in its general solution. Theorem 6.1.2 then tells us that they may be taken as $x(t_0)$, $y(t_0)$. Note that adding -1 times equation (17a) to (17b) gives
$$x' + 3y' = x - 5y + t, \quad (18a)$$
$$-5y' = 2x + 6y + e^t - t. \quad (18b)$$

If we then add $\frac{3}{5}$ of equation (18b) to (18a) and then multiply (18b) by $-\frac{1}{5}$, we get the explicit system
$$x' = \tfrac{11}{5}x - \tfrac{7}{5}y + \tfrac{2}{5}t + \tfrac{3}{5}e^t,$$
$$y' = -\tfrac{2}{5}x - \tfrac{6}{5}y - \tfrac{1}{5}(e^t - t),$$

as promised by Theorem 6.1.2. ∎

The consideration of first-order systems may seem to be restrictive, since we know, from Sections 3.16 and 3.17, that many electrical circuits and mechanical systems will lead to systems of second-order linear differential equations. However,

> Every system of differential equations can be rewritten as a first-order system of differential equations.

The easiest way to illustrate this fact is by examples.

Example 6.1.7 Rewrite

$$x'' + 3x' - y'' + y = \sin t,$$
$$x' - 4x + 5y'' - 6y' = \cos t, \tag{19}$$

as a first-order system.

Solution We introduce the two new variables $z = x'$, $w = y'$. Then (19) may be written as a first-order system in x, y, z, w:

$$z' + 3z - w' + y = \sin t,$$
$$z - 4x + 5w' - 6w = \cos t,$$
$$x' - z = 0,$$
$$y' - w = 0. \tag{20}$$

The resulting system (20) is not unique. For example, one could also use $x' - 4x + 5w' - 6y' = \cos t$ for the second equation in (20). ∎

In the case of a single-loop *RLC* circuit, this procedure amounts to thinking of the circuit as a first-order system in the charge and current rather than a second-order equation in the charge. For a mechanical system we have a first-order system in velocity and position rather than a second-order system in position.

Example 6.1.8 Rewrite

$$x''' + x' + y'' + y = t,$$
$$x + y'' - y = 1, \tag{21}$$

as a first-order system.

Solution The system (21) involves third-order derivatives of x and second-order derivatives of y. Accordingly, we introduce the variables $u = x'$, $v = x'' = u'$, $w = y'$. One way to rewrite (21) would be:

$$v' + u + w' + y = t,$$
$$x + w' - y = 1,$$
$$x' - u = 0,$$
$$u' - v = 0,$$
$$y' - w = 0. \blacksquare$$

Exercises

In Exercises 1 through 4, verify that the indicated functions form a solution of the system of differential equations.

1. $x' + 2x + y = 3e^t + e^{2t}$, $x = e^t$, $y = e^{2t}$
   $x' + y' + 3x = 4e^t + 2e^{2t}$

2. $x_1'' + x_1 + x_2 = \cos 2t$, $x_1 = \sin t$,
   $x_1' + x_2'' + 4x_2 = \cos t$, $x_2 = \cos 2t$

3. $x_1' = x_1 + x_2 + x_3 - 3t + 1$,
   $x_2' = x_1 - x_3$,
   $x_3' = x_1 + x_2 - 2t$
   $x_1 = t$, $x_2 = t + 1$, $x_3 = t - 1$

4. $x'' - x + y' + y - z = 3e^{2t} - e^{3t}$,
   $x' + x + y'' - z' = 2e^t + 4e^{2t} - 3e^{3t}$,
   $x' + y' - 2y + z' - 3z = e^t$
   $x = e^t$, $y = e^{2t}$, $z = e^{3t}$

In Exercises 5 through 10, determine whether or not the system is degenerate. If it is not degenerate, determine the number of arbitrary constants in the general solution.

5. $x' + x + y' - y = t$,
   $x' - x + 2y' + y = \sin t$

6. $x_1'' + x_1 + x_2'' = \sin 3t + 1$,
   $x_1' + x_2' = t^2$

7. $x'' + y'' + y = t$,
   $3x' + x + 4y' - y = \sin t$

8. $x_1'' + x_1 + x_2'' - x_2' = \cos t$,
   $x_1'' - x_1' - 2x_2'' + x_2 = t^3 - 2t$

9. $x' + x + y' = t$,
   $x'' + x' + y'' = t - 1$

10. $x'' - 2x + y'' - y = 0$,
    $2x' + x + 2y' + 3y = t$

In Exercises 11 through 14, determine whether or not the system is degenerate. If it is not degenerate, determine the number of arbitrary constants in the general solution. [*Note:* These exercises require a knowledge of determinants of 3×3 matrices (see Section 6.8).]

11. $x' + x + y + z = t$,
    $2x + y' = 2t$,
    $y + 3z' = t$

12. $x'' + x + y + z' = 1$,
    $x' + y' + z'' = t$,
    $x'' - x' + x - y' + y - z'' + z' = e^t$

13. $x_1' + x_1 + x_2' + x_2 + x_3 = t$,
    $x_1' + x_2' + 3x_3 = e^t$,
    $x_2' + x_3' = t$

14. $x'' + y'' + z'' = 0$,
    $x' + y' + y + z' + z = 1$,
    $x' + 3z = t$

In Exercises 15 through 20, rewrite the given system of differential equations as a first-order system of differential equations.

15. $x'' - x + y = t$,
    $x' + y' = 0$

16. $x''' + 3x' - 5x = \cos t$

17. $x_1'' + x_2'' = \sin t$,
    $x_1''' - x_1' + x_2 = t$

18. $x'''' + x'' + x = e^t$

19. $x_1'' = 3x_1' + 4x_2' - x_1$,
    $x_2'' = x_1' - x_2' + x_2$

20. $x_1'' + x_2'' + x_3 = t$,
    $x_3''' + x_1 - 6x_2 + x_1' = e^t$,
    $x_2' + x_1' + x_3 + x_3' = t^2$

In Exercises 21 through 24, determine (using Theorem 6.1.2) whether or not the system of differential equations can be put in explicit form using the same number of dependent variables by adding constant multiples of the equations. If it can be put in explicit form, do so.

21. $2x' + 3x + 5y' = t$,
    $x' - x + 2y' + y = \cos t$

22. $x' + x + 2y' - y = 0$,
    $2x' - 3x + 4y' - 7y = t$

23. $x' + y' = 6y + t$,
    $-3x' + 7y' = 3y - x$

24. $x' + x - 3y = t$,
    $6x' - 5y' - x + y = \cos t$

6.2 Elimination Methods

The algebraic system of equations

$$2x + 3y = 7, \tag{1a}$$
$$4x - 5y = 13, \tag{1b}$$

can be solved by adding constant multiples of one equation to another to obtain an equation in just one variable or, equivalently, by solving one equation, say (1a), for one of the variables, say x, and substituting this expression into the other equation, to obtain an equation in just one variable. A similar approach will be developed in this section for linear systems of differential equations with constant coefficients. This method is usually best suited for systems involving a small number of dependent variables, and will enable us to solve the differential equations that arise in the applications of Sections 6.4, 6.5, and 6.6. It is also possible to program this method on a computer that does symbolic calculus and algebra. However, a more common way to analyze and solve more complicated systems of linear differential equations is either numerically (Chapter 8) or by the matrix techniques developed later in this chapter.

To begin with, suppose we have two linear, constant-coefficient differential equations involving two dependent variables $x(t)$, $y(t)$. This system may be written as

$$L_1 x + L_2 y = f_1, \tag{2a}$$
$$L_3 x + L_4 y = f_2, \tag{2b}$$

where L_1, L_2, L_3, L_4 are differential operators (Section 3.6) and f_1, f_2 are known functions of t. We assume that f_1, f_2 are functions for which the method of undetermined coefficients (Section 3.12) is appropriate.

Example 6.2.1 The system of differential equations

$$x'' + 3x' + y' = t, \tag{3a}$$
$$x' + 3x + y'' + y' = e^{2t}, \tag{3b}$$

may be written in the form (2) as follows

$$(D^2 + 3D)x + D^2y = t \tag{4a}$$
$$(D + 3)x + (D^2 + D)y = e^{2t}. \tag{4b}$$

Here

$$L_1 = D^2 + 3D, \quad L_2 = D^2, \quad L_3 = D + 3, \quad L_4 = D^2 + D,$$

and

$$f_1 = t, \quad f_2 = e^{2t}. \;\blacksquare$$

In order to solve the system (2), we will derive a single equation in just one of the dependent variables. One way to do this is to "multiply" (2a) by L_3 and (2b) by L_1, to get

$$L_3 L_1 x + L_3 L_2 y = L_3 f_1 \tag{5a}$$
$$L_1 L_3 x + L_1 L_4 y = L_1 f_2. \tag{5b}$$

But $L_1 L_3 = L_3 L_1$, since L_3, L_1 have constant coefficients (Theorem 3.6.2). Thus, upon subtracting (5b) from (5a), we get a differential equation solely in terms of y:

$$(L_3 L_2 - L_1 L_4) y = L_3 f_1 - L_1 f_2.$$

If we apply this process to (4), we would "multiply" (4b) by $L_1 = D^2 + 3D$, and (4a) by $L_3 = D + 3$, to get

$$(D + 3)(D^2 + 3D)x + (D + 3)D^2 y = (D + 3)t$$
$$= 1 + 3t,$$
$$(D^2 + 3D)(D + 3)x + (D^2 + 3D)(D^2 + D)y = (D^2 + 3D)e^{2t}$$
$$= 4e^{2t} + 6e^{2t}$$
$$= 10e^{2t}.$$

Subtracting the second equation from the first yields

$$-(D^4 + 3D^3)y = 1 + 3t - 10e^{2t},$$

that is,

$$y'''' + 3y''' = 10e^{2t} - 1 - 3t.$$

This differential equation can be solved for y, for example, by the method of undetermined coefficients, to give

$$y = \frac{e^{2t}}{4} - \frac{t^4}{24} + c_1 + c_2 t + c_3 t^2 + c_4 e^{-3t}, \tag{6}$$

which has four arbitrary constants. This formula for y can be substituted into either of the original differential equations (3a) or (3b) to get a differential equation in x only. If (6) is substituted into (3b) for y, we get a first-order differential equation in x which, upon solution, provides one additional arbitrary constant, for a total of five arbitrary constants. On the other hand, if formula (6) is substituted for y in (3a), a second-order differential equation in x results, which, upon solution, gives us *two* additional arbitrary constants, for a total of six arbitrary constants. Thus, depending on how these calculations are done, we get either five or six arbitrary constants. But from Section 6.1, the correct total number of arbitrary constants is four, since $L_3 L_2 - L_1 L_4 = -D^4 - 3D^3$ is fourth-order.

To see how we could have avoided this problem, in this particular example, return to the system (4) and note that $D^2 + 3D = D(D + 3)$. Thus, to get the same coefficient in front of x in both equations, we could have merely multiplied (4b) by D to give the new system

$$(D^2 + 3D)x + D^2 y = t, \tag{7a}$$

$$D(D + 3)x + D(D^2 + D)y = De^{2t} = 2e^{2t}. \tag{7b}$$

Now if we subtract the first equation (7a) from (7b), we get

$$D^3 y = 2e^{2t} - t,$$

which can be solved by antidifferentiating three times to yield

$$y = \frac{e^{2t}}{4} - \frac{t^4}{24} + c_1 t^2 + c_2 t + c_3, \tag{8}$$

where c_1, c_2, c_3 are arbitrary constants. Substituting (8) into equation (3b) gives the first-order differential equation in x,

$$x' + 3x = \frac{-e^{2t}}{2} + \frac{t^3}{6} + \frac{t^2}{2} - 2c_1 t - (c_2 + 2c_1). \tag{9}$$

This differential equation (9) can be solved by the integrating-factor method, but it is probably quicker to use undetermined coefficients, as follows. The characteristic polynomial is $r + 3$, with a single root $r = -3$, so that

$$x_h = c_4 e^{-3t},$$

and x_p is of the form

$$x_p = A + Bt + Ct^2 + Et^3 + Fe^{2t}.$$

Substitute this form into (9) for x, to get

$$(B + 2Ct + 3Et^2 + 2Fe^{2t}) + 3(A + Bt + Ct^2 + Et^3 + Fe^{2t})$$
$$= -\frac{e^{2t}}{2} + \frac{t^3}{6} + \frac{t^2}{2} - 2c_1 t - (c_2 + 2c_1).$$

Equating the corresponding coefficients, we find

$$\begin{aligned}
1: \quad & B + 3A = -(c_2 + 2c_1), \\
t: \quad & 2C + 3B = -2c_1, \\
t^2: \quad & 3E + 3C = \frac{1}{2}, \\
t^3: \quad & 3E = \frac{1}{6}, \\
e^{2t}: \quad & 5F = -\frac{1}{2}.
\end{aligned}$$

In solving these equations, we treat A, B, C, E, F as the unknowns, and the c_1, c_2, c_3 as fixed, but unknown, constants. Thus, starting with the last equation and back-substituting yields:

$$F = -\frac{1}{10},$$

$$E = \frac{1}{18},$$

$$C = \frac{1}{6} - E = \frac{1}{9},$$

$$B = -\frac{2}{3}C - \frac{2}{3}c_1 = -\frac{2}{27} - \frac{2}{3}c_1,$$

$$\begin{aligned}
A = -\frac{1}{3}B - \frac{1}{3}(c_2 + 2c_1) &= \frac{2}{81} + \frac{2}{9}c_1 - \frac{1}{3}(c_2 + 2c_1) \\
&= \frac{2}{81} - \frac{4}{9}c_1 - \frac{1}{3}c_2.
\end{aligned}$$

The solution of (4) is thus

$$y = \frac{e^{2t}}{4} - \frac{t^4}{24} + c_1 t^2 + c_2 t + c_3, \tag{10a}$$

$$\begin{aligned}
x &= \left(\frac{2}{81} - \frac{4}{9}c_1 - \frac{1}{3}c_2\right) - \left(\frac{2}{27} + \frac{2}{3}c_1\right)t \\
&\quad + \frac{t^2}{9} + \frac{t^3}{18} - \frac{e^{2t}}{10} + c_4 e^{-3t},
\end{aligned} \tag{10b}$$

which has the correct number of arbitrary constants, four.

Caution If an arbitrary constant appears several places in the solution, as c_1 does in (10), we cannot change it in one location without changing it everywhere. For example, if we were to try to simplify (10b) by letting $\tilde{c}_1 = c_1/9$, we would have to replace c_1 by $9\tilde{c}_1$ everywhere in (10a) and (10b).

We shall summarize these observations and then carefully work additional examples.

Elimination Method

Elimination method for a system of two linear, constant-coefficient, differential equations in two unknowns:

1. First write the system in the form (2):

$$L_1 x + L_2 y = f_1, \tag{11a}$$
$$L_3 x + L_4 y = f_2. \tag{11b}$$

2. Multiply (11a) and/or (11b) by constant-coefficient differential operators, so that both equations have the same x or y coefficient. [One can always multiply (11a) by L_3 and (11b) by L_1, or (11a) by L_4 and (11b) by L_2.]
3. Subtract one equation from the other to get a differential equation in just one dependent variable, and solve this differential equation.
4. Substitute the answer for Part (3) into one of the equations in the original system, to get a differential equation in the other dependent variable. Solve this differential equation.
5. If the number of arbitrary constants is the same as the order of $L_1 L_4 - L_2 L_3$, we have the general solution.
6. If the number of arbitrary constants is greater than the order of $L_1 L_4 - L_2 L_3$, substitute the formulas for x, y from parts (3) and (4) into the original differential equation not used in part (4), in order to eliminate any extra constants.
7. If initial conditions are given, use them to eliminate remaining arbitrary constants.

Example 6.2.2 Using elimination, solve the system

$$4x' + 4x + y' = 1, \tag{12a}$$
$$3x' + y' - y = t. \tag{12b}$$

Solution (Step 1) First rewrite the system as

$$4(D + 1)x + Dy = 1, \tag{13a}$$
$$3Dx + (D - 1)y = t. \tag{13b}$$

(Step 2) We shall eliminate the y term. Multiply (13a) by $(D - 1)$ and (13b) by D, to get

$$4(D^2 - 1)x + (D - 1)Dy = (D - 1)1 = -1, \tag{14a}$$
$$3D^2 x + D(D - 1)y = Dt = 1. \tag{14b}$$

(Step 3) Subtract (14b) from (14a), to give

$$(D^2 - 4)x = -2$$

or

$$x'' - 4x = -2. \tag{15}$$

We shall solve this differential equation by the method of undetermined coefficients. The characteristic equation is $r^2 - 4 = 0$ with roots $r = \pm 2$. Thus $x_h = c_1 e^{-2t} + c_2 e^{2t}$. The particular solution has the form $x_p = A$, A a constant. Substituting this form into (15) gives $A'' - 4A = -2$ or $-4A = -2$ and $A = \frac{1}{2}$. Thus

$$x = \tfrac{1}{2} + c_1 e^{-2t} + c_2 e^{2t}. \tag{16}$$

(Step 4) This expression for x may be substituted into either (12a) or (12b), to find y. Since (12a) will give a slightly simpler equation for y, we substitute the formula (16) for x into (12a), to get

$$y' = 1 - 4x' - 4x$$

or

$$y' = -1 + 4c_1 e^{-2t} - 12c_2 e^{2t}.$$

Antidifferentiate both sides with respect to t, to find

$$y = -t - 2c_1 e^{-2t} - 6c_2 e^{2t} + c_3. \tag{17}$$

(Step 5) To check whether the number of arbitrary constants in (16) and (17) is correct, compute

$$\det \begin{bmatrix} 4D + 4 & D \\ 3D & D - 1 \end{bmatrix} = 4D^2 - 4 - 3D^2 = D^2 - 4,$$

which is *second*-order, so our answer (16) and (17) should have *two* arbitrary constants. Since we have three arbitrary constants, we proceed to Step 6.

(Step 6) Since (13a) was used to find y from x, we know (13a) holds. Thus, we use the other equation, (13b) or, equivalently (12b), to eliminate the extra constant. Substitute (16) and (17) into (12b), $3x' + y' - y = t$, to find

$$3(\tfrac{1}{2} + c_1 e^{-2t} + c_2 e^{2t})' + (-t - 2c_1 e^{-2t} - 6c_2 e^{2t} + c_3)'$$
$$- (-t - 2c_1 e^{-2t} - 6c_2 e^{2t} + c_3) = t,$$

or

$$-6c_1 e^{-2t} + 6c_2 e^{2t} - 1 + 4c_1 e^{-2t} - 12c_2 e^{2t} + t + 2c_1 e^{-2t} + 6c_2 e^{2t} - c_3 = t,$$

or, upon simplification,

$$-1 - c_3 = 0 \quad \text{and} \quad c_3 = -1.$$

Thus the general solution of (12) is

$$x = \tfrac{1}{2} + c_1 e^{-2t} + c_2 e^{2t},$$
$$y = -1 - t - 2c_1 e^{-2t} - 6c_2 e^{2t}. \ \blacksquare$$

Example 6.2.3

Using the elimination method, solve the initial-value problem

$$x' = x + 5y, \quad x(0) = 0, \tag{18a}$$
$$y' = -x - y, \quad y(0) = 1. \tag{18b}$$

Solution

Since x appears only once in (18b) and is undifferentiated, we could simply solve (18b) for $x = -y' - y$ and then substitute into (18a) to get $(-y' - y)' = (-y' - y) + 5y$. However we shall follow the general procedure. First, rewrite (18) as

$$(D - 1)x - 5y = 0, \tag{19a}$$
$$x + (D + 1)y = 0. \tag{19b}$$

Multiply the second equation by $(D - 1)$:

$$(D - 1)x - 5y = 0,$$
$$(D - 1)x + (D^2 - 1)y = 0,$$

and subtract the first from the second, to give an equation in only y:

$$(D^2 + 4)y = 0,$$

or $y'' + 4y = 0$. The characteristic polynomial is $r^2 + 4$ with roots of $\pm 2i$. Thus,

$$y = c_1 \cos 2t + c_2 \sin 2t. \tag{20a}$$

If this formula for y is substituted into (18a), a first-order differential equation in x results, whereas substituting into (18b) gives an algebraic equation for x. We substitute, then, into (18b);

$$x = -y' - y = -(c_1 \cos 2t + c_2 \sin 2t)' - (c_1 \cos 2t + c_2 \sin 2t),$$

so that

$$x = (-2c_2 - c_1) \cos 2t + (2c_1 - c_2) \sin 2t. \tag{20b}$$

The solution (20a), (20b) has the correct number of constants since we have $L_1 L_4 - L_2 L_3 = D^2 + 4$ is second-order. The initial conditions $x(0) = 0$, $y(0) = 1$ can now be applied to the general solution (20a), (20b),

$$0 = x(0) = -2c_2 - c_1,$$
$$1 = y(0) = c_1.$$

Thus $c_1 = 1$, $c_2 = -\frac{1}{2}$ and the solution of the original initial-value problem (18) is

$$x = \tfrac{5}{2} \sin 2t$$
$$y = \cos 2t - \tfrac{1}{2} \sin 2t. \blacksquare$$

The elimination method may also be used to solve systems with more than two equations.

Example 6.2.4 Using elimination, solve the system of differential equations

$$x' = x + y + z, \quad (21a)$$
$$y' = x + y + z, \quad (21b)$$
$$z' = x + y + z. \quad (21c)$$

Solution

This is a system of three equations in three unknowns. The procedure we follow is to use one equation to eliminate a variable from the other two equations. This gives us two equations in two unknowns. Using one of these equations we eliminate a dependent variable from the other giving an equation in one dependent variable, which we can then solve.

First, rewrite (21) as

$$(D-1)x - y - z = 0, \quad (22a)$$
$$-x + (D-1)y - z = 0, \quad (22b)$$
$$-x - y + (D-1)z = 0. \quad (22c)$$

We shall eliminate x from Eqs. (22b) and (22c). Multiply (22b) and (22c) by $(D-1)$,

$$(D-1)x - y - z = 0, \quad (23a)$$
$$-(D-1)x + (D-1)^2 y - (D-1)z = 0, \quad (23b)$$
$$-(D-1)x - (D-1)y + (D-1)^2 z = 0. \quad (23c)$$

Now add (23a) to both (23b) and (23c),

$$(D-1)x - y - z = 0, \quad (24a)$$
$$(D^2 - 2D)y - Dz = 0, \quad (24b)$$
$$-Dy + (D^2 - 2D)z = 0. \quad (24c)$$

Note that (24b) and (24c) give us two equations in the two unknowns y, z. Since $D^2 - 2D = D(D-2)$, we now multiply (24c) by $(D-2)$ and add (24b) to this (24c) to get the new (24c);

$$(D-1)x - y - z = 0, \quad (25a)$$
$$(D^2 - 2D)y - Dz = 0, \quad (25b)$$
$$(D^3 - 4D^2 + 3D)z = 0, \quad (25c)$$

resulting in a third-order differential equation in z. The characteristic polynomial is $r^3 - 4r^2 + 3r = r(r-3)(r-1)$, with roots of $r = 0, 1, 3$. Thus

$$z = c_1 + c_2 e^t + c_3 e^{3t}. \quad (26)$$

Substitute this expression for z into (25b), to get an equation in y:

$$(D^2 - 2D)y = Dz = c_2 e^t + 3c_3 e^{3t}.$$

Solving for y by the method of undetermined coefficients, and thinking of c_2, c_3 as fixed but unknown numbers, we find that

$$y_p = -c_2 e^t + c_3 e^{3t}, \qquad y_h = c_4 + c_5 e^{2t},$$

so that

$$y = c_4 + c_5 e^{2t} - c_2 e^t + c_3 e^{3t}. \tag{27}$$

The dependent variable x may be found directly from (22c), using (26) and (27),

$$x = z' - z - y = -(c_1 + c_4) + c_2 e^t + c_3 e^{3t} - c_5 e^{2t}. \tag{28}$$

Since (21) is in explicit form, according to Theorem 6.1.1, the original problem (21) should have three arbitrary constants instead of five. Since (22c) was used to find x, the other equations, (22a) and (22b) are used to eliminate the extra constants from (26), (27), (28). Substituting (26), (27), (28) for z, y, x in (22a) gives, after simplification,

$$-2c_5 e^{2t} = 0,$$

so that

$$c_5 = 0.$$

Now substituting (26), (27), and (28) for z, y, x in (22b) and simplifying gives

$$-c_2 e^t = c_2 e^t,$$

so that $c_2 = 0$. Thus, the solution of (21) is

$$x = -c_1 - c_4 + c_3 e^{3t},$$
$$y = c_4 + c_3 e^{3t},$$
$$z = c_1 + c_3 e^{3t}. \blacksquare$$

Later in this chapter we shall solve (21) using matrix theory. Even for the relatively simple problem (21), the added simplicity of the matrix method will be apparent.

Exercises

In Exercises 1 through 8, use the method of elimination to solve the differential equation. Find the general solution if no initial conditions are given.

1. $x' = -x + 3y,$
   $y' = -2x + 4y$

2. $x' = -x - y,$
   $y' = 2x + y$

3. $x' = -2x - 3y + t,$
   $y' = x + 2y + 3$

4. $x' = x + y + 1$, $x(0) = 1$,
   $y' = x + y + e^t$, $y(0) = 1$
5. $x' = x - 2y$, $x(0) = 0$,
   $y' = 2x + y$, $y(0) = 1$
6. $x' = 2x - y + t^2$,
   $y' = x$
7. $x' = 2x - y + \sin t$,
   $y' = 2x + 4y + \cos t$
8. $x' = x - y + 2t$, $x(0) = 3$,
   $y' = x - y - t$, $y(0) = 0$

In solving the algebraic equations (1), we could divide the first equation by 2 if we wanted to. However, we have carefully avoided dividing by a polynomial in $D = d/dt$. Exercises 9 through 11 examine why we have not done this.

9. If $(D - 1)x = 0$ for a function $x(t)$, then x need not be equal to zero. Find all x for which $(D - 1)x = 0$.

10. Suppose that $x(t)$, $y(t)$ are functions such that
$$(D^2 - 1)x + (D - 1)y = 0.$$
Note that $D^2 - 1$ and $D - 1$ are "divisible" by $D - 1$. Show that x, y are solutions of
$$(D + 1)x + y = c_1 e^t$$
for some constant c_1.

11. Suppose that L_1, L_2, L_3 are constant-coefficient differential operators. Show that if x, y are functions of t such that
$$L_1 L_2 x + L_1 L_3 y = 0,$$
then
$$L_2 x + L_3 y = f,$$
where f is a solution of $L_1 f = 0$.

In Exercises 12 through 25, solve the system of differential equations by the method of elimination. Find the general solution if no initial conditions are given.

12. $x'' + y' - y = 0$,
    $x' + y' + y = 0$

13. $x' + x + y' - y = 0$,
    $x' - x + 2y' + y = 0$

14. $x'' + x + y'' = 0$,
    $x' + y' = t$

15. $x'' - 2x + y'' - y = 1$,
    $2x' + x + 2y' + 2y = 2$

16. $x' = y + z$,
    $y' = 2x + y + 2z$,
    $z' = 3x + 3y + 2z$

17. $x' = 2x - y + z$, $x(0) = 0$,
    $y' = -x + 2y + z$, $y(0) = 1$,
    $z' = x + y$, $z(0) = 0$

18. $x' = 2x - y + z + 1$,
    $y' = 2x - y + 2z$,
    $z' = x - y + 2z + 2$

19. $x' = 3x - y + z$,
    $y' = -x + 3y + z$,
    $z' = x + y + z$

20. $x'' + 2x - y' - y = 0$,
    $-x' + x + y'' - y = 0$

21. $x'' = 8x - y - y'$,
    $y'' = x - x' + y$

22. $x'' + 2x' + x - 3y' - 3y = 0$,
    $2x + y' - 4y = 0$

23. $x'' + 2x' + 3y' = 1$,
    $x' + 2y' - y'' = 0$

24. $x'' - x' + 2y' = 0$,
    $-2x' + y'' - y' = 0$

25. $x'' - x + y'' - 2y' + y = e^{2t}$,
    $x' - x + 2y' + y = 0$

6.3 Solution by Laplace Transform

An alternative method to the elimination method for the solution of systems of linear differential equations with constant coefficients is to use the Laplace transform. This approach, sometimes in a somewhat disguised format, is especially popular in the electrical engineering and system and control literature. This method has the advantage that one can divide freely and not have to worry about having too many arbitrary constants. In order to show the similarities and differences of the two methods, we shall illustrate the method by reworking two of the examples from Section 6.2.

Example 6.3.1 Solve the system of differential equations

$$x' = x + 5y, \qquad x(0) = 0, \qquad (1a)$$
$$y' = -x - y, \qquad y(0) = 1, \qquad (1b)$$

using the Laplace transform (this is Example 6.2.3).

Solution Taking the Laplace transform of both sides of (1a) and (1b), we obtain:

$$sX - x(0) = X + 5Y,$$
$$sY - y(0) = -X - Y,$$

where $X(s) = \mathscr{L}[x(t)]$, $Y(s) = \mathscr{L}[y(t)]$. Using the initial conditions in (1) and rearranging terms gives

$$(s - 1)X - 5Y = 0, \qquad (2a)$$
$$X + (s + 1)Y = 1. \qquad (2b)$$

There are several ways to solve for X, Y. One could use Cramer's rule. We shall solve for X or Y in one equation and substitute into the other.

Solving the first equation (2a) for Y in terms of X we get:

$$Y = \frac{1}{5}(s - 1)X \qquad (3)$$

and, substituting this expression into (2b) for Y gives an equation for X:

$$X + \frac{1}{5}(s + 1)(s - 1)X = 1$$

or

$$X = \frac{5}{s^2 + 4} = \frac{5}{2}\frac{2}{s^2 + 4}, \qquad (4)$$

so that

$$x(t) = \frac{5}{2} \sin 2t. \qquad (5)$$

To find $y(t)$, we may either substitute formula (4) for $X(s)$ into (3) to find $Y(s)$ and then take the inverse Laplace transform, or we can substitute the formula

(5) for x(t) into (1a) and solve for y(t). In this example we shall do the latter. From (1a) and (5),

$$y(t) = \frac{1}{5}x' - \frac{1}{5}x = \cos 2t - \frac{1}{2}\sin 2t. \blacksquare$$

Comment Note that the system to be solved in (2) for X, Y is very similar to the system (19) of Section 6.2, where (1) is solved by elimination, except that now we have an s in place of D and the right-hand side is nonzero because of the initial conditions. Also, we could, if we wanted to, safely divide an equation by, say, $s - 1$, whereas "dividing" by $(D - 1)$ is to be avoided unless you have a very good understanding of what the operator notation means. (See Exercises 9 through 11 of Section 6.2).

Example 6.3.2 Solve the system of differential equations

$$4x' + 4x + y' = 1, \quad x(0) = 2, \tag{6a}$$

$$3x' + y' - y = t, \quad y(0) = -6, \tag{6b}$$

using Laplace transforms (this is Example 6.2.2 with initial conditions added).

Solution Taking the Laplace transform of both sides of equations (6a), (6b) gives

$$4(sX - x(0)) + 4X + sY - y(0) = \frac{1}{s},$$

$$3(sX - x(0)) + (sY - y(0)) - Y = \frac{1}{s^2}. \tag{7}$$

Using the initial conditions $x(0) = 2$, $y(0) = -6$, and combining terms, (7) becomes

$$4(s + 1)X + sY = \frac{1}{s} + 2 = \frac{1 + 2s}{s}, \tag{8a}$$

$$3sX + (s - 1)Y = \frac{1}{s^2}. \tag{8b}$$

Solve (8a) for Y,

$$Y = \frac{1 + 2s}{s^2} - \frac{4(s + 1)}{s}X, \tag{9}$$

and substitute into (8b) to yield an equation in X:

$$3sX + (s - 1)\left[\frac{1 + 2s}{s^2} - 4\frac{(s + 1)}{s}X\right] = \frac{1}{s^2}. \tag{10}$$

Simplifying (10), we obtain:

$$(-s^2 + 4)X = \frac{1}{s}(2 + s - 2s^2),$$

and, solving for X, we get

$$X = \frac{2s^2 - s - 2}{s(s^2 - 4)} = \frac{2s^2 - s - 2}{s(s - 2)(s + 2)}. \tag{11}$$

By partial fractions,

$$X = \frac{A}{s} + \frac{B}{s - 2} + \frac{C}{s + 2},$$

where

$$2s^2 - s - 2 = A(s - 2)(s + 2) + Bs(s + 2) + Cs(s - 2). \tag{12}$$

Since the three factors of the denominator are all linear factors, the simplest way to find A, B, C, is to evaluate (12) at the roots $s = 0$, $s = 2$, $s = -2$, of the factors in the denominator. This gives $A = \frac{1}{2}$, $B = \frac{1}{2}$, $C = 1$. Thus,

$$X(s) = \frac{1/2}{s} + \frac{1/2}{s - 2} + \frac{1}{s + 2},$$

so that

$$x(t) = \frac{1}{2} + \frac{1}{2}e^{2t} + e^{-2t}. \tag{13}$$

To find $y(t)$, we could substitute $X(s)$ from (11) into (9), get $Y(s)$, and calculate the inverse Laplace transform of $Y(s)$. It is somewhat quicker to substitute formula (13) for $x(t)$ into (6a) and then antidifferentiate to find y. We shall follow the second approach. From (6a),

$$y' = 1 - 4x' - 4x = -1 - 6e^{2t} + 4e^{-2t}.$$

Antidifferentiating gives

$$y = -t - 3e^{2t} - 2e^{-2t} + c_1.$$

The initial condition $y(0) = -6$ gives an equation for c_1:

$$-6 = y(0) = -3 - 2 + c_1,$$

so that $c_1 = -1$ and the solution of the differential equation (6) is

$$x = \frac{1}{2} + \frac{1}{2}e^{2t} + e^{-2t},$$
$$y = -1 - t - 3e^{2t} - 2e^{-2t}. \blacksquare$$

Exercises

In Exercises 1 through 18, solve the initial-value problem, using the Laplace transform. Except for the addition of initial conditions, Exercises 1 through 12 are Exercises 1 through 8, and 16 through 19 of Section 6.2.

1. $x' = -x + 3y,\quad x(0) = 2,$
   $y' = -2x + 4y,\quad y(0) = 1$
2. $x' = -x - y,\quad x(0) = 1,$
   $y' = 2x + y,\quad y(0) = -2$
3. $x' = -2x - 3y + t,\quad x(0) = 6,$
   $y' = x + 2y + 3,\quad y(0) = -5$
4. $x' = x + y + 1,\quad x(0) = -\frac{1}{2},$
   $y' = x + y + e^t,\quad y(0) = 1$
5. $x' = x - 2y,\quad x(0) = 0,$
   $y' = 2x + y,\quad y(0) = 1$
6. $x' = 2x - y + t^2,\quad x(0) = 4,$
   $y' = x,\quad y(0) = 5$
7. $x' = 2x - y + \sin t,\quad x(0) = \dfrac{23}{78},$
   $y' = 2x + 4y + \cos t,\quad y(0) = -\dfrac{67}{39}$
8. $x' = x - y + 2t,\quad x(0) = 3,$
   $y' = x - y - t,\quad y(0) = 0$
9. $x' = y + z,\quad x(0) = 2,$
   $y' = 2x + y + 2z,\quad y(0) = 0,$
   $z' = 3x + 3y + 2z,\quad z(0) = 4$
10. $x' = 2x - y + z,\quad x(0) = 0,$
    $y' = -x + 2y + z,\quad y(0) = -2,$
    $z' = x + y,\quad z(0) = -7$
11. $x' = 2x - y + z + 1,\quad x(0) = 3,$
    $y' = 2x - y + 2z,\quad y(0) = 6,$
    $z' = x - y + 2z + 2,\quad z(0) = 1$
12. $x' = 3x - y + z,\quad x(0) = 1,$
    $y' = -x + 3y + z,\quad y(0) = 3,$
    $z' = x + y + z,\quad z(0) = 5$
13. $x'' = -2x + y + y' + t,\quad x(0) = x'(0) = 0,$
    $y'' = -x + x' + y + 1,\quad y(0) = y'(0) = 0$
14. $x'' = 8x - y - y',\quad x(0) = 1,\quad x'(0) = -5,$
    $y'' = x - x' + y,\quad y(0) = -\dfrac{9}{4},\quad y'(0) = \dfrac{61}{4}$
15. $x'' + 2x' + 3y' = 1,\quad x(0) = 1,\quad x'(0) = 6,$
    $x' + 2y' - y'' = 0,\quad y(0) = 6,\quad y'(0) = -3$
16. $x'' + 2x' + x - 3y' - 3y = 0,$
    $2x + y' - 4y = 0,\quad y(0) = 0,\quad y'(0) = 0,$
    $y''(0) = 1$
    (Hint: compute $x(0)$, $x'(0)$)
17. $x'' - x + y'' - 2y' + y = -e^{2t},\quad x(0) = -\dfrac{47}{14},$
    $x' - x + 2y' + y = 0,\quad y(0) = \dfrac{43}{14},\quad y'(0) = -\dfrac{69}{7}$
    (Hint: compute $x'(0)$)
18. $x'' - x' + 2y' = 0,\quad x(0) = 0,\quad x'(0) = 1,$
    $-2x' + y'' - y' = 0,\quad y(0) = 1,\quad y'(0) = 0$

6.4 Mixing Problems

In this and the next two sections we will give several applications of systems of differential equations. Mixing problems will be discussed in this section. It is assumed the reader has read Section 2.10. At the end of this section, we shall

Figure 6.4.1
Picture for Example 6.4.1.

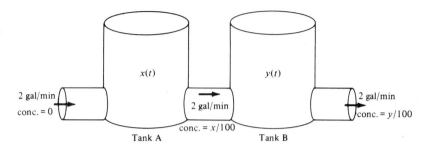

briefly discuss how a wide variety of problems can be viewed as mixing/flow problems.

In our problems we will have two or more tanks interconnected with pipes. The tanks contain a chemical, dissolved in a fluid, which we take as salt (NaCl) and water, respectively. At a given time t, we are interested in the amount of salt in each tank.

Example 6.4.1

We have two 100-gallon tanks. Tank A is initially full of salt water with a salt concentration of 0.5 lb/gal. Tank B is full of pure water. Pure water enters tank A at 2 gal/min. Water exits the well-mixed tank A at two gal/min and flows into tank B. Water also exits tank B at two gal/min and flows out of the system (down the drain?). Pictorially, we have Fig. 6.4.1.

Solution

We assume that the amount of water in the pipe between the tanks is negligible. Let $x(t)$ be the amount of salt in pounds in tank A at time t, and $y(t)$ the amount of salt in pounds in tank B at time t. Thus the concentrations in tanks A and B are $x/100$ and $y/100$, respectively (amount/volume). The equation describing the rate of change of the amount of salt in tank A is:

$$\frac{dx}{dt} = -\begin{bmatrix} \text{Outflow rate} \\ \text{of salt} \\ \text{from A} \end{bmatrix} = -\begin{bmatrix} \text{Flow rate} \\ \text{of} \\ \text{water} \end{bmatrix} \cdot \begin{bmatrix} \text{Conc. of} \\ \text{salt in} \\ \text{water} \end{bmatrix}$$

$$= -2\frac{x}{100}. \tag{1}$$

For tank B we have:

$$\frac{dy}{dt} = \begin{bmatrix} \text{Inflow rate} \\ \text{of salt} \\ \text{from A} \end{bmatrix} - \begin{bmatrix} \text{Outflow rate} \\ \text{of salt} \\ \text{from B} \end{bmatrix}$$

$$= 2\left(\frac{x}{100}\right) - 2\left(\frac{y}{100}\right). \tag{2}$$

Combining (1) and (2) and using our initial conditions, we have the linear system of differential equations

Figure 6.4.2
Graphs of (7).

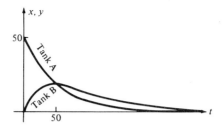

$$x' = -\frac{1}{50}x, \qquad x(0) = 50, \tag{3a}$$

$$y' = \frac{1}{50}x - \frac{1}{50}y, \qquad y(0) = 0. \tag{3b}$$

Since (3a) involves only x, we do not need to use elimination. Equation (3a) can be solved several ways. The quickest is to observe that the characteristic polynomial $\lambda + 1/50$ has only the single root $-1/50$. Thus, $x = c_1 e^{-t/50}$. The initial condition in (3a) implies that $c_1 = 50$ and

$$x = 50e^{-t/50}. \tag{4}$$

Substitute (4) for x in (3b) to get

$$y' + \frac{1}{50}y = e^{-t/50}. \tag{5}$$

This could be solved by undetermined coefficients. We shall use integrating factors (Section 2.3),

$$(e^{t/50}y)' = e^{-t/50}e^{t/50} = 1$$

and antidifferentiate

$$e^{t/50}y = t + c_2.$$

Solving for y and using the initial condition in (3b) to find $c_2 = 0$ gives

$$y = te^{-t/50}; \tag{6}$$

thus the solution of the system of differential equations (3) is

$$x = 50e^{-t/50}, \qquad y = te^{-t/50}. \tag{7}$$

This solution has the expected physical behavior. The amount of salt in tank A exponentially decays. In tank B there is an increase at first due to the inflow of salt from tank A and then the amount of salt in tank B decreases. This much we could have guessed. However, our solution (7) makes it possible to answer questions whose answer is not obvious. For example, is there ever more salt in tank B than tank A? The answer is "yes," as the graph of (7) in Fig. 6.4.2 shows. ∎

Example 6.4.2

Two well-mixed 100-liter tanks are initially full of pure water. Water containing salt at a concentration of 30 g/liter flows into tank A at a rate of 4 liters/min.

Figure 6.4.3
Picture for Example 6.4.2.

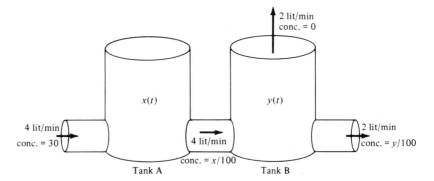

Water flows from tank A to tank B at 4 liters/min. Water evaporates from tank B at 2 liters/min and flows out at 2 liters/min. We want to find the amount of salt in both tanks as a function of time. Pictorially, we have Fig. 6.4.3.

Solution

Let $x(t)$ be the amount of salt in tank A at time t and $y(t)$ the amount of salt in tank B at time t; t is measured in minutes, and x and y are in grams. Note that the volume of both tanks is constant. The equations describing the change in the amounts of salt are

$$\frac{dx}{dt} = \begin{bmatrix} \text{Inflow rate of} \\ \text{water into} \\ \text{tank A} \end{bmatrix} \begin{bmatrix} \text{Conc. of} \\ \text{NaCl} \\ \text{in inflow} \end{bmatrix} - \begin{bmatrix} \text{Outflow rate of} \\ \text{water from} \\ \text{tank A} \end{bmatrix} \begin{bmatrix} \text{Conc. of} \\ \text{NaCl} \\ \text{in outflow} \end{bmatrix};$$

$$\frac{dy}{dt} = \begin{bmatrix} \text{Inflow rate of} \\ \text{water into} \\ \text{tank B} \end{bmatrix} \begin{bmatrix} \text{Conc. of} \\ \text{NaCl} \\ \text{in inflow} \end{bmatrix} - \begin{bmatrix} \text{Outflow rate of} \\ \text{water excluding} \\ \text{evaporation from} \\ \text{tank B} \end{bmatrix} \begin{bmatrix} \text{Conc. of} \\ \text{NaCl} \\ \text{in outflow} \end{bmatrix},$$

or

$$\frac{dx}{dt} = 4(30) - 4\frac{x}{100}, \quad x(0) = 0,$$

$$\frac{dy}{dt} = 4\frac{x}{100} - 2\frac{y}{100}, \quad y(0) = 0.$$

That is,

$$\frac{dx}{dt} + \frac{x}{25} = 120, \quad x(0) = 0, \tag{8a}$$

$$\frac{dy}{dt} + \frac{y}{50} - \frac{x}{25} = 0, \quad y(0) = 0. \tag{8b}$$

We could solve (8) in the same way we solved (3). Instead, we shall use the Laplace transform. Taking the Laplace transform of (8a), (8b), and using the initial conditions gives

$$sX + \frac{1}{25}X = \frac{120}{s}, \tag{9a}$$

$$sY + \frac{1}{50}Y - \frac{1}{25}X = 0. \tag{9b}$$

Solve (9a) for X,

$$X(s) = \frac{120}{s(s + 1/25)}. \tag{10}$$

Using partial fractions,

$$X = \frac{A}{s} + \frac{B}{s + 1/25},$$

where $120 = A(s + 1/25) + Bs$, so that $A = 25 \cdot 120 = 3000$ and $B = -3000$. Thus the inverse Laplace transform of $X(s)$ gives

$$x(t) = 3000 - 3000e^{-t/25}. \tag{11}$$

From (9b) and (10)

$$Y = \frac{1}{25(s + 1/50)}X = \frac{120}{25(s + 1/50)s(s + 1/25)}.$$

By partial fractions

$$Y = \frac{A}{s} + \frac{B}{s + 1/50} + \frac{C}{s + 1/25}, \tag{12}$$

where

$$\frac{120}{25} = A\left(s + \frac{1}{50}\right)\left(s + \frac{1}{25}\right) + Bs\left(s + \frac{1}{25}\right) + Cs\left(s + \frac{1}{50}\right).$$

Evaluating at $s = 0$, $s = -1/25$, $s = -1/50$ yields

$$A = 6000, \quad B = -12000, \quad C = 6000.$$

Taking the inverse Laplace transform in (12) and recalling (11) gives the solution of (8) as

$$\begin{aligned} x &= 3000 - 3000e^{-t/25}, \\ y &= 6000e^{-t/25} - 12000e^{-t/50} + 6000. \end{aligned} \tag{13}$$

The graphs of (13) are given in Fig. 6.4.4.

Both tanks approach an equilibrium, or fixed amount of salt. Initially the salt accumulates faster in tank A but then increases more rapidly in tank B

Figure 6.4.4
Graph of (13).

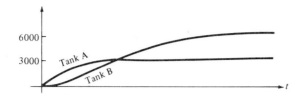

Figure 6.4.5
Picture for Example 6.4.3.

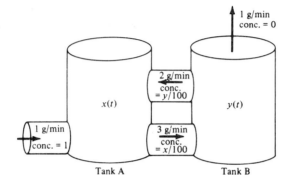

as more is passed on. If, instead of tanks, we think of a series of reservoirs in an arid region, each flowing into the next, this solution has interesting consequences. What do you think they are? ∎

Example 6.4.3 (Salt Lake.) There are two 100-gallon tanks both initially full of pure water. Water containing salt at a concentration of 1 lb/gal flows into tank A from an outside source at 1 gal/min. Water flows from tank A to tank B at 3 gal/min. Water evaporates from tank B at 1 gal/min and is also piped back to tank A at 2 gal/min. Find the amount of salt in each tank as a function of time. Pictorially, we have Fig. 6.4.5.

Solution Note that the amount of water in each tank is constant. Let x be the amount of salt in tank A at time t, and y the amount of salt in tank B at time t. The units are gal-lb-min. Then,

$$\frac{dx}{dt} = \begin{bmatrix} \text{Inflow rate} \\ \text{NaCl from} \\ \text{outside} \end{bmatrix} + \begin{bmatrix} \text{Inflow rate} \\ \text{NaCl from} \\ \text{tank B} \end{bmatrix} - \begin{bmatrix} \text{Outflow rate} \\ \text{NaCl from} \\ \text{tank A} \end{bmatrix},$$

$$\frac{dy}{dt} = \begin{bmatrix} \text{Inflow rate} \\ \text{NaCl from} \\ \text{tank A} \end{bmatrix} - \begin{bmatrix} \text{Outflow rate} \\ \text{NaCl from} \\ \text{tank B} \end{bmatrix},$$

or

$$\frac{dx}{dt} = 1 \cdot 1 + 2\frac{y}{100} - 3\frac{x}{100}, \quad x(0) = 0,$$

$$\frac{dy}{dt} = 3\frac{x}{100} - 2\frac{y}{100}, \quad y(0) = 0.$$

We have, then, the linear system of differential equations,

$$\frac{dx}{dt} = -\frac{3}{100}x + \frac{y}{50} + 1, \quad x(0) = 0, \tag{14a}$$

$$\frac{dy}{dt} = \frac{3}{100}x - \frac{y}{50}, \quad y(0) = 0. \tag{14b}$$

We shall solve (14) by elimination. First, we rewrite (14) in operator notation as

$$(100D + 3)x - 2y = 100, \tag{15a}$$

$$-3x + (100D + 2)y = 0. \tag{15b}$$

Multiply (15b) by $\frac{1}{3}(100D + 3)$ and add to (15a), to get a differential equation just in terms of y;

$$[\tfrac{1}{3}(100D + 3)(100D + 2) - 2]y = 100,$$

or

$$[100D^2 + 5D]y = 3, \tag{16}$$

that is, $100y'' + 5y' = 3$. The characteristic polynomial $100r^2 + 5r$ of (16) has roots $r = 0, -\frac{1}{20}$. Since 0 is a root, the method of undetermined coefficients says there will be a particular solution to (16) of the form $y_p = Ct$. Substituting this form for y_p into (16),

$$100(Ct)'' + 5(Ct)' = 3$$

implies that $5C = 3$ or $C = \frac{3}{5}$. Thus the general solution of (16) is

$$y = \frac{3}{5}t + c_1 + c_2 e^{-t/20}. \tag{17}$$

Then from (14b), $x = \frac{1}{3}[100y' + 2y]$, so that

$$x = 20 + \frac{2}{3}c_1 + \frac{2}{5}t - c_2 e^{-t/20}. \tag{18}$$

Equations (17), (18) give the general solution. The initial conditions $x(0) = 0$, $y(0) = 0$ applied to (17), (18) give

$$x(0) = 0 = 20 + \frac{2}{3}c_1 - c_2,$$

$$y(0) = 0 = c_1 + c_2,$$

and thus $c_1 = -12$, $c_2 = 12$. The solution of (14) is then

Figure 6.4.6
Graph of (19).

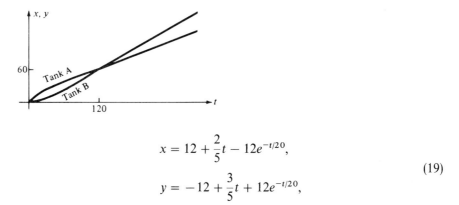

$$x = 12 + \frac{2}{5}t - 12e^{-t/20},$$
$$y = -12 + \frac{3}{5}t + 12e^{-t/20},$$
(19)

which is graphed in Fig. 6.4.6.

Both x and y have asymptotes and for large t, $x \approx 12 + \frac{2}{5}t$, $y \approx -12 + \frac{3}{5}t$, $y/x \approx \frac{3}{2}$. Thus while both tanks get steadily more "salty," tank B will approach being 1.5 times as "salty" as tank A. ∎

Comment While we have talked only of water tanks and salt, the modeling process used in our mixing problems is frequently used whenever the following situation occurs:

- There are several quantities of interest.
- The rate at which one quantity is changed into another quantity is proportional to a linear combination of the quantities.

In our examples, the quantities were the amount of salt in each tank. However, there are numerous other possibilities. Some of the more widely used ones are

i. A region is divided into several areas. The quantities of interest are the populations in each area. The populations change by birth, death, and migration. Birth and death rates are proportional to the number present. Migration rates may be proportional to the population of an area or to population differences.

ii. A species of animal is divided into several age groups. The quantities of interest are the number in each age group. Individuals die, or change age groups, at rates proportional to the number in the group. They are born at rates proportional to the numbers in the fertile age groups.

iii. An economy is divided into different sectors. The quantities of interest are the amount of goods of each sector. Goods from some sectors are either consumed, use goods from other sectors, or produce goods for other sectors, at rates that are linear combinations of the amount of goods in each sector.

iv. A lake is divided into regions on the basis of stable circulation patterns. The quantities of interest are the amount of a pollutant in the different regions.

Exercises

1. There are two jars initially full of pure water. Jar A is 0.5 liter and jar B is 0.25 liter. Water containing salt at a concentration of 100 g/liter is pumped into jar A at 0.5 liter/hr. Water flows from jar A to jar B at 0.5 liter/hr. Water flows out of jar B and down a drain at 0.5 liter/hr. Find the amounts of salt in each jar as a function of time t. Graph your solutions. (Figure 6.4.1 is applicable.)

2. There are two laboratory beakers. Beaker A contains 0.15 liter of pure water and beaker B contains 20 g of salt dissolved in 0.1 liter of water. Water containing salt at a concentration of 100 g/liter flows into beaker A at a rate of 0.3 liter/hr. Water flows from beaker A to beaker B at 0.3 liter/hr. Water flows from beaker B at 0.3 liter/hr and goes down the drain. Find the amount of salt in each beaker as a function of time, and graph your solution.

3. Water containing salt at a concentration of 1 lb/gal flows into a 2-gallon tank at 0.2 gal/min. Water flows out of the first tank and into a second 3-gallon tank at 0.2 gal/min. Both tanks are initially full of pure water. Pure water from a tap flows directly into the second tank at 0.1 gal/min. Water is piped out of the second tank and down a drain at 0.3 gal/min. Find the amount of salt in each tank as a function of time and graph your solution.

4. A tank initially contains 2 liters of water and 5 g of salt. Water containing salt at a concentration of 5 g/liter flows into this tank at 2 liters/hr. Water flows from this tank into a second tank at 2 liters/hr. The second tank contains one liter of fluid and initially contains 10 g of salt. Water evaporates from the second tank at 1 liter/hr and flows from the second tank and down a drain at 1 liter/hr. Find the amount of salt in each tank as a function of time, and graph the solutions.

5. There are two tanks. Tank A is a 100-gallon tank initially full of water containing salt at a concentration of 0.5 lb/gal. Tank B is a 200-gallon tank, initially full of water containing salt at a concentration of 0.1 lb/gal. Starting at time $t = 0$, water is pumped from tank A to tank B at 2 gal/min and from tank B to tank A at 2 gal/min. Find the amount of salt in each tank as a function of time, and graph the solutions.

6. There are two 100-gallon tanks full of water. Tank A contains salt at a concentration of 0.4 lb/gal, while tank B contains pure water. Pure water flows into tank A from an outside source at 2 gal/min. Water is pumped from tank A to tank B at 3 gal/min. Water evaporates from tank B at a rate of 2 gal/min. Water is also pumped from tank B to tank A at 1 gal/min. Find the amount of salt in each tank as a function of time, and graph your solution.

7. Two 2-liter jars are initially full of pure water. Water containing salt at a concentration of 10 g/liter is pumped into jar A at a rate of 2 liters/hr and into jar B at a rate of 2 liters/hr. Water is piped from jar A to jar B at 3 liters/hr and from jar B to jar A at 1 liter/hr. In addition, water flows from jar B down the drain at 4 liters/hr. Find the amount of salt in the jars as a function of time, and graph the solutions.

8. There are two tanks. Tank A contains 100 gallons of pure water while tank B contains 10 lb of salt dissolved in 200 gallons of water. Pure water enters tank A at 5 gal/min. Water is pumped from tank A to tank B at 8 gal/min and from tank B to tank A

at 3 gal/min. In addition, 5 gal/min of water is pumped out of tank B and sent out of the system. Find the amount of salt in each tank as a function of time, and graph the amounts.

9. Two large tanks initially each contain 100 gal of pure water. Water containing salt at a concentration of 0.5 lb/gal is pumped into tank A at 14 gal/min. Water is pumped from tank A to tank B at 9 gal/min. Water is pumped from tank B and sent down a drain at 6 gal/min. Find a differential equation for the amount of salt in each tank that is valid as long as the tanks are not full. You do not need to solve the differential equation. (*Note:* The volumes are not constant.)

10. Each of two large tanks in the desert initially contains 500 gallons of water with a salt concentration of 0.01 lb/gal. Water containing salt at a concentration of 0.2 lb/gal is pumped into tank A at a rate of 10 gal/hr. Water evaporates from tank A at a rate of 7 gal/hr and is pumped from tank A into tank B at 5 gal/hr. Water evaporates from tank B at a rate of 8 gal/hr. Water is also pumped out of tank B at 3 gal/hr. Write the differential equation for the amount of salt in each tank, which is valid until one of the tanks goes dry. You do not need to solve the differential equation.

11. There are three tanks each initially containing 100 gallons of pure water. Water containing salt at a concentration of 2 lb/gal flows into tank A at a rate of 5 gal/min. Water is pumped from tank A to tank B at 2 gal/min and from tank A to tank C at 3 gal/min. Water is pumped from tank C to tank B at 3 gal/min. Water is dumped out of tank B and then down a drain at a rate of 5 gal/min. Derive the differential equation for the amount of salt in each tank as a function of time. You need not actually solve the differential equation.

12. Three shallow ponds are out in the sun. Each is initially full of 5000 gallons of pure water. Water containing salt at a concentration of 0.2 lb/gal flows into pond A at a rate of 10 gal/hr. Water evaporates from pond A at 2 gal/hr. Water flows from pond A to pond B at 8 gal/hr. Water evaporates from pond B at 3 gal/hr. Water flows from pond B to pond C at 5 gal/hr. In pond C, water evaporates at 4 gal/hr and flows out at 1 gal/hr. Derive the differential equation for the amount of salt in each pond. You do not need to solve the differential equation.

13. Two tanks contain V_1 and V_2 gallons of water, respectively. Water containing salt at a concentration of δ lb/gal flows into the first tank at α gal/min, while β gal/min $(0 < \beta \leq \alpha)$ of water is pumped from the first tank into the second. Finally, γ gal/min of water is pumped out of the second tank $(0 \leq \gamma \leq \beta)$. Assume that evaporation rates for the tanks are such that V_1, V_2 are constant.

   i) Set up the differential equation for the amount of salt in each tank.

   ii) Using elimination, find a differential equation for the amount of salt in the second tank.

   iii) Find the roots of the characteristic equation from part (ii).

   iv) Determine for what values of the parameters there will be equilibrium solutions and determine the equilibrium solution if there is one.

(*Population Models*) Exercises 14 through 19 deal with the following situation. A species is divided into m groups which we take to be age groups. Suppose that w is the size of a group at time t. For that group we assume

   i) There is a loss due to death which is proportional to group size, $-\delta w$, with $\delta > 0$. (This could also represent harvesting for species such as trees.)

   ii) Individuals "graduate" from one group to the next at a rate proportional to group size, $-gw$, with $g > 0$.

   iii) Fertile groups give rise to offspring at a rate proportional to group size, βw, with $\beta > 0$.

14. Suppose that a population consists of two groups: adults and children. Let x be the number of children and y the number of adults. Assume that children cannot have offspring. Explain why the model

$$\frac{dx}{dt} = -(\delta_1 + g)x + \beta y, \quad x(t_0) = x_0 \geq 0$$
$$\frac{dy}{dt} = -\delta_2 y + gx, \quad y(t_0) = y_0 \geq 0 \qquad (20)$$

with δ_1, δ_2, g, β positive constants might be reasonable given assumptions (i) through (iii).

Exercises 15 through 17 illustrate the types of behavior possible for (20).

15. Suppose that (20) holds and $\delta_1 = g = 1$, $\delta_2 = 2$, and $\beta = 1$. Show that $\lim_{t \to \infty} x(t) = 0$, $\lim_{t \to \infty} y(t) = 0$, so that the species dies out.

16. Suppose that (20) holds and $\delta_1 = g = 1$, $\delta_2 = 2$, and $\beta = 4$.
  a) Show that (20) has nonzero equilibrium solutions $(\bar{x}, \bar{y})$.
  b) Show that every solution $x(t), y(t)$ converges to one of these equilibrium solutions.

17. Suppose that (20) holds and $\delta_1 = g = 1$, $\delta_2 = 2$, and $\beta = 9$. Show that for initial conditions $x(0) > 0$, $y(0) > 0$ we have $\lim_{t \to \infty} x(t) = \infty$, $\lim_{t \to \infty} y(t) = \infty$, and $\lim_{t \to \infty} \frac{x(t)}{y(t)}$ exists and is finite.

18. Suppose that (20) holds. Show that
  a) If $\delta_2(\delta_1 + g) - \beta g > 0$, then $\lim_{t \to \infty} x(t) = \lim_{t \to \infty} y(t) = 0$.
  b) If $\delta_2(\delta_1 + g) - \beta g = 0$, then there are nonzero equilibriums and every solution of (20) converges to one of these equilibriums as $t \to \infty$.
  c) If $\delta_2(\delta_1 + g) - \beta g < 0$, then $\lim_{t \to \infty} x(t) = \lim_{t \to \infty} y(t) = \infty$ for initial conditions $x(0) > 0$, $y(0) > 0$.
  d) Give a biological interpretation of parts (a) through (c) in terms of the effect of the birth rate constant β.

19. Suppose that the population consists of three stages which we shall call larva (x), pupa (y), and adult (z). Only adults can produce larva.
  a) Explain why a reasonable model might be
  $$\begin{aligned} x' &= -(\delta_1 + g_1)x + \beta z \\ y' &= -(\delta_2 + g_2)y + g_1 x \quad (21) \\ z' &= -\delta_3 z + g_2 y \end{aligned}$$
  with all constants positive and the initial conditions nonnegative.
  b) Discuss what other assumptions would probably need to be made for (21) to be an accurate model.

20. (*Interest*) Two investment accounts are set up with $1000 initially in account A and $2000 initially in account B. Account A, the long-term account, earns 10% a year compounded daily. Account B earns 5% a year compounded daily. Deposits are made into B at the rate of $10 a day. Every day the bank transfers money from B to A at an (annual) rate of 20% of the difference between B and $2000. Set up the differential equations that model this situation. (Interest is first discussed in Exercise 2.9.15.)

6.5 Mechanical Systems

In this section we will discuss how some mechanical systems can be modeled by linear systems of differential equations. This is a continuation of Sections 2.13, 3.3, and 3.16. In order to avoid nonlinear problems, we shall consider point masses, connected by springs in a linear array, undergoing small oscillations. Larger three-dimensional arrays can be used to model many physical structures and mechanical devices. We consider two configurations.

A Horizontal Array of Springs and Masses

Suppose we have two point masses of mass m_1, m_2 and three springs of lengths L_1, L_2, L_3 and spring constants k_1, k_2, k_3. The springs and masses are arranged as in Fig. 6.5.1.

The left end of spring 1 and the right end of spring 3 are attached to immovable surfaces. The masses are in contact with a surface whose coefficient

Figure 6.5.1
Spring–mass system at rest.

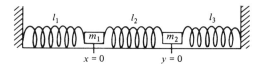

of friction (damping constant) is δ. We assume that the mass of the springs is negligible, friction does not affect the springs, Hooke's law is applicable to the springs, motion occurs only horizontally, and resistance is proportional to the velocity. As noted in the earlier sections, this will give us linear equations, and assumes small velocities, small displacements, and small masses.

It is not difficult to show that the configuration of Fig. 6.5.1 can be at equilibrium (at rest). Let l_1, l_2, l_3 be the lengths of the springs when the system is at rest (equilibrium). Let

$x =$ Distance of mass m_1 from its rest position,

$y =$ Distance of mass m_2 from its rest position.

A positive value indicates displacement to the right, a negative value is displacement to the left. Also let the difference between a spring's length at rest and its length be ΔL, so that

$$\Delta L_1 = l_1 - L_1,$$
$$\Delta L_2 = l_2 - L_2,$$
$$\Delta L_3 = l_3 - L_3.$$

We now use Newton's law to describe the motion of the masses m_1, m_2. Let F_{T_i} be the total force on mass i. Then

$$F_{T_1} = \begin{bmatrix} \text{Force from} \\ \text{Spring 1} \end{bmatrix} + \begin{bmatrix} \text{Force from} \\ \text{Spring 2} \end{bmatrix} + \begin{bmatrix} \text{Resistance} \\ \text{on } m_1 \end{bmatrix}, \qquad (1)$$

and

$$F_{T_2} = \begin{bmatrix} \text{Force from} \\ \text{Spring 2} \end{bmatrix} + \begin{bmatrix} \text{Force from} \\ \text{Spring 3} \end{bmatrix} + \begin{bmatrix} \text{Resistance} \\ \text{on } m_2 \end{bmatrix}. \qquad (2)$$

The lengths of the three springs are:

Spring 1: $L_1 + \Delta L_1 + x = l_1 + x$,
Spring 2: $L_2 + \Delta L_2 + y - x = l_2 + y - x$,
Spring 3: $L_3 + \Delta L_3 - y = l_3 - y$.

Thus $\Delta L_1 + x$, $\Delta L_2 + y - x$, $\Delta L_3 - y$ measure the amount of extension or compression of each spring. Keeping in mind that Spring 2 exerts an equal but opposite force on both m_1 and m_2, and that the force exerted by a spring is proportional to the amount by which its length differs from the rest length, we get that (1) and (2) are

$$(m_1 x')' = -k_1(\Delta L_1 + x) + k_2(\Delta L_2 + y - x) - \delta x' \qquad (3a)$$

and

$$(m_2 y')' = -k_2(\Delta L_2 + y - x) + k_3(\Delta L_3 - y) - \delta y'. \qquad (3b)$$

Since l_1, l_2, l_3 are the spring lengths at rest, the spring forces must balance at these lengths. Thus,

$$k_1 \Delta L_1 = k_2 \Delta L_2 = k_3 \Delta L_3,$$

and (3) simplifies to the system of two homogeneous, second-order linear differential equations,

$$m_1 x'' + \delta x' + (k_1 + k_2)x - k_2 y = 0, \qquad (4a)$$
$$m_2 y'' + \delta y' + (k_2 + k_3)y - k_2 x = 0. \qquad (4b)$$

If outside forces $f_1(t)$, $f_2(t)$ act horizontally on the masses m_1, m_2, we get the nonhomogeneous system

$$m_1 x'' + \delta x' + (k_1 + k_2)x - k_2 y = f_1, \qquad (5a)$$
$$m_2 y'' + \delta y' + (k_2 + k_3)y - k_2 x = f_2. \qquad (5b)$$

It is possible that there are different damping coefficients for m_1 and m_2. This could be caused, for example, by differences in the surfaces on which m_1, m_2 sit, or by properties of a mechanical device being modeled by this spring–mass system. In this event, $\delta x'$ is replaced by $\delta_1 x'$ in (5a) and $\delta y'$ is replaced by $\delta_2 y'$ in (5b). In our examples we assume that $\delta_1 = \delta_2$.

In order to determine the number of arbitrary constants in the solution of (4), we rewrite (4) in operator notation, as

$$[m_1 D^2 + \delta D + k_1 + k_2]x - k_2 y = 0, \qquad (6a)$$
$$-k_2 x + [m_2 D^2 + \delta D + k_2 + k_3]y = 0, \qquad (6b)$$

and take the determinant of the coefficients to get

$$[m_1 D^2 + \delta D + k_1 + k_2][m_2 D^2 + \delta D + k_2 + k_3] - k_2^2, \qquad (7)$$

which is a fourth-order polynomial in D. Thus, by Section 6.1, the system (4) has four arbitrary constants in its general solution. In fact, one can arbitrarily specify $x(0)$, $x'(0)$, $y(0)$, $y'(0)$ (although large values may not make physical sense, since the differential equation (4) is then no longer appropriate).

Example 6.5.1

Suppose that the spring–mass system of Fig. 6.5.1 has two 1-g masses and the spring constants are all 1 g/sec². Find the resulting motion if the first mass is initially at its equilibrium position with a velocity of 16 cm/sec to the right, while the second mass is initially at rest at its equilibrium position. Friction is assumed neligible.

Solution

We have $m_1 = m_2 = 1$, $k_1 = k_2 = k_3 = 1$, and $\delta = 0$. The system (4) is thus:

$$x'' + 2x - y = 0, \qquad x(0) = 0, \; x'(0) = 16, \qquad (8a)$$
$$y'' + 2y - x = 0, \qquad y(0) = y'(0) = 0. \qquad (8b)$$

We shall solve (8) by elimination. Rewriting the system (8) in operator notation, we obtain:

$$(D^2 + 2)x - y = 0, \qquad (9a)$$
$$-x + (D^2 + 2)y = 0. \qquad (9b)$$

Figure 6.5.2
Graph of (12b).

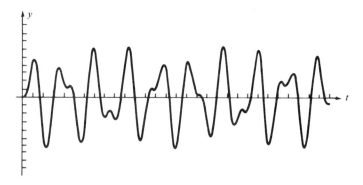

Add $(D^2 + 2)$ times the second equation (9b) to the first equation (9a) to give

$$[(D^2 + 2)(D^2 + 2) - 1]y = 0$$

or

$$(D^2 + 3)(D^2 + 1)y = 0. \tag{10}$$

Then the characteristic polynomial of this differential equation is $p(r) = (r^2 + 3)(r^2 + 1)$, which has roots $r = \pm\sqrt{3}i, \pm i$. Thus

$$y = c_1 \cos t + c_2 \sin t + c_3 \cos \sqrt{3}t + c_4 \sin \sqrt{3}t. \tag{11a}$$

Since $x = y'' + 2y$ from (9b) we use (11a) to give

$$x = c_1 \cos t + c_2 \sin t - c_3 \cos \sqrt{3}t - c_4 \sin \sqrt{3}t. \tag{11b}$$

The initial conditions of (8) applied to the general solution (11) yields

$$0 = x(0) = c_1 - c_3,$$
$$16 = x'(0) = c_2 - c_4\sqrt{3},$$
$$0 = y(0) = c_1 + c_3,$$
$$0 = y'(0) = c_2 + c_4\sqrt{3}.$$

Thus $c_1 = c_3 = 0$, and $c_2 = 8$, $c_4 = -8/\sqrt{3}$. The solution of the differential equation (8) is

$$x = 8 \sin t + \frac{8}{\sqrt{3}} \sin \sqrt{3}t, \tag{12a}$$

$$y = 8 \sin t - \frac{8}{\sqrt{3}} \sin \sqrt{3}t. \tag{12b}$$

Figure 6.5.2 shows a graph of y. ∎

This example is typical in the following respects. If there is no friction, ($\delta = 0$), then the polynomial (7) always has two distinct pairs of purely imaginary conjugate roots. The free response will then consist of the superposition of

Figure 6.5.3

two harmonic motions of different frequencies (periods). If the ratio of the frequencies is not a rational number, the free response will not be periodic except for very special initial conditions (see Exercises 12 through 14 at the end of the section). The functions (12) are not periodic.

Physical Implications Recall from Section 3.16 that resonance can occur if the forcing term has the same frequency as the free response. For Example 6.5.1, we would expect resonance with forcing terms of frequencies $\dfrac{1}{2\pi}$ or $\dfrac{\sqrt{3}}{2\pi}$ cycles/sec. Suppose now we had 50 masses arranged linearly, as in Fig. 6.5.3. If friction were negligible, we might expect (and in fact it's approximately true) that the free response would contain harmonic functions of 50 different frequencies. Thus, a forcing function containing terms at *any* of these 50 frequencies would cause resonance (excite the unstable modes). Now, imagine a large, highly flexible structure to be made of lightweight materials (for example, a solar collector or large antenna in space). The previous discussion suggests that such a structure might have a large number of frequencies at which it would exhibit resonance. The problem, then, is to build the structure so that these frequencies are not within the range of frequencies of the disturbances expected to act on the structure. Alternatively, we would need to design devices to counteract the disturbances at certain frequencies (active or passive controllers).

Example 6.5.2

Again we consider a spring–mass system arranged as in Fig. 6.5.1 with $m_1 = m_2 = 1$, $k_1 = k_2 = k_3 = 1$, but now we consider friction with a damping coefficient of $\delta = 2$ g/sec. Mass m_1 is displaced 1 cm to the left and released with a velocity of 11 cm/sec to the right, while at the same instant m_2 is displaced 1 cm to the right and released with a velocity of 9 cm/sec to the right.

Solution

The system of differential equations (4) is

$$x'' + 2x' + 2x - y = 0, \quad x(0) = -1,\ x'(0) = 11, \qquad (13a)$$
$$-x + y'' + 2y' + 2y = 0, \quad y(0) = 1,\ y'(0) = 9, \qquad (13b)$$

or, in operator notation,

$$(D^2 + 2D + 2)x - y = 0, \qquad (14a)$$
$$-x + (D^2 + 2D + 2)y = 0. \qquad (14b)$$

Using elimination to solve (14), we add $D^2 + 2D + 2$ times (14b) to (14a) and get

$$[(D^2 + 2D + 2)(D^2 + 2D + 2) - 1]y = 0. \qquad (15)$$

Now, in general, we have a fourth-degree polynomial, all of whose roots may be imaginary. The roots will usually be found numerically on a computer, although formulas exist. However, because of the choice of coefficients, we may solve the characteristic equation

$$(r^2 + 2r + 2)(r^2 + 2r + 2) - 1 = 0 \tag{16}$$

as follows. Equation (16) is

$$(r^2 + 2r + 2)^2 = 1.$$

Thus, taking the square root of both sides yields

$$r^2 + 2r + 2 = 1 \quad \text{or} \quad r^2 + 2r + 2 = -1,$$

that is,

$$r^2 + 2r + 1 = 0 \quad \text{or} \quad r^2 + 2r + 3 = 0.$$

Solving these quadratics gives the roots of (16) as

$$r = -1, -1 \quad \text{or} \quad r = -1 \pm i\sqrt{2}.$$

We have then that

$$y = c_1 e^{-t} + c_2 t e^{-t} + c_3 e^{-t} \cos\sqrt{2}t + c_4 e^{-t} \sin\sqrt{2}t. \tag{17a}$$

We may find x from (14b) as $x = y'' + 2y' + 2y$, so that (17a) gives (note shortcut below):

$$x = c_1 e^{-t} + c_2 t e^{-t} - c_3 e^{-t} \cos\sqrt{2}t - c_4 e^{-t} \sin\sqrt{2}t. \tag{17b}$$

Using the initial conditions from (13),

$$-1 = x(0) = c_1 - c_3,$$
$$11 = x'(0) = -c_1 + c_2 + c_3 - \sqrt{2}c_4,$$
$$1 = y(0) = c_1 + c_3,$$
$$9 = y'(0) = -c_1 + c_2 - c_3 + \sqrt{2}c_4,$$

we find that $c_1 = 0$, $c_3 = 1$, $c_2 = 10$, $c_4 = 0$, and the solution of (13) is

$$x = 10te^{-t} - e^{-t}\cos\sqrt{2}t,$$
$$y = 10te^{-t} + e^{-t}\cos\sqrt{2}t.$$

In this example we had two real roots and a complex conjugate pair. By changing δ, one may get several other types of solutions, ranging from four complex numbers to four distinct real numbers (see Exercise 9). ∎

A Shortcut

It is possible to sometimes reduce the amount of work in finding the second half of the solution when doing elimination. To see how this works, return to Example 6.5.2 and the derivation of y in (16) and (17). Since -1 is a repeated root of $r^2 + 2r + 1$, we know that (note Exercises 3.6.22 through 3.6.25)

$$(D^2 + 2D + 1)e^{-t} = 0, \quad (D^2 + 2D + 1)te^{-t} = 0. \tag{18}$$

Similarly, since $-1 \pm i\sqrt{2}$ are the roots of $r^2 + 2r + 3$, we have
$$(D^2 + 2D + 3)e^{-t}\cos\sqrt{2}t = 0, \qquad (D^2 + 2D + 3)e^{-t}\sin\sqrt{2}t = 0. \quad (19)$$
Now, to find x from (14b) and (17a), we have, from (14b), that $x = (D^2 + 2D + 2)y$. Now substitute in y, from (17a):
$$\begin{aligned}x = {} & c_1(D^2 + 2D + 2)e^{-t} + c_2(D^2 + 2D + 2)te^{-t} \\ & + c_3(D^2 + 2D + 2)e^{-t}\cos\sqrt{2}t \\ & + c_4(D^2 + 2D + 2)e^{-t}\sin\sqrt{2}t.\end{aligned} \quad (20)$$
But $D^2 + 2D + 2 = [D^2 + 2D + 1] + 1$ and $D^2 + 2D + 2 = [D^2 + 2D + 3] - 1$, so that, using (18) and (19) gives (20) as
$$\begin{aligned}x = {} & c_1(0 + 1e^{-t}) + c_2(0 + 1te^{-t}) + c_3(0 - 1e^{-t}\cos\sqrt{2}t) \\ & + c_4(0 - 1e^{-t}\sin\sqrt{2}t),\end{aligned}$$
which is (17b). If the solution involves two or four complex roots, this method can be much quicker than actually differentiating the expression for y twice and then combining terms. When real roots are involved,
$$p(D)e^{\alpha t} = p(\alpha)e^{\alpha t},$$
for any polynomial $p(\lambda)$. For example,
$$(D^3 - 3D^2 + D - 6)e^{5t} = (5^3 - 3\cdot 5^2 + 5 - 6)e^{5t} = 49\,e^{5t}.$$

A Vertical Array of Springs and Masses

Figure 6.5.4

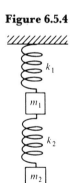

Now consider two masses hanging on springs in the configuration of Fig. 6.5.4. Again, let L_1, L_2 be the lengths of the springs, k_1, k_2 the spring constants, and l_1, l_2 the lengths of the springs at rest with the masses attached. Set $\Delta L_i = l_i - L_i$ for $i = 1, 2$. The resistive force acting on each mass will have damping coefficient δ_i. We may have $\delta_1 \neq \delta_2$ if the masses move in different media or if the spring–mass system is a model of a different physical structure and the δ_i represent internal damping. Let x, y measure the displacement of masses m_1, m_2 from equilibrium with the downward direction as positive. Then, from Newton's laws, ignoring external forces other than gravity,

$$(m_1 x')' = \begin{bmatrix}\text{Force from}\\ \text{Spring 1}\end{bmatrix} + \begin{bmatrix}\text{Force from}\\ \text{Spring 2}\end{bmatrix} + \begin{bmatrix}\text{Gravity on}\\ \text{Mass 1}\end{bmatrix} + \begin{bmatrix}\text{Friction on}\\ \text{Mass 1}\end{bmatrix},$$

$$(m_2 y')' = \begin{bmatrix}\text{Force from}\\ \text{Spring 2}\end{bmatrix} + \begin{bmatrix}\text{Gravity}\\ \text{On Mass 2}\end{bmatrix} + \begin{bmatrix}\text{Friction}\\ \text{On Mass 2}\end{bmatrix},$$

or
$$\begin{aligned}m_1 x'' &= -k_1(\Delta L_1 + x) + k_2(\Delta L_2 + y - x) + m_1 g - \delta_1 x',\\ m_2 y'' &= -k_2(\Delta L_2 + y - x) + m_2 g - \delta_2 y'.\end{aligned}$$

At equilibrium $k_1 \Delta L_1 = k_2 \Delta L_2 + m_1 g$ and $k_2 \Delta L_2 = m_2 g$, so that our system of differential equations is

$$m_1 x'' + \delta_1 x' + (k_1 + k_2)x - k_2 y = 0,$$
$$m_2 y'' + \delta_2 y' + k_2 y - k_2 x = 0. \tag{21}$$

Since (21) is so similar to (4), we shall not consider the configuration of Fig. 6.5.4 in the exercises.

Exercises

In Exercises 1 through 7, we have a spring–mass system arranged as in Fig. 6.5.1. Set up the differential equation that describes the dynamics, and solve the resulting differential equation. If friction is absent, tell for which frequencies a forcing term $\sin \alpha t$ would cause resonance. If you have access to a computer, you may wish to graph some of the solutions.

1. Two one-g masses are arranged in a spring–mass system, as in Fig. 6.5.1. The spring constants are $k_1 = 1$, $k_2 = 4$, and $k_3 = 1$ g/sec^2. Resistance is negligible. Initially, both masses are at their equilibrium positions and the first mass is at rest while the second mass is moving 1 cm/sec to the right.

2. Two two-g masses are arranged in a spring–mass system, as in Fig. 6.5.1. The spring constants are $k_1 = 4$, $k_2 = 6$, $k_3 = 4$ g/sec^2. Resistance is negligible. At time $t = 0$, both masses are released. Mass one is one cm to the left of its rest position; mass two is one cm to the right of its rest position.

3. Two two-g masses are arranged in a spring–mass system as in Fig. 6.5.1. The spring constants are $k_1 = 6$, $k_2 = 1$, $k_3 = 6$ g/sec^2. The same resistance acts on both masses. The damping constant is 8 g/sec. At time $t = 0$, both masses are at their rest positions. Mass one is initially moving at 1 cm/sec to the right; mass two is initially moving at 1 cm/sec to the left.

4. Two one-g masses are arranged in a spring–mass system as in Fig. 6.5.1. The spring constants are $k_1 = 2$, $k_2 = 1$, $k_3 = 2$ g/sec^2. The same resistance acts on both masses. The damping constant is 2 g/sec. Initially, both masses are at their rest positions but moving to the right at 3 cm/sec and 1 cm/sec, respectively.

5. Two one-g masses are arranged in a spring–mass system, as in Fig. 6.5.1. The spring constants are $k_1 = 4$, $k_2 = 6$, $k_3 = 4$ g/sec^2. Resistance is negligible. Initially, mass one is at its rest position and moving 2 cm/sec to the left while mass two is initially 2 cm to the right of its rest position and moving 6 cm/sec to the right.

6. Two one-g masses are arranged in a spring–mass system as in Fig. 6.5.1. The spring constants are $k_1 = 1$, $k_2 = 12$, $k_3 = 1$ g/sec^2. Resistance is negligible. Initially, the first mass is at its rest (equilibrium) position and moving to the right at 5 cm/sec. The second mass is initially 2 cm to the right of its rest position and moving 5 cm/sec to the left.

7. Two two-g masses are arranged in a spring–mass system as in Fig. 6.5.1. The spring constants are $k_1 = 10$, $k_2 = 3$, $k_3 = 10$ g/sec^2. The damping coefficient is $\delta = 12$ g/sec. Initially the first mass is one cm to the left of its rest position and moving to the right at 4 cm/sec. The second mass is initially 5 cm to the right of its rest position and moving to the left at 16 cm/sec.

8. Show that, if $\delta = 0$, then the characteristic polynomial for (7) always has two distinct pairs of purely imaginary conjugate complex roots.

9. Two one-g masses are arranged in a spring–mass system as in Fig. 6.5.1. The spring constants are $k_1 = k_3 = k$ and $k_2 = l$. The damping coefficient is δ. Thinking of k, l as unknown but fixed, we examine the effect of increasing the resistance.

Figure 6.5.5

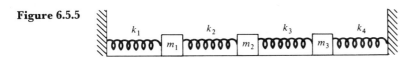

Figure 6.5.6

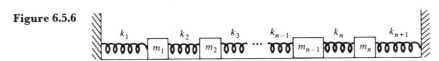

a) For what values of δ are all roots of the characteristic equation for (7) complex?

b) For what values of δ will there be two complex and two real roots for the characteristic equation of (7)?

c) For what values of δ will there be four distinct real roots for the characteristic equation of (7)?

10. Derive the system of differential equations that models the spring–mass system in Fig. 6.5.5 under the physical assumptions of this section. Include friction.

11. Derive the system of differential equations that models the spring–mass system with n masses and $(n + 1)$ springs in Fig. 6.5.6. Include friction.

12. Show that, if $A \neq 0$, $B \neq 0$ and $\alpha \neq 0$, $\beta \neq 0$, then
$$f(t) = A \sin \alpha t + B \sin \beta t$$
is periodic if and only if α/β is a rational number. (*Hint*: Use linear combinations of $f(t)$, $f''(t)$ to argue that $\sin \alpha t$, $\sin \beta t$ both have period τ if f has period τ.)

13. (Continuation of Exercise 12.) Show that if $\beta/\alpha = m/n$, where m, n are integers, then $A \sin \alpha t + B \sin \beta t$ has period $\tau = 2\pi n/\alpha = 2\pi m/\beta$.

14. (Continuation of Exercise 12.) Show that if $\beta/\alpha = m/n$, where m, n are integers, then $\tau = 2\pi n/\alpha$ is the smallest period of $A \sin \alpha t + B \sin \beta t$ with $A \neq 0$, $B \neq 0$, if and only if the rational number m/n is in reduced form (m, n have no common integer factors except 1 and hence are *relatively prime*).

Exercises 15 through 17 require a knowledge of matrix multiplication and differentiation. They may be assigned after Section 6.9 if the reader has not seen these concepts before.

15. Write the system (4) in the form
$$\mathbf{M}\mathbf{x}'' + \delta \mathbf{x}' + \mathbf{K}\mathbf{x} = \mathbf{0} \qquad (22)$$
where $\mathbf{M}$, $\mathbf{K}$ are 2×2 matrices and $\mathbf{x} = \begin{bmatrix} x \\ y \end{bmatrix}$.
Observe that (22) looks like the scalar equation $mx'' + \delta x' + kx = 0$ from Section 3.16 for a single mass and spring.

16. A 2×2 matrix is called *symmetric* if it can be written as
$$\begin{bmatrix} \alpha & \beta \\ \beta & \gamma \end{bmatrix}$$
for some scalars α, β, γ. Symmetric matrices have several useful properties and play an important role in many applications. Show that the $\mathbf{M}$, $\mathbf{K}$ from Exercise 15 are symmetric matrices.

17. A 3×3 matrix is called *symmetric* if its entries have the pattern
$$\begin{bmatrix} a & b & c \\ b & d & e \\ c & e & f \end{bmatrix}$$
for scalars a, b, c, d, e, f. Write the system of differential equations from Exercise 10 in the form (22), where $\mathbf{M}$, $\mathbf{K}$ are 3×3 matrices. Verify that $\mathbf{M}$, $\mathbf{K}$ are symmetric matrices.

18. Verify that, if $p(\lambda)$ is a polynomial, then
$$p(D)e^{\alpha t} = p(\alpha)e^{\alpha t}.$$

6.6 Multiloop Circuits

In Section 2.12 and 3.17 we discussed simple circuits. However, most circuits have more than one loop and may often be modeled by a system of differential equations. Unfortunately, the application of Kirchhoff's voltage and current laws usually leads to a system with too many variables. Algorithms based on graph theory exist for finding a reduced set of variables. However, we shall consider only two-loop circuits of the general form shown in (1).

$$\text{[circuit diagram with branches } A, B, C \text{ and currents } i_1, i_2, i_3\text{]} \tag{1}$$

This circuit has three *branches* denoted A, B, C with currents i_1, i_2, and i_3, respectively. On each branch we will allow some combination of linear resistors, inductors, capacitors, and independent voltage sources in series. Units are ohms, amps, coulombs, volts, etc., as discussed in Section 2.12.

The dynamics of (1) will be determined by three equations:

The voltage law applied to loop AB; (2)

The voltage law applied to loop BC; (3)

The current law at the top node, which is

$$i_1 - i_2 - i_3 = 0. \tag{4}$$

We also assume that at least one branch does not have a capacitor in it. Our procedure is then as follows.

Write down the voltage law for loops AB and BC, using the charge as the dependent variable if there is a capacitor. Otherwise use the current in the branch.

This gives two differential equations in three unknowns. Then,

Use the current law (4) to eliminate a current variable corresponding to a branch without a capacitor.

The result is a system of differential equations solvable by the methods of this chapter.

Example 6.6.1

Using this procedure, write the differential equation for (5).

(5)

Solution

Let i_1, i_2, i_3 be the currents in each branch and q_1, q_2 the charges in the capacitors in branch A and branch B. Then the voltage law applied to loops AB and BC gives

$$i_1 + 3i_2 + \tfrac{1}{4}q_2 + \tfrac{1}{6}q_1 + 5i_1' = e_1 \tag{6a}$$

and

$$2i_3 + 7i_3' - \tfrac{1}{4}q_2 - 3i_2 = e_2, \tag{6b}$$

respectively. Note that, since branch B is traversed in the opposite direction in loop BC, we reverse the signs of the corresponding voltages in (6b). Since there is no capacitor in branch C, use the current law $i_1 - i_2 - i_3 = 0$ to solve for i_3;

$$i_3 = i_1 - i_2,$$

which, upon substitution into (6), gives

$$i_1 + 3i_2 + \tfrac{1}{4}q_2 + \tfrac{1}{6}q_1 + 5i_1' = e_1, \tag{7a}$$
$$2i_1 - 2i_2 + 7i_1' - 7i_2' - \tfrac{1}{4}q_2 - 3i_2 = e_2. \tag{7b}$$

Now use the fact that $i_2 = q_2'$, $i_1 = q_1'$, to finally get

$$[5q_1'' + q_1' + \tfrac{1}{6}q_1] + [3q_2' + \tfrac{1}{4}q_2] = e_1(t),$$
$$[7q_1'' + 2q_1'] + [-7q_2'' - 5q_2' - \tfrac{1}{4}q_2] = e_2(t), \tag{8}$$

or, in the notation of Section 6.2,

$$[5D^2 + D + \tfrac{1}{6}]q_1 + [3D + \tfrac{1}{4}]q_2 = e_1(t),$$
$$[7D^2 + 2D]q_1 - [7D^2 + 5D + \tfrac{1}{4}]q_2 = e_2(t). \; \blacksquare \tag{9}$$

In multiloop circuits with several voltage sources, such as (5), we need to be careful about their polarity (which node is $+$). If the $+$ and $-$ were reversed on the right side of (5), then we would have $-e_2$ instead of e_2 on the right side of (6b).

Note that this procedure always leads to a system of the form

$$L_1 x + L_2 y = e_1(t),$$
$$L_3 x + L_4 y = e_2(t), \tag{10}$$

where L_1, L_2, L_3, L_4 are constant-coefficient linear differential operators of order at most two, and x, y are either a charge or current from two different branches.

Exercises

In Exercises 1 through 4, find the system of differential equations that model the circuit. Note that, if more than one branch is missing a capacitor, then there is more than one possible differential equation. Do not solve the resulting differential equations.

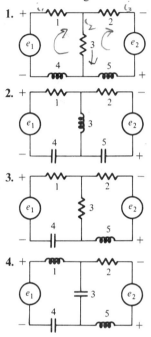

Exercises 5 through 8 refer to the circuit in (11).

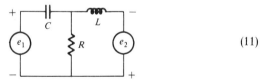

5. Suppose that $e_1 = e_2 = 0$, $L = 2$, $C = 2$, and $R = \frac{1}{2}$. Determine the free response of (11).

6. Take the same values of e_1, e_2, L, and C as in Exercise 5, but raise the resistance R to 1. Determine the free response to (11).

7. Take the same values of e_1, e_2, L, and C as in Exercise 5, but assume $R > \frac{1}{2}$. Show that the free response is made up of terms $e^{\alpha t} \cos \beta t$, $e^{\alpha t} \sin \beta t$, where $\alpha = -(4R)^{-1}$, $\beta = [4R^2 - 1]^{1/2}/(4R)$.

8. For large values of R, the circuit (11) would be expected to act like the circuit (12).

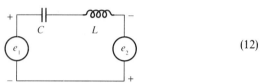

Determine the free response of (12) and compare to the results of Exercises 5, 6, and 7. In particular, show that as $R \to \infty$, the free response of (11) converges to that of (12).

Exercises 9 and 10 refer to the circuit (13). This example shows how the number of initial conditions can vary.

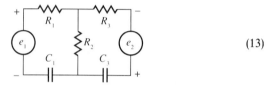

9. Take $e_1 = e_2 = 0$, $R_1 = R_3 = 0$, $R_2 = 1$, and $C_1 = C_2 = 1$ in (13). Derive the differential equation and determine the free response.

10. Take $e_1 = e_2 = 0$ in (13) and let $R_1 = 1$, $R_2 = 1$, $R_3 = 1$, $C_2 = 1$, $C_1 = 1$. Derive the differential equation and determine the free response.

Exercises 11 and 12 refer to the circuit (14).

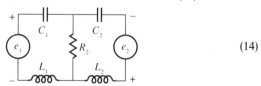

11. Show that the differential equation for (14) may be written as

$$\left(L_1 D^2 + R_2 D + \frac{1}{C_1}\right) q_1 - R_2 D q_2 = e_1,$$

$$-R_2 D q_1 + \left(L_2 D^2 + R_2 D + \frac{1}{C_2}\right) q_2 = e_2,$$

(15)

where q_1, q_2 are the charges on the capacitors.

12. Let $C_1 = C_2 = 1$, $R_2 = 1$, $L_1 = L_2 = 1$, $e_1 = e_2 = 0$, and (i) determine the free response; (ii) observe that a harmonic oscillation is present even though there is a resistor; (iii) determine a set of initial conditions so that $q_1(t) = \sin t$.

6.7 Matrices and Vectors

6.7.1 Introduction

In the preceding sections, we have come to realize that the modeling of many important problems naturally leads to systems of differential equations. Frequently, these systems either are linear or may be approximated by linear equations. In Sections 6.1, 6.2 and 6.3, we solved some fairly simple systems. The methods of those sections do not lend themselves well to many of the problems encountered in practice, where there are many more dependent variables. Neither do those techniques provide us with a sufficient theoretical understanding.

What is needed are some ideas and techniques from matrix theory and linear algebra. The remainder of this chapter is devoted to developing some of these ideas and applying them to systems of differential equations. Most students have probably seen matrices and vectors before. However, our presentation is self-contained and contains only that material which is needed for solving systems of differential equations of the types we shall study. The applications of matrix theory to differential equations begin in Section 6.9.

There is a fundamental pedagogical problem at this point. The ways in which these ideas are most easily presented are not the way they are usually computed in practice. Throughout the remainder of this chapter, we have tried to be as honest as possible. Where the method presented is primarily a textbook technique, we will so state. Unfortunately, to discuss how eigenvalues, eigenvectors, etc., are really found is well beyond the range of this book, and requires a separate course in numerical linear algebra.

6.7.2 Basic Terminology and Matrix Operations

A rectangular array is called a *matrix*. The *entries* of a matrix may be numbers, functions, or even other matrices. Unless stated otherwise, the entries in this chapter will be either *scalars* (real or complex numbers) or scalar-valued functions. If the matrix has m *rows* and n *columns*, it is called an $m \times n$ matrix (pronounced "m by n") and $m \times n$ is referred to as the *size* of the matrix.

Example 6.7.1 For example,

$$\begin{bmatrix} 3 \\ 1 \\ 0 \end{bmatrix} \text{ is } 3 \times 1, \qquad [t, t^2, 1] \text{ is } 1 \times 3,$$

$$\begin{bmatrix} i & -i \\ 2 & 0 \\ 1 & 4 \end{bmatrix} \text{ is } 3 \times 2, \qquad \text{and} \qquad [5] \text{ is } 1 \times 1. \blacksquare$$

An $m \times 1$ matrix is called an m-dimensional (*column*) *vector*, while a $1 \times n$ matrix is called an n-dimensional (*row*) *vector*. The first matrix in Example 1 is a column vector, while the second is a row vector. Unless stated otherwise, *vector* will mean a column vector. Matrices will be denoted by boldface capitals such as **A**, **B**, while row and column vectors will be denoted by boldface lower-case letters such as **u**, **v**, **x**.

The entry (or element) of **A** in the ith row and jth column is called the i,j-entry of **A** and is denoted a_{ij}. Thus, if **A** is 3×4, we would have

$$\mathbf{A} = \begin{bmatrix} a_{11} & a_{12} & a_{13} & a_{14} \\ a_{21} & a_{22} & a_{23} & a_{24} \\ a_{31} & a_{32} & a_{33} & a_{34} \end{bmatrix}.$$

If the number of rows equals the number of columns, **A** is called a *square matrix*. Two matrices **A**, **B** are *equal* if they are the same size and corresponding entries are equal. That is, $a_{ij} = b_{ij}$ for all values of i, j.

In order to add or subtract two matrices, they must be the same size. If **A**, **B** are both $m \times n$ matrices, then $\mathbf{A} + \mathbf{B}$ and $\mathbf{A} - \mathbf{B}$ are also $m \times n$ matrices. The i, j-entry of $\mathbf{A} + \mathbf{B}$ is $a_{ij} + b_{ij}$. That is, they are added entrywise. Similarly, the i, j-entry of $\mathbf{A} - \mathbf{B}$ is $a_{ij} - b_{ij}$.

Example 6.7.2 For example,

$$[1 \quad 2 \quad 0] + [3 \quad 4 \quad \pi] = [4 \quad 6 \quad \pi],$$

and

$$\begin{bmatrix} 3 \\ 6 \end{bmatrix} - \begin{bmatrix} 2 \\ 7 \end{bmatrix} = \begin{bmatrix} 1 \\ -1 \end{bmatrix},$$

while

$$\begin{bmatrix} t & t^2 \\ 1 & 0 \end{bmatrix} + \begin{bmatrix} 1 & -t \\ 0 & t \end{bmatrix} = \begin{bmatrix} t+1 & t^2 - t \\ 1 & t \end{bmatrix}. \blacksquare$$

If α is a number (also called a *scalar*), and **A** is an $m \times n$ matrix, then $\alpha \mathbf{A}$ is also an $m \times n$ matrix. The i, j-entry of $\alpha \mathbf{A}$ is αa_{ij}. For example,

$$3\begin{bmatrix} 1 & 2 \\ t & \pi \end{bmatrix} = \begin{bmatrix} 3 & 6 \\ 3t & 3\pi \end{bmatrix}.$$

If **r** is a $1 \times n$ row vector, and **c** is an $n \times 1$ column vector,

$$\mathbf{r} = [r_1, \ldots, r_n], \qquad \mathbf{c} = \begin{bmatrix} c_1 \\ \vdots \\ c_n \end{bmatrix},$$

then their *product* is the 1×1 matrix

$$\mathbf{rc} = r_1 c_1 + r_2 c_2 + \cdots + r_n c_n = \sum_{i=1}^{n} r_i c_i.$$

For example,

$$\begin{bmatrix} 1 & 2 & 3 \end{bmatrix} \begin{bmatrix} 4 \\ 5 \\ 6 \end{bmatrix} = 1 \cdot 4 + 2 \cdot 5 + 3 \cdot 6 = 32.$$

The expression $\sum_{i=1}^{n} r_i c_i$ is also called a *dot* or *inner product*.

Suppose, now, that **A** is $m \times n$ and **B** is $n \times p$. If $\mathbf{r}_1, \ldots, \mathbf{r}_m$ are the rows of **A** and $\mathbf{b}_1, \ldots, \mathbf{b}_p$ are the columns of **B**, then the *product* of **A** and **B**, denoted **AB**, is an $m \times p$ matrix. Its i,j-entry is $\mathbf{r}_i \mathbf{b}_j$. That is, $\sum_{k=1}^{n} a_{ik} b_{kj}$. Thus,

$$\mathbf{AB} = \begin{bmatrix} \mathbf{r}_1 \\ \mathbf{r}_2 \\ \vdots \\ \mathbf{r}_m \end{bmatrix} [\mathbf{b}_1, \ldots, \mathbf{b}_p] = \begin{bmatrix} \mathbf{r}_1 \mathbf{b}_1 & \mathbf{r}_1 \mathbf{b}_2 & \cdots & \mathbf{r}_1 \mathbf{b}_p \\ \mathbf{r}_2 \mathbf{b}_1 & \mathbf{r}_2 \mathbf{b}_2 & \cdots & \mathbf{r}_2 \mathbf{b}_p \\ \vdots & & & \vdots \\ \mathbf{r}_m \mathbf{b}_1 & \mathbf{r}_m \mathbf{b}_2 & \cdots & \mathbf{r}_m \mathbf{b}_p \end{bmatrix}. \qquad (1)$$

Note the pattern of the sizes

$$(m \times n)(n \times p) = m \times p,$$

and that **A** must have the same number of columns as **B** has rows.

Example 6.7.3 Find **AB** if

$$\mathbf{A} = \begin{bmatrix} 1 & 2 \\ 3 & 0 \\ 0 & 1 \end{bmatrix}, \qquad \mathbf{B} = \begin{bmatrix} 1 & 0 & 1 & 2 \\ -1 & 1 & 3 & 0 \end{bmatrix}.$$

Solution Note that **A** is 3×2 and **B** is 2×4, so that it is possible to form the product and **AB** will be 3×4. Then from (1),

$$\mathbf{AB} = \begin{bmatrix} [1\ 2]\begin{bmatrix}1\\-1\end{bmatrix} & [1\ 2]\begin{bmatrix}0\\1\end{bmatrix} & [1\ 2]\begin{bmatrix}1\\3\end{bmatrix} & [1\ 2]\begin{bmatrix}2\\0\end{bmatrix} \\ [3\ 0]\begin{bmatrix}1\\-1\end{bmatrix} & [3\ 0]\begin{bmatrix}0\\1\end{bmatrix} & [3\ 0]\begin{bmatrix}1\\3\end{bmatrix} & [3\ 0]\begin{bmatrix}2\\0\end{bmatrix} \\ [0\ 1]\begin{bmatrix}1\\-1\end{bmatrix} & [0\ 1]\begin{bmatrix}0\\1\end{bmatrix} & [0\ 1]\begin{bmatrix}1\\3\end{bmatrix} & [0\ 1]\begin{bmatrix}2\\0\end{bmatrix} \end{bmatrix}$$

$$= \begin{bmatrix} -1 & 2 & 7 & 2 \\ 3 & 0 & 3 & 6 \\ -1 & 1 & 3 & 0 \end{bmatrix}. \blacksquare$$

Comment Note that, to multiply two $n \times n$ matrices takes n^3 multiplications and $n^2(n-1)$ additions. For even moderately large n, this can become time-consuming. Many applied problems involve matrices where m, n are in the thousands. For this reason, in actually solving problems, one often tries to avoid computing matrix products whenever possible.

A very important special case of matrix multiplication is when $\mathbf{A}$ is a column vector and $\mathbf{B}$ is a row vector.

Example 6.7.4 Compute $\mathbf{uv}$ if

$$\mathbf{u} = \begin{bmatrix} 1 \\ 0 \\ 3 \end{bmatrix}, \quad \mathbf{v} = [1\ 0\ \pi\ 2].$$

Solution Since $\mathbf{u}$ is 3×1 and $\mathbf{v}$ is 1×4, the product will be 3×4, and it is

$$\mathbf{uv} = \begin{bmatrix} 1\cdot 1 & 1\cdot 0 & 1\cdot \pi & 1\cdot 2 \\ 0\cdot 1 & 0\cdot 0 & 0\cdot \pi & 0\cdot 2 \\ 3\cdot 1 & 3\cdot 0 & 3\cdot \pi & 3\cdot 2 \end{bmatrix} = \begin{bmatrix} 1 & 0 & \pi & 2 \\ 0 & 0 & 0 & 0 \\ 3 & 0 & 3\pi & 6 \end{bmatrix}. \blacksquare$$

It is also helpful to note that, if $\mathbf{A}$ is $m \times n$ and $\mathbf{B}$ is $n \times p$ and $\mathbf{b}_1, \ldots, \mathbf{b}_p$ are the columns of $\mathbf{B}$, then

$$\mathbf{AB} = \mathbf{A}[\mathbf{b}_1, \ldots, \mathbf{b}_p] = [\mathbf{Ab}_1, \mathbf{Ab}_2, \ldots, \mathbf{Ab}_p]. \tag{2}$$

That is,

Multiplying the matrix $\mathbf{A}$ times $\mathbf{B}$ is the same as multiplying each column of $\mathbf{B}$ by $\mathbf{A}$. (3)

Matrix multiplication is associative, $\mathbf{A}(\mathbf{BC}) = (\mathbf{AB})\mathbf{C}$, and distributive, $\mathbf{A}(\mathbf{B} + \mathbf{C}) = \mathbf{AB} + \mathbf{AC}$. However, in working with matrix products, it is important to keep in mind that, even if $\mathbf{AB}$ and $\mathbf{BA}$ are both defined, $\mathbf{AB}$ is usually *not* the same as $\mathbf{BA}$.

Example 6.7.5   Let

$$A = \begin{bmatrix} 0 & 1 \\ 0 & 0 \end{bmatrix}, \quad B = \begin{bmatrix} 1 & 0 \\ 0 & 0 \end{bmatrix}.$$

Then

$$AB = \begin{bmatrix} 0 & 0 \\ 0 & 0 \end{bmatrix}, \quad BA = \begin{bmatrix} 0 & 1 \\ 0 & 0 \end{bmatrix},$$

so that $AB \neq BA$. ∎

A matrix (vector) with all entries zero is called a *zero matrix* (*zero vector*) and is denoted **0**. Example 6.7.5 also shows that

$$AB = 0 \quad \text{need not imply that } B = 0 \text{ or } A = 0. \tag{4}$$

and similarly,

$$AB = AC \quad \text{need not imply that } B = C. \tag{5}$$

In order to understand (4) and (5) better, we define the *identity matrix* **I** to be a square matrix with its i, i-entries equal to one, and all the other entries zero. Examples of identity matrices are

$$[1], \quad \begin{bmatrix} 1 & 0 \\ 0 & 1 \end{bmatrix}, \quad \begin{bmatrix} 1 & 0 & 0 \\ 0 & 1 & 0 \\ 0 & 0 & 1 \end{bmatrix}.$$

Note that, for a given n, there is only one $n \times n$ identity matrix. If **I** is $m \times m$ and **A** is $m \times n$, then

$$IA = A,$$

while

$$AI = A,$$

if **I** is $n \times n$. The identity matrix is a special case of a diagonal matrix. An $n \times n$ matrix is a *diagonal matrix* if $a_{ij} = 0$ for $i \neq j$. Examples of diagonal matrices are

$$I, \quad \alpha I, \quad \begin{bmatrix} 1 & 0 \\ 0 & 2 \end{bmatrix}, \quad \begin{bmatrix} 1 & 0 & 0 \\ 0 & -3 & 0 \\ 0 & 0 & 1 \end{bmatrix}, \quad \begin{bmatrix} 2 & 0 & 0 \\ 0 & 0 & 0 \\ 0 & 0 & 0 \end{bmatrix}.$$

If **B** is a matrix such that

$$AB = BA = I,$$

then **B** is called the *inverse* of **A** and is denoted A^{-1}. Only a square matrix can have an inverse, although not all square matrices are *invertible* (have an inverse). The inverse, when it exists, is unique (Exercise 6.7.38). For example,

$$\begin{bmatrix} -2 & 1 \\ \frac{3}{2} & -\frac{1}{2} \end{bmatrix} \begin{bmatrix} 1 & 2 \\ 3 & 4 \end{bmatrix} = \begin{bmatrix} 1 & 0 \\ 0 & 1 \end{bmatrix} = \mathbf{I},$$

so that

$$\begin{bmatrix} -2 & 1 \\ \frac{3}{2} & -\frac{1}{2} \end{bmatrix} = \begin{bmatrix} 1 & 2 \\ 3 & 4 \end{bmatrix}^{-1} \quad \text{and} \quad \begin{bmatrix} 1 & 2 \\ 3 & 4 \end{bmatrix} = \begin{bmatrix} -2 & 1 \\ \frac{3}{2} & -\frac{1}{2} \end{bmatrix}^{-1}.$$

When matrices have inverses, we do not get the "unusual" behavior in (4) and (5).

Example 6.7.6 Show that if $\mathbf{A}$ has an inverse and $\mathbf{AB} = \mathbf{AC}$, then $\mathbf{B} = \mathbf{C}$.

Solution Suppose $\mathbf{A}$ has an inverse $\mathbf{A}^{-1}$ and

$$\mathbf{AB} = \mathbf{AC}.$$

Multiply both sides by $\mathbf{A}^{-1}$ on the left,

$$\mathbf{A}^{-1}\mathbf{AB} = \mathbf{A}^{-1}\mathbf{AC}.$$

Since $\mathbf{A}^{-1}\mathbf{A} = \mathbf{I}$ and matrix multiplication is associative, we have

$$\mathbf{IB} = \mathbf{IC}$$

or

$$\mathbf{B} = \mathbf{C}. \blacksquare$$

We shall return to the inverse at several points in this chapter and develop additional properties as needed.

6.7.3 Systems of Linear Equations

One of the most common problems in applications is to solve a system of m equations in n unknowns $x_1, \ldots, x_n$,

$$\begin{aligned} a_{11}x_1 + a_{12}x_2 + \cdots + a_{1n}x_n &= b_1, \\ a_{21}x_1 + a_{22}x_2 + \cdots + a_{2n}x_n &= b_2, \\ &\vdots \\ a_{m1}x_1 + a_{m2}x_2 + \cdots + a_{mn}x_n &= b_m. \end{aligned} \tag{6}$$

If we let

$$\mathbf{A} = \begin{bmatrix} a_{11} & a_{12} & \cdots & a_{1n} \\ \vdots & & & \vdots \\ a_{m1} & a_{m2} & \cdots & a_{mn} \end{bmatrix}, \quad \mathbf{x} = \begin{bmatrix} x_1 \\ \vdots \\ x_n \end{bmatrix}, \quad \mathbf{b} = \begin{bmatrix} b_1 \\ \vdots \\ b_m \end{bmatrix}, \tag{7}$$

then (6) may be written as

$$\mathbf{Ax} = \mathbf{b}. \tag{8}$$

The notation of (6), (7), and (8), including the size $m \times n$, will be used for the remainder of this chapter. Since we shall frequently need to solve systems such as (6) in the remainder of this chapter, the solution of (6) will be discussed fairly carefully. To solve $\mathbf{Ax} = \mathbf{b}$ for the unknowns $\mathbf{x}$, we must solve for variables, and substitute. However, it turns out to be much easier to set up an algorithm if we manipulate the equations by performing one of three operations:

i) Multiply an equation by a nonzero scalar (9a)

ii) Exchange two equations (9b)

iii) Add a multiple of one equation to another (9c)

Example 6.7.7 Solve the system of equations

$$\begin{aligned} y + 2z &= 1, \\ x + y + z &= 2, \\ 2x + 4y + 8z &= 3. \end{aligned} \tag{10}$$

Solution For reasons that will become clear shortly, we shall perform what appears to be a couple of unnecessary operations. Exchange Eqs. 1 and 2 in (10) to get

$$\begin{aligned} x + y + z &= 2, \\ y + 2z &= 1, \\ 2x + 4y + 8z &= 3. \end{aligned} \tag{11}$$

Add -2 times Eq. 1 to Eq. 3 in (11) to eliminate the x from the third equation:

$$\begin{aligned} x + y + z &= 2, \\ y + 2z &= 1, \\ 2y + 6z &= -1. \end{aligned} \tag{12}$$

The last two equations involve only y and z. Now add -2 times the second equation to the third, to get

$$\begin{aligned} x + y + z &= 2, \\ y + 2z &= 1, \\ 2z &= -3. \end{aligned} \tag{13}$$

The last equation now involves only z. Now perform the two operations of adding -1 times Eq. 3 to Eq. 2 in (13), and adding $-\frac{1}{2}$ times Eq. 3 to Eq. 1:

$$\begin{aligned} x + y &= 3\tfrac{1}{2}, \\ y &= 4, \\ 2z &= -3, \end{aligned} \tag{14}$$

which eliminates z from Eqs. 1 and 2. Finally, add -1 times Eq. 2 to Eq. 1 and multiply Eq. 3 by $\frac{1}{2}$, to get:

$$x = -\tfrac{1}{2},$$
$$y = 4, \qquad (15)$$
$$z = -\tfrac{3}{2}. \;\blacksquare$$

Looking back over this example, we see that, in the calculations we need not write down the variables x, y, z and the equal sign. Instead, we can proceed as follows:

Algorithm 6.7.1 **Solution of Ax = b by Gaussian elimination** Write down the *augmented matrix* [A | b]. This is A with an extra column b added. Perform the *elementary row operations* of:

$$\text{Multiplying a row by a nonzero scalar,} \qquad (16a)$$
$$\text{Exchanging two rows,} \qquad (16b)$$
$$\text{Adding a multiple of one row to another row,} \qquad (16c)$$

(which correspond to the operations (9) on the equations) according to the following pattern:

Part 1
a. Find a nonzero entry in the first column. If there is none, move to the next column. Repeat until a nonzero entry is found.
b. If this nonzero entry is not in the first row, exchange the row this entry is in with the first row.
c. Add multiples of this row to all the entries below it, so that all other entries in that column are zeroed out (made zero).

- Now ignore the first row and repeat (a), (b), (c) on the matrix formed by the rest of the rows.
- Now ignore the first two rows and repeat (a), (b), (c) on the matrix formed by the rest of the rows.
- Continue this pattern.

Part 1 terminates when we run out of rows; all the rest of the rows are identically zero, or there is a row whose only nonzero entry is the last one:

If there is a row whose only nonzero entry is the last one, the system of equations is *inconsistent*. That is, there is no solution.

If the system of equations is not inconsistent, then proceed to Part 2.

Part 2
Starting with the last nonzero row, find its first nonzero entry, and zero out all other entries in the column above. Then take the second to last nonzero row,

find its first nonzero entry, and zero out all other entries in the column above. Continue until reaching the first row. Often one also makes the leading nonzero entries in each nonzero row equal to one. The resulting matrix is called the *row echelon form* of the original matrix.

The columns that have the first nonzero entry of some row in them are called the *distinguished columns:*

If every column is distinguished except the last one, we have found a unique solution. (17)

If some columns other than the last one are *not* distinguished, then there are infinitely many solutions, and the variables corresponding to all undistinguished columns (except the last one) may be taken as arbitrary. The variables corresponding to the distinguished columns are found in terms of the arbitrary variables. (18)

Example 6.7.8 Consider again the system of Example 6.7.7:

$$\begin{aligned} y + 2z &= 1, \\ x + y + z &= 2, \\ 2x + 4y + 8z &= 3. \end{aligned} \quad (19)$$

Here

$$A = \begin{bmatrix} 0 & 1 & 2 \\ 1 & 1 & 1 \\ 2 & 4 & 8 \end{bmatrix}, \quad x = \begin{bmatrix} x \\ y \\ z \end{bmatrix}, \quad b = \begin{bmatrix} 1 \\ 2 \\ 3 \end{bmatrix},$$

and the augmented matrix is

$$[A \mid b] = \begin{bmatrix} 0 & 1 & 2 & \bigm| & 1 \\ 1 & 1 & 1 & \bigm| & 2 \\ 2 & 4 & 8 & \bigm| & 3 \end{bmatrix}. \quad (20)$$

We shall now apply the method of Gaussian elimination to (20) as described in Algorithm 6.7.1. By referring back to Example 6.7.7, the reader will see that the elementary row operations (16) on the augmented matrix (20) are equivalent to the earlier operations (9) on the system (10).

There are two nonzero entries in the first column of (20). To get a nonzero entry in the 1, 1 position, we exchange rows 1 and 2 to get:

$$\begin{bmatrix} 1 & 1 & 1 & \bigm| & 2 \\ 0 & 1 & 2 & \bigm| & 1 \\ 2 & 4 & 8 & \bigm| & 3 \end{bmatrix}. \quad (21)$$

(We could have exchanged rows 1 and 3, instead.)

Note that (21) is the augmented matrix of (11). To eliminate the rest of the entries in the first column, add -2 times the first row to the third row, to get

$$\begin{bmatrix} 1 & 1 & 1 & | & 2 \\ 0 & 1 & 2 & | & 1 \\ 0 & 2 & 6 & | & -1 \end{bmatrix}, \qquad (22)$$

which is the augmented matrix of (12). Leaving row 1 alone, we look at the remaining two rows. The 2, 2-entry is already nonzero, so we need to zero out all entries in the second column below it. To do this, add -2 times the second row to the third row

$$\begin{bmatrix} 1 & 1 & 1 & | & 2 \\ 0 & 1 & 2 & | & 1 \\ 0 & 0 & 2 & | & -3 \end{bmatrix}, \qquad (23)$$

which is the augmented matrix of (13). This completes Part 1. For Part 2, we use the 3, 3-entry to zero out all of the entries above it in the third column. This is done by adding -1 times the third row to the second row, and $-\frac{1}{2}$ times the third row to the first row:

$$\begin{bmatrix} 1 & 1 & 0 & | & \frac{7}{2} \\ 0 & 1 & 0 & | & 4 \\ 0 & 0 & 2 & | & -3 \end{bmatrix}, \qquad (24)$$

which is the augmented matrix of (14). Next, we go to the second row and use the 2, 2-entry to zero out all of the entries above it. This is done by adding -1 times row 2 to row 1:

$$\begin{bmatrix} 1 & 0 & 0 & | & -\frac{1}{2} \\ 0 & 1 & 0 & | & 4 \\ 0 & 0 & 2 & | & -3 \end{bmatrix}. \qquad (25)$$

Multiply row 3 by $\frac{1}{2}$ to get

$$\begin{bmatrix} 1 & 0 & 0 & | & -\frac{1}{2} \\ 0 & 1 & 0 & | & 4 \\ 0 & 0 & 1 & | & -\frac{3}{2} \end{bmatrix}. \qquad (26)$$

This is the augmented matrix of:

$$x = -\tfrac{1}{2},$$
$$y = 4,$$
$$z = -\tfrac{3}{2},$$

which gives the solution of (19) as desired. ∎

We now give two more examples illustrating Gaussian elimination (Algorithm 6.7.1).

Example 6.7.9 Find all solutions of

$$x_1 + 2x_2 = 3,$$
$$x_1 - x_2 + x_3 = 1, \tag{27}$$
$$x_1 + 5x_2 - x_3 = 10.$$

Solution The augmented matrix is

$$\begin{bmatrix} 1 & 2 & 0 & | & 3 \\ 1 & -1 & 1 & | & 1 \\ 1 & 5 & -1 & | & 10 \end{bmatrix}. \tag{28}$$

Since the 1, 1-entry is already nonzero, add -1 times row one to both rows 2 and 3 to zero the rest of the first column,

$$\begin{bmatrix} 1 & 2 & 0 & | & 3 \\ 0 & -3 & 1 & | & -2 \\ 0 & 3 & -1 & | & 7 \end{bmatrix}. \tag{29}$$

The 2, 2-entry is already nonzero. Zero the entry below it by adding row 2 to row 3,

$$\begin{bmatrix} 1 & 2 & 0 & | & 3 \\ 0 & -3 & 1 & | & -2 \\ 0 & 0 & 0 & | & 5 \end{bmatrix}. \tag{30}$$

The last row is [0 0 0 | 5], which is the equation

$$0x_1 + 0x_2 + 0x_3 = 5,$$

which has no solution. Thus the system (27) is inconsistent. There are no solutions. ∎

Example 6.7.10 Find all the solutions of:

$$x + y + z + 3w = 2,$$
$$x + y + w = 1,$$
$$-2x - 2y + z = -1, \tag{31}$$
$$-x - y + 2z + 3w = 1.$$

Solution The augmented matrix is

$$\begin{bmatrix} 1 & 1 & 1 & 3 & | & 2 \\ 1 & 1 & 0 & 1 & | & 1 \\ -2 & -2 & 1 & 0 & | & -1 \\ -1 & -1 & 2 & 3 & | & 1 \end{bmatrix}, \tag{32}$$

The 1, 1-entry is already nonzero. Use it to zero the rest of column 1 by adding -1 times row 1 to row 2, adding twice row 1 to row 3, and adding row 1 to row 4:

$$\begin{bmatrix} 1 & 1 & 1 & 3 & | & 2 \\ 0 & 0 & -1 & -2 & | & -1 \\ 0 & 0 & 3 & 6 & | & 3 \\ 0 & 0 & 3 & 6 & | & 3 \end{bmatrix}. \tag{33}$$

Ignoring the first row, the first nonzero entry appears in column 3. Since the 2, 3-entry is nonzero, we do not need to exchange rows. Now eliminate the entries below the 2, 3-entry by adding 3 times row 2 to rows 3 and 4:

$$\begin{bmatrix} 1 & 1 & 1 & 3 & | & 2 \\ 0 & 0 & -1 & -2 & | & -1 \\ 0 & 0 & 0 & 0 & | & 0 \\ 0 & 0 & 0 & 0 & | & 0 \end{bmatrix}. \tag{34}$$

This completes Part 1. Eliminate the entry above the 2, 3-entry by adding row 2 to row 1. Multiply row 2 by -1 to get each leading entry positive,

$$\begin{bmatrix} 1 & 1 & 0 & 1 & | & 1 \\ 0 & 0 & 1 & 2 & | & 1 \\ 0 & 0 & 0 & 0 & | & 0 \\ 0 & 0 & 0 & 0 & | & 0 \end{bmatrix}. \tag{35}$$

The matrix (35) is the row echelon form of (32). The distinguished columns are the first and third. Thus, by (18), the second and fourth variables are arbitrary:

$$y, w \quad \text{are arbitrary.} \tag{36}$$

From (35),

$$\begin{aligned} x + y + w &= 1, \\ z + 2w &= 1, \end{aligned} \tag{37}$$

or

$$\begin{aligned} x &= 1 - y - w, \\ z &= 1 - 2w, \end{aligned} \tag{38}$$

gives the other variables. It is very convenient for our later applications to write this solution (36) and (38) in vector notation

$$\mathbf{x} = \begin{bmatrix} x \\ y \\ z \\ w \end{bmatrix} = \begin{bmatrix} 1 - y - w \\ y \\ 1 - 2w \\ w \end{bmatrix} = \begin{bmatrix} 1 \\ 0 \\ 1 \\ 0 \end{bmatrix} + y \begin{bmatrix} -1 \\ 1 \\ 0 \\ 0 \end{bmatrix} + w \begin{bmatrix} -1 \\ 0 \\ -2 \\ 1 \end{bmatrix}, \tag{39}$$

where y, w are arbitrary constants. Two different people applying Algorithm 6.7.1 to a system of algebraic equations will always get the same answer, no matter in what order they do the calculations. If there is more than one solution and a choice of arbitrary variables different from that suggested in Algorithm 6.7.1 is made, the answer will appear different.

For example, in the previous problem, the original system (31) has been reduced to the equivalent one (37). In (37) one could take x, w arbitrary and then find y, z in terms of x and w,

$$y = 1 - x - w, \quad z = 1 - 2w,$$

so that the solution vectors have the form

$$\mathbf{X} = \begin{bmatrix} x \\ y \\ z \\ w \end{bmatrix} = \begin{bmatrix} x \\ 1 - x - w \\ 1 - 2w \\ w \end{bmatrix} = \begin{bmatrix} 0 \\ 1 \\ 1 \\ 0 \end{bmatrix} + x \begin{bmatrix} 1 \\ -1 \\ 0 \\ 0 \end{bmatrix} + w \begin{bmatrix} 0 \\ -1 \\ -2 \\ 1 \end{bmatrix}. \quad (40)$$

While (39) and (40) appear different, they actually describe the same set of vectors. ∎

Not just any pair of variables can be taken arbitrary in (37). For example, one cannot take z, w both arbitrary since $z + 2w = 1$. The beginning student is advised to follow Algorithm 6.7.1.

Algorithm 6.7.1 can be modified to find $\mathbf{A}^{-1}$ if it exists. The key is the observation that the *matrix equation* $\mathbf{AX} = \mathbf{B}$ can be solved by working with the augmented matrix $[\mathbf{A} \mid \mathbf{B}]$ and $\mathbf{A}^{-1}$ is the solution of $\mathbf{AX} = \mathbf{I}$.

Algorithm 6.7.2

For Computation of $\mathbf{A}^{-1}$, if A Is Square and $\mathbf{A}^{-1}$ Exists

1. Write down the augmented matrix $[\mathbf{A} \mid \mathbf{I}]$.
2. Perform elementary row operations to convert $\mathbf{A}$ to the identity, to give $[\mathbf{I} \mid \mathbf{C}]$.
3. Then $\mathbf{C} = \mathbf{A}^{-1}$.
4. If at any point in the calculations, the left side has a zero row, then $\mathbf{A}^{-1}$ does not exist.

Example 6.7.11

Find $\mathbf{A}^{-1}$, if possible, for

$$\mathbf{A} = \begin{bmatrix} 1 & 2 \\ 3 & 4 \end{bmatrix}.$$

Solution

Following Algorithm 6.7.2, we write down the augmented matrix

$$\begin{bmatrix} 1 & 2 & | & 1 & 0 \\ 3 & 4 & | & 0 & 1 \end{bmatrix}.$$

Adding -3 times row 1 to row 2 gives

$$\begin{bmatrix} 1 & 2 & | & 1 & 0 \\ 0 & -2 & | & -3 & 1 \end{bmatrix}.$$

Add row 2 to row 1,

$$\begin{bmatrix} 1 & 0 & | & -2 & 1 \\ 0 & -2 & | & -3 & 1 \end{bmatrix},$$

and multiply row 2 by $-\frac{1}{2}$,

$$\begin{bmatrix} 1 & 0 & | & -2 & 1 \\ 0 & 1 & | & \frac{3}{2} & -\frac{1}{2} \end{bmatrix}.$$

Since the left side is the identity, we have

$$\mathbf{A}^{-1} = \begin{bmatrix} -2 & 1 \\ \frac{3}{2} & -\frac{1}{2} \end{bmatrix}.$$

If the left side had a row of zeros, there would be no inverse. ∎

Comments If $\mathbf{Ax} = \mathbf{b}$ and $\mathbf{A}$ has an inverse, then multiplying both sides of $\mathbf{Ax} = \mathbf{b}$ by $\mathbf{A}^{-1}$ gives $\mathbf{x} = \mathbf{A}^{-1}\mathbf{b}$. Thus one could solve $\mathbf{Ax} = \mathbf{b}$ by computing $\mathbf{A}^{-1}$ (if it exists) and using $\mathbf{x} = \mathbf{A}^{-1}\mathbf{b}$. However, this is substantially more work then working directly with $[\mathbf{A} \mid \mathbf{b}]$. Inverses are usually not computed in practice. Also, while it can be shown that Algorithm 6.7.1 involves the least work of any general algorithm for solving $\mathbf{Ax} = \mathbf{b}$, it is not numerically implemented as stated, to avoid dividing by small numbers. Care must be used in choosing which rows to use to eliminate others (choice of pivots). For large systems and certain others that are numerically delicate, totally different techniques must be used.

Exercises

For Exercises 1 through 15, let

$$\mathbf{A} = \begin{bmatrix} 1 & 2 & 0 \\ 1 & -1 & 1 \end{bmatrix}, \quad \mathbf{B} = \begin{bmatrix} 1 & 3 \\ -1 & 4 \end{bmatrix},$$

$$\mathbf{u} = \begin{bmatrix} 1 \\ 2 \end{bmatrix}, \quad \mathbf{v} = \begin{bmatrix} 1 \\ 0 \\ 1 \end{bmatrix}, \quad \mathbf{w} = [3 \quad -4].$$

In each of Exercises 1 through 15, state whether or not it is possible to compute the given expression. Compute it if possible.

1. **AB**
2. **BA**
3. **uv**
4. **uw**
5. **wu**
6. **wv**
7. **2A**
8. **wAv**
9. **A + B**
10. **3I**   (**I** is 3 × 3)
11. **B − 2I**   (**I** is 2 × 2)
12. **3AI**   (**I** is 3 × 3)
13. **B − 2A**
14. **B²**   (**B² = BB**)
15. **A + 0**   (**0** is 2 × 3)
16. Let $\mathbf{A} = \begin{bmatrix} 1 & -1 \\ 1 & -1 \end{bmatrix}$ and compute $\mathbf{A}^2 = \mathbf{AA}$.

17. Let $A = \begin{bmatrix} 1 & -1 \\ 0 & -1 \end{bmatrix}$ and compute A^2.

For Exercises 18 through 29, solve the indicated system of algebraic equations, using augmented matrices and the method of Algorithm 6.7.1. If the system is inconsistent, so state. If there are infinitely many solutions, express them in vector form like (39) of Example 6.7.10.

18. $x + y = 1,$
    $x + 2y = 2$
19. $2x + 3y = 2,$
    $4x + 6y = 4$
20. $2x + 3y = 2,$
    $4x + 6y = 3$
21. $2x - y = 0,$
    $-6x + 3y = 0$
22. $x - y = 1,$
    $-2x + 2y = 3$
23. $x + y + z = 1,$
    $y + 2z = 0,$
    $2x + 3y + 4z = 0$
24. $x + 2y + 3z = 0,$
    $2x + 4y + 6z = 0,$
    $x + 2y + 3z = 0$
25. $x + y - z = 1,$
    $x - y - z = 2,$
    $2x + 3y - z = 0$
26. $x + y = 0,$
    $y + z = 0,$
    $x - z = 0$
27. $x + y = 1,$
    $y - z = 2,$
    $x - z = 3$
28. $x + y + z + w = 0,$
    $-x - y - z - w = 0,$
    $2x + 2y + 2z + 2w = 0,$
    $x + y + z + w = 0$

29. $x - y + z - w = 1,$
    $x - y + z - 3w = 7,$
    $2x - 2y + 2z + 2w = -10,$
    $-x + y + z + w = -4$

Exercises 30 through 38 provide some practice in manipulating matrix expressions.

30. Is $(A + 2I)^2 = A^2 + 4A + 4I$ for any square matrix A if A, I are the same size?
31. If A, B are square matrices of the same size, is $(A + B)^2 = A^2 + 2AB + B^2$?
32. Let u be $n \times 1$ and v be $1 \times n$ matrices such that $vu = 0$.
    a) Must $uv = 0$?
    b) Let $N = uv$. Show that $N^2 = 0$.
33. Suppose x_1, x_2 are two solutions of $Ax = 0$. Show $c_1 x_1 + c_2 x_2$ is also a solution of $Ax = 0$ for any scalars c_1, c_2.
34. If x_1 is a solution of $Ax = b$ and x_2 is a solution of $Ax = 0$, show that $x_1 + x_2$ is a solution of $Ax = b$.
35. If x_1 is a solution of $Ax = b_1$ and x_2 is a solution of $Ax = b_2$, show that $x_1 + x_2$ is a solution of $Ax = b_1 + b_2$.
36. Suppose that A is $n \times n$ and u is $n \times 1$. Show that if $u \neq 0$ and $Au = 0$, then A cannot be invertible.
37. Suppose that A, B are $n \times n$ invertible matrices. Show that AB is invertible and $(AB)^{-1} = B^{-1}A^{-1}$.
38. Suppose that A is invertible and B, C are both inverses of A. Show that $B = C$; that is, the inverse is unique.

In Exercises 39 through 42, compute A^{-1}, using Algorithm 6.7.2.

39. $A = \begin{bmatrix} 1 & 2 \\ 1 & 1 \end{bmatrix}$
40. $A = \begin{bmatrix} 1 & 3 \\ 2 & 1 \end{bmatrix}$
41. $A = \begin{bmatrix} 0 & 1 \\ 1 & 2 \end{bmatrix}$
42. $A = \begin{bmatrix} 2 & 1 \\ -1 & 0 \end{bmatrix}$

6.8 Determinants and Linear Independence

If $\mathbf{A}$ is a square matrix, then the *determinant* of $\mathbf{A}$, denoted det $\mathbf{A}$, is a number (scalar). There are several ways to define the determinant. We shall give two ways to compute it (in lieu of a definition).

a. If $\mathbf{A}$ is 1×1, then
$$\det[a_{11}] = a_{11}. \tag{1}$$

b. If $\mathbf{A}$ is 2×2, then
$$\det \begin{bmatrix} a_{11} & a_{12} \\ a_{21} & a_{22} \end{bmatrix} = a_{11}a_{22} - a_{21}a_{12}. \tag{2}$$

In general, if $\mathbf{A}$ is $n \times n$, then det $\mathbf{A}$ can be expressed in terms of n smaller $(n-1) \times (n-1)$ determinants, as follows:

Take any row or column of $\mathbf{A}$. Suppose the ith row is selected, so its entries are $a_{i1}, a_{i2}, \ldots, a_{in}$. Multiply each a_{ij} by $(-1)^{i+j}M_{ij}$, where M_{ij} is the determinant of the matrix obtained by deleting the ith row and jth column of $\mathbf{A}$. Adding the $a_{ij}(-1)^{i+j}M_{ij}$ gives det $\mathbf{A}$:

$$\det \mathbf{A} = \sum_{k=1}^{n} a_{ik}(-1)^{i+k}M_{ik}. \tag{3}$$

Thus, if $\mathbf{A}$ were, say, 5×5, formula (3) could be used to write det $\mathbf{A}$ as a sum of five 4×4 determinants. Then (3) could be used to express each of these 4×4 determinants as the sum of four 3×3 determinants each of which could then be expressed as three 2×2 determinants. The M_{ij} are called *minors*, and $A_{ij} = (-1)^{i+j}M_{ij}$ are called *cofactors*.

Example 6.8.1 Compute det $\mathbf{A}$, where

$$\mathbf{A} = \begin{bmatrix} 1 & 1 & 2 \\ 1 & 0 & -1 \\ 2 & 1 & 4 \end{bmatrix}. \tag{4}$$

Solution Since we can pick any row or column, it is a little less work to use one with a zero in it. If we select row 2, (3) becomes

$$\det \mathbf{A} = a_{21}(-1)^{2+1}M_{21} + a_{22}(-1)^{2+2}M_{22} + a_{23}(-1)^{2+3}M_{23}$$

$$= 1(-1)\det\begin{bmatrix} 1 & 2 \\ 1 & 4 \end{bmatrix} + 0(1)\det\begin{bmatrix} 1 & 2 \\ 2 & 4 \end{bmatrix} + (-1)(-1)\det\begin{bmatrix} 1 & 1 \\ 2 & 1 \end{bmatrix}.$$

The 2×2 determinants are evaluated using (2), to give

$$\det \mathbf{A} = -1(4-2) + 0 + 1(1-2) = -3. \blacksquare$$

This is obviously a lot of work for even fairly small matrices. In fact, one can show that, if $\mathbf{A}$ is $n \times n$, the number of multiplications is approximately $n!$. For this reason, evaluation of the determinant is usually avoided or computed by alternative means.* However, we will have no alternative to this method when we need to compute the characteristic polynomial two sections from now.

There are two ways to make the determinants in our problems easier to compute. The first is an alternative method for the 3 × 3 case.

Alternative Method for the 3 × 3 Determinant Take the 3 × 3 matrix

$$A = \begin{bmatrix} a & b & c \\ d & e & f \\ g & h & i \end{bmatrix},$$

and repeat the first two columns. Then multiply the elements down each of the six diagonals, as shown in (5).

$$\begin{bmatrix} a & b & c & a & b \\ d & e & f & d & e \\ g & h & i & g & h \end{bmatrix} \tag{5}$$

$$-gec - hfa - idb + aei + bfg + cdh = \det \mathbf{A}$$

Subtract the first three products from the sum of the last three products to get det **A**. This formula may be used even if the entries depend on another variable.

Example 6.8.2

Consider Example 6.8.1,

$$A = \begin{bmatrix} 1 & 1 & 2 \\ 1 & 0 & -1 \\ 2 & 1 & 4 \end{bmatrix},$$

and compute det **A** using (5).

Solution

From (5) we have

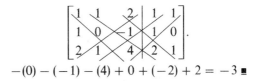

$$-(0) - (-1) - (4) + 0 + (-2) + 2 = -3 \blacksquare$$

However, probably the quickest way to compute det **A** is based on the following result.

---

*All of our discussion of numerical considerations assumes that the calculations are done *serially*. Different considerations apply on *parallel computers*.

Theorem 6.8.1 Suppose **A** is an $n \times n$ matrix. Then,

1. Exchanging two rows or columns of **A** changes the sign of det **A**.
2. Adding a multiple of one row (column) to another row (column) does not change det **A**.
3. If **A** is *upper triangular* (all entries below the principal diagonal are zero), then det **A** is the product of the diagonal entries.

This theorem can be used two ways. One can use parts (1) and (2) to get a row (or column) with only one nonzero entry, and then expand the determinant along that row or column, using (3). Alternatively, row operations may be performed on **A** following the Gaussian elimination pattern of the last section, and the determinant taken of the resulting upper triangular matrix.

Example 6.8.3 Find
$$\det \begin{bmatrix} 1 & 1 & 2 \\ 1 & 0 & -1 \\ 2 & 1 & 4 \end{bmatrix}.$$

Solution
$$\det \begin{bmatrix} 1 & 1 & 2 \\ 1 & 0 & -1 \\ 2 & 1 & 4 \end{bmatrix} \quad \text{adding } -1 \text{ times row 1 to row 2,} \\ \text{adding } -2 \text{ times row 1 to row 3,}$$

$$= \det \begin{bmatrix} 1 & 1 & 2 \\ 0 & -1 & -3 \\ 0 & -1 & 0 \end{bmatrix} \quad \text{adding } -1 \text{ times row 2 to row 3,}$$

$$= \det \begin{bmatrix} 1 & 1 & 2 \\ 0 & -1 & -3 \\ 0 & 0 & 3 \end{bmatrix}$$

$$= 1(-1)(3) = -3 \quad \text{by Part (3) of Theorem 6.8.1.} \blacksquare$$

For even small matrices such as 4×4 or 5×5, this method is much quicker than the expansion formula (3). Unfortunately, if the entries of **A** depend on a variable, then using row operations may become impractical or at least more difficult.

Example 6.8.4 Find det **A** if:

$$\mathbf{A}(\lambda) = \begin{bmatrix} \lambda & 2 & 1 \\ 1 & \lambda - 1 & 1 \\ 1 & 2 & \lambda \end{bmatrix}. \tag{6}$$

Solution We could evaluate this determinant by using the 3,1-entry to zero both entries above, and then expanding along the first column by (3) (Exercise 21). Instead, we shall use (3) directly and expand along the first row:

$$\det \mathbf{A}(\lambda) = \lambda(-1)^{1+1} \det \begin{bmatrix} \lambda - 1 & 1 \\ 2 & \lambda \end{bmatrix} + 2(-1)^{1+2} \det \begin{bmatrix} 1 & 1 \\ 1 & \lambda \end{bmatrix}$$
$$+ 1(-1)^{1+3} \det \begin{bmatrix} 1 & \lambda - 1 \\ 1 & 2 \end{bmatrix}$$
$$= \lambda[\lambda^2 - \lambda - 2] - 2(\lambda - 1) + (3 - \lambda)$$
$$= \lambda^3 - \lambda^2 - 5\lambda + 5. \ \blacksquare$$

6.8.1 Linear Independence

A set of r vectors $\{\mathbf{b}_1, \ldots, \mathbf{b}_r\}$ all the same size are *linearly independent* if

$$c_1 \mathbf{b}_1 + c_2 \mathbf{b}_2 + \cdots + c_r \mathbf{b}_r = \mathbf{0} \text{ for constants } c_1, c_2, \ldots, c_r \text{ implies that the only solution is } c_1 = c_2 = \cdots = c_r = 0. \tag{7}$$

The expression $c_1 \mathbf{b}_1 + c_2 \mathbf{b}_2 + \cdots + c_r \mathbf{b}_r$ is a *linear combination*. Intuitively, a set of vectors is linearly independent if no one of the vectors can be written as a linear combination of the others. If a set of vectors is not linearly independent, it is called *linearly dependent*.

Two vectors are linearly independent if neither is a multiple of the other. Three vectors are linearly independent if they do not lie in the same plane.

Example 6.8.5 Is the set of vectors

$$\left\{ \begin{bmatrix} 1 \\ 1 \\ 0 \end{bmatrix}, \begin{bmatrix} 1 \\ 0 \\ 1 \end{bmatrix}, \begin{bmatrix} 5 \\ 2 \\ 3 \end{bmatrix} \right\}$$

linearly independent?

Solution From (7) this is equivalent to asking whether

$$c_1 \begin{bmatrix} 1 \\ 1 \\ 0 \end{bmatrix} + c_2 \begin{bmatrix} 1 \\ 0 \\ 1 \end{bmatrix} + c_3 \begin{bmatrix} 5 \\ 2 \\ 3 \end{bmatrix} = \begin{bmatrix} 0 \\ 0 \\ 0 \end{bmatrix} \tag{8}$$

has $c_1 = c_2 = c_3 = 0$ as its only solution. The equation (8) is equivalent to

$$c_1 + c_2 + 5c_3 = 0,$$
$$c_1 + 2c_3 = 0, \tag{9}$$
$$c_2 + 3c_3 = 0.$$

If we apply Gaussian elimination to (9), we finally get an augmented matrix of

$$\begin{bmatrix} 1 & 0 & 2 & | & 0 \\ 0 & 1 & 3 & | & 0 \\ 0 & 0 & 0 & | & 0 \end{bmatrix}.$$

Thus we can take c_3 arbitrary and $c_1 = -2c_3$, $c_2 = -3c_3$. In particular, c_3 need not be zero, so the vectors are not linearly independent. ∎

Example 6.8.6   Is the set of vectors

$$\left\{ \begin{bmatrix} 1 \\ 2 \end{bmatrix}, \begin{bmatrix} 2 \\ 1 \end{bmatrix} \right\}$$

linearly independent?

Solution   From (7), this is equivalent to asking whether

$$c_1 \begin{bmatrix} 1 \\ 2 \end{bmatrix} + c_2 \begin{bmatrix} 2 \\ 1 \end{bmatrix} = \begin{bmatrix} 0 \\ 0 \end{bmatrix} \quad (10)$$

has as the only solution $c_1 = c_2 = 0$. Equation (10) is equivalent to

$$\begin{aligned} c_1 + 2c_2 &= 0, \\ 2c_1 + c_2 &= 0. \end{aligned} \quad (11)$$

Solving the system, we find $c_1 = 0$, $c_2 = 0$. Thus

$$\left\{ \begin{bmatrix} 1 \\ 2 \end{bmatrix}, \begin{bmatrix} 2 \\ 1 \end{bmatrix} \right\}$$

is a linearly independent set of vectors. ∎

The following theorem relates several of our concepts.

Theorem 6.8.2   Suppose $\mathbf{A}$ is an $n \times n$ matrix. Then the following are equivalent:

$$\text{The rows of } \mathbf{A} \text{ are linearly independent.} \quad (12)$$
$$\text{The columns of } \mathbf{A} \text{ are linearly independent.} \quad (13)$$
$$\det \mathbf{A} \neq 0. \quad (14)$$
$$\mathbf{A} \text{ has an inverse.} \quad (15)$$

This theorem gives us a way to check linear independence.

Algorithm 6.8.1   To check whether n vectors that are $n \times 1$ (or $1 \times n$) are linearly independent, make a matrix $\mathbf{A}$ with the vectors as either rows or columns. The vectors are linearly independent if and only if $\det \mathbf{A} \neq 0$.

Exercises

In Exercises 1 through 7, compute det **A** if

1. $\mathbf{A} = \begin{bmatrix} 1 & 2 \\ 3 & 4 \end{bmatrix}$

2. $\mathbf{A} = \begin{bmatrix} 1 & -1 \\ 2 & -6 \end{bmatrix}$

3. $\mathbf{A} = \begin{bmatrix} 1 & 1 & 2 \\ 0 & 1 & 0 \\ 2 & 1 & -1 \end{bmatrix}$

4. $\mathbf{A} = \begin{bmatrix} 1 & 1 & 1 \\ 2 & 2 & 2 \\ 3 & 3 & 3 \end{bmatrix}$

5. $\mathbf{A} = \begin{bmatrix} 1 & 1 & 0 \\ 0 & 1 & 1 \\ 1 & 0 & 1 \end{bmatrix}$

6. $\mathbf{A} = \begin{bmatrix} 1 & 2 & 3 & 4 \\ 0 & -1 & 2 & 6 \\ 0 & 0 & 3 & 5 \\ 0 & 0 & 0 & 9 \end{bmatrix}$

7. $\mathbf{A} = \begin{bmatrix} 1 & 2 & 0 \\ -1 & 1 & 1 \\ 1 & 2 & 3 \end{bmatrix}$

In Exercises 8 through 12, determine whether the given set of vectors is linearly independent.

8. $\left\{ \begin{bmatrix} 1 \\ 1 \end{bmatrix}, \begin{bmatrix} 1 \\ -1 \end{bmatrix} \right\}$

9. $\left\{ \begin{bmatrix} 1 \\ -1 \end{bmatrix}, \begin{bmatrix} -1 \\ 1 \end{bmatrix} \right\}$

10. $\left\{ \begin{bmatrix} 1 \\ 2 \\ 1 \\ 0 \end{bmatrix}, \begin{bmatrix} 0 \\ 1 \\ 1 \\ 1 \end{bmatrix}, \begin{bmatrix} 1 \\ 0 \\ 0 \\ 1 \end{bmatrix} \right\}$

11. $\left\{ \begin{bmatrix} 1 \\ 1 \\ 1 \end{bmatrix}, \begin{bmatrix} 1 \\ -1 \\ 1 \end{bmatrix}, \begin{bmatrix} -1 \\ 0 \\ 1 \end{bmatrix} \right\}$

12. $\left\{ \begin{bmatrix} 1 \\ 1 \\ 0 \end{bmatrix}, \begin{bmatrix} 0 \\ 1 \\ 1 \end{bmatrix}, \begin{bmatrix} 1 \\ 0 \\ -1 \end{bmatrix} \right\}$

13. a) Let **I** be the 2×2 identity. Verify that
$$\det(5\mathbf{I}) = 25.$$

b) Let **I** be the 3×3 identity. Verify that
$$\det(5\mathbf{I}) = 125.$$

c) Let **I** be the 4×4 identity. Verify that
$$\det(5\mathbf{I}) = 625.$$

14. (Note Exercise 13.) Show that if **I** is $n \times n$ and α is a scalar, then $\det(\alpha \mathbf{I}) = \alpha^n$.

In Exercises 15 through 17, compute $\det(\lambda \mathbf{I} - \mathbf{A})$, where **I** is 2×2 and λ is a parameter (variable) and:

15. $\mathbf{A} = \begin{bmatrix} 1 & 0 \\ 0 & 3 \end{bmatrix}$ **16.** $\mathbf{A} = \begin{bmatrix} 1 & 2 \\ 3 & 0 \end{bmatrix}$

17. $\mathbf{A} = \begin{bmatrix} 0 & 3 \\ 6 & 0 \end{bmatrix}$

In Exercises 18 through 20, compute $\det(\lambda \mathbf{I} - \mathbf{A})$, where **I** is 3×3 and λ is a parameter, and:

18. $\mathbf{A} = \begin{bmatrix} 3 & 6 & 7 \\ 0 & 0 & 5 \\ 0 & 0 & 2 \end{bmatrix}$ **19.** $\mathbf{A} = \begin{bmatrix} 1 & 0 & 1 \\ 0 & 1 & 0 \\ 1 & 0 & 1 \end{bmatrix}$

20. $\mathbf{A} = \begin{bmatrix} 0 & 0 & 1 \\ -1 & 3 & 2 \\ 2 & 0 & -1 \end{bmatrix}$

21. Let
$$\mathbf{A}(\lambda) = \begin{bmatrix} \lambda & 2 & 1 \\ 1 & \lambda - 1 & 1 \\ 1 & 2 & \lambda \end{bmatrix}.$$
Compute $\det \mathbf{A}(\lambda)$ by adding -1 times row 3 to

row 2, and $-\lambda$ times row 3 to row 1, and then expanding down the first column, using (3). Compare your answer to that of Example 6.8.4.

22. If **A**, **B** are $n \times n$ matrices, then it can be shown that $\det(\mathbf{AB}) = \det \mathbf{A} \det \mathbf{B}$. Use this to show that if **A** is invertible, then $\det(\mathbf{A}^{-1}) = (\det \mathbf{A})^{-1}$.

6.9 Differential Equations: Basic Theory

In this section we will begin to apply matrix theory to linear systems of differential equations in the explicit form

$$\begin{aligned}
x_1'(t) &= a_{11}(t)x_1(t) + \cdots + a_{1n}(t)x_n(t) + f_1(t), \\
x_2'(t) &= a_{21}(t)x_1(t) + \cdots + a_{2n}(t)x_n(t) + f_2(t), \\
&\vdots \qquad\qquad\qquad\qquad \vdots \\
x_n'(t) &= a_{n1}(t)x_1(t) + \cdots + a_{nn}(t)x_n(t) + f_n(t),
\end{aligned} \quad (1)$$

with initial conditions

$$x_1(t_0) = a_1, \ldots, x_n(t_0) = a_n. \quad (2)$$

In order to do so, we must briefly discuss matrix-valued functions (or equivalently, matrices whose entries are functions):

If $\mathbf{A}(t)$ is an $m \times n$ matrix whose entries depend on a variable t, then

$$\frac{d\mathbf{A}}{dt} = \mathbf{A}'(t) \quad (3)$$

is an $m \times n$ matrix whose i, j-entry is $a_{ij}'(t)$; that is, matrices are differentiated element-wise.

Of course, (3) requires the entries of $\mathbf{A}(t)$ to be differentiable functions.

Example 6.9.1 If

$$\mathbf{A}(t) = \begin{bmatrix} t & 1 & \sin t \\ 0 & 2 & \cos t \end{bmatrix}, \quad \text{then } \mathbf{A}'(t) = \begin{bmatrix} 1 & 0 & \cos t \\ 0 & 0 & -\sin t \end{bmatrix}. \blacksquare$$

Notational Comment The prime is also sometimes used in matrix theory to denote something called the transpose, so be careful when consulting other texts.

Differentiating matrix-valued functions works almost exactly like differentiating scalar functions:

$$\frac{d(\mathbf{A}(t) + \mathbf{B}(t))}{dt} = \frac{d\mathbf{A}(t)}{dt} + \frac{d\mathbf{B}(t)}{dt}; \tag{4}$$

$$\frac{d\mathbf{A}(t)}{dt} = 0 \text{ for all } t \text{ implies that } \mathbf{A}(t) \text{ is a constant matrix}; \tag{5}$$

$$\frac{d\mathbf{A}}{dt} = \frac{d\mathbf{B}}{dt} \text{ implies that } \mathbf{A}(t) = \mathbf{B}(t) + \mathbf{C}, \tag{6}$$

where $\mathbf{C}$ is a constant matrix of the same size as $\mathbf{A}$ and $\mathbf{B}$, and

$$\frac{d(\mathbf{A}(t)\mathbf{B}(t))}{dt} = \frac{d\mathbf{A}(t)}{dt}\mathbf{B}(t) + \mathbf{A}(t)\frac{d\mathbf{B}(t)}{dt}, \tag{7}$$

which is the *product rule*.

The only thing one has to be careful with is that the order of the products in the product rule cannot be changed. Thus, we cannot write $(\mathbf{AB})' = \mathbf{A}'\mathbf{B} + \mathbf{B}'\mathbf{A}$, which would be true with scalar functions.

If we introduce the notation

$$\mathbf{x}(t) = \begin{bmatrix} x_1(t) \\ \vdots \\ x_n(t) \end{bmatrix}, \mathbf{A}(t) = \begin{bmatrix} a_{11}(t) \ldots a_{1n}(t) \\ \vdots \qquad \vdots \\ a_{n1}(t) \ldots a_{nn}(t) \end{bmatrix}, \mathbf{f}(t) = \begin{bmatrix} f_1(t) \\ \vdots \\ f_n(t) \end{bmatrix}, \mathbf{a} = \begin{bmatrix} a_1 \\ \vdots \\ a_n \end{bmatrix}, \tag{8}$$

the nth-order linear system of differential equations (1) may be written as

$$\mathbf{x}'(t) = \mathbf{A}(t)\mathbf{x}(t) + \mathbf{f}(t), \qquad \mathbf{x}(t_0) = \mathbf{a}, \tag{9}$$

or, more simply, as

$$\mathbf{x}' = \mathbf{A}\mathbf{x} + \mathbf{f}, \qquad \mathbf{x}(t_0) = \mathbf{a}. \tag{10}$$

The vector $\mathbf{a}$ is called the *initial condition*, $\mathbf{A}$ is the *coefficient matrix*, $\mathbf{f}$ is the *input* or *forcing term*, and $\mathbf{x}$ is the *output* or *response*. A solution is a vector of functions $\mathbf{x}(t)$ (equivalently, a vector-valued function) which satisfies (10). We shall consider only systems where $\mathbf{A}$ is square. Theorem 6.1.1 can now be restated in matrix notation.

Theorem 6.9.1

Fundamental Theorem of Linear Systems Suppose that the entries of $\mathbf{A}$ and $\mathbf{f}$ are continuous on an interval I containing t_0. Let $\mathbf{a}$ be an arbitrary $n \times 1$ vector of scalars. Then

there exists a unique solution to $\mathbf{x}' = \mathbf{A}\mathbf{x} + \mathbf{f}, \mathbf{x}(t_0) = \mathbf{a}$, and this solution is defined on all of the interval I.

Just as in the scalar case, we get a breakdown of the solution into free response and forced response.

Theorem 6.9.2 Suppose that the entries of **A** and **f** are continuous on the interval I. Then

> Every solution of $\mathbf{x}' = \mathbf{A}\mathbf{x} + \mathbf{f}$ is of the form $\mathbf{x} = \mathbf{x}_p + \mathbf{x}_h$, where $\mathbf{x}_p$ is a *particular solution* of $\mathbf{x}' = \mathbf{A}\mathbf{x} + \mathbf{f}$, and $\mathbf{x}_h$ is a solution of the *associated homogeneous equation* $\mathbf{x}' = \mathbf{A}\mathbf{x}$.

The particular solution $\mathbf{x}_p$ is sometimes called the *forced response* to the input **f**, and $\mathbf{x}_h$ is then called the *free response*.

Thus, solving linear systems can again be broken down into two problems: solving the associated homogeneous equations and finding a particular solution. In the next two sections, we shall show how to do both when **A** is constant. The key facts about the solution of the associated homogeneous equation are covered in the next result.

Theorem 6.9.3 Suppose that the entries of the $n \times n$ matrix $\mathbf{A}(t)$ are continuous on the interval $t_0 \leq t \leq t_1$. Then the *general solution* of $\mathbf{x}'(t) = \mathbf{A}(t)\mathbf{x}(t)$ may be written as

$$\mathbf{x} = c_1 \mathbf{x}_1(t) + c_2 \mathbf{x}_2(t) + \cdots + c_n \mathbf{x}_n(t),$$

where $\{\mathbf{x}_1, \ldots, \mathbf{x}_n\}$ is any set of n solutions of $\mathbf{x}'(t) = \mathbf{A}(t)\mathbf{x}(t)$, such that the vectors $\{\mathbf{x}_1(t_0), \mathbf{x}_2(t_0), \ldots, \mathbf{x}_n(t_0)\}$ are linearly independent and the $c_1, \ldots, c_n$ are arbitrary constants. The set $\{\mathbf{x}_1(t), \ldots, \mathbf{x}_n(t)\}$ is called a *fundamental set of solutions* for $\mathbf{x}' = \mathbf{A}\mathbf{x}$.

A consequence of the Fundamental Theorem 6.9.1 on existence and uniqueness is that, if $\{\mathbf{x}_1, \ldots, \mathbf{x}_n\}$ are solutions of $\mathbf{x}' = \mathbf{A}\mathbf{x}$ and the set of vectors $\{\mathbf{x}_1(t), \ldots, \mathbf{x}_n(t)\}$ is linearly independent for one value of t, then they are linearly independent for all t.

Another way to express this is as follows:

> Suppose $\mathbf{A}(t)$ is $n \times n$. Let $\{\mathbf{x}_1, \ldots, \mathbf{x}_n\}$ be n solutions of $\mathbf{x}' = \mathbf{A}\mathbf{x}$ on the interval I. Let $\mathbf{X}(t)$ be the matrix with the $\mathbf{x}_i$ as columns. Let **c** be an $n \times 1$ vector of arbitrary constants. Then the following are equivalent:
>
> $\{\mathbf{x}_1, \ldots, \mathbf{x}_n\}$ is a fundamental set of solutions of $\mathbf{x}' = \mathbf{A}\mathbf{x}$. (11)
>
> $\det \mathbf{X}(t) \neq 0$ for some value of t in I. (12)
>
> $\det \mathbf{X}(t) \neq 0$ for all values of t in I. (13)
>
> $\mathbf{X}(t)\mathbf{c}$ is the general solution of $\mathbf{x}' = \mathbf{A}\mathbf{x}$.

Example 6.9.2

Verify that

$$\left\{ \begin{bmatrix} e^{2t} \\ e^{2t} \end{bmatrix}, \begin{bmatrix} 1 \\ -1 \end{bmatrix} \right\}$$

is a fundamental set of solutions of

$$\mathbf{x}' = \begin{bmatrix} 1 & 1 \\ 1 & 1 \end{bmatrix} \mathbf{x}. \tag{14}$$

Solution   We need to show that

$$\{\mathbf{x}_1, \mathbf{x}_2\} = \left\{ \begin{bmatrix} e^{2t} \\ e^{2t} \end{bmatrix}, \begin{bmatrix} 1 \\ -1 \end{bmatrix} \right\}$$

are both solutions and that they are linearly independent. First, we verify they are solutions: Substituting $\mathbf{x}_1$ into (14) gives

$$\begin{bmatrix} e^{2t} \\ e^{2t} \end{bmatrix}' \stackrel{?}{=} \begin{bmatrix} 1 & 1 \\ 1 & 1 \end{bmatrix} \begin{bmatrix} e^{2t} \\ e^{2t} \end{bmatrix}$$

or

$$\begin{bmatrix} 2e^{2t} \\ 2e^{2t} \end{bmatrix} = \begin{bmatrix} 2e^{2t} \\ 2e^{2t} \end{bmatrix},$$

so that $\mathbf{x}_1$ is a solution. Similarly,

$$\begin{bmatrix} 1 \\ -1 \end{bmatrix}' \stackrel{?}{=} \begin{bmatrix} 1 & 1 \\ 1 & 1 \end{bmatrix} \begin{bmatrix} 1 \\ -1 \end{bmatrix}$$

or

$$\begin{bmatrix} 0 \\ 0 \end{bmatrix} = \begin{bmatrix} 0 \\ 0 \end{bmatrix},$$

so that both $\mathbf{x}_1, \mathbf{x}_2$ are solutions. To verify that they are linearly independent, we compute that

$$\det[\mathbf{x}_1, \mathbf{x}_2] = \det \begin{bmatrix} e^{2t} & 1 \\ e^{2t} & -1 \end{bmatrix} = -2e^{2t} \neq 0. \blacksquare$$

Example 6.9.3

Given that we have found that

$$\{\mathbf{x}_1, \mathbf{x}_2\} = \left\{ \begin{bmatrix} e^{2t} \\ e^{2t} \end{bmatrix}, \begin{bmatrix} 1 \\ -1 \end{bmatrix} \right\}$$

is a fundamental set of solutions of (14),

a) Find the general solution of (14);

b) Find the solution of the initial-value problem

$$\mathbf{x}' = \begin{bmatrix} 1 & 1 \\ 1 & 1 \end{bmatrix} \mathbf{x}, \qquad \mathbf{x}(0) = \begin{bmatrix} 2 \\ 3 \end{bmatrix}. \tag{15}$$

Solution

a) From Theorem 6.9.3, the general solution will be

$$\mathbf{x} = \begin{bmatrix} x_1 \\ x_2 \end{bmatrix} = c_1 \begin{bmatrix} e^{2t} \\ e^{2t} \end{bmatrix} + c_2 \begin{bmatrix} 1 \\ -1 \end{bmatrix}, \tag{16}$$

or, equivalently,

$$x_1 = c_1 e^{2t} + c_2, \qquad x_2 = c_1 e^{2t} - c_2.$$

b) Applying the initial condition in (15) to the general solution in (16) gives

$$\mathbf{x}(0) = c_1 \begin{bmatrix} 1 \\ 1 \end{bmatrix} + c_2 \begin{bmatrix} 1 \\ -1 \end{bmatrix} = \begin{bmatrix} 2 \\ 3 \end{bmatrix}. \tag{17}$$

This is two equations in two unknowns,

$$c_1 + c_2 = 2,$$
$$c_1 - c_2 = 3.$$

The solution is $c_1 = \frac{5}{2}$, $c_2 = -\frac{1}{2}$, so that

$$\mathbf{x} = \frac{5}{2} \begin{bmatrix} e^{2t} \\ e^{2t} \end{bmatrix} - \frac{1}{2} \begin{bmatrix} 1 \\ -1 \end{bmatrix} = \begin{bmatrix} \frac{5}{2} e^{2t} - \frac{1}{2} \\ \frac{5}{2} e^{2t} + \frac{1}{2} \end{bmatrix},$$

is the solution of (15). ∎

Exercises

1. If $\mathbf{A}(t) = \begin{bmatrix} 1 & t \\ t^2 & 1 \end{bmatrix}$, compute $\mathbf{A}'(t)$.

2. Verify that $(\mathbf{ABC})' = \mathbf{A}'\mathbf{BC} + \mathbf{AB}'\mathbf{C} + \mathbf{ABC}'$.

3. a) Verify that

$$\left\{ \begin{bmatrix} e^{-t} \\ -e^{-t} \end{bmatrix}, \begin{bmatrix} e^{3t} \\ e^{3t} \end{bmatrix} \right\}$$

is a fundamental set of solutions of

$$\mathbf{x}' = \begin{bmatrix} 1 & 2 \\ 2 & 1 \end{bmatrix} \mathbf{x}. \tag{18}$$

b) Find the general solution of (18).

c) Find the solution of (18) that satisfies the initial condition

$$\mathbf{x}(0) = \begin{bmatrix} 2 \\ -1 \end{bmatrix}.$$

4. a) Verify that $\left\{ \begin{bmatrix} \cos t \\ \sin t \end{bmatrix}, \begin{bmatrix} -\sin t \\ \cos t \end{bmatrix} \right\}$ is a fundamental set of solutions for

$$\mathbf{x}' = \begin{bmatrix} 0 & -1 \\ 1 & 0 \end{bmatrix} \mathbf{x}. \tag{19}$$

b) Find the general solution of (19).

c) Find the solution of (19) for which

$$\mathbf{x}(0) = \begin{bmatrix} 5 \\ 7 \end{bmatrix}.$$

5. a) Verify that $\left\{ \begin{bmatrix} e^{3t} \\ e^{3t} \end{bmatrix}, \begin{bmatrix} 1 \\ -2 \end{bmatrix} \right\}$ is a fundamental set of solutions for

$$\mathbf{x}' = \begin{bmatrix} 2 & 1 \\ 2 & 1 \end{bmatrix} \mathbf{x}.$$

b) Find the general solution of this differential equation.

c) Find the solution for which
$$\mathbf{x}(0) = \begin{bmatrix} 7 \\ -7 \end{bmatrix}.$$

6. a) Verify that $\left\{ \begin{bmatrix} e^{4t} \\ e^{4t} \end{bmatrix}, \begin{bmatrix} 3e^{-t} \\ -2e^{-t} \end{bmatrix} \right\}$ is a fundamental set of solutions for
$$\mathbf{x}' = \begin{bmatrix} 1 & 3 \\ 2 & 2 \end{bmatrix} \mathbf{x}.$$

b) Find the general solution of this differential equation.

c) Find the solution for which
$$\mathbf{x}(0) = \begin{bmatrix} -11 \\ 4 \end{bmatrix}.$$

7. a) Verify that $\left\{ \begin{bmatrix} \cos 2t \\ \sin 2t \end{bmatrix}, \begin{bmatrix} -\sin 2t \\ \cos 2t \end{bmatrix} \right\}$ is a fundamental set of solutions for
$$\mathbf{x}' = \begin{bmatrix} 0 & -2 \\ 2 & 0 \end{bmatrix} \mathbf{x}.$$

b) Find the general solution of this differential equation.

c) Find the solution for which
$$\mathbf{x}(0) = \begin{bmatrix} 19 \\ -37 \end{bmatrix}.$$

8. a) Verify that $\left\{ \begin{bmatrix} e^{2t} \\ e^{2t} \end{bmatrix}, \begin{bmatrix} 3e^{t} \\ 2e^{t} \end{bmatrix} \right\}$ is a fundamental set of solutions of
$$\mathbf{x}' = \begin{bmatrix} -1 & 3 \\ -2 & 4 \end{bmatrix} \mathbf{x}.$$

b) Find the general solution of this differential equation.

c) Find the solution for which
$$\mathbf{x}(0) = \begin{bmatrix} 2 \\ \frac{3}{2} \end{bmatrix}.$$

9. Verify that a matrix $\mathbf{X}(t)$ satisfies the matrix differential equation
$$\mathbf{X}'(t) = \mathbf{A}(t)\mathbf{X}(t)$$
if and only if each column of $\mathbf{X}(t)$ satisfies $\mathbf{x}' = \mathbf{A}(t)\mathbf{x}$.

10. Suppose that $\mathbf{X}(t)$ is a matrix that satisfies $\mathbf{X}'(t) = \mathbf{A}(t)\mathbf{X}(t)$. Let $\mathbf{a}$ be a constant vector. Verify that $\mathbf{X}(t)\mathbf{a}$ is a solution of the vector differential equation $\mathbf{x}' = \mathbf{A}(t)\mathbf{x}$.

The theory of this section not only resembles that of Chapter 3; it includes it as a special case. To see this, consider the scalar linear homogeneous differential equation:
$$y''(t) + b(t)y'(t) + c(t)y(t) = 0, \qquad (20)$$
$$y(t_0) = a_0, \qquad y'(t_0) = a_1, \qquad (21)$$
where a_0, a_1 are constants.

11. Rewrite (20), (21) as a system in the form (recall Example 6.1.7) ($z = y'$)
$$\mathbf{x}'(t) = \mathbf{A}(t)\mathbf{x}(t), \qquad (22)$$
$$\mathbf{x}(t_0) = \mathbf{a}, \qquad (23)$$
with
$$\mathbf{x}(t) = \begin{bmatrix} y(t) \\ z(t) \end{bmatrix}.$$

12. (Same notation as Exercise 11.) Show that two scalar functions $y_1(t), y_2(t)$ are a fundamental set of solutions for the scalar equation $y'' + by' + cy = 0$ if and only if the vectors
$$\mathbf{x}_1(t) = \begin{bmatrix} y_1(t) \\ z_1(t) \end{bmatrix}, \qquad \mathbf{x}_2(t) = \begin{bmatrix} y_2(t) \\ z_2(t) \end{bmatrix}$$
are a fundamental set of solutions to the system (22).

13. (Same notation as Exercise 11, 12.) Show that the det $\mathbf{X}(t)$ in (12) for the system (22) is just the *Wronskian* of the solution $\{y_1, y_2\}$ of (20).

Usually when problems are modeled, for example, circuits, they are not initially in the *explicit* form
$$\mathbf{x}'(t) = \mathbf{A}(t)\mathbf{x}(t) + \mathbf{f}(t)$$
but rather in the *implicit* form
$$\mathbf{E}(t)\mathbf{x}'(t) + \mathbf{F}(t)\mathbf{x}(t) = \mathbf{g}(t). \qquad (24)$$
Several examples appear in Sections 6.1, 6.2, 6.3. Exercises 14 through 16 elaborate on some of the ideas behind Theorem 6.1.2.

14. Suppose that $\mathbf{E}(t), \mathbf{F}(t)$ are $n \times n$ matrices, and $\mathbf{E}(t)$ is invertible for each value of t. Show that (24) can be written in the form
$$\mathbf{x}'(t) = \mathbf{A}(t)\mathbf{x}(t) + \mathbf{f}(t) \qquad (25)$$
using the matrix $\mathbf{E}(t)^{-1}$.

15. (Uses Exercise 14.) a) Rewrite the system
$$\begin{aligned} x' + y' + 2x - 3y &= t, \\ x' - y' + 2x - y &= 1, \end{aligned} \qquad (26)$$
in the form $\mathbf{Ex}' + \mathbf{Fx} = \mathbf{g}(t)$.

b) Compute $\mathbf{E}^{-1}$ and use it to rewrite the system (26) in the form

16. a) Rewrite the system

$$x' + y' + x - y + z = \cos t,$$
$$y' - z' - x + 2y - 3z = 0, \quad (27)$$
$$x' + 2z' + 2x - 3y - z = 1$$

in the form $\mathbf{E}\mathbf{x}'(t) + \mathbf{F}\mathbf{x}(t) = \mathbf{g}(t)$.

b) Compute $\mathbf{E}^{-1}$ and use it to rewrite (27) in the form (25).

For Exercises 17 and 18 suppose that $\mathbf{A}(t)$ is 2×2 and continuous for all t. A solution of $\mathbf{x}'(t) = \mathbf{A}(t)\mathbf{x}(t)$ can be thought of as a point

$$\mathbf{x}(t) = \begin{bmatrix} x_1(t) \\ x_2(t) \end{bmatrix}$$

in the x_1, x_2 plane tracing out a curve (or *trajectory*).

17. Let

$$\mathbf{A}(t) = \frac{1}{1+t^2}\begin{bmatrix} t & 1 \\ -1 & t \end{bmatrix}$$

for $-\infty < t < \infty$.

a) Verify that

$$\mathbf{x}_1(t) = \begin{bmatrix} 1 \\ -t \end{bmatrix}, \quad \mathbf{x}_2(t) = \begin{bmatrix} t \\ 1 \end{bmatrix}$$

is a fundamental set of solutions for $\mathbf{x}' = \mathbf{A}\mathbf{x}$.

b) Graph the trajectories $\mathbf{x}_1(t), \mathbf{x}_2(t)$ for all t in the plane, and observe that the trajectories cross, since

$$\mathbf{x}_1(-1) = \mathbf{x}_2(1) = \begin{bmatrix} 1 \\ 1 \end{bmatrix}.$$

(This does not violate uniqueness, since $\mathbf{x}_1, \mathbf{x}_2$ never are at the same place at the same time.)

18. Suppose that $\mathbf{A}$ is a constant matrix and α is a scalar.

a) Show that, if $\mathbf{z}(t)$ is a solution of $\mathbf{x}'(t) = \mathbf{A}\mathbf{x}(t)$, then the function $\mathbf{w}(t) = \mathbf{z}(t - \alpha)$ is also a solution of $\mathbf{x}'(t) = \mathbf{A}\mathbf{x}(t)$.

b) Using the existence and uniqueness Theorem 6.9.1 and part (a), show that if $\mathbf{A}$ is a 2×2 constant matrix and the trajectories in the plane of two solutions $\mathbf{x}_1, \mathbf{x}_2$ ever intersect, then the solutions give the same trajectory and, in fact, there exists a scalar α so that $\mathbf{x}_1(t) = \mathbf{x}_2(t - \alpha)$ for all t. (*Hint*: Start by assuming $\mathbf{x}_1(t_0) = \mathbf{x}_2(t_1)$ for some fixed t_0, t_1).

19. Verify that, if $\mathbf{x}_1, \mathbf{x}_2$ are solutions of $\mathbf{x}' = \mathbf{A}(t)\mathbf{x}$, then $c_1\mathbf{x}_1 + c_2\mathbf{x}_2$ with c_1, c_2 constant is also a solution of $\mathbf{x}' = \mathbf{A}(t)\mathbf{x}$.

20. Suppose that $\mathbf{x}_p$ is a solution of $\mathbf{x}' = \mathbf{A}\mathbf{x} + \mathbf{f}$. Show that any other solution $\mathbf{x}$ of $\mathbf{x}' = \mathbf{A}\mathbf{x} + \mathbf{f}$ may be written as $\mathbf{x} = \mathbf{x}_p + \mathbf{x}_h$, where $\mathbf{x}_h$ is a solution of $\mathbf{x}' = \mathbf{A}\mathbf{x}$.

21. Show that, if $\mathbf{A}(t)$ is invertible for each t, then $(\mathbf{A}^{-1})' = -\mathbf{A}^{-1}\mathbf{A}'\mathbf{A}^{-1}$.

6.10 Homogeneous Systems with Constant Coefficients Using Eigenvectors

6.10.1 Introduction and Real Eigenvalues

This section will show how to solve $\mathbf{x}' = \mathbf{A}\mathbf{x}$ when $\mathbf{A}$ is a constant $n \times n$ matrix. One solution is $\mathbf{x}(t) = \mathbf{0}$. We wish to find nontrivial (nonzero) solutions. If $\mathbf{A}$ were a scalar, we know that a solution would be $x = e^{At}$. This suggests looking for solutions of the form

$$\mathbf{x} = e^{\lambda t}\mathbf{u}, \quad (1)$$

where **u** is a constant nonzero vector and λ is a scalar. Substituting $e^{\lambda t}\mathbf{u}$ for **x** in $\mathbf{x}' = \mathbf{A}\mathbf{x}$, we get

$$(e^{\lambda t}\mathbf{u})' = \mathbf{A}(e^{\lambda t}\mathbf{u})$$

or

$$\lambda e^{\lambda t}\mathbf{u} = e^{\lambda t}\mathbf{A}\mathbf{u},$$

and, upon division by $e^{\lambda t}$,

$$\lambda\mathbf{u} = \mathbf{A}\mathbf{u}. \tag{2}$$

In summary,

If λ is a scalar and **u** is a *nonzero vector* such that $\mathbf{A}\mathbf{u} = \lambda\mathbf{u}$, then $e^{\lambda t}\mathbf{u}$ is a nontrivial solution of $\mathbf{x}' = \mathbf{A}\mathbf{x}$. (3)

The scalar λ in (2) is called an *eigenvalue* of **A** and the vector **u** is called an *eigenvector* of **A** (*associated* with the eigenvalue λ). It is not obvious at first that such a λ, **u** need exist. However, by the end of this section we shall see that eigenvalues and eigenvectors are very helpful in solving linear systems of differential equations with constant coefficients.

Example 6.10.1   Let

$$\mathbf{A} = \begin{bmatrix} 1 & 2 \\ 2 & 1 \end{bmatrix}, \quad \mathbf{u} = \begin{bmatrix} 1 \\ 1 \end{bmatrix}, \quad \lambda = 3.$$

Then

$$\mathbf{A}\mathbf{u} = 3\mathbf{u} = \begin{bmatrix} 3 \\ 3 \end{bmatrix}.$$

Thus **u** is an eigenvector of **A** associated with the eigenvalue 3. ∎

We now turn to the problem of computing eigenvalues and eigenvectors. Unfortunately, we cannot, without a major digression, present any of the methods that are often used in practice. Instead, we shall present a method that, while generally applicable in principle, is actually used only for small examples like the ones we shall work in this section.

Notational Comment If **z** is an $n \times 1$ vector of variables, say in the equation $\mathbf{A}\mathbf{z} = \mathbf{0}$, we shall usually denote these variables by $\{z_1, \ldots, z_n\}$, so that

$$\mathbf{z} = \begin{bmatrix} z_1 \\ z_2 \\ \vdots \\ z_n \end{bmatrix}.$$

On the other hand, a collection of r vectors, for example, r solutions of $\mathbf{Az} = \mathbf{0}$, would be denoted $\{\mathbf{z}_1, \mathbf{z}_2, \ldots, \mathbf{z}_r\}$. We shall not need a notation to refer to the individual entries of a collection of vectors.

Note that $\mathbf{Au} = \lambda \mathbf{u}$ may be rewritten as

$$(\lambda \mathbf{I} - \mathbf{A})\mathbf{u} = \mathbf{0}. \tag{4}$$

If $\lambda \mathbf{I} - \mathbf{A}$ were invertible, we could multiply (4) by $(\lambda \mathbf{I} - \mathbf{A})^{-1}$ and conclude that $\mathbf{u} = \mathbf{0}$. But eigenvectors are always nonzero. Thus $\lambda \mathbf{I} - \mathbf{A}$ cannot be invertible if λ is an eigenvalue. But, from Theorem 6.8.2, $\lambda \mathbf{I} - \mathbf{A}$ is not invertible if and only if $\det(\lambda \mathbf{I} - \mathbf{A}) = 0$.

Theorem 6.10.1 Let $p(\lambda) = \det(\lambda \mathbf{I} - \mathbf{A})$, where $\mathbf{A}, \mathbf{I}$ are $n \times n$ matrices. Then $p(\lambda)$ is an nth-degree polynomial called the *characteristic* (or *auxiliary*) *polynomial* of $\mathbf{A}$. The eigenvalues of $\mathbf{A}$ are precisely the roots of the characteristic polynomial $p(\lambda)$.

Example 6.10.2 Let

$$\mathbf{A} = \begin{bmatrix} 1 & 2 \\ 2 & 1 \end{bmatrix}.$$

Find the characteristic polynomial and eigenvalues of $\mathbf{A}$.

Solution The characteristic polynomial is

$$p(\lambda) = \det(\lambda \mathbf{I} - \mathbf{A}) = \det\left(\begin{bmatrix} \lambda & 0 \\ 0 & \lambda \end{bmatrix} - \begin{bmatrix} 1 & 2 \\ 2 & 1 \end{bmatrix}\right)$$

$$= \det\begin{bmatrix} \lambda - 1 & -2 \\ -2 & \lambda - 1 \end{bmatrix}$$

$$= (\lambda - 1)^2 - 4 = \lambda^2 - 2\lambda - 3 = (\lambda - 3)(\lambda + 1).$$

The eigenvalues of $\mathbf{A}$ are the roots of the characteristic polynomial which, in this example, are $\lambda_1 = 3, \lambda_2 = -1$. ∎

Once the eigenvalues of $\mathbf{A}$ are found, we proceed as follows. To find the eigenvectors of $\mathbf{A}$, solve the system

$$(\lambda \mathbf{I} - \mathbf{A})\mathbf{u} = \mathbf{0} \tag{5}$$

for each eigenvalue. Equation (5) does not uniquely determine the eigenvectors. There will be one or more arbitrary constants in the solution. Once (5) is solved for a given eigenvalue, call it λ_0, we expand the solution in vector form, as in Example 6.7.10. Taking the vector times each arbitrary constant gives a set of linearly independent eigenvectors corresponding to the eigenvalue λ_0. The number of linearly independent eigenvectors is the same as the number of arbitrary constants in the solution of (5).

Example 6.10.3 Suppose that we solve (5) for a given eigenvalue, say $\lambda = 2$, and find that the solution is

$$\mathbf{u} = \begin{bmatrix} 3u_2 \\ -u_2 \\ u_2 \end{bmatrix},$$

or upon factoring,

$$\mathbf{u} = u_2 \begin{bmatrix} 3 \\ -1 \\ 1 \end{bmatrix}.$$

Then we would get

$$e^{2t} \begin{bmatrix} 3 \\ -1 \\ 1 \end{bmatrix}, \tag{6}$$

as a solution of $\mathbf{x}' = \mathbf{A}\mathbf{x}$. ∎

Example 6.10.4 Suppose that we solve (5) for a given eigenvalue, say $\lambda = -3$, and find that the solution of $(\lambda \mathbf{I} - \mathbf{A})\mathbf{u} = \mathbf{0}$ is:

$$\mathbf{u} = \begin{bmatrix} u_2 - u_3 \\ 2u_2 \\ 3u_3 - u_2 \end{bmatrix} = u_2 \begin{bmatrix} 1 \\ 2 \\ -1 \end{bmatrix} + u_3 \begin{bmatrix} -1 \\ 0 \\ 3 \end{bmatrix}.$$

Then

$$e^{-3t} \begin{bmatrix} 1 \\ 2 \\ -1 \end{bmatrix}, \quad e^{-3t} \begin{bmatrix} -1 \\ 0 \\ 3 \end{bmatrix} \tag{7}$$

will both be solutions of $\mathbf{x}' = \mathbf{A}\mathbf{x}$. ∎

Any nonzero multiple of an eigenvector may be used. For example, we could use

$$e^{-3t} \begin{bmatrix} 2 \\ 4 \\ -2 \end{bmatrix} \quad \text{instead of} \quad e^{-3t} \begin{bmatrix} 1 \\ 2 \\ -1 \end{bmatrix} \text{ in (7)}.$$

It turns out that the solutions of $\mathbf{x}' = \mathbf{A}\mathbf{x}$ generated in this manner are always linearly independent.

Theorem 6.10.2 If $\mathbf{A}$ is an $n \times n$ matrix and $\mathbf{A}$ has n *distinct* eigenvalues $\lambda_1, \ldots, \lambda_n$, then for each i, the solution of $(\lambda_i \mathbf{I} - \mathbf{A})\mathbf{u} = \mathbf{0}$ will have a solution of the form $c\mathbf{u}_i$, where c is an arbitrary constant. The set of vector functions
$$\{e^{\lambda_1 t}\mathbf{u}_1, e^{\lambda_2 t}\mathbf{u}_2, \ldots, e^{\lambda_n t}\mathbf{u}_n\}$$
is a fundamental set of solutions for $\mathbf{x}' = \mathbf{A}\mathbf{x}$.

Example 6.10.5 Find a fundamental set of solutions for
$$\begin{aligned} x' &= x + 2y, \\ y' &= 2x + y. \end{aligned} \tag{8}$$

Solution Let
$$\mathbf{A} = \begin{bmatrix} 1 & 2 \\ 2 & 1 \end{bmatrix}, \quad \mathbf{x} = \begin{bmatrix} x \\ y \end{bmatrix},$$
so that (8) can be written as $\mathbf{x}' = \mathbf{A}\mathbf{x}$. From Example 6.10.2 we know that $\mathbf{A}$ has characteristic polynomial
$$p(\lambda) = \det(\lambda \mathbf{I} - \mathbf{A}) = \lambda^2 - 2\lambda - 3 = (\lambda - 3)(\lambda + 1),$$
so that the eigenvalues are $\lambda_1 = 3$, $\lambda_2 = -1$. We now need to compute the eigenvectors using (5). For $\lambda_1 = 3$, $(\lambda \mathbf{I} - \mathbf{A})\mathbf{u} = \mathbf{0}$ becomes
$$\left(\begin{bmatrix} 3 & 0 \\ 0 & 3 \end{bmatrix} - \begin{bmatrix} 1 & 2 \\ 2 & 1 \end{bmatrix} \right) \mathbf{u} = \mathbf{0}$$
or
$$\begin{bmatrix} 2 & -2 \\ -2 & 2 \end{bmatrix} \begin{bmatrix} u_1 \\ u_2 \end{bmatrix} = \begin{bmatrix} 0 \\ 0 \end{bmatrix}.$$
Performing two row operations on the augmented matrix
$$\begin{bmatrix} 2 & -2 & | & 0 \\ -2 & 2 & | & 0 \end{bmatrix}$$
yields
$$\begin{bmatrix} 1 & -1 & | & 0 \\ 0 & 0 & | & 0 \end{bmatrix}.$$
Thus the eigenvectors are u_2 arbitrary, $u_1 - u_2 = 0$, or $u_1 = u_2$, that is,
$$\mathbf{u} = \begin{bmatrix} u_1 \\ u_2 \end{bmatrix} = \begin{bmatrix} u_2 \\ u_2 \end{bmatrix} = u_2 \begin{bmatrix} 1 \\ 1 \end{bmatrix}.$$
We take

$$\begin{bmatrix} 1 \\ 1 \end{bmatrix}$$

as our eigenvector (any nonzero multiple of it would do) and

$$e^{3t}\begin{bmatrix} 1 \\ 1 \end{bmatrix}$$

is one solution of (8). Similarly for the other eigenvalue $\lambda_2 = -1$, $(\lambda \mathbf{I} - \mathbf{A})\mathbf{u} = \mathbf{0}$ becomes

$$\left(\begin{bmatrix} -1 & 0 \\ 0 & -1 \end{bmatrix} - \begin{bmatrix} 1 & 2 \\ 2 & 1 \end{bmatrix}\right)\mathbf{u} = \mathbf{0}$$

or

$$\begin{bmatrix} -2 & -2 \\ -2 & -2 \end{bmatrix}\mathbf{u} = \mathbf{0}.$$

The solutions are $u_1 = -u_2$, u_2 arbitrary, or

$$\mathbf{u} = \begin{bmatrix} -u_2 \\ u_2 \end{bmatrix} = u_2 \begin{bmatrix} -1 \\ 1 \end{bmatrix}.$$

We may take

$$\begin{bmatrix} -1 \\ 1 \end{bmatrix}$$

as our eigenvector and

$$e^{-t}\begin{bmatrix} -1 \\ 1 \end{bmatrix}$$

is a second solution of (8). Since $\mathbf{A}$ is 2×2, and we have two linearly independent solutions of (8),

$$\left\{ e^{3t}\begin{bmatrix} 1 \\ 1 \end{bmatrix}, \; e^{-t}\begin{bmatrix} -1 \\ 1 \end{bmatrix} \right\}$$

is a fundamental set of solutions. The general solution of (8) would be

$$\mathbf{x} = c_1 e^{3t}\begin{bmatrix} 1 \\ 1 \end{bmatrix} + c_2 e^{-t}\begin{bmatrix} -1 \\ 1 \end{bmatrix}$$

or

$$\begin{aligned} x &= c_1 e^{3t} - c_2 e^{-t}, \\ y &= c_1 e^{3t} + c_2 e^{-t}. \end{aligned} \blacksquare \tag{9}$$

Example 6.10.6

If some eigenvalue is a repeated root of the characteristic polynomial, then the method we have just described may not find all solutions. How to find the other solutions in this case will be discussed in Section 6.10.3. In Section 6.10.2 we will discuss how to handle complex eigenvalues.

Solve the system of differential equations

$$x'_1 = -2x_1 + x_2 + x_3,$$
$$x'_2 = x_1 - 2x_2 + x_3, \quad (10)$$
$$x'_3 = x_1 + x_2 - 2x_3.$$

Solution Letting

$$\mathbf{x} = \begin{bmatrix} x_1 \\ x_2 \\ x_3 \end{bmatrix}, \quad \mathbf{A} = \begin{bmatrix} -2 & 1 & 1 \\ 1 & -2 & 1 \\ 1 & 1 & -2 \end{bmatrix},$$

we may rewrite (10) as $\mathbf{x}' = \mathbf{A}\mathbf{x}$. First, we find the characteristic polynomial of $\mathbf{A}$,

$$p(\lambda) = \det(\lambda \mathbf{I} - \mathbf{A}) = \det \begin{bmatrix} \lambda + 2 & -1 & -1 \\ -1 & \lambda + 2 & -1 \\ -1 & -1 & \lambda + 2 \end{bmatrix}.$$

Expanding along the first row, we obtain

$$p(\lambda) = (\lambda + 2)(-1)^{1+1} \det \begin{bmatrix} \lambda + 2 & -1 \\ -1 & \lambda + 2 \end{bmatrix}$$
$$+ (-1)(-1)^{1+2} \det \begin{bmatrix} -1 & -1 \\ -1 & \lambda + 2 \end{bmatrix}$$
$$+ (-1)(-1)^{1+3} \det \begin{bmatrix} -1 & \lambda + 2 \\ -1 & -1 \end{bmatrix}$$
$$= (\lambda + 2)[(\lambda + 2)^2 - 1] + (-\lambda - 2 - 1) - (1 + \lambda + 2)$$
$$= \lambda^3 + 6\lambda^2 + 9\lambda = \lambda(\lambda + 3)^2.$$

The roots of $p(\lambda)$ are $\lambda_1 = 0$ and $\lambda_2 = -3$. The *multiplicity* of an eigenvalue is its multiplicity as a root of the characteristic polynomial. In this example, $\lambda_1 = 0$ is an eigenvalue of multiplicity one, while λ_2 is an eigenvalue of multiplicity two. We now need to find the corresponding eigenvectors.

For $\lambda_1 = 0$, $(\lambda_1 \mathbf{I} - \mathbf{A})\mathbf{u} = \mathbf{0}$ is

$$\begin{bmatrix} 2 & -1 & -1 \\ -1 & 2 & -1 \\ -1 & -1 & 2 \end{bmatrix} \mathbf{u} = \mathbf{0}. \quad (11)$$

To solve (11) using Gaussian elimination, Algorithm 6.7.1, we consider the augmented matrix

$$\begin{bmatrix} 2 & -1 & -1 & | & 0 \\ -1 & 2 & -1 & | & 0 \\ -1 & -1 & 2 & | & 0 \end{bmatrix}. \tag{12}$$

Adding $\frac{1}{2}$ times row 1 to rows 2 and 3 yields

$$\begin{bmatrix} 2 & -1 & -1 & | & 0 \\ 0 & \frac{3}{2} & -\frac{3}{2} & | & 0 \\ 0 & -\frac{3}{2} & \frac{3}{2} & | & 0 \end{bmatrix}.$$

Now add row 2 to row 3 and then multiply row 2 by $\frac{2}{3}$:

$$\begin{bmatrix} 2 & -1 & -1 & | & 0 \\ 0 & 1 & -1 & | & 0 \\ 0 & 0 & 0 & | & 0 \end{bmatrix}.$$

This completes Part 1 of Algorithm 6.7.1. Now add row 2 to row 1 and multiply row 1 by $\frac{1}{2}$, to get the row echelon form

$$\begin{bmatrix} 1 & 0 & -1 & | & 0 \\ 0 & 1 & -1 & | & 0 \\ 0 & 0 & 0 & | & 0 \end{bmatrix}.$$

The distinguished columns are 1 and 2. Thus we may take u_3 arbitrary and

$$u_1 - u_3 = 0, \quad u_2 - u_3 = 0 \quad \text{gives us } u_1 = u_3, u_2 = u_3.$$

Thus the eigenvectors for $\lambda_1 = 0$ are

$$\mathbf{u} = \begin{bmatrix} u_3 \\ u_3 \\ u_3 \end{bmatrix} = u_3 \begin{bmatrix} 1 \\ 1 \\ 1 \end{bmatrix}.$$

We choose

$$\begin{bmatrix} 1 \\ 1 \\ 1 \end{bmatrix}$$

as our eigenvector, so that

$$\mathbf{x}_1 = e^{0t} \begin{bmatrix} 1 \\ 1 \\ 1 \end{bmatrix} = \begin{bmatrix} 1 \\ 1 \\ 1 \end{bmatrix} \tag{13}$$

is one solution of $\mathbf{x}' = \mathbf{A}\mathbf{x}$.

Now we must find the eigenvectors for $\lambda_2 = -3$. The eigenvectors **u** satisfy $(\lambda_2 \mathbf{I} - \mathbf{A})\mathbf{u} = \mathbf{0}$, or

$$\begin{bmatrix} -1 & -1 & -1 \\ -1 & -1 & -1 \\ -1 & -1 & -1 \end{bmatrix} \mathbf{u} = \mathbf{0}. \tag{14}$$

The augmented matrix for the system of equations (14) is

$$\begin{bmatrix} -1 & -1 & -1 & | & 0 \\ -1 & -1 & -1 & | & 0 \\ -1 & -1 & -1 & | & 0 \end{bmatrix},$$

which, after performing two row operations, is

$$\begin{bmatrix} 1 & 1 & 1 & | & 0 \\ 0 & 0 & 0 & | & 0 \\ 0 & 0 & 0 & | & 0 \end{bmatrix}.$$

Only the first column is distinguished. We take u_2, u_3 arbitrary and $u_1 + u_2 + u_3 = 0$ gives $u_1 = -u_2 - u_3$. Thus,

$$\mathbf{u} = \begin{bmatrix} u_1 \\ u_2 \\ u_3 \end{bmatrix} = \begin{bmatrix} -u_2 - u_3 \\ u_2 \\ u_3 \end{bmatrix} = u_2 \begin{bmatrix} -1 \\ 1 \\ 0 \end{bmatrix} + u_3 \begin{bmatrix} -1 \\ 0 \\ 1 \end{bmatrix}.$$

Corresponding to the eigenvalue $\lambda_2 = -3$, we take the two linearly independent eigenvectors

$$\begin{bmatrix} -1 \\ 1 \\ 0 \end{bmatrix}, \quad \begin{bmatrix} -1 \\ 0 \\ 1 \end{bmatrix},$$

which give us the solutions

$$e^{-3t} \begin{bmatrix} -1 \\ 1 \\ 0 \end{bmatrix}, \quad e^{-3t} \begin{bmatrix} -1 \\ 0 \\ 1 \end{bmatrix}. \tag{15}$$

The matrix **A** is 3×3, and (13) and (15) provide us with three linearly independent solutions, so that

$$\mathbf{x} = \begin{bmatrix} x_1 \\ x_2 \\ x_3 \end{bmatrix} = c_1 \begin{bmatrix} 1 \\ 1 \\ 1 \end{bmatrix} + c_2 e^{-3t} \begin{bmatrix} -1 \\ 1 \\ 0 \end{bmatrix} + c_3 e^{-3t} \begin{bmatrix} -1 \\ 0 \\ 1 \end{bmatrix},$$

or, equivalently,

$$x_1 = c_1 - c_2 e^{-3t} - c_3 e^{-3t},$$
$$x_2 = c_1 + c_2 e^{-3t},$$
$$x_3 = c_1 + c_3 e^{-3t},$$

is the general solution of (10). ∎

Helpful Check If **A** is $n \times n$, then the sum of the diagonal entries is the *trace* of **A**, denoted: trace(**A**). The trace is an easy-to-compute number that gives us two checks on our work.

$-$trace(**A**) is the coefficient of λ^{n-1} in the characteristic polynomial of **A**; (16)

trace(**A**) is the sum of all the eigenvalues of **A** if they are repeated according to their multiplicities. (17)

Example 6.10.7

If **A** is the matrix in Example 6.10.6,
$$\text{trace}(\mathbf{A}) = -2 + -2 + -2 = \boxed{-6}.$$

The characteristic polynomial was
$$p(\lambda) = \lambda^3 + \boxed{6}\lambda^2 + 9\lambda.$$

The eigenvalues, repeated according to multiplicity, were 0, -3, -3, and $0 + -3 + -3 = \boxed{-6}$. ∎

The discussion and examples of this section may be summarized as follows.

Summary of Method for Solving x' = Ax for a constant matrix A.

1. Compute the characteristic polynomial of **A**, $p(\lambda) = \det(\lambda \mathbf{I} - \mathbf{A})$. Use the trace as a check.
2. Find the roots of $p(\lambda)$; these are the eigenvalues $\lambda_1, \ldots, \lambda_m$ of **A**. Check if trace(**A**) = $\lambda_1 + \cdots + \lambda_m$.
3. If an eigenvalue is complex, see Section 6.10.2.
4. For each eigenvalue λ_i, solve the system $(\lambda_i \mathbf{I} - \mathbf{A})\mathbf{u} = \mathbf{0}$ to get a linearly independent set of eigenvectors.
5. If the number of linear independent eigenvectors for any eigenvalue is less than the multiplicity of that eigenvalue, see Section 6.10.3.
6. If (3) and (5) have not occurred, we now have a fundamental set of solutions of the form $e^{\lambda t}\mathbf{u}$.

We close this section with an example that illustrates what can sometimes happen with eigenvalues whose multiplicities are greater than one. A complete solution to this problem will be given in Section 6.10.3.

Example 6.10.8 Find all the solutions of

$$x'_1 = 3x_1 - x_2,$$
$$x'_2 = x_1 + x_2, \tag{18}$$

that are possible, using the eigenvalue/eigenvector method of this section.

Solution Let

$$\mathbf{x} = \begin{bmatrix} x_1 \\ x_2 \end{bmatrix}, \quad \mathbf{A} = \begin{bmatrix} 3 & -1 \\ 1 & 1 \end{bmatrix},$$

so that (18) is $\mathbf{x}' = \mathbf{A}\mathbf{x}$. First we compute the characteristic polynomial of $\mathbf{A}$,

$$p(\lambda) = \det[\lambda \mathbf{I} - \mathbf{A}] = \det \begin{bmatrix} \lambda - 3 & 1 \\ -1 & \lambda - 1 \end{bmatrix} = \lambda^2 - 4\lambda + 4.$$

Since $p(\lambda) = (\lambda - 2)^2$, there is a single eigenvalue $\lambda_1 = 2$ of multiplicity 2. To find the eigenvectors associated with $\lambda_1 = 2$, we solve $(\lambda_1 \mathbf{I} - \mathbf{A})\mathbf{u} = \mathbf{0}$ or

$$\begin{bmatrix} -1 & 1 \\ -1 & 1 \end{bmatrix} \mathbf{u} = \mathbf{0},$$

which is $-u_1 + u_2 = 0$. Thus we may take u_2 arbitrary, $u_1 = u_2$, and

$$\mathbf{u} = \begin{bmatrix} u_1 \\ u_2 \end{bmatrix} = \begin{bmatrix} u_2 \\ u_2 \end{bmatrix} = u_2 \begin{bmatrix} 1 \\ 1 \end{bmatrix}.$$

We have then

$$e^{2t} \begin{bmatrix} 1 \\ 1 \end{bmatrix}$$

is one solution. However, this method does not provide a second linearly independent solution. The reason, as will be shown in Section 6.10.3, is that the other solutions are not of the form $e^{\lambda t}\mathbf{u}$. ∎

Symmetric Matrices

A square matrix is *symmetric* if the i, j-entry equals the j, i-entry. For example,

$$\begin{bmatrix} 1 & 2 \\ 2 & 3 \end{bmatrix}, \quad \begin{bmatrix} 0 & 1 & 0 \\ 1 & 2 & 3 \\ 0 & 3 & 4 \end{bmatrix}, \quad \begin{bmatrix} 1 & 2 & 3 \\ 2 & 4 & 5 \\ 3 & 5 & 6 \end{bmatrix}$$

are all symmetric matrices. The $\mathbf{A}$ of Example 6.10.5 is also symmetric. Symmetric matrices arise in many applications (see Exercises 6.5.15 through 6.5.17).

Theorem 6.10.3 Suppose $\mathbf{A}$ is an $n \times n$ real symmetric matrix. Then

i. All eigenvalues of $\mathbf{A}$ are real.

ii. The number of linearly independent eigenvectors for each eigenvalue is the same as the multiplicity of the eigenvalue.

Thus all solutions of $\mathbf{x}' = \mathbf{A}\mathbf{x}$ can be found by the method of this section if $\mathbf{A}$ is symmetric.

Note that

$$\mathbf{A} = \begin{bmatrix} 3 & -1 \\ 1 & 1 \end{bmatrix}$$

of Example 6.10.8 is not symmetric, since the 1, 2-entry is -1 while the 2, 1-entry is 1.

Exercises

In Exercises 1 through 13, the method of Section 6.10.1 will provide a fundamental set of solutions and all eigenvalues have multiplicity one. Eigenvalues with multiplicity greater than one appear in Exercises 14 through 17.

In Exercises 1 through 17, solve the differential equation using the eigenvalue/eigenvector method. If no initial conditions are given, give the general solution. When checking your answers with those in the back of the book, keep in mind that, in the general solution, any nonzero multiple of the given eigenvectors may be used.

1. $x_1' = 3x_1 + x_2,$
   $x_2' = -2x_1$

2. $x_1' = x_1,$
   $x_2' = x_1 - x_2$

3. $x' = x - 2y, \quad x(0) = 1,$
   $y' = -2x + 4y, \quad y(0) = 2$

4. $x' = 3x + y,$
   $y' = -6x - 2y$

5. $x_1' = 6x_1 + 2x_2,$
   $x_2' = 2x_1 + 3x_2$

6. $x_1' = 3x_1 + x_2, \quad x_1(0) = 1,$
   $x_2' = -5x_1 - 3x_2, \quad x_2(0) = -2$

7. $v' = 2v - 2w,$
   $w' = -2v + 5w$

8. $v' = -3v + 3w, \quad v(0) = 0,$
   $w' = 3v + 5w, \quad w(0) = 2$

9. $x_1' = 2x_1 + x_3,$
   $x_2' = 2x_2 + x_3,$
   $x_3' = x_1 + x_2 + x_3$

10. $x' = x - 2y + z, \quad x(0) = 1,$
    $y' = -2x + 4y + z, \quad y(0) = -1,$
    $z' = 3z, \quad z(0) = 2$

11. $x_1' = x_1 + x_3,$
    $x_2' = 3x_2,$
    $x_3' = x_1 + x_3$

12. $u' = u + w, \quad u(0) = 0,$
    $v' = v - w, \quad v(0) = 0,$
    $w' = u - v, \quad w(0) = 2$

13. $x_1' = 2x_1 - 2x_2 - x_3,$
    $x_2' = -2x_1 + 2x_2 + x_3,$
    $x_3' = -x_1 + x_2 + 5x_3$

14. $x_1' = -2x_1 - x_2 + x_3,$
    $x_2' = -x_1 - 2x_2 - x_3,$
    $x_3' = x_1 - x_2 - 2x_3$

15. $x' = 2x + z,$
    $y' = y,$
    $z' = x + 2z$

16. $x_1' = x_1 + x_2 + 2x_3$,
    $x_2' = x_1 + x_2 + 2x_3$,
    $x_3' = 2x_1 + 2x_2 + 4x_3$
17. $x_1' = x_2 + x_4$,
    $x_2' = x_1 - x_3$,
    $x_3' = x_4$,
    $x_4' = x_3$
18. Theorem 6.8.1 can sometimes be used on matrices with variable coefficients. In Example 7.10.6,
$$\lambda I - A = \begin{bmatrix} \lambda+2 & -1 & -1 \\ -1 & \lambda+2 & -1 \\ -1 & -1 & \lambda+2 \end{bmatrix}.$$
Using Theorem 6.8.1, evaluate $\det(\lambda I - A)$ by using the 1,3-entry to zero out the 2,3- and 3,3-entries. Then expand the determinant down the third column.

19. Evaluate
$$\det \begin{bmatrix} \lambda & 2 & 1 \\ 1 & \lambda-1 & 1 \\ 1 & 2 & \lambda \end{bmatrix},$$
using Theorem 6.8.1 by using the 3,1-entry to zero out the 3,2- and 3,3-entry and then expanding the determinant along the bottom row.

20. Rewrite the differential equation $y'' + by' + cy = 0$, where b, c are constants, as a system $\mathbf{x}' = \mathbf{Ax}$ using
$$\mathbf{x} = \begin{bmatrix} y \\ y' \end{bmatrix}.$$
Show that the characteristic equation of $y'' + by' + cy$, as defined in Chapter 3, is the same as the characteristic equation of $\mathbf{x}' = \mathbf{Ax}$ defined in Section 6.10.1.

6.10.2 Complex Eigenvalues

As we saw in the applications of linear, constant-coefficient differential equations in Sections 3.16 and 3.17, solutions of the form $e^{\alpha t} \cos \beta t$, $e^{\alpha t} \sin \beta t$ are important physically; they occur when the characteristic equation has complex roots. The same is true for systems, as this section will show. (*Note:* Exercise 20 of Section 6.10.1 shows that both types of characteristic equation are actually the same.)

Assume, then, that $\mathbf{A}$ is an $n \times n$ *real matrix* (has real entries) and λ is a complex eigenvalue. Write λ as $\lambda = \alpha + i\beta$, where α, β are real numbers. Corresponding to λ there will be an eigenvector $\mathbf{u}$, so that $\mathbf{Au} = \lambda \mathbf{u}$. We may also break $\mathbf{u}$ into real and imaginary parts, $\mathbf{u} = \mathbf{a} + i\mathbf{b}$. For example,
$$\begin{bmatrix} 1-3i \\ 2+i \end{bmatrix} = \begin{bmatrix} 1 \\ 2 \end{bmatrix} + i\begin{bmatrix} -3 \\ 1 \end{bmatrix}.$$
The key fact we need is the following:

If $\mathbf{A}$ is an $n \times n$ real matrix with complex eigenvalue λ, and eigenvector $\mathbf{u}$,
$$\lambda = \alpha + i\beta, \qquad \mathbf{u} = \mathbf{a} + i\mathbf{b}, \tag{1}$$
then $\bar{\lambda}$ is another eigenvalue and $\bar{\mathbf{u}}$ its associated eigenvector, where
$$\bar{\lambda} = \alpha - i\beta, \qquad \bar{\mathbf{u}} = \mathbf{a} - i\mathbf{b}.$$

(Recall that $\bar{\lambda}$ is called the *complex conjugate* of λ.) That $\bar{\lambda}$ is also an eigenvalue follows from the fact that $p(\lambda) = \det(\lambda \mathbf{I} - \mathbf{A})$ has real coefficients if $\mathbf{A}$ is real, and thus the complex roots of the characteristic polynomial will occur in conjugate pairs. On the other hand, if $\mathbf{Au} = \lambda \mathbf{u}$, then taking conjugates of both sides gives

$$\overline{\mathbf{Au}} = \overline{\lambda \mathbf{u}}$$

or

$$\mathbf{A}\bar{\mathbf{u}} = \bar{\lambda}\bar{\mathbf{u}},$$

since $\mathbf{A}$ is real, so that $\bar{\mathbf{u}}$ is an eigenvector for the eigenvalue $\bar{\lambda}$.

Fact (1) is very helpful since it means that we actually have to compute only the eigenvectors for *one* of the eigenvalues in a conjugate pair.

Example 6.10.9 Find the eigenvalues and associated eigenvectors for

$$\mathbf{A} = \begin{bmatrix} 1 & -1 \\ 1 & 1 \end{bmatrix}.$$

Solution The characteristic polynomial is

$$p(\lambda) = \det(\lambda \mathbf{I} - \mathbf{A}) = \det \begin{bmatrix} \lambda - 1 & 1 \\ -1 & \lambda - 1 \end{bmatrix} = \lambda^2 - 2\lambda + 2.$$

The roots of the characteristic polynomial can be found using the quadratic formula

$$\lambda = \frac{2 \pm \sqrt{4 - 8}}{2} = 1 \pm i.$$

The eigenvalues are thus $\lambda_1 = 1 + i$, $\lambda_2 = 1 - i$. First, we find the eigenvectors for λ_1. Solving $(\lambda_1 \mathbf{I} - \mathbf{A})\mathbf{u} = \mathbf{0}$ gives

$$\begin{bmatrix} i & 1 \\ -1 & i \end{bmatrix} \mathbf{u} = \mathbf{0}.$$

The augmented matrix is

$$\begin{bmatrix} i & 1 & | & 0 \\ -1 & i & | & 0 \end{bmatrix}.$$

There are several ways to proceed. Exchange rows and multiply the first row by -1,

$$\begin{bmatrix} 1 & -i & | & 0 \\ i & 1 & | & 0 \end{bmatrix}.$$

Now add $-i$ times the first row to the second row:

$$\begin{bmatrix} 1 & -i & | & 0 \\ 0 & 0 & | & 0 \end{bmatrix}.$$

Thus $u_1 - iu_2 = 0$ or $u_1 = iu_2$, u_2 arbitrary, and

$$\mathbf{u} = \begin{bmatrix} iu_2 \\ u_2 \end{bmatrix} = u_2 \begin{bmatrix} i \\ 1 \end{bmatrix}.$$

We may take

$$\mathbf{u}_1 = \begin{bmatrix} i \\ 1 \end{bmatrix}$$

as the eigenvector associated with $\lambda_1 = 1 + i$. Then

$$\mathbf{u}_2 = \bar{\mathbf{u}}_1 = \overline{\begin{bmatrix} i \\ 1 \end{bmatrix}} = \begin{bmatrix} \bar{i} \\ \bar{1} \end{bmatrix} = \begin{bmatrix} -i \\ 1 \end{bmatrix}$$

will be the eigenvector associated with $\lambda_2 = 1 - i$. ∎

We still have $e^{\lambda t}\mathbf{u}$, $e^{\bar{\lambda} t}\bar{\mathbf{u}}$ as solutions but, as in Chapter 3, we wish to replace them by real solutions. Now, if $\lambda = \alpha + i\beta$, $\mathbf{u} = \mathbf{a} + i\mathbf{b}$, then

$$e^{\lambda t}\mathbf{u} = e^{(\alpha + i\beta)t}(\mathbf{a} + i\mathbf{b}) = e^{\alpha t}(\cos \beta t + i \sin \beta t)(\mathbf{a} + i\mathbf{b})$$
$$= e^{\alpha t}(\cos \beta t\, \mathbf{a} - \sin \beta t\, \mathbf{b}) + e^{\alpha t}i(\cos \beta t\, \mathbf{b} + \sin \beta t\, \mathbf{a}),$$

while

$$e^{\bar{\lambda} t}\bar{\mathbf{u}} = e^{(\alpha - i\beta)t}(\mathbf{a} - i\mathbf{b}) = e^{\alpha t}(\cos \beta t - i \sin \beta t)(\mathbf{a} - i\mathbf{b})$$
$$= e^{\alpha t}(\cos \beta t\, \mathbf{a} - \sin \beta t\, \mathbf{b}) - e^{\alpha t}i(\cos \beta t\, \mathbf{b} + \sin \beta t\, \mathbf{a}).$$

Since $e^{\lambda t}\mathbf{u}$, $e^{\bar{\lambda} t}\bar{\mathbf{u}}$ are solutions of $\mathbf{x}' = \mathbf{A}\mathbf{x}$, $c_1 e^{\lambda t}\mathbf{u} + c_2 e^{\bar{\lambda} t}\bar{\mathbf{u}}$ will also be a solution for any scalars c_1, c_2. Choosing

$$c_1 = c_2 = \frac{1}{2}$$

yields

$$e^{\alpha t}[\cos \beta t\, \mathbf{a} - \sin \beta t\, \mathbf{b}],$$

while

$$c_1 = \frac{1}{2i}, \quad c_2 = -\frac{1}{2i}$$

yields

$$e^{\alpha t}[\cos \beta t\, \mathbf{b} + \sin \beta t\, \mathbf{a}].$$

In summary,

6.10 Homogeneous Systems with Constant Coefficients Using Eigenvectors

Theorem 6.10.4

Real Solutions for Complex Eigenvalues If A is an $n \times n$ real matrix with complex eigenvalue $\lambda = \alpha + i\beta$ and associated eigenvector $\mathbf{u} = \mathbf{a} + i\mathbf{b}$, then $\bar{\lambda} = \alpha - i\beta$ is also an eigenvalue with associated eigenvector $\mathbf{a} - i\mathbf{b}$. Furthermore,

$$\mathbf{x}_1(t) = e^{\alpha t}[\cos \beta t \, \mathbf{a} - \sin \beta t \, \mathbf{b}] \qquad (2)$$

and

$$\mathbf{x}_2(t) = e^{\alpha t}[\cos \beta t \, \mathbf{b} + \sin \beta t \, \mathbf{a}] \qquad (3)$$

are two linearly independent solutions of $\mathbf{x}' = A\mathbf{x}$.

Example 6.10.10 Find a fundamental set of solutions of

$$\begin{aligned} x_1' &= x_1 - x_2, \\ x_2' &= x_1 + x_2. \end{aligned} \qquad (4)$$

Solution Let

$$\mathbf{x} = \begin{bmatrix} x_1 \\ x_2 \end{bmatrix}, \qquad A = \begin{bmatrix} 1 & -1 \\ 1 & 1 \end{bmatrix},$$

so that (4) is $\mathbf{x}' = A\mathbf{x}$. This A is the same as the A of Example 6.10.9, so that we have $\lambda = \alpha + \beta i = 1 + i$ is an eigenvalue and the associated eigenvector can be taken to be

$$\mathbf{u} = \mathbf{a} + i\mathbf{b} = \begin{bmatrix} i \\ 1 \end{bmatrix} = \begin{bmatrix} 0 \\ 1 \end{bmatrix} + i\begin{bmatrix} 1 \\ 0 \end{bmatrix}.$$

Thus,

$$\alpha = 1, \qquad \beta = 1, \qquad \mathbf{a} = \begin{bmatrix} 0 \\ 1 \end{bmatrix}, \qquad \mathbf{b} = \begin{bmatrix} 1 \\ 0 \end{bmatrix}.$$

The formulas (2) and (3) become

$$\mathbf{x}_1(t) = e^t \left[\cos t \begin{bmatrix} 0 \\ 1 \end{bmatrix} - \sin t \begin{bmatrix} 1 \\ 0 \end{bmatrix} \right] = \begin{bmatrix} -e^t \sin t \\ e^t \cos t \end{bmatrix},$$

$$\mathbf{x}_2(t) = e^t \left[\cos t \begin{bmatrix} 1 \\ 0 \end{bmatrix} + \sin t \begin{bmatrix} 0 \\ 1 \end{bmatrix} \right] = \begin{bmatrix} e^t \cos t \\ e^t \sin t \end{bmatrix},$$

and $\{\mathbf{x}_1(t), \mathbf{x}_2(t)\}$ is a fundamental set of solutions of (4). The general solution of (4) is $c_1 \mathbf{x}_1 + c_2 \mathbf{x}_2$ or:

$$\begin{aligned} x_1(t) &= -c_1 e^t \sin t + c_2 e^t \cos t, \\ x_2(t) &= c_1 e^t \cos t + c_2 e^t \sin t. \end{aligned} \quad \blacksquare$$

Review Fact (See also Appendix A.) In calculating with complex numbers, the following algebraic fact is helpful in eliminating complex numbers in the denominator of a fraction. The key is to multiply the numerator and denominator by the conjugate of the denominator:

$$\frac{1}{\alpha + i\beta} = \frac{1}{\alpha + i\beta} \frac{\alpha - i\beta}{\alpha - i\beta} = \frac{\alpha - i\beta}{\alpha^2 + \beta^2} = \frac{\alpha}{\alpha^2 + \beta^2} - i\left(\frac{\beta}{\alpha^2 + \beta^2}\right). \quad (5)$$

Example 6.10.11

$$\frac{1-i}{2-i} = \frac{1-i}{2-i}\frac{2+i}{2+i} = \frac{(1-i)(2+i)}{5}$$
$$= \frac{2 - 2i + i - i^2}{5} = \frac{3-i}{5} = \frac{3}{5} - \frac{i}{5}. \blacksquare$$

In general, of course, we expect to see both real and complex eigenvalues.

Example 6.10.12 Find the general solution of

$$x_1' = -x_2 + 2x_3,$$
$$x_2' = -x_1, \quad (6)$$
$$x_3' = -x_1 + x_2 - x_3.$$

Solution Let

$$\mathbf{x} = \begin{bmatrix} x_1 \\ x_2 \\ x_3 \end{bmatrix}, \quad \mathbf{A} = \begin{bmatrix} 0 & -1 & 2 \\ -1 & 0 & 0 \\ -1 & 1 & -1 \end{bmatrix},$$

so that (6) is $\mathbf{x}' = \mathbf{A}\mathbf{x}$. First we need the characteristic polynomial in order to find the eigenvalues:

$$p(\lambda) = \det(\lambda\mathbf{I} - \mathbf{A}) = \det\begin{bmatrix} \lambda & 1 & -2 \\ 1 & \lambda & 0 \\ 1 & -1 & \lambda + 1 \end{bmatrix},$$

(we continue by expanding down the third column)

$$= -2(-1)^{3+1} \det\begin{bmatrix} 1 & \lambda \\ 1 & -1 \end{bmatrix} + (\lambda + 1)(-1)^{3+3} \det\begin{bmatrix} \lambda & 1 \\ 1 & \lambda \end{bmatrix}$$
$$= -2[-1 - \lambda] + (\lambda + 1)(\lambda^2 - 1)$$
$$= (\lambda + 1)[2 + \lambda^2 - 1] = (\lambda + 1)(\lambda^2 + 1).$$

The roots are $\lambda_1 = -1, \lambda_2 = i, \lambda_3 = -i$. Now we need to find the eigenvectors. For $\lambda_1 = -1, (\lambda_1 \mathbf{I} - \mathbf{A})\mathbf{u} = \mathbf{0}$ is

$$\begin{bmatrix} -1 & 1 & -2 \\ 1 & -1 & 0 \\ 1 & -1 & 0 \end{bmatrix}\mathbf{u} = \mathbf{0}.$$

or, equivalently,

$$\begin{bmatrix} -1 & 1 & -2 & | & 0 \\ 1 & -1 & 0 & | & 0 \\ 1 & -1 & 0 & | & 0 \end{bmatrix}.$$

Exchange row 1 and row 2; then add the new row 1 to row 2 and -1 times row 2 to row 3, obtaining:

$$\begin{bmatrix} 1 & -1 & 0 & | & 0 \\ 0 & 0 & -2 & | & 0 \\ 0 & 0 & 0 & | & 0 \end{bmatrix}.$$

That is, $u_1 - u_2 = 0$, $-2u_3 = 0$. Thus, $u_3 = 0$, u_2 is arbitrary, and $u_1 = u_2$, so that

$$\mathbf{u} = \begin{bmatrix} u_2 \\ u_2 \\ 0 \end{bmatrix} = u_2 \begin{bmatrix} 1 \\ 1 \\ 0 \end{bmatrix},$$

and

$$\begin{bmatrix} 1 \\ 1 \\ 0 \end{bmatrix} \text{ is an eigenvector for } \lambda_1 = -1. \tag{7}$$

For $\lambda_2 = i$, $(\lambda_2 \mathbf{I} - \mathbf{A})\mathbf{u} = \mathbf{0}$ is:

$$\begin{bmatrix} i & 1 & -2 \\ 1 & i & 0 \\ 1 & -1 & i+1 \end{bmatrix} \mathbf{u} = \mathbf{0}.$$

We write down the augmented matrix, and, in order to reduce the complex arithmetic, exchange rows 1 and 2:

$$\begin{bmatrix} 1 & i & 0 & | & 0 \\ i & 1 & -2 & | & 0 \\ 1 & -1 & i+1 & | & 0 \end{bmatrix}.$$

Now add $-i$ times row 1 to row 2; then add -1 times row 1 to row 3:

$$\begin{bmatrix} 1 & i & 0 & | & 0 \\ 0 & 2 & -2 & | & 0 \\ 0 & -1-i & i+1 & | & 0 \end{bmatrix}.$$

Now multiply row 2 by $\tfrac{1}{2}$; then add $(1+i)$ times row 2 to row 3:

$$\begin{bmatrix} 1 & i & 0 & | & 0 \\ 0 & 1 & -1 & | & 0 \\ 0 & 0 & 0 & | & 0 \end{bmatrix}.$$

Now add $-i$ times row 2 to row 1 to obtain the row echelon form:

$$\begin{bmatrix} 1 & 0 & i & | & 0 \\ 0 & 1 & -1 & | & 0 \\ 0 & 0 & 0 & | & 0 \end{bmatrix}. \quad (8)$$

The solution of (8) is u_3 arbitrary, $u_2 = u_3$, $u_1 = -iu_3$. Thus,

$$\mathbf{u} = \begin{bmatrix} -iu_3 \\ u_3 \\ u_3 \end{bmatrix} = u_3 \begin{bmatrix} -i \\ 1 \\ 1 \end{bmatrix},$$

and we take

$$\begin{bmatrix} -i \\ 1 \\ 1 \end{bmatrix}$$

as the eigenvector associated with $\lambda_2 = i$. By Theorem 6.10.4, the eigenvector associated with $\lambda_3 = -i$ can be taken as

$$\begin{bmatrix} i \\ 1 \\ 1 \end{bmatrix}.$$

Thus, $\lambda = \alpha + i\beta = i$, so that $\alpha = 0$, $\beta = 1$, and

$$\begin{bmatrix} -i \\ 1 \\ 1 \end{bmatrix} = \begin{bmatrix} 0 \\ 1 \\ 1 \end{bmatrix} + i \begin{bmatrix} -1 \\ 0 \\ 0 \end{bmatrix} = \mathbf{a} + i\mathbf{b},$$

so that

$$\mathbf{a} = \begin{bmatrix} 0 \\ 1 \\ 1 \end{bmatrix}, \quad \mathbf{b} = \begin{bmatrix} -1 \\ 0 \\ 0 \end{bmatrix}.$$

From (2), (3), and (7) we get the real solutions of

$$e^{-t}\begin{bmatrix} 1 \\ 1 \\ 0 \end{bmatrix}, \quad \begin{bmatrix} \sin t \\ \cos t \\ \cos t \end{bmatrix}, \quad \begin{bmatrix} -\cos t \\ \sin t \\ \sin t \end{bmatrix},$$

which form a fundamental set of solutions for (6). The general solution is

$$\mathbf{x} = c_1 e^{-t} \begin{bmatrix} 1 \\ 1 \\ 0 \end{bmatrix} + c_2 \begin{bmatrix} \sin t \\ \cos t \\ \cos t \end{bmatrix} + c_3 \begin{bmatrix} -\cos t \\ \sin t \\ \sin t \end{bmatrix}. \blacksquare$$

Exercises

In Exercises 21 through 28, solve the system of differential equations by the eigenvalue-eigenvector method. If no initial conditions are given, find the general solution.

21. $x_1' = -4x_2$,
    $x_2' = x_1$
22. $x_1' = 3x_2$,
    $x_2' = -3x_1$
23. $x_1' = -x_1 - x_2$, $\quad x_1(0) = 1$,
    $x_2' = x_1 - x_2$, $\quad x_2(0) = -1$
24. $x_1' = 2x_1 - x_2$,
    $x_2' = 2x_1$
25. $x_1' = 3x_1 + 5x_2$,
    $x_2' = -x_1 + x_2$
26. $x_1' = 3x_1 - x_2$, $\quad x_1(0) = 0$,
    $x_2' = 5x_1 - x_2$, $\quad x_2(0) = 3$
27. $x_1' = 5x_1 + 2x_2$,
    $x_2' = -4x_1 + x_2$
28. $x_1' = x_1 - x_2$,
    $x_2' = 5x_1 - x_2$

29. Recall that if $\mathbf{A}$ is 2×2, then trace($\mathbf{A}$) = $a_{11} + a_{22}$ is the *trace* of $\mathbf{A}$. Suppose $\mathbf{A}$ has a complex conjugate pair of eigenvalues and $\mathbf{A}$ is a real 2×2 matrix. Show that:

   a) All solutions of $\mathbf{x}' = \mathbf{A}\mathbf{x}$ have $\lim_{t \to \infty} \mathbf{x}(t) = \mathbf{0}$ if and only if trace($\mathbf{A}$) < 0.

   b) Show that all nonzero solutions of $\mathbf{x}' = \mathbf{A}\mathbf{x}$ are periodic with the same period if trace($\mathbf{A}$) = 0.

   c) Show that, if trace($\mathbf{A}$) > 0, then there exist solutions whose entries are unbounded as $t \to \infty$.

30. Suppose that $\mathbf{A}$ is a real 2×2 matrix. Show that $\mathbf{A}$ has complex eigenvalues if and only if (trace($\mathbf{A}$))^2 < 4 det($\mathbf{A}$).

Problems involving complex numbers can often be reformulated as larger problems involving only real numbers. This is illustrated by Exercise 31.

31. Suppose that $\lambda = \alpha + i\beta$ is an eigenvalue of the real $n \times n$ matrix $\mathbf{A}$. Show that $\mathbf{u} = \mathbf{a} + i\mathbf{b}$ is an eigenvector for $\mathbf{A}$ corresponding to the eigenvalue $\lambda = \alpha + i\beta$ if and only if $\mathbf{a}, \mathbf{b}$ are solutions of the system of equations:

$$[\mathbf{A} - \alpha\mathbf{I}]\mathbf{a} + \beta\mathbf{b} = \mathbf{0},$$
$$\beta\mathbf{a} + [\alpha\mathbf{I} - \mathbf{A}]\mathbf{b} = \mathbf{0}, \quad (9)$$

or, equivalently,

$$\begin{bmatrix} \mathbf{A} - \alpha\mathbf{I} & \beta\mathbf{I} \\ \beta\mathbf{I} & \alpha\mathbf{I} - \mathbf{A} \end{bmatrix} \begin{bmatrix} \mathbf{a} \\ \mathbf{b} \end{bmatrix} = \begin{bmatrix} \mathbf{0} \\ \mathbf{0} \end{bmatrix}.$$

32. $\lambda_1 = 1 + i$ is an eigenvalue of $\mathbf{A} = \begin{bmatrix} 3 & -1 \\ 5 & -1 \end{bmatrix}$. Find the associated eigenvector $\mathbf{u} = \mathbf{a} + i\mathbf{b}$ by solving the system of equations (9). (*Note*: In solving (9) there will be two arbitrary constants. To find $\mathbf{a}, \mathbf{b}$ you may give these constants any values that are not both zero).

33. $\lambda_1 = 3 + 2i$ is an eigenvalue of $\mathbf{A} = \begin{bmatrix} 5 & 2 \\ -4 & 1 \end{bmatrix}$. Find the associated eigenvector $\mathbf{u} = \mathbf{a} + i\mathbf{b}$ by solving the system of equations (9). (See note for Exercise 32.)

34. Suppose that $\mathbf{A}$ is a 3×3 real matrix. Show $\mathbf{A}$ must have a real eigenvalue.

35. Show that if the $n \times n$ matrix $\mathbf{A}$ has more than one conjugate pair of eigenvalues, then $n \geq 4$.

For Exercises 36 through 37 find the general solution of the indicated system of differential equations.

36. $x_1' = 2x_1 - 2x_2 + 2x_3$
    $x_2' = -x_2 + 5x_3$
    $x_3' = -x_2 + x_3$
37. $x_1' = -x_2$
    $x_2' = 2x_1 + 2x_2$
    $x_3' = -2x_1 + 2x_3$

6.10.3 Eigenvalues of Higher Multiplicity

As was pointed out in Example 6.10.8, it is sometimes not possible to find all solutions of $\mathbf{x}' = \mathbf{A}\mathbf{x}$ by merely considering the eigenvalues and eigenvectors of $\mathbf{A}$. In this section we will discuss one method for obtaining the other solutions. An alternative method will be given in Section 6.12 on the matrix exponential. While all of the examples and exercises of this section deal only with real eigenvalues and eigenvector deficiencies of one or two, the approach may be generalized to cover the general case.

Suppose then that $\mathbf{A}$ is an $n \times n$ matrix and λ_0 is an eigenvalue of multiplicity m, but that, when we solve $(\lambda_0 \mathbf{I} - \mathbf{A})\mathbf{u} = \mathbf{0}$, we get only r linearly independent eigenvectors $\{\mathbf{u}_1, \ldots, \mathbf{u}_r\}$. Then we shall call $m - r$ the *deficiency* of the eigenvalue λ_0. Since λ_0 has multiplicity m, we need to provide m solutions of $\mathbf{x}' = \mathbf{A}\mathbf{x}$. However, we are getting only the r solutions $\{e^{\lambda_0 t}\mathbf{u}_1, \ldots, e^{\lambda_0 t}\mathbf{u}_r\}$. The procedure we give will be applied only to eigenvalues of deficiency 1 and 2. Additional matrix theory is needed to efficiently handle problems with higher deficiency.

Deficiency One

In Chapter 3 we discussed how to solve the scalar differential equation

$$y'' - 2y' + y = 0. \tag{1}$$

The characteristic polynomial $r^2 - 2r + 1 = (r - 1)^2$ has the root $r = 1$ of multiplicity 2. Thus a fundamental set of solutions for (1) is $\{e^t, te^t\}$. A similar result holds for systems.

If λ_0 is an eigenvalue of $\mathbf{A}$ of deficiency one, then there is a solution of $\mathbf{x}' = \mathbf{A}\mathbf{x}$ of the form

$$e^{\lambda_0 t}(\mathbf{u} + t\mathbf{v}) \tag{2}$$

with $\mathbf{v} \neq \mathbf{0}$.

Given the form (2) for a solution of $\mathbf{x}' = \mathbf{A}\mathbf{x}$, we may determine $\mathbf{u}, \mathbf{v}$ as follows: Assume that λ_0 is an eigenvalue of $\mathbf{x}' = \mathbf{A}\mathbf{x}$ of deficiency 1. Let

$$\mathbf{x} = e^{\lambda_0 t}(\mathbf{u} + t\mathbf{v}), \tag{3}$$

and substitute into $\mathbf{x}' = \mathbf{A}\mathbf{x}$:

$$[e^{\lambda_0 t}(\mathbf{u} + t\mathbf{v})]' = \mathbf{A}e^{\lambda_0 t}(\mathbf{u} + t\mathbf{v})$$

or

$$\lambda_0 e^{\lambda_0 t}(\mathbf{u} + t\mathbf{v}) + e^{\lambda_0 t}\mathbf{v} = e^{\lambda_0 t}(\mathbf{A}\mathbf{u} + t\mathbf{A}\mathbf{v}).$$

Divide by $e^{\lambda_0 t}$ and rearrange:

$$(\lambda_0 \mathbf{u} + \mathbf{v}) + t\lambda_0 \mathbf{v} = \mathbf{A}\mathbf{u} + t\mathbf{A}\mathbf{v}. \tag{4}$$

Since (4) holds for all t, we get the two equations

$$\lambda_0 \mathbf{u} + \mathbf{v} = \mathbf{A}\mathbf{u}, \tag{5a}$$
$$\lambda_0 \mathbf{v} = \mathbf{A}\mathbf{v}, \tag{5b}$$

or

$$(\lambda_0 \mathbf{I} - \mathbf{A})\mathbf{u} + \mathbf{v} = \mathbf{0}, \tag{6a}$$
$$(\lambda_0 \mathbf{I} - \mathbf{A})\mathbf{v} = \mathbf{0}. \tag{6b}$$

If $\mathbf{A}$ is $n \times n$, (5) or (6) gives $2n$ equations in the unknown components of $\mathbf{u}, \mathbf{v}$.

Comment Note that (5b) implies that $\mathbf{v}$ is an eigenvector. However, (5a) need not be consistent for every possible eigenvector $\mathbf{v}$. Thus, to be sure this method will *always* work, we must solve (5a), (5b) (or (6a), (6b)) at the same time. The exception is when λ_0 has multiplicity exactly 2 and deficiency one. Then we may solve (5b) first and then (5a). (See the special case after Algorithm 6.10.1.)

Example 6.10.13   Find a fundamental set of solutions of:

$$\begin{aligned} x_1' &= 3x_1 - x_2, \\ x_2' &= x_1 + x_2. \end{aligned} \tag{7}$$

Solution   This is Example 6.10.8. System (7) is of the form $\mathbf{x}' = \mathbf{A}\mathbf{x}$, where

$$\mathbf{x} = \begin{bmatrix} x_1 \\ x_2 \end{bmatrix}, \quad \mathbf{A} = \begin{bmatrix} 3 & -1 \\ 1 & 1 \end{bmatrix}.$$

The characteristic polynomial is

$$p(\lambda) = \lambda^2 - 4\lambda + 4 = (\lambda - 2)^2,$$

so that $\lambda = 2$ is an eigenvalue of multiplicity 2. However, all eigenvectors for $\lambda = 2$ are of the form

$$u_2 \begin{bmatrix} 1 \\ 1 \end{bmatrix},$$

so that we have only one linearly independent eigenvector. The deficiency is $2 - 1 = 1$, and we have one solution:

$$e^{2t} \begin{bmatrix} 1 \\ 1 \end{bmatrix}.$$

We shall illustrate the general procedure. To find additional solutions we solve (5). Note that (6) can be written:

$$\begin{bmatrix} \lambda_0 \mathbf{I} - \mathbf{A} & \mathbf{I} \\ \mathbf{0} & \lambda_0 \mathbf{I} - \mathbf{A} \end{bmatrix} \begin{bmatrix} \mathbf{u} \\ \mathbf{v} \end{bmatrix} = \begin{bmatrix} \mathbf{0} \\ \mathbf{0} \end{bmatrix},$$

or, letting

$$\lambda_0 = 2, \quad \mathbf{u} = \begin{bmatrix} u_1 \\ u_2 \end{bmatrix}, \quad \mathbf{v} = \begin{bmatrix} v_1 \\ v_2 \end{bmatrix},$$

we have
$$\begin{bmatrix} -1 & 1 & 1 & 0 \\ -1 & 1 & 0 & 1 \\ 0 & 0 & -1 & 1 \\ 0 & 0 & -1 & 1 \end{bmatrix} \begin{bmatrix} u_1 \\ u_2 \\ v_1 \\ v_2 \end{bmatrix} = \begin{bmatrix} 0 \\ 0 \\ 0 \\ 0 \end{bmatrix}.$$

Adding (-1) times row 1 to row 2, and -1 times row 3 to row 4, we get a new augmented matrix,
$$\left[\begin{array}{cccc|c} -1 & 1 & 1 & 0 & 0 \\ 0 & 0 & -1 & 1 & 0 \\ 0 & 0 & -1 & 1 & 0 \\ 0 & 0 & 0 & 0 & 0 \end{array}\right].$$

Now add row 2 to row 1 and add -1 times row 2 to row 3 to give:
$$\left[\begin{array}{cccc|c} -1 & 1 & 0 & 1 & 0 \\ 0 & 0 & -1 & 1 & 0 \\ 0 & 0 & 0 & 0 & 0 \\ 0 & 0 & 0 & 0 & 0 \end{array}\right], \tag{8}$$

or $-u_1 + u_2 + v_2 = 0$, $-v_1 + v_2 = 0$.

The first and third columns of (8) are distinguished, so we can take the variables corresponding to the second and fourth columns as arbitrary. That is, our solution is

$$u_2, v_2 \text{ arbitrary}, \qquad v_1 = v_2, \qquad u_1 = u_2 + v_2. \tag{9}$$

Thus all the solutions are of the form
$$\mathbf{x} = e^{\lambda t}(\mathbf{u} + t\mathbf{v})$$
$$= e^{2t}\left(\begin{bmatrix} u_1 \\ u_2 \end{bmatrix} + t \begin{bmatrix} v_1 \\ v_2 \end{bmatrix}\right),$$

which, from (9), is
$$= e^{2t}\left(\begin{bmatrix} u_2 + v_2 \\ u_2 \end{bmatrix} + t \begin{bmatrix} v_2 \\ v_2 \end{bmatrix}\right)$$
$$= u_2 e^{2t} \begin{bmatrix} 1 \\ 1 \end{bmatrix} + v_2 e^{2t} \begin{bmatrix} 1 + t \\ t \end{bmatrix}. \tag{10}$$

The solution
$$e^{2t} \begin{bmatrix} 1 \\ 1 \end{bmatrix}$$
had been found earlier. However,

$$e^{2t}\begin{bmatrix} 1+t \\ t \end{bmatrix}$$

is a new solution in the form (2). Finally, (10) with u_2, v_2 arbitrary constants is the general solution of (7) and

$$\left\{ e^{2t}\begin{bmatrix} 1 \\ 1 \end{bmatrix}, \ e^{2t}\begin{bmatrix} 1+t \\ t \end{bmatrix} \right\}$$

is a fundamental set of solutions. ∎

Comment If we are looking only for an additional (linearly independent) solution besides

$$e^{2t}\begin{bmatrix} 1 \\ 1 \end{bmatrix},$$

then in (9) we may take any values of u_2, v_2 for which $v_2 \neq 0$.

Algorithm 6.10.1 **Procedure for finding solutions of $\mathbf{x}' = \mathbf{Ax}$ for an eigenvalue λ_0 of deficiency one.**

1. Set up the system of equations

$$(\lambda_0 \mathbf{I} - \mathbf{A})\mathbf{u} + \mathbf{v} = \mathbf{0},$$
$$(\lambda_0 \mathbf{I} - \mathbf{A})\mathbf{v} = \mathbf{0}. \tag{11}$$

2. Solve (11) for $\mathbf{u}, \mathbf{v}$.
3. $\mathbf{x} = e^{\lambda_0 t}(\mathbf{u} + t\mathbf{v})$ gives all the solutions of $\mathbf{x}' = \mathbf{Ax}$ that correspond to the eigenvalue λ_0.

Special Case
If λ_0 is an eigenvalue of multiplicity 2 and deficiency 1, then (11) can be broken into the two steps:

1. Solve $(\lambda_0 \mathbf{I} - \mathbf{A})\mathbf{v} = \mathbf{0}$ for $\mathbf{v}$.
2. Solve $(\lambda_0 \mathbf{I} - \mathbf{A})\mathbf{u} = -\mathbf{v}$ to get $\mathbf{u}$.

Deficiency Two

Sometimes Algorithm 6.10.1 finds enough solutions even if the deficiency is greater than one. However, if λ_0 is an eigenvalue of deficiency two and Algorithm 6.10.1 does not find enough solutions, then we can get the needed number of solutions by considering solutions of the form

$$\mathbf{x} = e^{\lambda_0 t}(\mathbf{u} + t\mathbf{v} + t^2\mathbf{w}). \tag{12}$$

Substituting (12) into $\mathbf{x}' = \mathbf{Ax}$ gives

$$[e^{\lambda_0 t}(\mathbf{u} + t\mathbf{v} + t^2\mathbf{w})]' = \mathbf{A}e^{\lambda_0 t}(\mathbf{u} + t\mathbf{v} + t^2\mathbf{w})$$

or

$$\lambda_0 e^{\lambda_0 t}(\mathbf{u} + t\mathbf{v} + t^2\mathbf{w}) + e^{\lambda_0 t}(\mathbf{v} + 2t\mathbf{w}) = \mathbf{A}e^{\lambda_0 t}(\mathbf{u} + t\mathbf{v} + t^2\mathbf{w}).$$

Divide by $e^{\lambda_0 t}$ and equate like powers of t to get the following systems of algebraic equations:

$$\lambda_0 \mathbf{u} + \mathbf{v} = \mathbf{A}\mathbf{u},$$
$$\lambda_0 \mathbf{v} + 2\mathbf{w} = \mathbf{A}\mathbf{v},$$
$$\lambda_0 \mathbf{w} = \mathbf{A}\mathbf{w};$$

or equivalently,

$$(\lambda_0 \mathbf{I} - \mathbf{A})\mathbf{u} + \mathbf{v} = \mathbf{0}, \tag{13a}$$
$$(\lambda_0 \mathbf{I} - \mathbf{A})\mathbf{v} + 2\mathbf{w} = \mathbf{0}, \tag{13b}$$
$$(\lambda_0 \mathbf{I} - \mathbf{A})\mathbf{w} = \mathbf{0}. \tag{13c}$$

In matrix notation, (13) is

$$\begin{bmatrix} \lambda_0 \mathbf{I} - \mathbf{A} & \mathbf{I} & \mathbf{0} \\ \mathbf{0} & \lambda_0 \mathbf{I} - \mathbf{A} & 2\mathbf{I} \\ \mathbf{0} & \mathbf{0} & \lambda_0 \mathbf{I} - \mathbf{A} \end{bmatrix} \begin{bmatrix} \mathbf{u} \\ \mathbf{v} \\ \mathbf{w} \end{bmatrix} = \begin{bmatrix} \mathbf{0} \\ \mathbf{0} \\ \mathbf{0} \end{bmatrix}. \tag{14}$$

The system (13), or (14), is reasonably straightforward to solve on a computer. However, to keep the amount of algebra down, we shall consider only simple matrices $\mathbf{A}$.

Example 6.10.14

Solve the system of differential equations:

$$x_1' = x_1 + x_2,$$
$$x_2' = x_2 + x_3, \tag{15}$$
$$x_3' = x_3.$$

Solution

Let

$$\mathbf{x} = \begin{bmatrix} x_1 \\ x_2 \\ x_3 \end{bmatrix}, \quad \mathbf{A} = \begin{bmatrix} 1 & 1 & 0 \\ 0 & 1 & 1 \\ 0 & 0 & 1 \end{bmatrix}.$$

Then (15) is $\mathbf{x}' = \mathbf{A}\mathbf{x}$. The characteristic polynomial of $\mathbf{A}$ is

$$p(\lambda) = \det \begin{bmatrix} \lambda - 1 & -1 & 0 \\ 0 & \lambda - 1 & -1 \\ 0 & 0 & \lambda - 1 \end{bmatrix} = (\lambda - 1)^3,$$

so that 1 is an eigenvalue of multiplicity 3. We omit the calculation, but it turns out that the only eigenvectors are multiples of

$$\begin{bmatrix} 1 \\ 0 \\ 0 \end{bmatrix},$$

so that the deficiency is $3 - 1 = 2$. We look then for solutions of the form

$$\mathbf{x} = e^t(\mathbf{u} + t\mathbf{v} + t^2\mathbf{w}), \tag{16}$$

where

$$\mathbf{u}, \mathbf{v}, \mathbf{w} \quad \text{satisfy (14) for } \lambda_0 = 1.$$

That is,

$$\begin{bmatrix} 0 & -1 & 0 & 1 & 0 & 0 & 0 & 0 & 0 \\ 0 & 0 & -1 & 0 & 1 & 0 & 0 & 0 & 0 \\ 0 & 0 & 0 & 0 & 0 & 1 & 0 & 0 & 0 \\ 0 & 0 & 0 & 0 & -1 & 0 & 2 & 0 & 0 \\ 0 & 0 & 0 & 0 & 0 & -1 & 0 & 2 & 0 \\ 0 & 0 & 0 & 0 & 0 & 0 & 0 & 0 & 2 \\ 0 & 0 & 0 & 0 & 0 & 0 & -1 & 0 & 0 \\ 0 & 0 & 0 & 0 & 0 & 0 & 0 & -1 & 0 \\ 0 & 0 & 0 & 0 & 0 & 0 & 0 & 0 & 0 \end{bmatrix} \begin{bmatrix} u_1 \\ u_2 \\ u_3 \\ v_1 \\ v_2 \\ v_3 \\ w_1 \\ w_2 \\ w_3 \end{bmatrix} = \begin{bmatrix} 0 \\ 0 \\ 0 \\ 0 \\ 0 \\ 0 \\ 0 \\ 0 \\ 0 \end{bmatrix}. \tag{17}$$

Part 1 of Gaussian elimination is already done on (17). So add twice row 8 to row 6, add twice row 7 to row 5, and add row 5 to row 3. Again, dropping the zero rows, we now have the augmented matrix:

$$\left[\begin{array}{ccccccccc|c} 0 & -1 & 0 & 1 & 0 & 0 & 0 & 0 & 0 & 0 \\ 0 & 0 & -1 & 0 & 1 & 0 & 0 & 0 & 0 & 0 \\ 0 & 0 & 0 & 0 & -1 & 0 & 2 & 0 & 0 & 0 \\ 0 & 0 & 0 & 0 & 0 & -1 & 0 & 0 & 0 & 0 \\ 0 & 0 & 0 & 0 & 0 & 0 & 0 & -1 & 0 & 0 \\ 0 & 0 & 0 & 0 & 0 & 0 & 0 & 0 & -1 & 0 \end{array}\right].$$

Finally, add row 3 to row 2:

$$\left[\begin{array}{ccccccccc|c} 0 & -1 & 0 & 1 & 0 & 0 & 0 & 0 & 0 & 0 \\ 0 & 0 & -1 & 0 & 0 & 0 & 2 & 0 & 0 & 0 \\ 0 & 0 & 0 & 0 & -1 & 0 & 2 & 0 & 0 & 0 \\ 0 & 0 & 0 & 0 & 0 & -1 & 0 & 0 & 0 & 0 \\ 0 & 0 & 0 & 0 & 0 & 0 & 0 & -1 & 0 & 0 \\ 0 & 0 & 0 & 0 & 0 & 0 & 0 & 0 & -1 & 0 \end{array}\right]. \tag{18}$$

The 2, 3, 5, 6, 8, 9 columns are distinguished, so that the 1, 4, 7 variables u_1, v_1, w_1 are arbitrary, and the equations (18) tell us that

$$\begin{aligned} u_2 &= v_1, \\ u_3 &= 2w_1, \\ v_2 &= 2w_1, \\ -v_3 &= 0, \\ -w_2 &= 0, \\ -w_3 &= 0. \end{aligned} \quad (19)$$

Thus from (16),

$$\mathbf{x} = e^{\lambda_0 t}(\mathbf{u} + t\mathbf{v} + t^2 \mathbf{w})$$

$$= e^t \left(\begin{bmatrix} u_1 \\ u_2 \\ u_3 \end{bmatrix} + t \begin{bmatrix} v_1 \\ v_2 \\ v_3 \end{bmatrix} + t^2 \begin{bmatrix} w_1 \\ w_2 \\ w_3 \end{bmatrix} \right)$$

$$= e^t \left(\begin{bmatrix} u_1 \\ v_1 \\ 2w_1 \end{bmatrix} + t \begin{bmatrix} v_1 \\ 2w_1 \\ 0 \end{bmatrix} + t^2 \begin{bmatrix} w_1 \\ 0 \\ 0 \end{bmatrix} \right) \quad \text{[by (19)]}.$$

Rearranging gives

$$= e^t \left(u_1 \begin{bmatrix} 1 \\ 0 \\ 0 \end{bmatrix} + v_1 \begin{bmatrix} t \\ 1 \\ 0 \end{bmatrix} + w_1 \begin{bmatrix} t^2 \\ 2t \\ 2 \end{bmatrix} \right)$$

$$= u_1 e^t \begin{bmatrix} 1 \\ 0 \\ 0 \end{bmatrix} + v_1 e^t \begin{bmatrix} t \\ 1 \\ 0 \end{bmatrix} + w_1 e^t \begin{bmatrix} t^2 \\ 2t \\ 2 \end{bmatrix}.$$

where u_1, v_1, w_1 are arbitrary constants. ∎

This example has been worked to illustrate the general procedure. However, if the multiplicity is exactly 3 and the deficiency 2, then instead of solving (13) as a large system, one can solve (13c), then (13b), and finally (13a).

Exercises

In Exercises 38 through 45, the deficiency is one. Find the general solution of the differential equation.

38. $x'_1 = x_1 - x_2,$
$\quad x'_2 = 4x_1 - 3x_2$

39. $x'_1 = 3x_1 - 4x_2,$
$\quad x'_2 = x_1 - x_2$

40. $x'_1 = 3x_1 - x_2,$
$\quad x'_2 = 9x_1 - 3x_2$

41. $x' = -x + y,$
$\quad y' = x + y$

42. $x' = y,$
$\quad y' = -x + 2y$

43. $x_1' = -2x_1 + x_2,$
$x_2' = -16x_1 + 6x_2$

44. $x_1' = -2x_2 + x_3,$
$x_2' = x_1 - 3x_2 + x_3,$
$x_3' = x_1 - 2x_2$

45. $x_1' = 2x_1 - x_3,$
$x_2' = x_1 + x_2 - x_3,$
$x_3' = x_1$

In Exercises 46 through 49, the deficiency is two. Find the general solution of the differential equation.

46. $x_1' = 2x_1 - x_2,$
$x_2' = 2x_2 - 2x_3,$
$x_3' = 2x_3$

47. $x_1' = x_1,$
$x_2' = x_1 + x_2,$
$x_3' = x_1 + x_2 + x_3$

48. $x_1' = x_1 + x_2,$
$x_2' = 2x_2 + x_3,$
$x_3' = x_1 - x_2 + 3x_3$

49. $x_1' = -x_1 + 2x_2,$
$x_2' = x_2 + 2x_3,$
$x_3' = 2x_1 - 2x_2 + 3x_3$

6.11 Nonhomogeneous Systems (Undetermined Coefficients)

In the previous sections we have discussed how to solve the linear, homogeneous differential equation with constant coefficients,

$$\mathbf{x}'(t) = \mathbf{A}\mathbf{x}(t), \tag{1}$$

and find its general solution. In this section we will discuss how to solve the nonhomogeneous differential equation

$$\mathbf{x}'(t) = \mathbf{A}\mathbf{x}(t) + \mathbf{f}(t) \tag{2}$$

for a wide variety of forcing terms $\mathbf{f}(t)$ of interest. From Theorem 6.9.2, we know that the solution of (2) is of the form

$$\mathbf{x} = \mathbf{x}_p + \mathbf{x}_h,$$

where $\mathbf{x}_h$ is the general solution of the associated homogeneous equation (1) and $\mathbf{x}_p$ is a *particular solution* of (2). Since we know how to solve (1) we shall concentrate on how to find $\mathbf{x}_p$. The method presented in this section is a variation of the method of undetermined coefficients from Section 3.12.

Algorithm 6.11.1   This method is useful for finding a particular solution $\mathbf{x}_p$ of $\mathbf{x}' = \mathbf{A}\mathbf{x} + \mathbf{f}$, that is

$$x_1' = a_{11}x_1 + \cdots + a_{1n}x_n + f_1(t),$$
$$x_2' = a_{21}x_1 + \cdots + a_{2n}x_n + f_2(t),$$
$$\vdots \qquad \vdots$$
$$x_n' = a_{n1}x_1 + \cdots + a_{nn}x_n + f_n(t).$$
(3)

This method works if **A** is constant and each $f_i(t)$ is the type of function for which the method of undetermined coefficients of Section 3.12 works. That is, each $f_i(t)$ is a linear combination of functions $t^m e^{\alpha t} \cos \beta t$, $t^m e^{\alpha t} \sin \beta t$ for integers $m \geq 0$ and scalars α, β.

Step 1
Compute the characteristic polynomial $p(\lambda) = \det(\lambda \mathbf{I} - \mathbf{A})$ of **A**, and determine its roots (eigenvalues).

Step 2
Construct a form for the particular solution

$$\mathbf{x}_p = \begin{bmatrix} x_{1p} \\ x_{2p} \\ \vdots \\ x_{np} \end{bmatrix},$$

as follows. For terms that appear in *any* of the $f_i(t)$, add terms into *all* the x_{ip} according to the rules of Section 3.12, with the exception that, if we multiply by t^k for a root of multiplicity $k+1$ of $p(\lambda)$, we also include the terms multipled by the lower powers of t.

Step 3
Substitute the form for $\mathbf{x}_p$ into $\mathbf{x}_p' = \mathbf{A}\mathbf{x}_p + \mathbf{f}$ and solve for the constants in $\mathbf{x}_p$ by equating terms. The solution for the constants will be unique only if there are no terms in the form for $\mathbf{x}_p$ that also appear in the solution of the associated homogeneous equation (1).

We shall illustrate this algorithm with several examples.

Example 6.11.1   Find a particular solution of:

$$x_1' = 2x_1 + x_2 + t, \tag{4}$$
$$x_2' = x_1 + 2x_2 + 1 + e^{-t}. \tag{5}$$

Solution   *Step 1* The coefficient matrix

$$\mathbf{A} = \begin{bmatrix} 2 & 1 \\ 1 & 2 \end{bmatrix}$$

has characteristic polynomial

$$p(\lambda) = \det(\lambda \mathbf{I} - \mathbf{A}) = \det\begin{bmatrix} \lambda - 2 & -1 \\ -1 & \lambda - 2 \end{bmatrix} = \lambda^2 - 4\lambda + 3$$
$$= (\lambda - 3)(\lambda - 1),$$

so that the roots are $\lambda = 1, 3$.

Step 2 The forcing terms include 1, t, and e^{-t}. Since neither 0 nor -1 are roots of $p(\lambda)$, our particular solution will have the form:

$$x_{1p} = A + Bt + Ce^{-t}, \tag{6}$$
$$x_{2p} = E + Ft + Ge^{-t}. \tag{7}$$

Step 3 We substitute the form (6), (7) for $\mathbf{x}_p$ into the differential equations (4) and (5), to get

$$B - Ce^{-t} = 2(A + Bt + Ce^{-t}) + (E + Ft + Ge^{-t}) + t, \tag{8}$$
$$F - Ge^{-t} = (A + Bt + Ce^{-t}) + 2(E + Ft + Ge^{-t}) + 1 + e^{-t}. \tag{9}$$

Equating coefficients of like terms in (8) and then (9) yields equations on the coefficients:

$$\begin{array}{lll} 1: & B = 2A + E, & F = A + 2E + 1, \\ t: & 0 = 2B + F + 1, & 0 = B + 2F, \\ e^{-t}: & -C = 2C + G, & -G = C + 2G + 1. \end{array}$$

This is six equations in six unknowns, which could be solved using augmented matrices. However, it is quicker to solve the bottom two equations for C and G, then the middle two for B and F, and finally the top two equations for A and E to get:

$$A = -\tfrac{2}{9}, \quad B = -\tfrac{2}{3}, \quad C = \tfrac{1}{8}, \quad E = -\tfrac{2}{9}, \quad F = \tfrac{1}{3}, \quad G = -\tfrac{3}{8},$$

so that

$$\mathbf{x}_p = \begin{bmatrix} x_{1p} \\ x_{2p} \end{bmatrix} = \begin{bmatrix} -\tfrac{2}{9} - \tfrac{2}{3}t + \tfrac{1}{8}e^{-t} \\ -\tfrac{2}{9} + \tfrac{1}{3}t - \tfrac{3}{8}e^{-t} \end{bmatrix} \tag{10}$$

is a particular solution of (4) and (5).

Note This problem can also be worked in vector notation. That is, we can take

$$\mathbf{x} = \mathbf{a} + t\mathbf{b} + e^{-t}\mathbf{c}, \tag{11}$$

where $\mathbf{a}$, $\mathbf{b}$, $\mathbf{c}$ are unknown 2×1 constant vectors. Substituting (11) into $\mathbf{x}' = \mathbf{A}\mathbf{x} + \mathbf{f}$ and rewriting $\mathbf{f}$ gives

$$\mathbf{b} - e^{-t}\mathbf{c} = \mathbf{A}\mathbf{a} + t\mathbf{A}\mathbf{b} + e^{-t}\mathbf{A}\mathbf{c} + \begin{bmatrix} 0 \\ 1 \end{bmatrix} + t\begin{bmatrix} 1 \\ 0 \end{bmatrix} + e^{-t}\begin{bmatrix} 0 \\ 1 \end{bmatrix}.$$

Equating corresponding coefficients gives the vector equations

$$1: \quad \mathbf{b} = \mathbf{A}\mathbf{a} + \begin{bmatrix} 0 \\ 1 \end{bmatrix} \quad \text{or} \quad \mathbf{A}\mathbf{a} = \mathbf{b} - \begin{bmatrix} 0 \\ 1 \end{bmatrix}, \tag{12}$$

$$t: \quad \mathbf{0} = \mathbf{A}\mathbf{b} + \begin{bmatrix} 1 \\ 0 \end{bmatrix} \quad \text{or} \quad \mathbf{A}\mathbf{b} = -\begin{bmatrix} 1 \\ 0 \end{bmatrix}, \tag{13}$$

$$e^{-t}: \quad -\mathbf{c} = \mathbf{A}\mathbf{c} + \begin{bmatrix} 0 \\ 1 \end{bmatrix} \quad \text{or} \quad (\mathbf{A} + \mathbf{I})\mathbf{c} = -\begin{bmatrix} 0 \\ 1 \end{bmatrix}. \tag{14}$$

This gives us three systems. First, solve (14) for **c**, then (13) for **b**, and finally (12) for **a**. We leave the details to the interested reader (see Exercise 22 at the end of this section). ∎

Example 6.11.2   Find a particular solution of

$$x_1' = -x_1 + 4x_2 + e^t, \tag{15}$$
$$x_2' = x_1 - x_2 + 1. \tag{16}$$

Solution   *Step 1* The characteristic polynomial of

$$\mathbf{A} = \begin{bmatrix} -1 & 4 \\ 1 & -1 \end{bmatrix}$$

is

$$p(\lambda) = \det \begin{bmatrix} \lambda + 1 & -4 \\ -1 & \lambda + 1 \end{bmatrix} = \lambda^2 + 2\lambda - 3 = (\lambda + 3)(\lambda - 1),$$

with roots $\lambda = -3$, $\lambda = 1$.

Step 2 The functions 1 and e^t appear as forcing terms. Zero is not a root, while 1 is. Thus we take

$$x_{1p} = A + Be^t + Cte^t, \tag{17}$$
$$x_{2p} = E + Fe^t + Gte^t. \tag{18}$$

The inclusion of the Be^t, Fe^t terms in addition to the Cte^t, Gte^t terms is a major difference in the way the method of undetermined coefficients is applied to systems and the way it was applied in Section 3.12 to scalar equations. Substituting (17) and (18) into (15) and (16) gives us:

$$Be^t + C(e^t + te^t) = -(A + Be^t + Cte^t) + 4(E + Fe^t + Gte^t) + e^t,$$
$$Fe^t + G(e^t + te^t) = (A + Be^t + Cte^t) - (E + Fe^t + Gte^t) + 1.$$

Equating like coefficients we get the equations

$$1: \quad 0 = -A + 4E, \qquad\qquad 0 = A - E + 1, \tag{19}$$
$$e^t: \quad B + C = -B + 4F + 1, \qquad F + G = B - F, \tag{20}$$
$$te^t: \quad C = -C + 4G, \qquad\qquad G = C - G. \tag{21}$$

The first two equations (19) give $E = -\frac{1}{3}$, $A = -\frac{4}{3}$, while (21) implies only that $C = 2G$. Thus we can eliminate C in (20), and have left

$$2B + 2G - 4F = 1,$$
$$-B + G + 2F = 0.$$

This is only two equations in three unknowns, whose augmented matrix is

$$\begin{bmatrix} 2 & 2 & -4 & | & 1 \\ -1 & 1 & 2 & | & 0 \end{bmatrix} \quad \text{or} \quad \begin{bmatrix} 1 & 1 & -2 & | & 1/2 \\ -1 & 1 & 2 & | & 0 \end{bmatrix}.$$

Adding the first row to the second gives

$$\begin{bmatrix} 1 & 1 & -2 & | & \frac{1}{2} \\ 0 & 2 & 0 & | & \frac{1}{2} \end{bmatrix}.$$

Two more row operations yield the row echelon form

$$\begin{bmatrix} 1 & 0 & -2 & | & \frac{1}{4} \\ 0 & 1 & 0 & | & \frac{1}{4} \end{bmatrix}.$$

Thus $G = \frac{1}{4}$ and $B - 2F = \frac{1}{4}$ or $B = 2F + \frac{1}{4}$. Since we are looking for any particular solution of (15) and (16), we may set the arbitrary $F = 0$ and use $C = 2G$, to find

$$A = -\tfrac{4}{3}, \quad B = \tfrac{1}{4}, \quad C = \tfrac{1}{2}, \quad E = -\tfrac{1}{3}, \quad F = 0, \quad G = \tfrac{1}{4}.$$

Thus

$$x_{1p} = -\tfrac{4}{3} + \tfrac{1}{4}e^t + \tfrac{1}{2}te^t, \tag{22}$$
$$x_{2p} = -\tfrac{1}{3} + \tfrac{1}{4}te^t \tag{23}$$

is a particular solution of (15) and (16). ∎

Example 6.11.3 Find the general solution of

$$x_1' = -x_1 + 4x_2 + e^t, \tag{24}$$
$$x_2' = x_1 - x_2 + 1. \tag{25}$$

Solution The solution is $\mathbf{x} = \mathbf{x}_p + \mathbf{x}_h$, where $\mathbf{x}_p$ was computed in Example 6.11.2. We compute the homogeneous solution, using the eigenvector technique of Section 6.10. From Example 6.11.2, the eigenvalues of

$$\mathbf{A} = \begin{bmatrix} -1 & 4 \\ 1 & -1 \end{bmatrix}$$

are $\lambda = -3$, $\lambda = 1$. To find the eigenvectors, we must solve $(\lambda \mathbf{I} - \mathbf{A})\mathbf{x} = \mathbf{0}$. For $\lambda = 1$, we get

$$\begin{bmatrix} 2 & -4 & | & 0 \\ -1 & 2 & | & 0 \end{bmatrix} \quad \text{or} \quad \begin{bmatrix} 1 & -2 & | & 0 \\ 0 & 0 & | & 0 \end{bmatrix}.$$

That is,
$$u_1 = 2u_2 \quad \text{and} \quad \mathbf{u} = \begin{bmatrix} 2u_2 \\ u_2 \end{bmatrix} = u_2 \begin{bmatrix} 2 \\ 1 \end{bmatrix}.$$

For $\lambda = -3$, we get
$$\begin{bmatrix} -2 & -4 & | & 0 \\ -1 & -2 & | & 0 \end{bmatrix} \quad \text{or} \quad \begin{bmatrix} 1 & 2 & | & 0 \\ 0 & 0 & | & 0 \end{bmatrix}.$$

Thus, $u_1 = -2u_2$ and
$$\mathbf{u} = \begin{bmatrix} -2u_2 \\ u_2 \end{bmatrix} = u_2 \begin{bmatrix} -2 \\ 1 \end{bmatrix}.$$

Finally,
$$\mathbf{x}_h = c_1 e^t \begin{bmatrix} 2 \\ 1 \end{bmatrix} + c_2 e^{-3t} \begin{bmatrix} -2 \\ 1 \end{bmatrix} \tag{26}$$

is the general solution of the homogeneous equation. Combining (26), (22), and (23) yields the general solution of (24) and (25) as
$$\mathbf{x} = \mathbf{x}_p + \mathbf{x}_h$$
or
$$x_1 = -\tfrac{4}{3} + \tfrac{1}{4}e^t + \tfrac{1}{2}te^t + 2c_1 e^t - 2c_2 e^{-3t}, \tag{27}$$
$$x_2 = -\tfrac{1}{3} + \tfrac{1}{4}te^t + c_1 e^t + c_2 e^{-3t}, \tag{28}$$

with c_1, c_2 arbitrary constants. ∎

Example 6.11.4 Find the solution of
$$x_1' = -x_1 + 4x_2 + e^t, \quad x_1(0) = 0, \tag{29}$$
$$x_2' = x_1 - x_2 + 1, \quad x_2(0) = 1. \tag{30}$$

Solution First, we must find the general solution, which was done in the previous example. Then we apply the initial conditions in (29) and (30) to the general solution in (27) and (28), to obtain
$$0 = x_1(0) = -\frac{4}{3} + \frac{1}{4} + 2c_1 - 2c_2,$$
$$1 = x_2(0) = -\frac{1}{3} + c_1 + c_2.$$

That is,
$$\frac{13}{12} = 2c_1 - 2c_2,$$
$$\frac{4}{3} = c_1 + c_2$$

which has the solution $c_1 = \dfrac{15}{16}$, $c_2 = \dfrac{19}{48}$, and the solution of (25) and (26) is:

$$x_1 = -\frac{4}{3} + \frac{17}{8}e^t + \frac{1}{2}te^t - \frac{19}{24}e^{-3t},$$

$$x_2 = -\frac{1}{3} + \frac{1}{4}te^t + \frac{15}{16}e^t + \frac{19}{48}e^{-3t}. \blacksquare$$

We conclude with several additional examples of picking the form for $\mathbf{x}_p$.

Example 6.11.5 Give the form for $\mathbf{x}_p$ if

$$x_1' = x_1 + 2x_2 + \sin t,$$
$$x_2' = 3x_1 + 4x_2 + \cos 2t$$

is to be solved by the method of undetermined coefficients.

Solution Since $\sin t$, $\cos 2t$ are forcing terms, we must include $\sin t$, $\cos t$, $\sin 2t$, $\cos 2t$ in the form for $\mathbf{x}_p$. The characteristic polynomial of the coefficient matrix is

$$p(\lambda) = \det \begin{bmatrix} \lambda - 1 & -2 \\ -3 & \lambda - 4 \end{bmatrix} = \lambda^2 - 5\lambda - 2.$$

Since neither i nor $2i$ is a root of $p(\lambda)$, we do not need any additional terms, and the form of $\mathbf{x}_p$ is

$$x_{1p} = A \cos t + B \sin t + C \cos 2t + D \sin 2t,$$
$$x_{2p} = E \cos t + F \sin t + G \cos 2t + H \sin 2t. \blacksquare$$

Example 6.11.6 Give the form for $\mathbf{x}_p$ if

$$x_1' = x_1 + 5x_2 + e^{2t},$$
$$x_2' = -x_1 - x_2 + \sin 2t$$

is to be solved by the method of undetermined coefficients.

Solution Since e^{2t}, $\sin 2t$, are the forcing terms, we must include e^{2t}, $\sin 2t$, $\cos 2t$ in the particular solution. The characteristic polynomial is

$$p(\lambda) = \det \begin{bmatrix} \lambda - 1 & -5 \\ 1 & \lambda + 1 \end{bmatrix} = \lambda^2 + 4,$$

which has roots $\pm 2i$. Thus we need to include not only a $\sin 2t$, $\cos 2t$, but also $t \sin 2t$, $t \cos 2t$, and

$$x_{1p} = Ae^{2t} + B \sin 2t + C \cos 2t + Dt \sin 2t + Et \cos 2t,$$
$$x_{2p} = Fe^{2t} + G \sin 2t + H \cos 2t + Jt \sin 2t + Kt \cos 2t$$

is the form for $\mathbf{x}_p$. ∎

Example 6.11.7 Give the form for $\mathbf{x}_p$ if

$$x_1' = 3x_1 - 4x_2 + e^{3t},$$
$$x_2' = x_1 - x_2 + 6e^t$$

is to be solved by the method of undetermined coefficients.

Solution The characteristic polynomial is

$$p(\lambda) = \det \begin{bmatrix} \lambda - 3 & 4 \\ -1 & \lambda + 1 \end{bmatrix} = \lambda^2 - 2\lambda + 1 = (\lambda - 1)^2,$$

so that $\lambda = 1$ is a root of multiplicity two. Since 3 is not a root, the e^{3t} forcing term means that we include e^{3t} in the form for $\mathbf{x}_p$. Since 1 is a root of multiplicity two, the e^t term means we include e^t, te^t, $t^2 e^t$. Thus the form for $\mathbf{x}_p$ is

$$x_{1p} = Ae^{3t} + Be^t + Cte^t + Dt^2 e^t,$$
$$x_{2p} = Ee^{3t} + Fe^t + Gte^t + Ht^2 e^t. \quad ∎$$

Exercises

For Exercises 1 through 8, find a particular solution using the method of undetermined coefficients. Then find the general solution.

1. $x_1' = x_1 + x_2 + e^t,$
   $x_2' = x_1 + x_2 - e^{-t}$
2. $x_1' = x_1 + 3x_2 + t,$
   $x_2' = x_1 - x_2 - 1$
3. $x_1' = 2x_1 - x_2 + 2e^{-t},$
   $x_2' = 2x_1 - x_2$
4. $x_1' = 3x_1 - 2x_2 + e^t,$
   $x_2' = x_1$
5. $x_1' = 2x_1 + x_2 + \sin t,$
   $x_2' = x_1 + 2x_2 + 3 \cos t$
6. $x_1' = 2x_1 - 3x_2,$
   $x_2' = x_1 - 2x_2 + 2e^{-t}$
7. $x_1' = 2x_2 + e^{2t},$
   $x_2' = 2x_1$
8. $x_1' = x_2 + e^t,$
   $x_2' = -x_1$

In Exercises 9 through 20, give the form for $\mathbf{x}_p$. You need not actually solve for the arbitrary constants.

9. $x_1' = -x_1 - 2x_2 + \sin t,$
   $x_2' = x_1 + x_2 + \cos 2t$
10. $x_1' = x_1 - x_2 + t,$
    $x_2' = x_1 - x_2 - 3$
11. $x_1' = 3x_1 - 2x_2 + e^t,$
    $x_2' = 4x_1 - 3x_2$
12. $x_1' = x_1 + 2x_2 + e^{-t} \cos t,$
    $x_2' = x_1 + 3x_2 + \cos t$
13. $x_1' = 4x_1 - 2x_2 + t^2 e^{2t},$
    $x_2' = 4x_1 - 2x_2 - e^t$
14. $x_1' = 2x_1 - x_2 + e^t,$
    $x_2' = x_1$

15. $x_1' = 2x_1 - 4x_2 + t^2,$
    $x_2' = x_1 - 2x_2 + t - 1$
16. $x_1' = x_1 - x_2 + \sin 2t,$
    $x_2' = 5x_1 - x_2 - 3$
17. $x_1' = 2x_1 + 5x_2 + t \sin t,$
    $x_2' = -x_1 - 2x_2 + \cos t$
18. $x_1' = 2x_1 - x_2 + t^2 e^t,$
    $x_1' = 2x_1 + 3x_2 + e^{-4t}$
19. $x_1' = x_1 + 2x_2 + e^{-t} \cos 3t,$
    $x_2' = 3x_1 + 4x_2 + e^{-t}$
20. $x_1' = x_1 + 3x_2 + t \cos t,$
    $x_2' = 9x_1 + 13x_2 + \cos t$
21. Suppose that λ is an eigenvalue of the $n \times n$ matrix $\mathbf{A}$ of multiplicity one. Let $\mathbf{v}$ be a fixed vector. Show that finding a particular solution of $\mathbf{x}' = \mathbf{Ax} + e^{\lambda t}\mathbf{v}$ of the form $\mathbf{x}_p = e^{\lambda t}\mathbf{a} + te^{\lambda t}\mathbf{b}$ is equivalent to solving the system of algebraic equations

$$\mathbf{Ab} = \lambda \mathbf{b},$$
$$(\mathbf{A} - \lambda \mathbf{I})\mathbf{a} = \mathbf{b} - \mathbf{v},$$

(Note that this is a nonhomogeneous version of system (6) of the previous section.) or, equivalently,

$$\begin{bmatrix} \mathbf{A} - \lambda \mathbf{I} & \mathbf{0} \\ -\mathbf{I} & \mathbf{A} - \lambda \mathbf{I} \end{bmatrix} \begin{bmatrix} \mathbf{b} \\ \mathbf{a} \end{bmatrix} = \begin{bmatrix} \mathbf{0} \\ -\mathbf{v} \end{bmatrix}.$$

22. Solve the systems (14), (13), (12), and substitute the result into (11), to arrive at a particular solution to (4) and (5).

6.12 The Matrix Exponential

In Section 2.3, we found that the solution of $x' = ax$, a constant, was $x = e^{at}c$. The function e^{at} was then used in many of the calculations that followed. This section will show that similar notation is possible with the system $\mathbf{x}' = \mathbf{Ax}$.

The notation is very useful, both in working with constant-coefficient differential equations and in motivating the theory for linear systems with time-varying coefficients. If α is a scalar and $\mathbf{A}$ is a matrix, then we shall use the notation $\mathbf{A}\alpha$ for $\alpha \mathbf{A}$. This is standard practice, and allows our formulas to appear almost exactly like the formulas for scalar differential equations.

Recall that

$$e^{at} = \sum_{n=0}^{\infty} \frac{a^n t^n}{n!}.$$

Suppose that $\mathbf{A}$ is an $n \times n$ matrix. Then the *matrix exponential* may be defined as

$$e^{\mathbf{A}t} = \sum_{n=0}^{\infty} t^n \frac{\mathbf{A}^n}{n!} = \mathbf{I} + t\mathbf{A} + \frac{t^2}{2}\mathbf{A}^2 + \frac{t^3}{6}\mathbf{A}^3 + \cdots, \qquad (1)$$

where $\mathbf{A}^0 = \mathbf{I}$. This series converges to a matrix for all values of t and may be differentiated term by term, to show that

$$(e^{At})' = Ae^{At} = e^{At}A.$$

This leads to the following key theorem.

Theorem 6.12.1 Suppose that A is an $n \times n$ matrix. Then e^{At} is the unique solution of the matrix differential equation

$$X' = AX, \quad X(0) = I. \qquad (2)$$

Furthermore, if a is a constant $n \times 1$ vector, then the unique solution of

$$x' = Ax, \quad x(0) = a \qquad (3)$$

is

$$x = e^{At}a. \qquad (4)$$

Several other key facts about the matrix exponential are contained in the next theorem.

Theorem 6.12.2 Suppose that A is $n \times n$. Then

$$e^{At} \text{ is invertible} \quad \text{for all } t \quad \text{and} \quad (e^{At})^{-1} = e^{-At}, \qquad (5)$$
$$e^{A(t+s)} = e^{At}e^{As} \quad \text{for any scalars } s, t. \qquad (6)$$

However, in general, if B is an $n \times n$ matrix that does not commute with A (that is, $AB \neq BA$), then (see Exercises 10 and 11 at the end of this section),

$$e^{(A+B)t} \neq e^{At}e^{Bt}. \qquad (7)$$

An outline of a proof of these theorems appears in Exercises 25 through 27.

As we shall see, computing the matrix exponential is, in general, as difficult as solving the differential equation, and is avoided in practice whenever possible. This is quite similar to the useful notation $A^{-1}b$ for the solution of $Ax = b$. One rarely computes A^{-1}, and then $A^{-1}b$. Similarly, it is convenient to write $e^{At}a$ for the solution of

$$x' = Ax, \quad x(0) = a$$

even if the matrix exponential is not computed.

Before showing the usefulness of the matrix exponential notation, we shall compute a few examples. Occasionally it is possible to compute the series (1) directly.

Example 6.12.1 Let

$$A = \begin{bmatrix} 0 & 1 \\ 1 & 0 \end{bmatrix}.$$

Compute e^{At} from the series definition (1).

Solution

$$e^{At} = I + tA + \frac{t^2}{2}A^2 + \frac{t^3}{3!}A^3 + \frac{t^4}{4!}A^4 + \cdots$$

$$= I + t\begin{bmatrix} 0 & 1 \\ 1 & 0 \end{bmatrix} + \frac{t^2}{2}\begin{bmatrix} 1 & 0 \\ 0 & 1 \end{bmatrix} + \frac{t^3}{3!}\begin{bmatrix} 0 & 1 \\ 1 & 0 \end{bmatrix} + \frac{t^4}{4!}\begin{bmatrix} 1 & 0 \\ 0 & 1 \end{bmatrix} + \cdots$$

$$= I + \begin{bmatrix} 0 & t \\ t & 0 \end{bmatrix} + \begin{bmatrix} \frac{t^2}{2} & 0 \\ 0 & \frac{t^2}{2} \end{bmatrix} + \begin{bmatrix} 0 & \frac{t^3}{3!} \\ \frac{t^3}{3!} & 0 \end{bmatrix} + \begin{bmatrix} \frac{t^4}{4!} & 0 \\ 0 & \frac{t^4}{4!} \end{bmatrix} + \cdots$$

$$= \begin{bmatrix} 1 + \frac{t^2}{2!} + \frac{t^4}{4!} + \cdots, & t + \frac{t^3}{3!} + \frac{t^5}{5!} + \cdots \\ t + \frac{t^3}{3!} + \frac{t^5}{5!} + \cdots, & 1 + \frac{t^2}{2} + \frac{t^4}{4!} + \cdots \end{bmatrix} = \begin{bmatrix} \cosh t & \sinh t \\ \sinh t & \cosh t \end{bmatrix}. \blacksquare$$

However, it is usually easier to proceed in one of three ways.

1. Utilize some version of the Cayley–Hamilton Theorem (Exercises 23 and 24).
2. Compute eigenvectors and eigenvalues for **A** and use a change of coordinates (Exercises 13 through 21).
3. Solve the system of differential equations, $x' = Ax$ for several initial conditions.

We shall illustrate the third method of finding e^{At}, and leave the other two methods to the exercises.

Theorem 6.12.3 Suppose **A** is an $n \times n$ matrix. Let $X = e^{At}$. Let x_i be the ith column of e^{At}. Then x_i is the solution of

$$x' = Ax, \quad x(0) = e_i, \tag{8}$$

where e_i is the ith column of the $n \times n$ identity matrix.

Verification If $X = e^{At}$, then

$$X' = AX, \quad X(0) = I.$$

Let $X = [x_1, \ldots, x_n]$, where the x_i are the columns of **X**. Then $X' = AX$ becomes

$$[x'_1, \ldots, x'_n] = A[x_1, \ldots, x_n] = [Ax_1, \ldots, Ax_n],$$

while

$$X(0) = [x_1(0), \ldots, x_n(0)] = I = [e_1, e_2, \ldots, e_n].$$

Thus, $x'_i = Ax_i$, and $x_i(0) = e_i$. $\blacksquare$

Example 6.12.2   Let
$$\mathbf{A} = \begin{bmatrix} 0 & 1 \\ 1 & 0 \end{bmatrix}.$$

Find $e^{\mathbf{A}t}$, using Theorem 6.12.3.

Solution   First, we need to find the general solution of $\mathbf{x}' = \mathbf{A}\mathbf{x}$. The characteristic polynomial of $\mathbf{A}$ is

$$p(\lambda) = \det \begin{bmatrix} \lambda & -1 \\ -1 & \lambda \end{bmatrix} = \lambda^2 - 1,$$

so that the eigenvalues are $\lambda = \pm 1$. Computing the eigenvectors, we have an eigenvector

$$\begin{bmatrix} 1 \\ 1 \end{bmatrix}$$

associated with $\lambda = 1$, and an eigenvector

$$\begin{bmatrix} -1 \\ 1 \end{bmatrix}$$

associated with $\lambda = -1$. Thus,

$$\mathbf{x} = c_1 e^t \begin{bmatrix} 1 \\ 1 \end{bmatrix} + c_2 e^{-t} \begin{bmatrix} -1 \\ 1 \end{bmatrix} \tag{9}$$

is the general solution of $\mathbf{x}' = \mathbf{A}\mathbf{x}$. The columns of the matrix exponential are the solutions of

$$\mathbf{x}_1' = \mathbf{A}\mathbf{x}_1, \quad \mathbf{x}_1(0) = \begin{bmatrix} 1 \\ 0 \end{bmatrix}, \tag{10}$$

$$\mathbf{x}_2' = \mathbf{A}\mathbf{x}_2, \quad \mathbf{x}_2(0) = \begin{bmatrix} 0 \\ 1 \end{bmatrix}. \tag{11}$$

To find $\mathbf{x}_1$ we apply the initial condition (10) to the general solution (9), to get

$$\begin{aligned} c_1 - c_2 &= 1, \\ c_1 + c_2 &= 0, \end{aligned} \tag{12}$$

or $c_1 = \frac{1}{2}, c_2 = -\frac{1}{2}$. To find $\mathbf{x}_2$ we apply the initial condition (11) to the solution (9), to get

$$\begin{aligned} c_1 - c_2 &= 0, \\ c_1 + c_2 &= 1, \end{aligned} \tag{13}$$

or $c_1 = \frac{1}{2}, c_2 = \frac{1}{2}$. Thus,

$$e^{\mathbf{A}t} = [\mathbf{x}_1, \mathbf{x}_2] = \left[\begin{bmatrix} \frac{1}{2}e^t + \frac{1}{2}e^{-t} \\ \frac{1}{2}e^t - \frac{1}{2}e^{-t} \end{bmatrix}, \begin{bmatrix} \frac{1}{2}e^t - \frac{1}{2}e^{-t} \\ \frac{1}{2}e^t + \frac{1}{2}e^{-t} \end{bmatrix} \right]$$

$$= \begin{bmatrix} \cosh t & \sinh t \\ \sinh t & \cosh t \end{bmatrix}$$

as in Example 6.12.1. ∎

Note that in Example 6.12.2, the systems (12) and (13) had the same coefficients. This observation is closely related to the following fact.

Theorem 6.12.4 Suppose $\mathbf{A}$ is $n \times n$ and $\{\mathbf{x}_1, \ldots, \mathbf{x}_n\}$ is a fundamental set of solutions of $\mathbf{x}' = \mathbf{A}\mathbf{x}$. Then,

$$e^{\mathbf{A}t} = [\mathbf{x}_1, \ldots, \mathbf{x}_n][\mathbf{x}_1(0), \ldots, \mathbf{x}_n(0)]^{-1}. \tag{14}$$

Verification Let

$$\mathbf{Z}(t) = [\mathbf{x}_1, \ldots, \mathbf{x}_n][\mathbf{x}_1(0), \ldots, \mathbf{x}_n(0)]^{-1},$$

where $\{\mathbf{x}_1, \ldots, \mathbf{x}_n\}$ is a fundamental set of solutions of $\mathbf{x}' = \mathbf{A}\mathbf{x}$. Then

$$\mathbf{Z}' = [\mathbf{x}'_1, \ldots, \mathbf{x}'_n][\mathbf{x}_1(0), \ldots, \mathbf{x}_n(0)]^{-1}$$
$$= [\mathbf{A}\mathbf{x}_1, \ldots, \mathbf{A}\mathbf{x}_n][\mathbf{x}_1(0), \ldots, \mathbf{x}_n(0)]^{-1}$$
$$= \mathbf{A}[\mathbf{x}_1, \ldots, \mathbf{x}_n][\mathbf{x}_1(0), \ldots, \mathbf{x}_n(0)]^{-1} = \mathbf{A}\mathbf{Z},$$

and

$$\mathbf{Z}(0) = [\mathbf{x}_1(0), \ldots, \mathbf{x}_n(0)][\mathbf{x}_1(0), \ldots, \mathbf{x}_n(0)]^{-1} = \mathbf{I}.$$

Thus, $\mathbf{Z}(t) = e^{\mathbf{A}t}$ by Theorem 6.12.1. ∎

The invertibility of the matrix of initial values $[\mathbf{x}_1(0), \ldots, \mathbf{x}_n(0)]$ occurs because $\mathbf{x}_1(t), \ldots, \mathbf{x}_n(t)$ are a fundamental set of solutions.

Example 6.12.3 Given that we know, from (9), that

$$\mathbf{x}_1(t) = e^t \begin{bmatrix} 1 \\ 1 \end{bmatrix} = \begin{bmatrix} e^t \\ e^t \end{bmatrix}, \quad \mathbf{x}_2(t) = e^{-t} \begin{bmatrix} -1 \\ 1 \end{bmatrix} = \begin{bmatrix} -e^{-t} \\ e^{-t} \end{bmatrix} \tag{15}$$

are a fundamental set of solutions of $\mathbf{x}' = \mathbf{A}\mathbf{x}$, where

$$\mathbf{A} = \begin{bmatrix} 0 & 1 \\ 1 & 0 \end{bmatrix},$$

compute $e^{\mathbf{A}t}$ using Theorem 6.12.4.

Solution By (14),

$$e^{\mathbf{A}t} = [\mathbf{x}_1, \mathbf{x}_2][\mathbf{x}_1(0), \mathbf{x}_2(0)]^{-1}$$
$$= \begin{bmatrix} e^t & -e^{-t} \\ e^t & e^{-t} \end{bmatrix} \begin{bmatrix} 1 & -1 \\ 1 & 1 \end{bmatrix}^{-1}. \tag{16}$$

Using the technique of Section 6.7, we compute that

$$\begin{bmatrix} 1 & -1 \\ 1 & 1 \end{bmatrix}^{-1} = \begin{bmatrix} \frac{1}{2} & \frac{1}{2} \\ -\frac{1}{2} & \frac{1}{2} \end{bmatrix},$$

so that by (16)

$$e^{\mathbf{A}t} = \begin{bmatrix} e^t & -e^{-t} \\ e^t & e^{-t} \end{bmatrix} \begin{bmatrix} \frac{1}{2} & \frac{1}{2} \\ -\frac{1}{2} & \frac{1}{2} \end{bmatrix} = \begin{bmatrix} \dfrac{e^t + e^{-t}}{2} & \dfrac{e^t - e^{-t}}{2} \\ \dfrac{e^t - e^{-t}}{2} & \dfrac{e^t + e^{-t}}{2} \end{bmatrix}$$

$$= \begin{bmatrix} \cosh t & \sinh t \\ \sinh t & \cosh t \end{bmatrix}. \blacksquare$$

With these preliminaries out of the way, we may use the matrix exponential to solve linear, constant-coefficient differential equations. Consider, then, the nonhomogeneous problem

$$\mathbf{x}'(t) = \mathbf{A}\mathbf{x}(t) + \mathbf{f}(t)$$

or

$$\mathbf{x}' - \mathbf{A}\mathbf{x} = \mathbf{f}. \tag{17}$$

Proceeding as in Section 2.3 on first-order linear equations, multiply (17) by $e^{-\mathbf{A}t}$ on the left, to obtain

$$e^{-\mathbf{A}t}(\mathbf{x}' - \mathbf{A}\mathbf{x}) = e^{-\mathbf{A}t}\mathbf{f}, \tag{18}$$

which is

$$(e^{-\mathbf{A}t}\mathbf{x})' = e^{-\mathbf{A}t}\mathbf{f}. \tag{19}$$

Antidifferentiate both sides of (19), to yield

$$e^{-\mathbf{A}t}\mathbf{x} = \int e^{-\mathbf{A}s}\mathbf{f}(s)\, ds + \mathbf{c}, \tag{20}$$

where $\mathbf{c}$ is an arbitrary constant vector. Now multiply both sides of (20) by the inverse of $e^{-\mathbf{A}t}$, which is $e^{\mathbf{A}t}$, to get the following key result, which should be compared to that of Section 2.3.

Theorem 6.12.5 Suppose that $\mathbf{A}$ is an $n \times n$ constant matrix and $\mathbf{f}(t)$ is a continuous $n \times 1$ vector-valued function on the interval I containing t. (Equivalently, $\mathbf{f}$ is a vector of continuous functions.) Then the general solution of

$$\mathbf{x}' = \mathbf{A}\mathbf{x} + \mathbf{f}$$

6.12 The Matrix Exponential

is

$$\mathbf{x}(t) = e^{\mathbf{A}t} \int e^{-\mathbf{A}s}\mathbf{f}(s)\,ds + e^{\mathbf{A}t}\mathbf{c}$$

$$= \int e^{\mathbf{A}(t-s)}\mathbf{f}(s)\,ds + e^{\mathbf{A}t}\mathbf{c}, \tag{21}$$

where $\mathbf{c}$ is an arbitrary constant vector.

In particular, the solution of

$$\mathbf{x}' = \mathbf{A}\mathbf{x} + \mathbf{f}, \qquad \mathbf{x}(t_0) = \mathbf{a}$$

is

$$\mathbf{x}(t) = \int_{t_0}^{t} e^{\mathbf{A}(t-s)}\mathbf{f}(s)\,ds + e^{\mathbf{A}(t-t_0)}\mathbf{a}. \tag{22}$$

Example 6.12.4 Use Theorem 6.12.5 to solve the nonhomogeneous differential equation

$$\begin{aligned} x_1' &= x_2 + 1, & x_1(0) &= a_1, \\ x_2' &= x_1 + e^t, & x_2(0) &= a_2 \end{aligned} \tag{23}$$

on the interval $[0, \infty)$.

Solution We already have computed in Example 6.12.3 that

$$e^{\mathbf{A}t} = \begin{bmatrix} \cosh t & \sinh t \\ \sinh t & \cosh t \end{bmatrix}$$

so that

$$e^{-\mathbf{A}t} = e^{\mathbf{A}(-t)} = \begin{bmatrix} \cosh(-t) & \sinh(-t) \\ \sinh(-t) & \cosh(-t) \end{bmatrix} = \begin{bmatrix} \cosh t & -\sinh t \\ -\sinh t & \cosh t \end{bmatrix}.$$

By (22), the solution of (23) is

$$\mathbf{x} = e^{\mathbf{A}t} \int_0^t e^{-\mathbf{A}s}\mathbf{f}(s)\,ds + e^{\mathbf{A}t}\mathbf{a}$$

$$= e^{\mathbf{A}t} \int_0^t \begin{bmatrix} \cosh s & -\sinh s \\ -\sinh s & \cosh s \end{bmatrix} \begin{bmatrix} 1 \\ e^s \end{bmatrix} ds + e^{\mathbf{A}t}\mathbf{a}$$

$$= e^{\mathbf{A}t} \int_0^t \tfrac{1}{2}\begin{bmatrix} e^s + e^{-s} - e^{2s} + 1 \\ -e^s + e^{-s} + e^{2s} + 1 \end{bmatrix} ds + e^{\mathbf{A}t}\mathbf{a}$$

$$= \underbrace{\begin{bmatrix} \cosh t & \sinh t \\ \sinh t & \cosh t \end{bmatrix}}\cdot\underbrace{\tfrac{1}{2}\begin{bmatrix} e^t - e^{-t} - \dfrac{e^{2t}}{2} + t + \dfrac{1}{2} \\ -e^t - e^{-t} + \dfrac{e^{2t}}{2} + t + \dfrac{3}{2} \end{bmatrix}} + \underbrace{\begin{bmatrix} \cosh t & \sinh t \\ \sinh t & \cosh t \end{bmatrix}\begin{bmatrix} a_1 \\ a_2 \end{bmatrix}}$$
$$\tag{24}$$

$$= \tfrac{1}{4}\begin{bmatrix} 2te^t + e^t - e^{-t} \\ -4 + 2te^t + 3e^t + e^{-t} \end{bmatrix} + a_1\begin{bmatrix} \cosh t \\ \sinh t \end{bmatrix} + a_2\begin{bmatrix} \sinh t \\ \cosh t \end{bmatrix},$$

Exercises

In Exercises 1 through 6 write the system as $\mathbf{x}' = \mathbf{A}\mathbf{x} + \mathbf{f}$. Find $e^{\mathbf{A}t}$ using Theorem 6.12.3 or 6.12.4. Then solve the differential equation, using Theorem 6.12.5. If no initial conditions are given, find the general solution.

1. $x_1' = 3x_1 + x_2 + 1,$ $\quad x_1(0) = 1,$
   $x_2' = -2x_1 - 2,$ $\quad x_2(0) = 2$

2. $x_1' = x_1 + 2,$
   $x_2' = x_1 - x_2 + 3$

3. $x' = x - 2y + 6e^{-t},$
   $y' = -2x + 4y + e^t$

4. $x' = 3x + y + 1,$ $\quad x(0) = 0,$
   $y' = -6x - 2y + 1,$ $\quad y(0) = 0$

5. $x_1' = 6x_1 + 2x_2 + e^{2t},$
   $x_2' = 2x_1 + 3x_2$

6. $x_1' = 3x_1 + x_2,$
   $x_2' = -5x_1 - 3x_2 + e^t$

7. Let $\mathbf{A} = \begin{bmatrix} 0 & 1 \\ 0 & 0 \end{bmatrix}$.

   i) Compute $e^{\mathbf{A}t}$, using the series (1).
   ii) Compute $(e^{\mathbf{A}t})^{-1}$ and verify $(e^{\mathbf{A}t})^{-1} = e^{-\mathbf{A}t}$.

8. Let $\mathbf{A} = \begin{bmatrix} 0 & 1 & 0 \\ 0 & 0 & 1 \\ 0 & 0 & 0 \end{bmatrix}$.

   Compute $e^{\mathbf{A}t}$, using the series (1).

9. Let $\mathbf{A} = \begin{bmatrix} 0 & 1 \\ -1 & 0 \end{bmatrix}$.

   Compute $e^{\mathbf{A}t}$, using the series (1).

10. Let $\mathbf{A} = \begin{bmatrix} 0 & 1 \\ 0 & 0 \end{bmatrix}$, $\mathbf{B} = \begin{bmatrix} 0 & 0 \\ 1 & 0 \end{bmatrix}$,
    and $\mathbf{C} = \mathbf{A} + \mathbf{B} = \begin{bmatrix} 0 & 1 \\ 1 & 0 \end{bmatrix}$.

    Compute $e^{\mathbf{A}t}$, $e^{\mathbf{B}t}$, $e^{\mathbf{C}t}$, and verify $e^{\mathbf{A}t}e^{\mathbf{B}t} \neq e^{(\mathbf{A}+\mathbf{B})t}$.

11. Suppose that $e^{\mathbf{A}t}e^{\mathbf{B}t} = e^{(\mathbf{A}+\mathbf{B})t}$ for two $n \times n$ matrices $\mathbf{A}$, $\mathbf{B}$. Show that $\mathbf{A}\mathbf{B} = \mathbf{B}\mathbf{A}$. (*Hint:* Differentiate twice and evaluate at zero.)

12. Suppose $\mathbf{A}$ is an invertible $n \times n$ matrix. Verify that
    $$\int e^{\mathbf{A}t} \, dt = \mathbf{A}^{-1} e^{\mathbf{A}t} + \mathbf{C},$$
    where $\mathbf{C}$ is an arbitrary $n \times n$ constant matrix.

Exercises 13 through 21 provide an alternative way to compute $e^{\mathbf{A}t}$ and introduce the important concept of a *similarity transformation*.

13. Suppose $\mathbf{A}$ is a 2×2 matrix with distinct eigenvalues λ_1, λ_2 and nonzero eigenvectors $\mathbf{u}_1, \mathbf{u}_2$. Let $\mathbf{U} = [\mathbf{u}_1, \mathbf{u}_2]$ be the 2×2 matrix with $\mathbf{u}_1, \mathbf{u}_2$ as columns. Verify that
    $$\mathbf{A}[\mathbf{u}_1, \mathbf{u}_2] = [\mathbf{u}_1, \mathbf{u}_2] \begin{bmatrix} \lambda_1 & 0 \\ 0 & \lambda_2 \end{bmatrix}, \quad (25)$$
    or, equivalently,
    $$\mathbf{A}\mathbf{U} = \mathbf{U}\Lambda,$$
    where
    $$\Lambda = \begin{bmatrix} \lambda_1 & 0 \\ 0 & \lambda_2 \end{bmatrix}$$
    is the diagonal matrix with the eigenvalues as entries.

14. Suppose that $\mathbf{U}$ is an invertible $m \times m$ matrix and $\mathbf{B}$ is an $m \times m$ matrix. Verify that, for any integer $n \geq 0$,
$$(\mathbf{UBU}^{-1})^n = \mathbf{UB}^n\mathbf{U}^{-1}.$$
Then conclude from (1) that
$$e^{\mathbf{UBU}^{-1}t} = \mathbf{U}e^{\mathbf{B}t}\mathbf{U}^{-1}. \tag{26}$$

15. Show that, if
$$\boldsymbol{\Lambda} = \begin{bmatrix} \lambda_1 & & 0 \\ & \ddots & \\ 0 & & \lambda_n \end{bmatrix}$$
is a diagonal matrix, then
$$e^{\boldsymbol{\Lambda}t} = \begin{bmatrix} e^{\lambda_1 t} & & 0 \\ & \ddots & \\ 0 & & e^{\lambda_n t} \end{bmatrix}. \tag{27}$$

Exercises 13, 14, and 15 show that, if $\mathbf{A}$ is a 2×2 matrix and has two distinct eigenvalues λ_1, λ_2 with distinct eigenvectors $\mathbf{u}_1, \mathbf{u}_2$, then
$$e^{\mathbf{A}t} = \mathbf{U} \begin{bmatrix} e^{\lambda_1 t} & 0 \\ 0 & e^{\lambda_2 t} \end{bmatrix} \mathbf{U}^{-1}, \tag{28}$$
where $\mathbf{U}$ has for columns the eigenvectors $\mathbf{u}_1, \mathbf{u}_2$.

16. Let $\mathbf{A} = \begin{bmatrix} 3 & 1 \\ -2 & 0 \end{bmatrix}$.

Compute $e^{\mathbf{A}t}$, using (28).

17. Let $\mathbf{A} = \begin{bmatrix} 1 & -2 \\ -2 & 4 \end{bmatrix}$.

Compute $e^{\mathbf{A}t}$, using (28).

18. Let $\mathbf{A} = \begin{bmatrix} 3 & 1 \\ -6 & -2 \end{bmatrix}$.

Compute $e^{\mathbf{A}t}$, using (28).

19. Let $\mathbf{A} = \begin{bmatrix} 6 & 2 \\ 2 & 3 \end{bmatrix}$.

Compute $e^{\mathbf{A}t}$, using (28).

20. Let $\mathbf{A} = \begin{bmatrix} 3 & 1 \\ -5 & -3 \end{bmatrix}$.

Compute $e^{\mathbf{A}t}$, using (28).

21. Let $\mathbf{A} = \begin{bmatrix} -3 & 3 \\ 3 & 5 \end{bmatrix}$.

Compute $e^{\mathbf{A}t}$, using (28).

22. Show that if $\mathbf{A}$ is an $n \times n$ matrix with n distinct eigenvalues $\lambda_1, \ldots, \lambda_n$ with corresponding eigenvectors $\mathbf{u}_1, \ldots, \mathbf{u}_n$, then
$$e^{\mathbf{A}t} = \mathbf{U} \begin{bmatrix} e^{\lambda_1 t} & & 0 \\ & \ddots & \\ 0 & & e^{\lambda_n t} \end{bmatrix} \mathbf{U}^{-1}, \tag{29}$$
where $\mathbf{U} = [\mathbf{u}_1, \ldots, \mathbf{u}_n]$ has the eigenvectors as columns.

The *Cayley–Hamilton theorem* has as a consequence that if $\mathbf{A}$ is an $n \times n$ matrix, then
$$e^{\mathbf{A}t} = \alpha_0(t)\mathbf{I} + \alpha_1(t)\mathbf{A} + \cdots + \alpha_{n-1}(t)\mathbf{A}^{n-1} \tag{30}$$
for some scalar functions $\alpha_0(t), \ldots, \alpha_{n-1}(t)$. If $\mathbf{u}$ is an eigenvector for the eigenvalue λ, then
$$e^{\mathbf{A}t}\mathbf{u} = \alpha_0(t)\mathbf{I}\mathbf{u} + \alpha_1(t)\mathbf{A}\mathbf{u} + \cdots + \alpha_{n-1}(t)\mathbf{A}^n\mathbf{u}$$
or
$$e^{\lambda t}\mathbf{u} = (\alpha_0(t) + \alpha_1(t)\lambda + \cdots + \alpha_{n-1}(t)\lambda^n)\mathbf{u}. \tag{31}$$

23. If $\mathbf{A}$ is a 2×2 matrix with distinct eigenvalues λ_1, λ_2, then (30) becomes
$$e^{\mathbf{A}t} = \alpha_0(t)\mathbf{I} + \alpha_1(t)\mathbf{A}, \tag{32}$$
and (31) yields
$$\alpha_0(t) + \lambda_1\alpha_1(t) = e^{\lambda_1 t},$$
$$\alpha_0(t) + \lambda_2\alpha_1(t) = e^{\lambda_2 t},$$
or
$$\begin{bmatrix} 1 & \lambda_1 \\ 1 & \lambda_2 \end{bmatrix} \begin{bmatrix} \alpha_0(t) \\ \alpha_1(t) \end{bmatrix} = \begin{bmatrix} e^{\lambda_1 t} \\ e^{\lambda_2 t} \end{bmatrix}. \tag{33}$$

Thus, to find $e^{\mathbf{A}t}$ for a 2×2 matrix $\mathbf{A}$, one can find the eigenvalues λ_1, λ_2, solve (33) for α_0, α_1, and then use (32).

a) Let $\mathbf{A} = \begin{bmatrix} 3 & 1 \\ -2 & 0 \end{bmatrix}$.

Find $e^{\mathbf{A}t}$ by this method.

b) Let $\mathbf{A} = \begin{bmatrix} 1 & -2 \\ -2 & 4 \end{bmatrix}$.

Find $e^{\mathbf{A}t}$ by this method.

c) Let $\mathbf{A} = \begin{bmatrix} 3 & 1 \\ -6 & -2 \end{bmatrix}$.

Find $e^{\mathbf{A}t}$ by this method.

24. If $\mathbf{A}$ is a 3×3 matrix with distinct eigenvalues $\lambda_1, \lambda_2, \lambda_3$, then (30) is
$$e^{\mathbf{A}t} = \alpha_0(t)\mathbf{I} + \alpha_1(t)\mathbf{A} + \alpha_2(t)\mathbf{A}^2, \tag{34}$$
where $\alpha_0, \alpha_1, \alpha_2$ are the solutions of
$$\begin{bmatrix} 1 & \lambda_1 & \lambda_1^2 \\ 1 & \lambda_2 & \lambda_2^2 \\ 1 & \lambda_3 & \lambda_3^2 \end{bmatrix} \begin{bmatrix} \alpha_0 \\ \alpha_1 \\ \alpha_2 \end{bmatrix} = \begin{bmatrix} e^{\lambda_1 t} \\ e^{\lambda_2 t} \\ e^{\lambda_3 t} \end{bmatrix}. \tag{35}$$

Find $e^{\mathbf{A}t}$ for
$$\mathbf{A} = \begin{bmatrix} 1 & 1 & 0 \\ 0 & 0 & 1 \\ 0 & 0 & 2 \end{bmatrix}$$
by this method.

25. Verify $(e^{\mathbf{A}t})' = \mathbf{A}e^{\mathbf{A}t}$ by differentiating the series (1) with respect to t.

26. Verify that $(e^{At})^{-1} = e^{-At}$ by multiplying the series for e^{At} and e^{-At} to get I.

27. Let $F(t) = e^{A(t+s)}$, $G(t) = e^{At}e^{As}$ for a constant $n \times n$ matrix A and fixed but unknown scalar s. Verify that $F^{(n)}(0) = G^{(n)}(0) = A^n e^{As}$, and conclude, by the uniqueness of power series, that $e^{A(t+s)} = e^{At}e^{As}$.

Comment 1 The method of Exercises 23 and 24 seems so simple that we should probably explain why it is relegated to the exercises. It is not very practical for the larger-sized matrices that occur in real applications. There are two reasons for this. First, computing the powers of A becomes a lot of work. Surprisingly enough, computing A^2 involves about as much "work" as inverting A. Secondly, while the matrix

$$\begin{bmatrix} 1 & \lambda_1 & \cdots & \lambda_1^{n-1} \\ 1 & \lambda_2 & \cdots & \lambda_2^{n-1} \\ \vdots & & & \vdots \\ 1 & \lambda_n & \cdots & \lambda_n^{n-1} \end{bmatrix}$$

(known as the Vandermond matrix) is known to be invertible if the λ_i are distinct, it is also known to become very *ill-conditioned* as n increases. In other words, the equations for the α_i can be difficult to work with and are very susceptible to round-off error when they are solved numerically.

Comment 2 The eigenvalue/eigenvector approach of Exercises 13 through 21 is very important even though the actual exponential is usually not computed.

6.13 Fundamental Solution Matrices

For scalar constants a, the solution $x(t) = e^{at}c$ of $x' = ax$, $x(0) = c$ is just a special case of the solution

$$x(t) = e^{\int_0^t a(s)\,ds}c \tag{1}$$

of the linear, homogeneous, variable-coefficient problem

$$x'(t) = a(t)x(t), \qquad x(0) = c, \tag{2}$$

found in Section 2.3. Since $\mathbf{x} = e^{At}\mathbf{c}$ is the solution of $\mathbf{x}' = A\mathbf{x}$, $\mathbf{x}(0) = \mathbf{c}$, for a constant matrix, we might think that

$$\mathbf{x}(t) = e^{\int_0^t A(s)\,ds}\mathbf{c} \tag{3}$$

would be the solution of the linear, homogeneous, variable-coefficient system of differential equations

$$\mathbf{x}'(t) = A(t)\mathbf{x}(t). \tag{4}$$

Unfortunately, this is not generally the case. The difficulty is that, in general,

$$\frac{d}{dt}(e^{B(t)}) \neq e^{B(t)}B'(t),$$

unless $B(t)B'(t) = B'(t)B(t)$ for all t, and this last condition rarely holds in applications.

Example 6.13.1 Consider the linear, homogeneous system of differential equations

$$x_1' = x_1 + 2tx_2, \tag{5a}$$
$$x_2' = 0, \tag{5b}$$

or

$$\mathbf{x}' = \begin{bmatrix} 1 & 2t \\ 0 & 0 \end{bmatrix} \mathbf{x} = \mathbf{A}(t)\mathbf{x}. \tag{6}$$

Then

$$e^{\int_0^t \mathbf{A}(s)\,ds} = e^{\begin{bmatrix} t & t^2 \\ 0 & 0 \end{bmatrix}} = \begin{bmatrix} e^t & te^t - t \\ 0 & 1 \end{bmatrix}. \tag{7}$$

(This can be verified by using the Series (1) of Section 6.12.) Let

$$\mathbf{x} = e^{\int_0^t \mathbf{A}(s)\,ds} \begin{bmatrix} 1 \\ 1 \end{bmatrix} = \begin{bmatrix} e^t + te^t - t \\ 1 \end{bmatrix}. \tag{8}$$

To see that (8) is not a solution of (5), substitute into (5a):

$$(e^t + te^t - t)' = (e^t + te^t - t) + 2t$$

or

$$e^t + te^t + e^t - 1 = e^t + te^t + t,$$

which is not true. Thus (8) is not a solution of (5). ∎

We need something, however, to play the role for $\mathbf{x}'(t) = \mathbf{A}(t)\mathbf{x}(t)$ that $e^{\mathbf{A}t}$ plays for $\mathbf{x}' = \mathbf{A}\mathbf{x}$, so that we can derive a variation-of-constants formula (Theorem 6.12.5) for the variable-coefficient case. The key turns out to be the following:

Theorem 6.13.1 Suppose that $\mathbf{A}(t)$ is an $n \times n$ matrix of functions continuous on the interval I containing t_0. Let $\{\mathbf{w}_1(t), \ldots, \mathbf{w}_n(t)\}$ be a fundamental set of solutions of

$$\mathbf{x}'(t) = \mathbf{A}(t)\mathbf{x}(t), \tag{9}$$

as promised by Theorems 6.9.1 and 6.9.3. Let $\mathbf{W}(t) = [\mathbf{w}_1(t), \ldots, \mathbf{w}_n(t)]$ be the $n \times n$ matrix-valued function with the $\mathbf{w}_i$ as columns. Then $\mathbf{W}$ is called a *fundamental solution matrix* for (9). The fundamental matrix $\mathbf{W}$ has the following properties:

$$\mathbf{W}'(t) = \mathbf{A}(t)\mathbf{W}(t); \tag{10}$$
$$\mathbf{W}(t) \text{ is invertible for all } t; \tag{11}$$
$$\mathbf{W}(t)\mathbf{c}, \mathbf{c} \text{ an arbitrary constant vector, is the general solution of } \mathbf{x}' = \mathbf{A}(t)\mathbf{x}. \tag{12}$$

Verification Property (10) follows, since

$$\mathbf{W}' = [\mathbf{w}_1', \ldots, \mathbf{w}_n'] = [\mathbf{A}\mathbf{w}_1, \ldots, \mathbf{A}\mathbf{w}_n] = \mathbf{A}[\mathbf{w}_1, \ldots, \mathbf{w}_n] = \mathbf{A}\mathbf{W}.$$

Since $\mathbf{Wc} = c_1\mathbf{w}_1 + c_2\mathbf{w}_2 + \cdots + c_n\mathbf{w}_n$, (12) follows from (10), and the properties of fundamental sets of solutions described in Theorem 6.9.3. Property (11) follows from the fact that a fundamental set of solutions has values that are linearly independent for each t, or equivalently, by the uniqueness of solutions. ∎

If $\mathbf{W}(t)$ is a fundamental solution matrix for $\mathbf{x}'(t) = \mathbf{A}(t)\mathbf{x}(t)$ on the interval I define

$$\mathbf{Z}(t) = \mathbf{W}(t)\mathbf{W}^{-1}(t_0). \tag{13}$$

Theorem 6.13.2 If $\mathbf{Z}$ is defined by (13) for a fundamental solution matrix $\mathbf{W}$, then

$$\mathbf{Z}(t) \text{ is also a fundamental solution matrix,} \tag{14}$$
$$\mathbf{Z}'(t) = \mathbf{A}(t)\mathbf{Z}(t), \quad \mathbf{Z}(t_0) = \mathbf{I}, \tag{15}$$
$$\mathbf{x} = \mathbf{Z}(t)\mathbf{c} \text{ is the unique solution of } \mathbf{x}'(t) = \mathbf{A}(t)\mathbf{x}(t), \quad \mathbf{x}(t_0) = \mathbf{c}. \tag{16}$$

Verification From (13), $\mathbf{Z}(t_0) = \mathbf{W}(t_0)\mathbf{W}(t_0)^{-1} = \mathbf{I}$. Also,

$$\mathbf{Z}'(t) = \mathbf{W}'(t)\mathbf{W}(t_0)^{-1} = \mathbf{A}(t)\mathbf{W}(t)\mathbf{W}(t_0)^{-1} = \mathbf{A}(t)\mathbf{Z}(t).$$

Thus (14) and (15) hold, and (16) follows. ∎

The matrix $\mathbf{Z}(t)$ is a generalization of the matrix exponential.

Theorem 6.13.3 If $\mathbf{A}(t) = \mathbf{A}$ is a constant matrix, then $\mathbf{Z}(t)$ defined by (13) for a fundamental solution matrix $\mathbf{W}(t)$ of $\mathbf{x}' = \mathbf{A}\mathbf{x}$ is

$$\mathbf{Z}(t) = e^{\mathbf{A}(t-t_0)}.$$

Note that this is actually a restatement of Theorem 6.12.1

Example 6.13.2 Let us find $\mathbf{Z}(t)$ for Example 6.13.1 on the interval $[0, \infty)$. To do this, we need to find a fundamental set of solutions for (5), that is,

$$x_1' = x_1 + 2tx_2, \tag{17a}$$
$$x_2' = 0. \tag{17b}$$

From (17b) we get $x_2 = c_1$. Thus, (17a) is

$$x_1' - x_1 = 2tc_1.$$

The integrating factor for this first-order, linear differential equation is e^{-t}, and we have

$$(e^{-t}x_1)' = e^{-t}2tc_1$$

or

$$e^{-t}x_1 = -2(te^{-t} + e^{-t})c_1 + c_2.$$

Thus, the general solution of (17) is

$$\mathbf{x} = \begin{bmatrix} x_1 \\ x_2 \end{bmatrix} = \begin{bmatrix} (-2t-2)c_1 + c_2 e^t \\ c_1 \end{bmatrix} = c_1 \begin{bmatrix} -2t-2 \\ 1 \end{bmatrix} + c_2 \begin{bmatrix} e^t \\ 0 \end{bmatrix}.$$

A fundamental set of solutions would be

$$\{\mathbf{w}_1, \mathbf{w}_2\} = \left\{ \begin{bmatrix} -2t-2 \\ 1 \end{bmatrix}, \begin{bmatrix} e^t \\ 0 \end{bmatrix} \right\}.$$

Thus

$$\mathbf{W}(t) = \begin{bmatrix} -2t-2 & e^t \\ 1 & 0 \end{bmatrix}$$

is a fundamental solution matrix for (17). The matrix $\mathbf{Z}(t)$ is

$$\mathbf{Z}(t) = \mathbf{W}(t)\mathbf{W}(0)^{-1} = \begin{bmatrix} -2t-2 & e^t \\ 1 & 0 \end{bmatrix} \begin{bmatrix} -2 & 1 \\ 1 & 0 \end{bmatrix}^{-1}$$

$$= \begin{bmatrix} -2t-2 & e^t \\ 1 & 0 \end{bmatrix} \begin{bmatrix} 0 & 1 \\ 1 & 2 \end{bmatrix}$$

$$= \begin{bmatrix} e^t & -2t-2+2e^t \\ 0 & 1 \end{bmatrix}. \blacksquare$$

We now wish to use the fundamental solution matrix to solve nonhomogeneous linear systems of differential equations. The key is the following technical lemma. We let $\mathbf{W}^{-1}(t)$ denote the function $[\mathbf{W}(t)]^{-1}$.

Lemma 6.13.1 Suppose that $\mathbf{W}(t)$ is a fundamental solution matrix for $\mathbf{x}' = \mathbf{A}(t)\mathbf{x}$. Then,

$$(\mathbf{W}^{-1}(t))' = -\mathbf{W}^{-1}(t)\mathbf{A}(t). \tag{18}$$

If $\mathbf{h}(t)$ is any differentiable vector-valued function, then

$$(\mathbf{W}^{-1}(t)\mathbf{h}(t))' = \mathbf{W}^{-1}(t)\mathbf{h}'(t) - \mathbf{W}^{-1}(t)\mathbf{A}(t)\mathbf{h}(t). \tag{19}$$

Proof Suppose that $\mathbf{W}(t)$ is a fundamental solution matrix for $\mathbf{x}' = \mathbf{A}(t)\mathbf{x}$. Then $\mathbf{W}' = \mathbf{A}\mathbf{W}$ and $\mathbf{W}$ is invertible. By the definition of an inverse,

$$\mathbf{W}^{-1}\mathbf{W} = \mathbf{I}. \tag{20}$$

Differentiating both sides of (20) with respect to t yields, by the product rule,

$$(\mathbf{W}^{-1})'\mathbf{W} + \mathbf{W}^{-1}\mathbf{W}' = 0$$

or

$$(\mathbf{W}^{-1})'\mathbf{W} + \mathbf{W}^{-1}\mathbf{A}\mathbf{W} = 0. \tag{21}$$

Multiply (21) on the right by $\mathbf{W}^{-1}$ and use $\mathbf{W}\mathbf{W}^{-1} = \mathbf{I}$, to obtain

$$(\mathbf{W}^{-1})' + \mathbf{W}^{-1}\mathbf{A} = \mathbf{0},$$

or (18). Now, to verify (19), just differentiate,

$$(\mathbf{W}^{-1}\mathbf{h})' = (\mathbf{W}^{-1})\mathbf{h}' + (\mathbf{W}^{-1})'\mathbf{h}$$
$$= \mathbf{W}^{-1}\mathbf{h}' - \mathbf{W}^{-1}\mathbf{A}\mathbf{h} \qquad \text{(by (18))}. \blacksquare$$

Just as the matrix $\mathbf{W}$ (or $\mathbf{Z}$) plays the role of $e^{\mathbf{A}t}$, so $\mathbf{W}^{-1}$ (or $\mathbf{Z}^{-1}$) plays the role of $e^{-\mathbf{A}t}$. Thus, to solve

$$\mathbf{x}'(t) - \mathbf{A}(t)\mathbf{x}(t) = \mathbf{f}(t), \tag{22}$$

we multiply on the left by $\mathbf{W}^{-1}(t)$, to get:

$$\mathbf{W}^{-1}(t)\mathbf{x}'(t) - \mathbf{W}^{-1}(t)\mathbf{A}(t)\mathbf{x}(t) = \mathbf{W}^{-1}(t)\mathbf{f}(t).$$

By (19), with $\mathbf{x} = \mathbf{h}$, this becomes

$$(\mathbf{W}^{-1}(t)\mathbf{x}(t))' = \mathbf{W}^{-1}(t)\mathbf{f}(t).$$

Antidifferentiating both sides with respect to t yields

$$\mathbf{W}^{-1}(t)\mathbf{x}(t) = \int \mathbf{W}^{-1}(s)\mathbf{f}(s)\,ds + \mathbf{c}.$$

Now multiply by $\mathbf{W}(t)$ on the left to get the following theorem.

Theorem 6.13.4

Variation of Constants Suppose that $\mathbf{W}(t)$ is a fundamental solution matrix of $\mathbf{x}'(t) = \mathbf{A}(t)\mathbf{x}(t)$ on the interval I containing t_0. Then the general solution of $\mathbf{x}'(t) = \mathbf{A}(t)\mathbf{x}(t) + \mathbf{f}(t)$ is

$$\mathbf{x}(t) = \mathbf{W}(t)\int \mathbf{W}^{-1}(s)\mathbf{f}(s)\,ds + \mathbf{W}(t)\mathbf{c}, \tag{23}$$

where $\mathbf{c}$ is an arbitrary constant vector. If $\mathbf{Z}(t) = \mathbf{W}(t)\mathbf{W}^{-1}(t_0)$, then

$$\mathbf{x}(t) = \mathbf{Z}(t)\int_{t_0}^{t} \mathbf{Z}^{-1}(s)\mathbf{f}(s)\,ds + \mathbf{Z}(t)\mathbf{c} \tag{24}$$

is the unique solution of $\mathbf{x}'(t) = \mathbf{A}(t)\mathbf{x}(t) + \mathbf{f}(t)$, $\mathbf{x}(t_0) = \mathbf{c}$.

This theorem should be compared to Theorem 6.12.5. This method can require a great deal of calculating; it is, however, of considerable theoretical importance.

Example 6.13.3

Using Theorem 6.13.4, solve the initial-value problem

$$x_1' = x_1 + 2tx_2 + e^t, \qquad x_1(0) = 2, \tag{25a}$$
$$x_2' = 1, \qquad x_2(0) = 1. \tag{25b}$$

Solution

From Example 6.13.2, we already have that

$$Z(t) = \begin{bmatrix} e^t & -2t - 2 + 2e^t \\ 0 & 1 \end{bmatrix}.$$

A calculation then gives

$$Z^{-1}(t) = \begin{bmatrix} e^{-t} & 2te^{-t} + 2e^{-t} - 2 \\ 0 & 1 \end{bmatrix}.$$

Thus, (24) is

$$\begin{aligned}
\mathbf{x}(t) &= \begin{bmatrix} e^t & -2t - 2 + 2e^t \\ 0 & 1 \end{bmatrix} \int_0^t \begin{bmatrix} e^{-s} & 2se^{-s} + 2e^{-s} - 2 \\ 0 & 1 \end{bmatrix} \begin{bmatrix} e^s \\ 1 \end{bmatrix} ds \\
&\quad + \begin{bmatrix} e^t & -2t - 2 + 2e^t \\ 0 & 1 \end{bmatrix} \begin{bmatrix} 2 \\ 1 \end{bmatrix} \\
&= \begin{bmatrix} e^t & -2t - 2 + 2e^t \\ 0 & 1 \end{bmatrix} \int_0^t \begin{bmatrix} 2se^{-s} + 2e^{-s} - 1 \\ 1 \end{bmatrix} ds + \begin{bmatrix} 4e^t - 2t - 2 \\ 1 \end{bmatrix} \\
&= \begin{bmatrix} e^t & -2t - 2 + 2e^t \\ 0 & 1 \end{bmatrix} \begin{bmatrix} -2te^{-t} - 4e^{-t} - t + 4 \\ t \end{bmatrix} + \begin{bmatrix} 4e^t - 2t - 2 \\ 1 \end{bmatrix} \\
&= \begin{bmatrix} 8e^t + te^t - 2t^2 - 6t - 6 \\ t + 1 \end{bmatrix}. \blacksquare
\end{aligned}$$

Theorem 6.13.4 is actually a systems version of *variation of parameters* (Sections 3.14 and 3.15). To see this we shall rederive (23) by a variation of parameter argument.

Suppose that $\mathbf{W}(t)$ is a fundamental solution matrix for $\mathbf{x}' = \mathbf{A}\mathbf{x}$ so that $\{\mathbf{w}_1(t), \ldots, \mathbf{w}_n(t)\}$ is a fundamental set of solutions. To find a solution of the nonhomogeneous problem

$$\mathbf{x}' = \mathbf{A}\mathbf{x} + \mathbf{f}, \tag{26}$$

we shall look for a solution of the form

$$\mathbf{x}(t) = v_1(t)\mathbf{w}_1(t) + v_2(t)\mathbf{w}_2(t) + \cdots + v_n(t)\mathbf{w}_n(t), \tag{27}$$

where $v_1, \ldots, v_n$ are unknown scalar functions.

Let

$$\mathbf{v}(t) = \begin{bmatrix} v_1(t) \\ \vdots \\ v_n(t) \end{bmatrix},$$

so that (27) can be written as $\mathbf{x} = \mathbf{W}\mathbf{v}$. Substituting $\mathbf{x} = \mathbf{W}\mathbf{v}$ into (26) gives

$$(\mathbf{W}\mathbf{v})' = \mathbf{A}\mathbf{W}\mathbf{v} + \mathbf{f},$$

or, upon differentiating,

$$\mathbf{W}'\mathbf{v} + \mathbf{W}\mathbf{v}' = \mathbf{A}\mathbf{W}\mathbf{v} + \mathbf{f}. \tag{28}$$

But $\mathbf{W}' = \mathbf{A}\mathbf{W}$, so that (28) simplifies to

$$\mathbf{W}\mathbf{v}' = \mathbf{f}. \tag{29}$$

Multiply by $\mathbf{W}^{-1}$, to find

$$\mathbf{v}'(t) = \mathbf{W}^{-1}(t)\mathbf{f}(t).$$

Thus,

$$\mathbf{v}(t) = \int \mathbf{W}^{-1}(s)\mathbf{f}(s)\, ds + \mathbf{c},$$

and, from (27),

$$\mathbf{x}(t) = \mathbf{W}(t)\mathbf{v}(t) = \mathbf{W}(t)\int \mathbf{W}^{-1}(s)\mathbf{f}(s)\, ds + \mathbf{W}(t)\mathbf{c}, \tag{30}$$

which is exactly (23).

Exercises

For Exercises 1 and 2, you are given a linear system of differential equations $\mathbf{x}' = \mathbf{A}(t)\mathbf{x} + \mathbf{f}$ with variable coefficients. Find a fundamental solution matrix of the associated homogeneous equation $\mathbf{x}' = \mathbf{A}(t)\mathbf{x}$. Then solve the nonhomogeneous equation using Theorem 6.13.4.

1. $x' = x + e^{-t}y + e^{3t}$,
   $y' = y$.

2. $x' = x + e^t(1 - t)y + e^t$,
   $y' = -y$.

For Exercises 3 through 10, you are given a fundamental set of solutions and $\mathbf{f}(t)$ for a linear system $\mathbf{x}' = \mathbf{A}(t)\mathbf{x} + \mathbf{f}(t)$.
Find the coefficient matrix $\mathbf{A}(t)$. Then solve $\mathbf{x}' = \mathbf{A}\mathbf{x} + \mathbf{f}$ using Theorem 6.13.4 on the indicated interval.

3. $\left\{ \begin{bmatrix} e^t \\ 1 \end{bmatrix}, \begin{bmatrix} t \\ e^{-t} \end{bmatrix} \right\}$, $\mathbf{f}(t) = \begin{bmatrix} 0 \\ 1 - t \end{bmatrix}$, $t > 1$

4. $\left\{ \begin{bmatrix} t \\ t^2 \end{bmatrix}, \begin{bmatrix} 1 \\ t + 1 \end{bmatrix} \right\}$, $\mathbf{f}(t) = \begin{bmatrix} t \\ t^2 \end{bmatrix}$, $t > 0$

5. $\left\{ \begin{bmatrix} 1 \\ t \end{bmatrix}, \begin{bmatrix} t \\ 1 \end{bmatrix} \right\}$, $\mathbf{f}(t) = \begin{bmatrix} t - t^3 \\ 1 - t^2 \end{bmatrix}$, $-1 < t < 1$

6. $\left\{ \begin{bmatrix} e^t \\ 1 \end{bmatrix}, \begin{bmatrix} e^{2t} \\ e^{-t} \end{bmatrix} \right\}$, $\mathbf{f}(t) = \begin{bmatrix} e^t - e^{3t} \\ 3 - 3e^{2t} \end{bmatrix}$, $t > 0$

7. $\left\{ \begin{bmatrix} e^t \\ 1 \end{bmatrix}, \begin{bmatrix} 1 \\ 0 \end{bmatrix} \right\}$, $\mathbf{f}(t) = \begin{bmatrix} e^{2t} \\ 2 \end{bmatrix}$

8. $\left\{ \begin{bmatrix} 1 \\ e^t \end{bmatrix}, \begin{bmatrix} e^t \\ 0 \end{bmatrix} \right\}$, $\mathbf{f}(t) = \begin{bmatrix} 0 \\ e^{3t} \end{bmatrix}$

9. $\left\{ \begin{bmatrix} t \\ 1 \end{bmatrix}, \begin{bmatrix} 1 \\ 0 \end{bmatrix} \right\}$, $\mathbf{f}(t) = \begin{bmatrix} t^2 \\ t + 1 \end{bmatrix}$

10. $\left\{ \begin{bmatrix} t \\ e^t \end{bmatrix}, \begin{bmatrix} e^t \\ 0 \end{bmatrix} \right\}$, $\mathbf{f}(t) = \begin{bmatrix} e^{2t} \\ te^{2t} \end{bmatrix}$

11. Rewrite the second-order linear equation
$$a(t)x''(t) + b(t)x'(t) + c(t)x(t) = f(t)$$
as a linear system $\mathbf{x}'(t) = \mathbf{A}(t)\mathbf{x}(t) + \mathbf{f}(t)$, with
$$\mathbf{x} = \begin{bmatrix} x_1 \\ x_2 \end{bmatrix} = \begin{bmatrix} x \\ x' \end{bmatrix}.$$
Show that the system of equations (29) for $\mathbf{v}'$ is the same as the system (6) of Section 3.14.

12. Solve $y'' - 3y' + 2y = e^t$ by rewriting as a system as in Exercise 11, and then performing the variation of parameters (29) and (30).

13. Rewrite the nth-order linear equation
$$a_n(t)x^{(n)}(t) + \cdots + a_1(t)x'(t) + a_0(t)x(t) = f(t)$$
as a linear system $\mathbf{x}'(t) = \mathbf{A}(t)\mathbf{x}(t) + \mathbf{f}(t)$ with
$$\mathbf{x} = \begin{bmatrix} x_1 \\ \vdots \\ x_n \end{bmatrix} = \begin{bmatrix} x \\ x' \\ \vdots \\ x^{(n-1)} \end{bmatrix}.$$
Show that the system of equations (29) for $\mathbf{v}'$ is the same as the system (2) of Theorem 3.15.1.

14. *(Reduction of Order)* Suppose that $\{\mathbf{w}_1(t), \ldots, \mathbf{w}_n(t)\}$ are n linearly independent solutions of the associated homogeneous equation for the system
$$\mathbf{x}''(t) + \mathbf{A}(t)\mathbf{x}'(t) + \mathbf{B}(t)\mathbf{x}(t) = \mathbf{f}(t),$$
where $\mathbf{A}(t)$, $\mathbf{B}(t)$ are $n \times n$. Suppose also that $\mathbf{W}(t) = [\mathbf{w}_1, \ldots, \mathbf{w}_n]$ is invertible for all t. (This need not be true in general for second-order equations.) Let $\mathbf{x}(t) = \mathbf{W}(t)\mathbf{v}(t)$, and show that $\mathbf{v}'(t)$ satisfies the first-order linear system
$$\mathbf{u}'(t) = \mathbf{E}(t)\mathbf{u}(t) + \mathbf{g}(t),$$
where $\mathbf{E} = -\mathbf{W}^{-1}(2\mathbf{W}' + \mathbf{A}\mathbf{W})$, and $\mathbf{g} = \mathbf{W}^{-1}\mathbf{f}$.

7

Difference Equations

7.1 Introduction

As we have seen in previous chapters, differential equations play an important role in problems where there is a quantity y that depends on a continuous independent variable t. In many applications, however, the independent variable, denoted in this section by k, takes on only isolated or discrete values, and we are led to consider what is known as a *difference equation*. The increased use of digital circuits, computer simulations, and numerical methods (see Chapter 8) have made this topic even more important, since they all use difference equations as a fundamental concept.

There is a discrete (or difference) version of most differential-equation techniques. For example, there are difference-equation versions of undetermined coefficients, variation of parameters, resonance, the Laplace transform, and systems. We shall only touch on this theory here.

A difference equation relates different values of an unknown sequence $\{y_k\}$. Here k is an integer variable, and y is the quantity of interest. We usually assume that $k \geq M$ for some integer M. *Initial values* of the sequence are also sometimes specified.

Example 7.1.1

The temperature of an object is measured as 100°C. The temperature is measured each hour afterwards. It is observed that the amount the temperature changes each hour is -0.2 times the difference between the previous hour's temperature and the room temperature, which is 60°C. Model this problem as a difference equation.

Solution

Let T_k be the temperature at the end of k hours. Then $T_0 = 100$ is our initial value. Also, the third sentence is

$$\text{Change in temp} = -0.2[\text{Previous temp} - 60]. \tag{1}$$

If $k + 1$ is the current hour and k the previous hour, (1) is

$$T_{k+1} - T_k = -0.2[T_k - 60]. \tag{2}$$

Thus we have a difference equation with an initial condition; that is, an initial-value problem,

$$T_{k+1} = 0.8T_k + 12, \quad k \geq 0, \quad T_0 = 100. \ \blacksquare \tag{3}$$

Before developing methods for solving difference equations in later sections, we need some terminology.

A *solution* of a difference equation is a sequence that satisfies the difference equation (and any initial conditions) for all values of k.

Example 7.1.2

Verify that

$$T_k = (0.8)^k 40 + 60, \quad k \geq 0 \tag{4}$$

is the solution of the initial-value problem (3).

Solution

In the next section we shall show how to derive the formula (4). For now we will just verify that it is a solution. First, we must verify it satisfies the initial condition $T_0 = 100$. Letting $k = 0$ in (4) gives

$$T_0 = (0.8)^0 40 + 60 = 100,$$

as desired. Next we must verify that (4) satisfies the difference equation.

A sequence $\{T_k\}$ can also be thought of as a function of the integer variable k. From (4) we then have

$$T_{[k+1]} = (0.8)^{[k+1]} 40 + 60.$$

Thus, when we check (3), we find that

$$T_{k+1} - 0.8T_k = ((0.8)^{k+1} 40 + 60) - 0.8((0.8)^k 40 + 60)$$
$$= (0.8)^{k+1} 40 + 60 - (0.8)^{k+1} 40 - 48 = 12,$$

so that (4) satisfies the difference equation (3). $\blacksquare$

A *particular solution* of a difference equation is a solution with no arbitrary constants in it. The sequence (4) is a particular solution of the difference equation in (3).

A *general solution* of a linear difference equation is a sequence, usually with arbitrary constants. It is a solution for any value of the constants. Conversely, every solution is given by some value of the constants. For example,

$$T_k = C(0.8)^k + 60, \qquad k \geq 0,$$

is the general solution of the difference equation in (3). (This will be shown in the next section.)

The *order* of a difference equation is the difference between the largest and smallest subscript of the dependent variable that appears. For example,

$$y_{k+1} = 3y_k + \frac{2}{k+2}, \tag{5}$$

$$y_k = \sin(y_{k-1}), \tag{6}$$

are first-order, while

$$T_{k+2} - 3T_{k+1} + \frac{1}{k}T_k = k^2, \tag{7}$$

$$T_k = 3T_{k-2}, \tag{8}$$

are second-order, and

$$x_{k+3} - 3x_{k+1} + x_k^2 = 3 \tag{9}$$

is third-order.

The general first-, second-, and third-order *linear difference equations* in $\{y_k\}$ are those difference equations that can be written as

First-order: $\quad a_k y_{k+1} + b_k y_k = f_k;$ (10)

Second-order: $\quad a_k y_{k+2} + b_k y_{k+1} + c_k y_k = f_k;$ (11)

Third-order: $\quad a_k y_{k+3} + b_k y_{k+2} + c_k y_{k+1} + d_k y_k = f_k,$ (12)

respectively. The sequences a_k, b_k, c_k, d_k are the *coefficients* and are either constant or depend only on k. The sequence f_k is the *forcing* or *input* function and also depends only on k. For example, (5), (7), (8) are linear while (6), (9) are *nonlinear*. Of the linear equations, (5) and (8) have constant coefficients. We shall only discuss the solution of linear constant-coefficient difference equations.

It is possible to rewrite a difference equation in several ways. To illustrate, consider

$$T_{k+2} + kT_{k+1} + \frac{1}{k+1}T_k = k^2, \qquad k \geq 0, \tag{13}$$

which is second-order linear with nonconstant coefficients. Let $m = k + 2$. Then (13) is

$$T_m + (m-2)T_{m-1} + \frac{1}{m-1}T_{m-2} = (m-2)^2, \qquad m \geq 2. \tag{14}$$

Since the m, k in (13) and (14) are "dummy" variables, we could then rewrite (14) as

$$T_k + (k-2)T_{k-1} + \frac{1}{k-1}T_{k-2} = (k-2)^2, \qquad k \geq 2. \tag{15}$$

Another example is (3). If we had let k denote the current hour and $(k-1)$ the previous hour, then (2) would have been

$$[T_k - T_{k-1}] = -0.2[T_{k-1} - 60], \qquad k \geq 1, \qquad T_0 = 100,$$

or

$$T_k = 0.8T_{k-1} + 12, \qquad k \geq 1, \qquad T_0 = 100. \tag{16}$$

Even though they appear different, both (3) and (16) are the same difference equation, in that they have the same solution. For example, if $T_3 = 5$, then

$$k = 3 \text{ in (3) implies that } T_4 = 0.8T_3 + 12 = 16,$$

while

$$k = 4 \text{ in (16) implies that } T_4 = 0.8T_3 + 12 = 16,$$

which is the same. The existence of a solution is not a problem for most difference equations, since, given initial values, the difference equation can be used to *recursively calculate* the other values.

Example 7.1.3

Recursively find the first six values of $\{T_k\}$ in (3) and graph them.

Solution

We have $T_0 = 100$. Letting $k = 0$, the difference equation (3) is

$$T_1 = 0.8T_0 + 12 = 0.8(100) + 12 = 92.$$

Letting $k = 1$, and using T_1, the difference equation (3) gives

$$T_2 = 0.8T_1 + 12 = 0.8(92) + 12 = 85.6.$$

Letting $k = 2$, the difference equation (3) gives

$$T_3 = 0.8T_2 + 12 = 0.8(85.6) + 12 = 80.48.$$

Continuing, we find from (3) that

$$T_4 = 76.384,$$
$$T_5 = 73.1072,$$
$$T_6 = 70.48576.$$

When graphing $\{T_k\}$ we consider T to be a function of the integer variable k.

Figure 7.1.1
First 20 values of $\{T_k\}$ in (3).

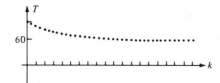

In the graph in Fig. 7.1.1, it appears that the T_k values are approaching a limit. (Intuitively, we expect it to be 60°, which is the room temperature.) To safely answer these kinds of problems, we need a formula for the solution directly in terms of k, such as (4). Techniques for developing these formulas are discussed in the next three sections. ∎

Exercises

For each of the difference equations in Exercises 1 through 8, state its order and whether or not it is linear. If it is linear, state whether or not it has constant coefficients.

1. $y_{k+1} = k y_k$
2. $3y_{k+2} - 2y_k = 0$
3. $T_{k+1} = \cos(T_{k-1})$
4. $T_k = k^2 T_{k-1} + k^3 T_{k-2}$
5. $y_{k+1} = y_k y_{k-1}$
6. $y_k - 3y_{k-1} = k^5$
7. $T_{k+3} = 5T_{k+2}$
8. $T_{k-1} - (T_{k-2})^2 = 3k$
9. Find the first six values of $y_{k+1} = (-1)y_k + 3$, given that $y_0 = 1$. Graph these solution values.
10. Find the first six values of $y_{k+1} = k y_k - k$, given that $y_0 = 2$. Graph these solution values.
11. Find the first six values of $y_k = y_{k-1} + 1/(k+1)$, given that $y_0 = 1$.
12. Let $y_0 = 1$, $y_1 = 2$. Find the first six values of the second-order difference equation $y_{k+2} = y_{k+1} + y_k$.
13. Let $y_0 = -1$, $y_1 = 0$. Find the first six values of the difference equation $y_{k+2} = y_{k+1} - y_k + 3$.

In Exercises 14 through 19, verify that the given sequence is a solution of the difference equation. Also verify that the sequence satisfies the initial conditions, if any.

14. $y_k = k^{-1}$; $y_{k+1} - y_k = -(k^2 + k)^{-1}$
15. $y_k = 2^k + 5^k$; $y_{k+1} - 2y_k = 3 \cdot 5^k$, $y_0 = 2$
16. $z_k = 3^k + 2 \cdot 5^k$; $z_{k+1} - 3z_k = 4 \cdot 5^k$, $z_0 = 3$
17. $z_k = k + (k+1)^{-1}$;
    $z_{k+2} - z_k = 2 - 2(k^2 + 4k + 3)^{-1}$
18. $z_k = 2^k + k^2$; $z_{k+2} - 4z_k = -3k^2 + 4k + 4$, $z_0 = 1$
19. $u_k = k + k2^k$; $u_{k+2} - 2u_{k+1} + u_k = (4+k)2^k$

Difference equations have many similarities to differential equations. However, uniqueness, for a difference equation, is different, in that solutions disagreeing at their starting values may later agree or have graphs that "cross."

20. For the difference equation $x_k = (2-k)x_{k-1}$, compute the first six values of the solution for which $x_0 = 0$ and of the solution for which $x_0 = 3$. Draw the graphs of both solutions on the same axis. Show that, in fact, these solutions agree for all $k \geq 2$.
21. For the difference equation $x_{k+1} = -x_k - 1$, verify that $x_k \equiv -\frac{1}{2}$ is an equilibrium solution, that is, it is a constant solution. Verify that $x_k = (-1)^k - \frac{1}{2}$ is another solution. Graph both solutions on the same graph for $0 \leq k \leq 8$. Observe that the solutions repeatedly "cross" each other.

7.2 First-Order Linear Difference Equations

For the first-order linear difference equation

$$a_k y_{k+1} + b_k y_k = f_k, \qquad k \geq 0, \tag{1}$$

we assume that $a_k \neq 0$ for all $k \geq 0$. Then y_0 may be taken arbitrarily and succeeding terms found recursively by

$$y_{k+1} = \frac{1}{a_k}[f_k - b_k y_k]. \tag{2}$$

Thus we have the following theorem.

Theorem 7.2.1 Suppose that $a_k \neq 0$ for all $k \geq 0$. Let z_0 be any number. Then there is a unique solution to the initial-value problem

$$a_k y_{k+1} + b_k y_k = f_k, \qquad y_0 = z_0.$$

As for first-order linear differential equations, we get the following theorem.

Theorem 7.2.2 The general solution of (1) can be written as

$$y_k = ch_k + p_k. \tag{3}$$

where c is an arbitrary constant, $\{ch_k\}$ is the general solution of the *associated homogeneous equation*

$$a_k y_{k+1} + b_k y_k = 0, \tag{4}$$

and $\{p_k\}$ is a particular solution of (1). We may take $\{h_k\}$ as any solution of (4) that is not identically zero.

We leave the verification of Theorem 7.2.2 to Exercises 24 and 25 and instead turn to finding the general solution of

$$ay_{k+1} + by_k = f, \tag{5}$$

where a, b, and f are constants. Methods of solving (1) with nonconstant $\{f_k\}$ are given in Section 7.4. Dividing by a and rearranging (5) gives

$$y_{k+1} = ry_k + e \tag{6}$$

with $r = -b/a$, $e = f/a$. Then

$$y_1 = ry_0 + e,$$
$$y_2 = ry_1 + e = r(ry_0 + e) + e$$
$$= r^2 y_0 + re + e,$$
$$y_3 = ry_2 + e = r^3 y_0 + r^2 e + re + e$$
$$\vdots \qquad \vdots \qquad \vdots$$

Thus,

$$y_k = r^k y_0 + (r^{k-1} + r^{k-2} + \cdots + 1)e. \tag{7}$$

If $r \neq 1$, then $(r-1)(r^{k-1} + r^{k-2} + \cdots + 1) = r^k - 1$. Thus we have the next theorem.

Theorem 7.2.3 The general solution of $y_{k+1} = ry_k + e$ is:

$$y_k = \begin{cases} r^k y_0 + \dfrac{(r^k - 1)}{r - 1} e, & \text{if } r \neq 1, \tag{8} \\ y_0 + ke, & \text{if } r = 1. \tag{9} \end{cases}$$

Note that in (8), $r^k y_0$ is the general solution of the associated homogeneous equation $y_{k+1} = ry_k$ and $p_k = \dfrac{r^k - 1}{r - 1} e$ is a particular solution of (6) if $r \neq 1$.

Example 7.2.1 Find the general solution of

$$T_{k+1} = 0.8 T_k + 12, \qquad T_0 = 100. \tag{10}$$

Solution We have $r = 0.8$, $e = 12$, so that, from (8),

$$T_k = (0.8)^k 100 + \left[\dfrac{(0.8)^k - 1}{0.8 - 1} \right] 12$$
$$= (0.8)^k 40 + 60.$$

This is the solution given in Example 7.1.2. ■

It is sometimes convenient to rewrite (8) if $r \neq 1$, to get

$$y_k = r^k \left[y_0 + \dfrac{e}{r - 1} \right] - \dfrac{e}{r - 1}$$

or

$$y_k = r^k c + \dfrac{e}{1 - r}, \tag{11}$$

where c is an arbitrary constant. Formula (11) is particularly convenient in some applications, since setting $c = 0$ in (11) shows that

$$y_k = \frac{e}{1-r} \quad (12)$$

is an *equilibrium* (or constant) solution.

Example 7.2.2   Find the general solution of

$$y_{k+1} + 2y_k - 3 = 0, \quad k \geq 0. \quad (13)$$

Solution   Rewriting (13) as

$$y_{k+1} = -2y_k + 3,$$

we have $r = -2$, $e = 3$, and (8) becomes

$$y_k = (-2)^k y_0 + (1 - (-2)^k). \quad \blacksquare$$

Exercises

In Exercises 1 through 9, find the solution of the initial-value problem if an initial condition is given. Otherwise find the general solution.

1. $y_{k+1} = 2y_k + 1, \quad y_0 = 1$
2. $y_{k+1} = -y_k + 3, \quad y_0 = 0$
3. $2y_{k+1} = 3y_k$
4. $4y_{k+1} - 2y_k = 8, \quad y_0 = 2$
5. $y_{k+1} - y_k = 3$
6. $3y_{k+1} = 7y_k, \quad y_0 = 2$
7. $2y_{k+1} + 2y_k = 6$
8. $3y_{k+1} - 2y_k = 9$
9. $y_{k+1} - 4y_k = 2, \quad y_0 = 6$
10. Show that $y_{k+1} = ry_k + e$ has a unique equilibrium if and only if $r \neq 1$, and in this case the equilibrium solution is $y_k = e/(1-r)$. Show that, if $r = 1$ and $e \neq 0$, then there is no equilibrium, and if $r = 1$ and $e = 0$, then every solution is an equilibrium solution.
11. Suppose that $r \neq 1$, so that, from Exercise 10, there is a unique equilibrium solution for $y_{k+1} = ry_k + e$. Assume that $y_0 \neq e/(1-r)$ and $\{y_k\}$ is a solution.

    a) Show that $\{y_k\}$ is *unbounded*, that is,
    $$\lim_{k \to \infty} |y_k| = \infty, \quad \text{if and only if } |r| > 1.$$

    b) show that $\{y_k\}$ is *asymptotic*, that is,
    $$\lim_{k \to \infty} y_k = \frac{e}{1-r}, \quad \text{if and only if } |r| < 1.$$

    c) Show that $\{y_k\}$ is *monotonic*, that is, the terms are increasing (or decreasing) if and only if $r > 0$.

    d) Show that $\{y_k\}$ oscillates about the equilibrium if $r < 0$.

    e) Show that the equilibrium is obtained in finite time, that is, there exists an integer $l > 0$ such that $y_l = e/(1-r)$, if and only if $r = 0$, in which case $l = 1$.

In Exercises 12 through 23, use Exercise 11, if possible, to determine whether the nonequilibrium solutions of the given difference equations are bounded, asymptotic, monotonic, or oscillating.

12. $y_{k+1} = 2y_k - 3$
13. $y_{k+1} = -3y_k + 6$
14. $2y_{k+1} - y_k = 8$

15. $3y_{k+1} + y_k = 9$
16. $y_{k+1} + y_k = 2$
17. $y_{k+1} - y_k = 3$
18. $2y_{k+1} - 2y_k = 3$
19. $y_{k+1} - 4y_k = 2$
20. $5y_{k+1} + y_k = 10$
21. $6y_{k+1} - 2y_k = 7$
22. $y_{k+1} + 7y_k = 3$
23. $5y_{k+1} + 5y_k = 15$

24. Assume that a_k is never zero. Let $\{p_k\}$ be a solution of $a_k y_{k+1} + b_k y_k = f_k$, $k \geq 0$. Show that, if $\{z_k\}$ is any other solution, then $z_k = p_k + w_k$, where $\{w_k\}$ is a solution of $a_k y_{k+1} + b_k y_k = 0$. Conversely, show that, if $\{p_k\}$ and $\{w_k\}$ are as described, then $z_k = p_k + w_k$ is a solution of $a_k y_{k+1} + b_k y_k = f_k$.

25. Assume that a_k is never zero. Let $\{h_k\}$ be a not identically zero solution of $a_k y_{k+1} + b_k y_k = 0$. Show that every other solution of $a_k y_{k+1} + b_k y_k = 0$ is of the form $\{ch_k\}$, c a constant.

7.3 Second-Order Homogeneous Difference Equations

The theory for second-order linear difference equations

$$a_k y_{k+2} + b_k y_{k+1} + c_k y_k = f_k, \qquad k \geq 0, \qquad (1)$$

is very similar to that for second-order linear differential equations. (The next theorem should be compared to Theorem 3.4.1.)

Theorem 7.3.1

Suppose $a_k \neq 0$ for all $k \geq 0$.

1. For any numbers z_0, z_1, there exists a unique solution to (1) such that $y_0 = z_0$, $y_1 = z_1$.
2. If $\{y_k\}$ is a solution of (1), then

$$y_k = p_k + z_k,$$

where $\{p_k\}$ is a particular solution of (1) and $\{z_k\}$ is a solution of the associated homogeneous difference equation

$$a_k y_{k+2} + b_k y_{k+1} + c_k y_k = 0. \qquad (2)$$

3. If $\{h_k\}$ and $\{q_k\}$ are solutions of (2) that are not identically zero and $\{h_k\}$ is not a constant multiple of $\{q_k\}$, then

$$z_k = \tilde{c}_1 h_k + \tilde{c}_2 q_k, \qquad (3)$$

with $\tilde{c}_1, \tilde{c}_2$ arbitrary constants, is the general solution of (2). $\{\{h_k\}, \{q_k\}\}$ is a *fundamental set* of solutions for (2).

4. If $\{p_k\}$, $\{h_k\}$, $\{q_k\}$ are as described in (2) and (3), then

$$y_k = p_k + \tilde{c}_1 h_k + \tilde{c}_2 q_k \qquad (4)$$

is the general solution of (1).

5. (*Superposition principle*) If $\{p_k\}$ is a solution of (1) and $\{w_k\}$ is a solution of $a_k y_{k+2} + b_k y_{k+1} + c_k y_k = g_k$, then $\{p_k + w_k\}$ is a solution of

$$a_k y_{k+2} + b_k y_{k+1} + c_k y_k = f_k + g_k.$$

Theorem 7.3.1 shows that we may reduce the solving of (1) to two problems. First, we need to solve the associated homogeneous equation (2). Then we need to find a particular solution of (1). This section will discuss how to solve the homogeneous equation (2) when there are constant coefficients.

7.3.1 Constant Coefficients

Assume that the homogeneous equation (2) has constant coefficients, so that we have

$$a y_{k+2} + b y_{k+1} + c y_k = 0, \qquad k \geq 0, \qquad (5)$$

with $a \neq 0$. To motivate what follows, recall that

$$ay' + by = 0$$

had a solution $e^{-(b/a)t}$. Thus when we wanted to solve

$$ay'' + by' + cy = 0,$$

we looked for solutions of the form $e^{\lambda t}$. Now, from Section 7.2, we know that $ay_{k+1} + by_k = 0$ has a solution $\{(-b/a)^k\}$. Thus, we begin by looking for solutions for (5) of the form λ^k with $\lambda \neq 0$. Suppose, then, that $\{\lambda^k\}$ is a nonzero solution of (5). Then we have

$$a\lambda^{k+2} + b\lambda^{k+1} + c\lambda^k = 0.$$

Dividing by λ^k yields

$$a\lambda^2 + b\lambda + c = 0. \qquad (6)$$

The polynomial $a\lambda^2 + b\lambda + c$ is called the *characteristic polynomial* of the difference equation (5). Equation (6) is called the *characteristic equation*.

If (6) has two distinct roots λ_1, λ_2, then $c_1(\lambda_1)^k + c_2(\lambda_2)^k$ will be the general solution of (5) unless one of these roots is zero. However, a root of zero can occur only if $c = 0$ in (6), and then (5) is a first-order equation for $k \geq 1$. We assume, then, that $c \neq 0$ in what follows.

Example 7.3.1 Find the general solution of

$$y_{k+2} - 5y_{k+1} + 6y_k = 0. \qquad (7)$$

Solution

The characteristic polynomial is $\lambda^2 - 5\lambda + 6 = (\lambda - 2)(\lambda - 3)$, which has the two distinct roots $\lambda_1 = 2, \lambda_2 = 3$. Thus $\{2^k\}, \{3^k\}$ are solutions of (7) and

$$y_k = c_1 2^k + c_2 3^k$$

is the general solution. ∎

Sometimes the characteristic equation (6) has a repeated root (of multiplicity two) so that $\lambda_1 = \lambda_2$. For differential equations, this would mean $e^{\lambda_1 t}, te^{\lambda_1 t}$ would be solutions. A similar situation applies here in that, if $\lambda_1 = \lambda_2$, then $\{\lambda_1^k\}$ and $\{k\lambda_1^k\}$ are both solutions of (5).

Example 7.3.2

Find the solution of

$$y_{k+2} - 4y_{k+1} + 4y_k = 0, \quad y_0 = 1, \quad y_1 = 3, \quad k \geq 0. \tag{8}$$

Solution

The characteristic polynomial is $\lambda^2 - 4\lambda + 4 = (\lambda - 2)^2$. Thus, $\lambda_1 = 2, \lambda_2 = 2$ are the roots. Two solutions of the difference equation are then $\{2^k\}, \{k2^k\}$. The general solution is

$$y_k = c_1 2^k + c_2 k 2^k. \tag{9}$$

Now we apply the initial conditions in (8) to (9):

$$1 = y_0 = c_1 2^0 + c_2 0 2^0 = c_1,$$
$$3 = y_1 = c_1 2^1 + c_2 1 \cdot 2^1 = 2c_1 + 2c_2,$$

to find $c_1 = 1, c_2 = \frac{1}{2}$ and

$$y_k = 2^k + \frac{1}{2} k 2^k = \left(1 + \frac{k}{2}\right) 2^k. \; ∎ \tag{10}$$

It is possible that there are two distinct roots that are both complex. In this case, $c_1 \lambda_1^k + c_2 \lambda_2^k$ is still the solution, but it involves complex numbers. We conclude by showing how to get solutions not expressed in terms of complex numbers.

Suppose then that a, b, c are real but that $a\lambda^2 + b\lambda + c$ has complex roots. Then they form a conjugate pair

$$\lambda_1 = \alpha + i\beta,$$
$$\lambda_2 = \alpha - i\beta.$$

We rewrite these roots in polar form (Appendix A) as:

$$\lambda_1 = re^{i\theta} = r(\cos\theta + i\sin\theta),$$
$$\lambda_2 = re^{-i\theta} = r(\cos\theta - i\sin\theta),$$

where $r = \sqrt{\alpha^2 + \beta^2}$ and θ is the angle the vector (α, β) makes with the x-axis. Then

$$\lambda_1^k = r^k e^{ik\theta} = r^k(\cos k\theta + i\sin k\theta),$$
$$\lambda_2^k = r^k e^{ik\theta} = r^k(\cos k\theta - i\sin k\theta).$$

Since $\{\lambda_1^k\}$ and $\{\lambda_2^k\}$ are solutions of the linear homogeneous problem, we have that

$$\frac{1}{2}\lambda_1^k + \frac{1}{2}\lambda_2^k = r^k \cos k\theta,$$

$$\frac{1}{2i}\lambda_1^k - \frac{1}{2i}\lambda_2^k = r^k \sin k\theta$$

are also solutions.

Example 7.3.3

Find the general solution of

$$y_{k+2} - 2y_{k+1} + 2y_k = 0. \tag{11}$$

Solution

The characteristic equation is $\lambda^2 - 2\lambda + 2 = 0$, which has roots $\lambda = 1 \pm i$. Thus $\alpha = 1$, $\beta = 1$ and $1 + i = \sqrt{2}e^{i(\pi/4)}$. Then $\left\{\sqrt{2}^k \cos\left(k\frac{\pi}{4}\right)\right\}$, $\left\{\sqrt{2}^k \sin\left(k\frac{\pi}{4}\right)\right\}$ are solutions of (11), and the general solution is

$$y_k = c_1 2^{k/2} \cos\left(\frac{k\pi}{4}\right) + c_2 2^{k/2} \sin\left(\frac{k\pi}{4}\right). \blacksquare$$

To summarize:

Theorem 7.3.2

Assume that $a \neq 0$, $c \neq 0$, and let λ_1, λ_2 be the roots of the characteristic equation $a\lambda^2 + b\lambda + c = 0$.

1. If $\lambda_1 \neq \lambda_2$, then

$$y_k = c_1 \lambda_1^k + c_2 \lambda_2^k$$

is the general solution of (5).

2. If $\lambda_1 = \lambda_2$, then

$$y_k = c_1 \lambda_1^k + c_2 k \lambda_1^k$$

is the general solution of (5).

3. If λ_1 is complex and a, b, c real, then $\lambda_1 = \alpha + \beta i$, $\lambda_2 = \alpha - \beta i$. Let $re^{i\theta}$ be the polar form of λ_1. Then

$$y_k = c_1 r^k \cos k\theta + c_2 r^k \sin k\theta$$

is the general solution of (5).

Exercises

In Exercises 1 through 15, solve the given difference equation. If there are no initial conditions, give the general solution.

1. $y_{k+2} + y_{k+1} - 6y_k = 0$
2. $y_{k+2} = 4y_k$
3. $y_{k+2} + y_{k+1} = 2y_k$, $\quad y_0 = 0, \quad y_1 = 3$
4. $x_{k+2} - x_k = 0$
5. $x_{k+2} - 6x_{k+1} + 9x_k = 0$, $\quad x_0 = 2, \quad x_1 = 15$
6. $x_{k+2} - 2x_{k+1} + x_k = 0$, $\quad x_0 = 3, \quad x_1 = 2$
7. $y_{k+2} = -y_k$
8. $4y_{k+2} + y_k = 0$
9. $y_{k+2} - 2y_{k+1} + 4y_k = 0$
10. $y_{k+2} - 4y_{k+1} - 5y_k = 0$
11. $y_{k+2} + 2y_{k+1} + 2y_k = 0$
12. $y_{k+2} + 16y_k = 0$
13. $9y_{k+2} + y_k = 0$
14. $y_{k+2} + 2y_{k+1} + 10y_k = 0$
15. $y_{k+2} + 10y_{k+1} + 25y_k = 0$

The next two exercises are important in that they are used to analyze error propagation in numerical procedures for solving differential equations. Let λ_1, λ_2 be the roots of $a\lambda^2 + b\lambda + c$, with $a \neq 0, c \neq 0$.

16. Show that $\lim_{k \to \infty} y_k = 0$ for every solution of $ay_{k+2} + by_{k+1} + cy_k = 0$ if and only if $|\lambda_1| < 1$, $|\lambda_2| < 1$.

17. Show that there is a solution of $ay_{k+2} + by_{k+1} + cy_k = 0$ such that $\lim_{k \to \infty} |y_k| = \infty$ if either
    a) $|\lambda_1| > 1$ or $|\lambda_2| > 1$,
    b) $\lambda_1 = \lambda_2$ and $|\lambda_1| = 1$.

18. Verify Part (2) of Theorem 7.3.1.
19. Verify that Eq. (3) is a solution of (2).
20. Verify that Eq. (4) is a solution of Eq. (1).
21. Verify the superposition principle, Part (5) of Theorem 7.3.1.
22. Suppose that $\lambda^2 + b\lambda + c = (\lambda - \lambda_1)^2$ and $c \neq 0$. Verify by a direct calculation that $\{\lambda_1^k\}$ and $\{k\lambda_1^k\}$ are solutions of $y_{k+2} + by_{k+1} + cy_k = 0$.
23. Suppose that $\lambda^3 + b\lambda^2 + c\lambda + d = (\lambda - \lambda_1)^3$ and $d \neq 0$. Verify by a direct calculation that $\{\lambda_1^k\}$, $\{k\lambda_1^k\}$, and $\{k^2\lambda_1^k\}$ are solutions of the third-order equation $y_{k+3} + by_{k+2} + cy_{k+1} + dy_k = 0$.

The ideas of this section extend readily to linear constant coefficient homogeneous difference equations with orders other than two. There are two major differences. First, there are n sequences in a fundamental set for an nth-order equation. Secondly, roots can be repeated more than twice. If λ_1 is a root of multiplicity m of the characteristic polynomial, then $\{\lambda_1^k\}$, $\{k\lambda_1^k\}$, ..., $\{k^{m-1}\lambda_1^k\}$ are solutions. (Note Exercises 22 and 23.) In Exercises 24 through 33 solve the given difference equation.

24. $y_{k+4} - 2y_{k+2} + y_k = 0$
25. $y_{k+4} - 8y_{k+2} + 16y_k = 0$
26. $y_{k+4} + 18y_{k+2} + 81y_k = 0$
27. $y_{k+4} + 2y_{k+2} + y_k = 0$
28. $y_{k+3} + 3y_{k+2} + 3y_{k+1} + y_k = 0$
29. $y_{k+3} - 3y_{k+2} + 3y_{k+1} - y_k = 0$
30. $y_{k+3} - 6y_{k+2} + 12y_{k+1} - 8y_k = 0$
31. $y_{k+3} + 6y_{k+2} + 12y_{k+1} + 8y_k = 0$
32. $y_{k+4} + 4y_{k+3} + 6y_{k+2} + 4y_{k+1} + y_k = 0$
33. $y_{k+4} - 4y_{k+3} + 6y_{k+2} - 4y_{k+1} + y_k = 0$

7.4 Nonhomogeneous Difference Equations

We now present a method for finding a particular solution of

$$ay_{k+2} + by_{k+1} + cy_k = f_k, \qquad k \geq 0, \tag{1}$$

with a, b, c constants for many f_k of interest. The method is a difference-equation version of the method of *undetermined coefficients* (Section 3.12). The general procedure is as follows.

Undetermined Coefficients

1. Calculate the roots λ_1, λ_2 of the characteristic polynomial $a\lambda^2 + b\lambda + c$.
2. Construct a form p_k for the particular solution using the rules that follow.
3. Substitute the form into (1), and equate corresponding functions of k, to find the constants (undetermined coefficients) in the form.

Rules

Suppose that f_k is a sum (linear combination) of terms of the form $r^k q(k)$, where r is a constant and $q(k)$ is a polynomial in k. In the rules to follow, the A_i are constants to be determined.

Rule 1 If a term r^k appears in f_k, then introduce a term

$$A_0 k^m r^k \tag{2}$$

into p_k, where m is the multiplicity of r as a root of the characteristic polynomial.

Rule 2 If a term $q(k)$, q an nth-degree polynomial, appears in f_k, then insert a term

$$k^m(A_0 + A_1 k + \cdots + A_n k^n) \tag{3}$$

into p_k, where m is the multiplicity of 1 as a root of the characteristic polynomial.

Rule 3 (Includes Rule 1 and Rule 2.) If a term $r^k q(k)$, q an nth-degree polynomial in k, appears in f_k, then include a term

$$k^m r^k (A_0 + A_1 k + \cdots + A_n k^n) \tag{4}$$

in p_k, where m is the multiplicity of r as a root of the characteristic polynomial.

Example 7.4.1 Find a particular solution of

$$y_{k+2} - 4y_k = 2 \cdot 3^k + 5^k, \qquad k \geq 0. \tag{5}$$

Solution The characteristic polynomial is $\lambda^2 - 4$ with roots $\lambda_1 = 2, \lambda_2 = -2$. In this case, f_k includes terms $3^k, 5^k$ (we ignore the constant 2). Since neither 3 nor 5 are roots of the characteristic equation, Rule 1 says we may take $\{p_k\}$ in the form

$$p_k = A_1 3^k + A_2 5^k.$$

Substitute this form into (5) to find

$$A_1 3^{k+2} + A_2 5^{k+2} - 4(A_1 3^k + A_2 5^k) = 2 \cdot 3^k + 5^k$$

or, grouping in terms of 3^k and 5^k,

$$(9A_1 - 4A_1)3^k + (25A_2 - 4A_2)5^k = 2 \cdot 3^k + 5^k.$$

Equating coefficients of the terms 3^k, 5^k gives

$$9A_1 - 4A_1 = 2,$$
$$25A_2 - 4A_2 = 1,$$

so that $A_1 = \frac{2}{5}$, $A_2 = \frac{1}{21}$, and the particular solution is

$$p_k = (\tfrac{2}{5})3^k + (\tfrac{1}{21})5^k. \quad \blacksquare \tag{6}$$

Example 7.4.2 Find the general solution of (5).

Solution From Example 7.4.1, we know that a particular solution p_k is given by (6). The solution of the associated homogeneous equation of (5) is $c_1 \lambda_1^k + c_2 \lambda_2^k = c_1 2^k + c_2(-2)^k$. Thus by Theorem 7.3.1 (Part 4), we have the general solution of (5) is

$$y_k = c_1 2^k + c_2(-2)^k + (\tfrac{2}{5})3^k + (\tfrac{1}{21})5^k. \quad \blacksquare$$

This procedure works on a difference equation of any order, as long as $\{f_k\}$ has the right form and the coefficients are constant. We illustrate this by showing how many standard series and sums may be expressed in closed form.

Example 7.4.3 Let

$$s_k = 1 + 2^2 + 3^2 + \cdots + k^2, \quad k \geq 1. \tag{7}$$

Find a formula for s_k in terms of k.

Solution From (7) we have

$$s_{k+1} = 1 + 2^2 + 3^2 + \cdots + k^2 + (k+1)^2. \tag{8}$$

Subtracting (7) from (8) gives $s_{k+1} - s_k = (k+1)^2$. Then $s_1 = 1$ implies $s_0 = 0$ and we have

$$s_{k+1} - s_k = (k+1)^2, \quad k \geq 0, \quad s_0 = 0. \tag{9}$$

This is a first-order linear constant-coefficient difference equation with $f_k = k^2 + 2k + 1$, which is a second-degree polynomial in k. The characteristic polynomial is $(\lambda - 1)$, which has the single root 1. Thus, by Rule 2, we look for a particular solution of the form ($m = 1$)

$$p_k = k(A_0 + A_1 k + A_2 k^2) = A_0 k + A_1 k^2 + A_2 k^3. \tag{10}$$

Substituting the form (10) into the difference equation (9) gives

$$p_{k+1} - p_k = k^2 + 2k + 1$$

or

$$[A_0(k+1) + A_1(k+1)^2 + A_2(k+1)^3] - [A_0 k + A_1 k^2 + A_2 k^3]$$
$$= k^2 + 2k + 1.$$

Multiplying out and combining terms:

$$A_0 + A_1(2k+1) + A_2(3k^2 + 3k + 1) = k^2 + 2k + 1.$$

Equating coefficients of 1, k, k^2 yields

$$1: \quad A_0 + A_1 + A_2 = 1,$$
$$k: \quad 2A_1 + 3A_2 = 2,$$
$$k^2: \quad 3A_2 = 1.$$

Thus $A_2 = \frac{1}{3}$, $A_1 = \frac{1}{2}$, $A_0 = \frac{1}{6}$ and $p_k = \frac{1}{6}k + \frac{1}{2}k^2 + \frac{1}{3}k^3$. The solution of the associated homogeneous equation is $c(1^k)$, so that the general solution of (9) is

$$s_k = c + \tfrac{1}{6}k + \tfrac{1}{2}k^2 + \tfrac{1}{3}k^3. \tag{11}$$

The initial condition $s_0 = 0$ applied to (11) implies that $c = 0$ and thus

$$s_k = \tfrac{1}{6}k + \tfrac{1}{2}k^2 + \tfrac{1}{3}k^3 \tag{12}$$

is the *closed-form expression* for (7). ∎

The same approach can be used to sum some infinite series.

Example 7.4.4 Find the value of

$$g(x) = \sum_{j=0}^{\infty} j x^j \tag{13}$$

if $|x| < 1$.

Solution Let

$$s_k = \sum_{j=0}^{k} j x^j \tag{14}$$

be the kth partial sum of the series (13). Then

$$g(x) = \lim_{k \to \infty} s_k. \tag{15}$$

The partial sums satisfy the difference equation

$$s_{k+1} - s_k = (k+1)x^{k+1} = (k+1)xx^k, \quad s_0 = 0. \tag{16}$$

Consider x as an unknown constant. The characteristic polynomial of (16) is $(\lambda - 1)$, with a root of 1. Since $x \neq 1$ by Rule 3 (with $r = x$, $n = 1$, and $m = 0$), a form for a particular solution will be

$$p_k = x^k(A_0 + A_1 k). \tag{17}$$

Substituting this form into the difference equation (16), we get

$$p_{k+1} - p_k = (k+1)xx^k$$

or

$$[x^{k+1}(A_0 + A_1(k+1))] - [x^k(A_0 + A_1 k)] = (k+1)xx^k. \quad (18)$$

The functions of k in (18) are x^k and kx^k. Multiplying out (18), we get

$$x^k x A_0 + kx^k x A_1 + xx^k A_1 - x^k A_0 - kx^k A_1 = xkx^k + xx^k.$$

Equating coefficients of x^k and kx^k yields

$$x^k: \quad xA_0 + xA_1 - A_0 = x,$$
$$kx^k: \quad xA_1 - A_1 = x,$$

so that, after some calculation,

$$A_1 = \frac{x}{x-1} \quad \text{and} \quad A_0 = \frac{-x}{(x-1)^2}.$$

Thus, from (17),

$$p_k = x^k \left[-\frac{x}{(x-1)^2} + \frac{xk}{(x-1)} \right].$$

The solution of the associated homogeneous equation is $c(1^k)$, so that the general solution of (16) is

$$s_k = c + x^k \left[-\frac{x}{(x-1)^2} + \frac{xk}{(x-1)} \right].$$

The initial condition implies that

$$0 = s_0 = c + 1 \left[-\frac{x}{(x-1)^2} + 0 \right]$$

and $c = x/(1-x)^2$. Thus,

$$s_k = \frac{x}{(1-x)^2} + x^k \left[-\frac{x}{(x-1)^2} + \frac{xk}{(x-1)} \right]. \quad (19)$$

Now since $|x| < 1$, we have

$$\lim_{k \to \infty} |x|^k = 0, \quad \lim_{k \to \infty} k|x|^k = 0 \quad (20)$$

[the second limit in (20) follows from l'Hôpital's rule], so that finally

$$\sum_{j=0}^{\infty} jx^j = \lim_{k \to \infty} s_k = \frac{x}{(1-x)^2}. \quad \blacksquare$$

If the forcing function $\{f_k\}$ in (1) has several terms, finding a particular solution can be reduced to several smaller problems by using the superposition principle [Theorem 7.3.1, Part (5)]. For example, in Example 7.4.1, we could have found a particular solution of (5) by finding a particular solution of

$$y_{k+2} - 4y_k = 2 \cdot 3^k$$

and

$$y_{k+2} - 4y_k = 5^k,$$

and then adding these particular solutions. Using the superposition principle in this manner does not alter the total amount of computation, but can reduce the length of the expressions being considered at any one time.

Exercises

In Exercises 1 through 7, find a closed-form expression for s_k, using the techniques of this section.

1. $s_k = 1 + 2 + \cdots + k$
2. $s_k = 1 + 3 + 3^2 + 3^3 + \cdots + 3^k$
3. $s_k = -1 + 2 - 3 + \cdots + (-1)^k k$
4. $s_k = 1 - 2^2 + 3^2 - + \cdots + (-1)^{k-1} k^2$
5. $s_k = 1 + 2^3 + 3^3 + \cdots + k^3$
6. $s_k = 1 - 1 + 1 - 1 + \cdots + (-1)^k$
7. $s_k = \underbrace{1 + 1 + \cdots + 1}_{(k+1)\text{terms}}$

In Exercises 8 through 14, sum the given series using the techniques of this section.

8. $\sum_{i=0}^{\infty} x^i$, $\quad |x| < 1$
9. $\sum_{i=0}^{\infty} (i+1) x^i$, $\quad |x| < 1$
10. $\sum_{i=0}^{\infty} (-2x)^i$, $\quad |2x| < 1$
11. $\sum_{i=0}^{\infty} 3^i x^i$, $\quad |x| < \frac{1}{3}$
12. $\sum_{i=0}^{\infty} i^2 x^i$, $\quad |x| < 1$
13. $\sum_{i=0}^{\infty} x^{2i}$, $\quad |x| < 1$
14. $\sum_{i=0}^{\infty} x^{3i}$, $\quad |x| < 1$

In Exercises 15 through 24, solve the difference equation. Give the general solution if no initial conditions are given.

15. $y_{k+2} - y_k = -4^k + 3 \cdot 7^k$
16. $y_{k+2} - 3y_{k+1} + 2y_k = k$, $\quad y_0 = 4$, $\quad y_1 = 6$
17. $y_{k+2} - 2y_{k+1} + y_k = k$
18. $y_{k+2} + 5y_{k+1} + 6y_k = k^2 - 1$
19. $y_{k+2} + 5y_{k+1} + 6y_k = 3(-2)^k$, $\quad y_0 = 6$, $y_1 = -13$
20. $y_{k+2} - 9y_k = k3^k - 2^k$
21. $y_{k+1} - 2y_k = k^2$
22. $y_{k+1} + 2y_k = k^2 + 1$
23. $y_{k+1} - 4y_k = 3^k$
24. $y_{k+1} + 3y_k = 2^k$

In Exercises 25 through 30, give the *form* of a particular solution. You need not actually find the particular solution.

25. $y_{k+2} - 2y_{k+1} + y_k = k^3 - 1$
26. $y_{k+2} - 4y_{k+1} + 4y_k = k^2 2^k + k^3$
27. $y_{k+2} - 5y_{k+1} + 6y_k = 3k2^k + 2k^2 3^k$
28. $y_{k+2} + 4y_{k+1} + 4y_k = 2 + 2 \cdot 2^k$
29. $y_{k+2} - 3y_{k+1} + 2y_k = k^3 + k - 1$
30. $y_{k+2} - 4y_{k+1} + 4y_k = 2 + 2 \cdot 2^k$
31. Suppose that $a\lambda^2 + b\lambda + c$ has roots $\lambda_1 = \alpha + \beta i$, $\lambda_2 = \alpha - \beta i$, that a, b, c are real, and that $\alpha + \beta i$

has polar form $\lambda_1 = re^{i\theta}$, $-\pi/2 < \theta \leq \pi/2$. Show that, if f_k includes a term of the form $h^k \sin(k\phi)$ or $h^k \cos(k\phi)$, with $-\pi/2 < \phi \leq \pi/2$, then the particular solution p_k includes

$$A_0 h^k \cos(k\phi) + A_1 h^k \sin(k\phi)$$

$$\text{if } \phi \neq \theta \quad \text{or} \quad h \neq r,$$

while, if $\phi = \theta$ and $h = r$, then p_k includes

$$k[A_0 h^k \cos(k\phi) + A_1 h^k \sin(k\phi)].$$

32. Using Exercise 31, find a particular solution of

$$y_{k+2} - y_k = \cos\left(k\frac{\pi}{3}\right).$$

33. Using Exercise 31, find a particular solution of

$$y_{k+2} + y_k = \sin\left(k\frac{\pi}{2}\right).$$

7.5 Applications

In Section 7.1, a cooling problem was modeled with a difference equation. Instead of the differential equation (10) in Section 2.9,

$$\frac{dT}{dt} = \alpha(T - T_{\text{ref}}), \tag{1}$$

we had

$$T_{k+1} - T_k = \beta[T_k - T_{\text{ref}}], \tag{2}$$

where T_{ref} is the room temperature and α, β are constants.

Note that the cooling problem can be modeled by either the differential-equation (1) or a difference equation (2). If one wants to solve the differential equation (1) on a computer, then the differential equation is usually replaced by a difference equation (the other alternative is using a symbolic language such as MCSYMA). This very important application of difference equations to the numerical solution of differential equations is discussed in Chapter 8.

In Section 7.4 we saw an additional important application of difference equations, the summing of series.

Since Chapter 8 will include a discussion of the conversion of differential-equation models to difference equations, in this section we will discuss only applications that are intrinsically discrete in the time variable. That is, they cannot be modeled as effectively by differential equations.

Probably the simplest and most intuitive are the *interest models.* They can be applied to a variety of growth and decay problems. In Chapter 2 we considered daily compounding which was approximated by continuous compounding. In this section, compounding is less frequent and a discrete model is more appropriate.

Example 7.5.1 You borrow $10,000 at 12% yearly interest compounded monthly. The monthly payments are $300. How long will it take to pay off the loan?

Solution

Let y_k be the amount owed at the end of k months. Thus $y_0 = \$10{,}000$. We think of the problem as a mixing-flow problem:

$$\begin{bmatrix} \text{Change in } y \text{ from} \\ \text{month } k \text{ to } (k+1) \end{bmatrix} = \begin{bmatrix} \text{Increase} \\ \text{due to} \\ \text{interest} \end{bmatrix} - \begin{bmatrix} \text{Decrease} \\ \text{due to} \\ \text{payment} \end{bmatrix}. \qquad (3)$$

The interest per month is $\dfrac{0.12}{12} = 0.01$, so that (3) is

$$y_{k+1} - y_k = (0.01)y_k - 300$$

or

$$y_{k+1} = (1.01)y_k - 300.$$

This is a first-order linear difference equation with constant coefficients. Since the forcing term -300 is constant, we have, by Theorem 7.2.3, that

$$y_k = (1.01)^k 10{,}000 + \frac{(1.01)^k - 1}{1.01 - 1}(-300)$$

or

$$y_k = -20{,}000(1.01)^k + 30{,}000. \qquad (4)$$

The loan will be paid off if $y_k = 0$. Setting $y_k = 0$ in (4) gives

$$0 = -20{,}000(1.01)^k + 30{,}000$$

or $(1.01)^k = 3/2$. Taking the natural logarithm and solving for k,

$$k = \frac{\ln(3/2)}{\ln(1.01)} = 40.75.$$

It will take about 41 months or $3\frac{5}{12}$ years to pay off the loan. Thus you will pay $12,300 for your $10,000 loan. It cost you $2,300 to borrow the money. ∎

Example 7.5.2

You are in charge of a research laboratory and must prepare the annual budget. This year your budget was one million dollars and your expenditures were $900,000. The company has decreed that each year your budget will be altered according to two rules. First, it will be increased by 10%. Secondly, it will be adjusted by 80% of the difference between your expenditures and the budgeted amount. If expenditures exceed the budget amount, your next budget adjustment is positive, while, if expenditures were less than the budgeted amount, the adjustment is negative. Suppose that, in fact, due to inflation, your expenditures increase by 14% a year. Compare your budget and expenditures over the next several years.

Solution

Let B_k be the amount budgeted for the kth year and E_k be the expenditures for the kth year. Taking our units as millions of dollars, we have the initial conditions

$$B_0 = 1, \qquad E_0 = 0.9. \tag{5}$$

The expenditures increase by 14% a year, so that

$$[\text{Next year's expenditures}] = \begin{bmatrix} \text{This year's} \\ \text{expenditures} \end{bmatrix} + \begin{bmatrix} 14\% \text{ of this} \\ \text{year's} \\ \text{expenditures} \end{bmatrix},$$

or

$$E_{k+1} = E_k + 0.14 E_k = (1.14) E_k. \tag{6}$$

From Theorem 7.2.3, we have the solution of (6) and (5) as

$$E_k = (1.14)^k E_0 = (1.14)^k (0.9). \tag{7}$$

On the other hand, the difference equation for B_k is

$$\begin{bmatrix} \text{Next year's} \\ \text{budget} \end{bmatrix} = \begin{bmatrix} \text{This year's} \\ \text{budget} \end{bmatrix} + 0.10 \begin{bmatrix} \text{This year's} \\ \text{budget} \end{bmatrix}$$
$$+ 0.8 \left(\begin{bmatrix} \text{This year's} \\ \text{expenditures} \end{bmatrix} - \begin{bmatrix} \text{This year's} \\ \text{budget} \end{bmatrix} \right),$$

or

$$B_{k+1} = B_k + 0.1 B_k + 0.8(E_k - B_k). \tag{8}$$

Combining (8) and (7) gives the difference equation for B_k as

$$B_{k+1} = 0.3 B_k + 0.72(1.14)^k, \qquad B_0 = 1. \tag{9}$$

This is a first-order difference equation. The characteristic polynomial is $\lambda - 0.3$. Since $1.14 \neq 0.3$, we have a particular solution (Rule 1, Section 7.4):

$$p_k = A(1.14)^k. \tag{10}$$

Substituting (10) in (9) gives

$$A(1.14)^{k+1} = 0.3 A(1.14)^k + 0.72(1.14)^k$$

or

$$0.84 A = 0.72 \quad \text{and} \quad A = \tfrac{6}{7};$$

hence

$$p_k = \tfrac{6}{7}(1.14)^k.$$

The homogeneous solution is $(0.3)^k c$ and thus

$$B_k = (0.3)^k c + \tfrac{6}{7}(1.14)^k.$$

Applying the initial condition $B_0 = 1$ gives $1 = B_0 = c + \tfrac{6}{7}$, and $c = \tfrac{1}{7}$. Thus

$$B_k = \tfrac{1}{7}(0.3)^k + \tfrac{6}{7}(1.14)^k. \tag{11}$$

To compare expenditures and budgets, note that, from (7) and (11),
$$B_k - E_k = \tfrac{1}{7}(0.3)^k - \tfrac{3}{70}(1.14)^k. \tag{12}$$
But $\lim_{k \to \infty} (0.3)^k = 0$, so that, for larger k,
$$B_k - E_k \approx -\tfrac{3}{70}(1.14)^k, \tag{13}$$
and from (11), $B_k \approx \tfrac{6}{7}(1.14)^k$. Thus, even though your budget grows at approximately 14% a year over the long run, (13) shows it cannot keep up with expenditures. ∎

7.5.1 Signal Processing

Many electronic devices, such as digital computers, and some image and signal-processing problems, are intrinsically discrete. To actually discuss these applications realistically would require discussing the discrete analogue of the Laplace transform (the z-transform) and would take us too far afield. However, we wish to give a very brief, cursory discussion of an elementary version of these ideas.

Suppose, then, that we have a *signal* $\{y_k\}$. We many think of y_k as the available input at time k. (However, in image processing, k is often a spatial rather than a time variable.)

It is convenient to consider our signal-processing device as made up of three types of objects governed by an *internal clock*. The first is a transmission line denoted

$$\xrightarrow{\;\;a\;\;} \tag{14}$$

The transmission is assumed to be instantaneous, and the signal is multiplied by the scalar a.

The second device is a *delay*. This is a storage device, which, at each value of k (at each cycle of the clock), transmits its current contents and stores the current input. It will be denoted by

$$\rightarrow \boxed{D} \rightarrow \tag{15}$$

In most texts z^{-1} is used instead of D.

The third device is an *adder* or summation. At each clock cycle, it accepts a number of inputs, sums them, and then transmits the sum. It is denoted by

$$\rightarrowtail\!\!\oplus\rightarrow \tag{16}$$

Example 7.5.3 Write a *signal-flow diagram* corresponding to the difference equation

$$y_{k+1} + 3y_k = f_k. \tag{17}$$

Solution We think of y_k as our input and f_k as the output. One diagram would be (18).

(18)

To see how this diagram is interpreted, consider what happens at the kth clock cycle.

1. The value y_k is sent to both the delay device and the adder.
2. The delay device stores the value y_k as its new contents and transmits its previous contents to the adder along a transmission line, which multiplies the previous contents by 3.
3. The adder accepts the input from two transmission lines and sends out their sum, which is denoted f_{k-1}.

Since the previous contents of the delay device was y_{k-1}, the adder receives y_k and $3y_{k-1}$. The action of the adder is given by

$$y_k + 3y_{k-1} = f_{k-1}, \tag{19}$$

which is the same as (17). ∎

The general constant-coefficient difference equation

$$ay_{k+1} + by_k = f_k \tag{20}$$

may be diagrammed as in (21), where the a refers only to the bottom transmission line and not to the transmission of y_k to the delay device. Note that in (20) we think of $\{y_k\}$ as the *input* and $\{f_k\}$ as the *output*, which is opposite from our other circuit applications.

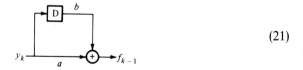

(21)

Similarly, the general second-order difference equation

$$ay_{k+2} + by_{k+1} + cy_k = f_k \tag{22}$$

could be represented as in either (23) or (24). (Note also Exercise 16.)

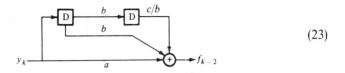
(23)

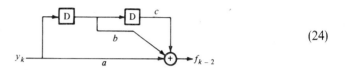
(24)

A device that takes a signal $\{y_k\}$ and changes it to give another signal is called a *filter*, so that (20) and (22) may be considered to be linear filters.

Now, in practice, one often wants to design the filter so that it affects certain ranges of frequencies a certain way. For example, it is often desirable to design low-pass filters. However, that again requires the z-transform and considerations of stability. Instead, we present the following simpler problem.

Suppose we have a signal $\{y_k\}$ made up of signals of two frequencies α, β. That is,

$$y_k = c_1 \sin 2\pi\alpha k + c_2 \cos 2\pi\alpha k + c_3 \sin 2\pi\beta k + c_4 \cos 2\pi\beta k, \qquad (25)$$

where c_1, c_2, c_3, and c_4 are some constants that are not currently important to us. We want to design a filter modeled by Eq. (22) such that the output has only the frequency α. That is, we want to filter out the component of frequency β from the signal. Since the signal $\{y_k\}$ is substituted into the difference equation to get the output $\{f_k\}$, we want $c_3 \sin 2\pi\beta k + c_4 \cos 2\pi\beta k$ to give zero output for all possible values of c_3, c_4. Thus, $c_3 \sin 2\pi\beta k + c_4 \cos 2\pi\beta k$ should be a solution of the associated homogeneous difference equation

$$ay_{k+2} + by_{k+1} + cy_k = 0.$$

From Theorem 7.3.2, Part (3), we know that the characteristic equation $a\lambda^2 + b\lambda + c = 0$ must then have roots $\lambda = e^{\pm i 2\pi\beta}$, so that the roots are

$$\lambda_1 = \cos 2\pi\beta + i \sin 2\pi\beta \qquad \text{and} \qquad \lambda_2 = \cos 2\pi\beta - i \sin 2\pi\beta.$$

Now if $\lambda_1 = z + iw$, $\lambda_2 = z - iw$, then λ_1, λ_2 are the roots of the polynomial

$$(\lambda - \lambda_1)(\lambda - \lambda_2) = \lambda^2 - 2z\lambda + z^2 + w^2.$$

Thus our characteristic polynomial would be

$$a[\lambda^2 - 2(\cos 2\pi\beta)\lambda + 1], \qquad (26)$$

and our difference equation, with $a = 1$, is

$$y_{k+2} - 2(\cos 2\pi\beta)y_{k+1} + y_k = f_k. \qquad (27)$$

To see why one might want to choose $a \neq 1$, suppose that $c_1 = 1$, $c_2 = 0$, so that (25) is

$$y_k = \sin 2\pi\alpha k + c_3 \sin 2\pi\beta k + c_4 \cos 2\pi\beta k. \tag{28}$$

Then substitute (28) into (27), to find

$$f_k = \sin(2\pi\alpha(k+2)) - 2\cos(2\pi\beta)\sin(2\pi\alpha(k+1)) + \sin(2\pi\alpha k),$$

which, by trigonometric identities, is

$$\begin{aligned} f_k &= \sin 2\pi\alpha k \cos 4\pi\alpha + \cos 2\pi\alpha k \sin 4\pi\alpha \\ &\quad - 2\cos 2\pi\beta[\sin 2\pi\alpha k \cos 2\pi\alpha + \cos 2\pi\alpha k \sin 2\pi\alpha] + \sin 2\pi\alpha k \\ &= \sin 2\pi\alpha k[\cos 4\pi\alpha - 2\cos 2\pi\beta \cos 2\pi\alpha + 1] \\ &\quad + \cos 2\pi\alpha k[\sin 4\pi\alpha - 2\cos 2\pi\beta \sin 2\pi\alpha] \\ &= \sin 2\pi\alpha k[\cos^2 2\pi\alpha - \sin^2 2\pi\alpha - 2\cos 2\pi\beta \cos 2\pi\alpha + 1] \\ &\quad + \cos 2\pi\alpha k[2\sin 2\pi\alpha \cos 2\pi\alpha - 2\cos 2\pi\beta \sin 2\pi\alpha] \\ &= \sin 2\pi\alpha k[2\cos^2 2\pi\alpha - 2\cos 2\pi\beta \cos 2\pi\alpha] \\ &\quad + \cos 2\pi\alpha k[2\sin 2\pi\alpha \cos 2\pi\alpha - 2\cos 2\pi\beta \sin 2\pi\alpha] \\ &= \sin 2\pi\alpha k[2\cos 2\pi\alpha]\{\cos 2\pi\alpha - \cos 2\pi\beta\} \\ &\quad + \cos 2\pi\alpha k[2\sin 2\pi\alpha]\{\cos 2\pi\alpha - \cos 2\pi\beta\}. \end{aligned}$$

Thus, by (11a) of Section 3.16, f_k can be written as

$$f_k = R \sin(2\pi\alpha k + \phi), \tag{29}$$

where R, after simplification, is

$$R = 2|\cos 2\pi\alpha - \cos 2\pi\beta|. \tag{30}$$

This has two consequences. First, if y_k has another component with frequency $\tilde{\beta}$ close to β, then its amplitude will be reduced by a factor of $2|\cos 2\pi\tilde{\beta} - \cos 2\pi\beta|$. However, if α is close to β, then the desirable component with frequency α will also have greatly reduced amplitude. By choosing $a > 1$ in (26), we will change the amplitude by aR instead of R.

Exercises

1. $10,000 is deposited in an account that earns 12% a year compounded quarterly. On the last day of each quarter, you deposit $200. How long does it take until you have $20,000?

2. You receive $100 when you are born, which is put into an account paying 10% interest. Each year after that on your birthday, a rich uncle deposits $10 for each year of age ($30 on your third birthday, $40 on your fourth, etc.). Find the formula, in terms of k, for the amount in your account at the end of k years. How much do you have on your twenty-first birthday?

3. You borrow $4000 at 12% interest compounded monthly.

a) Find the monthly payment required to pay the loan off in exactly 3 years.

b) Find the monthly payment required to pay the loan off in exactly 4 years.

c) What is the difference in the total interest paid between paying the amount in part a) and the amount in part b)?

4. You borrow $100,000 at 6% annual interest compounded monthly. You have a choice between paying $1000 a month or $2000 every other month. What would be the difference, after 10 years, in the amount owed between the two methods? Assume the first payment of the second plan is at the end of month two. (Note the second payment plan can be written $[1 + (-1)^n]\,2000$, $n \geq 1$)

5. On January 1, your uncle invests $10,000 in an account that earns 12% a year compounded yearly. At the same time, you invest $500 in an account that pays 10% a year compounded yearly. Each January 1 after that, your uncle gives you $\frac{1}{2}$ of the interest from his account. You immediately deposit it in yours. How much money do you have in your account after k years?

6. You have $100,000 in an account paying 12% a year compounded monthly. You wish to receive regular monthly payments P so that the account is empty after exactly T years. Find P as a function of T.

7. You run the purchasing department for an industrial concern. You need to know the amount A_k of a certain device to have delivered on day k. Of the shipment on day k, 80% pass inspection and are sent immediately to production. The other 20% need adjustment and are sent to the service department, where they are readjusted and sent to production the next day. Production wants to receive exactly 100 units on day 0 and then alternate between 200 and 100 units after that. How much should be ordered each day? [Hint: Note that $1 + (-1)^k$ takes on values of 2, 0, 2, 0, ..., and you need to compute the initial value.]

8. Same as Exercise 7, except that now the service department can fix only 80% of the devices sent to them. The remaining 20% are sent back to the factory, which repairs them and sends them back to production two days after they would have arrived if there had been no defects. How much should be ordered each day? (This problem involves more calculation than Exercise 7.)

9. Production of a certain memory chip began in 1981 with a first year's production of 200,000 chips. Each year after that the yearly chip production increases by 20%. Each year 90% of all available chips are bought. (That is, 90% of the new production *and* 90% of the chips remaining from the previous year.) Find the number of chips available at the end of each year.

10. You start an engineering firm, which initially needs 100 skilled employees. You need to add another 100 skilled employees each year afterwards. Each year you hire H_k new employees. Of this number, 80% are skilled and 20% require one year of on-the-job training. Most employees stay with the company but 10% of all employees leave at the end of their second year. Find the number of employees you need to hire each year, as a function of k.

11. A rare species of bean is being introduced into a country for general use. Initially there are 1000 seeds. Each seed, at harvest time, produces 45 new seeds. Two consecutive crops can be grown each year. The grower is allowed to eat 80% of each crop as a reward for using the seeds. How many seeds are there after 20 years?

12. A species of fish returns every 4 years to spawn and then dies. The survival rate of young is such that, on average, each spawning fish results in 107 fish returning to spawn 4 years later. At this time there are 500,000 fish returning to spawn. A fixed number H of fish are to be harvested out of the fish population as they return to spawn from this and each of the next five spawning runs. What should H be so that there are 2,000,000 fish spawning in 20 years?

13. Create a signal-flow graph for the third-order difference equation

$$ay_{k+3} + by_{k+2} + cy_{k+1} + dy_k = f_k. \qquad (31)$$

14. Show that (27), (29), and (30) are still valid if $\beta = 0$, and thus
$$y_{k+2} - 2y_{k+1} + y_k = f_k \qquad (32)$$
may be considered a filter of low frequencies, in that, if $y_k = \sin 2\pi\alpha k$, with α close to zero, then the output f_k is
$$R \sin(2\pi\alpha k + \phi),$$
where $R \approx 4\pi^2\alpha^2$.

15. Design a filter, $ay_{k+2} + by_{k+1} + cy_k = f_k$, so that it will eliminate signals of the form $\sin(\pi k/2)$.

16. One *cascades* devices by making the output of one the input of the other.

a) Find the difference equation (in terms of y_k) for the following cascade of two first-order filters.

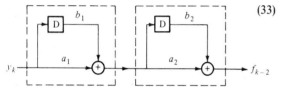
(33)

b) Show that the characteristic polynomial for part (a) is $(a_1\lambda + b_1)(a_2\lambda + b_2)$, where $a_1\lambda + b_1$, $a_2\lambda + b_2$ are the characteristic polynomials of the first and second filters.

17. (Continuation of Exercise 16.) For the cascade connection of two filters shown in the accompanying diagram (34),

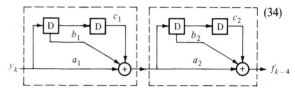
(34)

a) Find the difference equation for (34).

b) Show that the characteristic polynomial for part (a) is
$$(a_1\lambda^2 + b_1\lambda + c_1)(a_2\lambda^2 + b_2\lambda + c_2),$$
where $a_1\lambda^2 + b_1\lambda + c_1$, $a_2\lambda^2 + b_2\lambda + c_2$ are the characteristic polynomials of the first and second filters, respectively.

8

Numerical Methods

8.1 Introduction

Given a differential equation, there are several different ways we can begin to analyze it. In Chapters 2, 3, 4, and 6 we obtained explicit formulas for the solutions in terms of known functions. In Chapter 5 we obtained series solutions. These methods are often collectively referred to as the *analytic* solution of a differential equation. In Sections 2.1, 2.2, and Chapter 9 we try to determine *qualitative* properties of the solutions, such as boundedness and stability of equilibria.

This chapter will introduce the *numerical* approach. Here we are interested in obtaining estimates of the values of solutions at certain discrete points. With the advent of modern, widely accessible digital computers, many very complicated differential equations can now be solved numerically and a good understanding of numerical methods is increasingly important. These numerical procedures, however, have not changed the need for the other approaches. A qualitative analysis can show what types of numerical methods should be used, and help to determine whether the numerical results are reasonable.

On the other hand, plunging immediately into numerical methods can lead to unnecessary expense. We are aware of a computer program that was being used on a daily basis at a large corporation. Every time the program was used it numerically solved a system of differential equations several times. A mathematics major replaced the subroutine that solved the differential equation with a formula for the exact solution, obtained using the methods of Chapter 6. The result was a savings of several thousand dollars a year in computer time, and a pay raise for the programmer.

Comment on Computations It is not necessary to have access to a computer in order to utilize this chapter. All of the work may, in principle, be done on a pocket calculator, and some may be done by hand (but that is not advised except for the simplest problems). However, most of the homework is best done on at least a programmable pocket calculator. Such homework problems are denoted by an asterisk (*).

When comparing your answers with those given in the text, keep in mind that any of the following may affect the number of significant digits in your answers:

- Precision of the arithmetic on your machine.
- How round-off and floating-point operations are carried out.
- Order in which computation has been done (adding large numbers to small numbers and subtracting almost equal numbers will often give a loss of significance).
- Built-in functions, such as sin, cos, exp, log, a^x, may have less precision than arithmetic operations.
- How underflow and overflow are handled.

Unless noted otherwise, all of the numbers in this chapter were computed in double precision, even if only the first few (rounded) digits are given.

The basic problem of the first six sections of this chapter is:

Given the initial-value problem

$$y' = f(x, y), \qquad y(a) = y_0, \tag{1}$$

estimate the value of y at a later time $x = b$.

In some cases we are interested only in $y(b)$. In others, we also want some intermediate values of y. All of our estimates of the values of the solution of (1) will be arrived at using algorithms that follow the following pattern:

1. $N + 1$ values of the independent variable x will be chosen from a to b:

$$a = x_0 < x_1 < x_2 < \cdots < x_N = b.$$

2. Let $y_0 = y(x_0)$ be the initial condition.
3. Given $\{y_0, \ldots, y_n\}$, let the estimate y_{n+1} for $y(x_{n+1})$, the value of the solution at time x_{n+1}, be computed by some method. We now have $\{y_0, \ldots, y_n, y_{n+1}\}$. Repeat this step until we have computed $\{y_0, \ldots, y_N\}$.

4. Then y_N is an estimate of $y(x_N) = y(b)$. Because the estimates $y_1, \ldots, y_N$ are computed one after each other, we shall refer to each time (3) is done as one *step* of the method.

The amount the independent variable changes each time is $x_n - x_{n-1}$ and is called the *step size*. We shall use h to denote the step size, so that

$$h_n = x_n - x_{n-1}$$

Most numerical methods for solving Eq. (1), when actually used in practice, employ changes in step size. However, except in a couple of the homework exercises, we assume a *constant step size*.

Notation

The following notation will be used throughout this chapter. The differential equation is

$$y'(x) = f(x, y(x)), \qquad y(a) = y_0. \tag{2}$$

The points $x_0, \ldots, x_N$ are a *partition* (also a grid, or mesh) of the interval $[a, b]$:

$$a = x_0 < x_1 < \cdots < x_N = b.$$

The step size is h and

$$x_{n+1} = x_n + h.$$

The function $y(x)$ is the solution of the differential equation (2). The estimate of $y(x_n)$ provided by the numerical method is y_n. The error e_n at the nth step is the difference between the true solution $y(x_n)$ and our estimate y_n so that

$$e_n = y(x_n) - y_n.$$

How then do we compute the estimate y_{n+1} for the value of the solution y at time x_{n+1} if we already have the estimates $\{y_0, \ldots, y_n\}$? All the techniques described in subsequent sections are based on combinations of two ideas:

Taylor Series Estimate $y(x_{n+1}) = y(x_n + h)$, using the Taylor series of the solution y at time x_n.

Integral Formula Integrating Eq. (2) from x_n to x_{n+1} gives:

$$y(x_{n+1}) - y(x_n) = \int_{x_n}^{x_{n+1}} \frac{dy}{dx}\, dx = \int_{x_n}^{x_n+h} f(x, y)\, dx,$$

so that

$$y(x_{n+1}) = y(x_n) + \int_{x_n}^{x_n+h} f(x, y)\, dx. \tag{3}$$

Estimating the integral in Eq. (3) then gives an estimate y_{n+1} of $y(x_{n+1})$.

8.2 Euler's Method

We now develop our first method. Suppose that we wish to numerically solve

$$y' = f(x, y), \qquad y(a) = y_0, \tag{1}$$

and obtain an estimate for $y(b)$. We assume that f, f_y are continuous, so that the initial-value problem (1) has a unique solution. (See Section 2.1.) Subdivide the interval $[a, b]$ with $N + 1$ mesh points $x_0, \ldots, x_N$, with $x_0 = a$, $x_N = b$. Let $h = (b - a)/N$ be the step size, so that $x_n = a + hn$. In some problems h is given and $N = (b - a)/h$.

Let

$$y_0 = y(a) = y(x_0).$$

The first-order Taylor polynomial (Section 5.2) for the solution y of (1) at a point $\hat{x}$ is

$$y(x) \approx y(\hat{x}) + y'(\hat{x})(x - \hat{x}). \tag{2}$$

(This is the same as using the tangent line at $(\hat{x}, y(\hat{x}))$ as an approximation for $y(x)$ near $(\hat{x}, y(\hat{x}))$.) Thus, letting $x = \hat{x} + h$ in (2) gives

$$y(\hat{x} + h) \approx y(\hat{x}) + y'(\hat{x})h. \tag{3}$$

Pictorially, we have Fig. 8.2.1. But $y'(\hat{x}) = f(\hat{x}, y(\hat{x}))$, since y is assumed to be a solution of the differential equation (1). Thus (3) becomes

$$y(\hat{x} + h) \approx y(\hat{x}) + f(\hat{x}, y(\hat{x}))h. \tag{4}$$

This formula gives a way of estimating y at time x_{n+1}, given y at time x_n, by letting $\hat{x} = x_n$ in Eq. (4) and using our previous estimate y_n for $y(x_n)$. The resulting numerical method is called *Euler's method*.

Euler's Method

For the solution of $y' = f(x, y)$, $y(x_0) = y_0$,

1. Let $y_0 = y(a)$.
2. Recursively compute $y_1, \ldots, y_N$ by

$$y_{n+1} = y_n + hf(x_n, y_n) \tag{5}$$

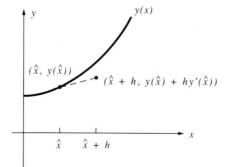

Figure 8.2.1

where
$$x_n = a + nh, \quad h = \frac{b-a}{N}.$$

3. y_N is an estimate of $y(b)$.

Note that (5) is a difference equation as discussed in Chapter 7.

Example 8.2.1 Let y be the solution of the differential equation
$$y' - xy = x, \quad y(0) = 1. \tag{6}$$
Estimate $y(1)$ by Euler's method using a step size of $h = 0.25$.

Solution The differential equation is
$$y' = x + xy, \quad y(0) = 1,$$
so that $f(x, y) = x + xy$. We have (from Eq. (6)) that
$$x_0 = 0, \quad y_0 = y(0) = 1.$$
The recursive relationship (5) is then
$$y_{n+1} = y_n + hf(x_n, y_n) = y_n + h(x_n + x_n y_n). \tag{7}$$
Thus we get

$$y_1 = y_0 + h(x_0 + x_0 y_0) = 1 + 0.25(0 + 0 \cdot 1) = 1;$$
$x_1 = x_0 + h = 0.25;$ $\quad y_2 = y_1 + h(x_1 + x_1 y_1)$
$$= 1 + 0.25(0.25 + 0.25 \cdot 1) = 1.125;$$
$x_2 = x_1 + h = 0.5;$ $\quad y_3 = y_2 + h(x_2 + x_2 y_2)$
$$= 1.125 + 0.25(0.5 + 0.5 \cdot 1.125) = 1.391,$$
$x_3 = x_2 + h = 0.75;$ $\quad y_4 = y_3 + h(x_3 + x_3 y_3)$
$$= 1.391 + 0.25(0.75 + 0.75 \cdot 1.391) = 1.839.$$

The estimate for $y(1)$ is $y_4 = 1.839$ (to three places). ∎

Equation (6) can be solved using separation of variables, to get the actual solution:
$$y = 2e^{x^2/2} - 1, \tag{8}$$
so that
$$y(1) = 2e^{0.5} - 1 \approx 2.2974425. \tag{9}$$
The error in our estimate of $y(1)$ is then

Figure 8.2.2

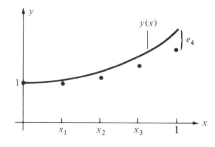

Figure 8.2.3

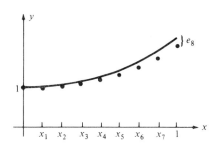

$$e_4 = y(1) - y(x_4) = 0.458. \tag{10}$$

These calculations are pictorially represented in Fig. 8.2.2.

It seems natural to try to get a smaller error by taking smaller steps.

Example 8.2.2 Again estimate $y(1)$ using Euler's method, where y is the solution of (6), but this time use a step size of $h = 0.125$.

Solution We have $a = 0$, $b = 1$, $h = 0.125$, and $N = (b - a)/n = 8$. Thus

$$x_0 = 0, \quad x_n = nh, \quad x_8 = 1.$$

Again use the difference relationship (7) and compute (x_n, y_n) for $n = 1, \ldots, 8$. The computed y_i, and the actual solution is graphed in Fig. 8.2.3. The estimate for $y(1)$ is

$$y(1) = y(x_8) \approx y_8 = 2.048. \blacksquare \tag{11}$$

Note that the error in the estimate (11) is

$$e_8 = y(1) - y_8 = 0.249. \tag{12}$$

Thus reducing the step size by a factor of one-half has reduced the error (10) by about one-half also. This in fact turns out to be a general property of Euler's method.

Table 8.2.1 Euler's Method for Eq. (6)

Step Size h	Steps N	Estimate y_N for $y(1)$	Error at $b = 1$ $e_N = y(1) - y_N$
0.1	10	2.09422	0.20322
0.01	100	2.27564	0.02180
0.001	1000	2.29525	0.00220
0.0001	10000	2.29722	0.00022

A method for solving (1) is said to be of *order r* (written $O(h^r)$) on the interval $[a, b]$ if there is a constant M depending on a, b, f, but not the step size h or the point $x_n \in [a, b]$, such that

$$|e_n| = |y(x_n) - y_n| \leq Mh^r. \tag{13}$$

Theorem 8.2.1 Euler's method is a first-order method. That is, if $x_n \in [a, b]$ for a given step size h, then

$$|e_n| = |y(x_n) - y_n| \leq Mh \tag{14}$$

for some constant M independent of n, h.

This result, which of course requires some technical assumptions on $f(x, y)$ and a careful consideration of the different types of error, will be proved in Section 8.3. Note that Eq. (14) says that, by taking h to be half as large, we should expect about half as much error, and that is what was observed in (10), (12). Theorem 8.2.1 also seems to suggest that, in order to get answers accurate to several significant figures using Euler's method, we would have to take very small steps and hence a large number of steps. Table 8.2.1 gives the result of solving Eq. (6) by Euler's method for several step sizes.

We can give a *heuristic* argument for Theorem 8.2.1 as follows. At any value of $\hat{x}$ we have, by Taylor's Theorem from calculus, that the error in using

$$y(\hat{x}) + hy'(\hat{x}) \quad \text{for } y(\hat{x} + h) \tag{15}$$

is [see Eq. (19) of Section 5.2.1]:

$$R_1(x) = \tfrac{1}{2} y''(\xi)(x - \hat{x})^2. \tag{16}$$

Since $x - \hat{x} = h$, the error induced at step $n + 1$ (*if* y_n *is exact*) can be estimated by

$$|R_1(x)| \leq \tfrac{1}{2} |y''(\xi)| h^2.$$

Let $\tilde{M}$ be the maximum of $\tfrac{1}{2}|y''(x)|$ on $[a, b]$. Then the error at each step is estimated to be $\tilde{M}h^2$. The number of steps is $N = (b - a)/h$. Thus the error could be estimated as the amount of error at each step times the number of steps, or

$$\tilde{M}h^2 N = \tilde{M}h^2 \frac{(b-a)}{h} = \tilde{M}(b-a)h.$$

Taking $M = \tilde{M}(b-a)$ would give Eq. (14).

This way of estimating the error is incorrect, however, since it overlooks the fact that we have to use an estimate y_n for $y(x_n)$ in Eq. (15) when $x = x_n$ and thus the error is *compounded*. It also overlooks roundoff error. However, it still remains a reasonable rule of thumb that, if the error on a given step is $O(h^{r+1})$ (sometimes called the *truncation* or *local error*), then one should expect the method to be $O(h^r)$ for h not so small that roundoff error becomes important.

The exercises that follow illustrate several aspects of Euler's method that are typical of numerical methods. Additional properties are developed in the (optional) Section 8.3.

Exercises

[Exercises marked with an asterisk are suggested only for programmable calculators and computers.] For Exercises 1 through 11, compute an estimate for the value of $y(b)$, using Euler's method, given the indicated step size.

1. $y' = y - x$, $\quad y(0) = 1$, $\quad b = 2$, $\quad h = 0.5$
2. $y' = xy$, $\quad y(0) = 1$, $\quad b = 1$, $\quad h = 0.2$
3. $y' = -y^2 x$, $\quad y(0) = 1$, $\quad b = 2$, $\quad h = 1$
4. $y' = 3y - 2x$, $\quad y(0) = 0$, $\quad b = 2$, $\quad h = 0.5$
5. $y' = 2y - 4x$, $\quad y(0) = 1$, $\quad b = 2$, $\quad h = 0.5$
6. $y' = xy - x$, $\quad y(0) = 1$, $\quad b = 4$, $\quad h = 0.2$
7. $y' = \sin y$, $\quad y(0) = 0$, $\quad b = 4$, $\quad h = 0.5$
8. $y' = \sin y$, $\quad y(0) = 1$, $\quad b = 4$, $\quad h = 0.2$
9. a) $y' = -20y$, $\quad y(0) = 1$, $\quad b = 2$, $\quad h = 0.2$
   b) On the same (x, y)-axis, plot the points (x_n, y_n) from part (a) and the true solution.
10. (Continuation of Exercise 9)
    a) $y' = -20y$, $\quad y(0) = 1$, $\quad h = 0.1$, $\quad b = 2$
    *b) On the same (x, y)-axis, plot the points (x_n, y_n) from part (a) and the true solution.
*11. (Continuation of Exercises 9, 10)
    a) $y' = -20y$, $\quad y(0) = 1$, $\quad h = 0.01$, $\quad b = 2$
    b) On the same (x, y)-axis, plot the point (x_n, y_n) from part (a) and the true solution.

Exercises 9, 10, 11 show that for some equations, Euler's method produces numerical solutions that resemble the actual solution only for small step sizes. In this example, this behavior is related to a property called *stiffness*, which is discussed in Section 8.3.

12. Estimate $y(1)$ for the solution of $y' = y^2$, $y(0) = 1$, using Euler's method with a step size of $h = 0.5$.
13. (Continuation of Exercise 12). Estimate $y(1)$ for the solution of $y' = y^2$, $y(0) = 1$, using Euler's method with step sizes of $h = 0.2$, $h = 0.1$, $h = 0.01$,* $h = 0.001$.*
14. (Continuation of Exercises 12, 13.) Explain what you observed about the estimates for $y(1)$ by solving the differential equation $y' = y^2$, $y(0) = 1$, by separation of variables.

Exercises 15 through 17 illustrate the point that, while analytically obtained solutions cannot cross equilibria if $f(x, y)$, $f_y(x, y)$ are continuous, the numerical solution can "jump" over an equilibrium.

15. Using Euler's method, calculate estimates for the solution of the differential equation
$$y' = 1 - 2y + y^2,$$
$$y(a) = -5, \qquad a \le x \le b, \qquad (17)$$
at points (x_n, y_n) where $a = 0$, $b = 2$, $h = 0.2$.

Graph these points (x_n, y_n) and sketch what you think the solution would look like.

16. (Continuation of Exercise 15). Sketch the solutions of $y' = 1 - 2y + y^2$ using the techniques of Section 2.2. Observe that $y = 1$ is an equilibrium solution. Compare this picture with that of Exercise 15.

17. (Continuation of Exercises 15, 16.) Solve Eq. (17) with $a = 0$, $b = 2$, using Euler's method with step of $h = 0.1$, plot* all values of (x_n, y_n), and compare to the analysis of Exercise 16.

Euler's method can also be derived using integration.

18. i) By antidifferentiating both sides of $y' = f(x, y)$ with respect to x, show that

$$y(\hat{x} + h) - y(\hat{x}) = \int_{\hat{x}}^{\hat{x}+h} f(x, y(x))\, dx. \quad (18)$$

ii) One estimate for an integral $\int_c^d g(x)\, dx$ (the *constant-slope estimate*) is $g(\xi)(d - c)$, where ξ is a point between c and d. Using $c = \hat{x}$, $d = \hat{x} + h$, and $\xi = \hat{x}$, obtain

$$y(\hat{x} + h) - y(\hat{x}) \approx f(\hat{x}, y(\hat{x}))h \quad (19)$$

from Eq. (18) using the constant-slope estimate.

iii) Let $\hat{x} = x_n$ and use the estimate y_n for $y(x_n)$ to derive Euler's formula (5) from Eq. (19).

Suppose that we are using an rth-order method with N_1 steps, so that

$$y_{N_1} - y(b) \approx Mh^r. \quad (20)$$

If we also use a step size that is one-half as big, we need to take $N_2 = 2N_1$ steps to reach b from a. Then we have

$$y_{N_2} - y(b) \approx M\left(\frac{h}{2}\right)^r. \quad (21)$$

From Eqs. (20) and (21) we find, after a bit of algebra, that we might expect that

$$\frac{y_{N_1} - y_{N_2}}{y_{N_2} - y_{N_3}} \approx 2^r, \quad (22)$$

where $N_3 = 2N_2$. This heuristic argument suggests that one can examine the order of the method by examining the left-hand side of Eq. (22).

19. Assume equality holds in Eqs. (20) and (21), and derive Eq. (22) with $\approx$ replaced by $=$, to heuristically justify Eq. (22).

20. Solve the differential equation $y' = y + x^2$, $y(0) = 0$, to obtain an estimate y_N for $y(2)$. Using each of the steps $h = 0.2$, $h = 0.1$, $h = 0.05$, $h = 0.025$, and $h = 0.0125$, compute the respective estimates y_{N_1}, y_{N_2}, y_{N_3}, y_{N_4}, and y_{N_5} for $y(2)$, where $N_1 = 10$, $N_2 = 20$, $N_3 = 40$, $N_4 = 80$, and $N_5 = 160$. Then compute

$$\frac{y_{N_1} - y_{N_2}}{y_{N_2} - y_{N_3}}, \quad \frac{y_{N_2} - y_{N_3}}{y_{N_3} - y_{N_4}}, \quad \frac{y_{N_3} - y_{N_4}}{y_{N_4} - y_{N_5}}.$$

If Euler's is a first-order method, from (22) with $r = 1$ we would expect these ratios to be approaching 2. Do they appear to be doing so?

8.3 An Analysis of Euler's Method

The other sections in this chapter present various numerical methods for solving ordinary differential equations. This section, which may be covered any time after Section 8.2, analyzes the Euler method of Section 8.2, and introduces several important concepts. Recall from Section 8.2 that $y(x)$ is the solution of

$$y' = f(x, y), \qquad y(a) = y_0, \quad (1)$$

where $x_0 = a$, $x_n = a + nh$, $x_N = b$, $h = (b - a)/N$, and Euler's method is

$$y_0 = y(a), \qquad y_{n+1} = y_n + hf(x_n, y_n), \qquad n \geq 0. \qquad (2)$$

The error in this estimate of $y(x_n)$ (ignoring round-off) is

$$e_n = y(x_n) - y_n. \qquad (3)$$

8.3.1 Discretization Error

In an actual computation we have two sources of error. One is due to the fact that the difference in Eq. (2) is only approximating the differential equation (1). This is called *discretization error*. The other major source is *rounding error* when doing the computations.

We consider the discretization error first. Let

$$E(h) = \max_{1 \leq n \leq N} |y_n - y(x_n)| = \max_{1 \leq n \leq N} |e_n|. \qquad (4)$$

Then $E(h)$ is the *global discretization* (or *truncation error*). For a given h, $E(h)$ is the largest error that occurs in our estimate at any mesh point. In order to estimate $E(h)$ for Euler's method, we need the following facts:

Fact 1
For constants α, β,

$$\lim_{h \to 0} (1 + h\alpha)^{\beta/h} = e^{\alpha\beta}. \qquad (5)$$

(This limit may be verified using l'Hôpital's rule; see Exercise 1 at the end of this section.)

Fact 2
Taylor's Theorem applied to the solution $y(x)$ gives (requires y being twice differentiable):

$$y(x + h) = y(x) + hy'(x) + R_1(x, h), \qquad (6)$$

where

$$R_1(x, h) = \tfrac{1}{2} y''(\xi) h^2 \qquad (7)$$

with ξ between x and $x + h$.

Fact 3
The Mean-Value Theorem (which is, of course, related to Taylor's Theorem) says

$$G(z) - G(w) = G'(\xi)(z - w), \qquad (8)$$

with ξ between z and w if $G(z)$ is continuously differentiable.

We are now ready to prove Theorem 8.2.1, which we restate here as Theorem 8.3.1. The proof is complete except that we do not carefully specify some of the sets over which we take maxima.

Theorem 8.3.1 Suppose that $f(x, y)$ is continuously differentiable with respect to both x and y. Suppose also that the solution y of

$$y' = f(x, y), \qquad y(a) = y_0, \tag{9}$$

exists on the interval $a \leq x \leq b$. Then there is a constant M, so that the global truncation error [Eqs. (3), (4)] for Euler's method satisfies

$$E(h) \leq Mh. \tag{10}$$

Thus Euler's method is a first-order method.

Proof Since $f(x, y)$ is continuously differentiable and $y(x)$ is continuously differentiable (Theorem 2.1.1), we see that y is *twice* differentiable, since differentiating (9) with respect to x gives

$$y'' = f_x(x, y) + f_y(x, y)y'. \tag{11}$$

Thus Facts 2 and 3 can be used. We now begin to estimate the error. Our immediate goal is to construct a difference equation to bound the error. From Eq. (3) we have

$$e_{n+1} = \underbrace{y(x_{n+1})} - \underbrace{y_{n+1}}. \tag{12}$$

Using Eq. (6) on $y(x_{n+1}) = y(x_n + h)$ and the difference equation (2) from Euler's method for y_{n+1}, Eq. (12) can be rewritten as

$$e_{n+1} = \underbrace{y(x_n) + hy'(x_n) + R_1(x_n, h)} - \underbrace{(y_n + hf(x_n, y_n))}$$
$$= y(x_n) - y_n + h(y'(x_n) - f(x_n, y_n)) + R_1(x_n, h)$$
$$= e_n + h(y'(x_n) - f(x_n, y_n)) + R_1(x_n, h). \tag{13}$$

We need to estimate the second and third terms in Eq. (13). Using the fact that $y(x)$ is a solution of the differential equation (9) we get, for the second term,

$$|y'(x_n) - f(x_n, y_n)| = |f(x_n, y(x_n)) - f(x_n, y_n)|. \tag{14}$$

But $f(x_n, z)$ is a function of z. Applying the mean-value theorem (8) to the right side of (14) gives

$$|y'(x_n) - f(x_n, y_n)| \leq M_1|y(x_n) - y_n| = M_1|e_n|, \tag{15}$$

where M_1 is an upper bound for $|f_y|$, taken over both x and y. We estimate $R_1(x_n, h)$ in (13) by using (7), to get

$$|R_1(x_n, h)| \leq M_2 h^2, \tag{16}$$

where $2M_2$ is the maximum of $|y''|$ on $[a, b]$.

Now, taking absolute values in (13),
$$|e_{n+1}| \leq |e_n| + h|y'(x_n) - f(x_n, y_n)| + |R_1(x_n, h)|$$
which by (15), (16) implies
$$|e_{n+1}| \leq |e_n| + hM_1|e_n| + h^2M_2. \tag{17}$$
Let ε_n be the solution of the difference equation
$$\varepsilon_{n+1} = \varepsilon_n + hM_1\varepsilon_n + h^2M_2, \qquad \varepsilon_0 = |e_0|. \tag{18}$$
It is easy to show by induction using (17), (18) that
$$\varepsilon_n \geq |e_n| \qquad \text{for all } n. \tag{19}$$
Thus, to estimate $|e_n|$ it suffices to solve the first-order difference equation (18). From Section 7.2 (or by direct verification), the solution of (18) is
$$\varepsilon_n = (1 + hM_1)^n \varepsilon_0 + \left(\sum_{i=0}^{n-1}(1 + hM_1)^i\right)h^2M_2. \tag{20}$$

Since from (18) we have $\varepsilon_n \leq \varepsilon_{n+1}$ it follows that the largest ε_n is the last one, ε_N, so that (4) and (19) give
$$E(h) \leq \varepsilon_N.$$
(It is not true, in general, that $E(h) \leq |e_N|$.) Thus to estimate $E(h)$ it suffices to estimate
$$\varepsilon_N = (1 + hM_1)^N \varepsilon_0 + \left(\sum_{i=0}^{N-1}(1 + hM_1)^i\right)h^2M_2. \tag{21}$$
The first term in Eq. (21) is easily handled by the fact that
$$(1 + hM_1)^N = (1 + hM_1)^{(b-a)/h},$$
which, by Eq. (5), converges to $e^{M_1(b-a)}$. In fact, it is monotonically increasing (see Exercise 2 at the end of this section) as $h \to 0^+$, so that
$$(1 + hM_1)^N \leq e^{M_1(b-a)}. \tag{22}$$
The second term in Eq. (21) is estimated by using the fact that, if $u \neq 1$, then
$$\sum_{i=0}^{n-1} u^i = \frac{u^n - 1}{u - 1},$$
so that
$$\left(\sum_{i=0}^{N-1}(1 + hM_1)^i\right)h^2M_2 = \left[\frac{(1 + hM_1)^N - 1}{(1 + hM_1) - 1}\right]h^2M_2$$
$$= \frac{(1 + hM_1)^{(b-a)/h} - 1}{hM_1}h^2M_2$$
$$\leq [e^{M_1(b-a)} - 1]h\left(\frac{M_2}{M_1}\right) \leq e^{M_1(b-a)}h\left(\frac{M_2}{M_1}\right).$$

This yields the final result

$$\varepsilon_N \le e^{M_1(b-a)} e_0 + e^{M_1(b-a)} \left(\frac{M_2}{M_1}\right) h. \tag{23}$$

If $e_0 = 0$, then

$$\varepsilon_N \le Mh,$$

where $M = e^{M_1(b-a)} M_2/M_1$, which is Theorem 8.3.1. ∎

As a bonus we get, from Eq. (23) an estimate of how much an error or change of the starting value y_0 affects the numerical solution (take $\varepsilon_0 = |e_0| \ne 0$).

8.3.2 Round-Off Error

Suppose that at each step there is an additional error ρ_i due to round-off and function evaluation. For convenience, assume that there is a bound ρ, so that

$$|\rho_i| \le \rho. \tag{24}$$

Note that ρ is independent of the step size and is at least as large as the machine round-off error. Let Y_n be the actual value calculated for y_n using Euler's method. Thus the values Y_n satisfy

$$Y_{n+1} = Y_n + hf(x_n, Y_n) + \rho_{n+1}. \tag{25}$$

The theoretical values of y_n satisfies

$$y_{n+1} = y_n + hf(x_n, y_n). \tag{26}$$

Let $E_n = Y_n - y_n$ be the difference between the computed and theoretical value of y_n. Subtracting (26) from (25) gives

$$E_{n+1} = E_n + h(f(x_n, Y_n) - f(x_n, y_n)) + \rho_{n+1}. \tag{27}$$

But $f(x_n, Y_n) - f(x_n, y_n) = f_y(x_n, \xi)(Y_n - y_n)$, with ξ between Y_n and y_n by the mean-value theorem (8). Thus, taking absolute values of (27) and using (24), we get

$$|E_{n+1}| \le |E_n| + hM_1|E_n| + \rho, \tag{28}$$

where M_1 is again an upper bound for $f_y(x, y)$. Again we solve the difference equation (see Exercise 5 at the end of this section) with equality in (28) to get

$$|E_N| \le e^{M_1(b-a)} |E_0| + e^{M_1(b-a)} \frac{\rho}{M_1 h}. \tag{29}$$

Note a significant difference in this new estimate from that of (23). As h decreases, the estimate for the round-off error *increases*, due to the fact that smaller steps require more calculations. The number ρ is usually quite small,

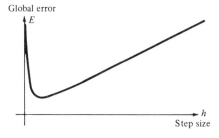

Figure 8.3.1
Graph of $\alpha h + \beta/h$ with $\alpha = 1, \beta = 0.01$.

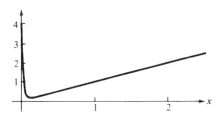

Figure 8.3.2
Graph of $y = x + 4e^{-40x}$.

between 10^{-8} and 10^{-17} on most machines. However, at some point the decreasing discretization error due to smaller steps is offset by increasing round-off error, and still smaller steps will only make things worse. Combining (23) and (29) with $e_0 = 0$, we see that if round-off error is included, then the global error for Euler's method is

$$\tilde{E}(h) = \max_{1 \le n \le N} |y(x_n) - Y_n| \approx \alpha h + \frac{\beta}{h},$$

which is graphed in Fig. 8.3.1.

In fact, Fig. 8.3.1 is appropriate for many numerical algorithms except that one gets $E(h) \approx \alpha h^r + \beta/h^s$, where r, s are constants depending on the method.

8.3.3 Stiffness

We conclude this section with a brief discussion of an important property called *stiffness*. This discussion will also illustrate that a theorem like Theorem 8.3.1 may require small h before it is applicable.

Consider the differential equation

$$y' = -40y + 40x + 1, \qquad y(0) = 4. \tag{30}$$

The solution of this linear differential equation is

$$y = x + 4e^{-40x}. \tag{31}$$

The solution (31) consists of two parts. A slowly varying portion, x, and a faster (transient) part, $4e^{-40x}$. The graph of Eq. (31) is given in Fig. 8.3.2.

These types of solution arise frequently in electrical circuits, in chemical reactions, and in "boundary layers" in fluid problems. Suppose that we attempt

Table 8.3.1 Solution of Eq. (30) with Euler's method, and $h = 0.1$.

x	n	Y_n	e_n	$E_n(0.1)$
0.0	0	4	0	0
0.5	5	-971.5	972	972
1.0	10	236197	236196	236196
1.5	15	-5.7×10^7	5.7×10^7	5.7×10^7
2.0	20	1.4×10^{10}	1.4×10^{10}	1.4×10^{10}

Table 8.3.2 Solution of Eq. (30) with Euler's method, and $h = 0.05$.

x	n	Y_n	e_n	$E_n(0.05)$
0.0	0	4.0	0	0
0.5	10	4.5	4	4.54
1.0	20	5.0	4	4.54
1.5	30	5.5	4	4.54
2.0	40	6.0	4	4.54

Table 8.3.3 Solution of Eq. (30) with Euler's method, and $h = 0.01$.

x	n	Y_n	e_n	$E_n(0.01)$
0.0	0	4.0	0	0
0.5	50	0.5	8.2×10^{-1}	0.357
1.0	100	1.0	3.4×10^{-17}	0.357
1.5	150	1.5	0	0.357
2.0	200	2.0	5.6×10^{-17}	0.357

to use Euler's method on Eq. (30) for $0 \leq x \leq 2$, and take a step size of $h = 0.1$. Let

$$E_n(h) = \max_{0 \leq i \leq n} |e_i|.$$

The result is Table 8.3.1.

This is not even close. Trying again with $h = 0.05$, we get Table 8.3.2. The error is no longer increasing unboundedly, but it still gives a poor estimate. Letting $h = 0.01$, we get Table 8.3.3. This gives excellent estimates for $x \geq 0.5$ but less accuracy on the first few steps, since $E(0.01) = 0.357$.

Table 8.3.4

x	Y_n	e_n
0.0	4	0
0.5	0.496	0.004
1.0	1.960	0.960
1.5	−231.664	233.164
2.0	56660	56658

This is somewhat frustrating. We are forced to use small steps just to approximate a function that is quickly almost the straight line $y = x$.

A natural thing to attempt is to use smaller steps when the solution is changing rapidly and large steps later. If we take $h = 0.01$ for $0 \leq x \leq 0.2$ and $h = 0.1$ for $0.2 \leq x \leq 2$, we get the result in Table 8.3.4. Again we have a poor numerical estimate.

To understand what is happening, we need to examine Euler's method more carefully. Euler's method applied to Eq. (30) gives

$$y_{n+1} = y_n + h(-40y_n + 40nh + 1), \qquad y_0 = 4, \qquad (32)$$

or

$$y_{n+1} = (1 - h40)y_n + 40nh^2 + h, \qquad y_0 = 4; \qquad (33)$$

the solution of Eq. (33) from Section 7.4 is

$$y_n = (1 - h40)^n y_0 + hn. \qquad (34)$$

It is true that $\lim_{h \to 0^+} (1 - h40)^n = 1$ for fixed n, but if h is not very small, then $(1 - h40)^n$ can be quite large for large n. In fact, if $h = 0.1$, then $(1 - h40)^n y_0 = (-3)^n y_0$ oscillates with growing amplitude as n increases. Only if $0 < h < 1/20$ is $|1 - h40| < 1$, so that the numerical solution resembles the real solution.

Backward Euler

Seemingly small differences in a numerical method can greatly affect its behavior. Euler's method for $y' = f(x, y)$ may be written as

$$\frac{y_{n+1} - y_n}{h} = f(x_n, y_n), \qquad (35)$$

where $(y_{n+1} - y_n)/h$ is an approximation to y'. The *backward* or *implicit Euler* uses a backward difference to approximate y', and is given by

$$\frac{y_n - y_{n-1}}{h} = f(x_n, y_n),$$

or, equivalently,

$$\frac{y_{n+1} - y_n}{h} = f(x_{n+1}, y_{n+1}). \qquad (36)$$

Table 8.3.5 Backward Euler on Eq. (30), with $h = 0.1$.

x	n	Y_n	e_n	$E_n(0.1)$
0.0	0	4	0	0.727
0.5	5	0.501	1.2×10^{-3}	0.727
1.0	10	1.000	4.1×10^{-7}	0.727
1.5	15	1.500	1.3×10^{-10}	0.727
2.0	20	2.000	4.2×10^{-14}	0.727

It is more effort to solve for y_{n+1} in (36) than in (35), since y_{n+1} appears inside f. Frequently a numerical method like Newton's must be used to solve for y_{n+1}. To illustrate (36), we solve Eq. (30) by the backward Euler's (36). This time the difference equation is

$$\frac{y_{n+1} - y_n}{h} = -40 y_{n+1} + 40(n+1)h + 1,$$

or

$$(1 + 40h) y_{n+1} = y_n + 40(n+1)h^2 + h,$$

and finally

$$y_{n+1} = \frac{1}{1 + 40h} y_n + h + \frac{40nh^2}{1 + 40h}. \tag{37}$$

When this difference equation is solved, one gets a $(1 + 40h)^{-n}$ term. But

$$\left| \frac{1}{1 + 40h} \right| < 1 \quad \text{for any } h > 0,$$

so we expect better convergence. The result of using the backward Euler method (37) on (30) with a step size of $h = 0.1$ is given in Table 8.3.5.

Comparing Table 8.3.5 to Table 8.3.1, we see that changing from a forward difference to a backward difference has greatly improved our estimate for $y(2)$ in this example.

Exercises

1. Using l'Hôpital's rule, verify that $\lim_{h \to 0} (1 + h\alpha)^{\beta/h} = e^{\alpha \beta}$, as stated in Eq. (5).

2. Show that if $\alpha, \beta > 0$, then, as $h \to 0$, $(1 + h\alpha)^{\beta/h}$ is monotonically increasing. (*Hint*: Show that, if $g(h) = (1 + h\alpha)^{\beta/h}$, then $g'(h) < 0$ for all $h > 0$).

3. Let $|e_n|$, ε_n be defined as in Section 8.3. Using Eqs. (17) and (18), verify by induction that $\varepsilon_n \geq |e_n|$ for all $n \geq 0$ and thus Eq. (19) holds.

4. Verify that the sequence (20) is a solution of the difference equation (18).

5. Let $T_{n+1} = T_n + hM_1 T_n + \rho$, with $M_1 > 0$, $T_0 \geq 0$. Show that

$$T_N \leq e^{M_1(b-a)} T_0 + e^{M_1(b-a)} \frac{\rho}{M_1 h},$$

where $N = (b-a)/h$. This verifies Eq. (29).

6. Note that Eq. (33) can be written as $y_{n+1} = \alpha y_n + \beta n + \gamma$. Using the theory of Section 7.4, solve this difference equation. Use this solution to derive the solution (34) of Eq. (33).

The function $g(h) = \alpha h + \beta/h$, with $\alpha > 0$, $\beta > 0$, and $h > 0$, attains its minimum at $\hat{h} = \sqrt{\beta/\alpha}$. In the error estimate (23) we usually have $\alpha > 1$ so that $\hat{h} \geq \sqrt{\beta}$. Since β is at least as big as machine precision, we have that round-off error can become a major factor in Euler's method by the time the step size nears the square root of machine precision.

*7. For a given step size h, solve

$$y' = 2y + x, \quad y(0) = 1, \quad 0 \leq x \leq 1, \quad (38)$$

by Euler's method. At each step compute the difference between the computed Y_n and the true $y(x_n)$. Let $\hat{E}(h)$ be the largest of these h's. Compute $\hat{E}(h)$ for progressively smaller h's, starting with $h = 0.1$, $h = 0.01$, etc. You should observe that $\hat{E}(h)$ decreases for a while and then increases. For what value of h was $\hat{E}(h)$ smallest? How does that compare to the square root of your machine precision? (It is recommended that you print out only the $\hat{E}(h)$ and not all the y_n values.)

8. Verify Eq. (15).

*9. a) Solve the differential equation

$$y' = -30y + 30x + 31, \quad y(0) = 4, \quad (39)$$

on the interval $0 \leq x \leq 2$ using both Euler's method and the Backward Euler's method with $h = 0.1$.

b) Find the formula for the solution of (39).

c) Graph the values of (x_n, y_n) from both methods in part (a) and the solution from part (b) on the same axis.

*10. Newton's method for solving $g(z) = 0$, given an initial guess, $\hat{z}$, is to let $z_0 = \hat{z}$ and consider the iteration

$$z_{m+1} = z_m - g'(z_m)^{-1} g(z_m). \quad (40)$$

The iteration (40) is repeated for M steps, and z_M is taken as the estimate of a solution of $g(z) = 0$. Depending on the application, M can range from 1 to a large number. This exercise is to solve

$$y' = 30 \cos y, \quad y(0) = 1, \quad 0 \leq x \leq 2 \quad (41)$$

using the Backward Euler method (36) with $h = 0.1$. This will require solving

$$y_{n+1} - y_n - h30 \cos y_{n+1} = 0 \quad (42)$$

for y_{n+1}, given y_n. Do this using Newton's method. This entails, at each n, having an initial guess for y_{n+1}, denoted $\hat{y}_{n+1}$, and a rule for when to stop the Newton iteration. Experiment with these choices for $\hat{y}_{n+1}$;

i) $\hat{y}_{n+1} = y_n$

ii) $\hat{y}_{n+1} = y_n + h30 \cos y_n$ (Euler predictor),

and the following rules for M;

i) Stop after one iteration ($M = 1$)

ii) Stop after two iterations ($M = 2$)

iii) Stop when $|z_{m+1} - z_m| \leq 0.01$

11. (Continuation of Exercise 10.)
a) Solve (41) using Euler's method with $h = 0.1$. Compare to what you observed in Exercise 10.
b) Using the techniques of Section 2.2, sketch the solutions of (41) and compare to the values of the two computed approximations.

8.4 Second-Order Methods

As noted in Sections 8.2 and 8.3, in order to get sufficient accuracy with Euler's method we may have to take a large number of steps with a small step size. Such calculations can be time-consuming, even on a computer, and present difficulties caused by the accumulation of round-off error. One way to circumvent this difficulty is to use higher-order methods. This section will present two second-order methods.

Perhaps the most natural way to improve on Euler's method is to replace the first-order Taylor polynomial used in Euler's method with a second-order Taylor polynomial.

If y is the solution of

$$y' = f(x, y), \quad y(x_0) = y_0, \quad (1)$$

then the second-order Taylor approximation (Section 5.2) for y at x is

$$y(x + h) = y(x) + hy'(x) + \frac{h^2}{2} y''(x). \quad (2)$$

From (1) we have $y'(x) = f(x, y)$. Differentiating both sides of (1) with respect to x yields

$$y'' = f_x(x, y) + f_y(x, y)y' = f_x(x, y) + f_y(x, y)f(x, y). \quad (3)$$

8.4.1 Second-Order Taylor Method

Substituting (1) and (3) into (2) gives the *Second-Order Taylor Method* for solving $y' = f(x, y)$, $y(a) = y_0$ with a fixed step size of h.

1. Let y_0 be as in (1), $x_0 = a$.
2. Compute x_n, y_n for $n \geq 1$ by $x_{n+1} = x_n + h$:

$$y_{n+1} = y_n + hf(x_n, y_n) + \frac{h^2}{2} [f_x(x_n, y_n) + f_y(x_n, y_n)f(x_n, y_n)]. \quad (4)$$

Example 8.4.1 If we are going to solve

$$y' = xy + x, \quad (5)$$

using this second-order Taylor method, we have

$$f(x, y) = xy + x, \quad f_x(x, y) = y + 1, \quad f_y(x, y) = x.$$

Thus (4) becomes

$$y_{n+1} = y_n + h(x_n y_n + x_n) + \frac{h^2}{2} [(y_n + 1) + x_n(x_n y_n + x_n)]. \quad (6)$$

These calculations will be carried out in Example 8.4.3. ∎

Unfortunately, the Taylor method requires the calculation of two partials and three function evaluations. In many applications, these partials are not easily available and the function evaluations are computationally expensive. The next approach is closer in spirit to those usually used in practice.

To motivate this next method, integrate (1) from x_n to x_{n+1} to get

$$\int_{x_n}^{x_{n+1}} y'\, dx = \int_{x_n}^{x_{n+1}} f(x, y)\, dx \tag{7}$$

or

$$y(x_{n+1}) - y(x_n) = \int_{x_n}^{x_n + h} f(x, y(x))\, dx. \tag{8}$$

Estimate this integral by using the trapezoid rule with just two mesh points, x_n, x_{n+1}. Equation (8) becomes

$$y(x_{n+1}) - y(x_n) \approx \frac{h}{2}[f(x_n, y(x_n)) + f(x_{n+1}, y(x_{n+1}))]$$

or

$$y(x_{n+1}) \approx y(x_n) + \frac{h}{2}[f(x_n, y(x_n)) + f(x_{n+1}, y(x_{n+1}))]. \tag{9}$$

The only difficulty with (9) is that we do not know $y(x_{n+1})$ when evaluating the right-hand side, since $y(x_{n+1})$ is what y_{n+1} is supposed to estimate. This difficulty can be circumvented by estimating $y(x_{n+1})$ in the right-hand side of (9), using Euler's method. The result is the modified *Euler method*.

8.4.2 Modified Euler Method

To solve $y' = f(x, y)$, $y(a) = y_0$ with step size h, by the Modified Euler Method:

1. Let y_0 be as in (1), $x_0 = a$.
2. For $n \geq 0$, define x_n, y_n by the recursion

$$x_{n+1} = x_n + h,$$
$$z_{n+1} = y_n + hf(x_n, y_n), \tag{10}$$
$$y_{n+1} = y_n + \frac{h}{2}[f(x_n, y_n) + f(x_{n+1}, z_{n+1})]. \tag{11}$$

The modified Euler method is also called the *Heun* (or *improved Euler*) method. It is an example of a *second-order Runge–Kutta method* (the reasons for this will be mentioned in Section 8.5). The modified Euler method illustrates the idea of a *predictor-corrector* method. Given the values x_n, y_n, we *predict* the value of $y(x_{n+1})$. This prediction is denoted by z_{n+1} in (10). Then in (11) the prediction z_{n+1} is *corrected* to give the final estimate y_{n+1} for $y(x_{n+1})$. In the modified Euler method, the "predictor" is Euler's method. The "corrector" is integrating (8) by the trapezoid rule.

Example 8.4.2 If we are to solve $y' = xy + x$, $y(0) = 1$, by the modified Euler method with step size h, we have

$$x_n = nh,$$
$$z_{n+1} = y_n + h(x_n y_n + x_n), \tag{12}$$
$$y_{n+1} = y_n + \frac{h}{2}[x_n y_n + x_n + x_{n+1} z_{n+1} + x_{n+1}]. \tag{13}$$

In this case $x_0 = 0$, $y_0 = 1$. Thus $x_1 = h$. We shall calculate y_1 using $h = 0.1$. [Additional calculations appear in Example 8.4.3.] From (12),

$$z_1 = y_0 + h(x_0 y_0 + x_0) = 1 + (0.1)0 = 1.$$

Then, from (13),

$$y_1 = y_0 + \frac{h}{2}[x_0 y_0 + x_0 + x_1 z_1 + x_1] = 1 + \frac{0.1}{2}[0 + 0 + (0.1)1 + 0.1]$$
$$= 1.01. \blacksquare$$

The Taylor, Euler, and modified Euler methods are examples of *one-step methods*. That is, to compute the estimate y_{n+1} for $y(x_{n+1})$, we need only know the values x_n, y_n from *one* step earlier. For example, in Example 8.4.2, Eqs. (12) and (13) may be combined to express y_{n+1} directly in terms of y_n, as

$$y_{n+1} = y_n + \frac{h}{2}[x_n y_n + x_n + x_{n+1}[y_n + h(x_n y_n + x_n)] + x_{n+1}]. \tag{14}$$

We note without proof that, under the appropriate smoothness assumptions on $f(x, y)$:

Theorem 8.4.1 Both the Taylor method (4) and the modified Euler method are second-order methods on a finite interval $[a, b]$. That is, there exists a constant M depending on the method, $f(x, y)$, the interval $[a, b]$, and the initial condition, but not on the step size h, such that the global error,

$$E(h) = \max_{1 \leq n \leq (b-a)/h} |y(x_n) - y_n| = \max_{1 \leq n \leq (b-a)/h} |e_n| \tag{15}$$

satisfies

$$E(h) \leq Mh^2. \tag{16}$$

Table 8.4.1 Error $|e_N|$ in estimate of $y(1)$ for (17).

h	Euler	Second-Order Taylor	Modified Euler
0.1	0.203	0.008602	0.001680
0.01	0.022	0.000095	0.000015

Example 8.4.3   Table 8.4.1 gives the errors that result in solving

$$y' = xy + x, \quad y(0) = 1, \quad 0 \le x \le 1, \qquad (17)$$

for $y(1)$, using a Euler's, second-order Taylor, and modified Euler method with a step size of $h = 0.1$ and $h = 0.01$.

The dramatic reduction in error for only slightly more computational effort shows why higher-order methods are almost always chosen over Euler's method ∎

Exercises

In Exercises 1 through 7, compute the solution on the interval [0, 1] with the given step size using Euler's, second-order Taylor, and modified Euler's methods. For Exercises 1 through 5, compare your result to the true solution. Values for y_N and e_N are given in the answer section at the end of this book.

1. $y' = 2y, \quad y(0) = 1, \quad h = 0.1$
*2. $y' = 2y, \quad y(0) = 1, \quad h = 0.01$
3. $y' = -xy, \quad y(0) = 1, \quad h = 0.2$
4. $y' = 1 - y^2, \quad y(0) = 0, \quad h = 0.1$
*5. $y' = 1 - y^2, \quad y(0) = 0, \quad h = 0.01$
6. $y' = \cos y, \quad y(0) = 0, \quad h = 0.1$
*7. $y' = \cos y, \quad y(0) = 0, \quad h = 0.01$
8. Let λ be a constant.
    i) Show that the second-order Taylor and modified Euler methods give the same result when applied to the linear constant-coefficient problem

    $$y' = \lambda y, \quad y(a) = y_0. \qquad (18)$$

    ii) Show that both these methods applied to (18) give the recursion relationship

    $$y_{n+1} = \left[1 + h\lambda + \frac{h^2\lambda^2}{2}\right] y_n. \qquad (19)$$

    iii) Show that $1 + h\lambda + h^2\lambda^2/2$ are the first three terms in a series expansion for $e^{h\lambda}$.

9. Take x_n, y_n as given numbers.
    i) Show that the difference relations in solving Eq. (5) of Example 8.4.1 by the second-order Taylor (6) and the modified Euler (14) methods give different values of y_{n+1}.
    ii) Show that these two estimates for y_{n+1} differ by terms all of which have an h^3 in them.

10. Derive a third-order Taylor method for solving $y' = f(x, y), y(a) = y_0$ similar to the way (4) was derived.

Using the *third-order Taylor* method defined in Exercise 10, estimate $y(1)$ for the following differential equations with the indicated step sizes:

11. $y' = 2y, \quad y(0) = 1, \quad h = 0.2, 0 \le x \le 1$
12. $y' = xy, \quad y(0) = 1, \quad h = 0.1, 0 \le x \le 1$
13. $y' = xy + x^2, \quad y(0) = 1,$
    $h = 0.1, 0 \le x \le 1$

14. It can be shown that if round-off error is considered, then the global error from either the modified Euler or second-order Taylor method has an estimate of the form

$$\hat{E}(h) = Mh^2 + \frac{\beta}{h}. \tag{20}$$

Assume that $M = 1$, and β is a small number approximately equal to machine precision.

a) Show that $\hat{E}(h)$ attains its minimum at about $h = \sqrt[3]{\beta}$.

b) Let $\beta = 10^{-16}$. Compute the value of $\hat{E}(h)$ at the minimum from part (a).

c) Accepting that the minimum error from Euler's method is approximately $\sqrt{\beta}$, can you (heuristically) conclude that the second-order methods not only can be expected to give greater accuracy at less computational cost but are capable of attaining greater accuracy before the effect of round-off error dominates the calculation?

8.5 Fourth-Order Runge–Kutta

In some problems the numerical (or actual) cost of evaluating $f(x, y)$ can be high. In other problems we must worry about rapid transients (see Section 8.3). Finally, many modern numerical packages for solving differential equations not only automatically vary the step size but also change the order, and even the type of method being used. However, among the more popular general purpose methods used are the Runge–Kutta methods. These are one-step methods and can thus easily vary the step size from one step to another to better control errors. One feature of Runge–Kutta methods is that they often compute intermediate values of the independent variable.

There are actually several fourth-order Runge–Kutta methods. One of the more popular is given below.

Fourth-Order Runge–Kutta

Fourth-Order Runge–Kutta for solving

$$y' = f(x, y), \qquad y(a) = y_0 \tag{1}$$

with step size h proceeds as follows.

1. Let $x_0 = a$, $x_{n+1} = x_n + h$, $y_0 = y(a)$.
2. For $n \geq 0$ define y_{n+1} in terms of y_n by:

   a) Compute

   $$F_1(n) = hf(x_n, y_n), \tag{2}$$

   $$F_2(n) = hf\left(x_n + \frac{h}{2}, y_n + \frac{F_1}{2}\right), \tag{3}$$

$$F_3(n) = hf\left(x_n + \frac{h}{2}, y_n + \frac{F_2}{2}\right), \tag{4}$$

$$F_4(n) = hf(x_n + h, y_n + F_3), \tag{5}$$

b) Let

$$y_{n+1} = y_n + \tfrac{1}{6}(F_1(n) + 2F_2(n) + 2F_3(n) + F_4(n)) \tag{6}$$

Example 8.5.1   To illustrate, again consider

$$y' = xy + x, \qquad y(0) = 1, \tag{7}$$

so that $f(x, y) = xy + x$ and $x_0 = 0$, $y_0 = 1$. We shall compute y_1 using the fourth-order Runge–Kutta method [Eqs. (2) through (6)] with $h = 0.1$. From (2),

$$F_1 = 0.1(x_0 y_0 + x_0) = 0.1(0 \cdot 1 + 0) = 0,$$

so that

$$F_2 = 0.1f\left(x_0 + \frac{h}{2}, y_0 + \frac{F_1}{2}\right) = 0.1f(0.05, 1)$$

$$= 0.1(0.05(1) + 0.05) = 0.01$$

$$F_3 = 0.1f\left(x_0 + \frac{h}{2}, y_0 + \frac{F_2}{2}\right) = 0.1f(0.05, 1.005)$$

$$= 0.1(0.05(1.005) + 0.05) = 0.010025,$$

$$F_4 = 0.1f(x_0 + h, y_0 + F_3) = 0.1f(0.1, 1.010025)$$

$$= 0.1(0.1(1.010025) + 0.1) = 0.02010025,$$

and then from (6)

$$y_1 = 1 + \frac{1}{6}(0 + 0.02 + 0.02005 + 0.02010025) = 1.010025. \blacksquare$$

It is interesting to compare this result with the solution of (7) by the methods of Sections 8.2 and 8.4. For (7) we have

> True rounded value of $y(0.1) = 1.010025042$;
> y_1 from Euler's method $= 1$,
> y_1 from modified Euler $= 1.01$,
> y_1 from second-order Taylor $= 1.01$,
> y_1 from Fourth-Order Runge–Kutta $= 1.010025$.

The fourth-order Runge–Kutta method can be understood much as the modified Euler was. Again, antidifferentiate both sides of (1) from x_n to $x_n + h$, to give

$$y(x_{n+1}) = y(x_n) + \int_{x_n}^{x_n + h} f(x, y)\, dx. \tag{8}$$

In the modified Euler we estimate the integral using the trapezoid rule. In the fourth-order Runge–Kutta a variation of Simpson's rule is used to estimate the integral in (8).

The Runge–Kutta methods may also be derived using a Taylor-series argument.

Exercises

For Exercises 1 through 8, estimate $y(1)$ using a fourth-order Runge–Kutta with the indicated step size. Values of y_1 and y_N are given in the appendices.

1. $y' = xy + x$, $y(0) = 1$, $h = 0.1$
*2. $y' = xy + x$, $y(0) = 1$, $h = 0.01$
3. $y' = 1 - y^2$, $y(0) = 0.5$, $h = 0.1$
4. $y' = xy + x^2$, $y(0) = 0$, $h = 0.1$
5. $y' = \cos y$, $y(0) = 1$, $h = 0.1$
6. $y' = y - xy^2$, $y(0) = 1$, $h = 0.1$
7. $y' = -40y + 40x + 1$, $y(0) = 4$, $h = 0.1$
   (See the discussion of stiffness in Section 8.3.)
*8. $y' = -40y + 40x + 1$, $y(0) = 4$, $h = 0.01$
9. a) Show that if the fourth-order Runge–Kutta method of this section is applied to $y' = \lambda y$, $y(0) = 1$, λ a constant, then
$$y_{n+1} = \left[1 + h\lambda + \frac{(h\lambda)^2}{2} + \frac{(h\lambda)^3}{3!} + \frac{(h\lambda)^4}{4!}\right] y_n.$$
   b) Observe that
$$1 + h\lambda + \frac{(h\lambda)^2}{2} + \frac{(h\lambda)^3}{3!} + \frac{(h\lambda)^4}{4!}$$
   is the first five terms in an expansion of $e^{\lambda h}$ and the Runge–Kutta method is *fourth*-order. Compare to Exercise 8 at the end of Section 8.4.

(Runge–Kutta methods are also used to compute starting values in some of the Exercises of Section 8.6.)

8.6 Multistep Methods

In all of the methods examined so far for solving

$$y' = f(x, y), \quad y(a) = y_0, \tag{1}$$

higher accuracy was obtained by performing additional evaluations of the function $f(x, y)$. The Euler method required one evaluation, the modified Euler required two, and the fourth-order Runge–Kutta required four evaluations per step.

In many problems, computing these values of f requires a great deal of computational effort. One way to circumvent this difficulty is by using earlier values of f rather than more function evaluations of f to get higher-order methods. One such family of methods is "The Adams Family" for solving (1). Three of these methods are the:

8.6.1 Second-Order Adams–Bashforth

Given y_0, y_1, then for $n \geq 1$,

$$y_{n+1} = y_n + \frac{h}{2}(3f(x_n, y_n) - f(x_{n-1}, y_{n-1})). \tag{2}$$

8.6.2 Third-Order Adams–Bashforth

Given y_0, y_1, y_2, then for $n \geq 2$,

$$y_{n+1} = y_n + \frac{h}{12}(23f(x_n, y_n) - 16f(x_{n-1}, y_{n-1}) + 5f(x_{n-2}, y_{n-2})). \tag{3}$$

8.6.3 Fourth-Order Adams–Bashforth

Given y_0, y_1, y_2, y_3, then for $n \geq 3$,

$$y_{n+1} = y_n + \frac{h}{24}(55w_n - 59w_{n-1} + 37w_{n-2} - 9w_{n-3}), \tag{4}$$

where

$$w_n = f(x_n, y_n).$$

Note that while (4), for example, involves four evaluations, only one of them, $f(x_n, y_n) = w_n$, has to be computed at each step. The other three, $w_{n-1}, w_{n-2}, w_{n-3}$, have been computed on previous steps.

It is important that the starting values be as accurate as the method is expected to be. Thus, in using the fourth-order method (4), we would probably use a fourth-order one-step method, such as a fourth-order Runge–Kutta, to compute y_1, y_2, y_3 from y_0. Other options are to use a Euler method with step size h^4 or a modified Euler with step size h^2.

Example 8.6.1 Estimate $y(1)$ for the solution of

$$y' = xy + x, \qquad y(0) = 1, \tag{5}$$

using the second-order Adams–Bashforth method (2) with step size $h = 0.1$. Estimate y_1 using a modified Euler method.

Solution   Let
$$w_n = f(x_n, y_n) = x_n y_n + x_n.$$
Then (2) becomes
$$y_{n+1} = y_n + \frac{h}{2}(3w_n - w_{n-1}). \tag{6}$$

The initial conditions
$$x_0 = 0, \quad y_0 = 1$$
imply that $w_0 = 0$. We compute y_1 using a modified Euler (see Example 8.4.2), so that
$$x_1 = 0.1, \quad y_1 = 1.01.$$
Thus $w_1 = 0.201$.

We can now use (6) to compute the other y_n. First
$$y_2 = y_1 + \frac{0.1}{2}(3w_1 - w_0) = 1.04015.$$
Then
$$x_2 = 0.2, \quad w_2 = x_2(y_2 + 1) = 0.40803.$$
Continuing
$$y_3 = y_2 + \frac{0.1}{2}(3w_2 - w_1) = 1.0913045,$$
and finally after seven more iterations of this process,
$$y_{10} = 2.27768.$$
The actual value of $y(1)$ is approximately $y(1) = 2.2974$. ∎

8.6.4 Second-Order Adams–Moulton

The starting value y_1 is computed by a second-order one-step method. Then given y_n, y_{n-1}, the predicted value of y_{n+1} is given by the second-order Adams–Bashforth,
$$z_{n+1} = y_n + \frac{h}{2}(3f(x_n, y_n) - f(x_{n-1}, y_{n-1})). \tag{7}$$

The prediction z_{n+1} is corrected using the trapezoid rule,

$$y_{n+1} = y_n + \frac{h}{2}(f(x_n, y_n) + f(x_{n+1}, z_{n+1})). \qquad (8)$$

The Adams–Moulton method (7), (8) requires two function evaluations at each step. However, the method can sometimes have numerical advantages. (Note Exercise 5.)

Exercises

In Exercises 1 through 4, estimate $y(1)$ using a second-order Adams–Bashforth with a step size of 0.1. Calculate the second starting value y_1 using a modified Euler method.

1. $y' = 3y + 1$, $\quad y(0) = 1$
2. $y' = 1 - y^2$, $\quad y(0) = 2$
3. $y' = \cos y$, $\quad y(0) = 0$
4. $y' = xy - x^2$, $\quad y(0) = 1$

*5. Find the actual solution of
$$y' = -40y + 40x + 1, \quad y(0) = 4.$$
Using step sizes of $h = 0.1, 0.05, 0.04, 0.01$, estimate $y(1)$, using both a second-order Adams–Bashforth and a second-order Adams–Moulton. Compare the result to the actual answer. [This is the stiffness example of Section 8.3.]

6. Let y be the solution of $y' = f(x, y)$, $y(x_0) = y_0$. Using the Taylor expansion of $y(x)$ centered at x_n, show that
$$y(x_n + h) - y(x_n - h) = 2y'(x_n)h + O(h^3). \qquad (9)$$
The resulting two-step method
$$y_{n+1} = y_{n-1} + 2hf(x_n, y_n) \qquad (10)$$
is called the *centered-difference* method.

7. Estimate $y(1)$ for $y' = -xy^2$, $y(0) = 1$, by the centered-difference method with a step size of $h = 0.1$. [Find y_1 using a modified Euler's.]

*8. For the differential equation
$$y' = \sin(xy) + \cos(y^2) + e^{-x^2}, \quad y(0) = 1,$$
using a step size of $h = 0.01$, estimate $y(10)$ using both:

 i) A fourth-order Runge–Kutta.

 ii) A fourth-order Adams–Bashforth with starting values, y_1, y_2, y_3, determined by a fourth-order Runge–Kutta.

 Compare the amount of computer time each method took.

*9. Verify that $y_{10} = 2.27768$ for Example 8.6.1.

*10. Estimate $y(1)$ for (5) using the fourth-order Adams–Bashforth method (4) with $h = 0.01$. Find starting values with a fourth-order Runge–Kutta method. Find the actual solution to (5) and determine $e_{100} = y(1) - y_{100}$.

8.7 Systems

There are at least three ways a computer can be used to solve systems of differential equations. One is by use of *symbolic* languages or programs such as *MACSYMA*.[1] The second, if considering linear constant-coefficient differential

---

[1] R. H. Rand, *Computer Algebra in Applied Mathematics: An Introduction to MACSYMA*. Pitman, 1984.

equations, is to compute eigenvalues, and eigenvectors as in Chapter 6. Numerical routines for doing this may be found in such packages as MATLAB, EISPACK, and they even appear as primitives in such languages as APL, version 2. The third method, and the one we shall briefly discuss, is by extending the numerical methods of Sections 8.1 through 8.6 to systems. This turns out to be notational. Given a system of m equations;

$$\begin{aligned} y_1'(t) &= f_1(t, y_1, \ldots, y_m), & y_1(a) &= y_{10}, \\ y_2'(t) &= f_2(t, y_1, \ldots, y_m), & y_2(a) &= y_{20}, \\ &\vdots & &\vdots \\ y_m'(t) &= f_m(t, y_1, \ldots, y_m), & y_m(a) &= y_{m0}, \end{aligned} \quad (1)$$

we adopt the vector notation

$$\mathbf{y} = \begin{bmatrix} y_1 \\ \vdots \\ y_m \end{bmatrix}, \quad \mathbf{f} = \begin{bmatrix} f_1 \\ \vdots \\ f_m \end{bmatrix}, \quad \mathbf{y}_0 = \begin{bmatrix} y_{10} \\ \vdots \\ y_{m0} \end{bmatrix}. \quad (2)$$

Then (1) becomes

$$\mathbf{y}' = \mathbf{f}(t, \mathbf{y}), \quad \mathbf{y}(a) = \mathbf{y}_0. \quad (3)$$

All of the previous techniques may be applied to (3) by using boldface* type for the y and f wherever they appear. For example, the Euler method of Section 8.2 is

$$\mathbf{y}_{n+1} = \mathbf{y}_n + h\mathbf{f}(t_n, \mathbf{y}_n) \quad \text{for } n \geq 0. \quad (4)$$

[Note that the n in $\mathbf{y}_n$ denotes the nth step whereas y_n in (1) is the nth function in the system. To avoid this notational confusion, we shall consider only small-order systems and give each dependent function a different letter.]

Example 8.7.1 Estimate $x(1)$, $y(1)$ for the solution of the system of differential equations

$$\begin{aligned} x'(t) &= x(t) + 3y(t) - 1, & x(0) &= 1, \\ y'(t) &= x(t) - 2y(t) + t, & y(0) &= 3, \end{aligned} \quad (5)$$

using a modified Euler method with step size $h = 0.1$.

Solution We have $t_0 = 0$ and $t_n = nh = 0.1n$. Also,

$$N = \frac{b-a}{h} = \frac{2}{0.1} = 20.$$

---

* In handwritten calculations, other conventions, such as underlining, are used to denote vectors.

Let
$$\mathbf{y} = \begin{bmatrix} x \\ y \end{bmatrix}, \quad \mathbf{f}(t, \mathbf{y}) = \begin{bmatrix} x + 3y - 1 \\ x - 2y + t \end{bmatrix}. \tag{6}$$

We also need the predictor variable $\mathbf{z}$ which we denote
$$\mathbf{z} = \begin{bmatrix} u \\ v \end{bmatrix}. \tag{7}$$

The modified Euler method in vector form is then:
$$\mathbf{z}_{n+1} = \mathbf{y}_n + h\mathbf{f}(t_n, \mathbf{y}_n), \tag{8}$$
$$\mathbf{y}_{n+1} = \mathbf{y}_n + \frac{h}{2}[\mathbf{f}(t_n, \mathbf{y}_n) + \mathbf{f}(t_{n+1}, \mathbf{z}_{n+1})]. \tag{9}$$

In terms of (6) and (7), the equations (8) and (9) become:
$$\begin{aligned} u_{n+1} &= x_n + h(x_n + 3y_n - 1), \\ v_{n+1} &= y_n + h(x_n - 2y_n + t_n), \end{aligned} \tag{10}$$

and
$$\begin{aligned} x_{n+1} &= x_n + \frac{h}{2}[(x_n + 3y_n - 1) + (u_{n+1} + 3v_{n+1} - 1)] \\ y_{n+1} &= y_n + \frac{h}{2}[(x_n - 2y_n + t_n) + (u_{n+1} - 2v_{n+1} + t_{n+1})], \end{aligned} \tag{11}$$

respectively.

For example, given $x_0 = 1$, $y_0 = 3$, $h = 0.1$, we get from (10)
$$u_1 = x_0 + (0.1)(x_0 + 3y_0 - 1) = 1 + 0.1(1 + 9 - 1) = 1.9,$$
$$v_1 = y_0 + (0.1)(x_0 - 2y_0 + 0) = 3 + 0.1(1 - 6) = 2.5.$$

Then from (11)
$$x_1 = 1 + \frac{0.1}{2}[(1 + 3(3) - 1) + (1.9 + 3(2.5) - 1)] = 1.87,$$
$$y_1 = 3 + \frac{0.1}{2}[(1 - 2(3) + 0) + (1.9 - 2(2.5) + 0.1)] = 2.6.$$

Given (x_1, y_1) we may now compute (x_2, y_2), (x_3, y_3), etc., recursively, from (10) and (11). The estimates for $x(1)$, $y(1)$ are $x_{10} = 14.630$, $y_{10} = 4.361$.
∎

Table 8.7.1 Solution of Eq. (13) by Modified Euler.

n	t_n	x_n	y_n	Error in x $x(t_n) - x_n$	Error in y $y(t_n) - y_n$
0	0	0	1	0	0
1	0.1	−0.40000	0.98000	0.00267	0.00007
5	0.5	−1.69171	0.53528	0.00877	0.00502
10	1	−1.81109	−0.42894	−0.00750	0.01280

If a programming language with vector variables and vector operators is used, then the previous methods on systems are programmed almost exactly as they were for a single equation.

Higher-Order Equations

Our numerical methods may be applied to higher-order equations by rewriting them as systems. (See Chapter 6.)

Example 8.7.2 Find $y(1)$ for the solution of

$$y''(t) + 4y(t) = 0, \quad y(0) = 1, \quad y'(0) = 0, \quad (12)$$

using a modified Euler with step size of 0.1.

Solution Let $x(t) = y'(t)$. Then (12) becomes $x' + 4y = 0$, so that we have the first-order linear system

$$\begin{aligned} x' &= -4y, & x(0) &= 0, \\ y' &= x, & y(0) &= 1. \end{aligned} \quad (13)$$

Let $t_0 = 0$, $t_n = n(0.1)$. If we let

$$\mathbf{z}_n = \begin{bmatrix} u_n \\ v_n \end{bmatrix}.$$

be the predictor variable, the predictor equations are

$$\begin{aligned} u_n &= x_n + h(-4y_n), \\ v_n &= y_n + h x_n, \end{aligned} \quad (14)$$

and the corrector equations are

$$\begin{aligned} x_{n+1} &= x_n + \frac{h}{2}[-4y_n + (-4v_{n+1})], \\ y_{n+1} &= y_n + \frac{h}{2}[x_n + u_{n+1}]. \end{aligned} \quad (15)$$

The actual solution is $x = -2 \sin 2t$, $y = \cos 2t$. Table 8.7.1 gives the computed values and their error at several steps. The actual solution is periodic with period π and traces out an ellipse. The numerical solution, however, is not periodic. ∎

Exercises

For the systems in Exercises 1 through 6, estimate $x(1)$, $y(1)$, using both a Euler and a modified Euler method.

1. $x' = x + y + 1$, $\quad h = 0.2$,
   $y' = x + 3y + t^2$, $\quad x(0) = 0$, $\quad y(0) = 1$
2. $x' = xy$, $\quad h = 0.2$,
   $y' = -xy$, $\quad x(0) = 1$, $\quad y(0) = 1$
3. $x' = tx - ty$, $\quad h = 0.1$,
   $y' = x - y + t$, $\quad x(0) = 1$, $\quad y(0) = 0$
4. $x' = ty$, $\quad h = 0.1$,
   $y' = -tx$, $\quad x(0) = 1$, $\quad y(0) = 1$
5. $x' = -y^2$, $\quad h = 0.1$,
   $y' = x$, $\quad x(0) = y(0) = 1$
6. $x' = x - y - t$, $\quad h = 0.1$,
   $y' = y + 3x + t$, $\quad x(0) = y(0) = 0$

In Exercises 7 through 10, find the exact solution of the differential equation. Then compute a numerical estimate for $y(1)$, using a Euler and a modified Euler. Find the error in your estimate by comparing to the true solution.

7. $y'' + y = 0$, $\quad y(0) = 0$, $\quad y'(0) = 1$, $\quad h = 0.1$
8. $y'' - y = t$, $\quad y(0) = y'(0) = 1$, $\quad h = 0.1$
9. $y'' - 5y' + 6y = 0$, $\quad y(0) = 1$,
   $y'(0) = 2$, $\quad h = 0.1$
10. $y'' + y' = 1$, $\quad y(0) = y'(0) = 1$, $\quad h = 0.1$
11. Estimate $y(1)$ for Exercise 9, using a fourth-order Runge–Kutta, and compare to the true solution.

9

Qualitative Analysis of Nonlinear Equations in the Plane

9.1 Introduction

In Section 2.2 we introduced the qualitative analysis of a single differential equation. This chapter will discuss the qualitative analysis of systems of *autonomous differential equations* of the form

$$\frac{dx}{dt} = f(x, y),$$
$$\frac{dy}{dt} = g(x, y). \tag{1}$$

The system (1) is called autonomous (time-independent) since the equations (1) for the velocities

$$\frac{dx}{dt} \quad \text{and} \quad \frac{dy}{dt}$$

depend explicitly only on the values of x and y and not on the time t. In a qualitative analysis, one is interested in questions such as:

- Are there any equilibria (constant solutions)?
- Are these equilibria stable or unstable?

- Are solutions bounded or unbounded?
- Are there oscillations (periodic solutions)?

The answers to these questions can be important not only to verify numerical results, but also to explain physical behavior when (1) cannot actually be solved or when only forms of f and g are known. Such an analysis underlies many physical theories.

Before beginning this analysis in the next section, we shall give several examples of nonlinear systems, some of which will be examined more carefully later in this chapter.

As noted before, a single second-order differential equation,

$$\frac{d^2x}{dt^2} + f\left(x, \frac{dx}{dt}\right) = 0, \qquad x(0) = x_0, \qquad x'(0) = y_0, \tag{2}$$

may be rewritten as a first-order system.

$$\begin{aligned} \frac{dx}{dt} &= y, & x(0) &= x_0, \\ \frac{dy}{dt} &= -f(x, y), & y(0) &= y_0. \end{aligned} \tag{3}$$

Electrical Circuits

(Note Section 2.12.) In many devices, such as diodes, the voltage v and current i satisfy a nonlinear relationship. That is, the device has a nonlinear v–i characteristic. Assuming the device is current-controlled, that is, the voltage drop is a function of current, we get circuit models of the form (4)

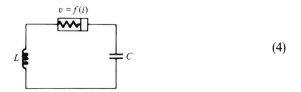

(4)

which, by Kirchhoff's voltage law, leads to the differential equation

$$L\frac{di}{dt} + f(i) + \frac{1}{C}q = 0, \tag{5}$$

or, equivalently, the system

$$\frac{di}{dt} = -\frac{1}{L}f(i) - \frac{1}{LC}q,$$

$$\frac{dq}{dt} = i,$$
(6)

where i is the current in the loop and q is the charge on the capacitor.
Additional electrical examples will be given later.

Mechanical Systems

(Note Section 2.13.) If we no longer assume small velocities and small displacements, we expect the resistance and spring force to vary in a nonlinear manner. ("Linear behavior is only for small variations" is a heuristic physical version of Taylor's theorem.) This would lead to a spring-mass system with differential equation

$$m\frac{d^2x}{dt^2} + f\left(\frac{dx}{dt}\right) + g(x) = 0,$$

or the equivalent system

$$\frac{dx}{dt} = y,$$

$$\frac{dy}{dt} = -\frac{1}{m}f(y) - \frac{1}{m}g(x).$$

Chemical Reactions

Chemical reactions involving two substances with concentrations x and y, which not only can combine but also disassociate, lead to differential equations of the form

$$\frac{dx}{dt} = ax + bxy + cy + d,$$

$$\frac{dy}{dt} = ex + fxy + gy + h.$$

where a, b, c, d, e, f, g, h are constants. (A special case is Eq. (11) of Section 2.13.6.)

A closely related idea is that of populations (which could be nonbiological) that are either competing or feeding on each other. Under the appropriate modeling assumptions, these often lead to models of the general form

$$\frac{dx}{dt} = ax + bxy + cx^2,$$

$$\frac{dy}{dt} = dy + exy + fy^2.$$
(7)

with a, b, c, d, e, f constants.

9.2 The Phase Plane

This section will present some basic results for systems of the form

$$x' = f(x, y),$$
$$y' = g(x, y), \tag{1}$$

and introduce the important concept of the *phase plane* of (1). The basic existence and uniqueness theorem for (1) is:

Theorem 9.2.1 Suppose that f, g, f_x, f_y, g_x, g_y are continuous in a region R (a connected open set) of the x,y-plane containing the point (x_0, y_0) in its interior. Let t_0 be a fixed value of t. Then there is a unique solution to (1) in the region R such that $x(t_0) = x_0, y(t_0) = y_0$.

A solution $(x(t), y(t))$ of (1) traces out a curve in the x,y-plane. This parameterized curve, along with an indication of the direction the solution moves along the curve, is called a *trajectory* or *orbit*. The term *solution curve* is also sometimes used. (The curve without a sense of direction is also referred to as an *integral curve*.) The set of trajectories in the x,y-plane, together with an indication of the solutions' directions along them, is the *phase plane*. Usually only a few representative solutions are drawn. This sketch is sometimes called a *phase portrait*.

The fact that (1) is autonomous has several implications for the phase portrait if Theorem 9.2.1 holds. First, observe that if (x, y) is a solution of (1) defined for $t \in (\alpha, \beta)$, then $\hat{x}(t) = x(t - c), \hat{y}(t) = y(t - c)$ is a solution for $t \in (\alpha + c, \beta + c)$ that gives rise to the same trajectory. Conversely, every solution giving rise to the same trajectory is of the form $x(t - c), y(t - c)$ for some constant c.

A curve $\phi(x, y) = C$ is an *invariant curve* for the autonomous system (1) if for any point (x_0, y_0) on the curve, and for any t_0, the solution to (1) satisfying $x(t_0) = x_0, y(t_0) = y_0$, stays on the curve. That is, solutions starting on the curve, stay on the curve. The points on a trajectory are an invariant curve. In general, an invariant curve is the union of the points making up one or more trajectories.

To find the invariant curves of (1), we can use the fact that if $x(t), y(t)$ is a solution of (1), then

$$\frac{dy}{dx} = \frac{dy/dt}{dx/dt} = \frac{g(x, y)}{f(x, y)} \tag{2}$$

or

$$f(x, y)\, dy - g(x, y)\, dx = 0.$$

Theorem 9.2.2 A curve $\phi(x, y) = C$ is an invariant curve for the autonomous system (1) if and only if $\phi(x, y) = C$ is a solution of $f(x, y)\, dy - g(x, y)\, dx = 0$.

Figure 9.2.1
Phase portrait for system (3).

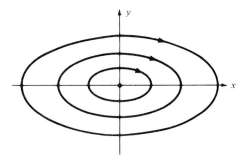

Example 9.2.1 Find the invariant curves and draw the phase portrait for the system

$$\frac{dx}{dt} = 4y,$$
$$\frac{dy}{dt} = -x. \quad (3)$$

Solution The invariant curves satisfy

$$\frac{dy}{dx} = \frac{dy/dt}{dx/dt} = \frac{-x}{4y}.$$

Solving this differential equation by separation of variables, we get the invariant curves are

$$x^2 + 4y^2 = C.$$

If $C < 0$, there is no curve. If $C = 0$, the curve is a point representing the constant solution $x = 0$, $y = 0$ of (3). If $C > 0$, the invariant curve is an ellipse. Suppose $C > 0$. Then

$$\mathbf{v} = \left[\frac{dx}{dt}, \frac{dy}{dt}\right] = [4y, -x]$$

is the *velocity vector* and is tangent to the curve. The length of the velocity vector is the *speed* $s(t)$. Since

$$s(t) = \sqrt{\left(\frac{dx}{dt}\right)^2 + \left(\frac{dy}{dt}\right)^2} = \sqrt{16y^2 + x^2} > \sqrt{4y^2 + x^2} = \sqrt{C},$$

we see that the solutions keep moving along the ellipse without ever slowing down below $\sqrt{C}$. Thus they repeatedly traverse the ellipse. (A more sophisticated technique to reach this same conclusion is given in Section 9.5.) The direction of movement can be determined by noting that, if $x > 0$, $y > 0$, then the velocity vector points down and to the right. Thus the ellipses are traversed clockwise. The phase portrait is then Fig. 9.2.1. ∎

While informative, some information is lost in Fig. 9.2.1 because of the supression of the t variable. For example, Fig. 9.2.1 does not indicate how fast solutions go around the ellipses.

In many applications, an invariant curve $\phi(x, y) = C$ is referred to as a *conservation law* since the quantity $\phi(x, y)$ remains constant. Conservation of energy is one such law.

In drawing these phase planes, it is helpful to realize that, if f, g, f_x, f_y, g_x, g_y are continuous for all x, y, as they are for most of our problems, then the only way a solution can fail to exist for all t is if (x, y) goes to infinity. Also, under these same circumstances, two distinct solution curves can never cross, since to do so would violate the basic existence and uniqueness Theorem 9.2.1. On the other hand, suppose that a trajectory given by the solution $(x(t), y(t))$ crosses itself at time t_1. Let t_0 be the previous time the solution was at this same point, so that $(x(t_0), y(t_0)) = (x(t_1), y(t_1))$. But then $(x(t), y(t))$ and $(x(t + t_0 - t_1), y(t + t_0 - t_1))$ are both solutions of (1) that satisfy the same initial condition at t_1. By Theorem 9.2.1, we have $x(t) = x(t + t_0 - t_1)$, $y(t) = y(t + t_0 - t_1)$ and the solution $(x(t), y(t))$ is periodic with period $t_0 - t_1$.

This means that in our problems there are three possibilities for a solution curve:

- A point (constant solution); (Fig 9.2.1, $x = 0$, $y = 0$)
- A simple closed curve (like an ellipse but not like a figure eight); (Figure 9.2.1)
- A curve that at each end either goes to infinity or does not include its endpoint. (Examples of this appear in the next section.) (Figure 9.2.2b)

If an equilibrium lies on an invariant curve, then we get several trajectories on the invariant curve as shown in the next example.

Example 9.2.2

Note that

$$x^4 - y^4 = 0 \tag{4}$$

is one of the invariant curves of

$$\begin{aligned}\frac{dx}{dt} &= y^3, \\ \frac{dy}{dt} &= x^3.\end{aligned} \tag{5}$$

The graph of (4) is shown in Fig. 9.2.2a. The origin $(0, 0)$ is a constant solution of (5). Each of the line segments that are left is a separate trajectory. Checking the direction on each we get Fig. 9.2.2b. For example, if $x > 0$, $y > 0$,

Figure 9.2.2

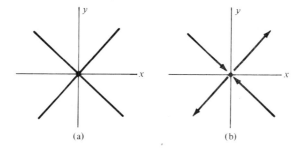

(a) (b)

then (5) says that $x' > 0$, $y' > 0$. Thus the trajectory in the first quadrant moves up ($y' > 0$) and to the right ($x' > 0$) away from the origin. Similarly, if $x > 0$, $y < 0$, then from (5), $x' < 0$, $y' > 0$ and the trajectory in the fourth quadrant moves up ($y' > 0$) and to the left ($x' < 0$). ∎

Note The precise definitions of solution curve and trajectory varies slightly in some texts. However, all treatments lead to the same phase portrait (Figure 9.2.1 for (4)).

Since $[f(x_0, y_0), g(x_0, y_0)]$ is the tangent vector to the solution at (x_0, y_0), we can also draw vector fields as in Section 2.2. Care must be taken, however. If, for example, the tangent vectors are normalized, we are really sketching tangents to invariant curves, rather than trajectories. The resulting sketch, while helpful, may not indicate an equilibrium. It is helpful to indicate the direction of each tangent vector on the sketch.

Exercises

For Exercises 1 through 11, (a) find and sketch the invariant curves, (b) determine the equilibrium point, and (c) find the direction traveled by the trajectories.

1. $\dfrac{dx}{dt} = x$, $\dfrac{dy}{dt} = -y$

2. $\dfrac{dx}{dt} = y$, $\dfrac{dy}{dt} = x$

3. $\dfrac{dx}{dt} = xy$, $\dfrac{dy}{dt} = x$

4. $\dfrac{dx}{dt} = xy$, $\dfrac{dy}{dt} = xy^3 + xy$

5. $\dfrac{dx}{dt} = 3$, $\dfrac{dy}{dt} = 1$

6. $\dfrac{dx}{dt} = y^3$, $\dfrac{dy}{dt} = -x^3$

7. $\dfrac{dx}{dt} = 2 - 2y$, $\dfrac{dy}{dt} = 2x - 2$

8. $\dfrac{dx}{dt} = y$, $\dfrac{dy}{dt} = xy + y$

9. $\dfrac{dx}{dt} = x^2$, $\dfrac{dy}{dt} = 3yx$

10. $\dfrac{dx}{dt} = -4x$, $\dfrac{dy}{dt} = x$

11. $\dfrac{dx}{dt} = -x^2$, $\dfrac{dy}{dt} = yx$

12. Rewrite the second-order differential equation
$$x'' + f(x) = 0$$
as a first-order system using $y = x'$ and show that the integral curves can always, in principle, be found by separation of variables.

For Exercises 13 through 20, use Exercise 12 to rewrite the second-order nonlinear equation as a first-order system, and then find its invariant curves.

13. $x'' + \sin x = 0$ **14.** $x'' + e^x = 0$

15. $x'' + x^2 = 0$ **16.** $x'' + \dfrac{1}{1+x^2} = 0$

17. $x'' + \dfrac{x}{1+x^2} = 0$ **18.** $x'' + x^4 + 1 = 0$

19. $x'' - x^2 = 0$ **20.** $x'' + x^3 = 0$

21. a) Show that $x^2 + y^2 = 1$ is an invariant curve of
$$\frac{dx}{dt} = -y^2,$$
$$\frac{dy}{dt} = xy$$
(6)

b) Find all equilibriums of (6) on $x^2 + y^2 = 1$
c) Sketch the trajectories of (6) on $x^2 + y^2 = 1$.

22. a) Show that $x^2 + y^2 = 1$ is an invariant curve of
$$\frac{dx}{dt} = -xy^2,$$
$$\frac{dy}{dt} = x^2 y$$
(7)

b) Find all equilibriums of (7) on $x^2 + y^2 = 1$.
c) Sketch the trajectories of (7) on $x^2 + y^2 = 1$.

23. a) Show that $x^2 + y^2 = 1$ is an invariant curve of
$$\frac{dx}{dt} = -y(y-1),$$
$$\frac{dy}{dt} = x(y-1)$$
(8)

b) Find all equilibriums of (8) on $x^2 + y^2 = 1$.
c) Sketch the trajectories of (8) on $x^2 + y^2 = 1$.

24. Use Theorem 9.2.1 to show that if a solution stays in the open region R, then the solution exists for all time.

25. Show that if $(x(t), y(t))$ is a solution of (1) and the speed is zero at any time t_0, then (x, y) is a constant solution.

26. Prove Theorem 9.2.2.

In Exercises 27 through 30, find equilibria for the given system. Then sketch the vector field indicating the direction of each vector. (You may wish to experiment with variable length also.)

27. $\dfrac{dx}{dt} = -y$, $\dfrac{dy}{dt} = x$

28. $\dfrac{dx}{dt} = y$, $\dfrac{dy}{dt} = x$

29. $\dfrac{dx}{dt} = y^2$, $\dfrac{dy}{dt} = x^2$

30. $\dfrac{dx}{dt} = 2 - 2y$, $\dfrac{dy}{dt} = 2x - 2$

9.3 Linear Systems

One of the keys to drawing phase portraits for nonlinear systems turns out to be understanding linear systems. In this section we shall derive almost all possible phase portraits for linear homogeneous systems in the form:

$$\frac{dx}{dt} = ax + by,$$
$$\frac{dy}{dt} = cx + dy,$$
(1)

where a, b, c, d are constants. Recall from Chapter 6 that

$$p(\lambda) = \det \begin{bmatrix} \lambda - a & -b \\ -c & \lambda - d \end{bmatrix} = \lambda^2 - (a+d)\lambda + (ad - bc)$$

is the *characteristic polynomial* of (1). The roots of the characteristic polynomial are denoted λ_1, λ_2 and are called *eigenvalues* of the matrix

$$\begin{bmatrix} a & b \\ c & d \end{bmatrix}.$$

For some questions, just knowing λ_1, λ_2 is sufficient. However, we shall point out how extra information, such as the eigenvectors (Section 6.10) of

$$\begin{bmatrix} a & b \\ c & d \end{bmatrix}$$

can be useful.

It turns out, for our later purposes, that if either or both eigenvalues are zero, our analysis cannot proceed. (Note Exercises 17 and 18.) Thus we assume

$$\lambda_1 \neq 0, \quad \lambda_2 \neq 0. \tag{2}$$

An *equilibrium* is a constant solution $x(t) \equiv C_1$, $y(t) \equiv C_2$. Substituting into (1), we find that C_1, C_2 must satisfy:

$$\begin{aligned} 0 &= aC_1 + bC_2, \\ 0 &= cC_1 + dC_2. \end{aligned} \tag{3}$$

By solving (3) we can show that (2) implies that $C_1 = C_2 = 0$ (see Exercise 19 at the end of this section, or Theorem 6.8.2). Thus the only equilibrium is the origin. Note that the curve $e^{\lambda t}\mathbf{u}$ for a scalar λ and vector $\mathbf{u}$ is a ray from the origin parallel to $\mathbf{u}$ in the (x, y)-plane.

Case 1

If $\lambda_1 > \lambda_2 > 0$, then the nonzero solutions involve $e^{\lambda_1 t}, e^{\lambda_2 t}$, and all solutions except the equilibrium tend to infinity as $t \to \infty$. The resulting phase portrait is Fig. 9.3.1. The origin is an *unstable* or *repelling* equilibrium, since all the other

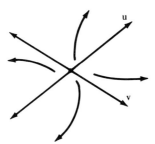

Figure 9.3.1

trajectories move away from it. The two straight trajectories are in the directions of the eigenvectors of

$$\begin{bmatrix} a & b \\ c & d \end{bmatrix}$$

Figure 9.3.2

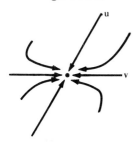

with **u** being the eigenvector for λ_1 and **v** the eigenvector for λ_2. (See Exercises 20 and 21 at the end of this section.) If $\lambda_1 = \lambda_2$ and $\lambda_1 > 0$, then all nonzero trajectories still go to infinity but they may bend differently.

Case 2

If $\lambda_1 < \lambda_2 < 0$, then the solutions involve $e^{\lambda_1 t}$, $e^{\lambda_2 t}$ which now go to zero as $t \to \infty$. Thus, all solutions approach the equilibrium and it is called an *attractor*, or an *asymptotically stable equilibrium*. The phase portrait is given in Fig. 9.3.2. The trajectories that are approaching the equilibrium solution (origin) get arbitrarily close to the origin but never reach it. Again **u** is in the direction of an eigenvector corresponding to λ_1 and **v** an eigenvector corresponding to λ_2. If $\lambda_1 = \lambda_2$ but both are negative, the curves may bend differently but they still approach the origin.

Figure 9.3.3

Case 3

If $\lambda_1 < 0 < \lambda_2$, the equilibrium is a *saddle point*. It has solutions with both decaying and increasing terms. Case 3 is illustrated in Fig. 9.3.3. Again **u**, **v** are directions of eigenvectors corresponding to λ_1 and λ_2, respectively.

Before considering what happens with complex eigenvalues, we shall give an example with real eigenvalues.

Example 9.3.1 Draw the phase portrait for the linear system

$$x' = -7x + 6y,$$
$$y' = 6x + 2y.$$

Solution This is $\mathbf{x}' = \mathbf{A}\mathbf{x}$ with

$$\mathbf{x} = \begin{bmatrix} x \\ y \end{bmatrix} \quad \text{and} \quad \mathbf{A} = \begin{bmatrix} -7 & 6 \\ 6 & 2 \end{bmatrix}.$$

The characteristic polynomial of **A** is

$$p(\lambda) = \det \begin{bmatrix} \lambda + 7 & -6 \\ -6 & \lambda - 2 \end{bmatrix} = \lambda^2 + 5\lambda - 50 = (\lambda - 5)(\lambda + 10).$$

Thus the eigenvalues are $\lambda_1 = -10$, $\lambda_2 = 5$, which have opposite sign so that we are in Case 3, and will have a saddle point at the origin. Using the techniques of Chapter 6, we compute eigenvectors of

$$\mathbf{u} = \begin{bmatrix} -2 \\ 1 \end{bmatrix} \quad \text{for } \lambda_1 = -10 \quad \text{and} \quad \mathbf{v} = \begin{bmatrix} 1 \\ 2 \end{bmatrix} \quad \text{for } \lambda_2 = 5.$$

Figure 9.3.4

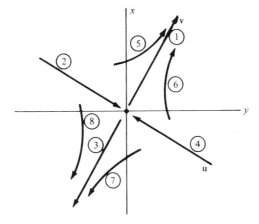

We know that (0, 0) is an equilibrium. The trajectories labeled 1 through 4 in Fig. 9.3.4 are the solutions

$$e^{5t}\mathbf{v}, \qquad e^{-10t}\mathbf{u}, \qquad -e^{5t}\mathbf{v}, \qquad -e^{-10t}\mathbf{u},$$

respectively. Any other solution is of the form

$$c_1 e^{5t}\mathbf{v} + c_2 e^{-10t}\mathbf{u}$$

for constants c_1, c_2. As t increases, we have

$$c_2 e^{-10t}\mathbf{u} \to \mathbf{0} \qquad \text{and} \qquad c_1 e^{5t}\mathbf{v} + c_2 e^{-10t}\mathbf{u} \approx c_1 e^{5t}\mathbf{v}.$$

This is illustrated by trajectories 5, 6, 7, and 8 in Fig. 9.3.4. ∎

Case 4

Figure 9.3.5

$\lambda_1 = \beta i$, $\lambda_2 = -\beta i$ with β a real nonzero number. The solutions involve $\sin \beta t$, $\cos \beta t$, and are "skewed ellipses." Motion is periodic and the axes are again eigenvectors. The equilibrium is *stable*, since nearby solutions do not move very far away. However, since the solutions do not approach the equilibrium, the equilibrium is not asymptotically stable. See Fig. 9.3.5.

Case 5

$\lambda_1 = \alpha + \beta i$, $\lambda_2 = \alpha - \beta i$, $\beta \neq 0$, $\alpha \neq 0$. The solutions of (1) now involve $e^{\alpha t} \cos \beta t$, $e^{\alpha t} \sin \beta t$. The result is a spiral centered at the origin. The solution moves out from the origin if $\alpha > 0$ (since $e^{\alpha t}$ increases if $\alpha > 0$) and inward if $\alpha < 0$. The solutions may also spiral either clockwise or counterclockwise

Figure 9.3.6

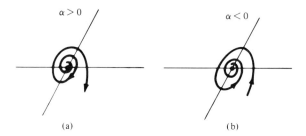

depending on the eigenvectors but that is not usually an important consideration. See Fig. 9.3.6.

In Fig. 9.3.6 one should visualize an infinite number of spirals both inside and outside of those drawn. If $\alpha > 0$, then the equilibrium is unstable, or a *repeller*. If $\alpha < 0$, the equilibrium is asymptotically stable, or an *attractor*.

Example 9.3.2   Determine the phase portrait for

$$x' = 2x + y,$$
$$y' = -x + 2y. \qquad (4)$$

Solution   The characteristic polynomial is

$$p(\lambda) = \det\begin{bmatrix} \lambda - 2 & -1 \\ 1 & \lambda - 2 \end{bmatrix} = \lambda^2 - 4\lambda + 5.$$

The roots are $\lambda_1 = 2 + i, \lambda_2 = 2 - i$. Since $\alpha = 2 > 0$ we have an outward spiral. Taking a nonzero point, say $x_0 = 1, y_0 = 0$, we get the tangent vector

$$[x', y'] = [2, -1],$$

which points down and to the right from $(1, 0)$. Thus the spiral is clockwise and we have precisely Fig. 9.3.6a. ∎

In concluding this section, note that the invariant curves for these linear homogeneous differential equations always satisfy

$$\frac{dy}{dx} = \frac{cx + dy}{ax + by},$$

which is a homogeneous equation, in the sense of Section 2.8, and may be solved by the methods of that section.

Exercises

In Exercises 1 through 16, determine the behavior of the solutions near the equilibrium $x = 0, y = 0$, and sketch the phase portrait.

1. $x' = x, \qquad y' = x + 2y$
2. $x' = 2x - y, \qquad y' = 3x - 2y$

3. $x' = -x - 5y, \quad y' = x + y$
4. $x' = 2x - y, \quad y' = 2x + 5y$
5. $x' = x - y, \quad y' = x + y$
6. $x' = -2x + 2y, \quad y' = -x$
7. $x' = -5x - 4y, \quad y' = 2x + y$
8. $x' = x + 5y, \quad y' = -2x - y$
9. $x' = y, \quad y' = 2x + y$
10. $x' = -x - 2y, \quad y' = 2x - y$
11. $x' = -5x - y, \quad y' = 3x - y$
12. $x' = x + 2y, \quad y' = -4x - 3y$
13. $x' = -x + 4y, \quad y' = -4x - y$
14. $x' = 3x + 2y, \quad y' = -2x + 3y$
15. $x' = 4x + 3y, \quad y' = 3x + 4y$
16. $x' = 2x + 3y, \quad y' = 3x + 2y$

In Exercises 17 through 19, λ_1, λ_2 are the roots of the characteristic equation for system (1).

17. Show that, if $\lambda_1 = \lambda_2 = 0$, then either all solutions of (3) are constants or constants and straight lines.

18. Show that, if $\lambda_1 = 0$ but $\lambda_2 \neq 0$, then there is a line of equilibrium points and
    a) all other solutions tend toward this line if $\lambda_2 < 0$;
    b) all other solutions go away from this line if $\lambda_2 > 0$.

19. Show that if $\lambda_1 \neq 0, \lambda_2 \neq 0$, then the only equilibrium of (1) is $x = 0, y = 0$.

20. Show that if λ is a nonzero scalar and $\mathbf{u} \neq \mathbf{0}$ is a two-vector, then the curve
$$\begin{bmatrix} x \\ y \end{bmatrix} = e^{\lambda t}\mathbf{u}$$
in the x, y-plane is a ray extending from (but not including) the origin.

21. (Continuation of Exercise 20.) Show that if $\lambda > 0$, then (x, y) moves away from the origin along the ray, while if $\lambda < 0$, then (x, y) moves toward the origin along the ray.

In Exercises 22 through 24 find the invariant curves. Exercises 23 and 24 use Section 2.8 on homogeneous equations.

22. $x' = -x - 5y$ (Exercise 3)
    $y' = x + y$
23. $x' = -x - 2y$ (Exercise 10)
    $y' = 2x - y$
24. $x' = x - y$ (Exercise 5)
    $y' = x + y$

9.4 Equilibria of Nonlinear Systems

In this section we return to studying the nonlinear autonomous systems
$$\begin{aligned} x' &= f(x, y), \\ y' &= g(x, y). \end{aligned} \tag{1}$$

If $x(t), y(t)$ is an *equilibrium*, or constant solution, then $x(t) = r, y(t) = s$ for constants r, s. Substituting into (1) we find that r, s must satisfy
$$\begin{aligned} 0 &= f(r, s), \\ 0 &= g(r, s). \end{aligned} \tag{2}$$

Conversely, if r, s are constants satisfying (2), then $x(t) = r, y(t) = s$ is a solution of (1). That is,

The equilibrium solutions $x = r$, $y = s$ of (1) are exactly the solutions (r, s) of the system of nonlinear algebraic equations (2).

Example 9.4.1 Find all equilibria of

$$x' = -x + xy,$$
$$y' = -y + 2xy. \quad (3)$$

Solution Let $x(t) = r$, $y(t) = s$, where r, s are constants. Then (3) becomes

$$0 = -r + rs = r(-1 + s), \quad (4a)$$
$$0 = -s + 2rs = s(-1 + 2r). \quad (4b)$$

Equation (4a) implies that $r = 0$ or $s = 1$. Then,

If $r = 0$, Eq. (4b) implies that $s = 0$, while
If $s = 1$, Eq. (4b) implies that $-1 + 2r = 0$, so that $r = \frac{1}{2}$.

Thus

$$x = 0, y = 0 \quad \text{and} \quad x = \tfrac{1}{2}, y = 1 \quad (5)$$

are the only two equilibria of (3). ∎

We now wish to determine the behavior of solutions of (1) near an equilibrium. Suppose that $x(t) = r$, $y(t) = s$ is an equilibrium. If we use the two-dimensional version of Taylor's approximations learned in calculus we have

$$f(x, y) = f(r, s) + f_x(r, s)(x - r) + f_y(r, s)(y - s) + \theta_1,$$
$$g(x, y) = g(r, s) + g_x(r, s)(x - r) + g_y(r, s)(y - s) + \theta_2, \quad (6)$$

where θ_1, θ_2 involve quadratic terms in $x - r$, $y - s$ and are thus "small" if x is close to r and y is close to s. But $f(r, s) = 0$, $g(r, s) = 0$ by (2). This suggests that, near the equilibrium, the solutions of (1) should resemble these of the *linearized differential equation*.

$$x' = f_x(r, s)(x - r) + f_y(r, s)(y - s),$$
$$y' = g_x(r, s)(x - r) + g_y(r, s)(y - s). \quad (7)$$

Letting $z = x - r$, $w = y - s$ (which translates the equilibrium point r, s to the origin $z = 0$, $w = 0$), (7) becomes

$$z' = az + bw,$$
$$w' = cz + dw, \quad (8a)$$

where a, b, c, and d are constants given by

$$a = f_x(r, s), \quad b = f_y(r, s), \quad c = g_x(r, s), \quad d = g_y(r, s). \quad (8b)$$

But this is a linear homogeneous system like that discussed in the last section. This motivates, but does not prove, the following.

Theorem 9.4.1

Suppose that (r, s) is an equilibrium point of (1). Define a, b, c, d by (8b). Let λ_1, λ_2 be the roots of the characteristic polynomial of (8a),

$$\lambda^2 - (a + d)\lambda + ad - bc.$$

Then

1. If $\lambda_1, \lambda_2 > 0$, the equilibrium is unstable and the local phase portrait resembles Fig. 9.3.1.
2. If $\lambda_1, \lambda_2 < 0$, the equilibrium is asymptotically stable and the local phase portrait resembles Fig. 9.3.2.
3. If λ_1, λ_2 are nonzero and of opposite signs, the equilibrium is unstable and the local phase portrait resembles Fig. 9.3.3 (is a saddle point).
4. If $\lambda_1 = \alpha + \beta i$, $\lambda_2 = \alpha - \beta i$ with $\beta \neq 0$, then $\alpha > 0$ means the equilibrium is unstable and a spiral repeller (Fig. 9.3.6a) while $\alpha < 0$ means that the equilibrium is asymptotically stable and a spiral attractor (Fig. 9.3.6b).
5. If $\lambda_i = 0$, this procedure gives no immediate information. If $\lambda_1 = \beta i$, β real, then trajectories *may* spiral or be "ellipse like."

Local means the picture is valid only near the equilibrium point. *Resembles* means the axis can be bent and the picture somewhat distorted.

Example 9.4.2

Determine the phase portrait of (3) near the equilibria (5) found in Example 9.4.1.

Solution

We have that

$$f_x = -1 + y, \qquad f_y = x, \qquad g_x = 2y, \qquad g_y = -1 + 2x. \tag{9}$$

Consider first the equilibrium $x = 0, y = 0$. By Theorem 9.4.1 the behavior of (3) near $(0, 0)$ is that of (8a) which is

$$\begin{aligned} z' &= (-1)z + 0w, \\ w' &= 0z + (-1)w. \end{aligned} \tag{10}$$

The characteristic polynomial is $(\lambda + 1)^2 = \lambda^2 + 2\lambda + 1$, with roots $-1, -1$. Thus $(0, 0)$ is an attractor. Near $(0, 0)$ the phase portrait of (3) will resemble that of (10), which is shown in Fig. 9.4.1a.

Near the other equilibrium point $x = \frac{1}{2}, y = 1$, the behavior of (3) is that of (8a) with $x = \frac{1}{2}, y = 1$, or

$$\begin{aligned} z' &= (-1 + 1)z + \tfrac{1}{2}w = \tfrac{1}{2}w, \\ w' &= (2 \cdot 1)z + (-1 + 2 \cdot \tfrac{1}{2})w = 2z. \end{aligned}$$

Figure 9.4.1

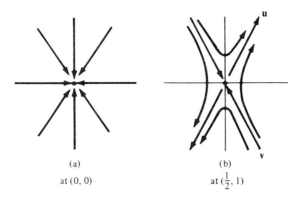

(a) at $(0, 0)$ (b) at $(\frac{1}{2}, 1)$

Figure 9.4.2
Phase portrait for (3).

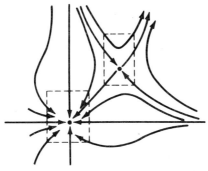

The characteristic polynomial is $\lambda^2 - 1$ with roots ± 1 so that we have a saddle point. The eigenvectors for $\lambda_1 = 1, \lambda_2 = -1$ of

$$\begin{bmatrix} 0 & \frac{1}{2} \\ 2 & 0 \end{bmatrix}$$

are

$$\mathbf{u} = \begin{bmatrix} 1 \\ 2 \end{bmatrix} \quad \text{and} \quad \mathbf{v} = \begin{bmatrix} -1 \\ 2 \end{bmatrix},$$

leading to the local picture in Fig. 9.4.1b. ∎

The actual full phase portrait for (3) is given in Fig. 9.4.2.

It is important to keep in mind that Theorem 9.4.1 describes only the behavior near the equilibria as indicated in Fig. 9.4.2 by the dashed rectangles. The extra detail in Fig. 9.4.2, such as the trajectories first approaching $(\frac{1}{2}, 1)$ and

Figure 9.4.3

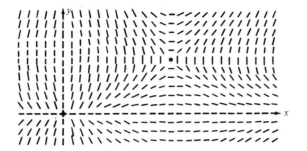

then veering off toward $(0, 0)$, come from a more sophisticated analysis and the use of more advanced theorems. Some of these theorems and facts will be presented in the next section. Alternatively, the techniques of Section 2.2 can be used to help sketch the phase portrait of (3). For example, the invariant curves of (3) satisfy

$$\frac{dy}{dx} = \frac{-y + 2xy}{-x + xy}. \tag{11}$$

The slope portrait for (11) is shown in Fig. 9.4.3. Figure 9.4.3 should be compared to Fig. 9.4.2.

Exercises

In Exercises 1 through 12, determine all equilibria and the behavior of solutions near them.

1. $x' = x + xy, \quad y' = y - 2xy$
2. $x' = -x + xy, \quad y' = -2y + 4xy$
3. $x' = x - xy, \quad y' = y - xy$
4. $x' = 1 - x^2, \quad y' = y + 1$
5. $x' = x^2 - y^2, \quad y' = x - xy$
6. $x' = y^3 + 1, \quad y' = x^2 + y$
7. $x' = 1 - y^2, \quad y' = 1 - x^2$
8. $x' = x(1 - y^2), \quad y' = x + y$
9. $x' = x - y + x^2, \quad y' = x + y$
10. $x' = 2x - y - xy, \quad y' = x + 2y$
11. $x' = -x - 2y, \quad y' = 2x - y + xy^2$
12. $x' = -2x + y, \quad y' = -x - 2y + y^3$

Exercises 13 through 17 refer to

$$x' = x - xy + \gamma x^2,$$
$$y' = -y + xy.$$

In each case find all equilibrium points and determine the behavior near each equilibrium using Theorem 9.4.1.

13. $\gamma = -8$
14. $\gamma = -\frac{1}{3}$
15. $\gamma = \frac{1}{3}$
16. $\gamma = 1$
17. $\gamma = 8$

Exercises 18 through 22 show that, if the roots of the characteristic polynomial are pure imaginary, then a nonlinear system is different from a linear system in that there need not be a periodic orbit which is not an equilibrium. First consider the system

$$x' = -x^3 - y,$$
$$y' = x - y^3. \tag{12}$$

18. Show that $x = 0, y = 0$ is the only equilibrium of

system (12), and that the characteristic polynomial at this equilibrium is $\lambda^2 + 1$ with roots $\pm i$.

19. Let $x(t)$, $y(t)$ be a nonzero trajectory of (12). Define
$$d(t) = (x(t)^2 + y(t)^2)^{1/2},$$
which is the distance from $(x(t), y(t))$ to $(0, 0)$. Differentiate $d(t)$ with respect to t and use (12) for x', y', to conclude that $d'(t) < 0$ for all t. Thus the trajectory cannot be periodic since it always moves toward the origin.

Now consider
$$\begin{aligned} x' &= x^3 - y, \\ y' &= x + y^3. \end{aligned} \qquad (13)$$

20. Show that $x = 0$, $y = 0$ is the only equilibrium of (13) and that the characteristic polynomial at this equilibrium is $\lambda^2 + 1$ with roots $\pm i$.

21. Let $(x(t), y(t))$ be a nonzero trajectory of (13). Define
$$d(t) = (x(t)^2 + y(t)^2)^{1/2},$$
which is the distance from $(x(t), y(t))$ to the origin at time t. Differentiate $d(t)$ with respect to t and use (13) for x', y' to conclude that $d'(t) > 0$ for all t. Thus the trajectory cannot be periodic since it always moves away from the origin.

22. Let d be as in Exercise 21. Show that
$$\lim_{t \to \infty} d(t) = +\infty.$$

The function $d(t)$ in Exercises 19 through 22 is an example of a *Liapunov function*. Liapunov functions are very important in studying the stability of nonlinear systems, but we will not go into the topic in much greater detail since the construction of Liapunov (also sometimes called Lyapunov) functions takes some experience. (See Section 9.5.)

Let $x' = y$ and rewrite the second-order equation $x'' = g(x, x')$ as the system
$$\begin{aligned} x' &= y, \\ y' &= g(x, y). \end{aligned} \qquad (14)$$

23. **a)** Show that the equilibria of $x'' = g(x, x')$ are those r such that $g(r, 0) = 0$ and that in this case $(r, 0)$ is an equilibrium also of (14). Conversely, show that if (r, s) is an equilibrium of (13), then $s = 0$ and r is an equilibrium of $x'' = g(x, x')$.

b) Let $(r, 0)$ be an equilibrium of (14). Show that the characteristic polynomial at this equilibrium point is
$$\lambda^2 - g_y(r, 0)\lambda - g_x(r, 0).$$

24. Using Exercise 23, show that if $x'' = g(x)$, then Theorem 9.4.1 either guarantees the equilibrium $(r, 0)$ is a saddle point of (14) or fails to determine the behavior of trajectories near the equilibrium.

If a point (r, s) is not an equilibrium of (1), one may still use (6) to get a linear differential equation that approximates (1) as long as trajectories stay near (r, s). For each of the following systems, determine the linear system that approximates it near the given point.

25. $x' = x^2 + xy^4$, $\quad r = 1, \quad s = 1$
    $y' = x - y^3$

26. $x' = x^2 + xy^4$, $\quad r = 0, \quad s = 0$
    $y' = x - y^3$

27. $x' = \sin(x + y)$, $\quad r = 0, \quad s = 0$
    $y' = \cos\left(x + y + \dfrac{\pi}{2}\right)$

28. $x' = -xy$, $\quad r = 1, \quad s = -1$
    $y' = x^2 + y^2$

9.5 Periodic Solutions

Oscillations are of interest in many applications. In the case of linear autonomous systems (Section 9.3), the existence of pure imaginary roots guaranteed the existence of periodic trajectories. Unfortunately, as noted in Theorem 9.4.1,

9.5 Periodic Solutions

and Exercises 9.4.18 through 9.4.21, this technique breaks down for nonlinear systems. Several theorems have been developed to overcome this difficulty. We shall discuss one of these.

Assumptions

Suppose that $x(t)$, $y(t)$ is a solution of

$$\begin{aligned} x' &= f(x, y), \\ y' &= g(x, y), \end{aligned} \tag{1}$$

and that $(x(t), y(t))$ is always in the interior of a *bounded* region R of the x, y-phase plane. Assume also that f, g, f_x, f_y, g_x, g_y are continuous on an open set containing R.

With these assumptions what can the trajectory $x(t)$, $y(t)$ do? It will exist for all time. Intuitively, there are two possibilities. It will approach an equilibrium or wander about the region R. It turns out that in the second case it must eventually become almost periodic. The formal statement is:

Theorem 9.5.1 (**Poincaré–Bendixson**) Under the assumptions just described there are three possibilities for a trajectory $(x(t), y(t))$:

i. There exists an equilibrium point toward which the trajectory approaches arbitrarily close ($(0, 0)$ in Fig. 9.4.2 or Fig. 9.3.6b)
ii. The trajectory is periodic (Fig. 9.5.1)
iii. The trajectory approaches a periodic orbit asymptotically (Fig. 9.5.2).

There are two ways to use this theorem. The simplest is the following.

Theorem 9.5.2 Suppose that $F(x, y) = C$ gives an invariant curve for (1) that does not contain any equilibrium points. Then, for each connected piece of $F(x, y) = C$, there is a solution of (1) with that trajectory. In particular, if a connected piece of $F(x, y) = C$ is a closed curve, then that trajectory will be periodic.

By *connected*, we mean that the set can be continuously parameterized by a variable t on a connected interval. This is sometimes referred to as being *arcwise connected*.

Example 9.5.1 Consider the nonlinear system

$$\begin{aligned} x' &= y^3, \\ y' &= -x^3. \end{aligned} \tag{2}$$

The only equilibrium of (2) is $x = 0$, $y = 0$ and Theorem 9.4.1 gives no information about the behavior near these points since the roots of the characteristic polynomial λ^2 are 0, 0.

The invariant curves satisfy

$$\frac{dy}{dx} = \frac{dy/dt}{dx/dt} = \frac{-x^3}{y^3}, \tag{3}$$

which can be solved by separation of variables as

$$\frac{y^4}{4} + \frac{x^4}{4} = C. \tag{4}$$

For any $C > 0$, the equation (4) defines a closed curve about the origin. (Note that $|x| \leq 2\sqrt[4]{C}$, $|y| \leq 2\sqrt[4]{C}$.) The only equilibrium $(0, 0)$ is not on the curve. Thus by Theorem 9.5.2 we have that (4) with $C > 0$ defines a periodic trajectory. Figure 9.5.1 gives the phase portrait. ∎

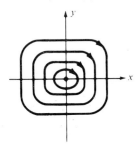

Figure 9.5.1 Phase portrait for system (2).

9.5.1 Liapunov Functions (Optional)

It is not always possible to find invariant curves as in Example 9.5.1. In that case, one can use a different approach in order to apply Theorem 9.5.1.

A set S in the phase plane is an *invariant set* for the autonomous system (1) if for any solution of (1) such that $x(t_0) = x_0$, $y(t_0) = y_0$, and $(x_0, y_0) \in S$, then $(x(t), y(t)) \in S$ for all $t \geq t_0$. That is, solutions that start in S, or enter S, stay in S. An invariant curve is a special example of an invariant set.

Theorem 9.5.3 Suppose that the assumptions of Theorem 9.5.1 hold and that (1) has a *closed* invariant set $S \subseteq R$. If S does not contain any equilibrium points, then (1) has a periodic orbit in S.

To illustrate, consider the system

$$x' = y + x^3 \cos(x^2 + y^2), \tag{5a}$$
$$y' = -x + y^3 \cos(x^2 + y^2). \tag{5b}$$

In order to concentrate on the application of Theorem 9.5.1, we shall leave some of the details to the exercises.

First, note that $x = 0$, $y = 0$ is the only equilibrium (see Exercise 16 at the end of this section) and that the roots there are $\pm i$.

Let $d(t) = x^2 + y^2$ be the bounded square of the distance of (x, y) from the origin at time t. Then

$$d' = 2xx' + 2yy' = 2(x^4 + y^4)\cos(d) \tag{6}$$

from (5).

Let R be the bounded region in the x,y-plane;

$$R = \{(x, y): d = x^2 + y^2 < \pi\}. \tag{7}$$

Figure 9.5.2

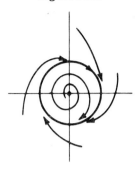

Figure 9.5.3

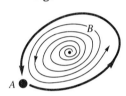

This is a disk around the origin of radius $\sqrt{\pi}$. If (x, y) is close to the origin, say $d < \pi/2$, then (6) shows that $d' > 0$, and the trajectory moves farther away from the equilibrium point. On the other hand, if $\pi/2 < d < \pi$, then $d' < 0$ so that (x, y) moves closer to the origin.

Let $S = \left\{(x, y) \,\Big|\, \dfrac{\pi}{2} \le d = x^2 + y^2 \le \dfrac{3}{4}\pi\right\}$. Then S is invariant and closed.

Thus S contains a periodic orbit for (5) by Theorem 9.5.3. Alternatively, we can argue directly that the trajectory can neither leave R nor approach the equilibrium. Thus by Theorem 9.5.1, there is a periodic orbit. Pictorially we have Fig. 9.5.2.

The function $d(t)$ in this example is a special case of a *Liapunov function*. In general, if $d(t)$ is a function of x, y, and $d(t) = c$ is a level curve of this function, and $d'(t) < 0$ on this level curve, then the trajectories can never cross $d(t) = c$ in the direction that would increase $d(t)$. Those level curve can then often be used to either define R or an invariant set S.

Comment Possibility i of Theorem 9.5.1 does not say the trajectory has the equilibrium point as its limit. The trajectory may just come closer and closer to the equilibrium point (trajectory B and equilibrium A in Fig. 9.5.3).

Exercises

In Exercises 1 through 13, find all equilibria and determine their local behavior. Then use Theorem 9.5.2 to show the existence of periodic trajectories. Note that some of these are linear and could be done by Section 9.3, but these exercises provide practice in using Theorem 9.5.2.

1. $x' = 8y, \quad y' = -2x$
2. $x' = -2y, \quad y' = 8x$
3. $x' = 4y - 8, \quad y' = -2x + 2$
4. $x' = -2y + 6, \quad y' = 2x - 4$
5. $x' = 4y^3 - 12y^2 + 12y - 4,$
   $y' = -4x^3 + 12x^2 - 12x + 4$
6. $x' = -6y^5, \quad y' = 6x^5$
7. $x' = -4x + 10y, \quad y' = -16x + 4y$
8. $x' = 4y^3, \quad y' = -2x$
9. $x' = -4y, \quad y' = 4x^3$
10. $x' = -8y^3, \quad y' = 2x + 4x^3$
11. $x' = 8y + 32y^3, \quad y' = -5x^3$
12. $x' = 7y + 13y^3, \quad y' = -9x$
13. $x' = 6y, \quad y' = -3x - 5x^3$

To illustrate it is important to know not only the integral curves but also the equilibria, consider

$$x' = -xy,$$
$$y' = x^2. \quad (8)$$

14. Show that $(0, y_0)$ is an equilibrium of (8) for any $y_0 \ne 0$.

15. Show that (8) has integral curves that are circles but that on every circle there are two equilibrium points. Conclude that there are no nonconstant periodic trajectories. Sketch the phase portrait of (8).

16. Verify that $(0, 0)$ is the only equilibrium of (5).

17. Verify that $x = a \sin t$, $y = a \cos t$ is a solution of (5) for any a such that $\cos a^2 = 0$. With this added information, sketch the phase portrait for $|x|, |y| \le 3\sqrt{\pi}$.

Autonomous systems in other coordinate systems are also important. The remaining exercises concern systems

$$\frac{dr}{dt} = f(r, \theta), \qquad (8)$$
$$\frac{d\theta}{dt} = g(r, \theta)$$

with the trajectory $(r(t), \theta(t))$ being in polar coordinates. The existence and uniqueness theorems still hold if f, g satisfy their assumptions. However, it is now possible for trajectories to cross, since different values of (r, θ) can be coordinates of the same point.

18. Find the invariant curves for
$$r' = -\sin \theta,$$
$$\theta' = 1 \qquad (9)$$
Sketch the curves through $(1.5, 0), (2.3, 0), (3.1, 0)$. (Graph with solutions in the solutions section at the end of this book).

19. Show that if $r(0) > 2$, then the trajectory for (9) starting at $(r(0), 0)$ is periodic (in polar coordinates).

20. Show that if $0 < r(0) < 2$, then the trajectory for (9) starting at $(r(0), 0)$ crosses itself (in polar coordinates).

9.6 Population Models

One of the many interesting applications of systems has been in developing population models for interacting species. Actually the species could be chemical, economic, or other nonbiological quantities. However, we shall consider two species and speak of them as *biological species*.

Assume that there are two species whose populations at time t are given by $x(t), y(t)$. In order to consider x, y as continuous functions, we must assume that the populations are fairly large. It is also often more appropriate to measure x, y in *mass units* rather than to count individuals. There are two reasons for this. First, the *biomass* of a species sometimes varies more continuously. Secondly, what is important for a predator is not how *many* things it eats but how *much* it eats. Nonetheless, we shall talk of the numbers of a given species. We also assume that:

> At time t, the rates of changes of the populations of x and y depend only on the number of x and y at time t and on no other factors.

Thus we get the first-order system;

$$\frac{dx}{dt} = f(x, y), \qquad (1a)$$

$$\frac{dy}{dt} = g(x, y). \qquad (1b)$$

We refer to f and g as the *growth rates* of x and y. It should be stressed that merely writing down (1) has ruled out any effects due to time (seasonal variations), delay effects, and external factors such as a fluctuating food supply. However, models like (1) do provide some insight into the dynamics of populations. Two special cases will illustrate how a model might be developed.

Case 1: Two Prey Species

In this example, consider two populations that are competing for the same limited food and shelter. In this situation, it is often reasonable to make the following assumptions (we assume $x \geq 0$, $y \geq 0$):

Assumption 1 If one species is absent, there can be no change in that population. Thus,

$$f(0, y) = 0, \qquad g(x, 0) = 0 \qquad \text{for all } x, y. \tag{2}$$

Assumption 2 Increasing the numbers of either species decreases the growth rate of the other species. Since f_y is the rate of change in the growth rate of x with respect to a change in y (and similarly for g_x), this says

$$f_y(x, y) < 0, \qquad g_x(x, y) < 0 \qquad \text{if } x > 0, y > 0. \tag{3}$$

Assumption 3 If the other species is absent, and the population is small, the first species will grow. Thus

$$f(x, 0) > 0, \qquad g(0, y) > 0 \qquad \text{for small } x, y. \tag{4}$$

Assumption 4 If the population of either species is too large, the growth rate of that species is negative. That is, there is a maximum number of each species that can be supported by the available food and shelter. Thus,

$$\begin{aligned} f(x, y) < 0 & \quad \text{for large } x, \\ g(x, y) < 0 & \quad \text{for large } y. \end{aligned} \tag{5}$$

Ideally, we should now consider (1) using only these assumptions and not relying on any special choice of f, g. However, to simplify the analysis, we shall look for functions f, g that are quadratics in x, y and satisfy Eqs. (2) through (5). Intuitively, we are taking second-order approximations to $f(x, y)$, $g(x, y)$. Thus suppose

$$\begin{aligned} f(x, y) &= a_1 + a_2 x + a_3 y + a_4 x^2 + a_5 xy + a_6 y^2, \\ g(x, y) &= b_1 + b_2 x + b_3 y + b_4 x^2 + b_5 xy + b_6 y^2. \end{aligned}$$

Assumption (2) implies that

$$a_1 + a_3 y + a_6 y^2 = 0, \qquad b_1 + b_2 x + b_4 x^2 = 0 \qquad \text{for all } y, x.$$

Thus $a_1 = a_3 = a_6 = b_1 = b_2 = b_4 = 0$. Now

$$f_y = a_5 x, \qquad g_x = b_5 y.$$

Thus Assumption (3) implies that $a_5 < 0$, $b_5 < 0$. Similarly (4) implies that $a_2 > 0$, $b_3 > 0$, while (5) implies $a_4 < 0$, $b_6 < 0$. This leads to the nonlinear system:

$$x' = ax - bxy - cx^2, \quad (6a)$$
$$y' = qy - rxy - sy^2, \quad (6b)$$

where a, b, c, q, r, s are positive constants. The equilibria of (6) are given by

$$x = 0 \quad \text{or} \quad a - by - cx = 0$$

and

$$y = 0 \quad \text{or} \quad q - rx - sy = 0,$$

so that the equilibria are

$$(0, 0), \quad \left(0, \frac{q}{s}\right), \quad \left(\frac{a}{c}, 0\right), \quad (\alpha, \beta),$$

where α, β is the solution of

$$cx + by = a,$$
$$rx + sy = q.$$

(We assume that $cs - br \neq 0$, so that the point (α, β) exists and is unique.) By Cramer's rule (or some algebra),

$$\alpha = \frac{as - bq}{cs - br}, \quad \beta = \frac{cq - ar}{cs - br}. \quad (7)$$

Note that if $x(t)$ satisfies

$$x' = ax - cx^2, \quad (8)$$

then $(x, 0)$ is a solution of (6), and if $y(t)$ satisfies

$$y' = qy - sy^2, \quad (9)$$

then $(0, y)$ is a solution of (6). Thus the x,y-axis contains trajectories. The equations (8) and (9) are sometimes called *logistics equations*. Since the x- and y-axes contain trajectories, trajectories beginning with $x(0) > 0$ and $y(0) > 0$ must stay in the first quadrant. The only portion of the phase plane of physical interest is $x \geq 0$, $y \geq 0$. Depending on the values of the constants, the equilibrium (α, β) may not be of physical interest.

Example 9.6.1 Suppose that $b = q = r = s = 1$, and $c = 3$, $a = 2$. Then (6) is

$$\begin{aligned} x' &= 2x - xy - 3x^2, \\ y' &= y - xy - y^2. \end{aligned} \quad (10)$$

The equilibria are

$$(0, 0), \quad (0, 1), \quad (\tfrac{2}{3}, 0), \quad \text{and} \quad (\tfrac{1}{2}, \tfrac{1}{2}).$$

From Section 9.4 the characteristic polynomial is

$$p(\lambda) = (\lambda - f_x)(\lambda - g_y) - f_y g_x$$
$$= \lambda^2 - (2 - y - 6x + 1 - x - 2y)\lambda$$
$$+ [(2 - y - 6x)(1 - x - 2y) - xy].$$

We determine the behavior near each equilibrium using Theorem 9.4.1.

1. $(0, 0)$ gives $p(\lambda) = \lambda^2 - 3\lambda + 2 = (\lambda - 1)(\lambda - 2)$, so that
$$\lambda_1 = 1 > 0, \qquad \lambda_2 = 2 > 0,$$
and the equilibrium is a repellor.

2. $(0, 1)$ gives $p(\lambda) = \lambda^2 - 1$, so that $\lambda = \pm 1$ and the equilibrium is a saddle.

3. $(\frac{2}{3}, 0)$ gives $p(\lambda) = (\lambda + 2)(\lambda - \frac{1}{3})$ with roots $-2, \frac{1}{3}$, and the equilibrium is a saddle.

4. $(\frac{1}{2}, \frac{1}{2})$ gives $p(\lambda) = \lambda^2 + 2\lambda + \frac{1}{2}$, so that $\lambda = -1 \pm 1/\sqrt{2}$ and the equilibrium is an attractor.

By checking the directions of the different trajectories, we get the phase portrait in Fig. 9.6.1. (Again, techniques besides Theorem 9.4.1 are needed to determine trajectories away from the equilibria.)

This model suggests that, for these species, the populations should approach a nonzero equilibrium. This model runs counter to the *niche* theory, which says that, given two essentially similar species, one will eventually replace the other. The reason for this difference is that the terms $-cx^2$, $-sy^2$ act as limitations on the population sizes and prevent either species from running the other one off. What happens if $c = s = 0$ is covered in the exercises. ∎

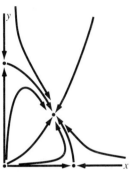

Figure 9.6.1
Phase portrait for (10).

Case 2: Predator–Prey Models

Now suppose that we have a *prey* species x that is eaten by a *predator* species y. We again assume (1), (2), and $x \geq 0$, $y \geq 0$ so that

$$f(0, y) = 0, \qquad g(x, 0) = 0 \qquad \text{for all } x, y.$$

However, we now assume the following:

Assumption 5 Increasing the prey population increases the growth rate of the predator (more food yields faster growth):

$$g_x(x, y) > 0 \qquad \text{for all } x, y. \tag{11}$$

Assumption 6 Increasing the predator population decreases the growth rate of the prey:

$$f_y(x, y) < 0 \qquad \text{for all } x, y. \tag{12}$$

Assumption 7 If either population is large enough, increasing it further decreases its growth rate (there's a finite amount of food for prey and a finite amount of territory/shelter for predators):

$$f_x(x, y) < 0, \qquad g_y(x, y) < 0 \qquad \text{for large } x, y \text{ respectively.} \tag{13}$$

Assumption 8 If there are no predators and a few prey, the prey increases:

$$f(x, 0) > 0 \qquad \text{for small } x. \tag{14}$$

Assumption 9 If there are no prey, the predator decreases:

$$g(0, y) < 0 \qquad \text{for all } y. \tag{15}$$

If we now seek a quadratic equation for f, g satisfying these assumptions, we get, after some calculation (see Exercise 17 at the end of this section), the model

$$\begin{aligned} x' &= ax - bxy - cx^2, \\ y' &= -qy + rxy - sy^2, \end{aligned} \tag{16}$$

with a, b, c, q, r, s positive constants. Again the equilibria are given by

$$x = 0 \qquad \text{or} \qquad a - by - cx = 0,$$

and

$$y = 0 \qquad \text{or} \qquad -q + rx - sy = 0,$$

which yields

$$(0, 0), \qquad \left(0, -\frac{q}{s}\right), \qquad \left(\frac{a}{c}, 0\right), \qquad \text{and} \qquad (\alpha, \beta),$$

where α, β satisfy

$$\begin{aligned} c\alpha + b\beta &= a, \\ r\alpha - s\beta &= q, \end{aligned}$$

so that

$$\alpha = \frac{as + bq}{cs + br}, \qquad \beta = \frac{ra - cq}{cs + br},$$

assuming that $cs + br \neq 0$.

Example 9.6.2 Suppose that $b = q = r = s = 1, c = 2, a = 3$, so that (16) is

$$\begin{aligned} x' &= 3x - xy - 2x^2, \\ y' &= -y + xy - y^2. \end{aligned}$$

The equilibria are

$$(0, 0), \quad (0, -1), \quad (\tfrac{3}{2}, 0), \quad (\tfrac{4}{3}, \tfrac{1}{3}).$$

The $(0, -1)$ equilibrium is nonphysical because of the negative y value. The characteristic polynomial for the linearized differential equation at each point (x, y) is

$$\lambda^2 - [(3 - y - 4x) + (-1 + x - 2y)]\lambda \\ + (3 - y - 4x)(-1 + x - 2y) - (-x)(y).$$

Thus, by Theorem 9.4.1, at each equilibrium we have:

1. $(0, 0)$ gives $p(\lambda) = \lambda^2 - 2\lambda - 3 = (\lambda - 3)(\lambda + 1)$ and $\lambda_1 = -1, \lambda_2 = 3$, so that $(0, 0)$ is a saddle.
2. $(\tfrac{3}{2}, 0)$ gives $p(\lambda) = \lambda^2 + \tfrac{5}{2}\lambda - \tfrac{3}{2}$, so that $\lambda_1 = \tfrac{1}{2}, \lambda_2 = -3$, which also gives a saddle.
3. $(\tfrac{4}{3}, \tfrac{1}{3})$ gives $p(\lambda) = \lambda^2 + 3\lambda + \tfrac{4}{3}$ so that $\lambda_1 \simeq -0.54, \lambda_2 \simeq -2.45$, and the equilibrium is an attractor.

A complete phase portrait is Fig. 9.6.2. ∎

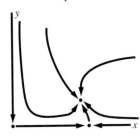

Figure 9.6.2
Phase portrait of Example 9.6.2.

As will be shown in the exercises, if all resource limitations are removed from (16) (take $c = 0$, $s = 0$), then periodic oscillations exist.

In the predator–prey model (16) there is a tendency by some to refer to the xy-term as the *predation term*. However, it is really not exactly a predation term. Suppose that $-bxy$ was the loss rate of (say) rabbits and $+sxy$ was a gain (in weight) for foxes. This would imply that, given a *fixed* number (or weight) of foxes and any number of rabbits, that doubling the number of rabbits would double the amount of predation. This is clearly nonsense. A given number of foxes can eat only so many rabbits per day. Thus, as x increases, the loss of rabbits due to just predation would approach a multiple of y, the number of foxes. Typical functions used in the biological literature are

$$-b\frac{x}{1+x}y, \quad -b\frac{x^2}{1+x^2}y.$$

These functions are called *uptake functions*. The effect of trying to actually model predation is examined in Exercises 12 through 16 at the end of this section.

The key thing to remember is that one is often better off thinking of $f(x, y)$, $g(x, y)$ as *representative* functions satisfying the right sort of change conditions [such as Assumptions 1 through 4 under Case 1, Eqs. (2) through (5)] rather than trying to impart special biological significance directly to each term.

Exercises

In Exercises 1–17, the x- and y-axes contain trajectories and you are only to consider the trajectories for $x \geq 0$, $y \geq 0$.

Two-Prey Models

In Exercises 1 through 3, determine the behavior near all equilibria.

1. In Example 9.6.1, decrease the limitation term on x by making $c = 2$, so that
$$a = c = 2, \quad q = r = s = b = 1.$$

2. In Example 9.6.1, increase the effect of competition on x by changing b to 2. Thus
$$a = b = 2, \quad c = 3, \quad q = r = s = 1.$$
This makes the competition between x and y hurt x more than y.

3. In Example 9.6.1, reduce the inherent growth rate of x by changing a to 1, increase the effect of competition on x by changing b to 2, and increase the resource limitation on x by changing c to 4, so that $b = 2$, $c = 4$, and $a = q = r = s = 1$.

4. (Continuation of Exercise 3.) Assuming that, in fact, $(0, 1)$ is an attractor for the entire open first quadrant $(x > 0, y > 0)$, sketch the phase portrait. Explain in biological terms what this represents.

5. In the system (6), if there is ample food and shelter, then one might set $c = 0$, $s = 0$. Show that there are only two equilibria, $(0, 0)$ and $(q/r, a/b)$, which are a repellor and a saddle, respectively.

*6. (Continuation of Exercise 5.) Sketch the phase portrait for Exercise 5, and explain why this is compatible with the niche theory. Considering Exercises 1, 4, 5, and Example 9.6.1, what do you think are the limitations and possible truth in the niche theory?

Predator-Prey Problems

In Exercises 7 through 10, determine the behavior near all physically meaningful equilibria.

7. In Example 9.6.2, effectively decrease the available prey by increasing c to 6, so that $q = r = s = b = 1$, $a = 3$, $c = 6$.

8. In Example 9.6.2, remove the limitation on prey population by setting $c = 0$, so that $q = r = s = b = 1$, $a = 3$, $c = 0$.

9. In Example 9.6.2, remove the limitation on prey and predator populations by setting $c = s = 0$, so that $q = r = b = 1$, $a = 3$. Show that periodic trajectories exist.

10. In Example 9.6.2, increase the effect of the predator on the prey by increasing b to 6, so that $q = r = s = 1$, $a = 3$, $c = 2$, $b = 6$.

11. In Exercises 7 through 10, express the change in the phase portrait in biological terms.

Consider a predator–prey model in which we assume that there is unlimited food and shelter available so that, left to itself, the prey will grow exponentially. Assume that, without the prey, the predator will die out slowly. [Perhaps there is an alternative food supply, but it lacks the proper nutrients.] Instead of a constant multiple of xy, use the more realistic uptake function discussed in this section, so that the model is

$$x' = ax - b\left(\frac{x}{1+x}\right)y,$$
$$y' = -qy + r\left(\frac{x}{1+x}\right)y. \quad (17)$$

Note that again the x,y-axes contain trajectories and we need only consider $x \geq 0$, $y \geq 0$.

12. Show that if $q = r$, then $(0, 0)$ is the only equilibrium of (17). Show that if $q \neq r$, then
$$(0, 0) \quad \text{and} \quad \left(\frac{q}{r-q}, \frac{ar}{b(r-q)}\right)$$
are equilibria of (17).

13. Show that, if $q > r$ in (17), then the predator y dies out and the prey population x goes to infinity if $x(0) > 0$, $y(0) > 0$.

For Exercises 14 through 16, determine the physically meaningful equilibria, and the behavior of solutions near them, and sketch the phase portrait.

14. Let $a = b = 1$, $q = 2$, $r = 1$ in (17).

15. Lower the predator death rate in (17) by setting $a = b = 1$, $q = 0.5$, $r = 1$.

16. Increase the predator uptake rate in (17) by setting $a = b = 1$, $q = 1$, $r = 2$.

17. Suppose $f(x, y)$, $g(x, y)$ are quadratic functions. Verify that Eqs. (11) through (15) implies that (1) has the form of (16).

All of the systems examined in this section have had the x- and y-axes as invariant curves. This need not always be the case, as the remaining exercises in this section show. Consider a species which has an immature and an adult phase. Let x be the number of immature organisms and y the number of adults. Assume each group suffers a death rate that is proportional to its size. Also assume that each group produces members of the other group at a rate proportional to its own size (immatures grow up and adults give birth to immatures). Suppose also that limitations on food and shelter only affect the adults.

18. Explain why

$$x' = -ax + by,$$
$$y' = cx - dy - ey^2 \qquad (18)$$

with a, b, c, d, e positive constants might be a reasonable model for this species made up of immatures and adults.

19. a) Show that the coordinate axes are not invariant curves for (18).
    b) Show that $R = \{(x, y): x \geq 0, y \geq 0\}$ and $\tilde{R} = \{(x, y): x > 0, y > 0\}$ are invariant sets for (18).
    c) Explain why (a) and (b) are biologically reasonable.

In Exercises 20 through 22, find all physical ($x \geq 0$, $y \geq 0$) equilibrium and sketch the phase portrait for $x \geq 0$, $y \geq 0$. Note that these exercises differ only in the adult death rate.

20. Let $b = 2$, $a = c = e = d = 1$ in (18).
21. Let $d = b = 2$, $a = c = e = 1$ in (18).
22. Let $b = 2$, $d = 3$, $a = c = e = 1$ in (18).

9.7 Nonlinear Circuits

As noted earlier, many electrical devices such as diodes, transitors, etc., are inherently nonlinear. Often they can be modeled by a combination of capacitors, inductors, and nonlinear current-controlled resistors. (See Section 2.12.) As pointed out in Section 9.1, the circuit in Fig. 9.7.1 may be modeled by the system

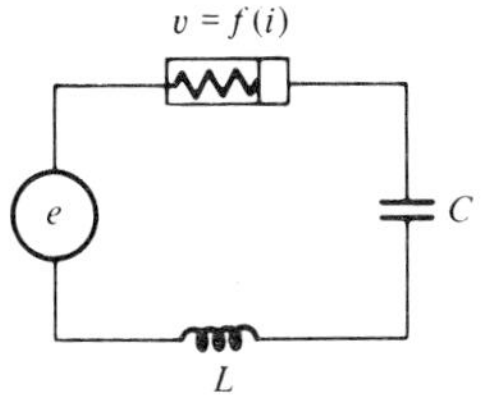

$$q' = i,$$
$$i' = \frac{-q}{CL} - \frac{f(i)}{L} + \frac{e}{L}, \qquad (1)$$

where we assume that $f(i)$ is a differentiable function of i and e is a *constant* voltage source. (Note Exercises 21 and 22 at the end of this section). Since Fig. 9.7.1 is equivalent to Fig. 9.7.2, we shall assume also throughout this section that

$$f(0) = 0. \qquad (2)$$

If we take $\hat{q} = q - Ce$, then system (1) becomes

$$\hat{q}' = i = h(\hat{q}, i),$$
$$i' = -\frac{\hat{q}}{CL} - \frac{f(i)}{L} = k(\hat{q}, i). \qquad (3)$$

Hence the effect of the voltage source is merely to shift the entire phase portrait

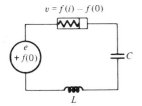

Figure 9.7.1

Figure 9.7.2

over Ce units in the q direction. Thus we may assume for the moment that $e = 0$. We are led then to consider (3) or, equivalently, (1) with $e = 0$. The only equilibrium of (3) is

$$\hat{q} = 0, \quad i = 0. \tag{4}$$

To determine the behavior near this equilibrium, by the method of Section 9.4, compute the eigenvalues of the coefficient matrix of the linearization of (3) at $(0, 0)$, that is

$$\begin{bmatrix} h_{\hat{q}}(0, 0) & h_i(0, 0) \\ k_{\hat{q}}(0, 0) & k_i(0, 0) \end{bmatrix} = \begin{bmatrix} 0 & 1 \\ -\dfrac{1}{CL} & -\dfrac{f'(0)}{L} \end{bmatrix}.$$

The characteristic polynomial,

$$p(\lambda) = \lambda \left(\lambda + \frac{f'(0)}{L} \right) + \frac{1}{CL} = \lambda^2 + \frac{f'(0)\lambda}{L} + \frac{1}{CL} \tag{5}$$

has roots

$$\lambda = -\frac{f'(0)}{2L} \pm \sqrt{\frac{f'(0)^2}{4L^2} - \frac{1}{CL}}. \tag{6}$$

From (6) and Theorem 9.4.1, we have the following theorem.

Theorem 9.7.1   Let $v = f(i)$ be the v-i characteristic for the nonlinear resistor in Fig. 9.7.1. Then there is one equilibrium given by (4), which is $(0, 0)$ and it is

i. An attractor if $f'(0) > 0$,
ii. A repellor if $f'(0) < 0$.

Furthermore, solutions near the equilibrium spiral about the equilibrium if $f'(0)^2 < 4L/C$. They do not spiral if $f'(0)^2 > 4L/C$. If $f'(0) = 0$, the solution near the origin will either spiral while near the origin, or be periodic around the origin.

It is of interest to see if periodic solutions can exist. To do so, we shall use a *Liapunov function* (Section 9.5) since it is too difficult, in general, to find invariant curves. Those who skipped Liapunov functions may wish to move directly to the exercises.

Suppose that (i, q) is a solution of (1) and assume $f(0) = 0$, $e = 0$. Let

$$F = i^2 + \frac{1}{CL} q^2. \tag{7}$$

Using (1) we find that

$$\frac{dF}{dt} = -2 \frac{if(i)}{L}. \tag{8}$$

We shall now verify the following theorem.

Theorem 9.7.2   Let $f(i)$ be as in Theorem 9.7.1 with $f(0) = 0, e = 0$.

a. If $if(i) \geq 0$ for i near zero, then the equilibrium is stable. That is, solutions starting near it stay close to it. If $if(i) > 0$ for i near zero and $i \neq 0$, then there is a region about the equilibrium that contains no periodic trajectories.

b. If $if(i) < 0$ for i near zero and $i \neq 0$, then there is a region about the equilibrium that contains no periodic trajectories.

Verification   Note that if K is a nonzero constant, then $F = K$ defines a family of ellipses in the i,q-plane about the equilibrium. If $if(i) \geq 0$ for i near zero, we have that $F' \leq 0$ along any trajectory near the equilibrium. Thus the trajectory must either stay on that ellipse or move to an ellipse closer to the equilibrium.

If $if(i) < 0$, then the solutions must move out to a further ellipse since $F' > 0$ and thus the solutions cannot be periodic. ∎

Example 9.7.1   Suppose that

$$f(i) = \begin{cases} 0 & \text{if } i \leq 0, \\ i^3 & \text{if } i \geq 0. \end{cases}$$

Determine the phase portrait for (1) if $C = 1, L = 1, e = 0$.

Solution   The graph of $f(i)$, given in Fig. 9.7.3, is similar to the characteristic curve of some diodes. The equilibrium is $q = 0, i = 0$. From (6), the roots are $\pm i$ since $f'(0) = 0$. Thus, by Theorem 9.7.1, near the equilibrium the trajectories will move around the equilibrium.

Figure 9.7.4

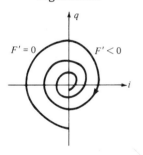

Now if F is as in (7), then, by (8),

$$F'(t) = \begin{cases} 0 & \text{if } i \leq 0, \\ -2i^4 & \text{if } i \geq 0. \end{cases}$$

Thus, while the solution has $i \leq 0$, it keeps F constant and hence lies on an ellipse, while if $i > 0$, then F decreases. The result is that $(0, 0)$ is a type of spiraling attractor, as shown in Fig. 9.7.4. In fact, Fig. 9.7.4 represents the *global behavior*. That is, solutions spiral into the origin from any point in the i,q-plane. ∎

Exercises

In Exercises 1 through 16, determine the behavior near the equilibrium $i = 0, q = 0$ for the circuit (1) with $C = 1, L = 1, e = 0$, and the indicated v-i characteristic.

1. $f(i) = 2i$
2. $f(i) = -3i$
3. $f(i) = \sin i$
4. $f(i) = \sin 3i$
5. $f(i) = 1 - e^i$
6. $f(i) = e^{-i} - 1$

7. $f(i) = \sinh 4i$
8. $f(i) = 6i + i^3$
9. $f(i) = 7i^3 - 4i$
10. $f(i) = i^3$
11. $f(i) = 3i^3 + i^5$
12. $f(i) = -2i + i^4$
13. $f(i) = i^5$
14. $f(i) = -i^5$
15. $f(i) = \begin{cases} 0 & \text{if } i \geq 0, \\ i^2 & \text{if } i \text{ is } < 0 \end{cases}$
16. $f(i) = \tan^{-1} i$

If $f(i) = \alpha i^2 + \beta$ for constants α, β, then it is possible to determine invariant curves by the integrating-factor method of Section 2.7. In Exercises 17 through 20, determine invariant curves for (1) if $e = 0$, $C = 1$, $L = 1$, and:

17. $f(i) = i^2$
18. $f(i) = -i^2$
19. $f(i) = 3i^2$
20. $f(i) = -3i^2$

Exercises 21 and 22 show how the analysis of equilibria can be used to study the dynamic case. The idea is as follows: Suppose that e is a constant in (1) and that there is an equilibrium (q_e, i_e) of (1), which is an attractor. (The subscript e is not denoting a partial derivative here.) If we change e by a little bit and hold it constant, then (q, i) should go to the new equilibrium. Intuitively, if $e(t)$ varies slowly, then a solution $(i(t), q(t))$ should stay close to the points $(q_{e(t)}, i_{e(t)})$. To simplify the exposition consider

$$\begin{aligned} x' &= -x + u(t), \\ y' &= -2y + u(t). \end{aligned} \quad (9)$$

Of course, (9) can be solved if $u(t)$ is known, but the ideas involved work for a general system. Suppose that $u(t)$ is constant; $u(t) \equiv u_0$. Then (9) has a single equilibrium $x_u = u_0$, $y_u = u_0/2$.

21. Show that the phase portrait for (9) at the equilibrium is given by Fig. 9.7.5 if $u(t) \equiv u_0$.
22. Suppose now that we consider (9) for $0 \leq t \leq 20$ with $x(0) = 1$, $y(0) = 0.5$, and

$$u(t) = \begin{cases} -1, & 0 \leq t < 10, \\ 1, & 10 \leq t \leq 20. \end{cases}$$

For $0 \leq t < 10$, (x, y) moves toward the "equilibrium" $(-1, -0.5)$. However, for $10 \leq t < 20$, the solution (x, y) moves toward the equilibrium $(1, 0.5)$. On the basis of Fig. 9.7.5, we would expect the resulting motion to look like the trajectory in Fig. 9.7.6. Verify this by solving (9) explicitly and plotting the resulting trajectory in the x, y-plane.

Figure 9.7.5
$\hat{x} = x - 1$,
$\hat{y} = y - 0.5$

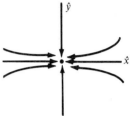

Figure 9.7.6

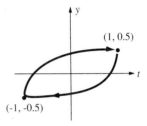

23. Theorem 9.7.2 is closely related to Theorem 9.7.1. Suppose that f is continuous and $f(0) = 0$.
   a) Show that $if(i) \geq 0$ for i near zero and f differentiable implies $f'(0) \geq 0$.
   b) Show that $if(i) \leq 0$ for i near zero and f differentiable implies $f'(0) \leq 0$.
   c) Show $f'(0) > 0$ implies $if(i) \geq 0$ for i near zero.
   d) Show $f'(0) < 0$ implies $if(i) \leq 0$ for i near zero.

9.8 Mechanical Systems

Most mechanics problems are linear only if one assumes that there are "small displacements" or "limited variations in velocity." In this section we shall analyze one nonlinear problem. An additional problem is considered in the exercises at the end of the section. The problem to be considered is a rigid

Figure 9.8.1

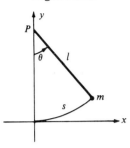

pendulum of length l. (See Fig. 9.8.1.) It is free to rotate around the point P: $(0, l)$. We suppose that the mass of the arm of the pendulum is negligible with respect to the mass m at the end of the pendulum. The mass moves along a circle, which we denote C.

Let (x, y) be the location of the mass at time t, s the distance along the circle C from the origin to the point (x, y), and θ the angle the pendulum makes with the y-axis. At a given s, the velocity along C is ds/dt, and is in a direction tangential to the circle C, which the mass is following. The component of the force of gravity in this direction is $-mg \sin \theta$. As in Section 3.16, we assume that resistance is linear and in the opposite direction to the velocity, so that

$$\text{Force total} = \text{Gravity} + \text{damping},$$

or

$$ms'' = -mg \sin \theta - \beta s'. \tag{1}$$

If θ is measured in radians, then $s = l\theta$ and equation (1) becomes $ml\theta'' = -mg \sin \theta - \beta l\theta'$ or

$$ml\theta'' + \beta l\theta' + mg \sin \theta = 0. \tag{2}$$

If θ is small, then $\sin \theta \approx \theta$, and (2) is often approximated by the linear constant coefficient equation

$$ml\theta'' + \beta l\theta' + mg\theta = 0, \tag{3}$$

which we studied in Chapter 3. For $\beta = 0$ we find that (3) predicts oscillations, as shown in Section 3.16. We shall analyze (2). Let $\gamma = \beta/m$, $\delta = g/l$, $\phi = \theta'$; then (2) can be rewritten as the system

$$\phi' = -\gamma\phi - \delta \sin \theta, \tag{4a}$$
$$\theta' = \phi. \tag{4b}$$

The frictionless case $\beta = \gamma = 0$ will be analyzed here. The $\beta \neq 0$ case is covered in the exercises at the end of this section. If $\beta = 0$, the equations (4) become:

$$\phi' = -\delta \sin \theta = f(\phi, \theta), \tag{5a}$$
$$\theta' = \phi \qquad = g(\phi, \theta). \tag{5b}$$

The equilibria are $\phi = 0$, $\sin \theta = 0$, or

$$\phi = 0, \qquad \theta = n\pi, \qquad n = 0, \pm 1, \pm 2, \dots \tag{6}$$

Note that, in (4), the change of variables $\hat{\phi} = \phi$, $\hat{\theta} = \theta + 2\pi$, again produces the same system of differential equations. Thus, the phase portrait will repeat itself every 2π units in θ. In particular, the equilibria (ϕ_0, θ_0) and $(\phi_0, \theta_0 + 2n\pi)$ will look the same. To use Theorem 9.4.1 to determine the behavior near the equilibria, we need the eigenvalues of the linearized coefficient matrix

$$\begin{bmatrix} f_\phi & f_\theta \\ g_\phi & g_\theta \end{bmatrix} = \begin{bmatrix} 0 & -\delta \cos \theta \\ 1 & 0 \end{bmatrix};$$

that is, the roots of the characteristic polynomial

$$\lambda^2 + \delta \cos \theta. \tag{7}$$

There are two cases:

Case 1

$\phi = 0, \theta = (2m + 1)\pi, m$ an integer. Then the characteristic polynomial is $\lambda^2 - \delta$ with real roots $\pm\sqrt{\delta}$. These equilibria, which correspond to the pendulum sticking straight up, are unstable (saddles).

Case 2

$\phi = 0, \theta = 2m\pi, m$ an integer. Then the characteristic polynomial is $\lambda^2 + \delta$. The roots are $\pm i\sqrt{\delta}$. This suggests the possibility of, but does not prove that, there are periodic oscillations. As noted in Theorem 9.4.1, it does show that there are trajectories that circle the equilibrium for a while. To determine what these equilibria look like, we shall determine the invariant curves. From (4) with $\beta = 0$,

$$\frac{d\phi}{d\theta} = -\frac{\delta \sin \theta}{\phi}.$$

Solving by separation of variables yields the invariant curves

$$\frac{\phi^2}{2} = \delta \cos \theta + c. \tag{8}$$

If $(\phi(t), \theta(t))$ is a trajectory near the origin $(0, 0)$, we have at $t = 0$

$$c = \frac{\phi^2(0)}{2} - \delta \cos (\theta(0))$$

and c is negative if $\phi^2(0) < 2\delta \cos \theta(0)$. Suppose then that $c = -K^2$, where $K \neq 0$.

To show that the solutions passing through these $\theta(0), \phi(0)$ are periodic, we shall use Theorem 9.5.2, and show that the invariant curve is a simple closed curve not containing an equilibrium. We already know that it does not contain an equilibrium since $K^2 > 0$. Suppose, then, that we have

$$\frac{\phi^2}{2} = \delta \cos \theta - K^2$$

and $K^2/\delta < 1$. Then

$$\phi = 0 \quad \text{for } \theta = \cos^{-1}\left(\frac{K^2}{\delta}\right).$$

Figure 9.8.2

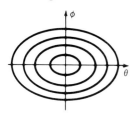

Figure 9.8.3

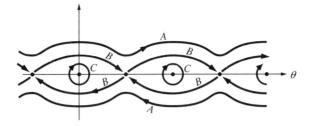

As θ increases, there are two values of ϕ,
$$\phi = \pm(2\delta \cos\theta - 2K^2)^{1/2},$$
until θ is the next solution of $\cos\theta = K^2/\delta^2$, when $\phi = 0$ again. The resulting curves, shown in Fig. 9.8.2, are simple closed curves. Thus every trajectory near enough to the origin is periodic, by Theorem 9.5.2. The actual phase portrait is shown in Fig. 9.8.3.

Trajectories A in Fig. 9.8.3 correspond to the situation when the pendulum is given enough initial velocity to keep swinging around the pivot point P. Note that it slows down as the mass approaches the top $[\theta = (2n + 1)\pi]$ and speeds up as the mass swings down. The trajectories B correspond to when the conditions are just right so that the pendulum approaches the unstable equilibrium. The trajectories B separate the trajectories C where the pendulum oscillates about the stable equilibrium from those like A, where it keeps circling the pivot point.

Exercises

For Exercises 1 through 4, suppose, in the pendulum problem, that there is a resistive force of $-\beta s'$. In Exercises 1 through 3, determine the nature of the equilibria under the given conditions.

1. $l = 32$ ft, $m = 1$ slug, $\beta = 1$ slug/sec.
2. $l = 32$ ft, $m = \frac{1}{2}$ slug, $\beta = 1$ slug/sec.
3. $l = 32$ ft, $m = \frac{1}{3}$ slug, $\beta = 1$ slug/sec.
4. Determine the values of β for which $(0, 0)$ is:
   i) A spiral attractor;
   ii) An attractor that does not spiral;
   iii) Interpret the answers to (i) and (ii) in terms of underdamping and overdamping.

Rather than thinking of resistance as due to the movement of the mass through some medium, let us consider the pendulum as being attached at point P (perhaps by a coupling with ball bearings) to a shaft. Suppose that the resistance is proportional to the angular velocity of the pendulum relative to the shaft. If the shaft is not moving, then the resistance term is again a constant multiple of θ'. Suppose that the shaft is rotating at a fixed velocity of r rad/sec. Then our model is

$$ml\theta'' + k(\theta' - r) + mg \sin\theta = 0. \qquad (9)$$

5. Rewrite (9) as a system with variables θ, $\phi = \theta'$.

In Exercises 6 through 8, determine the equilibria and phase portrait for (9) under the given conditions. Try to physically interpret your results. Units are ft-slug-rad-sec.

6. $k = 2$, $\quad l = 2$, $\quad m = 1$, $\quad r = 8$
7. $k = 2$, $\quad l = 2$, $\quad m = 1$, $\quad r = 16$
8. $k = 2$, $\quad l = 2$, $\quad m = 1$, $\quad r = 32$

(*Hint*: Show that for any initial condition, eventually $15 \leq \phi \leq 49$ and $\theta \to \infty$.)

This section considered an equation of the form $x'' + \gamma x' + f(x) = 0$ while Section 9.7 considered $x'' + f(x') + \gamma x = 0$. More complicated circuits and mechanical systems can take the more general form

$$x'' + f(x, x') = 0. \qquad (10)$$

9. Let $y = x'$ and rewrite (10) as a system in (x, y).

a) Show that the only equilibriums of this system are of the form $(c, 0)$ where $x(t) \equiv c$ is an equilibrium of (10).

b) Show that the characteristic polynomial of the linearized coefficient matrix at an equilibrium $(c, 0)$ is $\lambda^2 + \lambda f_y(c, 0) + f_x(c, 0)$.

Appendix A

Complex Numbers

As mankind has considered more difficult mathematical problems, it has been necessary to expand the number system. Thus to the integers 1, 2, 3, ..., it has been necessary to add zero, negative numbers, fractions (rational numbers), algebraic numbers ($\sqrt{2}$, $\sqrt[3]{5}$, etc.) and transcendental numbers (e, π, etc.). Some problems require additional entities called *complex numbers*. The complex numbers are formed by adding to the real numbers the number i, where i has the property $i^2 = -1$. The number i is one of the two square roots of -1 and is also denoted by

$$i = \sqrt{-1}.$$

The other square root of -1 is $-i$. All complex numbers z may be written as

$$z = a + ib, \qquad (1)$$

where a and b are real numbers; a is called the *real part* of z, and b is called the *imaginary part* of z. The real and imaginary parts are also denoted

$$a = \text{Re}(z), \qquad b = \text{Im}(z).$$

This section is not intended as a rigorous development of complex numbers but rather as a review of basic facts and as an introduction to some required notation.

Example A.1 Find the real and imaginary parts of $z_1 = 3 - 5i$, $z_2 = 6i$, $z_3 = 13$.

Solution

$z_1 = 3 - 5i = 3 + (-5)i$ has $\text{Re}(z_1) = 3$, $\text{Im}(z_1) = -5$.
$z_2 = 6i = 0 + 6i$ has $\text{Re}(z_2) = 0$, $\text{Im}(z_2) = 6$,
$z_3 = 13 = 13 + 0i$ has $\text{Re}(z_3) = 13$, $\text{Im}(z_3) = 0$. ∎

Complex numbers can be multiplied using the usual laws for real numbers and the added fact that $i^2 = -1$.

Example A.2 Multiply $3 - 2i$ times $-1 + 4i$.

Solution

$$(3 - 2i)(-1 + 4i) = -3 + 14i - 8i^2$$
$$= -3 + 14i - 8(-1) = 5 + 14i. \blacksquare$$

To each complex number $z = a + bi$, there corresponds a second complex number

$$\bar{z} = a - bi,$$

called its *conjugate*. For example,

$$\overline{3 + i} = 3 - i, \quad \bar{5} = 5, \quad \text{and} \quad \overline{7i} = -7i.$$

Conjugates may be used to express the real and imaginary parts of a complex number since if $z = a + bi$, then

$$a = \text{Re}(z) = \frac{z + \bar{z}}{2}, \quad b = \text{Im}(z) = \frac{z - \bar{z}}{2i}. \quad (2)$$

Example A.3 Verify (2) for $z = 2 - 5i$.

Solution If $z = 2 - 5i$, then $\bar{z} = 2 + 5i$. Thus

$$\frac{z + \bar{z}}{2} = \frac{2 - 5i + 2 + 5i}{2} = 2 = \text{Re}(z),$$

$$\frac{z - \bar{z}}{2i} = \frac{2 - 5i - (2 + 5i)}{2i} = \frac{-10i}{2i} = -5 = \text{Im}(z). \blacksquare$$

For any complex number $z = a + bi$,

$$z\bar{z} = (a + ib)(a - ib) = a^2 + b^2. \quad (3)$$

Note that $z\bar{z}$ is always a real number and $z\bar{z} \neq 0$ if $z \neq 0$.
The absolute value of $z = a + bi$, denoted $|z|$, is defined by

$$|z| = (z\bar{z})^{1/2} = \sqrt{a^2 + b^2}. \quad (4)$$

The number $|z|$ is the same as the Euclidean length of the vector (a, b) in the plane. Sometimes $|z|$ is also called the *magnitude*, or *modulus*, of z.

The conjugate is also useful in division. If $z = a + ib$ and $w = c + id$ are two complex numbers, then

$$\frac{a + ib}{c + id} = \frac{z}{w} = \frac{z\bar{w}}{w\bar{w}} = \frac{(a + ib)(c - id)}{(c + id)(c - id)}$$

$$= \frac{(ac + bd) + i(bc - da)}{c^2 + d^2}$$

$$= \frac{ac + bd}{c^2 + d^2} + i\frac{(bc - da)}{c^2 + d^2}.$$

Example A.4 Compute $(1 - i)/(-1 + 2i)$.

Solution We multiply numerator and denominator by

$$\overline{-1 + 2i} = -1 - 2i,$$

obtaining

$$\frac{1 - i}{-1 + 2i} = \frac{1 - i}{-1 + 2i}\frac{(-1 - 2i)}{(-1 - 2i)} = \frac{-1 + i - 2i + 2i^2}{1 - 4i^2}$$

$$= \frac{-3 - i}{1 + 4} = -\frac{3}{5} - \frac{1}{5}i. \blacksquare$$

In working with differential equations we shall need to consider the exponential of a complex number. If x is a real number, we define

$$e^{ix} = \cos x + i \sin x. \tag{5}$$

Note One way to motivate formula (5) is to use the series for e^z and assume that it is valid for complex z:

$$e^{ix} = \sum_{n=0}^{\infty} \frac{(ix)^n}{n!} = \sum_{m=0}^{\infty} \frac{(ix)^{2m}}{(2m)!} + \sum_{m=0}^{\infty} \frac{(ix)^{2m+1}}{(2m + 1)!}$$

$$= \sum_{m=0}^{\infty} i^{2m}\frac{x^{2m}}{(2m)!} + \sum_{m=0}^{\infty} ii^{2m}\frac{x^{2m+1}}{(2m + 1)!}$$

$$= \sum_{m=0}^{\infty} (-1)^m\frac{x^{2m}}{(2m)!} + i\sum_{m=0}^{\infty} (-1)^m\frac{x^{2m+1}}{(2m + 1)!} = \cos x + i \sin x.$$

For a general complex number $z = a + ib$, we may use the law of exponents:

$$e^{a+bi} = e^a e^{bi} = e^a(\cos b + i \sin b)$$
$$= e^a \cos b + i e^a \sin b. \qquad (6)$$

As a consequence of (6) we have

$$(e^{a+bi})^n = e^{an+bni} = e^{an}(\cos bn + i \sin bn).$$

The special case

$$(\cos b + i \sin b)^n = (e^{bi})^n = \cos bn + i \sin bn$$

is known as *de Moivre's formula*.

Note that

$$e^{a-bi} = e^a e^{-bi} = e^a \cos(-b) + i e^a \sin(-b)$$
$$= e^a \cos b - i e^a \sin b, \qquad (7)$$

so that

$$e^{\bar{z}} = \overline{e^z}. \qquad (8)$$

Polar Form

It is sometimes helpful to have an alternative to the $a + bi$ notation for a complex number. For this purpose, we view the complex number $z = a + bi$ as a vector (a, b) in the plane, as shown in Fig. A.1, where the horizontal axis is the *real axis* and the vertical axis is the *imaginary axis*. By trigonometry (Fig. A.2),

$$z = a + ib = r \cos \theta + ir \sin \theta = re^{i\theta},$$

where

$$r = \sqrt{a^2 + b^2} = |z| \sim \text{modulus of } z$$

is the *modulus* of z, and θ is the angle z makes with the real axis.

The angle θ is often called the *argument* of z. We shall measure θ in radians (rad). The expression $z = re^{i\theta}$ is called the *polar form* of the complex number z since r and θ are polar coordinates in the plane. The argument is not unique, since θ and $\theta + 2n\pi$, n an integer, describe the same angle. It is often easier to multiply and divide complex numbers when they are in polar form.

If

$$z_1 = r_1 e^{i\theta_1}, \qquad z_2 = r_2 e^{i\theta_2},$$

then

$$z_1 z_2 = r_1 r_2 e^{i(\theta_1 + \theta_2)} \qquad (9)$$

and

$$\frac{z_1}{z_2} = \frac{r_1 e^{i\theta_1}}{r_2 e^{i\theta_2}} = \frac{r_1}{r_2} e^{i(\theta_1 - \theta_2)}. \qquad (10)$$

The product (9) may be restated as:

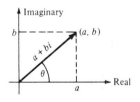

Figure A.1
Vector representation of $z = a + bi$.

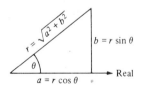

Figure A.2

Complex multiplication multiplies moduli and adds arguments.

Example A.5   Write $z = 1 + i$ in polar form.

Solution   Let $z = 1 + i$. Then the modulus $r = \sqrt{1^2 + 1^2} = \sqrt{2}$ and the argument $\theta = 45°$ or $\pi/4$ rad. Thus

$$1 + i = \sqrt{2}e^{(\pi/4)i}. \blacksquare$$

Example A.6   Compute $(1 + i)^3$.

Solution   From Example A.5, $1 + i = \sqrt{2}e^{(\pi/4)i}$. Thus

$$\begin{aligned}(1 + i)^3 &= (\sqrt{2}e^{(\pi/4)i})^3 = (\sqrt{2})^3 (e^{(\pi/4)i})^3 \\ &= 2^{3/2} e^{(3\pi/4)i} \\ &= 2^{3/2}\left(\cos\frac{3\pi}{4} + i\sin\frac{3\pi}{4}\right) \\ &= 2^{3/2}\left(\frac{-1}{\sqrt{2}} + i\frac{1}{\sqrt{2}}\right) = -2 + 2i. \blacksquare\end{aligned}$$

Exercises

In Exercises 1 through 6, express the result in the form $a + bi$.

1. $(1 + 3i)(6 - 2i)$
2. $(2 + 3i)^2$
3. $\dfrac{1 + i}{2 + 3i}$
4. $\dfrac{i}{-1 + i}$
5. $\dfrac{2 + 5i}{-3i}$
6. $\dfrac{7}{2i}$

In Exercises 7 through 12, sketch the given complex number z_k in vector form (see Fig. A.1) and express z_k in polar form.

7. $z_1 = 1 - i$
8. $z_2 = 1 + \sqrt{3}i$
9. $z_3 = -\sqrt{3} - i$
10. $z_4 = -i$
11. $z_5 = -3$
12. $z_6 = 2i$

In Exercises 13 through 19, compute the indicated complex number using the polar form obtained in Exercises 7 through 10 for z_1, z_2, z_3, z_4.

13. $\dfrac{z_1}{z_2}$
14. $z_1 z_2$
15. $\dfrac{z_3}{z_4}$
16. $(z_1)^3$
17. $z_1 z_3$
18. With z_2 as in Exercise 8, find a fourth root of z_2; that is, a number z such that $z^4 = z_2$.
19. Find a complex number z such that $z^3 = z_1$.
20. If z is a complex number, verify that $e^z \neq 0$ for every z.

21. Verify that $e^{-z} = 1/e^z$.

22. Verify that $f(z) = e^z$ is a periodic function with period $2\pi i$.

23. Show that
$$r^{1/m} \exp\left[i\left(\frac{\theta}{m} + \frac{2\pi}{m}n\right)\right], \quad n = 0, 1, \ldots, m-1$$
are m distinct mth roots of $z = re^{i\theta}$ if $r \neq 0$.

24. (Uses Exercise 23.) Find the five fifth roots of $1 + i$ and sketch the results.

25. (Uses Exercise 23.) Find the seven seventh roots of $-i$ and sketch the results.

Appendix B

Review of Partial Fractions

Partial fractions is an important algebraic technique. In this book it will be used both in the evaluation of integrals, a procedure with which you are familiar, and as a means of evaluating inverse Laplace transforms (Chapter 4). In both applications the idea is the same. A complicated ratio of polynomials is rewritten as a sum of simpler terms that are easier to work with. We first describe the general procedure for partial fractions and then consider several examples.

The key fact is the following:

Suppose that $p(x)/q(x)$ is a ratio of two polynomials and that the degree of $q(x)$ is greater than the degree of $p(x)$. Then $p(x)/q(x)$ may be written as a sum of terms of the form

$$\frac{A}{(x-a)^m} \quad \text{and} \quad \frac{Bx+C}{(x^2+ax+b)^m}. \tag{1}$$

Application of Partial Fractions

We employ partial fractions to express a fraction of two polynomials $r(x)/q(x)$ as the sum of a polynomial and terms in the form (1) by using the following steps:

1. If $q(x)$ does not have a higher degree than $r(x)$, divide $q(x)$ into $r(x)$ to get

$$\frac{r(x)}{q(x)} = h(x) + \frac{p(x)}{q(x)},$$

where $h(x)$ is a polynomial, and $p(x)$ is a polynomial of lower degree than $q(x)$.

The rest of the procedure expresses $p(x)/q(x)$ as a sum of terms in the form (1).

2. Factor $q(x)$ into *linear* (first-degree) and *irreducible quadratic* (second-degree) factors:

$$q(x) = a_0(x - a_1)^{m_1} \cdots (x - a_r)^{m_r}$$
$$\times (x^2 + b_1 x + c_1)^{n_1} \cdots (x^2 + b_s x + c_s)^{n_s}. \qquad (2)$$

Here $a_1, \ldots, a_r$ are the distinct real roots of $q(x)$. The integers $m_1, \ldots, m_r$ are their *multiplicities* (i.e., how often they are repeated). The polynomials $x^2 + b_i x + c_i$, $i = 1, \ldots, s$, have no real roots and are distinct.

3. If the linear factor $(x - a)$ appears exactly m times in the factorization of $q(x)$, then the terms

$$\frac{C_1}{(x-a)} + \frac{C_2}{(x-a)^2} + \cdots + \frac{C_m}{(x-a)^m} \qquad (3)$$

are put into the formula for $p(x)/q(x)$, where $C_1, \ldots, C_m$ are constants to be determined.

4. If the irreducible quadratic $(x^2 + bx + c)$ appears exactly n times in the factorization of the denominator $q(x)$, then

$$\frac{E_1 x + F_1}{(x^2 + bx + c)} + \frac{E_2 x + F_2}{(x^2 + bx + c)^2} + \cdots + \frac{E_n x + F_n}{(x^2 + bx + c)^n} \qquad (4)$$

is put into the formula for $p(x)/q(x)$, where $E_1, F_1, \ldots, E_n, F_n$ are constants to be determined.

Once the ratio $p(x)/q(x)$ has been written as a sum of terms such as in (3) and (4), it is necessary to solve for the constants. In general, we must:

5. Multiply both sides by $q(x)$ and rewrite the right-hand side as a polynomial in x.
6. Equate the coefficients of corresponding powers of x.
7. Solve the resulting system of equations for the constants.

Step 6 may sometimes be considerably simplified.

6'. If all factors of the denominator $q(x)$ are distinct linear factors, then evaluate the result of step 5 at the roots of $q(x)$ to solve for the constants.

If not all the factors are distinct linear factors, then a combination of steps 6 and 6' may often be helpful.

Appendix B: Review of Partial Fractions

Before illustrating the entire procedure of partial fractions, four examples of steps 2, 3, and 4 will be given. Underlining brackets are used to show corresponding terms.

Example B.1

$$\frac{x^2 + 1}{x(x - 1)(x + 3)} = \frac{A}{x} + \frac{B}{x - 1} + \frac{C}{x + 3}.$$

This example has three distinct linear factors: x, $x - 1$, $x + 3$, each of multiplicity 1. ∎

Example B.2

$$\frac{x^2 - 3x + 1}{x^3(x - 1)(x + 3)^2} = \frac{A}{x} + \frac{B}{x^2} + \frac{C}{x^3} + \frac{D}{x - 1} + \frac{E}{(x + 3)} + \frac{F}{(x + 3)^2}.$$

This example also has three distinct linear factors: x, $x - 1$, $x + 3$, but x has multiplicity 3 and $x + 3$ has multiplicity 2. ∎

Example B.3

$$\frac{x^3 + 1}{(x^2 + 1)(x^2 + x + 1)} = \frac{Ax + B}{x^2 + 1} + \frac{Cx + D}{x^2 + x + 1}.$$

This example has two distinct quadratic factors: $x^2 + 1$ and $x^2 + x + 1$, each of multiplicity 1. ∎

Example B.4

$$\frac{x^3}{(x^2 + 1)^3 x (x^2 + x + 2)}$$

$$= \frac{Ax + B}{x^2 + 1} + \frac{Cx + D}{(x^2 + 1)^2} + \frac{Ex + F}{(x^2 + 1)^3} + \frac{G}{x} + \frac{Hx + I}{x^2 + x + 2}.$$

This example has two distinct quadratic factors: $x^2 + 1$ and $x^2 + x + 2$, and one linear factor x. The factor $x^2 + 1$ has multiplicity 3, while x and $x^2 + x + 2$ each have multiplicity 1. ∎

The next two examples illustrate the general procedure of partial fractions.

Example B.5

Expand $x/[(x^2 + 1)(x - 1)]$, using partial fractions.

Solution

Since the polynomial in the denominator has larger degree than the polynomial x in the numerator, we omit step 1. In this example

$$q(x) = (x^2 + 1)(x - 1)$$

has one linear factor $(x - 1)$ and one irreducible quadratic factor $(x^2 + 1)$. Thus from steps 3 and 4,

$$\frac{x}{(x^2 + 1)(x - 1)} = \frac{Ax + B}{x^2 + 1} + \frac{C}{x - 1}.$$

Multiplying both sides by $q(x) = (x^2 + 1)(x - 1)$ gives

$$x = (Ax + B)(x - 1) + C(x^2 + 1) = (A + C)x^2 + (B - A)x + (-B + C).$$

Since $x = 0 \cdot 1 + 1 \cdot x + 0 \cdot x^2$, equating the coefficients of powers of x on both sides of the equals sign gives:

Power of x	Left-hand side	=	Right-hand side
1	0	=	$-B + C$
x	1	=	$B - A$
x^2	0	=	$A + C$

We must now solve this system of equations for A, B and C. (The reader familiar with solving systems of linear equations using an augmented matrix and row operations is encouraged to do so; see Section 6.7.) The result is $B = \frac{1}{2}$, $A = -\frac{1}{2}$, $C = \frac{1}{2}$. Thus

$$\frac{x}{(x^2 + 1)(x - 1)} = \frac{-\frac{1}{2}x + \frac{1}{2}}{x^2 + 1} + \frac{\frac{1}{2}}{x - 1}. \blacksquare$$

Example B.6 Expand $(x^2 + 1)/[x(x - 1)(x + 3)]$, using partial fractions.

Solution Since $q(x) = x(x - 1)(x + 3)$ has three distinct linear factors, x, $x - 1$, and $x + 3$, we write

$$\frac{x^2 + 1}{x(x - 1)(x + 3)} = \frac{A}{x} + \frac{B}{x - 1} + \frac{C}{x + 3}.$$

Multiplying both sides by $q(x) = x(x - 1)(x + 3)$ gives

$$x^2 + 1 = A(x - 1)(x + 3) + Bx(x + 3) + Cx(x - 1). \tag{5}$$

Following step 6' and evaluating (5) at the roots of $x(x - 1)(x + 3)$, which are $x = 0, 1, -3$, we have

$x = 0$: $\quad 1 = A(-1)(3) + 0 + 0 \quad$ so $\quad A = -\frac{1}{3}$,
$x = 1$: $\quad 2 = 0 + B(1)(4) + 0 \quad$ so $\quad B = \frac{1}{2}$,
$x = -3$: $\quad 10 = 0 + 0 + C(-3)(-4) \quad$ so $\quad C = \frac{5}{6}$.

Hence
$$\frac{x^2+1}{x(x-1)(x+3)} = \frac{-\frac{1}{3}}{x} + \frac{\frac{1}{2}}{x-1} + \frac{\frac{5}{6}}{x+3}. \blacksquare$$

Exercises

Apply the technique of partial fractions to each of the following quotients of polynomials.

1. $\dfrac{2x^2 - 3x - 1}{x^3 - x}$

2. $\dfrac{1}{x^2 - 5x - 6}$

3. $\dfrac{1}{(x^2+1)^2}$

4. $\dfrac{x^3 + 3x^2 + x + 12}{(x^2+1)(x^2+4)}$

5. $\dfrac{x^3 + x^2 + x}{(x^2+1)^2}$

6. $\dfrac{3}{(x+2)^3}$

7. $\dfrac{2x+7}{(x+2)^3}$

8. $\dfrac{2}{x(x^2+4)}$

9. $\dfrac{x^2+x}{x+3}$

10. $\dfrac{x^2+5}{x^3 + 2x^2 + 5x}$

Appendix C

Existence and Uniqueness

In this appendix we shall prove the basic existence and uniqueness theorem for first-order initial value problems given in Section 2.1 (Theorem 2.1.1). The proof we give may be modified to prove the other existence and uniqueness theorems in this book, such as those for systems and higher-order equations. The proof will also introduce several ideas, such as iterative methods and integral equations, which are of great importance in applied mathematics.

Given the first-order initial value problem,

$$\frac{dy}{dx} = f(x, y), \qquad y(x_0) = y_0, \qquad (1)$$

how are we to prove the existence and uniqueness of a solution? There are two key steps. The first is to rewrite (1) as an *integral equation* by integrating both sides of (1)

$$\int_{x_0}^{x} \frac{dy}{ds} ds = \int_{x_0}^{x} f(s, y(s)) ds,$$

evaluating the integral on the left side

$$\int_{x_0}^{x} \frac{dy}{ds} ds = y(x) - y(x_0) = y(x) - y_0,$$

and solving for y to get an integral equation for y

$$y(x) = \int_{x_0}^{x} f(s, y(s))\, ds + y_0. \tag{2}$$

Let $I = [\alpha, \beta]$ be an interval containing x_0. Note that if $y(x)$ is a differentiable function defined on I that satisfies the differential equation (1), then it satisfies the integral equation (2). Conversely, if $y(x)$ is a continuous function defined on I which satisfies the integral equation (2) for $x \in I$, and if $f(s, y(s))$ is continuous for $s \in I$, then by the fundamental theorem of calculus, the right-hand side of (2) is continuously differentiable. Thus $y(x)$ is continuously differentiable, and $y(x)$ satisfies (1). We have established that:

> Finding a differentiable solution of the initial value problem (1) is equivalent to finding a continuous solution of the integral equation (2).

The second key in establishing existence and uniqueness is to try and solve the integral equation (2) by the *method of successive approximation*, or *Picard iteration*. [Emile Picard was a French mathematician (1856–1941) who extensively developed the idea of successive approximations. The method had been used earlier by Liouville.] In the method of successive approximations, one solves an equation of the form

$$y = G(y) \tag{3}$$

for y by making an initial guess ϕ_0 for y and then defining a sequence of approximations by

$$\begin{aligned} \phi_1 &= G(\phi_0) \\ \phi_2 &= G(\phi_1) \\ &\vdots \\ \phi_{n+1} &= G(\phi_n) \\ &\vdots \end{aligned} \tag{4}$$

If the approximations ϕ_n converge, say $\phi_n \to \phi$, and if $G(\phi_n) \to G(\phi)$, then (4) implies that $\lim_{n \to \infty} \phi_{n+1} = \lim_{n \to \infty} G(\phi_n)$ so that $\phi = G(\phi)$ and ϕ is a solution of (3).

In our problem, (3) is the integral equation (2) and y is an unknown function, so that the iterates ϕ_n will also be functions. We take ϕ_0 to be the constant function y_0;

$$\phi_0(x) = y_0. \tag{5}$$

The successive approximations are then

$$\phi_1(x) = \int_{x_0}^{x} f(s, \phi_0(s))\, ds + y_0 \tag{6}$$

$$\phi_2(x) = \int_{x_0}^{x} f(s, \phi_1(s))\, ds + y_0 \tag{7}$$

$$\vdots$$

$$\phi_{n+1}(x) = \int_{x_0}^{x} f(s, \phi_n(s))\, ds + y_0 \tag{8}$$

$$\vdots$$

In order for this approach to work, there are several requirements that must be met:

1. For each $\phi_n(x)$ it must be possible, at least theoretically, to evaluate the integral in (8) to get the next iterate $\phi_{n+1}(x)$.
2. The iterates $\phi_n(x)$ must converge to a function $\phi(x)$.
3. $\int_{x_0}^{x} f(s, \phi_n(s))\, ds + y_0$ must converge to $\int_{x_0}^{x} f(s, \phi(s))\, ds + y_0$ as $n \to \infty$.
4. $\phi(x)$ must be continuous on an interval containing x_0.

Before showing how to satisfy all these requirements, and thereby prove existence and uniqueness for (1), we will give a specific example of the iteration process.

Example C.1

Find the first four iterates on $[0, 2]$ for the initial value problem

$$\frac{dy}{dx} = y^2, \quad y(0) = 1 \tag{9}$$

using the successive approximations (5) through (8).

Solution

In this example we have $x_0 = 0$, $y_0 = 1$, and $f(s, y(s)) = y(s)^2$. Thus the equations (5) through (8) defining the iterates become

$$\phi_0(x) = 1$$

$$\phi_1(x) = \int_0^x \phi_0(s)^2\, ds + 1 = \int_0^x ds + 1 = x + 1$$

$$\phi_2(x) = \int_0^x \phi_1(s)^2\, ds + 1 = \int_0^x (1 + s)^2\, ds + 1 = 1 + x + x^2 + \frac{1}{3}x^3$$

$$\phi_3(x) = \int_0^x \phi_2(s)^2\, ds + 1 = \int_0^x \left(1 + s + s^2 + \frac{1}{3}s^3\right)^2 ds + 1$$

$$= 1 + x + x^2 + x^3 + \frac{2}{3}x^4 + \frac{1}{3}x^5 + \frac{1}{9}x^6 + \frac{1}{63}x^7.$$

Additional iterates could be computed in a similar fashion. ∎

It is instructive to examine this example more carefully. The actual solution of (9) can be found by separation of variables to be

Figure C.1

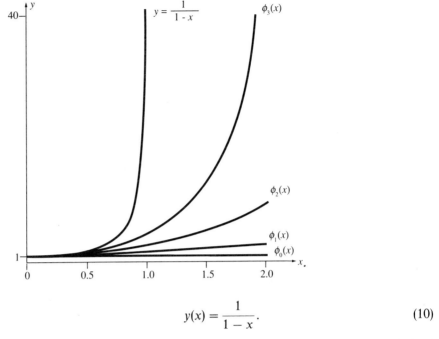

$$y(x) = \frac{1}{1-x}. \tag{10}$$

Notice that while the iterates $\phi_n(x)$ are defined on $[0, 2]$ [in fact, they are defined on $(-\infty, \infty)$ for all n], the solution (10) is only defined on $[0, 1)$ [or more generally $(-\infty, 1)$]. This shows that in our proof we will have to make restrictions on the interval of definition in order to obtain convergence of the $\phi_n(x)$ on all of that interval. Next notice that the solution (10) has the series expansion on $[0, 1)$,

$$y(x) = \frac{1}{1-x} = \sum_{i=0}^{\infty} x^i = 1 + x + x^2 + x^3 + x^4 + \cdots \tag{11}$$

Looking at the $\phi_n(x)$ we computed in Example C.1, we see that $\phi_0(x)$ gives the first term of the expansion (11), $\phi_1(x)$ agrees with the first two terms, $\phi_2(x)$ has the same first three terms, and $\phi_3(x)$ has the same first four terms. The solution (10) and $\phi_0(x), \phi_1(x), \phi_2(x), \phi_3(x)$ are graphed in Fig. C.1.

We now prove the following existence and uniqueness theorem.

Theorem C.1 Suppose that $f(x, y)$ and $f_y(x, y)$ are continuous at and near (x_0, y_0). Then there is an interval I containing x_0 on which there is a unique solution to the initial value problem $dy/dx = f(x, y), y(x_0) = y_0$.

In the proof, we shall be more precise about the sets involved. To begin, let R be a rectangular region centered at (x_0, y_0),

$$R = \{(x, y): |x - x_0| \le a, \quad |y - y_0| \le b\}, \quad a > 0, b > 0. \tag{12}$$

We shall assume that:

Assumption 1 $f(x, y)$ is continuous on the rectangular region R.

Assumption 2 $f_y(x, y)$ is continuous on the rectangular region R.

From advanced calculus we know that a continuous function on a closed and bounded set, such as R, is bounded. Thus there exists constants M, L such that

$$|f(x, y)| \le M \quad \text{all } (x, y) \in R, \tag{13}$$
$$|f_y(x, y)| \le L \quad \text{all } (x, y) \in R. \tag{14}$$

Before giving the technical version of Theorem 1 to be proven, we need one more estimate which will be discussed more fully in the exercises. If (x, y_1), (x, y_2) are both in R and Assumptions 1 and 2 hold, then we may use the mean value theorem and (14) on $f(x, y_1) - f(x, y_2)$ to get

$$|f(x, y_1) - f(x, y_2)| = |f_y(x, \hat{y})||y_1 - y_2| \le L|y_1 - y_2|, \tag{15}$$

where $\hat{y}$ is a number between y_1 and y_2. We are now ready to prove our technical version of Theorem 1.

Theorem C.2

Technical Version of Theorem 1 Let $f(x, y)$ satisfy Assumptions 1 and 2. Let $\bar{a}$ be the smaller of a and b/M where M is from (13). Then there is a unique solution to $dy/dx = f(x, y)$, $y(x_0) = y_0$, defined for $|x - x_0| \le \bar{a}$.

Proof

We shall use Picard iteration and show that Requirements 1 through 4 are met. Let

$$\bar{R} = \{(x, y): \quad |x - x_0| \le \bar{a}, |y - y_0| \le b\}$$

and define

$$\phi_0(x) = y_0 \quad \text{for } |x - x_0| \le \bar{a}.$$

1. Notice that the graph of $\phi_0(x)$ lies in $\bar{R}$ and $\phi_0(x)$ is continuous. We shall show that if $\phi_n(x)$ is a continuous function defined on $|x - x_0| \le \bar{a}$, whose graph lies in $\bar{R}$, then $\phi_{n+1}(x)$ has these same properties and hence, by induction, the sequence of iterates $\{\phi_n(x)\}$ is well defined. Suppose that $\phi_n(x)$ is a continuous function defined on $|x - x_0| \le \bar{a}$ whose graph lies in $\bar{R}$. Clearly, from (8) we have that $\phi_{n+1}(x)$ will be continuous and defined on $|x - x_0| \le \bar{a}$. Thus we need to show only that the graph of $\phi_{n+1}(x)$ lies in $\bar{R}$. But if $|x - x_0| \le \bar{a}$, then from the definition (8) of the sequence $\{\phi_n(x)\}$, we have

$$|\phi_{n+1}(x) - y_0| = \left|\int_{x_0}^x f(s, \phi_n(s))\, ds\right| \leq \left|\int_{x_0}^x |f(s, \phi_n(s))|\, ds\right|$$

$$\leq |x - x_0| \max_{(x,y) \in \bar{R}} |f(x, y)|$$

$$= M|x - x_0| \leq M\frac{b}{M} = b, \tag{16}$$

which shows the graph of $\phi_{n+1}(x)$ also lies in $\bar{R}$.

2. We now show that $\lim_{n \to \infty} \phi_n(x) = \phi(x)$ for a continuous function $\phi(x)$. This is done by viewing $\phi_n(x)$ as the partial sum of a series. Observe that

$$\phi_n(x) - \phi_0(x)$$
$$= [\phi_n(x) - \phi_{n-1}(x)] + [\phi_{n-1}(x) - \phi_{n-2}(x)] + \cdots + [\phi_1(x) - \phi_0(x)]$$
$$= \sum_{i=1}^n [\phi_i(x) - \phi_{i-1}(x)]. \tag{17}$$

We need to obtain estimates for the terms in (17). First, we compute as in (16) that

$$|\phi_1(x) - \phi_0(x)| = |\phi_1(x) - y_0|$$
$$= \left|\int_{x_0}^x f(s, \phi_0(s))\, ds\right|$$
$$\leq M|x - x_0|. \tag{18}$$

For notational convenience we shall do the remaining calculations for $x_0 \leq x \leq x_0 + \bar{a}$ but they can be carried out for $-\bar{a} + x_0 \leq x \leq x_0$ with only some sign changes. Using (15), (18), and the definition of the sequence of iterates (8), we compute that:

$$|\phi_2(x) - \phi_1(x)| = \left|\int_{x_0}^x f(s, \phi_1(s)) - f(s, \phi_0(s))\, ds\right|$$
$$\leq \int_{x_0}^x |f(s, \phi_1(s)) - f(s, \phi_0(s))|\, ds$$
$$\leq \int_{x_0}^x L|\phi_1(s) - \phi_0(s)|\, ds \qquad \text{[by (15)]}$$
$$\leq L\int_{x_0}^x M|s - x_0|\, ds \qquad \text{[by (18)]}$$
$$= ML\frac{|x - x_0|^2}{2}.$$

We are now ready to show by induction that

$$|\phi_n(x) - \phi_{n-1}(x)| \leq ML^{n-1}\frac{|x - x_0|^n}{n!} \tag{19}$$

for $n = 1, 2, \ldots$. We have already shown (19) holds for $n = 1, 2$. Suppose that (19) holds for $n = 1, 2, \ldots, k - 1$. We shall show that it holds for $n = k$.

$$\begin{aligned}
|\phi_k(x) - \phi_{k-1}(x)| &= \left| \int_{x_0}^x f(s, \phi_{k-1}(s)) - f(s, \phi_{k-2}(s)) \, ds \right| \\
&\leq \int_{x_0}^x |f(s, \phi_{k-1}(s)) - f(s, \phi_{k-2}(s))| \, ds \\
&\leq \int_{x_0}^x L|\phi_{k-1}(s) - \phi_{k-2}(s)| \, ds \quad \text{[by (15)]} \\
&\leq L \int_{x_0}^x ML^{k-2} \frac{|s - x_0|^{k-1}}{(k-1)!} \, ds \quad \text{[by (19)]} \\
&= ML^{k-1} \frac{|x - x_0|^k}{k!}, \quad (20)
\end{aligned}$$

which is (19). Using the estimate (19) we now show that the series $\sum_{i=1}^\infty [\phi_i(x) - \phi_{i-1}(x)]$ converges absolutely since the partial sums of the absolute values are bounded:

$$\begin{aligned}
\sum_{i=1}^n |\phi_i(x) - \phi_{i-1}(x)| &\leq \sum_{i=1}^n ML^{i-1} \frac{|x - x_0|^i}{i!} \\
&\leq \frac{M}{L} \sum_{i=1}^\infty \frac{L^i |x - x_0|^i}{i!} \\
&= \frac{M}{L} (e^{L|x - x_0|} - 1). \quad (21)
\end{aligned}$$

The absolute convergence of this series has several consequences. First, from calculus, we get that the partial sums, which by (17) are $\phi_n(x) - \phi_0(x)$, converge to a function $\psi(x)$ for all $|x - x_0| \leq \bar{a}$. Since $\phi_0(x)$ does not depend on n, we have $\phi_n(x) \to \phi(x) = \psi(x) + \phi_0(x)$. Secondly, we can modify estimate (20) to show that

$$\phi_n(x) \text{ converges uniformly to } \phi(x) \text{ on } |x - x_0| \leq \bar{a} \quad (22)$$

so that, from advanced calculus, $\phi(x)$ is a continuous function. This is Requirement 4. (Uniform convergence is discussed in Exercises 18 and 19.)

3. There remains only to show that

$$\lim_{n \to \infty} \int_{x_0}^x f(s, \phi_n(s)) \, ds + y_0 = \int_{x_0}^x f(s, \phi(s)) \, ds + y_0. \quad (23)$$

But

$$\left| \int_{x_0}^{x} f(s, \phi_n(s))\, ds - \int_{x_0}^{x} f(s, \phi(s))\, ds \right| \leq \int_{x_0}^{x} |f(s, \phi_n(s)) - f(s, \phi(s))|\, ds$$

$$\leq L \int_{x_0}^{x} |\phi_n(s) - \phi(s)|\, ds \quad \text{[by (14)]}$$

$$\leq L\bar{a} \max_{|x - x_0| \leq \bar{a}} |\phi_n(x) - \phi(x)|, \quad (24)$$

which goes to zero since $\phi_n(x)$ converges to $\phi(x)$ uniformly on $|x - x_0| \leq \bar{a}$.

Since Requirements 1 through 4 have been met, we have established the existence of a solution to (1). To show uniqueness of this solution, suppose that $y(x)$, $z(x)$ are two solutions of the initial value problem (1) on $|x - x_0| \leq \bar{a}$. Let

$$d(x) = (y(x) - z(x))^2.$$

We shall show that $y(x) = z(x)$ by showing that $d(x) \equiv 0$ on $|x - x_0| \leq \bar{a}$.

First note that

$$d(x_0) = (y(x_0) - z(x_0))^2 = (y_0 - y_0)^2 = 0$$

since $y(x)$ and $z(x)$ satisfy the same initial condition. Also, since $y(x)$, $z(x)$ are solutions of (1),

$$d'(x) = 2(y(x) - z(x))(y'(x) - z'(x))$$
$$= 2(y(x) - z(x))(f(x, y(x)) - f(x, z(x)))$$

so that by (15),

$$|d'(x)| \leq 2|y(x) - z(x)|L|y(x) - z(x)| = 2Ld(x).$$

Hence

$$d'(x) \leq 2Ld(x), \qquad (25)$$

so that

$$d'(x) - 2Ld(x) \leq 0.$$

Multiplying this last inequality by e^{-2Lx} gives

$$e^{-2Lx}d'(x) - e^{-2Lx}2Ld(x) \leq 0$$

or

$$(e^{-2Lx}d(x))' \leq 0. \qquad (26)$$

Taking the definite integral from x_0 to x of both sides of (26) yields

$$e^{-2Lx}d(x) - e^{-2Lx_0}d(x_0) \leq 0$$

or $d(x) \leq e^{2L(x-x_0)}d(x_0)$. But $d(x_0) = 0$ and $d(x) \geq 0$ implies that $d(x) \equiv 0$ as desired. ∎

This completes our proof of Theorem C.2. Additional discussion and examples can be found in the exercises at the end of this appendix.

Exercises

Exercises 1 through 6 illustrate the use of iteration on algebraic equations. These exercises are best done with a programmable calculator or computer.

As mentioned in the introduction to this appendix, iteration can sometimes be used to solve algebraic equations, $h(y) = 0$, by rewriting them as $y = G(y)$, making an initial guess ϕ_0, and then using the iteration $\phi_{n+1} = G(\phi_n)$ to generate a sequence of approximations $\{\phi_n\}$ to the solution of $y = G(y)$. There are many ways to get $y = G(y)$ from $h(y) = 0$. For example,

$$y^3 + 4y - 1 = 0 \tag{27}$$

could be rewritten as

$$y = \frac{1}{y^2 + 4} \tag{28}$$

or

$$y = \frac{1}{4}(1 - y^3). \tag{29}$$

Since the polynomial $y^3 + 4y - 1$ is continuous on $[0, 1]$ and changes sign in going from 0 to 1, it is clear that the polynomial has a root in the interval $[0, 1]$.

1. Take ϕ_0 in $[0, 1]$ and find a solution of (27) using iteration on (28).

2. Take ϕ_0 in $[0, 1]$ and find a solution of (27) using iteration on (29).

3. Using iteration and ϕ_0 in $[0, 1]$, find a solution of $y^5 + y^3 + 9y - 1 = 0$.

4. Using iteration and ϕ_0 in $[0, 1]$, find a solution of $y^5 + 2y^3 + 14y - 1 = 0$.

5. Observe that $y^5 - 1 = 0$ has 1 as a solution and may be rewritten as $y = G(y) = y^{-4}$. Take any $\phi_0 \neq 1$ and observe that the iteration $\phi_{n+1} = G(\phi_n)$ does not converge to 1.

6. Observe that $y^5 - 1 = 0$ has 1 as a solution and may be rewritten as $y = G(y) = y^5 + y - 1$. Take any $\phi_0 \neq 1$ and observe that the iteration $\phi_{n+1} = G(y_n)$ does not converge to 1.

In Exercises 7 through 16,

a) Rewrite the given initial value problem as an integral equation.

b) Using Picard iteration, find the first m approximations to the solution of the given initial value problem.

c) Find the exact solution of the given initial value problem.

d) Graph the m approximations from part (b) and the exact solution from part (c) on the same set of coordinates. (Best done with a computer.)

7. $\dfrac{dy}{dx} = 3y$, $\quad y(0) = 1$, $\quad m = 4$

8. $\dfrac{dy}{dx} = -y$, $\quad y(0) = 2$, $\quad m = 4$

9. $\dfrac{dy}{dx} = y + x$, $\quad y(0) = 0$, $\quad m = 4$

10. $\dfrac{dy}{dx} = -y + x$, $\quad y(0) = 0$, $\quad m = 4$

11. $\dfrac{dy}{dx} = 2xy$, $\quad y(0) = 1$, $\quad m = 4$

12. $\dfrac{dy}{dx} = -xy$, $\quad y(0) = 2$, $\quad m = 4$

13. $\dfrac{dy}{dx} = \dfrac{1}{y}$, $\quad y(0) = 1$, $\quad m = 3$

14. $\dfrac{dy}{dx} = y^2 - 1$, $\quad y(0) = 0$, $\quad m = 3$

15. $\dfrac{dy}{dx} = y^2 - 2y + 1$, $\quad y(1) = 0$, $\quad m = 3$

16. $\dfrac{dy}{dx} = \dfrac{x}{y^2}$, $\quad y(1) = 1$, $\quad m = 2$

17. The proof of Theorem 2 can be modified to yield estimates for the error in using $\phi_n(x)$ as an approximation for the solution $\phi(x)$.

a) Show that for $|x - x_0| \leq \bar{a}$,

$$|\phi_n(x) - \phi(x)| = \left| \sum_{i=n+1}^{\infty} [\phi_i(x) - \phi_{i-1}(x)] \right|.$$

b) Show that for $|x - x_0| \leq \bar{a}$,

$$|\phi_n(x) - \phi(x)| \leq \frac{M}{L} \sum_{i=n+1}^{\infty} \frac{L^i |x - x_0|^i}{i!}$$

$$= \frac{M}{L} \left(e^{L|x - x_0|} - \sum_{i=0}^{n} \frac{L^i |x - x_0|^i}{i!} \right).$$

c) Show that there is a constant P, which depends on $\bar{a}$ and f, such that
$$|\phi_n(x) - \phi(x)| \le \frac{M}{L} P \frac{|x - x_0|^{n+1}}{(n+1)!}.$$

18. By definition, $\{\phi_n(x)\}$ converges uniformly to $\phi(x)$ on the interval I if for every $\varepsilon > 0$, there is an N so that $|\phi_n(x) - \phi(x)| \le \varepsilon$ for all $n \ge N$ and all x in I. Verify that (21) holds.

19. Uniform convergence of the $\{\phi_n(x)\}$ was needed to insure that the limit $\phi(x)$ was continuous. To see how a limit function may fail to be continuous, let
$$\psi_n(x) = \frac{x^n}{1 + x^n}, \quad \text{for } 0 \le x \le 2.$$
   a) Show that there is a function $\psi(x)$ such that $\lim_{n \to \infty} \psi_n(x) = \psi(x)$ exists for every x in $[0, 2]$.
   b) Find $\psi(x)$ and show that $\psi(x)$ is not continuous.

20. Show that $\phi_n(x) = x^n$ converges to 0 for every x in $[0, 1)$, but that the convergence is not uniform.

A function $g(y)$ is *Lipschitz continuous* on the interval I if there is a constant L such that $|g(y_1) - g(y_2)| \le L|y_1 - y_2|$ for all $y_1, y_2 \in I$. L is called the *Lipschitz constant*. A function $g(x, y)$ is *uniformly Lipschitz continuous* on a region R if there is a constant L so that $|g(x, y_1) - g(x, y_2)| \le L|y_1 - y_2|$ for all (x, y_1) and (x, y_2) in R. Equation (15) showed that Assumption 2 implied $f(x, y)$ was uniformly Lipschitz continuous on a rectangular region containing (x_0, y_0).

21. Show that Theorem 2 is still true if Assumption 2 is replaced by the assumption that $f(x, y)$ is uniformly Lipschitz continuous on the rectangular region R with Lipschitz constant L.

22. Show that $f(y) = y^{1/3}$ is continuous on $[-1, 1]$ but not Lipschitz continuous. (Hint: Show that if f is Lipschitz continuous on $[-1, 1]$ with Lipschitz constant L, then $|f'(\hat{y})| \le L$ for all $\hat{y} \in [-1, 1]$ where $f'(\hat{y})$ is defined.)

23. Show that $f(y) = |y|$ is Lipschitz continuous on $[-1, 1]$, but not differentiable everywhere on $[-1, 1]$, so that assuming $f(x, y)$ is uniformly Lipschitz continuous in y is a weaker assumption than assuming $f_y(x, y)$ is continuous.

As noted in the text, a stronger existence and uniqueness result is possible for first-order linear equations such as
$$\frac{dy}{dx} + p(x)y = q(x), \quad y(x_0) = y_0. \tag{30}$$

Existence and uniqueness can be proven for (30) by using the formulas derived in Section 2.3. Exercises 24 and 25 discuss how to establish existence and uniqueness for (30) using the ideas of this appendix.

24. Suppose that $p(x)$, $q(x)$ are continuous on a fixed interval $[\alpha, \beta]$. Show that one can prove Theorem C.2 for (30) for any point $x_0 \in [\alpha, \beta]$, with the solution existing and being unique on $|x - x_0| \le \tilde{a}$ where $\tilde{a} = \min\{\bar{a}, |\alpha - x_0|, |\beta - x_0|\}$ and $\bar{a}, M, L$ do not depend on x_0.

25. (Continuation of Exercise 24.) Show that under the assumptions of Exercise 24, the solutions of (30) are unique and exist on all of $[\alpha, \beta]$.

The linear result giving existence of the solution on all of $[\alpha, \beta]$ in Exercises 24 and 25 can be extended to cover some nonlinear systems.

26. a) Show that if $f(x, y)$, $f_y(x, y)$ are continuous for all (x, y) and are bounded functions, then for any (x_0, y_0), the solution of $dy/dx = f(x, y)$, $y(x_0) = y_0$ exists for all x.
   b) Verify that $dy/dx = (1 + x^2 + y^2)^{-1}$ satisfies the assumptions of part (a).

27. a) Show that if $f(x, y)$, $f_y(x, y)$ are continuous for all (x, y) and are bounded on sets of the form $|x - x_0| \le n$, y arbitrary, where the bound may depend on n, then solutions of $dy/dx = f(x, y)$ exist for all x.
   b) Show that $dy/dx = \sin(xy)$ satisfies the assumptions of part (a) of this exercise, but not of part (a) of Exercise 26.

Solutions to Selected Exercises

Chapter 1

Section 1.1

1. Ordinary, first-order linear
3. Ordinary, first-order nonlinear
5. Ordinary, second-order nonlinear
7. Ordinary, first-order linear
9. Ordinary, first-order linear
11. Ordinary, first-order nonlinear
13. $c = e^{-2}$, linear
15. $c_1 = -3$, $c_2 = -2$, linear
17. $c = 16$, nonlinear
19. $c = 2e$, linear
21. $c = 3e$, linear
23. $r = -2$
25. $r = -3$
27. $r = 1, -1$
29. $r = 1, 3$
31. $r = 0, -1$
33. $y = \cos x + c_1 \dfrac{x^2}{2} + c_2 x + c_3$
35. $c_1 = -5c_2$
37. $c_1 = 3c_3$, $c_2 = -4c_4$
39. $c_1 = c_3$, $c_2 + 3 = c_3 + c_4$
47.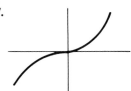
49. a) $y = \tfrac{1}{2}x^2 + c$ 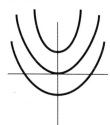 b) $y = -\tfrac{1}{2}x^2 + c$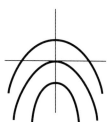
51. $c_1 = 3$, $c_2 = -4$
53. $y = c_2 \sin t$, c_2 arbitrary
55. $\dfrac{d^2 x}{dt^2} = k(x - \cos t)$
57. $\dfrac{d^2 x}{dt^2} = \dfrac{k}{x^3 - x^2}$

Chapter 2

Section 2.1
1. All (x_0, y_0)
3. All (x_0, y_0)
5. $x_0 + y_0 \neq 0$
7. $y_0 \neq 1$
9. $x_0^2 + 2y_0^2 < 1$
11. $f(x, y) = x^{-1/3}$, $f_y = 0$ (if $x \neq 0$). Both continuous if $x \neq 0$.
13. $y_0 = 0$
15. $y_0 = 0$
17. $x_0 = \dfrac{\pi}{2} + n\pi$, $n = 0, \pm 1, \pm 2, \ldots$
19. None
21. $y = 1$ and $y = \left(\dfrac{4}{5}x - \dfrac{4}{5}x_0\right)^{5/4} + 1$, f_y not continuous at $y_0 = 1$.

Section 2.2

1.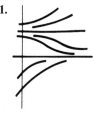
3.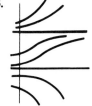
5.
7.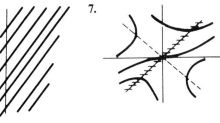

9.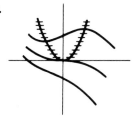
11.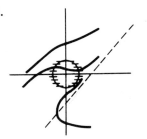
13.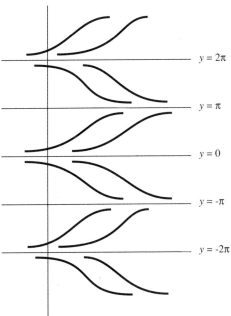

15.

17. $y'' = f_x + f_y y'$
19. $g(c) = 0, g'(c) < 0 \Rightarrow g(y) > 0$ for $y < c$, but near c, and $g(y) < 0$ for $y > c$, but near c.
21. Implies $g(y) > 0$ for y near c but not equal to c.
23. $y = 0$ stable, $y = 1$ unstable
25. $y = 1$ unstable, $y = 2$ semistable
27. $y = 2n\pi$ stable, $y = (2n+1)\pi$ unstable, n an integer

31.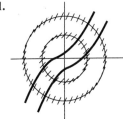

33.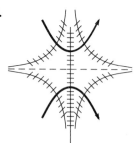

35. (Note $y = x - 1$ is also a solution.)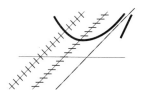

Section 2.3

1. $y = x^{-1}e^x + (1 - e)x^{-1}$

3. $y = 3e^x + c$ (*Note* This can be done by just antidifferentiating both sides.)

5. $y = x(x^2 + 1)^{-1} + C(x^2 + 1)^{-1}$

7. $y = \dfrac{x}{4} - \dfrac{1}{16} + \dfrac{1}{16}e^{-4x}$

9. $y = 4x^2$

11. $y = x^{-1} \ln x + Cx^{-1}$

13. $y = \dfrac{1}{4}x + Cx^{-3}$. One solution continuous at $(0, 0)$. For the rest, $|y| \to \infty$ as $x \to 0$.

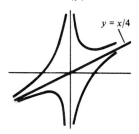

15. $y = x^2 + Cx$. All solutions continuous, and pass through $(0, 0)$.

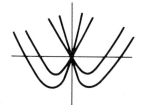

17. $x = -y - 1 + Ce^y$

19. $y = \begin{cases} 2e^{-2x}, & 0 \le x \le 1, \\ 2e^{-1}e^{-x} = 2e^{-(x+1)} & 1 \le x \le 2 \end{cases}$

21. $y = \begin{cases} x, & 0 \le x \le 1, \\ 2 - x, & 1 \le x \le 2 \end{cases}$

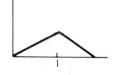

23. $y = \begin{cases} 2e^{-x}, & 0 \le x \le 1, \\ x + 2e^{-1} - 1, & 1 \le x \le 2 \end{cases}$

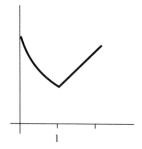

25. *Hint:* $e^{\int p(x)dx + C_1} = e^{\int p(x)dx}e^{C_1} = ue^{C_1}$, so the new integrating factor is just a constant times the old one.

27. Differentiate uy and use $u(y' + py) = (uy)'$ to get $u'/u = p$. Antidifferentiate both sides with respect to x to get $\ln |u| = \int p(x) \, dx + C$. Exponentiate both sides to show that $u = e^{\int p(x)dx}\tilde{C}$.

29. $\frac{dy}{dx} = 1$, $(\sin y = \sin(\pi/2) + \cos(\pi/2)(y - (\pi/2)) + \cdots)$

31. $\frac{dy}{dx} = 4y + 2$, $(y^3 + y = (-1)^3 + (-1) + (3(-1)^2 + 1)(y - -1) + \cdots)$

35. $y = e^{-x^2} \int e^{x^2} dx + Ce^{-x^2}$

37. $y = e^{-x^3/3} \int e^{x^3/3} x \, dx + Ce^{-x^3/3}$

39. $y = e^{-e^x} \int 3e^{e^x} dx + Ce^{-e^x}$

Section 2.4

1. $y = (\frac{1}{3} + Ce^{-3x})^{-1}$ and $y = 0$
3. $y = (C - 3x)^{-1/3}$ and $y = 0$
5. $y = (-3x^2 + Cx)^{-1}$ and $y = 0$
7. $y = (1 + Ce^{x^2})^{-1/2}$ and $y = 0$
9. $y = (-1 + Ce^{x/2})^2$ and $y = 0$
11. $y = (-\frac{4}{3}x + C)^{-3/4}$ and $y = 0$
13. $y = (-\frac{1}{4}x^3 + Cx)^{-2}$ and $y = 0$
15. $y = (-1 + Ce^{-4x})^{-1/4}$ and $y = 0$
17. $\frac{dy}{dx} = y - \frac{y^3}{6}$
19. $\frac{dw}{dx} = 3w + 3w^2$
21. $z' = az^2 + (b + 2ar)z$
23. $z' = z^2$, $y = -1 + (-x + C)^{-1}$ and $y = -1$
25. $z' = z^2 - z$, $y = 1 + (1 + Ce^x)^{-1}$ and $y = 1$ (or $z' = z^2 + z$, $y = 2 + (-1 + Ce^{-x})^{-1}$, and $y = 2$)
27. Show $v' + (1 - n)pv = (1 - n)q$, $v(x_0) = 0$, has a solution v that is nonzero near x_0.

Section 2.5

1. $y = Cx - 1$
3. $y = e^x + C$
5. $y = Ce^{(x^2/2)+4x} - 3$
7. $y = 3x + C$
9. $y = (9 - 4x)^{-1/4}$
11. $u^3 + 12u = t^3 + 3t + 13$
13. $y = \tan\left(\frac{x^3}{3} + x + \tan^{-1} 2\right)$
15. $-\ln|y| + \ln|y - 1| = x + C$; $y = (1 - Ke^x)^{-1}$ and $y = 0$
17. $\ln|y - 1| - \ln|y - 2| - (y - 2)^{-1} = x + C$; $y = 1$, $y = 2$
19. $xe^x ye^y = \tilde{C}$
21. $\left[\frac{z-3}{z+3}\right] = \tilde{C}\left[\frac{t+2}{t-2}\right]^{3/2}$
23. $-e^{-x} - \frac{e^{-4y}}{4} = C$; $y = -\frac{1}{4}\ln(-4e^{-x} + K)$
25. $z' = bf(z) + a$
27. $\frac{1}{2}\tan^{-1}(2x + 8y - 2) = x + C$; $y = \frac{1}{4}(1 - x + \frac{1}{2}\tan(2x + K))$
29. $(x + y + 1)e^{-x-y} = -x + C$

Section 2.6

1. $x^2 + xy + y^2 = C$
3. $y \ln x + y^3 = C$
5. Not exact
7. $x^{-2} + e^{x^2}y + y^2 = C$
9. $u^2 + u\theta + u^2\theta^2 + \theta^3 = C$
11. Not exact
13. $x^2 + x^3 y^2 - y^5 = C$
15. $x^2 e^{xy} + y^3 = C$
17. $\ln(x + y) + y^3 = C$
19. $\sin(x + y) + \cos(y^2) = C$
23. Need $F(x, y) = F(x_0, y_0)$ for all (x, y) on the path followed.
25. Exact $\Leftrightarrow (-F_y)_y = (F_x)_x$

Section 2.7
1. $x^2 + yx + y^{-2} = C, u(y) = y^{-3}$
3. $x^{-2} + xy + 3y^2 = C, u(x) = x^{-3}$
5. $ye^x + x = C, u(x) = e^x$
7. $x^2 e^{-y} + 3y^2 = C, u(y) = e^{-y}$
9. $x^3 y^3 + y^5 = C, u(y) = y^2$
11. $x^4 y^5 + y^6 = C, u(y) = y^4$
13. $x^3 y^3 + y^2 + x^2 = C, u(y) = y^2$
15. The method of this section gives the solution as
$$e^{\int p(x)dx} y - \int e^{\int p(x)dx} g(x)\, dx = C.$$
19. $y^2 e^x + e^x = C, u(x) = e^x, x + \ln(y^2 + 1) = C, \tilde{u}(y) = (y^2 + 1)^{-1}$
21. $u = xy^2, x^2 y^3 + x^3 y^4 = C$
23. $u = x^{-3} y^{-3}, x^{-3} y^{-3} + x^{-2} y = C$

Section 2.8
7. $e^{-y/x} = -\ln|x| + C$ or $y = -x \ln(C - \ln|x|)$

9. $\ln\left|\dfrac{y}{x} - 1\right| - 2\left(\dfrac{y}{x} - 1\right)^{-1} = -\ln|x| + C$

11. $\dfrac{1}{2}\ln\left|\left(\dfrac{y}{x}\right)^2 + 2\dfrac{y}{x} + 3\right| = -\ln|x| + C$ or $y^2 + 2xy + 3x^2 = K$

13. $-\ln\left|\dfrac{y}{x}\right| + \dfrac{3}{2}\ln\left|\dfrac{y}{x} - 1\right| + \dfrac{1}{2}\ln\left|\dfrac{y}{x} + 1\right| = -\ln|x| + C$ or $(y - x)^3(y + x) = \tilde{C}y^2$

15. $\ln\left|\dfrac{y}{x} - 1\right| + \dfrac{2}{\sqrt{3}} \tan^{-1}\left(\dfrac{2}{\sqrt{3}}\left(\dfrac{y}{x} + \dfrac{1}{2}\right)\right) = -\ln|x| + C$

17. $y = x(C - 3\ln|x|)^{-1/3}$

19. $y = x \tan(\ln x + C)$

21. a) $(ax + by) = m(dx + ey)$ for all x, y implies that $a = md$ and $b = me$.
   b) k, l are the solution of $\begin{cases} ak + bl = c, \\ dk + el = f \end{cases}$

23. $x = u + 1, y = w - 3; \dfrac{1}{2}\ln\left|\dfrac{y+3}{x-1} - 4\right| + \dfrac{1}{2}\ln\left|\dfrac{y+3}{x-1} - 2\right| = -\ln|x - 1| + C, (y - 4x + 7)(y + 5 - 2x) = \tilde{C}$

25. $x = u, y = w - 2; \dfrac{1}{2}\ln\left[\left(\dfrac{y+2}{x} + 1\right)^2 + 2\right] = -\ln|x| + C, y^2 + 4y + 4 + 2yx + 4x + 3x^2 = K$

Section 2.9
1. 110.04 g

3. a) 31.063 years
   b) 75.018 years

5. a) $\dfrac{10 \ln 17}{\ln 17 - \ln 15} \approx 226.4$ minutes

7. $k = -\dfrac{1}{10} \ln 3 = -0.1099$

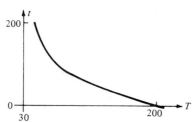

b) $t(T) = \dfrac{10}{\ln 15 - \ln 17} \ln\left(\dfrac{T - 30}{170}\right)$

9. a) $\dfrac{dT}{dt} = k(T - T_0) = k(T - (20 + 10t))$

b) $T = 10 + 10t + 30e^{-t}$

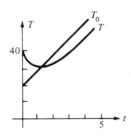

11. a) $\dfrac{dh}{dt} = \dfrac{9}{\pi} kh^{-1}$, $h(t) = (36 - 27t)^{1/2}$

b) $t = 4/3$ minutes

13. $h = (-2t + 18\pi)^2 (9\pi)^{-2}$, $v = 9\pi h$

15. a) $\dfrac{dA}{dt} = rA$, $A = e^{rt} A(0)$

b) $e^{0.06} - 1 \approx 0.0618$

c) $R = e^r - 1$

17. a) $\dfrac{dA}{dt} = 0.06(A - 500)$, $A(0) = 2000$

b) $A = 500 + 1500 e^{0.06t}$, $\$3233.18$

c) $\$411.06$

19. 9.97 yr

Section 2.10

1. a) $Q(t) = 120 - 60 e^{-t/150}$ lb,

$c(t) = \dfrac{Q}{300} = \dfrac{2}{5} - \dfrac{1}{5} e^{-t/150}$ lb/gallon

b) $t = -150 \ln(0.5) \approx 104$ minutes

3. a) $Q(t) = 25 - e^{-t/25}\, 15$,

$c(t) = 0.25 - e^{-t/25}\, 0.15$

b) $\lim_{t \to \infty} c(t) = 0.25 > 0.2$

5. a) $Q(t) = \begin{cases} 10\left(\dfrac{100}{100 + t}\right)^{1/5}, & \text{if } 0 \le t \le 100, \\ 10 e^{3(100-t)/500}(2^{-1/5}), & 100 \le t \end{cases}$

$c(t) = \begin{cases} \dfrac{Q(t)}{500 + 5t}, & 0 \le t \le 100 \\ \dfrac{Q(t)}{1000}, & 100 \le t \end{cases}$

b)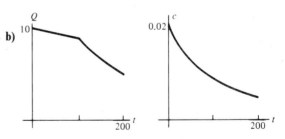

7. $V \equiv 2000$, $Q = 12 \times 10^4 (1 - e^{-t/200})$

$c = \dfrac{Q}{2000} = 60(1 - e^{-t/200})$

9. a) No, $\dfrac{dV}{dt} = 0$ when $h = 1$, $V' > 0$ if $h < 1$, $V' < 0$ if $h > 1$, $V = \dfrac{1}{3}\pi$ m^3

b) $\dfrac{h^2 dh}{dt} = 2 - h - h^2$, $h - \dfrac{4}{3} \ln|h + 2| + \dfrac{1}{3} \ln|h - 1| = -t - \dfrac{4}{3} \ln 2$

11. $Q' = 0.08 Q + 365 B$, $Q(0) = 1000$

a) $B = \$3.79$

b) $B(x) = (800 - 80 e^{0.08x})(365 e^{0.08x} - 365)^{-1}$

13. $Q' = 0.2Q + 400 \cos 2\pi t$, $Q(0) = 100$

a) $Q = \dfrac{1}{\pi^2 + 0.01}[-20 \cos 2\pi t + 200\pi \sin 2\pi t] + \left[100 + \dfrac{20}{\pi^2 + 0.01}\right]e^{0.2t}$

b)

15. $Q' = k_1 Q - k_2 Q + k$

Section 2.11

1. $y = -x + c_2$

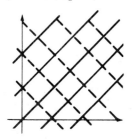

3. $y^2 - x^2 = c_2$

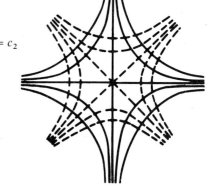

4. Uses Section 2.8 (Homogeneous)

5. $2y^2 \ln|y| - y^2 = x^2 - 2x^2 \ln|x| + c_2$

7. $y = -\dfrac{\ln|x|}{2} + C_2$

9. $y = \pm\dfrac{4}{3}x^{3/2} + C_2$

11. $y + \dfrac{y^3}{3} = -x + C_2$

13. $y^2 = 2\ln|\sin x| + C_2$

15. $2y^2 = -x^2 + C_2$

17. $y = [C - \tfrac{3}{2}\ln|x|]^{-1}$

19. $y' + p(x)y = q(x)$ with $p(x) = 0$

21. Always

23. $(N_y + M_x)/M$, a function of x, or $(M_x + N_y)/N$, a function of y

Section 2.12

1. $\dfrac{di}{dt} + 2i - 1 = 0$. Equilibrium is $i = 1/2$

$i = \tfrac{1}{2} + Ce^{-2t} = \tfrac{1}{2} + (i(0) - \tfrac{1}{2})e^{-2t}$

3. $\dfrac{di}{dt} + i - \sin t = 0$, $i = \dfrac{1}{2}\sin t - \dfrac{1}{2}\cos t + \left(i(0) + \dfrac{1}{2}\right)e^{-t}$

5. $L\dfrac{di}{dt} + Ri = e$

a) $\lim\limits_{t \to \infty}\left[\dfrac{e}{R} + \left(i(0) - \dfrac{e}{R}\right)e^{-Rt/L}\right] = \dfrac{e}{R}$

b) L arbitrary, $R = 8.5/4.2 = 2.02$

7. $\dfrac{di}{dt} + 2i = e(t)$, $i(t) = \begin{cases} 0 & 0 \le t \le 10 \\ 0.75(1 - e^{20-2t}) & 10 \le t \le 20 \end{cases}$

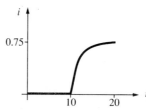

9. $\dfrac{dq}{dt} + q = 3 \sin t, q(0) = 1,$

$q = \dfrac{3}{2}(\sin t - \cos t) + \dfrac{5}{2}e^{-t}$

11. $i\dfrac{di}{dt} + i = 2t$ or $\dfrac{di}{dt} = \dfrac{2t - i}{i}$ (Homogeneous, Section 2.8)

$\dfrac{2}{3}\ln\left|\dfrac{i}{t} + 2\right| + \dfrac{1}{3}\ln\left|\dfrac{i}{t} - 1\right| = -\ln t + \dfrac{2}{3}\ln 4$ or $(i + 2t)^2(i - t) = 16$

12. (Bernoulli, Section 2.4)

Section 2.13

1. **a)** $v = 1960(1 - e^{-t/2})$ **b)** $v = 1947$ **c)** 1960 3. 14416 cm/sec

5. **a)** $v = \sqrt{32}\dfrac{(1 + Ke^{-2\sqrt{32}t})}{(1 - Ke^{-2\sqrt{32}t})}$, $K = (1000 - \sqrt{32})(1000 + \sqrt{32})^{-1}$

   b) $v = \sqrt{32}$

7. **a)** $x = \alpha\beta(1 - e^{(\alpha - \beta)kt})(\beta - \alpha e^{(\alpha - \beta)kt})^{-1}$, $a = \alpha - x$

   b) $\lim_{t \to \infty} x = \alpha$, $\lim_{t \to \infty} a = 0$, $\lim_{t \to \infty} b = \beta - \alpha$

9. **a)** $-\tfrac{1}{2}\ln|x - 1| + \ln|x - 2| - \tfrac{1}{2}\ln|x - 3| = t + \ln 2 - \tfrac{1}{2}\ln 3$

   or $\dfrac{|x - 2|}{|x - 1|^{1/2}|x - 3|^{1/2}} = \dfrac{2}{\sqrt{3}}e^t$ **b)** $\lim_{t \to \infty} x = 1$ **c)** 0.7318 sec

11. $R(v) \approx R(0) + R'(0)v = R'(0)v$ 13. $Q = (1 + e^{-t}C)^{-1}$, $C = Q(0)^{-1} - 1$

15. $Q = \dfrac{r}{k}(1 + e^{-rt}C)^{-1}$, $C = \dfrac{r}{k}Q(0)^{-1} - 1$

17. $f(0) = a = 0$, $f'(Q) = b + 2cQ$, $f'(Q) < 0$ large $Q \Rightarrow c < 0$ $f'(Q) > 0$ for any Q in $0 < Q < \dfrac{|b|}{4|c|} \Rightarrow b > 0$

Chapter 3

Section 3.1

1. All (x_0, y_0, z_0)
5. $z_0^2 + y_0^2 \neq 0$
9. $x_0 \neq 0$ and $z_0 \neq -1$
13. $y = x^2 + Cx$

3. All (x_0, y_0, z_0)
7. All (x_0, y_0, z_0)
11. $y_0 \neq 0$ and $z_0 \neq -1$
15. $y = x^2 + C$

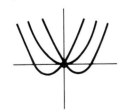

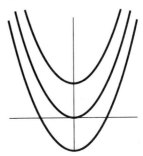

17. $y = \dfrac{x^3}{6} + \dfrac{x^2}{2} + Cx + 1$

19. $y = \dfrac{x^3}{6} + \dfrac{x^2}{2} - x + C$

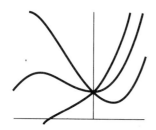
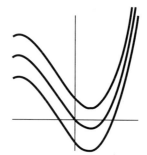

21. $f(x, y, z) = \dfrac{xz}{y}$, not continuous at $(0, 0, 0)$.

23. Theorem 3.1.1 implies that for any (x_0, y_0, z_0), there is a solution y such that $(x_0, y_0, z_0) = (x, y(x), y'(x))$ when $x = x_0$.

Section 3.2

1. $y = \dfrac{2x}{3} - \dfrac{C_1}{3}e^{-3x} + C_2 = \dfrac{2x}{3} + \tilde{C}_1 e^{-3x} + C_2$

3. $y = \cos x$

5. $y = x + 3$

7. $y = C_1 \tan^{-1} x + C_2$

9. $\dfrac{y}{C_1} + \dfrac{1}{C_1^2}\ln|-1 + C_1 y| = x + C_2$ and $\dfrac{-y^2}{2} = x + C_2$ and $y = C_1$

11. $y = -\tan^{-1} x$

13. $\ln|e^{-y} - C_1| = -x + C_2$ and $y = $ constant, or $y = -\ln(C_1 + Ke^{-x})$

15. $y = C_2 - \ln|C_1 - x|$, $y = C$

17. $y = (C_1 + x)\ln(C_1 + x) - x + C_2$

19. Get $\dfrac{dy}{dx} = \dfrac{1}{-x^2 - K}$

21. $dv = f(y)\,dy$, $dy = [\int f(y)\,dy]\,dx$

Section 3.3

1. $20x'' + 980x = 9800$, $x = 10\sin(7t + C) + 10$, $C = \sin^{-1}(-1) = -\pi/2$;
$x = 10 - 10\cos 7t$.

3. $(20x')' = 980x + 9800$,
$\ln|x + 10 + \sqrt{x^2 + 20x}| = \ln 10 + 7t$,
$x = -10 + 5(e^{7t} + e^{-7t}) = -10 + 10\cosh 7t$

5. $(\delta x x')' = F - \delta x 980$
$x = \dfrac{3}{1960}\left\{\dfrac{F}{\delta} - \left[\dfrac{980}{3}t - \sqrt{\dfrac{F}{\delta}}\right]^2\right\}$
$= \dfrac{3F}{1960\delta}\left\{1 - \left[\dfrac{980\sqrt{\delta}}{3\sqrt{F}}t - 1\right]^2\right\}$

7. $x = 4\ln|y + \sqrt{y^2 - 16}| - 4\ln 4$ or
$y = 4\cosh\left(\dfrac{x}{4}\right)$

Section 3.4

1. $y = c_1 \sin x + c_2 \cos x + 1$, $y = 1 - \cos x$

3. $y = c_1 e^x + c_2 e^{2x} + x + \tfrac{3}{2}$, $y = -\tfrac{1}{2}e^{2x} + x + \tfrac{3}{2}$

5. $y = c_1 e^{-x} \cos x + c_2 e^{-x} \sin x + 3$, $y = -2e^{-x}\cos x - e^{-x}\sin x + 3$

7. $y = c_1 \cos x + c_2 \sin x + x \sin x$, $y = \cos x - \sin x + x \sin x$
9. b) $\tilde{y} = 2e^x + e^{2x} + e^{-x} = e^x + e^{2x} + 2\cosh x = y$
11. $y = c_1 \cos x + c_2 \sin x - \frac{1}{3} \sin 2x - \frac{1}{3} \cos 2x$
13. $a_2 \neq 0 \Rightarrow p, q, f$ continuous
15. $\int \frac{y_1'}{y_1} dx = \int \frac{y_2'}{y_2} dx \Rightarrow \ln|y_1| = \ln|y_2| + \tilde{c}$
17. Let $x = -1, x = 1, x = 2$

Section 3.5
1. Yes, $W[\sin x, \cos x] = -1 \neq 0$.
3. x, no; only one solution.
5. a) $W[y_1, y_2](1) = -1 \neq 0$ b) $y_3 = 2y_1 + 2y_2$
7. $W[y_1, y_2](x_0) = 1$
9. Manipulate definition (1)
11. $\tan x = 1$, no

Section 3.6
1. $6x^2 + x^4$
3. $3e^{-x} - 3xe^{-x}$
5. e^{-x}
7. $L_1 L_2 = xD^2 + D$, $L_2 L_1 = xD^2$
9. $D^3 + 3D^2 + D + 3$
11. $D^3 - D$
13. $L_1 L_2 = 2 \cos x D^2 + \sin x D^3$, $L_2 L_1 = \sin x (D^3 + D)$
15. $L_2 L_1(y) = L_2(f) = 0$
17. $L_1(y) = 3 \Rightarrow y_1 = c_1 e^{-x} + 3$, $L_2(y_2) = y_1 \Rightarrow y_2' - y_2 = c_1 e^{-x} + 3$
$\Rightarrow y_2 = c_2 e^x - \frac{c_1}{2} e^{-x} - 3$
21. $h = f - g$, $L(h) = L(f) - L(g) = 0$.
23. $p(D)f = m(D)q(D)f + r(D)f = m(D)0 + r(D)f$
$= r(D)f$
25. b) $(D^5 + 3D^2 + 1) = (D^3 - 2D^2 + 3D - 1)(D^2 + 2D + 1) - D + 2$
$\Rightarrow (D^5 + 3D^2 + 1)xe^{-x} = (-D + 2)xe^{-x} = -e^{-x} + 3xe^{-x}$

Section 3.7
1. e^{2x}, $y = e^x + c_2 e^{2x} + c_1 x e^{2x}$
3. e^x (or e^{2x}), $y = -xe^x + c_1 e^{2x} + c_2 e^x$
5. e^x (or e^{-x}), $y = \frac{xe^x}{2} + c_1 e^{-x} + c_2 e^x$
7. e^{2x} (or e^{-2x}), $y = \frac{xe^{2x}}{4} + \tilde{c}_1 e^{-2x} + c_2 e^{2x}$
9. $\left\{ \frac{\ln x}{x}, \frac{1}{x} \right\}$, $y = c_1 \frac{\ln x}{x} + \frac{c_2}{x}$
11. $y_1 = x^{-2}$, $y = \frac{x^6}{64} + c_1 x^{-2} \ln x + c_2 x^{-2}$
13. $y_1 = x^3$, $y = x^3 \left(\frac{\ln x}{2} \right)^2 + c_1 x^3 \ln x + c_2 x^3$
15. $y = -c_1(x + 1) + c_2 e^x = \tilde{c}_1(x + 1) + c_2 e^x$
17. $y = c_1 x + c_2 \left[1 - \frac{x}{2} \ln \left| \frac{1 + x}{1 - x} \right| \right]$

Section 3.8
1. $y = c_1 e^{2x} + c_2 e^{-3x}$
3. $y = c_1 \cos x + c_2 \sin x$
5. $y = c_1 e^{-2x} \cos x + c_2 e^{-2x} \sin x$
7. $y = c_1 e^x + c_2 e^{2x}$
9. $y = c_1 + c_2 e^x$
11. $y = c_1 + c_2 x$
13. $y = c_1 e^{-x} + c_2 e^{x/3}$
15. $y = \frac{1}{3} e^x - \frac{1}{3} e^{-2x}$
17. $y = c_1 \cos \sqrt{2} x + c_2 \sin \sqrt{2} x$
19. $y = -e^{-3x} + 3e^{-x}$
21. $y = c_1 e^{-5x} + c_2 x e^{-5x}$
23. $y = c_1 e^{7x} + c_2 x e^{7x}$
25. $y = c_1 e^{3x} \cos 4x + c_2 e^{3x} \sin 4x$
27. $W[e^{\alpha x} \cos \beta x, e^{\alpha x} \sin \beta x] = \beta e^{2\alpha x} \neq 0$ if $\beta \neq 0$
29. $y'' + 3y' + 2y = 0$
31. $y'' - 6y' + 9y = 0$
33. $y'' + 16y = 0$
35. $y'' + 9y = 0$
37. $y'' = 0$
39. $y'' - 2y' + 2y = 0$

Section 3.9
1. Yes
3. b) $\tilde{y} = y_1 + \frac{1}{3} y_3$
5. b) $y = x + c_1 + c_2 e^x + c_3 x e^x$
7. b) $y = -e^{-x} + c_1 e^x + c_2 x e^x + c_3 x^2 e^x$
9. $[-1, 1]$
11. x_0 not singular point $\Rightarrow W(x)$ always or never zero on an interval containing x_0.

Section 3.10
1. $y = c_1 e^{2x} + c_2 x e^{2x} + c_3 x^2 e^{2x}$
3. $y = c_1 e^x + c_2 e^{-x} + c_3 e^{2x} + c_4 e^{-2x}$
5. $y = c_1 + c_2 e^x + c_3 e^{-2x}$
7. $y = c_1 e^{-x} + c_2 x e^{-x} + c_3 x^2 e^{-x} + c_4 x^3 e^{-x}$
9. $y = c_1 e^{3x}$
11. $y = c_1 e^{-x} \cos x + c_2 e^{-x} \sin x + c_3 e^x$
13. $y = c_1 e^{-x} \cos x + c_2 e^{-x} \sin x + c_3 x e^{-x} \cos x + c_4 x e^{-x} \sin x$
15. $y = c_1 e^x + c_2 x e^x + c_3 x^2 e^x + c_4 x^3 e^x$
17. $y = c_1 e^x + c_2 x e^x + c_3 x^2 e^x + c_4 e^{-x} + c_5 x e^{-x} + c_6 x^2 e^{-x}$
19. $y = c_1 e^x + c_2 e^{-x} + c_3 \cos x + c_4 \sin x$
21. $y = c_1 \cos 5x + c_2 \sin 5x + c_3 x \cos 5x + c_4 x \sin 5x$
23. $y = c_1 e^x + c_2 e^{-2x} \cos x + c_3 e^{-2x} \sin x$
25. $y''' - 4y'' + 4y' = 0; r(r-2)^2$
27. $y'''' + 50y'' + 625y = 0; (r^2 + 25)^2 = (r + 5i)^2 (r - 5i)^2$
29. $y''' + 6y'' + 12y' + 8y = 0; (r + 2)^3$
31. $y'''' - 4y''' + 8y'' - 8y' + 4y = 0; (r^2 - 2r + 2)^2$
33. $y'''' + 2y'' + y = 0; (r^2 + 1)^2$

Section 3.11
1. $y = c_1 x + c_2 x^{-1}$
3. $y = c_1 x^{-1} + c_2 x^{-1} \ln x$
5. $y = c_1 x^{-1/2} + c_2 x^{-1/2} \ln x$
7. $y = 2x^{-1} - x^{-2}$

9. i) $x\dfrac{dy}{dx} = x\dfrac{dy}{ds}\dfrac{ds}{dx} = ks\dfrac{dy}{ds}\dfrac{1}{k} = s\dfrac{dy}{ds}$ (etc.)

11. $y = c_1 x^{-2} + c_2 x^{-2} \ln x$

13. $y = c_1 x^{-1} + c_2 x^{-3}$

15. $y = c_1 + c_2 \ln x$

19. $y = c_1 \cos(\ln x) + c_2 \sin(\ln x) + c_3 (\ln x) \cos(\ln x) + c_4 (\ln x) \sin(\ln x)$

21. $y = c_1 + c_2 x + c_3 x^{-2}$

Section 3.12

1. Yes

3. No; $\ln|x|$

5. No; $\dfrac{\sin x}{\cos x}$ not allowed

7. No; not constant coefficients

9. No; negative power of x

11. Yes, $\sinh 3x = \dfrac{1}{2} e^{3x} - \dfrac{1}{2} e^{-3x}$

13. $y = -2xe^x + c_1 e^{2x} + c_2 e^x$

15. $y_p = A_1 \sin x + A_2 \cos x$, $y = \dfrac{3}{5} \sin x - \dfrac{3}{10} \cos x + c_1 e^{-x} \cos 2x + c_2 e^{-x} \sin 2x$

17. $y_p = A_1 \sin 2x + A_2 \cos 2x$, $y = -\dfrac{1}{3} \cos 2x + \dfrac{1}{3} \cos x + \sin x$

19. $y_p = (A_0 x + A_1 x^2) e^x$, $y = -\dfrac{1}{4} x e^x - \dfrac{1}{4} x^2 e^x + c_1 e^{3x} + c_2 e^x$

21. $y_p = A_1 e^x \cos 3x + A_2 e^x \sin 3x$, $y = -\dfrac{1}{5} e^x \cos 3x + c_1 e^x \cos 2x + c_2 e^x \sin 2x$

23. $y_p = x(A_0 + A_1 x + A_2 x^2) + A_3 e^x$, $y = \dfrac{3}{8} x - \dfrac{3}{4} x^2 + x^3 + \dfrac{1}{5} e^x + \dfrac{29}{32} - \dfrac{17}{160} e^{-4x}$

25. $y_p = A x^2 e^{-x}$, $y = \dfrac{3}{2} x^2 e^{-x} + c_1 e^{-x} + c_2 x e^{-x}$

27. $y_p = A_1 e^x + A_2 x e^x + A_3 e^{3x}$, $y = e^x + x e^x + 2 e^{3x} + c_1 e^{2x} + c_2 x e^{2x}$

29. $y_p = A_0 + A_1 x + A_2 x^2 + A_3 \cos 2x + A_4 \sin 2x$, $y = 6 - 5x + 2x^2 + \dfrac{1}{5} \sin 2x + c_1 e^{-x} + c_2 e^{-4x}$

31. $y_p = A_0 + A_1 x + A_2 x^2$, $y = \dfrac{11}{27} - \dfrac{2}{9} x + \dfrac{1}{3} x^2 + c e^{-3x}$

33. $y_p = A_0 x \sin 2x + B_0 x \cos 2x$, $y = -\dfrac{1}{4} x \cos 2x + c_1 \cos 2x + c_2 \sin 2x$

35. $y_p = (A_1 + A_2 x) e^x$, $y = -3 e^x + x e^x + c e^{2x/3}$

37. $y_p = A_1 x e^x \sin 2x + A_2 x e^x \cos 2x$

39. $y_p = x^2 (A_0 + A_1 x + A_2 x^2 + A_3 x^3 + A_4 x^4) e^{3x}$

41. $y_p = (A_1 + A_2 x) e^{-x} \sin 2x + (A_3 + A_4 x) e^{-x} \cos 2x$

43. $y_p = A_1 x e^{-x} \cos x + A_2 x e^{-x} \sin x + A_3 e^x \cos x + A_4 e^x \sin x$

45. $y_p = x(A_0 + A_1 x + A_2 x^2) e^{2x} + A_3 e^{5x}$

47. $y_p = x(A_0 + A_1 x) \cos 4x + x(A_2 + A_3 x) \sin 4x + A_4 e^{-x} \sin 4x + A_5 e^{-x} \cos 4x + A_6 e^{-4x}$

49. $y_p = (A_0 + A_1 x + A_2 x^2 + A_3 x^3) e^{-x} \sin x + (A_4 + A_5 x + A_6 x^2 + A_7 x^3) e^{-x} \cos x + A_8 x e^x \sin x + A_9 x e^x \cos x$

Section 3.13

1. $y = e^{2x} + c_1 e^x + c_2 x e^x + c_3 x^2 e^x$

3. $y = \frac{1}{2}\cos x + c_1 + c_2 e^x + c_3 e^{-x}$

5. $y_p = Ax\cos x + Bx\sin x + Cx$, $y = -x\cos x + 3x - 2\sin x$

7. $y = -\frac{4}{45}e^x - \frac{1}{3}xe^x + c_1 e^{2x} + c_2 e^{-2x} + c_3 \cos 2x + c_4 \sin 2x$

9. $y = -\frac{1}{40}e^{3x} + \frac{1}{12}xe^{2x} + c_1 e^x + c_2 e^{-x} + c_3 e^{2x} + c_4 e^{-2x}$

11. $y_p = x^3(A_0 + A_1 x + A_2 x^2)e^x$

13. $y_p = x^4(A_0 + A_1 x + A_2 x^2 + A_3 x^3)e^x + (A_4 + A_5 x + A_6 x^2)e^{-x}$

15. $y_p = x[A_0 e^{-x}\cos x + B_0 e^{-x}\sin x]$

17. $r^4 + 4r^3 + 8r^2 + 8r + 4 = (r^2 + 2r + 2)^2$, $y_p = x^2[A_0 e^{-x}\cos x + B_0 e^{-x}\sin x]$

(*Note* In the solutions for Exercises 19 through 25, subscript on constants may vary.)

19. $(D^2 + 1)$, $y_p = c_3 x\sin x + c_4 x\cos x$

21. $(D^2 + 1)^2$, $y_p = c_3 \sin x + c_4 x\sin x + c_5 \cos x + c_6 x\cos x$

23. $D^2 + 2D + 5$, $y_p = c_3 xe^{-x}\cos 2x + c_4 xe^{-x}\sin 2x$

25. $D^3(D + 1)$, $y_p = c_1 + c_2 x + c_3 x^2 + c_4 e^{-x}$

Section 3.14

1. $y = \frac{1}{3}e^{2x} + c_1 e^x + c_2 e^{-x}$, yes

3. $y = (\sin x)\ln|\sin x| - x\cos x + c_1 \sin x + c_2 \cos x$, no

5. $y = (-\ln|\sec x + \tan x| + \sin x)\cos x + (-\cos x)\sin x + c_1 \cos x + c_2 \sin x$
$= -(\cos x)\ln|\sec x + \tan x| + c_1 \cos x + c_2 \sin x$, no

7. $y = -\frac{1}{2}e^{-2x}\ln(1 + e^{2x}) + e^{-x}\tan^{-1}e^x + c_1 e^{-2x} + c_2 e^{-x}$, no

9. $y = -\frac{2}{7}x^{7/2}e^{3x} + \frac{2}{5}x^{5/2}xe^{3x} + c_1 e^{3x} + c_2 xe^{3x}$

$= \frac{4}{35}x^{7/2}e^{3x} + c_1 e^{3x} + c_2 xe^{3x}$, no

11. $y = -\frac{\ln x}{4}e^{-x/2} - \frac{x^{-1}}{4}xe^{-x/2} + c_1 e^{-x/2} + c_2 xe^{-x/2}$

$= -\frac{\ln x}{4}e^{-x/2} + \tilde{c}_1 e^{-x/2} + c_2 xe^{-x/2}$, no

13. $y = \left(-\frac{1}{25}e^{-5x}\right)e^{3x} - \frac{1}{5}xe^{-2x} + c_1 e^{3x} + c_2 e^{-2x} = -\frac{1}{5}xe^{-2x} + c_1 e^{3x} + \tilde{c}_2 e^{-2x}$, yes

15. $y = \frac{x^3}{2} + c_1 x + c_2 x^2$

17. $y = -x^{-1} - x^{-1}\ln x + c_1 + c_2 x^{-1} = -x^{-1}\ln x + c_1 + \tilde{c}_2 x^{-1}$

19. Use (9) with $v_i = \int_0^x v_i'(s)\,ds$ and $y_p = v_1 y_1 + v_2 y_2$

21. $\int_0^x \sinh(x - s)e^{-s^2}\,ds$

23. $-\int_0^x [e^{3(s-x)} - e^{2(s-x)}]\frac{1}{s+1}\,ds$

25. $Kf = \int_0^x f(s)\,ds$

27. $g = K(Lf) \Rightarrow Lg = Lf \Rightarrow g = f + y_h$ and $g(0) = 0$, $g'(0) = 0$. Solve for constants in y_h.

Section 3.15

1. $y = \left(-\dfrac{e^{2x}}{2}\right)1 + \left(\dfrac{e^x}{2}\right)e^x + \left(\dfrac{e^{3x}}{6}\right)e^{-x} + c_1 + c_2 e^x + c_3 e^{-x}$

$= \dfrac{e^{2x}}{6} + c_1 + c_2 e^x + c_3 e^{-x}$

3. $y = \left(-\dfrac{x^4}{8} - \dfrac{x^2}{2} + \dfrac{x^3}{3}\right)1 + \left(\dfrac{x^3}{3} - \dfrac{x^2}{2}\right)x - \dfrac{x^2}{4} \cdot x^2$

$\quad - (x+1)e^{-x}e^x + c_1 + c_2 x + c_3 x^2 + c_4 e^x$

$= \dfrac{-x^4}{24} - \dfrac{x^3}{6} + \tilde{c}_1 + \tilde{c}_2 x + \tilde{c}_3 x^2 + c_4 e^x$

7. $y = -\dfrac{x}{2}e^x - \dfrac{1}{3}e^{-x}e^{2x} + \dfrac{e^{2x}}{12}e^{-x} + c_1 e^x + c_2 e^{2x} + c_3 e^{-x}$

$= -\dfrac{x}{2}e^x + \tilde{c}_1 e^x + c_2 e^{2x} + c_3 e^{-x}$

9. $y = -\dfrac{1}{6}e^{-x}e^x - \dfrac{1}{2}e^x e^{-x} + \dfrac{1}{6}e^{2x}e^{-2x} + c_1 e^x + c_2 e^{-x} + c_3 e^{-2x}$

$= -\dfrac{1}{2} + c_1 e^x + c_2 e^{-x} + c_3 e^{-2x}$

11. $y = \dfrac{x}{8}e^x - \dfrac{1}{8}e^{2x}e^{-x} + \dfrac{1}{32}e^{4x}e^{-3x} + c_1 e^x + c_2 e^{-x} + c_3 e^{-3x}$

$= \dfrac{1}{8}xe^x + \tilde{c}_1 e^x + c_2 e^{-x} + c_3 e^{-3x}$

15. $y = \dfrac{1}{2}\ln x\, e^{-x} + x^{-1}(xe^{-x}) - \dfrac{1}{4}x^{-2}x^2 e^{-x} + c_1 e^{-x} + c_2 xe^{-x} + c_3 x^2 e^{-x}$

$= \dfrac{1}{2}\ln x\, e^{-x} + \tilde{c}_1 e^{-x} + c_2 xe^{-x} + c_3 x^2 e^{-x}$

17. $y = \dfrac{1}{13}x^{13/2}e^{2x} - \dfrac{2}{11}x^{11/2}xe^{2x} + \dfrac{1}{9}x^{9/2}x^2 e^{2x} + c_1 e^{2x} + c_2 xe^{2x} + c_3 x^2 e^{2x}$

$= \dfrac{8}{1287}x^{13/2}e^{2x} + c_1 e^{2x} + c_2 xe^{2x} + c_3 x^2 e^{2x}$

19. $y = \dfrac{x^3}{12}e^x - \dfrac{x^2}{4}xe^x + \dfrac{x}{4}x^2 e^x + c_1 e^x + c_2 xe^x + c_3 x^2 e^x$

$= \dfrac{x^3}{12}e^x + c_1 e^x + c_2 xe^x + c_3 x^2 e^x$

21. $y = \dfrac{x^4}{2}1 + \dfrac{2}{3}x^3 x - \dfrac{8}{7}x^{7/2}x^{1/2} + c_1 + c_2 x + c_3 x^{1/2}$

$= \dfrac{1}{42}x^4 + c_1 + c_2 x + c_3 x^{1/2}$

Section 3.16

1. $30x'' + 1470x = 0$, $x(0) = 10$, $x'(0) = 0$, $x = 10 \cos 7t$

3. $2x'' + 128x = 0$, $x(0) = 2$, $x'(0) = 1$, $x = 2 \cos 8t + \frac{1}{8} \sin 8t$

5. $k = 1000 \pi^2$ ft/sec^2

7. $x = A \cos 2t + B \sin 2t$, $A = \sqrt{A^2 + B^2} \sin \phi = 1.732$
   $B = \sqrt{A^2 + B^2} \cos \phi = 1$, $x(0) = A = 1.732$, $x'(0) = 2B = 2$

9. a) $x = 10 \sqrt{\frac{m}{k}} \sin \sqrt{\frac{k}{m}} t$ b) Amplitude $= 10 \sqrt{\frac{m}{k}}$ c) Decreases d) Increases

11. a) $x = \cos \sqrt{\frac{k}{m}} t + \sqrt{\frac{m}{k}} \sin \sqrt{\frac{k}{m}} t$ b) Amplitude $= \sqrt{1 + \frac{m}{k}}$

    c) Increasing m increases amplitude, increasing k decreases amplitude toward 1.

    d) $\phi = \sin^{-1}\left[\left(1 + \frac{m}{k}\right)^{-1/2}\right]$

    e) As m increases, ϕ decreases toward zero. As k increases, ϕ increases toward $\pi/2$.

13. $\int 0 \, dt = \int mx''x' + kxx' \, dt \Rightarrow c = \frac{1}{2} m(x')^2 + \frac{1}{2} kx^2$

15. $10x'' + 40x' + 30x = 0$, $x(0) = 3$, $x'(0) = -5$, $x = e^{-3t} + 2e^{-t}$, overdamped

17. $x'' + 49x = 0$, $x(0) = 1$, $x'(0) = 7$, $x = \sin 7t + \cos 7t$, $x = \sqrt{2} \sin\left(7t + \frac{\pi}{4}\right)$, harmonic

19. $x'' + 7x' + 12x = 0$, $x(0) = -1$, $x'(0) = 1$, $x = 2e^{-4t} - 3e^{-3t}$, overdamped.

21. $x'' + 4x' + 5x = 0$, $x(0) = 2$, $x'(0) = 0$, $x = 4e^{-2t} \sin t + 2e^{-2t} \cos t$, damped oscillation, $x = e^{-2t} \sqrt{20} \sin(t + \phi)$, $\phi = \sin^{-1}(1/\sqrt{5}) = 0.464$ radians.

23. Differentiate x, set equal to zero, and show that there is at most one solution.

25. $0 < m < \delta^2/4k$, overdamped; $m = \delta^2/4k$, critically damped; $\frac{\delta^2}{4k} < m$ is damped oscillation.

    As $m \to +\infty$, solution decays slower, period $\to \infty$, frequency $\to 0$.

31. $m = (240\pi^2)^{-1}$ g

34. $x = \frac{1}{0.84}(\cos 2t - \cos 2.2t) = \frac{1}{0.42} \sin(0.1t) \sin(2.1t)$ [See figure below.]

35. $x = \dfrac{1}{13} + \dfrac{38}{13}e^{-3t}\cos 2t + \dfrac{57}{13}e^{-3t}\sin 2t$; steady state $= \dfrac{1}{13}$, transient $= \dfrac{38}{13}e^{-3t}\cos 2t + \dfrac{57}{13}e^{-3t}\sin 2t$

37. **a)** $x = \dfrac{1}{10}\sin t - \dfrac{3}{10}\cos t - \dfrac{1}{5}e^{-2t} + \dfrac{1}{2}e^{-t}$ **b)** $-\dfrac{1}{5}e^{-2t} + \dfrac{1}{2}e^{-t}$

39. **a)** $x = \dfrac{3}{25} - \dfrac{1}{17}e^{-2t} + \dfrac{393}{425}e^{-3t}\sin 4t + \dfrac{399}{425}e^{-3t}\cos 4t$ **b)** all but $\dfrac{3}{25}$

41. **a)** $x = -\dfrac{1}{4}t\sin t - \dfrac{1}{4}t^2\cos t$ **b)** none

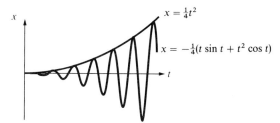

45. $mx'' = mg - \delta x' - k(x + \Delta L - h)$

Section 3.17

1. $q = \dfrac{6}{13}\cos t + \dfrac{9}{13}\sin t - \dfrac{1}{2}e^{-t} + \dfrac{1}{26}e^{-5t}$,

   $i = -\dfrac{6}{13}\sin t + \dfrac{9}{13}\cos t + \dfrac{1}{2}e^{-t} - \dfrac{5}{26}e^{-5t}$

3. $q = 3 + 6e^{-0.5t} - 7e^{-t}$

   $i = -3e^{-0.5t} + 7e^{-t}$

5. $L < 10(60\pi)^{-2}$ or $L > 10(40\pi)^{-2}$

7. $q(t) = \begin{cases} \dfrac{1}{2} - \dfrac{1}{2}e^{-t}\cos t - \dfrac{1}{2}e^{-t}\sin t, & 0 \le t \le \pi \\ -\dfrac{1}{2}(e^{\pi} + 1)e^{-t}(\cos t + \sin t), & \pi \le t \le 2\pi \end{cases}$

9. $\dfrac{dE}{dt} = Lii' + \dfrac{1}{c}qq' = \left(Li' + \dfrac{1}{c}q\right)i = 0i = 0$

11. $q(0) = -\dfrac{4}{35}$, $i(0) = 0$

17. Output $= \dfrac{1}{C}q = \dfrac{1}{C}(A\sin\omega t + B\cos\omega t) = \dfrac{1}{C}\sqrt{A^2 + B^2}\sin(\omega t + \phi)$

19. $M = (2305)^{-1/2}$, $\phi = 4.5$

Section 3.18

1. $y_p = \dfrac{1}{\sqrt{5}}\sin(t - 1.107)$ 3. $y_p = \dfrac{1}{\sqrt{10}}\sin(t - 1.249)$ 5. $y_p = \tfrac{1}{2}\sin(t - \pi/2)$ 7. $b = 2^{-5/2}$, $c - 4a = 2^{-3/2}$

11. May be done directly since $\sin(\omega t + \psi) = \operatorname{Im}(e^{i\psi}e^{i\omega t})$. Let $f = e^{i\psi}e^{i\omega t}$ and proceed as in the text, to compute y_p, then $\operatorname{Im}(y_p)$. Alternatively, use trigonometric identity $\sin(\omega t + \psi) = \sin\omega t\cos\psi + \sin\psi\cos\omega t$. Then, by Exercise 10, we have $y_p = \sin(\omega t + \psi)\cos\psi + \sin\psi\cos(\omega t + \psi) = \sin(\omega t + \psi + \phi)$.

13. $y_2 = \dfrac{1}{5}\sin(t - 2.214)$

15. $\sin\phi < 0 \Rightarrow -\pi < \phi < 0$

17. Get output $R^{1/3}\sin\left(\omega t + \dfrac{\phi}{3}\right)$. Then cascade twice with itself.

Chapter 4

Section 4.1

1. $\lim_{t \to 2^+} f(t)$ does not exist.

3. Continuous except at $t = 1$; $\lim_{t \to 1^+} f(t)$, $\lim_{t \to 1^-} f(t)$ exist.

5. $(1 - e^{-s})/s$

7. $(1 - 2e^{-s} + e^{-2s})/s^2$

13. $\dfrac{3s}{s^2 - 4}$

15. $\dfrac{30}{s^2 + 36}$

17. $-\dfrac{1}{s^2} + \dfrac{3}{s}$

19. $\dfrac{2}{s} + \dfrac{s}{s^2 + 25}$

21. $\dfrac{3}{s^2 - 9}$

23. $\dfrac{2}{s+1} + \dfrac{6}{s-3}$

25. $\dfrac{3}{s^2} - \dfrac{1}{s} + \dfrac{s}{s^2 - 4}$

27. $t - 1$

29. $\dfrac{1}{3} \sin 3t$

31. $1 + t$

33. $3 - 7t + 19 \sin t$

35. $3 \cos 4t + \dfrac{7}{4} \sin 4t$

37. $\cosh t + 2 \sinh t$ or $\dfrac{3}{2} e^t - \dfrac{1}{2} e^{-t}$

39. $1 + e^t$

41. $-\dfrac{1}{3} e^{-2t} - \dfrac{2}{3} e^t$

43. $-\dfrac{4}{5} e^{-3t} + \dfrac{4}{5} e^{2t}$

45. $\dfrac{1}{2} + \dfrac{1}{2} e^{2t} - e^t$

47. $-\dfrac{2}{7} + \dfrac{1}{6} e^{-t} + \dfrac{47}{42} e^{-7t}$

49. $-3 + 2e^t + e^{-t}$

53. Yes; $e^{\sqrt{t}} \leq e^t$ if $t \geq 1$.

Section 4.2

1. $\dfrac{2}{3} \sin 3t$

3. $e^t \cos 5t + \dfrac{1}{5} e^t \sin 5t$

5. $\dfrac{4}{5!} t^5 e^{-3t} = \dfrac{1}{30} t^5 e^{-3t}$

7. $\dfrac{1}{6} e^{3t} - \dfrac{1}{6} e^{-3t}$ if we use (T2) and partial fractions; $\dfrac{1}{3} \sinh 3t$ if we use (T7).

9. $2e^{3t} \cos 3t + \dfrac{11}{3} e^{3t} \sin 3t$

11. $3e^{-5t} \cos t - 17 e^{-5t} \sin t$

13. $\dfrac{3}{9!} t^9$

15. $\dfrac{1}{8} \cosh 2t - \dfrac{1}{8} \cos 2t = \dfrac{1}{16} e^{2t} + \dfrac{1}{16} e^{-2t} - \dfrac{1}{8} \cos 2t$

17. $\dfrac{1}{24} \sin 3t + \dfrac{9}{8} \cos 3t - \dfrac{1}{8} \sin t - \dfrac{1}{8} \cos t$

19. $\dfrac{4}{3} \cosh 2t - \dfrac{1}{3} \cosh t = \dfrac{2}{3} e^{2t} + \dfrac{2}{3} e^{-2t} - \dfrac{1}{6} e^t - \dfrac{1}{6} e^{-t}$

21. $\dfrac{3}{\sqrt{14}} \sin \sqrt{\dfrac{7}{2}} t$

23. $\dfrac{1}{3! \, 3^5} t^4 e^{-t/3}$

25. $-\dfrac{1}{4} + \dfrac{3}{8} e^{2t} - \dfrac{1}{8} e^{-2t}$

27. e^t

29. $\dfrac{1}{6} + \dfrac{5}{2} e^{-2t} - \dfrac{5}{3} e^{-3t}$

31. $\dfrac{1}{2} + \dfrac{3}{2} e^{2t} - 2e^t$

33. $\dfrac{1}{10} e^{-t} + \dfrac{7}{10} \sin 3t + \dfrac{9}{10} \cos 3t$

35. $\dfrac{2}{13} + \dfrac{22}{39} e^{-2t} \sin 3t + \dfrac{11}{13} e^{-2t} \cos 3t$

37. $-1 + \dfrac{9}{4} e^t - \dfrac{1}{4} e^{-t} + 2 \cos t + \dfrac{5}{2} \sin t$

39. $t - e^t$

41. $\dfrac{1}{2} t^2 e^{-t} + t e^{-t}$

43. $\dfrac{1}{2} e^{-t} \cos t + \dfrac{1}{2} e^{-t} \sin t - \dfrac{1}{2} e^{-2t}$

45. Use change of variables $\tau = ct$.

49. If $F(s) = 1/(s - a)$, $F^{(n)}(s) = (-1)^n \dfrac{n!}{(s-a)^{n+1}}$

51. $-\dfrac{d}{ds}\left(\dfrac{s}{s^2+a^2}\right) = \dfrac{s^2-a^2}{(s^2+a^2)^2}$

53. $\dfrac{2s+1}{(s^2+1)^2} = \dfrac{2s}{(s^2+1)^2} + \dfrac{1}{2}\dfrac{2}{(s^2+1)^2}$, $y = t\sin t + \dfrac{1}{2}(\sin t - t\cos t)$

55. $\dfrac{s^2-s}{(s^2+4)^2} = \dfrac{1}{s^2+4} - \dfrac{s+4}{(s^2+4)^2} = \dfrac{1}{2}\dfrac{2}{s^2+4} - \dfrac{1}{4}\dfrac{2\cdot 2s}{(s^2+4)^2} - \dfrac{1}{2}\dfrac{2\cdot 4}{(s^2+4)^2}$,

$$y = \dfrac{1}{2}\sin 2t - \dfrac{1}{4}t\sin 2t - \dfrac{1}{2}\left(\dfrac{1}{2}\sin 2t - t\cos 2t\right) = \dfrac{1}{4}\sin 2t - \dfrac{1}{4}t\sin 2t + \dfrac{1}{2}t\cos 2t$$

57. $\dfrac{s^3-1}{(s^2+9)^2} = \dfrac{s^3+9s-9s-1}{(s^2+9)^2} = \dfrac{s}{s^2+9} - \dfrac{9s+1}{(s^2+9)^2} = \dfrac{s}{s^2+9} - \dfrac{3}{2}\dfrac{2\cdot 3s}{(s^2+9)^2} - \dfrac{1}{18}\dfrac{2\cdot 9}{(s^2+9)^2}$,

$$y = \cos 3t - \dfrac{3}{2}t\sin 3t - \dfrac{1}{18}\left(\dfrac{1}{3}\sin 3t - t\cos 3t\right)$$

59. $\dfrac{1}{2}(\sin t - t\cos t)$ **61.** $\dfrac{5}{8}t\sin 4t + \cos 4t + \dfrac{1}{4}\sin 4t$

Section 4.3

1. $f(t) = 3[H(t-2) - H(t-5)] + tH(t-5)$
$ = 3H(t-2) + (t-3)H(t-5)$

3. $f(t) = \sin t[1 - H(t-\pi)] + \sin t[H(t-2\pi) - H(t-3\pi)]$

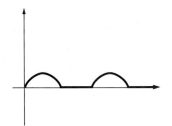

5. $f(t) = 1[1 - H(t-1)] + 1[H(t-2) - H(t-3)] = 1 - H(t-1) + H(t-2) - H(t-3)$

7. $f(t) = (t-1)[H(t-1) - H(t-2)] + [H(t-2) - H(t-3)]$
$ + (4-t)[H(t-3) - H(t-4)]$
$ = (t-1)H(t-1) + (2-t)H(t-2) + (3-t)H(t-3) - (4-t)H(t-4)$

9. $f(t) = (2-t)[H(t-1) - H(t-3)] + (t-4)[H(t-3) - H(t-5)]$

11. $Y(s) = e^{-2s}\left[\dfrac{1}{s^2} + \dfrac{2}{s}\right]$ **13.** $Y(s) = e^{-s}\left[\dfrac{6}{s^4} + \dfrac{6}{s^3} + \dfrac{3}{s^2} + \dfrac{2}{s}\right]$

15. $Y(s) = e^{-\pi s}\left[-\dfrac{1}{s^2+1}\right]$ **17.** $Y(s) = e^{-2\pi s}\left[\dfrac{s}{s^2+1}\right]$

19. $Y(s) = e^{-3s}\left[\dfrac{e^6}{s-2}\right]$

21. $Y(s) = e^{-2s}e^{10}\left[\dfrac{1}{(s-5)^2} + \dfrac{2}{s-5}\right]$

23. $y(t) = [-e^{-(t-2)} + 1]H(t-2) + [-e^{-(t-3)} + 1]H(t-3)$
$= [1 - e^{2-t}]H(t-2) + [1 - e^{3-t}]H(t-3)$

25. $y(t) = e^{-(t-3)}\sin(t-3)H(t-3)$

27. $y(t) = \sin(t-1)H(t-1) - \dfrac{1}{2}\sin(2(t-2))H(t-2)$

29. $[4 + 6\cos(3(t-1))]H(t-1)$

31. $y(t) = [1 - \cos(t-2)]H(t-2) - [1 - \cos(t-5)]H(t-5)$

33. $y(t) = e^{3t} + \left[-\dfrac{1}{9} - \dfrac{t}{3} + \dfrac{1}{9}e^{3t}\right] - 2\left[-\dfrac{1}{9} - \dfrac{(t-1)}{3} + \dfrac{1}{9}e^{3(t-1)}\right]H(t-1)$
$+ \left[-\dfrac{1}{9} - \dfrac{(t-2)}{3} + \dfrac{1}{9}e^{3(t-2)}\right]H(t-2)$

35. $y(t) = \left[\dfrac{1}{10} + \dfrac{9}{10}e^{-t}\cos 3t + \dfrac{3}{10}e^{-t}\sin 3t\right]$
$- \left[-\dfrac{1}{10}e^{-(t-3)}\cos(3t-9) - \dfrac{1}{30}e^{-(t-3)}\sin(3t-9) + \dfrac{1}{10}\right]H(t-3)$

37. $y(t) = \dfrac{1}{4}[1 - \cos 2t] - \dfrac{1}{4}[1 - \cos(2(t-1))]H(t-1)$
$+ \dfrac{1}{4}[1 - \cos(2(t-2))]H(t-2) - \dfrac{1}{4}[1 - \cos(2(t-3))]H(t-3)$

39. $y(t) = \dfrac{1}{4}[1 - \cos 2t] + \dfrac{1}{2}\sum_{n=1}^{\infty}(-1)^n[1 - \cos(2(t-n))]H(t-n)$

Section 4.4

1. $G(s) = \dfrac{1 + e^{-\pi s}}{(s^2 + 1)(1 - e^{-\pi s})}$

3. $G(s) = \left[\dfrac{2}{s^3} - \left(\dfrac{4}{s^3} + \dfrac{4}{s^2}\right)e^{-s} + \left(\dfrac{2}{s^3} + \dfrac{4}{s^2} + \dfrac{2}{s}\right)e^{-2s}\right]\dfrac{1}{1 - e^{-2s}}$

5. $G(s) = \dfrac{e^{1-s} - 1}{(1 - s)(1 - e^{-s})}$

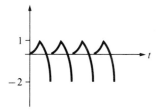

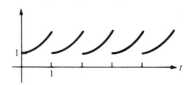

7. $G(s) = \left[\dfrac{1}{s} - \dfrac{2}{s}e^{-s} + \dfrac{1}{s}e^{-2s}\right]\dfrac{1}{1-e^{-2s}}$

9. $\displaystyle\sum_{n=0}^{\infty}\left[(t-n) + \dfrac{(t-n)^2}{2}\right]H(t-n)$

11. $\displaystyle\sum_{n=0}^{\infty}\cos(2(t-n))H(t-n)$

13. $\displaystyle\sum_{n=0}^{\infty}\left[(t-2n) + \dfrac{(t-1-2n)^2}{2}H(t-1-2n)\right]H(t-2n)$

$= \displaystyle\sum_{n=0}^{\infty}(t-2n)H(t-2n) + \dfrac{(t-1-2n)^2}{2}H(t-1-2n)$

15. $\displaystyle\sum_{n=0}^{\infty}\left[1 + \sin\left(t - \dfrac{\pi}{2} - n\pi\right)H\left(t - \dfrac{\pi}{2} - n\pi\right)\right]H(t-n\pi)$

$= \displaystyle\sum_{n=0}^{\infty}H(t-n\pi) + \sin\left(t - \dfrac{\pi}{2} - n\pi\right)H\left(t - \dfrac{\pi}{2} - n\pi\right)$

17. $\displaystyle\sum_{n=0}^{\infty}(-1)^n\left[\dfrac{1}{2}(t-5n)^2 H(t-5n) + \dfrac{1}{6}(t-5n-2)^3 H(t-5n-2)\right]$

19. $G(s) = \dfrac{1}{s^2}\operatorname{csch} s = \dfrac{2}{s^2(e^s - e^{-s})} = \dfrac{2e^{-s}}{s^2(1 - e^{-2s})}$

$g(t) = \displaystyle\sum_{n=0}^{\infty} 2(t-2n-1)H(t-2n-1)$

23. $\dfrac{1}{3}\displaystyle\sum_{n=0}^{\infty}\left\{\left[2\sin\left(t - \dfrac{n\pi}{2}\right) - \sin(2t - n\pi)\right]H\left(t - \dfrac{n\pi}{2}\right)\right.$

$\left. + \left[2\sin\left(t - \dfrac{\pi}{2} - \dfrac{n\pi}{2}\right) - \sin(2t - \pi - n\pi)\right]H\left(t - \dfrac{\pi}{2} - \dfrac{n\pi}{2}\right)\right\}$

25. $e^{-t} + \displaystyle\sum_{n=0}^{\infty}\sinh(t-n)H(t-n) - e\sinh(t-n-1)H(t-n-1)$

27. $\dfrac{1}{4}\displaystyle\sum_{n=0}^{\infty}[\cosh(2(t-2n)) - 1]H(t-2n) - 2[\cosh(2(t-2n-1)) - 1]H(t-2n-1)$

$+ [\cosh(2(t-2n-2)) - 1]H(t-2n-2)$

29. $\displaystyle\sum_{n=0}^{\infty}\dfrac{1}{s^{n+1}} = \dfrac{1}{s}\displaystyle\sum_{n=0}^{\infty}\dfrac{1}{s^n} = \dfrac{1}{s\left(1 - \dfrac{1}{s}\right)} = \dfrac{1}{s-1}.$

Section 4.5

1. $y(t) = e^{-8(t-1)}H(t-1) + e^{-8(t-2)}H(t-2) = e^{8-8t}H(t-1) + e^{16-8t}H(t-2)$

3. $y(t) = \dfrac{1}{10}e^{-3(t-1)}\sin(10(t-1))H(t-1) - \dfrac{1}{10}e^{-3(t-7)}\sin(10(t-7))H(t-7)$

5. $y(t) = \dfrac{1}{3}(1 - e^{-3t}) + \dfrac{1}{2}[e^{-(t-3)} - e^{-3(t-3)}]H(t-3)$

7. $y(t) = 1 + \sin(t - 2\pi)H(t-2\pi) = 1 + \sin t\, H(t-2\pi)$

9. (a) The following are graphs of the solutions of Exercises 1 through 4.

(1)

(2)

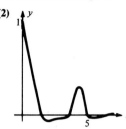

(3)

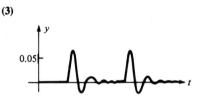

(4)

11. $\int_0^\infty \delta^{(n)}(t-a)e^{-st}\,dt = (-1)^n \left[\dfrac{d^n(e^{-st})}{dt^n}\right]\bigg|_{t=a} = (-1)^n(-s)^n e^{-as}$

13. d), e). Note that $\int_{-\infty}^\infty g(t)f_n(t)\,dt = n\int_0^{1/n} g(t)\,dt.$

Section 4.6

1. $t*e^t = -t - 1 + e^t$

3. t

5. $t - \sin t$

9. $X(s) = \dfrac{1}{s^2 - s + 1}$, $x(t) = \dfrac{2}{\sqrt{3}} e^{t/2} \sin\left(\dfrac{\sqrt{3}}{2}t\right)$

11. $X(s) = \dfrac{2(s+1)}{s^2}$, $x(t) = 2 + 2t$

13. $X(s) = \dfrac{s^2 + 4}{s^2(s^2+1)}$, $x(t) = 4t - 3\sin t$

15. $x(t) = 3 - \dfrac{3}{2}t^2$

17. $|F(s)| \le \int_0^\infty e^{-st}|f(t)|\,dt \le \int_0^\infty e^{-st}Me^{\alpha t}\,dt$

$= \dfrac{M}{s-\alpha}$ for $s > \alpha.$

Section 4.7

1. $s + 1$

3. $\dfrac{s}{s+1}$

5. $\dfrac{s+1}{[s(s+1)(s^2+1) - 1]}$

7. $s + \dfrac{1}{s}$

9. $\dfrac{1}{s}V_1 = V_0 \Rightarrow v_0 = \int_0^t v_1(\tau)\,d\tau \Rightarrow v_1 = v_0'$

13. $\dfrac{(s^2 + 16\pi^2)(s^2 + 64\pi^2)}{(s^2 + 5184\pi^2)}$

15. $T(s) = T_0(s) + T_1(s)$, where $T_0(s)$ is a polynomial, and $T_1(s)$ has numerator of lower degree than the denominator.

$$v_1(s) = T_0(s)v_0(s) + T_1(s)v_0(s).$$

$T_0(s)v_0(s)$ gives derivatives of $v_0(t)$.

Chapter 5

Section 5.2

1. Yes

3. Yes

5. Yes (all but the first coefficient are zero)

7. No (negative power)

9. Yes

11. No (negative power)

13. Yes

15. Fails

17. $r = \infty, (-\infty, \infty)$

19. $r = 0, \{0\}$

21. $e^{2x} = \sum_{n=0}^{\infty} \frac{2^n e^2}{n!} (x-1)^n$

23. $\cos x = 1 - \frac{x^2}{2} + \frac{x^4}{4!} - \cdots = \sum_{n=0}^{\infty} (-1)^n \frac{x^{2n}}{(2n)!}$

25. $x^3 = x^3$

27. $x = x$

29. $\frac{1}{1+x^2} = \sum_{n=0}^{\infty} (-1)^n x^{2n}$

31. $e^{x^3} = \sum_{n=0}^{\infty} \frac{(x^3)^n}{n!} = \sum_{n=0}^{\infty} \frac{x^{3n}}{n!}$

33. $\ln(1+x) = \sum_{n=0}^{\infty} (-1)^n \frac{x^{n+1}}{n+1}$

37. $p_2(x) = e^2 + 2e^2(x-1) + 2e^2(x-1)^2$

39. $p_5(x) = 1 - \frac{1}{2}\left(x - \frac{\pi}{2}\right)^2 + \frac{1}{24}\left(x - \frac{\pi}{2}\right)^4$

41. $p_2(x) = x$

Section 5.3

1. All x; $(-\infty, \infty)$

3. $x \neq 0$ and $x \neq 1$; $(-\infty, 0) \cup (0, 1) \cup (1, \infty)$

5. $x \neq 2$, $x \neq 3$, and $x \neq n\pi$ for $n = 0, \pm 1, \pm 2, \ldots$

7. $r \geq 1$

9. $r \geq 3$

11. $r \geq 1$ (i a root of $x^2 + 1$), $|0 - i| = 1$

13. $\min\{|3.3 - 2|, |5 - 3.3|\} = 1.3$

Section 5.4

1. $y = 2x - \frac{x^3}{6} + \cdots$

3. $y = 2 + x - \frac{2x^3}{3} + \cdots$

5. $y = (x-1) + \frac{(x-1)^3}{3} + \cdots$

7. $y = y_0 + y_1 x - y_0 \frac{x^2}{2} - y_1 \frac{x^3}{6} + y_0 \frac{x^4}{24} + \cdots = y_0 \left[1 - \frac{x^2}{2} + \frac{x^4}{24} + \cdots\right]$

$+ y_1 \left[x - \frac{x^3}{6} + \cdots\right], (y_p = 0)$

9. $y = y_0 + y_1 x + \frac{x^2}{2} + \frac{x^3}{6} + (1 - 2y_0)\frac{x^4}{24} + \cdots$

$= \left[\frac{x^2}{2} + \frac{x^3}{6} + \frac{x^4}{24} + \cdots\right] + y_0 \left[1 - \frac{x^4}{12} + \cdots\right] + y_1 [x + \cdots]$

11. $y = y_0 + y_1\left(x - \frac{\pi}{2}\right) + y_0 \frac{(x - (\pi/2))^2}{2} + (y_1 - 1)\frac{(x - (\pi/2))^3}{6}$

$= \left[-\frac{(x - (\pi/2))^3}{6} + \cdots\right] + y_0 \left[1 + \frac{(x - (\pi/2))^2}{2} + \cdots\right]$

$+ y_1 \left[(x - (\pi/2)) + \frac{(x - (\pi/2))^3}{6} + \cdots\right]$

13. $y = y_0 + (1 - 2y_0)(x - 1) + (-2 + 6y_0)\dfrac{(x-1)^2}{2} + (8 - 24y_0)\dfrac{(x-1)^3}{6} + (-40 + 120y_0)\dfrac{(x-1)^4}{24}$

$= \left[(x-1) - (x-1)^2 + \dfrac{4}{3}(x-1)^3 - \dfrac{5}{3}(x-1)^4 + \cdots\right]$

$\quad + y_0[1 - 2(x-1) + 3(x-1)^2 - 4(x-1)^3 + 5(x-1)^4 + \cdots]$

15. $y = y_0 + y_1 x + y_0 \dfrac{x^2}{2} + (y_0 + y_1)\dfrac{x^3}{6} + (3y_0 + 2y_1 + 2)\dfrac{x^4}{24} + \cdots$

$= \left[\dfrac{x^4}{12} + \cdots\right] + y_0\left[1 + \dfrac{1}{2}x^2 + \dfrac{1}{6}x^3 + \dfrac{1}{8}x^4 + \cdots\right]$

$\quad + y_1\left[x + \dfrac{1}{6}x^3 + \dfrac{1}{12}x^4 + \cdots\right]$

Section 5.5

1. $(n+1)c_{n+1} + 3c_n = 0$ for $n \geq 0$, $y = 2 - 6x + 9x^2 - 9x^3 + \dfrac{27}{4}x^4 + \cdots$

   (Note $y = 2e^{-3x}$.)

3. $c_{n+1} + c_n = 0$ for $n \geq 0$, $y = 2 - 2(x-1) + 2(x-1)^2 - 2(x-1)^3 + 2(x-1)^4 + \cdots$, $y = 2x^{-1}$

5. $c_2 = \tfrac{1}{2}$, $(n+2)(n+1)c_{n+2} + c_{n-1} = 1$ for $n \geq 1$,

   $y = \dfrac{x^2}{2} + \dfrac{1}{6}x^3 + \dfrac{1}{12}x^4 + \cdots$

7. $2c_2 + 6c_0 = 0$, $(n+2)(n+1)c_{n+2} + (6 - 2n)c_n = 0$ for $n \geq 1$, $y = x - \tfrac{2}{3}x^3$

9. $c_0 = 0$, $c_1 = 1$, $c_2 = \tfrac{1}{2}$, $(n+1)c_{n+1} + 3c_{n-2} = \dfrac{1}{n!}$ for $n \geq 2$

   $y = x + \dfrac{1}{2}x^2 + \dfrac{1}{6}x^3 + \cdots$

11. $c_0 = 0$, $c_1 = 0$, $c_2 = 0$, $c_3 = 1/6$, $(n+2)(n+1)c_{n+2} + c_{n-2} = \begin{cases} 0 & \text{if } n \text{ even} \\ \dfrac{(-1)^{(n-1)/2}}{n!} & \text{if } n \text{ odd} \end{cases}$

13. $C_0 = 2$, $c_1 + c_0 = 0$, $2c_2 + c_1 = 0$, $3c_3 + c_2 + c_0 = 1$, $(n+1)c_{n+1} + c_n + c_{n-2} = 0$ if $n > 2$;
    $c_0 = 2$, $c_1 = -2$, $c_2 = 1$, $c_3 = -2/3$

15. $y = c_0\left[1 + \dfrac{x^2}{2} + \dfrac{x^4}{4!} + \cdots\right] + c_1\left[x + \dfrac{x^3}{3!} + \cdots\right] = c_0 \cosh x + c_1 \sinh x$

17. $y = c_0\left[1 - \dfrac{x^2}{2} + \dfrac{x^4}{8} + \cdots\right] + c_1\left[x - \dfrac{x^3}{6} + \cdots\right]$; recursion is $(n+2)(n+1)c_{n+2} + (n^2 - n + 1)c_n = 0$.

19. $y = c_0\left[1 - \dfrac{x^3}{3} + \cdots\right] + c_1\left[x - \dfrac{x^4}{6} + \cdots\right]$; recursion is $c_2 = 0$, $(n+2)(n+1)c_{n+2} + 2c_{n-1} = 0$ for $n \geq 1$.

21. $y = c_0\left[1 - \dfrac{x^6}{2} + \dfrac{x^{12}}{8} + \cdots\right]$; recursion is $c_1 = \cdots = c_5 = 0$, $(n+1)c_{n+1} + 3c_{n-5} = 0$, $n \geq 5$.

23. Let $h(x)$ be the solution of $ay'' + by' + cy = 0$, $y(0) = 1$, $y'(0) = 0$ and $g(x)$ be the solution of $ay'' + by' + cy = 0$, $y(0) = 0$, $y'(0) = 1$. Solutions are unique and series expansions are unique.

Section 5.6

1. $P_4(x) = \dfrac{1}{8}(3 - 30x^2 + 35x^4)$

2. $P_5(x) = \dfrac{63}{8}x^5 - \dfrac{35}{4}x^3 + \dfrac{15}{8}x$

7. $T_0(x) = 1,\ T_1(x) = x,\ T_2(x) = 2x^2 - 1,\ T_3(x) = 4x^3 - 3x$

9. $H_0(x) = 1,\ H_1(x) = 2x,\ H_2(x) = -2 + 4x^2,\ H_3(x) = -12x + 8x^3$

11. Let $x = 1$ in (1) to get $-2p'(1) + (v^2 + v)p(1) = 0$.

    Then $p(1) = 0 \Rightarrow p'(1) = 0$. Thus $p \equiv 0$ by (3).

Section 5.7

1. Yes

3. Yes (divide by x)

5. No; limits exist but $x^{2/3}$ is not analytic at zero.

7. Yes; $\lim\limits_{x \to 0}\left[\dfrac{\sin x}{x}\right] = 1$, $Q(x) = \dfrac{\sin x}{x}$ is analytic if you let $Q(0) = 1$

9. $2^n n!$

11. $3^{n+1}\dfrac{\Gamma(n + 4/3)}{\Gamma(1/3)}$

13. $4^{n+1}\dfrac{\Gamma(n + 3/2)}{\Gamma(1/2)}$

15. $\dfrac{1}{5^{n+1}}\dfrac{\Gamma(6 + n)}{\Gamma(5)}$

19. $r_1 = \tfrac{1}{2},\ r_2 = 0,\ y_1 = x^{1/2}\sum\limits_{n=0}^{\infty}(-1)^n\dfrac{x^n}{(2n+1)!},\ y_2 = \sum\limits_{n=0}^{\infty}(-1)^n\dfrac{x^n}{(2n)!}$

    $y_1 = \sin(x^{1/2}),\ y_2 = \cos(x^{1/2})$

21. $r_1 = \tfrac{1}{3},\ r_2 = 0,\ y_1 = x^{1/3}\sum\limits_{n=0}^{\infty}\dfrac{(-1)^n \Gamma(4/3)}{3^n n!\, \Gamma(n + 4/3)}x^n$,

    $y_2 = \sum\limits_{n=0}^{\infty}\dfrac{(-1)^n \Gamma(2/3)}{3^n n!\, \Gamma(n + 2/3)}x^n$.

23. $r_1 = \tfrac{1}{2},\ r_2 = 0,\ y_1 = x^{1/2}\sum\limits_{n=0}^{\infty}x^n = \dfrac{x^{1/2}}{1-x},\ y_2 = \sum\limits_{n=0}^{\infty}x^n = \dfrac{1}{1-x}$

25. $r_1 = r_2 = 0,\ y_1 = 1,\ y_2 = \ln x + \sum\limits_{n=0}^{\infty}\dfrac{x^n}{n} = \ln x - \ln(1 - x)$

    (Note $\tilde{C}_0$ arbitrary, take $\tilde{C}_0 = 0$.)

27. $r_1 = 0,\ r_2 = -1,\ y_1 = \sum\limits_{n=0}^{\infty}\dfrac{x^n}{n+1} = x^{-1}\ln(1 - x),\ y_2 = x^{-1}$

    (Note $\tilde{C}_0,\ \tilde{C}_1$ arbitrary, take $\tilde{C}_0 = 1,\ \tilde{C}_1 = 0 \Rightarrow \tilde{C}_n = 0,\ n \geq 1$.)

29. $C_n = \dfrac{(n-1+\alpha)(n-1+\beta)}{n(n-1+\gamma)} C_{n-1} \Rightarrow C_n = \dfrac{\Gamma(n+\alpha)}{\Gamma(\alpha)} \dfrac{\Gamma(n+\beta)}{\Gamma(\beta)} \dfrac{\Gamma(\gamma)}{\Gamma(n+\gamma)n!} C_0$

31. $p_0(x) = 1, p_1(x) = 1 - x, p_2(x) = 1 - 2x + \tfrac{1}{2}x^2$

Section 5.8

1. May be verified like (25).

4. Subtract (29) from (28).

5. $J_{3/2}(x) = \sqrt{\dfrac{2}{\pi x}}\left[\dfrac{\sin x}{x} - \cos x\right]$

6. $J_{5/2}(x) = \sqrt{\dfrac{2}{\pi}}[3x^{-5/2}\sin x - 3x^{-3/2}\cos x - x^{-1/2}\sin x]$

8. $J_{3/2}(x) = 0 \Leftrightarrow \tan x = x$

10.

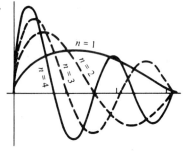

11. $q_5 = 0.673, q_{10} = 0.626, q_{20} = 0.602, q_{30} = 0.593, q_{50} = 0.587, q_{100} = 0.582,$
$q_{200} = 0.579, q_{300} = 0.578$

Chapter 6

Section 6.1

5. Two

7. Three

9. Degenerate

11. Three

13. Two

15. $z' - x + y = t,$
$y' + z = 0,$
$x' - z = 0$

17. $x_3' + x_5' = \sin t,$
$x_4' - x_3 + x_2 = t,$
$x_1' - x_3 = 0,$
$x_3' - x_4 = 0,$
$x_2' - x_5 = 0$

19. $x_3' = 3x_3 + 4x_4 - x_1,$
$x_4' = x_3 - x_4 + x_2,$
$x_1' = x_3,$
$x_2' = x_4$

21. Yes; $x' = 11x - 5y - 2t + 5\cos t,$
$y' = -5x + 2y + t - 2\cos t$

23. Yes; $x' = \dfrac{1}{10}x + \dfrac{39}{10}y + \dfrac{7}{10}t$
$y' = -\dfrac{1}{10}x + \dfrac{21}{10}y + \dfrac{3}{10}t$

Section 6.2

1. $x = \dfrac{3}{2}c_1 e^t + c_2 e^{2t}, \ y = c_1 e^t + c_2 e^{2t}$

3. $x = 8 + 2t - c_1 e^t - 3c_2 e^{-t}, \ y = -6 - t + c_1 e^t + c_2 e^{-t}$

5. (General solution):

   $x = c_2 e^t \cos 2t - c_1 e^t \sin 2t;$

   $y = c_1 e^t \cos 2t + c_2 e^t \sin 2t;$

   $x = -e^t \sin 2t, \ y = e^t \cos 2t$

7. $x = c_1 e^{3t} \cos t + c_2 e^{3t} \sin t - \dfrac{4}{13} \sin t - \dfrac{8}{39} \cos t,$

   $y = -(c_1 + c_2)e^{3t} \cos t + (c_1 - c_2)e^{3t} \sin t - \dfrac{4}{39} \cos t + \dfrac{7}{39} \sin t$

9. $x = c_1 e^t$

11. Let $z = L_2 x + L_3 y$ and note that $L_1 z = 0$

13. $x = c_1 - \dfrac{3}{2}c_2 e^{-5t}, \ y = c_1 + c_2 e^{-5t}$

15. $x = -\dfrac{4}{3} + c_1 e^{3t}, \ y = \dfrac{5}{3} - \dfrac{7}{8}c_1 e^{3t} + c_2 e^{-t}$

17. $x = c_1 e^{3t} + c_2 e^{2t} + c_3 e^{-t}, \ y = -c_1 e^{3t} + c_2 e^{2t} + c_3 e^{-t},$

    $z = c_2 e^{2t} - 2c_3 e^{-t}, \ c_1 = -\dfrac{1}{2}, \ c_2 = \dfrac{1}{3}, \ c_3 = \dfrac{1}{6}$

19. $x = -\dfrac{c_1}{2} + c_2 e^{3t} - c_3 e^{4t}, \ y = -\dfrac{c_1}{2} + c_2 e^{3t} + c_3 e^{4t}, \ z = c_1 + c_2 e^{3t}$

21. $x = c_1 e^t + c_2 e^{3t} + c_3 e^{-3t}, \ y = \dfrac{7}{2}c_1 e^t - \dfrac{1}{4}c_2 e^{3t} + \dfrac{1}{2}c_3 e^{-3t} + c_4 e^{-t}$

23. $x = -1 + 2t - 3c_2 + 2c_1 - c_3 e^t - 3c_4 e^{-t} = -1 + 2t + \tilde{c}_2 - c_3 e^t - 3c_4 e^{-t}$

    $y = -t + 2c_2 - c_1 + c_3 e^t + c_4 e^{-t} = -t + \tilde{c}_1 + c_3 e^t + c_4 e^{-t}$

25. $x = c_1 + \tfrac{5}{14} e^{2t} - \tfrac{3}{2} c_2 e^{-5t} + c_3 e^t$

    $y = c_1 + c_2 e^{-5t} - \tfrac{1}{14} e^{2t}$

Section 6.3

1. $x = 3e^t - e^{2t}, \ y = 2e^t - e^{2t}$

3. $x = 8 + 2t - \dfrac{1}{2}e^t - \dfrac{3}{2}e^{-t}, \ y = -6 - t + \dfrac{1}{2}e^t + \dfrac{1}{2}e^{-t}$

5. $x = -e^t \sin 2t, \ y = e^t \cos 2t$

7. $x = \dfrac{1}{2}e^{3t} \cos t + \dfrac{87}{78}e^{3t} \sin t - \dfrac{4}{13} \sin t - \dfrac{8}{39} \cos t$

   $y = -\dfrac{63}{39}e^{3t} \cos t + \dfrac{8}{13}e^{3t} \sin t + \dfrac{7}{39} \sin t - \dfrac{4}{39} \cos t$

9. $x = e^{5t} + e^{-t}$, $y = 2e^{5t} - 2e^{-t}$, $z = 3e^{5t} + e^{-t}$

11. $x = 2 + te^t + e^t$, $y = 6 + 2te^t$, $z = 1 + te^t$

13. $x = \frac{1}{2}e^t + \frac{1}{2}\cos t - \frac{3}{2}\sin t - 1 + t$, $y = \frac{3}{4}e^t + \cos t - \frac{1}{2}\sin t + \frac{5}{4}e^{-t} - 3 + t$

15. $x = 3 + 2t + e^t - 3e^{-t}$
    $y = -t + 6 - e^t + e^{-t}$

17. $x = 1 - \frac{5}{14}e^{2t} - 3e^{-5t} - e^t$

   $y = 1 + 2e^{-5t} + \frac{1}{14}e^{2t}$

Section 6.4

1. $x' = -x + 50$, $x(0) = 0$,
   $y' = x - 2y$, $y(0) = 0$;
   $x = 50 - 50e^{-t}$,
   $y = 25 - 50e^{-t} + 25e^{-2t}$

3. $x' = -0.1x + 0.2$, $x(0) = 0$,
   $y' = 0.1x - 0.1y$, $y(0) = 0$,
   $x = 2 - 2e^{-0.1t}$,
   $y = 2 - 2e^{-0.1t} - \frac{1}{5}te^{-0.1t}$

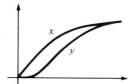

5. $x' = -\frac{x}{50} + \frac{y}{100}$, $x(0) = 50$,

   $y' = \frac{x}{50} - \frac{y}{100}$, $y(0) = 20$,

   $x = \frac{70}{3} + \frac{80}{3}e^{-3t/100}$

   $y = \frac{140}{3} - \frac{80}{3}e^{-3t/100}$

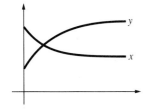

7. $x' = -\frac{3}{2}x + \frac{1}{2}y + 20$, $x(0) = 0$,

   $y' = \frac{3}{2}x - \frac{5}{2}y + 20$, $y(0) = 0$,

   $x = y = 20 - 20e^{-t}$

9. $x' = -9\dfrac{x}{100 + 5t} + 7$, $x(0) = 0$,

$y' = 9\dfrac{x}{100 + 5t} - 6\dfrac{y}{100 + 3t}$, $y(0) = 0$

11. $x' = -5\dfrac{x}{100} + 10$, $\quad x, y, z$ amounts in tanks A, B, C, $x(0) = y(0) = z(0) = 0$.

$y' = 2\dfrac{x}{100} + 3\dfrac{z}{100} - 5\dfrac{y}{100}$,

$z' = 3\dfrac{x}{100} - 3\dfrac{z}{100}$

13. $x' = \alpha\delta - \dfrac{\beta}{V_1}x$,

$y' = \beta\dfrac{x}{V_1} - \gamma\dfrac{y}{V_2}$

Let $\tau = \dfrac{\beta}{V_1}, k = \dfrac{\gamma}{V_2}$,

$x' = -\tau x + \alpha\delta$,

$y' = \tau x - ky$

ii) $y'' + (\tau + k)y' + \tau k y = \alpha\delta\tau$

iii) $-\tau, -k$

iv) $\gamma \neq 0$ and $\beta \neq 0$ produces equilibrium (some outflow from both tanks).

v) Equilibrium $x = \dfrac{\alpha\delta}{\tau}, y = \dfrac{\alpha\delta}{k}$

15. $x = c_1 e^{-t} + c_2 e^{-3t}$, $\quad y = c_1 e^{-t} - c_2 e^{-3t}$

17. $x = 3c_1 e^t - 3c_2 e^{-5t}$, $\quad y = c_1 e^t + c_2 e^{-5t}$

$x(0) > 0, y(0) > 0 \Rightarrow c_1 > 0$. $\lim\limits_{t \to \infty} \dfrac{x}{y} = 3$.

20. $a' = 0.1a + 0.2(b - 2000)$, $\quad a(0) = 1000$

$b' = 0.05b + 3650 - 0.2(b - 2000)$, $\quad b(0) = 2000$

Section 6.5

1. $x = \tfrac{1}{2} \sin t - \tfrac{1}{6} \sin 3t$,

$y = \tfrac{1}{2} \sin t + \tfrac{1}{6} \sin 3t$;

Resonance at frequencies of $1/2\pi$, $3/2\pi$ cycles/sec.

3. $x = te^{-2t}, y = -te^{-2t}$

5. $x = \cos 2t + \sin 2t - \cos 4t - \sin 4t$,

$y = \cos 2t + \sin 2t + \cos 4t + \sin 4t$;

Resonance at $1/\pi$, $2/\pi$ cycles/sec.

7. $x = e^{-t} - e^{-2t} - 2e^{-4t} + e^{-5t}$,
   $y = e^{-t} + e^{-2t} + 2e^{-4t} + e^{-5t}$

9. a) $0 \le \delta^2 < 4k$;
   b) $4k \le \delta^2 < 4k + 8l$;
   c) $4k + 8l < \delta^2$

11. $m_1 x_1'' + \delta_1 x_1' + (k_1 + k_2)x_1 - k_2 x_2 = 0$,
    $m_i x_i'' + \delta_i x_i' + (k_i + k_{i+1})x_i - k_i x_{i-1} - k_{i+1} x_{i+1} = 0, i = 2, \ldots, n-1$
    $m_n x_n'' + \delta_n x_n' + (k_n + k_{n+1})x_n - k_n x_{n-1} = 0$

15. $\mathbf{M} = \begin{bmatrix} m_1 & 0 \\ 0 & m_2 \end{bmatrix}, \mathbf{K} = \begin{bmatrix} k_1 + k_2 & -k_2 \\ -k_2 & k_2 + k_3 \end{bmatrix}$

17. $\mathbf{M} = \begin{bmatrix} m_1 & 0 & 0 \\ 0 & m_2 & 0 \\ 0 & 0 & m_3 \end{bmatrix}, \mathbf{K} = \begin{bmatrix} k_1 + k_2 & -k_2 & 0 \\ -k_2 & k_2 + k_3 & -k_3 \\ 0 & -k_3 & k_3 + k_4 \end{bmatrix}$

Section 6.6

1. $(4D + 4)i_1 - 3i_3 = e_1$,
   $-3i_1 + (5D + 5)i_3 = e_2$

3. $(4D + \frac{1}{4})q_1 - 3i_3 = e_1$,
   $-3Dq_1 + (5D + 5)i_3 = e_2$

5. $i_3 = C_1 e^{-t/2} + C_2 t e^{-t/2}$,
   $q_1 = (2C_1 - 4C_2)e^{-t/2} + 2C_2 t e^{-t/2}$

7. $\alpha + \beta i$ is root of $2RD^2 + D + R/2$.

9. $q_1 = c_1 e^{-t/2}, q_3 = -c_1 e^{-t/2}$

Section 6.7

1. Not possible

3. Not possible

5. -5

7. $\begin{bmatrix} 2 & 4 & 0 \\ 2 & -2 & 2 \end{bmatrix}$

9. Not possible

11. $\begin{bmatrix} -1 & 3 \\ -1 & 2 \end{bmatrix}$

13. Not possible

15. $\begin{bmatrix} 1 & 2 & 0 \\ 1 & -1 & 1 \end{bmatrix}$

17. $\begin{bmatrix} 1 & 0 \\ 0 & 1 \end{bmatrix} = \mathbf{I}$

19. y arbitrary, $x = 1 - \frac{3}{2}y$; $\mathbf{x} = \begin{bmatrix} 1 - \frac{3}{2}y \\ y \end{bmatrix} = \begin{bmatrix} 1 \\ 0 \end{bmatrix} + y \begin{bmatrix} -\frac{3}{2} \\ 1 \end{bmatrix}$

21. $x = \frac{1}{2}y$, y arbitrary; $\mathbf{x} = \begin{bmatrix} y/2 \\ y \end{bmatrix} = y \begin{bmatrix} \frac{1}{2} \\ 1 \end{bmatrix}$

23. Inconsistent

25. $x = 0, y = -1/2, z = -3/2$

27. $x = 1, y = 0, z = -2$

29. $\mathbf{x} = \begin{bmatrix} -1/2 \\ 0 \\ -3/2 \\ -3 \end{bmatrix} + y \begin{bmatrix} 1 \\ 1 \\ 0 \\ 0 \end{bmatrix}$, y arb.

31. Not generally; $(\mathbf{A} + \mathbf{B})^2 = (\mathbf{A} + \mathbf{B})(\mathbf{A} + \mathbf{B}) = \mathbf{A}^2 + \mathbf{BA} + \mathbf{AB} + \mathbf{B}^2$.

33. $\mathbf{A}(c_1 \mathbf{x}_1 + c_2 \mathbf{x}_2) = c_1 \mathbf{A}\mathbf{x}_1 + c_2 \mathbf{A}\mathbf{x}_2 = c_1 \mathbf{0} + c_2 \mathbf{0} = \mathbf{0}$

Chapter 6

35. $A(x_1 + x_2) = Ax_1 + Ax_2 = b_1 + b_2$

37. $B^{-1}A^{-1}(AB) = B^{-1}IB = B^{-1}B = I$

39. $A^{-1} = \begin{bmatrix} -1 & 2 \\ 1 & -1 \end{bmatrix}$

41. $A^{-1} = \begin{bmatrix} -2 & 1 \\ 1 & 0 \end{bmatrix}$

Section 6.8

1. -2

3. -5

5. 2

7. 9

9. No

11. Yes, det $= -4$

15. $\lambda^2 - 4\lambda + 3$

17. $\lambda^2 - 18$

19. $\lambda^3 - 3\lambda^2 + 2\lambda$

21. $\lambda^3 - \lambda^2 - 5\lambda + 5$

Section 6.9

1. $A'(t) = \begin{bmatrix} 0 & 1 \\ 2t & 0 \end{bmatrix}$

3. b) $x = c_1 \begin{bmatrix} e^{-t} \\ -e^{-t} \end{bmatrix} + c_2 \begin{bmatrix} e^{3t} \\ e^{3t} \end{bmatrix}$

  c) $x = \begin{bmatrix} \frac{3}{2}e^{-t} + \frac{1}{2}e^{3t} \\ -\frac{3}{2}e^{-t} + \frac{1}{2}e^{3t} \end{bmatrix}$

5. b) $x = c_1 \begin{bmatrix} e^{3t} \\ e^{3t} \end{bmatrix} + c_2 \begin{bmatrix} 1 \\ -2 \end{bmatrix}$

  c) $x = \begin{bmatrix} \frac{7}{3}e^{3t} + \frac{14}{3} \\ \frac{7}{3}e^{3t} - \frac{28}{3} \end{bmatrix}$

7. b) $x = c_1 \begin{bmatrix} \cos 2t \\ \sin 2t \end{bmatrix} + c_2 \begin{bmatrix} -\sin 2t \\ \cos 2t \end{bmatrix}$

  c) $x = \begin{bmatrix} 19\cos 2t + 37\sin 2t \\ 19\sin 2t - 37\cos 2t \end{bmatrix}$

11. $A(t) = \begin{bmatrix} 0 & 1 \\ -c(t) & -b(t) \end{bmatrix}$, $x(t) = \begin{bmatrix} y(t) \\ z(t) \end{bmatrix}$, $a = \begin{bmatrix} y_0 \\ y_1 \end{bmatrix}$

13. Note that $z_i = y_i'$

15. a) $E = \begin{bmatrix} 1 & 1 \\ 1 & -1 \end{bmatrix}$, $F = \begin{bmatrix} 2 & -3 \\ 2 & -1 \end{bmatrix}$, $g = \begin{bmatrix} t \\ 1 \end{bmatrix}$

  b) $E^{-1} = \frac{1}{2}\begin{bmatrix} 1 & 1 \\ 1 & -1 \end{bmatrix}$, $A = \begin{bmatrix} -2 & 2 \\ 0 & 1 \end{bmatrix}$, $f = \frac{1}{2}\begin{bmatrix} t+1 \\ t-1 \end{bmatrix}$

17.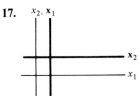

19. $(c_1\mathbf{x}_1 + c_2\mathbf{x}_2)' = c_1\mathbf{x}_1' + c_2\mathbf{x}_2' = c_1\mathbf{A}\mathbf{x}_1 + c_2\mathbf{A}\mathbf{x}_2 = \mathbf{A}(c_1\mathbf{x}_1 + c_2\mathbf{x}_2)$

21. Differentiate $\mathbf{A}\mathbf{A}^{-1} = \mathbf{I}$

Section 6.10

1. $\mathbf{x} = c_1 e^t \begin{bmatrix} -\frac{1}{2} \\ 1 \end{bmatrix} + c_2 e^{2t} \begin{bmatrix} -1 \\ 1 \end{bmatrix}$

3. $\mathbf{x} = \frac{4}{5}\begin{bmatrix} 2 \\ 1 \end{bmatrix} + \frac{6}{5} e^{5t}\begin{bmatrix} -1/2 \\ 1 \end{bmatrix}$

5. $\mathbf{x} = c_1 e^{2t} \begin{bmatrix} -1/2 \\ 1 \end{bmatrix} + c_2 e^{7t} \begin{bmatrix} 2 \\ 1 \end{bmatrix}$

7. $\mathbf{x} = c_1 e^t \begin{bmatrix} 2 \\ 1 \end{bmatrix} + c_2 e^{6t} \begin{bmatrix} -1/2 \\ 1 \end{bmatrix}$

9. $\mathbf{x} = c_1 \begin{bmatrix} -\frac{1}{2} \\ -\frac{1}{2} \\ 1 \end{bmatrix} + c_2 e^{2t} \begin{bmatrix} -1 \\ 1 \\ 0 \end{bmatrix} + c_3 e^{3t} \begin{bmatrix} 1 \\ 1 \\ 1 \end{bmatrix}$

11. $\mathbf{x} = c_1 \begin{bmatrix} -1 \\ 0 \\ 1 \end{bmatrix} + c_2 e^{2t} \begin{bmatrix} 1 \\ 0 \\ 1 \end{bmatrix} + c_3 e^{3t} \begin{bmatrix} 0 \\ 1 \\ 0 \end{bmatrix}$

13. $\mathbf{x} = c_1 \begin{bmatrix} 1 \\ 1 \\ 0 \end{bmatrix} + c_2 e^{3t} \begin{bmatrix} 1 \\ -1 \\ 1 \end{bmatrix} + c_3 e^{6t} \begin{bmatrix} -\frac{1}{2} \\ \frac{1}{2} \\ 1 \end{bmatrix}$

15. $\mathbf{x} = c_1 e^t \begin{bmatrix} -1 \\ 0 \\ 1 \end{bmatrix} + c_2 e^t \begin{bmatrix} 0 \\ 1 \\ 0 \end{bmatrix} + c_3 e^{3t} \begin{bmatrix} 1 \\ 0 \\ 1 \end{bmatrix}$

17. $\mathbf{x} = c_1 e^t \begin{bmatrix} 1 \\ 0 \\ 1 \\ 1 \end{bmatrix} + c_2 e^t \begin{bmatrix} 1 \\ 1 \\ 0 \\ 0 \end{bmatrix} + c_3 e^{-t} \begin{bmatrix} -1 \\ 0 \\ -1 \\ 1 \end{bmatrix} + c_4 e^{-t} \begin{bmatrix} -1 \\ 1 \\ 0 \\ 0 \end{bmatrix}$

19. $p(\lambda) = \lambda^3 - \lambda^2 - 5\lambda + 5$

21. $\lambda = \pm 2i$, $\mathbf{x} = c_1 \begin{bmatrix} -2\sin 2t \\ \cos 2t \end{bmatrix} + c_2 \begin{bmatrix} 2\cos 2t \\ \sin 2t \end{bmatrix}$

23. $\lambda = -1 \pm i$, $\mathbf{x} = e^{-t}\left(c_1 \begin{bmatrix} -\sin t \\ \cos t \end{bmatrix} + c_2 \begin{bmatrix} \cos t \\ \sin t \end{bmatrix}\right)$, $c_1 = -1$, $c_2 = 1$

25. $\lambda = 2 \pm 2i$, $\mathbf{x} = c_1 e^{2t} \begin{bmatrix} -\cos 2t + 2\sin 2t \\ \cos 2t \end{bmatrix} + c_2 e^{2t} \begin{bmatrix} -2\cos 2t - \sin 2t \\ \sin 2t \end{bmatrix}$

27. $\lambda = 3 \pm 2i$, $\mathbf{x} = c_1 e^{3t} \begin{bmatrix} -\frac{\cos 2t}{2} + \frac{\sin 2t}{2} \\ \cos 2t \end{bmatrix} + c_2 e^{3t} \begin{bmatrix} -\frac{\cos 2t}{2} - \frac{\sin 2t}{2} \\ \sin 2t \end{bmatrix}$

29. $\lambda = \alpha \pm \beta i$ eigenvalues, trace $(\mathbf{A}) = (\alpha + \beta i) + (\alpha - \beta i) = 2\alpha$. Solutions involve functions of the form $e^{\alpha t} \cos \beta t$.

31. $\mathbf{A}(\mathbf{a} + i\mathbf{b}) = (\alpha + i\beta)(\mathbf{a} + i\mathbf{b}) \Rightarrow \mathbf{A}\mathbf{a} + i\mathbf{A}\mathbf{b} = (\alpha\mathbf{a} - \beta\mathbf{b}) + i(\beta\mathbf{a} + \alpha\mathbf{b})$
$\Rightarrow \mathbf{A}\mathbf{a} = \alpha\mathbf{a} - \beta\mathbf{b}$ and $\mathbf{A}\mathbf{b} = \beta\mathbf{a} + \alpha\mathbf{b}$

33. $\begin{bmatrix} 2 & 2 & 2 & 0 & | & 0 \\ -4 & -2 & 0 & 2 & | & 0 \\ 2 & 0 & -2 & -2 & | & 0 \\ 0 & 2 & 4 & 2 & | & 0 \end{bmatrix} \rightarrow \begin{bmatrix} 1 & 0 & -1 & -1 & | & 0 \\ 0 & 1 & 2 & 1 & | & 0 \\ 0 & 0 & 0 & 0 & | & 0 \\ 0 & 0 & 0 & 0 & | & 0 \end{bmatrix}, \mathbf{a} = \begin{bmatrix} a_1 \\ a_2 \end{bmatrix}, \mathbf{b} = \begin{bmatrix} b_1 \\ b_2 \end{bmatrix}$

Choose b_1, b_2 arbitrary but not both zero. Let $a_1 = b_1 + b_2, a_2 = -2b_1 - b_2$.

35. If $\mathbf{A}$ has more than one conjugate pair of eigenvalues, $\mathbf{A}$ has *at least four* eigenvalues, so $\mathbf{A}$ is at least 4×4.

37. $\lambda_1 = 2, \mathbf{u} = \begin{bmatrix} 0 \\ 0 \\ 1 \end{bmatrix}, \lambda_2 = 1 + i, \mathbf{u} = \begin{bmatrix} \frac{1}{2} - \frac{1}{2}i \\ -1 \\ 1 \end{bmatrix}$

$\mathbf{x} = c_1 e^{2t} \begin{bmatrix} 0 \\ 0 \\ 1 \end{bmatrix} + c_2 e^t \begin{bmatrix} \frac{1}{2}\cos t + \frac{1}{2}\sin t \\ -\cos t \\ \cos t \end{bmatrix} + c_3 e^t \begin{bmatrix} -\frac{1}{2}\cos t + \frac{1}{2}\sin t \\ -\sin t \\ \sin t \end{bmatrix}$

39. $\lambda = 1, \mathbf{x} = e^t \left(\begin{bmatrix} 2u_2 + v_2 \\ u_2 \end{bmatrix} + t \begin{bmatrix} 2v_2 \\ v_2 \end{bmatrix} \right)$

$= u_2 e^t \begin{bmatrix} 2 \\ 1 \end{bmatrix} + v_2 e^t \begin{bmatrix} 1 + 2t \\ t \end{bmatrix}$

41. $\lambda = 0, \mathbf{x} = e^{0t} \left(\begin{bmatrix} u_2 - v_2 \\ u_2 \end{bmatrix} + t \begin{bmatrix} v_2 \\ v_2 \end{bmatrix} \right) = u_2 \begin{bmatrix} 1 \\ 1 \end{bmatrix} + v_2 \begin{bmatrix} -1 + t \\ t \end{bmatrix}$

43. $\lambda = 2, \mathbf{x} = e^{2t} \left(\begin{bmatrix} \frac{1}{4}u_2 - \frac{1}{16}v_2 \\ u_2 \end{bmatrix} + t \begin{bmatrix} \frac{1}{4}v_2 \\ v_2 \end{bmatrix} \right)$

$= u_2 e^{2t} \begin{bmatrix} \frac{1}{4} \\ 1 \end{bmatrix} + v_2 e^{2t} \begin{bmatrix} -\frac{1}{16} + \frac{1}{4}t \\ t \end{bmatrix}$

45. $\lambda = 1, \mathbf{x} = e^t \left(\begin{bmatrix} u_3 + v_3 \\ u_2 \\ u_3 \end{bmatrix} + t \begin{bmatrix} v_3 \\ v_3 \\ v_3 \end{bmatrix} \right)$

$= u_2 e^t \begin{bmatrix} 0 \\ 1 \\ 0 \end{bmatrix} + u_3 e^t \begin{bmatrix} 1 \\ 0 \\ 1 \end{bmatrix} + v_3 e^t \begin{bmatrix} 1 + t \\ t \\ t \end{bmatrix}$

47. $\lambda = 1, \mathbf{x} = e^t \left(\begin{bmatrix} 2w_3 \\ v_3 - 2w_3 \\ u_3 \end{bmatrix} + t \begin{bmatrix} 0 \\ 2w_3 \\ v_3 \end{bmatrix} + t^2 \begin{bmatrix} 0 \\ 0 \\ w_3 \end{bmatrix} \right)$

$= u_3 e^t \begin{bmatrix} 0 \\ 0 \\ 1 \end{bmatrix} + v_3 e^t \begin{bmatrix} 0 \\ 1 \\ t \end{bmatrix} + w_3 e^t \begin{bmatrix} 2 \\ 2t - 2 \\ t^2 \end{bmatrix}$

49. $\lambda = 1, \mathbf{x} = e^t \left(\begin{bmatrix} u_2 - \frac{1}{2}v_2 + \frac{1}{2}w_2 \\ u_2 \\ \frac{1}{2}v_2 \end{bmatrix} + \begin{bmatrix} v_2 - w_2 \\ v_2 \\ w_2 \end{bmatrix} t + \begin{bmatrix} w_2 \\ w_2 \\ 0 \end{bmatrix} t^2 \right)$

$= u_2 e^t \begin{bmatrix} 1 \\ 1 \\ 0 \end{bmatrix} + v_2 e^t \begin{bmatrix} -\frac{1}{2} + t \\ t \\ \frac{1}{2} \end{bmatrix} + w_2 e^t \begin{bmatrix} \frac{1}{2} - t + t^2 \\ t^2 \\ t \end{bmatrix}$

Section 6.11

1. $x_{1p} = Ae^t + Be^{-t}$,

$x_{2p} = Ce^t + De^{-t}$;

$$\mathbf{x} = \mathbf{x}_p + \mathbf{x}_h = \begin{bmatrix} -\frac{1}{3}e^{-t} \\ -e^t + \frac{2}{3}e^{-t} \end{bmatrix} + c_1 \begin{bmatrix} -1 \\ 1 \end{bmatrix} + c_2 e^{2t} \begin{bmatrix} 1 \\ 1 \end{bmatrix}$$

3. $x_{1p} = Ae^{-t}$,

$x_{2p} = Be^{-t}$;

$$\mathbf{x} = \mathbf{x}_p + \mathbf{x}_h = \begin{bmatrix} 0 \\ 2e^{-t} \end{bmatrix} + c_1 \begin{bmatrix} \frac{1}{2} \\ 1 \end{bmatrix} + c_2 e^t \begin{bmatrix} 1 \\ 1 \end{bmatrix}$$

5. $x_{1p} = A \sin t + B \cos t$,

$x_{2p} = C \sin t + D \cos t$;

$$\mathbf{x} = \mathbf{x}_p + \mathbf{x}_h = \begin{bmatrix} -\sin t \\ \sin t - \cos t \end{bmatrix} + c_1 e^t \begin{bmatrix} 1 \\ -1 \end{bmatrix} + c_2 e^{3t} \begin{bmatrix} 1 \\ 1 \end{bmatrix}$$

7. $x_{1p} = Ae^{2t} + Bte^{2t} \quad B = D = \frac{1}{2}$

$x_{2p} = Ce^{2t} + Dte^{2t} \quad A - C = \frac{1}{4}$ (let $A = 0$);

$$\mathbf{x} = \begin{bmatrix} \frac{1}{2}te^{2t} \\ -\frac{1}{4}e^{2t} + \frac{1}{2}te^{2t} \end{bmatrix} + c_1 e^{2t} \begin{bmatrix} 1 \\ 1 \end{bmatrix} + c_2 e^{-2t} \begin{bmatrix} -1 \\ 1 \end{bmatrix}$$

9. $x_{1p} = A \sin t + B \cos t + Ct \sin t + Dt \cos t + E \sin 2t + F \cos 2t$,

$x_{2p} = G \sin t + H \cos t + It \sin t + Jt \cos t + K \sin 2t + L \cos 2t$

11. $x_{1p} = Ae^t + Bte^t$,

$x_{2p} = Ce^t + Dte^t$

13. $x_{1p} = Ae^{2t} + Bte^{2t} + Ct^2 e^{2t} + Dt^3 e^{2t} + Fe^t$,

$x_{2p} = Ge^{2t} + Hte^{2t} + It^2 e^{2t} + Jt^3 e^{2t} + Ke^t$

15. $x_{1p} = A + Bt + Ct^2 + Dt^3 + Et^4$

$x_{2p} = F + Gt + Ht^2 + It^3 + Jt^4$

17. $x_{1p} = A \sin t + B \cos t + Ct \sin t + Dt \cos t + Et^2 \sin t + Ft^2 \cos t$

$x_{2p} = G \sin t + H \cos t + It \sin t + Jt \cos t + Kt^2 \sin t + Lt^2 \cos t$

19. $x_{1p} = Ae^{-t} + Be^{-t} \cos 3t + Ce^{-t} \sin 3t$

$x_{2p} = De^{-t} + Ee^{-t} \cos 3t + Fe^{-t} \sin 3t$

Section 6.12

1. $e^{At} = \begin{bmatrix} -e^t + 2e^{2t} & -e^t + e^{2t} \\ 2e^t - 2e^{2t} & 2e^t - e^{2t} \end{bmatrix}$, $\mathbf{x} = \begin{bmatrix} -1 - 2e^t + 4e^{2t} \\ 2 + 4e^t - 4e^{2t} \end{bmatrix}$

3. $e^{At} = \frac{1}{5}\begin{bmatrix} 4 + e^{5t} & 2 - 2e^{5t} \\ 2 - 2e^{5t} & 1 + 4e^{5t} \end{bmatrix}$, $\mathbf{x} = \begin{bmatrix} -5e^{-t} + \frac{1}{2}e^t + \frac{22}{5} + \frac{1}{10}e^{5t} \\ -2e^{-t} + \frac{11}{5} - \frac{1}{5}e^{5t} \end{bmatrix} + e^{At}\mathbf{c}$

5. $e^{At} = -\frac{1}{5}\begin{bmatrix} -e^{2t} - 4e^{7t} & 2e^{2t} - 2e^{7t} \\ 2e^{2t} - 2e^{7t} & -4e^{2t} - e^{7t} \end{bmatrix}$, $\mathbf{x} = \frac{1}{25}\begin{bmatrix} 5te^{2t} - 4e^{2t} + 4e^{7t} \\ -10te^{2t} - 2e^{2t} + 2e^{7t} \end{bmatrix} + e^{At}\mathbf{c}$

7. $e^{At} = \begin{bmatrix} 1 & t \\ 0 & 1 \end{bmatrix}$. (Note: $\mathbf{A}^m = \mathbf{0}$ if $m \geq 2$.) $e^{-At} = \begin{bmatrix} 1 & -t \\ 0 & 1 \end{bmatrix}$.

9. $e^{At} = \begin{bmatrix} \cos t & \sin t \\ -\sin t & \cos t \end{bmatrix}$

11. $A^2 + AB + BA + B^2 = (A + B)^2 \Rightarrow AB = BA$

13. $A[u_1, u_2] = [Au_1, Au_2] = [\lambda_1 u_1, \lambda_2 u_2] = [u_1, u_2]\begin{bmatrix} \lambda_1 & 0 \\ 0 & \lambda_2 \end{bmatrix}$

15. Use (1).

17. $e^{At} = \frac{1}{5}\begin{bmatrix} 4 + e^{5t} & 2 - 2e^{5t} \\ 2 - 2e^{5t} & 1 + 4e^{5t} \end{bmatrix}$

19. $e^{At} = -\frac{1}{5}\begin{bmatrix} -e^{2t} - 4e^{7t} & 2e^{2t} - 2e^{7t} \\ 2e^{2t} - 2e^{7t} & -4e^{2t} - e^{7t} \end{bmatrix}$

21. $e^{At} = \frac{1}{10}\begin{bmatrix} e^{6t} + 9e^{-4t} & 3e^{6t} - 3e^{-4t} \\ 3e^{6t} - 3e^{-4t} & 9e^{6t} + e^{-4t} \end{bmatrix}$

23. a) $e^{At} = \begin{bmatrix} -e^t + 2e^{2t} & -e^t + e^{2t} \\ 2e^t - 2e^{2t} & 2e^t - e^{2t} \end{bmatrix}$

b) $e^{At} = \frac{1}{5}\begin{bmatrix} 4 + e^{5t} & 2 - 2e^{5t} \\ 2 - 2e^{5t} & 1 + 4e^{5t} \end{bmatrix}$

c) $e^{At} = \begin{bmatrix} -2 + 3e^t & -1 + e^t \\ 6 - 6e^t & 3 - 2e^t \end{bmatrix}$

Section 6.13

1. $Z(t) = \begin{bmatrix} e^t & e^t - 1 \\ 0 & e^t \end{bmatrix}$, $x = \begin{bmatrix} \frac{1}{2}(e^{3t} - e^t) \\ 0 \end{bmatrix} + Z(t)c$

Note Form of answer in Ex. 3 through 9 depends on choice of $W(t)$ and c.

3. $A(t) = \frac{1}{1-t}\begin{bmatrix} 0 & e^t - te^t \\ e^{-t} & -1 \end{bmatrix}$, $x = \begin{bmatrix} -\frac{t^2}{2}e^t + te^t \\ -\frac{t^2}{2} + 1 \end{bmatrix} + \begin{bmatrix} e^t & t \\ 1 & e^{-t} \end{bmatrix}\tilde{c}$

5. $A(t) = \frac{1}{1-t^2}\begin{bmatrix} -t & 1 \\ 1 & -t \end{bmatrix}$, $x = \begin{bmatrix} t^2 - \frac{t^4}{3} \\ t - \frac{t^3}{3} \end{bmatrix} + \begin{bmatrix} 1 & t \\ t & 1 \end{bmatrix}c$

7. $A(t) = \begin{bmatrix} 0 & e^t \\ 0 & 0 \end{bmatrix}$, $x = \begin{bmatrix} 2te^t + \frac{1}{2}e^{2t} - 2e^t + \frac{3}{2} \\ 2t \end{bmatrix} + \begin{bmatrix} e^t & 1 \\ 1 & 0 \end{bmatrix}c$

9. $A(t) = \begin{bmatrix} 0 & 1 \\ 0 & 0 \end{bmatrix}$, $x = \begin{bmatrix} \frac{t^3}{2} + \frac{t^2}{2} \\ \frac{t^2}{2} + t \end{bmatrix} + \begin{bmatrix} t & 1 \\ 1 & 0 \end{bmatrix}c$

Chapter 7

Section 7.1
1. First-order linear
3. Second-order nonlinear
5. Second-order nonlinear
7. First-order linear constant coefficient
9. 1, 2, 1, 2, 1, 2
11. 1, 3/2, 11/6, 25/12, 137/60, 147/60
13. $-1, 0, 4, 7, 6, 2$
15. $2^{k+1} + 5^{k+1} - 2[2^k + 5^k] = 3 \cdot 5^k, 2^0 + 5^0 = 2$
17. $k + 2 + \dfrac{1}{k+3} - \left[k + \dfrac{1}{k+1}\right] = 2 - 2[k^2 + 4k + 3]^{-1}$
19. $k + 2 + (k+2)2^{k+2} - 2[k + 1 + (k+1)2^{k+1}] + k + k2^k = (4+k)2^k$

Section 7.2
1. $y_k = 2 \cdot 2^k - 1 = 2^{k+1} - 1$
3. $y_k = (\tfrac{3}{2})^k y_0$
5. $y_k = y_0 + 3k$
7. $y_k = \tfrac{3}{2} + (-1)^k(y_0 - \tfrac{3}{2}) = \tfrac{3}{2} + (-1)^k c$
9. $y_k = -\tfrac{2}{3} + \tfrac{20}{3}(4)^k$
11. Use Eqs. (8) and (9).
13. Unbounded, oscillating, $r = -3, r < 0, |r| = 3 > 1$
15. Oscillating, bounded, asymptotic, $r = -\tfrac{1}{3} < 0, |r| < 1$
17. Monotonic, unbounded, $r = 1$
19. Unbounded, monotonic, $r = 4, r > 1$
21. Monotonic, bounded, asymptotic, $r = \tfrac{1}{3} < 1, r > 0$
23. Oscillating, bounded, $r = -1$

Section 7.3
1. $y_k = c_1 2^k + c_2(-3)^k$
3. $y_k = c_1 + c_2(-2)^k, c_1 = 1, c_2 = -1$
5. $x_k = c_1 3^k + c_2 k 3^k, c_1 = 2, c_2 = 3$
7. $y_k = c_1 \cos\left(\dfrac{\pi k}{2}\right) + c_2 \sin\left(\dfrac{\pi k}{2}\right)$
9. $y_k = c_1 2^k \cos\left(\dfrac{\pi k}{3}\right) + c_2 2^k \sin\left(\dfrac{\pi k}{3}\right)$
11. $y_k = c_1 2^{k/2} \cos\left(\dfrac{3\pi}{4}k\right) + c_2 2^{k/2} \sin\left(\dfrac{3\pi}{4}k\right)$
13. $y_k = c_1 3^{-k} \cos\left(\dfrac{\pi k}{2}\right) + c_2 3^{-k} \sin\left(\dfrac{\pi k}{2}\right)$
15. $y_k = c_1(-5)^k + c_2 k(-5)^k$
17. $\lim\limits_{k \to \infty} |\lambda|^k = \infty$ iff $|\lambda| > 1$, $\lim\limits_{k \to \infty} k|\lambda|^k = \infty$ iff $|\lambda| \geq 1$
19. Substitute and rearrange.
25. $\lambda = \pm 2, \pm 2, y_k = c_1 2^k + c_2 k 2^k + c_3(-2)^k + c_4 k(-2)^k$
27. $\lambda = \pm i, \pm i, y_k = c_1 \cos\left(\dfrac{\pi k}{2}\right) + c_2 \sin\left(\dfrac{\pi k}{2}\right) + c_3 k \cos\left(\dfrac{\pi k}{2}\right) + c_4 k \sin\left(\dfrac{\pi k}{2}\right)$

29. $\lambda = 1, 1, 1, y_k = c_1 + c_2 k + c_3 k^2$

31. $\lambda = -2, -2, -2, y_k = c_1(-2)^k + c_2 k(-2)^k + c_3 k^2(-2)^k$

33. $\lambda = 1, 1, 1, 1, y_k = c_1 + c_2 k + c_3 k^2 + c_4 k^3$

Section 7.4

1. $s_k = \dfrac{1}{2}k^2 + \dfrac{1}{2}k = \dfrac{k(k+1)}{2}$

3. $s_k = (-1)^k\left(\dfrac{1}{2}k + \dfrac{1}{4}\right) - \dfrac{1}{4}$

5. $s_k = \dfrac{1}{4}k^4 + \dfrac{1}{2}k^3 + \dfrac{1}{4}k^2 = \dfrac{k^2(k+1)^2}{4}$

7. $s_k = k + 2$

9. $\dfrac{1}{(1-x)^2}$

11. $\dfrac{1}{1-3x}$

13. $\dfrac{1}{1-x^2}$

15. $y_k = -\dfrac{1}{15}4^k + \dfrac{1}{16}7^k + c_1 + c_2(-1)^k$

17. $y_k = -\dfrac{1}{2}k^2 + \dfrac{1}{6}k^3 + c_1 + c_2 k$

19. $y_k = -\dfrac{3}{2}k(-2)^k + c_1(-2)^k + c_2(-3)^k$, $c_1 = 2, c_2 = 4$

21. $y_k = -3 - 2k - k^2 + c2^k$

23. $y_k = -3^k + c4^k$

25. $p_k = k^2[A_0 + A_1 k + A_2 k^2 + A_3 k^3]$

27. $p_k = k[A_0 + A_1 k]2^k + k[A_2 + A_3 k + A_4 k^2]3^k$

29. $p_k = k[A_0 + A_1 k + A_2 k^2 + A_3 k^3]$

31. (One option) Let $\hat{f}_k = h^k e^{ik\phi}$, so that $\text{Im}(\hat{f}_k) = f_k$. Using Rule 3, solve $ay_{k+2} + by_{k+1} + cy_k = \hat{f}_k = (he^{i\phi})^k$ with $r = he^{i\phi}$. Then show $\text{Im}(p_k)$ is a solution of $ay_{k+2} + by_{k+1} + cy_k = f_k$.

33. $\lambda_1 = e^{i\pi/2}, p_k = k\left[A_0 \cos\left(k\dfrac{\pi}{2}\right) + A_1 \sin\left(k\dfrac{\pi}{2}\right)\right]$, $p_k = -\dfrac{1}{2}k \sin\left(k\dfrac{\pi}{2}\right)$

Section 7.5

1. $y_{k+1} = (1.03)y_k + 200$, $y_0 = 10{,}000$; $k = 15.9$ quarters ≈ 4 years

3. $y_{k+1} = (1.01)y_k - b$, $y_0 = 4000$
  a) \$132.86   b) \$105.33   c) \$272.88

5. $y_{k+1} = (1.1)y_k + (0.06)[(1.06)^k 10{,}000]$
 $= (1.1)y_k + 600(1.06)^k$, $y_0 = 500$
 $y_k = 15{,}500(1.1)^k - 15{,}000(1.06)^k$ dollars

7. $0.8A_k + 0.2A_{k-1} = 150 + 50(-1)^{k-1}, k \geq 1, 0.8A_0 = 100$,

$A_k = \dfrac{175}{3}\left(-\dfrac{1}{4}\right)^k + 150 - \dfrac{250}{3}(-1)^k$

9. Production during kth year $= (1.2)^k 200{,}000$. Available chips a_k at end of kth year.
$a_1 = 20{,}000$, $a_k = 0.1a_{k-1} + 0.1(1.2)^k 200{,}000$, for $k \geq 2$, $a_k = -\dfrac{680{,}000}{11}(0.1)^k + \dfrac{240{,}000}{11}(1.2)^k$ for $k \geq 1$.

11. $s_0 = 1000$, $s_{k+1} = 9s_k$, $s_{40} = 9^{40}(1000)$

13.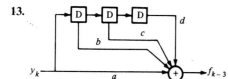

15. $a\lambda^2 + b\lambda + c = a\left(\lambda^2 - 2\cos\left(\dfrac{\pi}{2}\right)\lambda + 1\right) \Rightarrow [y_{k+2} + y_k] = f_k$ ($a = 1$ for convenience)

17. a) $a_1 a_2 y_k + (a_1 b_2 + b_1 a_2) y_{k-1} + (c_1 a_2 + b_1 b_2 + a_1 c_2) y_{k-2}$
$+ (b_1 c_2 + c_1 b_2) y_{k-3} + c_1 c_2 y_{k-4} = f_{k-4}$

Chapter 8

Section 8.2

1. 3
3. 0
5. 5
7. 0
9. a) 59049
11. a) 4.15×10^{-20}
13. 4.109, 6.129, 30.390, 193.137
15. $y(2) = 551{,}627$. Appears to be going from -5 to infinity.

16.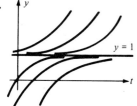

17. $y_N = y_{20} = 0.601$ with $h = 0.1$

19. $\dfrac{y_{N_1} - y_{N_2}}{y_{N_2} - y_{N_3}} = \dfrac{2^r - 1}{1 - (\frac{1}{2})^r} \approx 2^r$

20. 1.7106, 1.8411, 1.9165 (The next ratio is 1.957.)

Section 8.3

9. b) $y = x + 1 + 3e^{-30x}$

11. Numerical solution oscillates between approximately 0.07 and 3.06. Actual nonequilibrium solutions are monotonic.

Section 8.4

	Euler	Modified Euler	Second-order Taylor
1. y_N	6.1917	7.3046	7.3046
$\|e_N\|$	1.1973	0.0844	0.0844
3. y_N	0.6529	0.60698	0.60141
$\|e_N\|$	0.0463	0.00044	0.00513
5. y_N	0.7634	0.761582	0.761598
$\|e_N\|$	0.0018	0.000012	0.000004
7. y_N	0.8672	0.865763	0.865775
e_N	0.0014	0.000006	0.000006

Note In Ex. 7, $y = \sin^{-1}\left(\dfrac{e^{2x}-1}{e^{2x}+1}\right)$.

11. $y_5 = 7.3604$, $e_5 = 0.0286$ **13.** $y_{10} = 2.0588$

Section 8.5
1. $y_1 = 1.010025042$, $y_{10} = 2.297442014$ **3.** $y_1 = 0.571202424$, $y_{10} = 0.913669564$
5. $y_1 = 1.051795725$, $y_{10} = 1.355751179$ **7.** $y_1 = 20.1$, $y_{10} = 39{,}062{,}499$
8. $y_1 = 2.6916$, $y_{100} = 1.000000$

Section 8.6
1. $y_{10} = 24.30623$ **3.** $y_{10} = 0.86706$
5. $y(1) = 1$ (to 18 digits)

	$h = .1$	$h = .05$	$h = .04$	$h = .01$
y_N for Adams-Bashforth	-6×10^7	-3×10^7	1.7×10^6	1
y_N for Adams-Moulton	4×10^9	5	0.98880	1

7. 0.66547 ($y(1) = 0.66666\ldots$)

Section 8.7
1. Euler, $x_5 = 5.813$, $y_5 = 12.983$;
Modified Euler, $x_5 = 10.8926$, $y_5 = 25.2104$.

3. Euler, $x_{10} = 1.20608$, $y_{10} = 1.05994$;
Modified Euler, $x_{10} = 1.21620$, $y_{10} = 1.07288$.

5. Euler, $x_{10} = -0.68137$, $y_{10} = 1.33303$;
Modified Euler, $x_{10} = -0.61452$, $y_{10} = 1.24624$.

	Euler y_{10}	Error in Euler	Modified Euler y_{10}	Error in Modified Euler
7.	0.8825	-0.041	0.84247	-0.001
9.	6.1917	1.197	7.30463	0.0844

11. $y_{10} = 7.38889$, $e_{10} = 0.00017$

Chapter 9

Section 9.2

1. **a)** $xy = C$ **b)** $x = y = 0$,

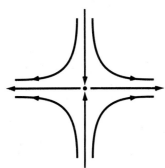

3. **a)** $\dfrac{y^2}{2} = x + C$ **b)** $x = 0, y = C$ (whole y axis are equilibrium points)

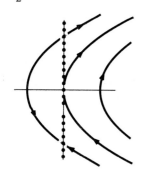

5. **a)** $y = \dfrac{1}{3}x + C$ **b)** no equilibria **c)** upward

7. $x^2 - 2x + y^2 - 2y = C$ (counterclockwise ellipses)

9. **a)** $y = Cx^3$ **b)** y axis is equilibria

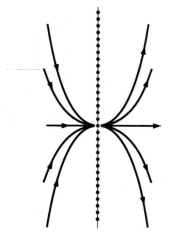

11. a) $yx = C$ **b)** y axis is equilibria

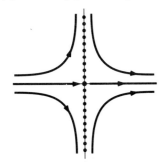

13. $y^2 = 2\cos x + C$
15. $y^2 = -2x^3/3 + C$
17. $y^2 = -\ln(1 + x^2) + C$
19. $y^2 = 2x^3/3 + C$
21. b) $y = 0, x = \pm 1$ **c)**

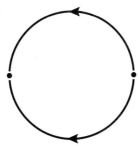

23. b) $y = 1, x = 0$ **c)**

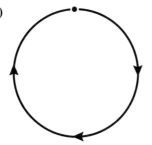

25. $(x(t_0), y(t_0))$ is a constant solution so that $x(t) \equiv x(t_0)$, $y(t) \equiv y(t_0)$ by Theorem 9.2.1.

27.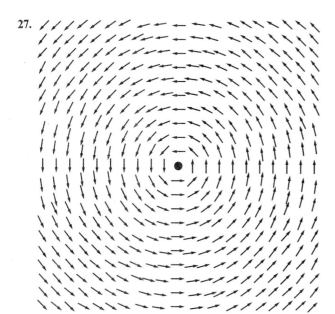

29.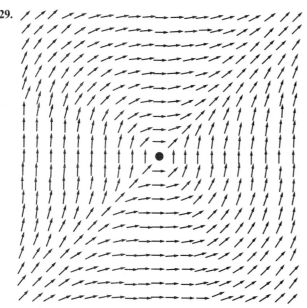

Section 9.3

1. $\lambda = 1, 2$; unstable; repeller
3. $\lambda = \pm 2i$; stable; ellipses traversed counterclockwise
5. $\lambda = 1 \pm i$; unstable; outward spiral counterclockwise
7. $\lambda = -1, -3$; attractor
9. $\lambda = 2, -1$; saddle
11. $\lambda = 2, -4$; attractor

13. $\lambda = -1 \pm 4i$; stable, inward spiral clockwise

15. $\lambda = 1, 7$; unstable; repeller

23. $\tan^{-1} \frac{y}{x} + \ln\left(\frac{y^2}{x^2} + 1\right) = -2 \ln x + C$

Section 9.4

1. $(0, 0)$, repeller; $(\frac{1}{2}, -1)$, saddle

3. $(0, 0)$, repeller; $(1, 1)$, saddle

5. $(0, 0)$, no info; $(-1, 1)$, saddle; $(1, 1)$, saddle

7. $(1, 1)$ and $(-1, -1)$ are saddles; $(1, -1)$ and $(-1, 1)$ pure imaginary roots $\Rightarrow$ circles equilibrium, long-run behavior unclear

9. $(0, 0)$, spiral repeller; $(-2, 2)$, saddle.

11. $(0, 0)$, spiral attractor

13. $(0, 0)$, saddle; $(\frac{1}{8}, 0)$, attractor; $(1, -7)$, saddle

15. $(0, 0)$, saddle; $(-3, 0)$, attractor; $(1, \frac{4}{3})$, spiral repeller

17. $(0, 0)$, saddle; $(-\frac{1}{8}, 0)$, attractor; $(1, 9)$, repeller

In 25 through 27, $z = x - r, w = y - s$.

25. $z' = 3z + 4w + 2,$
$w' = z - 3w$

27. $z' = z + w,$
$w' = -z - w$

Section 9.5

The equilibrium, and simple closed invariant curves not containing the equilibrium are:

1. $(0, 0)$; $\quad x^2 + 4y^2 = C, C > 0$

3. $(1, 2)$; $\quad x^2 - 2x + 1 + 2y^2 - 8y = C, C > -8$

5. $(1, 1)$; $\quad (x - 1)^4 + (y - 1)^4 = C, C > 0$

7. $(0, 0)$; $\quad 8x^2 - 4xy + 5y^2 = C, C > 0$

9. $(0, 0)$; $\quad 2y^2 + x^4 = C, C > 0$

11. $(0, 0)$; $\quad 4y^2 + 8y^3 + \frac{5}{4}x^4 = C, C > 0$

13. $(0, 0)$; $\quad 3y^2 + \frac{3}{2}x^2 + \frac{5}{4}x^4 = C, C > 0$

15.

17.

18. $r = \cos \theta + C$

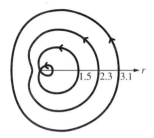

19. $r(t) = \cos \theta(t) + (r(0) - 1)$, $\theta = t$
(Note figure for Exercise 18.)

Section 9.6

1. $(0, 0)$, $(0, 1)$, $(1, 0)$; are repeller, saddle, and test fails.

3. $(0, 0)$, $(0, 1)$, $(\frac{1}{4}, 0)$; are repeller, attractor, saddle.

4. Extinction of x. Equilibrium value of y is great enough so that the $-2xy$ term eventually dominates the growth term for x.

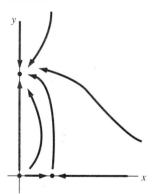

5, 6.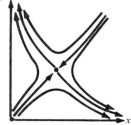

7. $(0, 0)$, $(\frac{1}{2}, 0)$; are saddle, and attractor (predators die out).

9. (0, 0), is saddle. Invariant curves are $3 \ln|y| + \ln|x| = y + x + c$. Apply Theorem 9.5.2. Periodic around (1, 3).

13. (0, 0) is a saddle. Use $y' = -qy + r\dfrac{x}{1+x} y \le (r-q)y$, $x' = ax - b\dfrac{x}{1+x} y \ge ax - by$.

15. (0, 0), (1, 2); are saddle, spiral repeller.

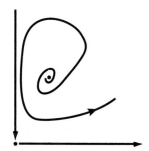

19. a, c. Velocity vectors (x', y') point into R from positive x and y axes.

21. (0, 0); test fails

Section 9.7

1. $f'(0) = 2$, stable, attractor
3. $f'(0) = 1$, spiral attractor
5. $f'(0) = -1$, spiral repeller
7. $f'(0) = 4$, stable, attractor
9. $f'(0) = -4$, repeller
11. $f'(0) = 0$, $if(i) = 3i^4 + i^6 \ge 0$, stable
13. $f'(0) = 0$, $if(i) = i^6 \ge 0$, stable
15. $f'(0) = 0$, $if(i) \ge 0$, stable
17. $2i^2 + 2q - 1 = ce^{-2q}$
19. $18i^2 + 6q - 1 = ce^{-6q}$

23. a) $f(i) \le 0$ for $i \le 0$ and $f(i) \ge 0$ for $i \ge 0 \Rightarrow f'(0) \ge 0$.
   b) $f(i) \ge 0$ for $i < 0$ and $f(i) \le 0$ for $i > 0 \Rightarrow f'(0) \le 0$.
   c) $f'(0) > 0$, $f(0) = 0 \Rightarrow f(i) < 0$ for $i < 0$ (i near 0), $f(i) > 0$ for $i > 0$ (i near 0) $\Rightarrow if(i) \ge 0$.
   d) $f'(0) < 0$, $f(0) = 0 \Rightarrow f(i) > 0$ for $i < 0$, $f(i) < 0$ for $i > 0$ (i near 0).

Section 9.8

1. $(0, 2n\pi)$ are spiral attractors, $(0, (2n+1)\pi)$ are saddles

3. $(0, 2n\pi)$ are attractors, $(0, (2n+1)\pi)$ are saddles

5. $\theta' = \phi, \phi' = [-mg \sin\theta - k\phi + rk]/ml$

6. Then

$(\phi, \theta) = \left(0, \dfrac{\pi}{6} + 2n\pi\right)$ spiral attractor (n integer),

$(\phi, \theta) = \left(0, \dfrac{5\pi}{6} + 2n\pi\right)$ unstable, saddle

7. $\phi = 0, \theta = \pi/2 + 2n\pi$, one zero eigenvalue, test fails

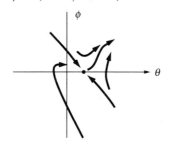

Appendix A

1. $12 + 16i$

3. $\dfrac{5}{13} - \dfrac{i}{13}$

5. $-\dfrac{5}{3} + \dfrac{2}{3}i$

7. $z_1 = \sqrt{2}e^{-(\pi/4)i}$ (or $\sqrt{2}e^{(7/4)\pi i}$)

9. $z_3 = 2e^{(7\pi/6)i}$ (or $2e^{-(5/6)\pi i}$)

11. $z_5 = 3e^{\pi i}$ (or $3e^{-\pi i}$)

13. $\dfrac{\sqrt{2}}{2}e^{(-(\pi/4)-(\pi/3))i} = \dfrac{1}{\sqrt{2}}e^{-(7/12)\pi i}$

15. $e^{((7/6)\pi - -(\pi/2))i} = 2e^{(5/3)\pi i}$

17. $\sqrt{2}\,2e^{(-(\pi/4)+(7\pi/6))i} = 2^{3/2}e^{(11/12)\pi i}$

19. There are three. One is $2^{1/6}e^{-(\pi/12)i}$.

21. $\dfrac{1}{e^z} = \dfrac{1}{e^a(\cos b + i \sin b)} = \dfrac{\cos b - i \sin b}{e^a} = e^{-a}(\cos b - i \sin b) = e^{-z}$

23. $[r^{1/m}e^{i((\theta/m)+(2\pi/m)n)}]^m = re^{i(\theta + 2\pi n)} = re^{i\theta}e^{i2\pi n} = re^{i\theta}$

25. $e^{i((3\pi/14)+(2\pi/7)n)}$; $n = 0, 1, 2, 3, 4, 5, 6$

Appendix B

1. $\dfrac{1}{x} - \dfrac{1}{x-1} + \dfrac{2}{x+1}$

3. $\dfrac{1}{(x^2+1)^2}$ (already reduced)

5. $\dfrac{x+1}{x^2+1} - \dfrac{1}{(x^2+1)^2}$

7. $\dfrac{2}{(x+2)^2} + \dfrac{3}{(x+2)^3}$

9. $x - 2 + \dfrac{6}{x+3}$

Appendix C

1. 0.2462661722

3. 0.1109574581

7. **a)** $y(x) = \displaystyle\int_0^x 3y(s)\,ds + 1$ **b)** $\phi_0(x) = 1$, $\phi_1(x) = 1 + 3x$, $\phi_2(x) = 1 + 3x + \tfrac{9}{2}x^2$,
$\phi_3(x) = 1 + 3x + \tfrac{9}{2}x^2 + \tfrac{9}{2}x^3$ **c)** $y(x) = e^{3x}$

9. **a)** $y(x) = \displaystyle\int_0^x (y(s) + s)\,ds$ **b)** $\phi_0(x) = 0$, $\phi_1(x) = \tfrac{1}{2}x^2$, $\phi_2(x) = \tfrac{1}{2}x^2 + \tfrac{1}{6}x^3$,
$\phi_3(x) = \tfrac{1}{2}x^2 + \tfrac{1}{6}x^3 + \tfrac{1}{24}x^4$ **c)** $y(x) = -1 - x + e^x$

11. **a)** $y(x) = \displaystyle\int_0^x 2sy(s)\,ds + 1$, **b)** $\phi_0(x) = 1$, $\phi_1(x) = 1 + x^2$, $\phi_2(x) = 1 + x^2 + \tfrac{1}{2}x^4$,
$\phi_3(x) = 1 + x^2 + \tfrac{1}{2}x^4 + \tfrac{1}{6}x^6$, **c)** $y(x) = e^{x^2}$

13. **a)** $y(x) = \displaystyle\int_0^x y(s)^{-1}\,ds + 1$, **b)** $\phi_0(x) = 1$, $\phi_1(x) = x + 1$, $\phi_2(x) = \ln(x+1) + 1$ **c)** $y(x) = \sqrt{2x+1}$

15. **a)** $y(x) = \displaystyle\int_1^x y(s)^2 - 2y(s) + 1\,ds$ **b)** $\phi_0(x) = 0$, $\phi_1(x) = x - 1$, $\phi_2(x) = \tfrac{1}{3}x^3 - 2x^2 + 4x - \tfrac{7}{3}$ **c)** $y = x(x+1)^{-1}$

17. **a)** and **b)** as in proof of Theorem 2. **c)** $|\phi_n(x) - \phi(x)| \le \dfrac{ML^n|x - x_0|^{n+1}}{(n+1)!} H(x)$ where $H(x) = \displaystyle\sum_{i=0}^{\infty} \dfrac{L^i|x - x_0|^i i!}{i!(n+1+i)!}$.
Show that $H(x)$ is continuous on $|x - x_0| \le \tilde{a}$. Let $P = \max_{|x - x_0| \le \tilde{a}} |H(x)|$.

19. $\psi(x) = \begin{cases} 0 & \text{if } 0 \le x < 1 \\ \tfrac{1}{2} & \text{if } x = 1 \\ 1 & \text{if } 1 < x \le 2 \end{cases}$

23. $||y_1| - |y_2|| \le 1|y_1 - y_2|$ so Lipschitz continuous with $L = 1$.

25. Take $x_0 \in [\alpha, \beta]$. Apply Theorem 2 using Exercise 24. If $x_0 + \tilde{a} < \beta$, then apply Theorem 2 and Exercise 24 at $x_0 + \tilde{a}$ to extend solution to smaller of $x_0 + 2\tilde{a}$ and β. If $x_0 + 2\tilde{a} < \beta$, repeat process until solution defined on $[x_0, \beta]$. If $\alpha < x_0 - \tilde{a}$, similar argument extends solution to all of $[\alpha, \beta]$.

Index

Abel's formula, 122
Absolute value of a complex number, 528
Adams-Bashforth methods, 484–485
Adams-Moulton method, 485–486
Adder, 453
Ambient temperature, 59
Amplitude, 183
   time varying, 187
Analytic approach, 2
Analytic function, 264
Annihilators and undetermined coefficients, 168–171
Approximation
   linear 34, 97. *See also* Linearization
   nonlinear, 37
   quadratic, 37
   second-order, 513
Arbitrary constant, 312
Asymptotically stable equilibrium, 500
Attractor, 23, 500, 502
Augmented matrix, 360
Auxiliary polynomial, 382

Basic existence and uniqueness theorem, 14–18, 538–547

Bernoulli equation, 35–37
Bessel's equation, 302–303
Bessel's functions, 302–308
Boundary conditions, 59
Boundary value problem, 59
Branch, 82
Buoyancy, 200

Capacitance, 83
Capacitor, 83
Cascade, 207, 458
Catenary, 111
Cayley-Hamilton theorem, 423
Centered difference, 486
Characteristic equation, 136, 392, 441
Characteristic polynomial, 136, 155–162, 382, 441, 499
Chemical reactions, 93–95, 493
Circuits, 82–90, 200–202, 234, 242–246, 252–256, 310, 350–352, 492–493, 519–522
Clock, 453
Closed form, 447
Coefficients, 4, 112, 313, 434
Column, 353
Complex numbers, 527–532

Compounding of interest, 68
Concavity, 21
Conjugate, 528
Conservation law, 496,
Conservation of mechanical energy, 186
Convergence of a series, 258
   absolute, 259
   conditional, 260
Convergence set, 261
Convolution, 246–252
Cooling, 59–61
Cramer's rule, 176, 178
Current source, 84

Damping
   constant, 182
   critical, 188
   over, 189
   under, 186
Dashpot, 182
Deficiency of eigenvalues, 400–406
Degenerate system, 312
Degree of a polynomial, 55
Delay, 453
Delta function, 245
De Moirre's formula, 530

Determinant, 120, 368–372
Diagonal matrix, 357, 422–423
Difference equation, 432–458
Differential equation
  autonomous, 491
  Euler's 149–154
  exact, 43–49
  explicit, 13–14
  homogeneous, 53–56
  implicit, 9, 13
  linear, 4, 26–35, 111–431, 498–503
  nonlinear, 4
  ordinary, 2
  partial, 2
  separable, 37–43
  system, 2, 309–431, 486–490
Differentiator, 255
Direction, field, 18–24
Discretization error, 468–471
Distinguished column, 361
Distributions, 242–246
Divergence of series, 258
Doubling time, 69

Eigenvalue, 380–407, 498–500
Eigenvector, 380–407, 499–500
Elementary row operation, 360
Elimination method, 318–327
Equidimensionality, 153
Equilibrium, 20, 60, 88, 439, 491, 499, 503
  semistable, 23, 25, 499
  stable, 23, 25, 499
  unstable, 23, 25, 499
Equipotentials, 49
Equivalent system, 311
Error, 148, 468–472
Escape velocity, 111
Euler constant, 306
Euler method, 462–476, 487
  backward, 474–475
  implicit, 474
  improved, 478
  modified, 478–480, 489
Euler's equation, 149–154, 288
Evaporation, 62–64
Existence, 14–18, 538–547
Exponential decay, 58
Exponential matrix, 415–424
Exponential order, 211

Family, 11
Feedback circuit, 254
Filter, 455
Flow problems, 64–78, 331–339
Flux lines, 79
Forced response, 154, 191–198, 376
Forcing function, 112, 313, 434
Fourier series, 163–164, 302

Free response, 182–189, 376
Frequency domain, 252
Friction, 186–189
Frobenius method, 287–302
Fundamental set of solutions, 113, 122, 376, 419
Fundamental solution matrix, 424–431

Gamma function, 219, 296–301
Gaussian elimination, 360–365
Generalized function, 244
Geometric series, 259
Global behavior, 521
Gravity, 181–182
Grid, 461
Growth rates, 512

Half-life, 58
Harmonic functions, 49
Harmonic motion, 183–185
Heaviside function, 227–237
Hermite equation and polynomial, 287
Heun method, 478
Homogeneous equations
  associated linear, 34, 112, 141, 376
  linear, 135–146
  nonlinear, 53–56
Homogeneous polynomial, 55
Hooke's law, 181
Hypergeometric equation, 302

Identity matrix, 357
Ill-conditioned, 424
Imaginary part, 527
Impulses, 242–246
Inconsistent system, 312, 360
Indicial equation and polynomial, 290
Inductor and inductance, 84
Initial conditions, 8, 432
Initial value problem, 8
Input, 191, 203
Input voltage, 202
Integral curves, 494, 509
Integral equations, 249–251
Integrating factor, 28, 49–52
Integrator, 255
Interest, 68, 76, 450
Interlacing property, 308
Invariant curve, 494
Invariant set, 510
  closed, 510
Inverse Laplace transform, 212
Inverse matrix, 357–358, 365–366, 372
Isobars, 79
Isoclines, 26
Isotherms, 79

Jump discontinuity, 211

Kirchoff's laws, 82, 84–85, 200

Laguerre's equation and polynomial, 302
Laplace's equation, 49, 302
Laplace transform, 209–256, 328–331
Legendre equation, 135
Legendre polynomials, 284–287
Leontief model, 339
Leslie model, 339
Level curves, 49
Liapunov function, 508, 510–511, 520
Linear approximation, 34, 97. See also Linearization
Linear combination, 113, 122, 371
Linear factor, 534
Linearization, 34, 503–504, 520
Linearly dependent, 120–125, 371
Linearly independent, 120–125, 141–145, 371
Linear transformation, 212
Line integrals, 49
Local, 505
Logistic equation, 97, 514

Machine precision, 460, 471–472, 476
Magnitude, 529
Mass action law, 94
Matrix operations, 353–356
Matrix product, 355
Mean value theorem, 468
Measurements as integrals, 252
Mechanical systems, 309, 341–349, 522–526
Mechanics, 91–93, 105–111, 180–200, 493
Mesh, 461
Mixing problems, 69–78, 331–341
Modulus, 529
Momentum, 91
Multiplicity
  roots, 147, 534
  eigenvalues, 386
Multistep methods, 483–486

Natural response, 182–189
Newton's law
  of cooling, 2, 59
  of motion, 91, 105, 180–181
Newton's method, 148, 476
Niche theory, 515
Node, 82
Numerical approach, 3, 459–490

Operator
  constant coefficient, 128–136
  differential, 126–131, 168–171, 177, 318–327
  integral, 177
  linear, 127

Orbit, 494
Order
　difference equation, 434
　differential equation, 4
　numerical method, 465–466, 469, 479, 483
Ordinary point, 268
Orthogonal family, 49, 79
Orthogonal trajectories, 79–82
Output, 191, 203
Output voltage, 202

Partial fractions, 533–537
Partial sum, 258
Partition, 461
Pendulum, 522–525
Periodic functions, 237–242
Periodic solution, 484, 508–512
Phase amplitude form, 203
Phase angle, 183
Phase plane, 494–498
Phase portrait, 494
Phasor, 207
Piecewise continuous, 211, 227–237
Poincaré-Bendixson, 509
Polar decomposition. See Polar form
Polar form, 205, 442, 530–31
Poles, 255
Population models, 340, 512–519
Potential, 49
Predation term, 517
Predator-prey model, 515
Predictor-corrector, 478, 488
Pressure, 64–66
Prime, 255
Product rule, 375

Qualitative approach, 3

Radioactive decay, 57–58
Radius of convergence, 261, 267–269
Rate constant, 94
Ratio test, 260–261
Real part, 527
Recursion, 435
Reduction of order, 131–135, 431
Regular singular point, 288–289
Repeller, 502
Resembles, 505
Resistance, 525
　electrical, 83
　mechanical, 93, 182
Resistor, 83
Resonance, 196, 345
Rodrigues's formula, 286
Roots, 136
　characteristic polynomial, 164, 441
　of complex numbers, 532
　repeated, 136
Root test formula, 261
Row, 353
Round-off error, 148, 468, 471–472
Row echelon form, 361
Runge-Kutta
　second-order, 478
　fourth-order, 481–483

Saddle point, 500, 505
Scalar, 353
Series
　alternating, 260
　expansion of, 264
　Maclurin's, 260
　summation of, 445–449
Shifting theorem, 217
Signal, 453
Signal-flow diagram, 454
Signal processing, 453–456
Similarity transformation, 422
Simpson's rule, 483
Singular points, 27, 112, 141, 248
Sketching of solutions, 18–26
Solution curves, 494
Solution of difference equation
　general, 433–434
　particular, 433, 445
　unbounded, 439
　asymptotic, 439
　monotonic, 439
Solution of differential equation
　existence of, 14–15, 27, 99, 112–113, 314, 375
　explicit, 6, 40
　general, 7, 28–29, 376
　implicit, 6, 40
　particular, 7, 315, 376
　revised definition, 32–33
　singular, 12, 38–39
　uniqueness of, 14–15, 27, 99, 112–113, 314, 375
Speed, 495
Spiral, 501, 505
Spiral attractor, 521
Spring-mass system, 181, 310, 341–349
Square matrix, 354
Stable, 25, 501, 505
State variable model, 13–14, 252
Steady-state operating points, 88
Step size, 461
Stiffness, 472–475
Stream lines, 79
Substitutions, 35, 100–105
Superposition principle, 113, 130, 163, 441

Symbolic language, 225, 450, 486
Symmetric matrix, 349, 390–391
Systems
　of algebraic equations, 358–367
　explicit, 313, 375, 379
　implicit, 379
　numerical, 486–490
　solution, 310

Taylor methods
　second-order, 477
　third-order, 480
　series, 270–271
Taylor's theorem, 34, 37, 468
Taylor series, 260–265
Tchebycheff polynomials and equation, 287
Trace, 389, 399
Trajectory, 380, 494
Transfer functions, 252–256
　proper, 256
　stable, 255
Transient, 191
Trapezoid rule, 478, 485
Two-port, 252

Undetermined coefficients, 154–171, 275, 407–415, 445–449
Uniqueness, 14–18, 538–547
Units
　centimeter/grams/seconds (cgs), 92
　electrical, 84
　foot/pound/second (fps), 92
Unit step function, 227
Unstable, 499–500, 505
Upper triangular matrix, 370
Uptake function, 517

Vandermond matrix, 424
Variable mass, 106–108
Variation of constants, 428
Variation of parameters, 171–180, 429
Vector, 354
Vector field, 19
Velocity vector, 495
v–i characteristic, 83
Volume and surface area, 65

Wave equation, 302
Weight, 91
Weighted averages, 252
Work, 49
Wronskian, 121–122, 144, 173, 176, 379

Zeroes of transfer function, 255
Zero matrix, 357